Candida albicans: Cellular and Molecular Biology

Rajendra Prasad
Editor

Candida albicans: Cellular and Molecular Biology

Second Edition

Springer

Editor
Rajendra Prasad
Amity Institute of Integrative Sciences
and Health and Amity Institute
of Biotechnology
Amity University Haryana, Amity Education
Valley
Gurgaon, Haryana
India

ISBN 978-3-319-50408-7 ISBN 978-3-319-50409-4 (eBook)
DOI 10.1007/978-3-319-50409-4

Library of Congress Control Number: 2016963194

Printed on acid-free paper

This Springer imprint is published by Springer Nature
The registered company is Springer International Publishing AG
The registered company address is: Gewerbestrasse 11, 6330 Cham, Switzerland

Contents

Contributors

David R. Andes Departments of Medicine and Medical Microbiology and Immunology, University of Wisconsin, Madison, WI, USA

Marie-Elisabeth Bougnoux Unité Biologie et Pathogénicité Fongiques, Institut Pasteur, INRA, Paris, France; Unité de Parasitologie-Mycologie, Service de Microbiologie Clinique, Hôpital Necker-Enfants-Malades, Assistance Publique Des Hôpitaux de Paris (APHP), Paris, France

Gordon D. Brown Aberdeen Fungal Group, MRC Centre for Medical Mycology, Institute of Medical Sciences, University of Aberdeen, Aberdeen, UK

Richard D. Cannon Sir John Walsh Research Institute, Faculty of Dentistry, University of Otago, Dunedin, New Zealand

Neeraj Chauhan Public Health Research Institute, New Jersey Medical School, Rutgers, The State University of New Jersey, Newark, NJ, USA; Department of Microbiology, Biochemistry and Molecular Genetics, New Jersey Medical School, Rutgers, The State University of New Jersey, Newark, NJ, USA

Arunaloke Chakrabarti Department of Medical Microbiology, Postgraduate Institute of Medical Education and Research, Centre of Advance Research in Medical Mycology & WHO Collaborating Centre, Chandigarh, India

Christophe d'Enfert Unité Biologie et Pathogénicité Fongiques, Institut Pasteur, INRA, Paris, France

Patrick Van Dijck VIB Department of Molecular Microbiology, KU Leuven, Louvain, Belgium

Eddie G. Dominguez Departments of Medicine and Medical Microbiology and Immunology, University of Wisconsin, Madison, WI, USA

John E. Edwards Los Angeles Biomedical Research Institute, Harbor-UCLA, Torrance, CA, USA

Joachim F. Ernst Department Biologie, Molekulare Mykologie, Heinrich-Heine-Universität Düsseldorf, Düsseldorf, Germany

Saranna Fanning Whitehead Institute for Biomedical Research, Cambridge, MA, USA

Adeline Feri Unité Biologie et Pathogénicité Fongiques, Institut Pasteur, INRA, Paris, France; Cellule Pasteur, Univ. Paris Diderot, Sorbonne Paris Cité, Paris, France

Neil A.R. Gow Aberdeen Fungal Group, Institute of Medical Sciences, University of Aberdeen, Aberdeen, UK

Chibuike Ibe Aberdeen Fungal Group, Institute of Medical Sciences, University of Aberdeen, Aberdeen, UK

Mary Ann Jabra-Rizk University of Maryland School of Dentistry, Baltimore, MD, USA

David Kadosh Department of Microbiology and Immunology, University of Texas Health Science Center at San Antonio, San Antonio, TX, USA

Mikhail V. Keniya Department of Oral Sciences, University of Otago, Dunedin, New Zealand

Afshin Khan Tulane University School of Medicine, New Orleans, LA, USA

Erwin Lamping Sir John Walsh Research Institute, Faculty of Dentistry, University of Otago, Dunedin, New Zealand

Hee Ji Lee Sir John Walsh Research Institute, Faculty of Dentistry, University of Otago, Dunedin, New Zealand

Mélanie Legrand Unité Biologie et Pathogénicité Fongiques, Institut Pasteur, INRA, Paris, France

Raphaël Loll-Krippleber Unité Biologie et Pathogénicité Fongiques, Institut Pasteur, INRA, Paris, France; Cellule Pasteur, Univ. Paris Diderot, Sorbonne Paris Cité, Paris, France

Jose Luis Lopez Ribot Department of Biology, University of Texas San Antonio, San Antonio, TX, USA

Golnoush Madani Sir John Walsh Research Institute, Faculty of Dentistry, University of Otago, Dunedin, New Zealand

Timea Marton Unité Biologie et Pathogénicité Fongiques, Institut Pasteur, INRA, Paris, France

Corinne Maufrais Institut Pasteur, Centre d'Informatique pour la Biologie, Paris, France

Sławomir Milewski Department of Pharmaceutical Technology and Biochemistry, Gdańsk University of Technology, Gdańsk, Poland

Aaron P. Mitchell Department of Biological Sciences, Carnegie Mellon University, Pittsburgh, PA, USA

Brian C. Monk Department of Oral Sciences, University of Otago, Dunedin, New Zealand

Joachim Morschhäuser Institut für Molekulare Infektionsbiologie, Universität Würzburg, Würzburg, Germany

Carol A. Munro Aberdeen Fungal Group, Institute of Medical Sciences, University of Aberdeen, Aberdeen, UK

Remya Nair Membrane Biology Laboratory, School of Life Sciences, Jawaharlal Nehru University, New Delhi, India

Krishnamurthy Natarajan Laboratory of Eucaryotic Gene Regulation, School of Life Sciences, Jawaharlal Nehru University, New Delhi, India

Masakazu Niimi Sir John Walsh Research Institute, Faculty of Dentistry, University of Otago, Dunedin, New Zealand; Mycology Unit, Department of Microbiology, Faculty of Medicine, Chulalongkorn University, Bangkok, Thailand

David S. Perlin Public Health Research Institute, New Jersey Medical School – Rutgers Biomedical and Health Sciences, Newark, NJ, USA

Attilio Di Pietro Institute of Protein Biology and Chemistry, MMSB, UMR 5086 CNRS-University of Lyon, Lyon, France

Rajendra Prasad Amity Institute of Integrative Sciences and Health and Institute of Biotechnology, Amity University Haryana, Gurgaon, Haryana, India

Jeanne Ropars Unité Biologie et Pathogénicité Fongiques, Institut Pasteur, INRA, Paris, France

Dominique Sanglard Institute of Microbiology, University of Lausanne and University Hospital Center, CH-1011 Lausanne, Switzerland

Kaustuv Sanyal Molecular Mycology Laboratory, Molecular Biology and Genetics Unit, Jawaharlal Nehru Centre for Advanced Scientific Research, Jakkur, Bangalore, India

Natacha Sertour Unité Biologie et Pathogénicité Fongiques, Institut Pasteur, INRA, Paris, France

Sudhanshu Shukla Amity Institute of Biotechnology, Amity University, Haryana, India

Ashutosh Singh Department of Biochemistry, Lucknow University, Lucknow, India

Rachna Singh Department of Microbial Biotechnology, Panjab University, Chandigarh, India

Emilie Sitterlé Unité Biologie et Pathogénicité Fongiques, Institut Pasteur, INRA, Paris, France; Cellule Pasteur, Univ. Paris Diderot, Sorbonne Paris Cité, Paris, France

Lakshmi Sreekumar Molecular Mycology Laboratory, Molecular Biology and Genetics Unit, Jawaharlal Nehru Centre for Advanced Scientific Research, Jakkur, Bangalore, India

Manjit Kumar Srivastav Laboratory of Eucaryotic Gene Regulation, School of Life Sciences, Jawaharlal Nehru University, New Delhi, India

Mark H.T. Stappers Aberdeen Fungal Group, MRC Centre for Medical Mycology, Institute of Medical Sciences, University of Aberdeen, Aberdeen, UK

Marc Swidergall Division of Infectious Diseases, Los Angeles Biomedical Research Institute, Harbor-UCLA Medical Center, Torrance, CA, USA

Priya Uppuluri Los Angeles Biomedical Research Institute, Harbor-UCLA, Torrance, CA, USA

Neha Varshney Molecular Mycology Laboratory, Molecular Biology and Genetics Unit, Jawaharlal Nehru Centre for Advanced Scientific Research, Jakkur, Bangalore, India

Louise A. Walker Aberdeen Fungal Group, Institute of Medical Sciences, University of Aberdeen, Aberdeen, UK

Chapter 1
Introduction

Rajendra Prasad

Almost 25 years have passed since the first version of this book was released (Prasad 1991). It is noteworthy that within this span of more than two decades, the importance of *Candida albicans* as a model system for studying pathogenic fungi has only multiplied. The advent of modern age scientific tools and techniques has made the pathogenic fungi readily amenable for genetic exploitations. Further, exploring the intricacies within its genomic landscape has also become much easier with the boon of Next-Gen sequencing methods. This has culminated in an increased understanding of the pathogen's genetic makeup and biology, virulence mechanisms and interaction with the host. The phenomenon of multidrug resistance in *C. albicans* and the related pathogenic species has taken toll on the clinicians because the management of fungal diseases has become extremely difficult. In order to explore alternate drug targets and develop modern age drugs and vaccines, thorough understanding of the pathogen's biology has become vital.

Considering the explosive growth in *Candida* research, it was a long time desire to collect almost all recent scientific developments made on the molecular and cellular aspects of this opportunistic human pathogen. Hopefully, this multiauthored book would be one of the most comprehensive and sought after compilation on the organism written by leading experts in areas encompassing cell biology, drug resistance, infection biology, host–pathogen interactions and drug discovery. The book will not only be of interest to mycologists but also to medical professionals and pharmacologists with interest in development of novel antifungals.

Two chapters, one on currents trends in candidiasis (Chap. 2) and the other with focus on invasive candidiasis with respect to southeast Asia (Chap. 3) lay the foundation for this compilation. In recent years, a general shift has been observed with regard to the species, which are commonly recovered from patients suffering

R. Prasad (✉)
Amity Institute of Integrative Sciences and Health and Institute of Biotechnology,
Amity University Haryana, Gurgaon, Haryana, India
e-mail: rprasad@ggn.amity.edu

R. Prasad (ed.), *Candida albicans: Cellular and Molecular Biology*,
DOI 10.1007/978-3-319-50409-4_1

from candidiasis. For instance, non albicans species such as *C. krusei* and *C. glabrata* have been found to be frequently associated with life threatening infections. Such a shift from azole susceptible to azole resistant strains is a cause of deep concern. When one considers the Southeast Asian region, *C. tropicalis* accounts for the maximum share. Interestingly, a new species *C. auris* has been in the news for outbreaks across the globe (Pfaller and Diekema 2007). Both the chapters discuss its status as well. Apart from focusing on the species spectrum, the chapters also deliberate on the diagnostic procedures and disease management options.

C. albicans has the ability to alter its morphology and undergo transition from yeast to the hyphal form. This property of the microbe has been found to be important for the overall biology of the pathogen as well as its pathogenesis. Chapter 4 focuses on the morphological attributes as well as the factors and pathways which govern it. *C. albicans* is present in the form of multicellular structures known as biofilms which can develop both on abiotic as well as biotic surfaces. These structures are much more resistant to antifungals and the immune system than the planktonic forms. While Chap. 5 discusses the characteristic features of *C. albicans* biofilm formation and the regulatory modules associated with it, the two subsequent chapters (#6 and 7) highlight the resilience of *Candida* biofilm capable of withstanding extraordinarily high antifungal concentrations, and discuss the animal models for studying it. *C. albicans* and *Staphylococcus aureus* form mix biofilms which is beneficial for both microbes as it leads to enhancement in their virulence potential and makes them much more resistant to the antimicrobials (Harriott and Noverr 2009). There are other examples as well where the interaction between fungi and bacteria is actually antagonistic (Mallick and Bennet 2013). Next chapter (# 8) particularly highlights the factors governing polymicrobial interactions.

It is of utmost importance to understand a disease from the perspective of both pathogen and the host to devise therapeutic strategies. Chapter 9 sheds light on the entities and mechanisms through which our innate and adaptive immune system recognizes and tackles the onslaught of *C. albicans* infections. This chapter also stresses on counter strategies adopted by the pathogen to evade the immune response. Next chapter underscores an account of the antimicrobial peptide (AMP) response of host to *C. albicans* infection and discusses their immunomodulatory and antifungal actions.

The easy access to high-throughput genomic and transcriptomic data is now changing the landscape of our understanding which was not feasible till recently. Chapters 11 and 12 highlight our understanding of the pathogen in the context of its genome. While the former focuses on the genetic diversity and genome dynamics in general, the latter reveals chromosomal elements responsible for the genome plasticity. The latter, however, is not only restricted to *C. albicans* but for a better perspective, includes discussion on other species as well.

A well-knitted regulatory network is required by a pathogen to thrive inside the host. The technologies available today for transcriptomic analyses are summarized in Chap. 13 and presents the information gathered so far with respect to gene expression and remodeling happening upon *C. albicans* infection. Signaling pathways play a central role in every biological process from survival to death. Pathogenesis too requires a dynamic signaling cascade to take place. Chapter 14 discusses two-component signal transduction pathway which plays a pivotal role in the pathogen's virulence mechanism.

Inappropriate usage of fungistatic (antifungals) compounds have led to the development (emergence) of multidrug resistance (MDR) phenomenon amongst pathogenic fungi. Their dynamic genome too has exacerbated the situation such that the need for newer antifungals is ever increasing (Morschhauser 2016). Chapter 15 throws light on the mechanisms of drug resistance in *C. albicans* and also describes the concept of antifungal tolerance mechanisms.

Contrary to the earlier belief, which considered cell walls to be simply the protective structure, it is now clear that cell wall of *C. albicans* plays a key role in its biology and pathogenesis. Apart from involvement in morphological growth, it also contains factors which trigger immune responses in the host. Chapter 16 brings the cell wall into focus highlighting its importance in drug resistance and also underlines it as an interesting target for improving therapeutic strategies.

Iron as we all know is an essential micronutrient which has recently been realized to play important role in fungal pathogenesis (Noble et al. 2010). Chapter 17 underscores the iron uptake systems, transcriptional circuitry associated with it and the role (connection) of iron homeostasis with multidrug resistance and lipid homeostasis in *Candida* cells. The chapter also presents an atlas enlisting genes involved in the iron acquisition, their expression profiles under iron depletion and association with virulence of the pathogen.

One of the key mechanism by which *C. albicans* develops multidrug resistance is through the over expression of membrane transporters encoding genes belonging to the ABC and MFS superfamilies. These promiscuous pump proteins, thus act as important pharmacological targets. However, their ability to accommodate a vast array of structurally and functionally unrelated molecules makes it difficult to develop inhibitors against them. Next chapter presents a detailed overview of the two prominent multidrug transporters Cdr1p and Mdr1p of *C. albicans* compiling all the data that has recently been gathered from mutational analyses while Chap. 19 summarizes the transcriptional control of *MDR1*. The close association of lipid biosynthetic pathways with that of the MDR and the fact that most of the available antifungal drugs target the lipid biosynthetic pathways, make them highly sought after biomolecules. Chap. 20 features the role of lipids in morphogenesis, virulence and trafficking/localization of MDR pump proteins. The chapter also underlines the current technologies deployed for generation of high-throughput lipidomics data. Chapter 21 provides a detailed description of the new targets that have been the hot spots for development of drugs. It also sheds light on the targets, which have shown potential to some extent, but needs deeper investigation to qualify for pharmacological interest.

The last three chapters are dedicated to the management of fungal infections. Chapter 22 provides a detailed account of the major antifungal drug classes and their key representatives, while Chaps. 23 and 24 focus on the development of inhibitors of MDR pump proteins which is of importance considering the fact that the major ploy of fungal pathogens is to over express these proteins for rapidly extruding out the administered drugs.

Together this multiauthored book compilation provides most recent advances in different dimensions of *C. albicans* research. Thus, it may serve as prospective handbook (reference book) for anyone and everyone interested in fungal pathogens. The book is also hoped to function as a base for searching alternative strategies for management of life threatening fungal infections.

Acknowledgements I am indebted to AT (Atanu Banerjee) for his unqualified help, but for him it would have been difficult to complete this project. I would like to thank all my contributors who readily accepted my invitation and for their forbearance whilst the final chapters were awaited.

References

Harriott MM, Noverr MC (2009) *Candida albicans* and *Staphylococcus aureus* form polymicrobial biofilms: effects on antimicrobial resistance. Antimicrob Agents Chemother 53:3914–3922

Mallick EM, Bennett RJ (2013) Sensing of the microbial neighborhood by *Candida albicans*. PLoS Pathog 9:e1003661

Morschhauser J (2016) The development of fluconazole resistance in Candida albicans—an example of microevolution of a fungal pathogen. J Microbiol 54(3): 192–201

Noble SM, French S, Kohn LA, Chen V, Johnson AD (2010) Systematic screens of a Candida albicans homozygous deletion library decouple morphogenetic switching and pathogenicity. Nat Genet 42(7):590–598

Pfaller MA, Diekema DJ (2007) Epidemiology of invasive candidiasis: a persistent public health problem. Clin Microbiol Rev 20:133–163

Prasad (1991) Introduction in *Candida albicans*: cellular and molecular biology, Prasad R (ed), pp 1–4

Chapter 2
Current Trends in Candidiasis

Priya Uppuluri, Afshin Khan and John E. Edwards

Abstract In the 1940s, coincident with the introduction of antibacterial antibiotics into therapeutic strategies, a gradual increase in the number of reports of serious, deep organ candidal infections began to appear in the medical literature. Since then, there has been an abrupt increase in the incidence of hematogenously disseminated candidiasis and its complications. Candidiasis now represents the third-to-fourth most frequent nosocomial infection in hospitals in the US and worldwide. An important reason for this is that *Candida* species have a propensity to infect devices, and there has been a highly significant increase in the implantation of devices including indwelling catheters, artificial hips, knees and shoulders and prosthetic heart components, in recent years. The gravity of the situation deepens with the emergence of novel multidrug resistant species of *Candida*, such as *C. auris*, and the inability of existing diagnostic techniques to identify such species. Here, we discuss the current trends in the field of Candidiasis: epidemiology, most common or unusual clinical manifestations, and latest development in the areas of immunology and diagnostic testing. While this chapter is not meant to be exhaustive in content related to *Candida*, we have attempted to inform most recent updates in the field.

2.1 Introduction

In the 1940s, coincident in timing with the introduction of antibacterial antibiotics into modern medical therapeutic strategies, a gradual increase in the number of reports of serious, deep organ candidal infections began to appear in the medical

P. Uppuluri (✉) · J.E. Edwards
Los Angeles Biomedical Research Institute, Harbor-UCLA, 1124 W. Carson St., Torrance 90502, CA, USA
e-mail: puppuluri@labiomed.org

J.E. Edwards
e-mail: jedwards@labiomed.org

A. Khan
Tulane University School of Medicine, 1430 Tulane Avenue, New Orleans 70112, LA, USA
e-mail: afshin.khan.2007@gmail.com

R. Prasad (ed.), *Candida albicans: Cellular and Molecular Biology*,
DOI 10.1007/978-3-319-50409-4_2

literature. It was during this time that widespread use of antibiotics was introduced, and since then, there has been an abrupt increase in the incidence of hematogenously disseminated candidiasis and its complications. Currently, there are approximately 2000 references per year on *Candida* and infections caused by the species. One important reason for this increase in infections is that *Candida* species have a propensity to infect devices, and there has been a highly significant increase in the implantation of devices including indwelling catheters, artificial hips, knees and shoulders and, prosthetic heart components.

Candidiasis now represents the third-to-fourth most frequent nosocomial infection in hospitals in the US and worldwide (Beck-Sague and Jarvis 1993; Edmond et al. 1999; Wisplinghoff et al. 2014; Wright and Wenzel 1997). The incidence of systemic candidiasis in the US is approximately 20 cases per 100,000 people (or about 60,000 cases per year) and in high-risk hospitalized patients this incidence increases by a factor of 50. Of note, these rates represent a 20-fold increase compared with just two decades ago, mostly as a result of an expanding population of immunocompromised patients (Edmond et al. 1999; Hajjeh et al. 2004; Viudes et al. 2002; Wright and Wenzel 1997). Disseminated candidiasis carries unacceptably high mortality rates, about 40–60%, even with treatment using antifungal agents. This high mortality may be due to poor diagnosis, inappropriate disease management, associated septic shock, or the general critical condition of the patient. The total estimated direct cost of candidiasis to the US health care system was ~ $2–4 billion yearly in the year 2000 (Wilson et al. 2002; Zaoutis et al. 2005b).

Deep organ, or disseminated candidiasis, is more likely to occur when the skin or gastrointestinal barriers are disrupted. The skin and mucosal infections may be more superficial and mild such as intertrigo, esophageal, and oropharyngeal candidiasis. In more serious situations, the superficial infections can advance toward an invasion of the bloodstream (blood stream infections) and dissemination to different organs in the body (hematogenously disseminated candidiasis). Invasive candidiasis (IC) itself includes severe complications such as endophthalmitis, meningitis, peritonitis, pancreatitis, endocarditis, arthritis, central nervous system infections, and osteomyelitis, besides others elaborated in their respective sections of this chapter.

Risk factors for invasive candidiasis include but are not limited to surgery, prolonged stay in an intensive care unit, severe burns, prior administration of broad-spectrum antibiotics and immunosuppressive agents, organ transplantation (especially liver), total parenteral nutrition, hemodialysis, antineoplastic chemotherapy, and catheter use (Pappas 2006; Bouza and Muñoz 2008; Playford et al. 2008). In addition to these risk factors, neonates and children are also susceptible if they are born premature, have low birth weight or any congenital malformations, or have low APGAR (American Pediatric Gross Assessment) score (Simonsen et al. 2014). An overall comprehensive picture of the pathogenesis of invasive candidiasis is explained elegantly in a review article by Kullberg and Arendrup (2015), and summarized in Fig. 2.1.

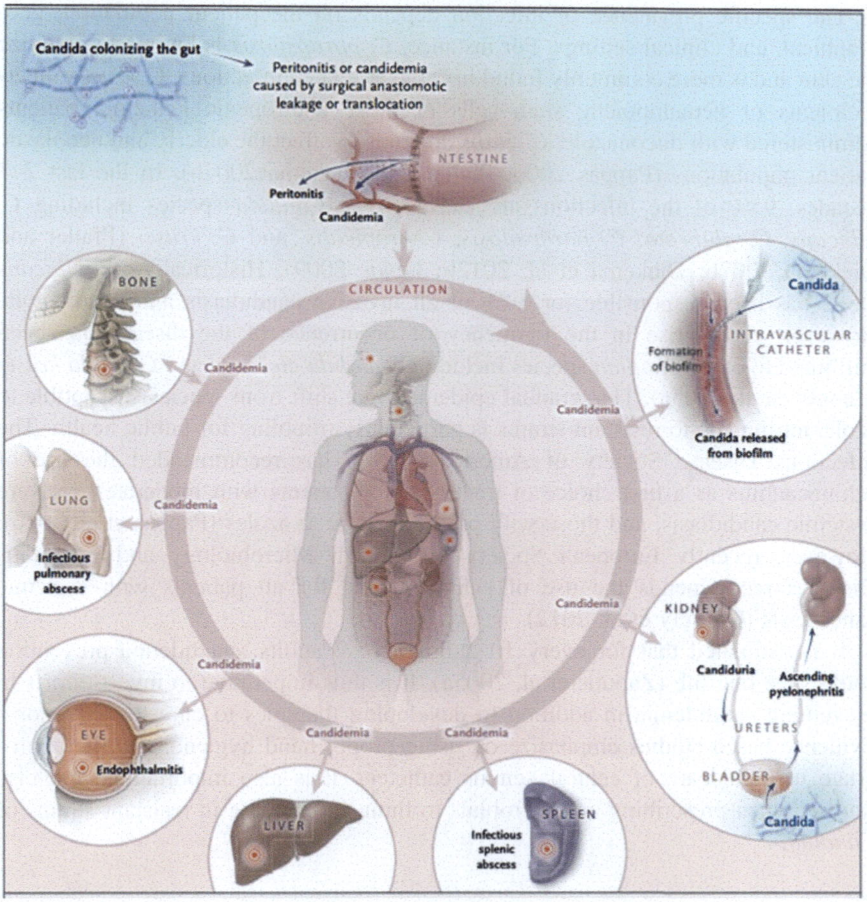

Fig. 2.1 Pathogenesis of invasive candidiasis (Courtesy Dr. Maiken C. Arendrup) (Kullberg and Arendrup 2015)

2.2 Epidemiology

C. albicans isolates have been recovered from soil, animals, hospital environments and food, and non-albicans spp may also be found in animal environments. Of the 150 known species of *Candida*, only 15 have caused infections in humans. These pathogens include *Candida albicans*, *Candida glabrata*, *Candida tropicalis*, *Candida parapsilosis*, *Candida krusei*, *Candida guilliermondii*, *Candida lusitaniae*, *Candida dubliniensis*, *Candida pelliculosa*, *Candida kefyr*, *Candida lipolytica*, *Candida famata*, *Candida inconspicua*, *Candida rugosa*, and *Candida norvegensis* (Pfaller et al. 2005). Of significance is the newly isolated species *C. auris* (http://www.cdc.gov/fungal/diseases/candidiasis/candida-auris-alert.html), which will be discussed below.

The specific prevalence of infection depends on the patient population, geographical, and clinical settings. For instance, *C. parapsilosis* is known to colonize the skin and is more commonly found in catheter related infections. *C. krusei* affects recipients of hematopoietic stem cells or those neutropenic leukemia patients administered with fluconazole. *C. glabrata* tends to affect the elderly and neoplastic patient populations (Pappas 2006; Pfaller and Diekema 2007b). In the last 2–3 decades, 95% of the infections are caused by 5 *Candida* species including *C. albicans*, *C. glabrata*, *C. parapsilosis*, *C. tropicalis*, and *C. krusei* (Pfaller and Diekema 2007b; Diekema et al. 2012b; Lewis 2009). Historically, *C. albicans* alone has been responsible for 50% of all invasive candidiasis and candidemia. However, recent rise in the frequency of occurrence of the disease has been attributed to other *Candida* species including *Candida glabrata* and *Candida krusei* (Zaoutis et al. 2005a). This gradual epidemiologic shift from species susceptible to azole antifungals to resistant strains is particularly troubling for public health. The Infectious Disease Society of America (IDSA) has recommended the use of echinocandins as a first choice of treatment for patients with moderate to severe systemic candidiasis, and those with prior exposure to azoles (Pappas et al. 2016). However, recently European Society of Clinical Microbiology and Infectious Diseases recommends the use of echinocandins for all patients with systemic candidiasis (Cornely et al. 2012).

It is estimated that for every 10 children or 7 adults, candidemia prevention could save one life (Zaoutis et al. 2005a). It is thus imperative to invest efforts in preventing candidemia in addition to developing therapies to cure the infections. Evidence-based studies emphasize on using proper hand hygiene, and the appropriate use and care of central venous catheters. It is also important to exercise caution when prescribing antimicrobials to limit the increase in resistant strains of *Candida*.

2.3 Emergence of *Candida auris*

The Centers for Disease Control and Prevention (CDC) has received reports from international healthcare facilities that *Candida auris*, an emerging multidrug-resistant (MDR) yeast, is causing invasive healthcare-associated infections with high mortality. *Candida auris* is a newly emerging form of the yeast, first described in 2009, when it was isolated from the ear discharge of a Japanese patient (Satoh et al. 2009). Since then, the organism has been reported across four continents in countries including India, Pakistan, South Korea, South Africa, Kuwait, UK, Venezuela, and Colombia (Pfaller and Diekema 2007a). It is not known why the species has emerged but it is likely that there are new selection pressures posed by humans, animals, or the environment for *C. auris* to evolve and increase its incidence over the years. Infections caused due to *C. auris* commonly occur in hospitals, several weeks into a patient's hospital stay. This species of fungus has been reported to cause bloodstream infections, wound infections, and ear infections

(Lee et al. 2011). While it has also been cultured from urine and the respiratory tract, it is unknown if isolation from these sites represented infection verses colonization. *C. auris* has been documented to cause infections in patients of all ages. Co-infection with other *Candida* spp. and detection of *C. auris* while the patient was being treated with antifungals have also been reported (Lee et al. 2011).

Diagnostic tests using traditional biochemical methods, API strips and VITEK-2 cannot differentiate *C. auris* from other *Candida* species and these isolates have been reported as "other *Candida* spp." http://www.cdc.gov/fungal/diseases/candidiasis/candida-auris-alert.html. The CDC suggests that the clinical, state, and public health laboratories should be aware of this organism and of the limitations in its identification.

C. auris is also extremely drug resistant. Although no established minimum inhibitory concentration (MIC) breakpoints exist for *C. auris*, resistance testing of an international collection of isolates conducted by CDC demonstrated that nearly all isolates are highly resistant to fluconazole, based on breakpoints established for other *Candida* spp. More than half of *C. auris* isolates were resistant to voriconazole, 1/3rd were resistant to amphotericin B (AmB; MIC greater than or equal to 2), and a few were resistant to echinocandins. Some isolates have demonstrated elevated MICs to all three major antifungal classes, including azoles, echinocandins, and polyenes, indicating that treatment options would be limited.

2.4 Recent Trends in Anatomical Distribution of Candidal Infections

As the frequency of diseases due to *Candida* has increased, a relatively large number of manifestations, that were previously either not recognized or extremely infrequent, have become well documented. We have discussed them in the following sections.

2.4.1 Intra-Abdominal Candidiasis (IAC)

Our knowledge about intra-abdominal candidiasis (IAC) is limited, as most epidemiological studies, diagnostic studies, and antifungal clinical trials have focused primarily on candidemia. Since IAC occurs in different clinical presentations, has poor clinical diagnosis coupled with the fact that there are no standardized definitions for the disease, the scope of the research in this field has been limited (Bassetti et al. 2013; Blot et al. 2007; Montravers et al. 2015). Some studies define IAC to be comprised of primary peritonitis, secondary peritonitis, biliary infections, intra-abdominal abscesses, or infected pancreatic necrosis. (Bassetti et al. 2013; Blot et al. 2007; Lamme et al. 2006).

IAC has poor prognosis and afflicts 35–40% of patients who have had gastrointestinal surgery, necrotizing pancreatitis or recurrent gastrointestinal perforation (De Ruiter et al. 2009). After candidemia, IAC is the most prevalent type of invasive candidiasis, especially patients in intensive care units (Leroy et al. 2009). In postoperative patients suffering from IAC, *C. albicans* is the most frequently isolated pathogen (65–82%) followed by *C. glabrata* (Dupont et al. 2002; Sandven et al. 2002). Associated mortality rates for IAC range between 25 and 60%, while 10% of all peritonitis cases are caused by IAC (Dupont et al. 2002; Montravers et al. 2006). Among all causes of IAC, the highest mortality rates are attributed to primary and secondary peritonitis (88 and 75% respectively) (Vergidis et al. 2016). High mortality of IAC is due to difficulties in diagnosing the disease, since cultures have low sensitivity and specificity. Substantial delays in obtaining results, both before and after IAC, further complicates the diagnosis. In recent years, there is increasing evidence to suggest that IAC could act as a hidden reservoir for echinocandin resistant *Candida* (Shields et al. 2014).

Clinical evidence for the use of antifungal therapy for patients with suspected intra-abdominal invasive candidiasis is limited. Source control with adequate drainage and/or debridement is an important part of therapy of intra-abdominal candidiasis (Pappas et al. 2016). Preferred empiric therapy for this infection is an echinocandin, especially for patients likely to be infected by *C. glabrata* or *C. krusei*. Studies suggest that initiating treatment is more important than the choice of antifungal therapy (Morrell et al. 2005; Garey et al. 2006; Clancy and Nguyen 2012). It is thus essential to perform blood-based non-culture techniques such as β-D-glucan and polymerase chain reaction assays to quickly diagnose patients with IAC as compared to conventional cultures of intra-abdominal specimens (Nguyen et al. 2012; Clancy and Nguyen 2013, 2014). Needless to say, there is a need for the infectious diseases community to invest its efforts in developing standardized protocols for managing IAC, besides carrying out early diagnostic tests, and conducting research studies to gain insights about IAC.

2.4.2 Hepatosplenic Disease

Hepatosplenic candidiasis (HSC) is also called chronic disseminated candidiasis. It mainly involves liver and spleen. The disease is more likely to occur in patients with severe and prolonged neutropenia, especially those suffering from acute leukemia. The incidence of the disease ranges from 3 to 29%, but in recent years, there has been a steady decline (Rammaert et al. 2012; Masood and Sallah 2005). This decrease in disease incidence is attributed to the use of prophylactic antifungal therapy. With the use of newer antifungal agents, mortality has reduced, from 74 to 21% overall (De Castro et al. 2012). Lipid formulations of AmB have demonstrated better efficacy, perhaps due to superior tissue concentrations (Gokhale et al. 1993; Masood and Sallah 2005; Sallah et al. 1999). A prophylactic treatment with fluconazole predisposes the patient to an increased risk of infection with a fluconazole-resistant organism. In such populations, a broader spectrum azole, or an echinocandin is considered more appropriate therapy (Cornely et al. 2007; De Castro et al. 2012; Lehrnbecher et al. 2010; Ostrosky-Zeichner et al. 2003; Poon et al. 2009; Rammaert et al. 2012). In recent years, corticosteroids have been used in addition to antifungal therapy. It is important to diagnose and treat HSC early as, any delays could result in negative patient outcomes (Legrand et al. 2008).

2.4.3 Neonatal Candidiasis

Close to four million neonates are born every year in the United States. About 11.4% of these babies are born preterm, 8% have low birth weight (LBW) and 1.4% are of very low birth weight (VLBW), bringing the cumulative sum of these delicate, susceptible population to 20.8% (~750,000 cases/yr) (Martin et al. 2015).

Of infants admitted to the NICU, 75% are colonized with *Candida* by the first month (Bendel 2005). Infection is acquired by; (1) vertical transmission during vaginal delivery; (2) postnatally from contact with maternal skin or the skin of direct care providers; or (3) direct transmissions via contaminated equipment or intravenous catheters. *Candida albicans* remains the most prominent pathogen in neonates, followed by significant cases due to *C. parapsilosis.* (Trofa et al. 2008; Hoffmann-Santos et al. 2013; Leibovitz et al. 2013) *Candida* infections are responsible for ~10–12% of nosocomial sepsis in VLBW (<1500 g) infants, with a collective incidence of up to 4% among all NICU admissions. (Botero-Calderon et al. 2015) In fact, *Candida* is the 3rd most frequently isolated organisms (after coagulase negative *Staphylococcus* spp. and *S. aureus*) in late onset sepsis in VLBW infants. (Bendel 2005) Despite empirical antifungal therapy, mortality related to the disease remains considerably high (20–30%), with even higher rates (59–73%) of long-term neurodevelopmental impairment in survivors. (Bendel 2005; Botero-Calderon et al. 2015).

Dosing of antifungal agents is substantially different for neonates than it is for older children and adults. Numerous studies examining fluconazole prophylaxis for the prevention of invasive candidiasis in neonates have consistently demonstrated efficacy and possibly reduced mortality (Kaufman et al. 2001; Manzoni et al. 2006, 2007) (for a complete list of these studies refer reference # (Pappas et al. 2016)). Enteral/orally administered nystatin has been shown to be effective in reducing invasive candidiasis in preterm infants (Howell et al. 2009; Violaris et al. 2010). Another antifungal drug, AmB deoxycholate (dose of 1 mg/kg daily) is also well tolerated in neonates without a high risk for nephrotoxicity (Benson and Nahata 1989). The duration of therapy is based primarily on adult and pediatric data, and there are no data to guide duration specifically in neonate (Pappas et al. 2016).

2.4.4 Genitourinary Candidiasis

The most frequent manifestations of genitourinary candidiasis include vulvovaginal candidiasis (VVC) in women, balanitis and balanoposthitis in men, and candiduria in both sexes. These diseases are remarkably common but occur in different populations, immunocompetent as well as immunocompromised. While VVC affects mostly healthy women, candiduria is commonly diagnosed in immunocompromised patients or neonates. In the majority of women, a diagnosis of VVC is made at least once during their childbearing years (Sobel et al. 1998). VVC is the second most common genital infection, after bacterial vaginosis and is diagnosed in up to 40% of women with vaginal complaints in the primary care setting (Anderson et al. 2004). *Candida* is also the most common infectious agent causing inflammation of the glans penis (Edwards 1996). In contrast to genital manifestations of candidiasis, candiduria is usually diagnosed in elderly hospitalized patients.

Over a decade ago, VVC was classified into uncomplicated (sporadic and infrequent) and complicated (recurrent and severe) cases, a classification that has been internationally accepted and adapted (Pappas et al. 2016; Sobel et al. 1998; Centers for Disease et al. 2006). Long-term suppressive antifungal therapy is commonly required to control complicated cases. However, recurrence often occur at rates of up to 50% after discontinuation of suppressive therapy (Sobel et al. 2004).

Candidal balanitis is defined as inflammation of the glans penis, often involving the prepuce (balanoposthitis), in the presence of *Candida* spp. and the absence of other infectious etiology. Candidal balanitis is generally sexually acquired and is often associated with the presences of diabetes (Edwards 1996). Diagnosis is based mostly on clinical appearance alone but should be confirmed by microscopy and/or culture if other differential diagnoses are considered.

A variety of topical and systemic oral agents are available for treatment of genital candidiasis. Uncomplicated infections can be effectively treated with either single-dose fluconazole or short-course fluconazole for 3 days, both of which achieve >90% response (Sobel et al. 1995; Watson et al. 2002). Complicated candidiasis requires that therapy be administered intravaginally with topical agents for 5–7 days or orally with fluconazole for the same duration (Sobel et al. 1998, 2001).

Women with type 2 diabetes mellitus (T2DM) are at increased risk for vaginal *Candida* colonization, perhaps because of glucosuria. Sodium glucose co-transporter 2 (SGLT2) inhibitors, in development for the treatment of T2DM, improve glycemic control by increasing urinary glucose excretion. Invokana (Canagliflozin) is one such drug. However, treatment with this drug was associated with an increase in genital candidiasis in both women and men suffering from T2DM (Nyirjesy et al. 2012).

2.4.5 Catheter-Associated Disseminated Candidiasis

Central venous catheters and other intravascular devices are important risk factors in the development and persistence of candidemia in non-neutropenic patients (Diekema et al. 2012a; Pfaller et al. 2012; Wagner et al. 2011). The rate of candidemia in such patients harboring central venous catheters (CVC) is between 50 and 70% (Diekema et al. 2012a; Pfaller et al. 2012; Wagner et al. 2011; Kett et al. 2011). The relationship of candidemia to CVCs has been assumed on the basis of observation, clinical experience, and an understanding of the role of biofilm in the genesis of bloodstream infections (Ruiz et al. 2013; Tumbarello et al. 2012). That candidemia in non-neutropenic patients is commonly due to contaminated CVCs is undeniable, but it is still unclear how best to distinguish a catheter-associated candidemia from one that is related to another source, such as the gastrointestinal tract. There have been no prospective clinical studies designed to examine CVC management as a primary measurement related to outcome. No prospective studies have been done on the relationship of early CVC removal and survival benefit to patients who have candidemia. However, several studies have demonstrated a shorter duration of candidemia and/or a trend toward improved outcomes (Garnacho-Montero et al. 2013; Lai et al. 2012; Kollef et al. 2012), also refer (Pappas et al. 2016), for a complete list of other related references.

2.5 Host Defense Against Candidiasis

A primary defense mechanism against *Candida* is intact skin and mucosal membranes. Any process causing skin disruption or mucosal harm leaves the involved site susceptible to *Candida* invasion, even in healthy individuals. In recent years, the importance of the dendritic cell for maintaining skin and mucosal integrity and as an immune effector cell has been recognized (d'Ostiani et al. 2000).

Figure 2.2, modified from Lionakis (Lionakis and Netea 2013), summarizes current topics regarding the cell surface pattern recognition receptors involved in the recognition of *Candida* by the innate immune system, and findings related to STAT1, Dectin-1, and CARD9.

STAT1p is a signal transducer protein that modulates cellular response to a variety of cytokines and growth factors (Lionakis 2012). Individuals having a deficiency in the production of the STAT1 protein display a proclivity for having a defect in their defense against mucocutaneous candidiasis. This deficiency results in insufficient production of interferon-γ, IL-17, and IL-22,—important regulators of a variety of inflammatory cells. Another population of patients with a propensity for recurrent episodes of mucocutaneous candidiasis are patients with Job Syndrome, who have varying degrees of deficiency in the STAT3 gene product and have cutaneous infections of both *Candida* and *Staphyloccus* (Chandesris et al. 2012).

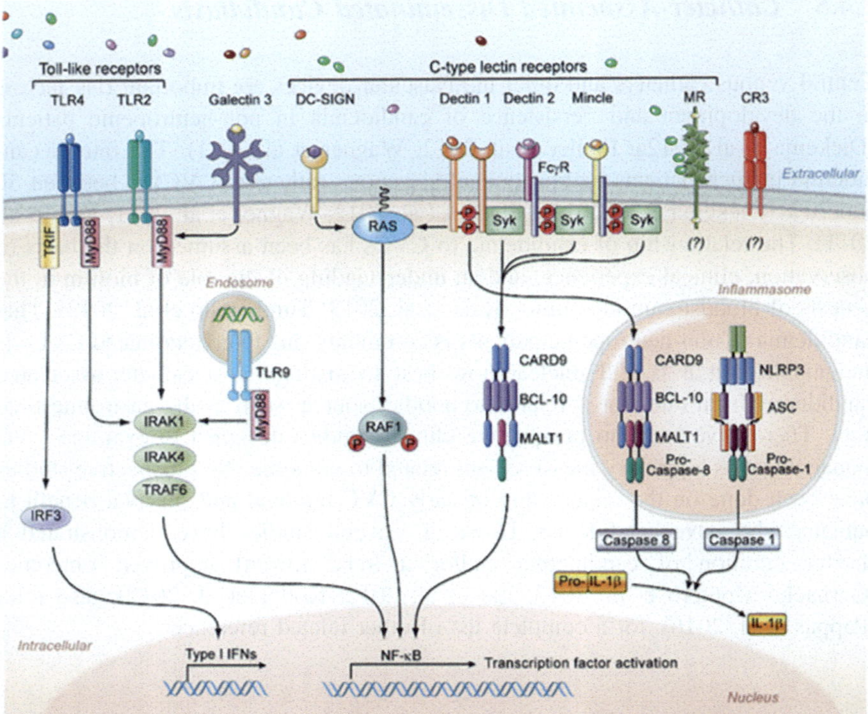

Fig. 2.2 Pathways involved in immune defense against *Candida* (*Courtesy* Dr. Mihailis Lionakis)

Dectin-1 has been found to be a major recognition factor for β-glucan on *Candida* and participates in the CARD9 pathway (Gow et al. 2007) (Fig. 2.2). Defective surface expression of Dectin-1 results in impaired cytokine response in monocytes and macrophages but not neutrophils. Homozygous mutations in CARD9 result in chronic mucocutaneous candidiasis, *Candida* brain abscess and deep dermatophytosis (Glocker et al. 2009).

Another area of intense evaluation has been the role of Th1 and Th17 cells in defense against invasion by *Candida* (Hernandez-Santos and Gaffen 2012). Several reports reveal that these cells play a role in defense against both mucosal and disseminated disease (Conti et al. 2009; Hernandez-Santos and Gaffen 2012; van de Veerdonk et al. 2011). In fact, vaccination with the N-terminus of *C. albicans* cell wall protein Als3, has been shown to increase IL-17 levels in humans (Schmidt et al. 2012; Spellberg et al. 2006).

Numerous other cells and components of the immune system (cytokines, defensins, platelets, complement components, macrophages and monocytes, Treg and natural killer (NK) cells, extracellular traps, Th2 cells and antibodies) play critical roles in defense against *Candida* damage (Edwards 2015; Romani 2000, 2004; Shoham and Levitz 2005). A discussion of these entities will be found in other chapters of this publication.

2.6 Diagnosis

Diagnosis of candidiasis cultures of blood or other samples collected under sterile conditions have long been considered diagnostic gold standards for invasive candidiasis. Nonculture diagnostic tests, such as antigen, antibody, or β-D-glucan detection assays, and polymerase chain reaction (PCR) are now entering clinical practice as adjuncts to cultures. If used and interpreted judiciously, these tests can identify more patients with invasive candidiasis and better direct antifungal therapy. To fully realize the benefits of combining culture and nonculture tests, however, clinicians must carefully consider the types of invasive candidiasis, understand the strengths and limitations of each assay, and interpret test results in the context of the clinical setting.

2.6.1 Use of Cultures for Diagnosis of Systemic Candidiasis

Invasive candidiasis encompasses 3 entities: candidemia in the absence of deep-seated candidiasis, candidemia associated with deep-seated candidiasis, and deep-seated candidiasis in the absence of candidemia (Clancy and Nguyen 2013). The overall sensitivity of blood cultures for diagnosing invasive candidiasis is roughly 50% (Clancy and Nguyen 2013). While blood cultures are positive during active *Candida* bloodstream infections, they may be negative in cases of extremely low-level or intermittent candidemia, or deep-seated candidiasis in the absence of candidemia. Cultures of tissues or fluid recovered from infected sites during deep-seated candidiasis also exhibit poor sensitivity.

2.6.2 Antigen and Antibody Detection

Candida antigen and anti-*Candida* antibody detection has gained greater acceptance in Europe, where it is approved for use. The best-studied test is a combined mannan/antimannan antibody assay (Platelia *Candida* Ag and Ab; BioRad). In a meta-analysis of 14 studies, the sensitivity/specificity for the diagnosis of invasive candidiasis of mannan and antimannan IgG individually were 58/93 and 59/83%, respectively (Mikulska et al. 2010). Values for the combined assay were 83 and

86%, with best performances for *C. albicans*, *C. glabrata*, and *C. tropicalis* infections. This assay is not used widely in the United States, and its role in the diagnosis and management of invasive candidiasis is unclear.

2.6.3 β-D-Glucan Detection

β-D-glucan is a cell wall constituent of *Candida* species, *Aspergillus* species, *Pneumocystis jiroveci*, and several other fungi. A serum β-D-glucan assay (Fungitell; Associates of Cape Cod, East Falmouth, Massachusetts) has been approved by the FDA as an adjunct to cultures for the diagnosis of invasive fungal infections. True-positive results are not specific for invasive candidiasis, but rather suggest the possibility of an invasive fungal infection. β-D-glucan detection can identify cases of invasive candidiasis days to weeks prior to positive blood cultures, and shorten the time to initiation of antifungal therapy. The major concern about β-D-glucan detection is the potential for poor specificity and false positivity, which may be particularly problematic in the patient populations for which nonculture diagnostics would be most helpful. For example, false-positive results are rare in healthy controls, but decidedly more common among patients in an ICU. For example, a routine surveillance β-D-glucan testing in a recent study of lung transplant recipients revealed sensitivity/specificity and positive/negative predictive values of 64/9 and 14/50%, respectively (Alexander et al. 2010).

2.6.4 Polymerase Chain Reaction

Candida PCR shares many of the potential benefits and shortcomings of β-D-glucan detection. Compared to cultures, PCR assays of various blood fractions have been shown to shorten the time to diagnosis of invasive candidiasis and initiation of antifungal therapy (Avni et al. 2011; McMullan et al. 2008). The pooled sensitivity and specificity of PCR for suspected invasive candidiasis in a recent meta-analysis were 95 and 92%, respectively (Avni et al. 2011). In probable invasive candidiasis, sensitivity of PCR and blood cultures was 85 and 38%, respectively. A major limitation of PCR studies is the lack of standardized methodologies and multicenter validation of assay performance. A multicenter US study assessing the performance of a self-contained instrument that amplifies and detects *Candida* DNA by PCR and T2 magnetic resonance (T2 Biosystems, Lexington, Massachusetts), respectively, has been completed (Mylonakis et al. 2015). This assay is FDA approved, but its role in the early diagnosis and management of candidemia remains unclear until more data are available. PCR has potential advantages over β-D-glucan or antigen-antibody assays, including the capacity for species identification, detection of molecular markers for drug resistance, and multiplex formatting. Furthermore, the role of PCR in testing samples other than blood is not established.

2.7 Summary

In summary, the incidence of nearly all forms of candidal infections is predicted to increase, due partially to the ever increasing number of devices being implanted. *Candida* infections are predominantly becoming healthcare-related infections, with immunocompromised milieu as primary populations suffering from these diseases. Newer approaches to the management of diabetes will likely result in an increase in genital/urinary tract *Candida* infections.

Predicted also is an increasing risk of resistance, not only of *C. albicans*, but also other species of *Candida*. It is likely that the intense interest in the molecular mechanisms of resistance, coupled with increased stewardship will result in strategies to minimize its development. Of recent concern is the emergence of *C. auris*, which is resistant to all antifungals.

In recent years, there has been an increase in abdominal/peritoneal candidiasis, which poses complex diagnostic challenges. For reasons that are not entirely clear, there has been a significant reduction in the presence of hepatosplenic candidiasis.

Substantial efforts have been sustained in developing diagnostic tests for hematogenously disseminated candidiasis. Presently, their value is most significant for their strong negative predictive value, resulting in the withdrawal of unnecessary antifungal treatment. The development of hematogenous *Candida* endophthalmitis is an important physical finding in establishing a 90% likelihood of the presence of microabscesses in the brain, heart and kidney as well, and in that sense, is a valuable diagnostic tool.

In general, *Candida* species have a very high propensity to spread hematogenously to the eye. The organism is likely the most common organism causing endopthalmitis in hospitalized patients. Its presence not only is a helpful diagnostic tool (indicating widespread dissemination, but also necessitates careful observation due to its capability of causing irreversible blindness. Therefore, the IDSA (Infectious Disease Society of America) currently recommends that every non-neutropenic patient with candidemia have a dilated ophthalmoscopic examination (Pappas et al. 2016). However, we would recommend that every patient with candidemia have a dilated ophthamoscopic exam. The typical lesion is an off-white, cotton ball looking lesion, projecting out into the vitreous, and frequently accompanied by a vitreal haze (Edwards 2015).

A significant roadblock in the elimination of *Candida* infections is the presence of drug resistant biofilms. Fungal cells within the biofilms display resistance to azoles and polyenes (Taff et al. 2013) and echinocandins likely achieve better results against *Candida* biofilms. Recently, some compounds with known anti-inflammatory properties have been investigated for their antifungal activity.

The interplay between fungus and host, i.e., immune system and inflammatory milieu, is crucial in determining the tolerance or the disease status (Romani 2004). Drugs displaying dual activity, antifungal and anti-inflammatory, could thus represent novel approaches to treat biofilm-related infections. After a significant hiatus, the pharmaceutical industry has renewed efforts to develop newer antifungals, including drugs in new classes. These efforts are in early stage clinical trials, and will be discussed in subsequent chapters on drug resistance in this book.

A vaccine to prevent or ameliorate mortality from *Candida* sepsis is currently in clinical trials. This vaccine is based on the recombinant N-terminis of Als3p. It has shown safety and a signal of efficacy for recurrent *Candida* vaginitis in early phase 1b/2a clinical trials (Edwards 2016). Currently there are no vaccines approved by regulatory agencies for any fungal infections.

References

Alexander BD, Smith PB, Davis RD, Perfect JR, Reller LB (2010) The (1,3){beta}-D-glucan test as an aid to early diagnosis of invasive fungal infections following lung transplantation. J Clin Microbiol 48:4083–4088

Anderson MR, Klink K, Cohrssen A (2004) Evaluation of vaginal complaints. JAMA 291:1368–1379

Avni T, Leibovici L, Paul M (2011) PCR diagnosis of invasive candidiasis: systematic review and meta-analysis. J Clin Microbiol 49:665–670

Bassetti M, Marchetti M, Chakrabarti A, Colizza S, Garnacho-Montero J, Kett DH, Munoz P, Cristini F, Andoniadou A, Viale P (2013) A research agenda on the management of intra-abdominal candidiasis: results from a consensus of multinational experts. Intensive Care Med 39:2092–2106

Beck-Sague C, Jarvis WR (1993) Secular trends in the epidemiology of nosocomial fungal infections in the United States, 1980-1990. National Nosocomial Infections Surveillance System. J Infect Dis 167:1247–1251

Bendel CM (2005) Nosocomial neonatal candidiasis. Pediatr Infect Dis J 24:831–832

Benson JM, Nahata MC (1989) Pharmacokinetics of amphotericin B in children. Antimicrob Agents Chemother 33:1989–1993

Blot SI, Vandewoude KH, de Waele JJ (2007) Candida peritonitis. Current opin crit care 13:195–199

Botero-Calderon L, Benjamin DK, Cohen-Wolkowiez M (2015) Advances in the treatment of invasive neonatal candidiasis. Expert Opin Pharmacother 16:1035–1048

Bouza E, Muñoz P (2008) Epidemiology of candidemia in intensive care units. Int J Antimicrob agents 32: S87–S91, 0924–8579

Chandesris MO, Melki I, Natividad A, Puel A, Fieschi C, Yun L, Thumerelle C, Oksenhendler E, Boutboul D, Thomas C, Hoarau C, Lebranchu Y, Stephan JL, Cazorla C, Aladjidi N, Micheau M, Tron F, Baruchel A, Barlogis V, Palenzuela G, Mathey C, Dominique S, Body G, Munzer M, Fouyssac F, Jaussaud R, Bader-Meunier B, Mahlaoui N, Blanche S, Debre M, le Bourgeois M, Gandemer V, Lambert N, Grandin V, Ndaga S, Jacques C, Harre C, Forveille M, Alyanakian MA, Durandy A, Bodemer C, Suarez F, Hermine O, Lortholary O, Casanova JL, Fischer A, Picard C (2012) Autosomal dominant STAT3 deficiency and hyper-IgE syndrome: molecular, cellular, and clinical features from a French national survey. Medicine (Baltimore) 91:e1–19

Clancy CJ, Nguyen MH (2012) The end of an era in defining the optimal treatment of invasive candidiasis. Clinical infectious diseases, cis023

Clancy CJ, Nguyen MH (2013) Finding the "missing 50%" of invasive candidiasis: how nonculture diagnostics will improve understanding of disease spectrum and transform patient care. Clin Infect Dis 56:1284–1292

Clancy CJ, Nguyen MH (2014) Undiagnosed invasive candidiasis: incorporating non-culture diagnostics into rational prophylactic and preemptive antifungal strategies. Expert review of anti-infective therapy 12:731–734

Conti HR, Shen F, Nayyar N, Stocum E, Sun JN, Lindemann MJ, Ho AW, Hai JH, Yu JJ, Jung JW, Filler SG, Masso-Welch P, Edgerton M, Gaffen SL (2009) Th17 cells and IL-17 receptor signaling are essential for mucosal host defense against oral candidiasis. J Exp Med 206:299–311

Cornely O, Bassetti M, Calandra T, Garbino J, Kullberg B, Lortholary O, Meersseman W, Akova M, Arendrup M, Arikan-Akdagli S (2012) ESCMID* guideline for the diagnosis and management of Candida diseases 2012: non-neutropenic adult patients. Clin Microbiol Infect 18:19–37

Cornely OA, Lasso M, Betts R, Klimko N, Vazquez J, Dobb G, Velez J, Williams-Diaz A, Lipka J, Taylor A, Sable C, Kartsonis N (2007) Caspofungin for the treatment of less common forms of invasive candidiasis. J Antimicrob Chemother 60:363–369

D'Ostiani CF, del Sero G, Bacci A, Montagnoli C, Spreca A, Mencacci A, Ricciardi-Castagnoli P, Romani L (2000) Dendritic cells discriminate between yeasts and hyphae of the fungus Candida albicans. Implications for initiation of T helper cell immunity in vitro and in vivo. J Exp Med 191:1661–1674

de Castro N, Mazoyer E, Porcher R, Raffoux E, Suarez F, Ribaud P, Lortholary O, Molina JM (2012) Hepatosplenic candidiasis in the era of new antifungal drugs: a study in Paris 2000–2007. Clin Microbiol Infect 18:E185–187

de Ruiter J, Weel J, Manusama E, Kingma W, van der Voort P (2009) The epidemiology of intra-abdominal flora in critically ill patients with secondary and tertiary abdominal sepsis. Infection 37:522–527

Diekema D, Arbefeville S, Boyken L, Kroeger J, Pfaller M (2012a) The changing epidemiology of healthcare-associated candidemia over three decades. Diagn Microbiol Infect Dis 73:45–48

Diekema D, Arbefeville S, Boyken L, Kroeger J, Pfaller M (2012b) The changing epidemiology of healthcare-associated candidemia over three decades. Diagn Microbiol Infect Dis 73: 45–48, 0732-8893

Dupont H, Paugam-Burtz C, Muller-Serieys C, Fierobe L, Chosidow D, Marmuse J-P, Mantz J, Desmonts J-M (2002) Predictive factors of mortality due to polymicrobial peritonitis with Candida isolation in peritoneal fluid in critically ill patients. Arch Surg 137:1341–1346

Edmond MB, Wallace SE, McClish DK, Pfaller MA, Jones RN, Wenzel RP (1999) Nosocomial bloodstream infections in United States hospitals: a three-year analysis. Clin Infect Dis 29:239–244

Edwards S (1996) Balanitis and balanoposthitis: a review. Genitourin Med 72:155–159

Garey KW, Rege M, Pai MP, Mingo DE, Suda KJ, Turpin RS, Bearden DT (2006) Time to initiation of fluconazole therapy impacts mortality in patients with candidemia: a multi-institutional study. Clin Infect Dis 43:25–31

Garnacho-Montero J, Diaz-Martin A, Garcia-Cabrera E, Ruiz Perez De Pipaon M, Hernandez-Caballero C, Lepe-Jimenez JA (2013) Impact on hospital mortality of catheter removal and adequate antifungal therapy in *candida* spp. bloodstream infections. J Antimicrob Chemother 68:206–213

Glocker EO, Hennigs A, Nabavi M, Schaffer AA, Woellner C, Salzer U, Pfeifer D, Veelken H, Warnatz K, Tahami F, Jamal S, Manguiat A, Rezaei N, Amirzargar AA, Plebani A, Hannesschlager N, Gross O, Ruland J, Grimbacher B (2009) A homozygous CARD9 mutation in a family with susceptibility to fungal infections. N Engl J Med 361:1727–1735

Gokhale PC, Barapatre RJ, Advani SH, Kshirsagar NA, Pandya SK (1993) Successful treatment of disseminated candidiasis resistant to amphotericin B by liposomal amphotericin B: a case report. J Cancer Res Clin Oncol 119:569–571

Gow NA, Netea MG, Munro CA, Ferwerda G, Bates S, Mora-Montes HM, Walker L, Jansen T, Jacobs L, Tsoni V, Brown GD, Odds FC, van der Meer JW, Brown AJ, Kullberg BJ (2007) Immune recognition of Candida albicans beta-glucan by dectin-1. J Infect Dis 196:1565–1571

Hajjeh RA, Sofair AN, Harrison LH, Lyon GM, Arthington-Skaggs BA, Mirza SA, Phelan M, Morgan J, Lee-Yang W, Ciblak MA, Benjamin LE, Sanza LT, Huie S, Yeo SF, Brandt ME, Warnock DW (2004) Incidence of bloodstream infections due to Candida species and in vitro susceptibilities of isolates collected from 1998 to 2000 in a population-based active surveillance program. J Clin Microbiol 42:1519–1527

Hernandez-Santos N, Gaffen SL (2012) Th17 cells in immunity to Candida albicans. Cell Host Microbe 11:425–435

Hoffmann-Santos HD, Paula CR, Yamamoto AC, Tadano T, Hahn RC (2013) Six-year trend analysis of nosocomial candidemia and risk factors in two intensive care hospitals in Mato Grosso, midwest region of Brazil. Mycopathologia 176:409–415

Howell A, Isaacs D, Halliday R (2009) The Australian study group for neonatal infections. Oral nystatin prophylaxis and neonatal fungal infections. Arch Dis Child Fetal Neonatal Ed 94: F429–33

John E. Edwards Jr (2015) *Candida* species. In: Bennett JE, Dolin R, Blaser MJ (eds) Principles and Practice of Infectious Diseases, 8th edn. Elsevier, Amsterdam

John E. Edwards Jr (2016) Candida vaccine trial for recurrent candida vaginitis exploratory 1b/2a trial preliminary 6 Month interim analysis. ASM Candida Candidiasis Meet, Seattle

Kaufman D, Boyle R, Hazen KC, Patrie JT, Robinson M, Donowitz LG (2001) Fluconazole prophylaxis against fungal colonization and infection in preterm infants. N Engl J Med 345:1660–1666

Kett DH, Azoulay E, Echeverria PM, Vincent JL (2011) For the extended prevalence of infection in the ICU study (EPIC II) group of investigators. Candida bloodstream infections in intensive care units: analysis of the extended prevalence of infection in intensive care unit study. Crit Care Med 39:665–670

Kollef M, Micek S, Hampton N, Doherty JA, Kumar A (2012) Septic shock attributed to Candida infection: importance of empiric therapy and source control. Clin Infect Dis 54:1739–1746

Kullberg BJ, Arendrup MC (2015) Invasive Candidiasis. N Engl J Med 373:1445–1456

Lai YC, Huang LJ, Chen TL, Yang YW, Hsiao LT, Teng HW, Fung CP, Chiou TJ, Tzeng CH, Liu CY (2012) Impact of Port-A-Cath device management in cancer patients with candidaemia. J Hosp Infect 82:281–285

Lamme B, Mahler CW, van Ruler O, Gouma DJ, Reitsma JB, Boermeester MA (2006) Clinical predictors of ongoing infection in secondary peritonitis: systematic review. World J Surg 30:2170–2181

Lee WG, Shin JH, Uh Y, Kang MG, Kim SH, Park KH, Jang HC (2011) First three reported cases of nosocomial fungemia caused by Candida auris. J Clin Microbiol 49:3139–3142

Legrand F, Lecuit M, Dupont B, Bellaton E, Huerre M, Rohrlich P-S, Lortholary O (2008) Adjuvant corticosteroid therapy for chronic disseminated candidiasis. Clin Infect Dis 46:696–702

Lehrnbecher T, Attarbaschi A, Duerken M, Garbino J, Gruhn B, Kontny U, Luer S, Phillips R, Scholz J, Wagner HJ, Wiesel T, Groll AH (2010) Posaconazole salvage treatment in paediatric patients: a multicentre survey. Eur J Clin Microbiol Infect Dis 29:1043–1045

Leibovitz E, Livshiz-Riven I, Borer A, Taraboulos-Klein T, Zamir O, Shany E, Melamed R, Rimon OF, Bradenstein R, Chodick G, Golan A (2013) A prospective study of the patterns and dynamics of colonization with candida spp. in very low birth weight neonates. Scand J Infect Dis 45:842–848

Leroy O, Gangneux J-P, Montravers P, Mira J-P, Gouin F, Sollet J-P, Carlet J, Reynes J, Rosenheim M, Regnier B (2009) Epidemiology, management, and risk factors for death of

invasive Candida infections in critical care: a multicenter, prospective, observational study in France (2005–2006). Crit Care Med 37:1612–1618
Lewis RE (2009) Overview of the changing epidemiology of candidemia. Curr Med Res Opin 25:1732–1740
Lionakis MS (2012) Genetic susceptibility to fungal infections in humans. Curr Fungal Infect Rep 6:11–22
Lionakis MS, Netea MG (2013) Candida and host determinants of susceptibility to invasive candidiasis. PLoS Pathog 9:e1003079
Manzoni P, Arisio R, Mostert M, Leonessa M, Farina D, Latino MA, Gomirato G (2006) Prophylactic fluconazole is effective in preventing fungal colonization and fungal systemic infections in preterm neonates: a single-center, 6-year, retrospective cohort study. Pediatrics 117:e22–32
Manzoni P, Stolfi I, Pugni L, Decembrino L, Magnani C, Vetrano G, Tridapalli E, Corona G, Giovannozzi C, Farina D, Arisio R, Merletti F, Maule M, Mosca F, Pedicino R, Stronati M, Mostert M, Gomirato G (2007) For the Italian task force for the study and prevention of neonatal fungal infections and the Italian society of neonatology. A multicenter, randomized trial of prophylactic fluconazole in preterm neonates. N Engl J Med 356: 2483–2495
Martin JA, Hamilton BE, Osterman MJ, Curtin SC, Matthews TJ (2015) Births: final data for 2013. Natl Vital Stat Rep 64:1–65
Masood A, Sallah S (2005) Chronic disseminated candidiasis in patients with acute leukemia: emphasis on diagnostic definition and treatment. Leuk Res 29:493–501
McMullan R, Metwally L, Coyle PV, Hedderwick S, McCloskey B, O'Neill HJ, Patterson CC, Thompson G, Webb CH, Hay RJ (2008) A prospective clinical trial of a real-time polymerase chain reaction assay for the diagnosis of candidemia in nonneutropenic, critically ill adults. Clin Infect Dis 46:890–896
Mikulska M, Calandra T, Sanguinetti M, Poulain D, Viscoli C (2010) Third European Conference on Infections in Leukemia Group. The use of mannan antigen and anti-mannan antibodies in the diagnosis of invasive candidiasis: recommendations from the Third European Conference on Infections in Leukemia. Crit Care 14:R222
Montravers P, Dupont H, Gauzit R, Veber B, Auboyer C, Blin P, Hennequin C, Martin C (2006) Candida as a risk factor for mortality in peritonitis. Crit Care Med 34:646–652
Montravers P, Leroy O, Eckmann C (2015) Intra-abdominal candidiasis: it's still a long way to get unquestionable data. Intensive Care Med 41:1682–1684
Morrell M, Fraser VJ, Kollef MH (2005) Delaying the empiric treatment of Candida bloodstream infection until positive blood culture results are obtained: a potential risk factor for hospital mortality. Antimicrob Agents Chemother 49:3640–3645
Mylonakis E, Clancy CJ, Ostrosky-Zeichner L, Garey KW, Alangaden GJ, Vazquez JA, Groeger JS, Judson MA, Vinagre YM, Heard SO, Zervou FN, Zacharioudakis IM, Kontoyiannis DP, Pappas PG (2015) T2 magnetic resonance assay for the rapid diagnosis of candidemia in whole blood: a clinical trial. Clin Infect Dis 60:892–899
Nguyen MH, Wissel MC, Shields RK, Salomoni MA, Hao B, Press EG, Shields RM, Cheng S, Mitsani D, Vadnerkar A (2012) Performance of Candida real-time polymerase chain reaction, β-D-glucan assay, and blood cultures in the diagnosis of invasive candidiasis. Clin Infect Dis 54:1240–1248
Nyirjesy P, Zhao Y, Ways K, Usiskin K (2012) Evaluation of vulvovaginal symptoms and Candida colonization in women with type 2 diabetes mellitus treated with canagliflozin, a sodium glucose co-transporter 2 inhibitor. Curr Med Res Opin 28:1173–1178
Ostrosky-Zeichner L, Oude Lashof AM, KULLBERG BJ, Rex JH (2003) Voriconazole salvage treatment of invasive candidiasis. Eur J Clin Microbiol Infect Dis 22:651–655
Pappas PG (2006) Invasive candidiasis. Infect Dis Clin North Am 20:485–506
Pappas PG, Kauffman CA, Andes DR, Clancy CJ, Marr KA, Ostrosky-Zeichner L, Reboli AC, Schuster MG, Vazquez JA, Walsh TJ, Zaoutis TE, Sobel JD (2016) Clinical practice guideline for the management of candidiasis: 2016 update by the infectious diseases society of america. Clin Infect Dis 62:e1–50

Pfaller M, Neofytos D, Diekema D, Azie N, Meier-Kriesche HU, Quan SP, Horn D (2012) Epidemiology and outcomes of candidemia in 3648 patients: data from the Prospective Antifungal Therapy (PATH Alliance(R) registry, 2004–2008. Diagn Microbiol Infect Dis 74:323–331

Pfaller MA, Boyken L, Hollis RJ, Messer SA, Tendolkar S, Diekema DJ (2005) In vitro activities of anidulafungin against more than 2500 clinical isolates of candida spp, including 315 isolates resistant to fluconazole. J clin microbiol 43: 5425–5427, 0095–1137

Pfaller MA, Diekema DJ (2007a) Epidemiology of invasive candidiasis: a persistent public health problem. Clin Microbiol Rev 20:133–163

Pfaller MA, Diekema DJ (2007b) Epidemiology of invasive candidiasis: a persistent public health problem. Clinical microbiology reviews 20: 133–163, 0893–8512

Playford EG, Marriott D, Nguyen Q, Chen S, Ellis D, Slavin M, Sorrell TC (2008) Candidemia in nonneutropenic critically ill patients: risk factors for non-albicans candida spp. Crit care med, 36: 2034–2039, 0090–3493

Poon LM, Chia HY, Tan LK, Liu TC, Koh LP (2009) Successful intensive chemotherapy followed by autologous hematopoietic cell transplantation in a patient with acute myeloid leukemia and hepatosplenic candidiasis: case report and review of literature. Transpl Infect Dis 11:160–166

Rammaert B, Desjardins A, Lortholary O (2012) New insights into hepatosplenic candidosis, a manifestation of chronic disseminated candidosis. Mycoses 55:e74–84

Romani L (2000) Innate and adaptive immunity in Candida albicans infections and saprophytism. J Leukoc Biol 68:175–179

Romani L (2004) Immunity to fungal infections. Nat Rev Immunol 4:1–23

Ruiz LS, Khouri S, Hahn RC, da Silva EG, de Oliveira VK, Gandra RF, Paula CR (2013) Candidemia by species of the Candida parapsilosis complex in children's hospital: prevalence, biofilm production and antifungal susceptibility. Mycopathologia 175:231–239

Sallah S, Semelka RC, Sallah W, Vainright JR, Philips DL (1999) Amphotericin B lipid complex for the treatment of patients with acute leukemia and hepatosplenic candidiasis. Leuk Res 23:995–999

Sandven P, Qvist H, Skovlund E, Giercksky KE (2002) Significance of Candida recovered from intraoperative specimens in patients with intra-abdominal perforations. Crit Care Med 30:541–547

Satoh K, Makimura K, Hasumi Y, Nishiyama Y, Uchida K, Yamaguchi H (2009) Candida auris sp. nov., a novel ascomycetous yeast isolated from the external ear canal of an inpatient in a Japanese hospital. Microbiol Immunol 53:41–44

Schmidt CS, White CJ, Ibrahim AS, Filler SG, Fu Y, Yeaman MR, Edwards JE, Hennessey JP (2012) NDV-3, a recombinant alum-adjuvanted vaccine for Candida and Staphylococcus aureus, is safe and immunogenic in healthy adults. Vaccine 30:7594–7600

Shields RK, Nguyen MH, Press EG, Clancy CJ (2014) Abdominal candidiasis is a hidden reservoir of echinocandin resistance. Antimicrob Agents Chemother 58:7601–7605

Shoham S, Levitz SM (2005) The immune response to fungal infections. Br J Haematol 129:569–582

Simonsen KA, Anderson-Berry AL, Delair SF, Davies HD (2014) Early-onset neonatal sepsis. Clin Microbiol Rev 27:21–47

Sobel JD, Brooker D, Stein GE, Thomason JL, Wermeling DP, Bradley B, Weinstein L (1995) Single oral dose fluconazole compared with conventional clotrimazole topical therapy of Candida vaginitis. Fluconazole vaginitis study group. Am J Obstet Gynecol 172:1263–1268

Sobel JD, Faro S, Force RW, Foxman B, Ledger WJ, Nyirjesy PR, Reed BD, Summers PR (1998) Vulvovaginal candidiasis: epidemiologic, diagnostic, and therapeutic considerations. Am J Obstet Gynecol 178:203–211

Sobel JD, Kapernick PS, Zervos M, Reed BD, Hooton T, Soper D, Nyirjesy P, Heine MW, Willems J, Panzer H, Wittes H (2001) Treatment of complicated Candida vaginitis: comparison of single and sequential doses of fluconazole. Am J Obstet Gynecol 185:363–369

Sobel JD, Wiesenfeld HC, Martens M, Danna P, Hooton TM, Rompalo A, Sperling M, Livengood C, Horowitz B, Von Thron J, Edwards L, Panzer H, Chu TC (2004) Maintenance fluconazole therapy for recurrent vulvovaginal candidiasis. N Engl J Med 351:876–883

Spellberg BJ, Ibrahim AS, Avanesian V, Fu Y, Myers C, Phan QT, Filler SG, Yeaman MR, Edwards JE (2006) Efficacy of the anti-Candida rAls3p-N or rAls1p-N vaccines against disseminated and mucosal candidiasis. J Infect Dis 194:256–260

Taff HT, Mitchell KF, Edward JA, Andes DR (2013) Mechanisms of candida biofilm drug resistance. Future Microbiol 8:1325–1337

Trofa D, Gacser A, Nosanchuk JD (2008) Candida parapsilosis, an emerging fungal pathogen. Clin Microbiol Rev 21:606–625

Tumbarello M, Fiori B, Trecarichi EM, Posteraro P, Losito AR, de Luca A, Sanguinetti M, Fadda G, Cauda R, Posteraro B (2012) Risk factors and outcomes of candidemia caused by biofilm-forming isolates in a tertiary care hospital. PLoS ONE 7:e33705

van de Veerdonk FL, Joosten LA, Shaw PJ, Smeekens SP, Malireddi RK, van der Meer JW, Kullberg BJ, Netea MG, Kanneganti TD (2011) The inflammasome drives protective Th1 and Th17 cellular responses in disseminated candidiasis. Eur J Immunol 41:2260–2268

Vergidis P, Clancy CJ, Shields RK, Park SY, Wildfeuer BN, Simmons RL, Nguyen MH (2016) Intra-abdominal candidiasis: the importance of early source control and antifungal treatment. PLoS ONE 11:e0153247

Violaris K, Carbone T, Bateman D, Olawepo O, Doraiswamy B, Lacorte M (2010) Comparison of fluconazole and nystatin oral suspensions for prophylaxis of systemic fungal infection in very low birthweight infants. Am J Perinatol 27:73–78

Viudes A, Peman J, Canton E, Ubeda P, Lopez-Ribot JL, Gobernado M (2002) Candidemia at a tertiary-care hospital: epidemiology, treatment, clinical outcome and risk factors for death. Eur J Clin Microbiol Infect Dis 21:767–774

Wagner M, Bonhoeffer J, Erb TO, Glanzmann R, Hacker FM, Paulussen M, Weibel D, Heininger U (2011) Prospective study on central venous line associated bloodstream infections. Arch Dis Child 96:827–831

Watson MC, Grimshaw JM, Bond CM, Mollison J, Ludbrook A (2002) Oral versus intra-vaginal imidazole and triazole anti-fungal agents for the treatment of uncomplicated vulvovaginal candidiasis (thrush): a systematic review. BJOG 109:85–95

Wilson LS, Reyes CM, Stolpman M, Speckman J, Allen K, Beney J (2002) The direct cost and incidence of systemic fungal infections. Value Health 5:26–34

Wisplinghoff H, Ebbers J, Geurtz L, Stefanik D, Major Y, Edmond MB, Wenzel RP, Seifert H (2014) Nosocomial bloodstream infections due to candida spp. in the USA: species distribution, clinical features and antifungal susceptibilities. Int J Antimicrob Agents 43:78–81

Workowski KA, Berman SM (2006) Centers for disease control and prevention. Sexually transmitted diseases treatment guidelines. MMWR Recomm Rep 55: 1–94

Wright WL, Wenzel RP (1997) Nosocomial candida. epidemiology, transmission, and prevention. Infect Dis Clin North Am 11:411–425

Zaoutis TE, Argon J, Chu J, Berlin JA, Walsh TJ, Feudtner C (2005a) The epidemiology and attributable outcomes of candidemia in adults and children hospitalized in the United States: a propensity analysis. Clinical infectious diseases 41: 1232–1239, 1058–4838

Zaoutis TE, Argon J, Chu J, Berlin JA, Walsh TJ, Feudtner C (2005b) The epidemiology and attributable outcomes of candidemia in adults and children hospitalized in the United States: a propensity analysis. Clin Infect Dis 41:1232–1239

Chapter 3
Invasive Candidiasis in the Southeast-Asian Region

Rachna Singh and Arunaloke Chakrabarti

Abstract Invasive fungal infections (IFIs) are being reported at a rising frequency worldwide. Amongst the opportunistic mycoses, invasive candidiasis, aspergillosis and mucormycosis are the common infections reported from Asia. Although the precise epidemiology of IFIs in Asia remains largely unknown and the magnitude of the problem is not available for all countries, the existing literature suggests a unique epidemiology with respect to the spectrum of agents, risk factors and disease pattern. The present chapter focuses on distinct peculiarities of invasive candidiasis in South-Asian region, highlighting epidemiology, diagnosis and management modalities. The projected prevalence of invasive candidiasis in South-Asian countries (1–12 cases/1000 hospital admissions) is nearly 20–30 times higher than the developed world. Several reasons have been predicted for this high prevalence. The compromise in health care in overpopulated public sector hospitals is the one of the possible reasons, as *Candida* hand carriage rates are significantly high among health-care workers. For the same reason, nosocomial outbreaks have been reported due to rare yeasts like *Kodamaea ohmeri* and *Pichia anomala.* Candidemia remains the predominant clinical manifestation. Intra-abdominal candidiasis and *Candida* pancreatitis are emerging problems. Nearly 70–90% of the invasive candidiasis are caused by non-*albicans Candida* species, with *Candida tropicalis* being the most common (35–40%) in the tropical regions. A new multidrug-resistant clonal *Candida auris* has emerged recently in India, causing infection in critically ill patients. The rising azole resistance in so-called susceptible species, *C. albicans* and *C. tropicalis*, is a cause of concern. Echinocandin resistant *C. glabrata* is also

R. Singh
Department of Microbial Biotechnology, Panjab University, Chandigarh 160014, India
e-mail: rachna.singh@pu.ac.in

A. Chakrabarti (✉)
Department of Medical Microbiology, Postgraduate Institute of Medical Education and Research, Centre of Advance Research in Medical Mycology & WHO Collaborating Centre, Chandigarh 160012, India
e-mail: arunaloke@hotmail.com

R. Prasad (ed.), *Candida albicans: Cellular and Molecular Biology*,
DOI 10.1007/978-3-319-50409-4_3

reported. Diagnosis and management of invasive candidiasis remains a challenge in South-Asian settings, owing to high *Candida* colonization rates, late presentation to the hospitals, limited availability of newer diagnostic techniques and poor affordability of antifungal agents. Although echinocandin is the preferred first-line drug for invasive candidiasis, fluconazole and amphotericin B deoxycholate are prescribed in majority of critically ill and unstable patients owing to cost constraints.

Keywords Invasive candidiasis · Candidemia · India · South Asia · Southeast Asia

3.1 Introduction

Fungal infections are a public health concern worldwide, and are being increasingly reported from both developed and developing countries. Patients with invasive fungal infections (IFIs) have high morbidity and mortality rates. The environment, host as well as social factors contribute to the high prevalence of fungal infections in Asian countries. Majority of South-Asian countries lie in the tropical/subtropical zone, where fungi thrive easily in the environment and patients are exposed to high burden of fungi both in hospital and home. Socially, the population is bifurcated in two groups: (i) a large, economically weaker section with improper hygiene, sanitation and poor access to health-care facilities and (ii) a limited privileged section with access to modern medical management including transplants and intensive cancer therapy. Both these groups are prone to invasive fungal infections for different reasons. Furthermore, the rampant drug abuse, misuse of steroids and broad-spectrum antibiotics and spurious health-care providers (quacks) play additional roles in making the population susceptible to fungal diseases. Unabated construction activities in the hospital without protective coverage from the patient area, climate change and natural disasters also contribute to the rise of fungal diseases in this region. Although the exact burden of IFIs in South Asia is largely unknown due to lack of awareness and diagnostic facilities, the limited available literature suggests a high incidence and a unique epidemiology of systemic mycoses with respect to the population at risk, disease pattern and the spectrum of fungal species involved (Chakrabarti and Singh 2011; Chakrabarti 2013a).

IFIs are broadly classified into endemic and opportunistic mycoses. Histoplasmosis, penicilliosis, blastomycosis and sporotrichosis are the endemic mycoses present in South Asia. While histoplasmosis (pocket distribution) and penicilliosis (endemic in Southeast and Eastern Asia) are predominantly associated with AIDS patients, blastomycosis is relatively rare though the fungus has been isolated from patients and bats in India. Sporotrichosis is reported from few countries of South Asia. Coccidioidomycosis (few cases) has been reported as imported mycoses in India (Khan et al. 1982; Randhawa et al. 1985; Padhye et al.

1994; Chakrabarti 2013a; Chakrabarti and Padhye 2013). Amongst the opportunistic IFIs, candidiasis, aspergillosis and mucormycosis are predominant diseases. Despite availability of antiretroviral therapy, the incidence of cryptococcosis and pneumocystosis has not come down significantly in AIDS patients. Scedosporiosis, fusariosis and pythiosis are reported occasionally (Chakrabarti and Singh 2011; Chakrabarti 2013a).

The prevalence of candidiasis in South Asia ranges from 0.16 to 4.53 cases per 1000 discharges in hospitals and 11.7 per 1000 discharges in intensive care units (ICUs), a rate which is nearly 20–30 times higher than the western world (Chakrabarti 2013a, b; Tan et al. 2015). Alike other IFIs, certain peculiarities have been observed in the epidemiology of invasive candidiasis in South Asia. Nearly 70–90% of these infections are caused by non-*albicans Candida* (NAC) species, with *Candida tropicalis* being the most common species (35–40%) in the tropical regions. *Candida* hand carriage rates are significantly high among health-care workers (45–80%). Intra-abdominal candidiasis and *Candida* pancreatitis are emerging problems. Nosocomial outbreaks have been reported due to rare yeasts like *Kodamaea ohmeri* (teleomorph of *Candida guilliermondii* var. *membranaefaciens*), *Pichia anomala* (anamorph of *Candida pelliculosa*) and *Candida auris* (Chakrabarti 2013a, b; Chen 2013; Chowdhary et al. 2013). The present chapter focuses on these peculiarities of invasive candidiasis in the South-Asian region, with an emphasis on the risk factors, disease pattern, spectrum of *Candida* species involved and challenges in diagnosis as well as management.

3.2 Disease Spectrum

Candida species are common commensals of the human body. These fungi inhabit the mucosa of gastrointestinal tract, upper respiratory tract and vagina, and occasionally the skin. Invasive candidiasis is often endogenous in origin, unlike most of the other mycoses. Candidiasis ranges from superficial infection, encompassing the skin and mucous membrane, to invasive disease, involving the bloodstream and/or other body organs either through secondary dissemination or direct inoculation of the fungus at sterile body sites. The major clinical manifestations of invasive candidiasis include candidemia, abdominal candidiasis, endocarditis, endophthalmitis, meningitis, urinary tract infection, osteomyelitis, hepatosplenic candidiasis and other disseminated forms (Kullberg and Arendrup 2015; Pappas et al. 2016).

Candidemia remains the most common clinical entity in hospitalized patients throughout the world. It is the fourth leading cause of nosocomial blood stream infections in the USA hospitals and the third most common in its ICUs, accounting for 9% of all infections with high crude mortality (42%). In Europe, candidemia due to *C. albicans* and NAC species accounted for 40 and 42% mortality, respectively, across hospitals, and 34 and 28% attributable mortality, respectively, in ICUs

(Vincent et al. 2006; Pfaller and Diekema 2007; Todi 2013). Candidemia is also the most frequent form of invasive candidiasis in South Asia, with an estimated 30-day crude and attributable mortality of 44.7 and 19.8%, respectively, in the Indian ICUs (Chakrabarti et al. 2015). *Candida* accounts for 8.2% of device-related infections (Todi 2013).

Abdominal candidiasis, especially pancreatitis, is an emerging infection in South Asia. It generally results from the invasion of *Candida* across gastrointestinal tract mucosa during abdominal surgery or chemotherapy. Additional risk factors include severe injury to the pancreas and prophylactic fluconazole. In an Indian study, nearly 12% of the patients with acute pancreatitis (41 out of 335 cases) were infected with *Candida*. True infection (isolation from pancreatic necrotic tissue) was observed in 22 (6.6%) and possible infection in 19 (5.7%) patients. Nineteen patients with a clinical diagnosis of acute pancreatitis were positive for *Candida* blood cultures (Chakrabarti et al. 2007).

Candida central nervous system infections, such as meningitis and brain abscess, occur in patients at extremes of age. Meningitis is more frequent in infants than older patients. Rarely, *Candida* meningitis may occur in an otherwise healthy person. The development of meningitis or mass lesion in the brain due to *Candida* spp. depends on multiple factors that control local proliferation and access of *Candida* to the central nervous system. One such example is pregnancy, when *Candida* vaginal populations flourish. When a mother with dense *Candida* vaginal growth delivers a premature infant, especially with defects in neural tube, rapid infection of the meninges may occur. Neurosurgery has also been linked with an increased risk of *Candida* meningitis (Chakrabarti 2007; Sundaram et al. 2006; Taneja et al. 2009).

Candida spp. also account for nearly 50% of the cases of fungal endocarditis, a relatively uncommon infection, which is increasing in recent years partly due to the increased use of prosthetic intravascular devices. In a single-institutional Indian study over a 14-year period, 14 cases of *Candida* endocarditis were reported. Twelve of these cases were health-care associated, and the infection was predominantly right-sided valvular infection (Vaideeswar 2015).

Candiduria is common among hospitalized patients in South Asia, especially those with indwelling bladder catheters (Jain et al. 2011; Yashavanth et al. 2013). However, a true invasive *Candida* infection of the lower and/or upper urinary tract needs to be distinguished from mere colonization of lower urinary tract (Singla et al. 2012). Asymptomatic candiduria often does not require antifungal treatment except in case of very low-birth-weight infants, high-risk groups, e.g. neutropenic patients, or those undergoing urological manipulation. The major manifestations of invasive candidiasis in urinary tract that require active management are cystitis and ascending pyelonephritis (Pappas et al. 2016).

Ocular infections, such as keratitis and endophthalmitis, are primarily caused by filamentous-fungi like *Aspergillus* (54.4%) in South Asia. Yeasts are less common (24.6%), with nearly 9% of these being *C. tropicalis* (Chakrabarti et al. 2008).

3.3 Epidemiology

3.3.1 *Prevalence*

In western countries, the incidence of invasive candidiasis is either stabilized or coming down in most of the settings. However, the disease continues to be reported at an increasing rate from many of the South-Asian countries. The prevalence rate is nearly 20–30 times higher in South-Asian countries than the western world: 1–12 cases/1000 admissions in India, compared with 0.05–0.36/1000 admissions in Australia, 0.8/1000 discharges in the USA and 0.2–0.5/1000 discharges in European countries (Hajjeh et al. 2004; Almirante et al. 2005; Arendrup et al. 2005; Chen et al. 2006; Chakrabarti 2013b). Several outbreaks of invasive candidiasis have also been described from South Asia, especially in the resource-limited settings (Chowdhary et al. 2003; Juyal et al. 2013; Roy et al. 1993).

Amongst the Southeast-Asian countries, the epidemiology of invasive candidiasis varies significantly amongst different geographical locations (Table 3.1), though the data is not available for all countries (Tan et al. 2015). The overall prevalence of candidemia is the highest in India (Table 3.1). A recent prospective, multi-centric observational study carried out at 27 medical and surgical ICUs across India (study identifier NCT01281345), reported candidemia at an average burden of 6.51 cases/1000 admissions, with the prevalence varying widely amongst institutions and private versus public sector hospitals. The highest burden of candidemia was noted in ICUs of North India (8.91/1000 admissions), and lowest in the West India (3.61/1000 admissions). The true burden of candidemia is expected to be even larger, as only 33% of the cases occur in ICU and blood culture positivity rate in invasive candidiasis does not exceed 50%. This high prevalence of invasive candidiasis has been linked to sub-optimal hospital care practices, especially in public sector hospitals due to heavy patient load, poor affordability for disposables in health-care practice and the high yeast carriage rate (46–82%) amongst health-care workers (Chakrabarti et al. 2015). Even neonatal candidemia has been reported at a

Table 3.1 Prevalence of candidemia in Southeast-Asian countries (data from Tan et al. 2015)

Country/region	Setting/hospital	Prevalence (range)
India	Hospital based survey (3 hospitals) 27 intensive care units	1.94 per 1000 discharges (0.30–4.53) 6.51 per 1000 discharges
Singapore	Hospital based survey (1 hospital)	2.93 per 1000 discharges (1.99–3.89)
Thailand	Hospital based survey (3 hospitals)	1.31 per 1000 discharges
Asia	Hospital based survey (25 hospitals in 6 countries/regions)	1.22 per 1000 discharges (0.16–4.53)
	Intensive care units (25 hospitals in 6 countries/regions)	11.7 per 1000 discharges

very high rate in India. A study from a tertiary-care centre reported 46 cases of neonatal candidemia per 1000 admissions, a rate which is nearly three times the prevalence reported in the USA (Chakrabarti 2013b).

Out of the different hospital services (Tan et al. 2015), medical wards generally report the highest number of cases whereas outpatient units report the lowest rates in Southeast Asia (Table 3.2). The prevalence is also higher amongst patients with acute rather than chronic leukemias (Table 3.3).

Table 3.2 Distribution of patients with candidemia by hospital services (modified form Tan et al. 2015)

S. No.	Hospital service	% cases
1	Medical wards	30.30
2	Surgical wards	23.74
3	Intensive care units	23.11
4	Haematology/oncology	10.14
5	Paediatric wards	6.46
6	Emergency department	5.57
7	Outpatient units	0.68

Table 3.3 Frequency of invasive fungal infections and invasive candidiasis amongst patients with malignancies and transplants in Asia (data from Chakrabarti 2013b; Malhotra 2013)*, rare in Asia

S. No.	Risk group	Invasive fungal infections (IFIs; %)	Invasive candidiasis (%) amongst IFIs
1	Hematopoietic stem cell transplants	3–20	30–70
2	Kidney transplants	0–20	50
3	Liver transplants*	5–40	70
4	Lung transplants*	8–35	20–25
5	Heart transplants*	5–20	50–60
6	Small bowel transplants*	12–60	80
7	Pancreas transplants*	3–35	75
8	Acute myeloid leukaemia	43	44
9	Acute lymphoblastic leukaemia	6.5	29
10	Lymphomas	1.4	43
11	Chronic lymphocytic leukaemia	0.5	20
12	Chronic myeloid leukaemia	2.5	8
13	Multiple myeloma	0.5	40

* It was used as asymbol to indicate rare prevalence in India

3.3.2 Risk Factors

Invasive candidiasis is largely considered a disease of medical progress, involving both host-related and health-care associated risk factors (Pappas et al. 2009). The disease is generally preceded by prior colonization with the infecting species. Amongst the host-related risk factors, underlying diseases such as diabetes and renal failure, Acute Physiology and Chronic Health Evaluation II (APACHE II) score greater than 20, burns, premature birth, extremes of age and neutropenia due to any disease are the major factors involved. In the health-care associated risk factors, colonization at multiple sites, broad-spectrum antibiotics, major abdominal surgery, total parenteral nutrition, central venous catheters, hemodialysis, multiple transfusions, immunosuppressive therapy and ICU stay are implicated (Chakrabarti 2013b; Faksri et al. 2014).

In the Indian candidemia network study of ICUs (Chakrabarti et al. 2015), the common underlying disease and risk factors were found to be respiratory illness (25%), renal disease (22.9%), malignancies (12.8%, with 82.9% being solid organ and 17.1% being haematological malignancies), surgery (37.3%, particularly gastrointestinal, hepatobiliary and pancreatic), central venous catheters (74%), parenteral nutrition (13.4%), antibiotic therapy (93%) and corticosteroids (18%). Additionally, the patients were considerably younger (mean 49.7 years), comparatively less serious (mean APACHE II score 17.2), and acquired candidemia early (median 8 days of ICU admission compared to 23 days in developed nations). This different epidemiology of ICU-acquired candidemia in India may be linked to high exposure to broad-spectrum antibiotics and steroids, and compromised health care.

One study from Pakistan documented the use of health-care devices (85.4% of cases) and admission to special care units (75%) as predominant risk factors for invasive candidiasis. Notably, 66.7% of their patients with invasive candidiasis were not immunosuppressed (Farooqi et al. 2013). Differences have also been noted in the predominant risk factors for *C. albicans* versus NAC species. A study from Thailand reported APACHE II score greater than 15 and central venous catheters, but not, immunocompromised conditions, as the predominant risk factors for acquiring bloodstream infections due to *C. albicans* compared with NAC species (Apisarnthanarak et al. 2009).

3.3.3 Spectrum of Candida Species

Historically, *C. albicans* accounted for 70–80% cases of invasive candidiasis, but the trend has gradually shifted towards NAC species in the past few decades. The shift is much more in South-Asian countries where NAC species were reported in 70–90% of cases (Chakrabarti et al. 2015). The spectrum of *Candida* species causing this disease is also broad in these countries (Table 3.4). In the Indian multi-centric ICU study, 31 species of *Candida* were isolated from candidemia

Table 3.4 Distribution of different *Candida* species amongst candidemia patients in Asia (adapted from Tan et al. 2015)

S. No.	*Candida* species	% isolation
1	*C. albicans*	41.36
2	*C. tropicalis*	25.45
3	*C. glabrata*	13.93
4	*C. parapsilosis*	12.15
5	*C. guilliermondii*	2.25
6	*C. krusei*	1.94
7	*C. famata*	0.89
8	*C. pelliculosa*	0.73
9	*C. haemulonii*	0.37
10	*C. intermedia*	0.26
11	*C. lusitaniae*	0.26
12	*C. sake*	0.21
13	*C. dubliniensis*	0.10
14	*C. pararugosa*	0.05
15	*C. catenulate*	0.05

cases, compared with the western world where top five *Candida* species, namely *C. albicans*, *C. glabrata, C. tropicalis*, *C. parapsilosis* and *C. krusei* are implicated in majority of the reports (Pfaller and Diekema 2007; Chakrabarti 2013a, b; Chen 2013; Chakrabarti et al. 2015; Tan et al. 2015).

Even different geographical locations within Asia vary in their species distribution. In tropical Asian countries (India, Thailand and Singapore), *C. tropicalis* is the leading cause of invasive candidiasis (35–46%), compared with the other Asian regions where *C. albicans* still remains the commonest *Candida* species isolated (41%) though *C. tropicalis* is rated as the second most common *Candida* species (18.9%) (Chakrabarti 2013b; Tan et al. 2015). This picture is in contrast to the developed nations, where *C. albicans* is the most common species (45–74%) and *C. glabrata* (12–22.6%) is the next common species (Silva et al. 2012; Chen 2013). The rise in number of cases with *C. glabrata* infection is associated with fluconazole prophylaxis. Though fluconazole resistance in *C. tropicalis* is on the rise in South-Asian countries (2.6–6.5%) (Chakrabarti et al. 2015; Pfaller et al. 2010), it could not be the main reason for the high incidence of *C. tropicalis* candidemia in those countries. Possibly, the increasing incidence of *C. tropicalis* in South Asia is linked to hand colonization in health-care workers and compromise of infection-control practices, as nearly 82% of health-care workers in an Indian tertiary-care hospital were colonized with *Candida*, out of which 80% was *C. tropicalis* (Chakrabarti et al. 2009, 2015).

The shift observed in species distribution over the past years has important epidemiological and therapeutic implications, considering that many of the *Candida* species are associated with distinct risk groups, pathogenicity and antifungal susceptibility. *C. glabrata* is less susceptible to fluconazole while *C. krusei* is inherently resistant. Though *C. parapsilosis* is considered less virulent to *C. albicans*,

C. tropicalis and *C. glabrata*, it flourishes in clinical settings and leads to nosocomial outbreaks, owing to its ability to colonize indwelling medical devices and human skin. *C. glabrata* and *C. tropicalis* are commonly seen in patients with malignancies and neutropenia; *C. krusei* in hematopoietic stem cell transplant recipients; and *C. dubliniensis* in AIDS patients (Kuhn et al. 2002; Kullberg and Arendrup 2015; Polke et al. 2015; Silva et al. 2012). *C. tropicalis* is frequently isolated from hemato-oncology wards in Southeast Asia and is common in neonates, though *C. parapsilosis* is the predominant species causing neonatal invasive candidiasis associated with parenteral nutrition and central lines in western world (Farooqi et al. 2013; Polke et al. 2015; Tan et al. 2015). The time to positivity (TTP) of *C. tropicalis* blood cultures is significantly shorter than others, and septic shock plus skin emboli are common findings; both these features are associated with poor prognosis (Lai et al. 2012; Chen 2013).

The spectrum of *Candida* species also differs amongst private/corporate and public sector hospitals in South-Asian countries. *C. auris* and *C. rugosa* were reportedly more common in public sector Indian ICUs (8.2 and 5.6%, respectively) compared with private/corporate sector ICUs (3.9 and 1.5%, respectively). This difference is again linked to a compromise in health care and horizontal transmission in public sector hospitals due to huge patient load. Around 12% of these isolates were also azole resistant (Chakrabarti et al. 2015).

In recent years, there has been an increase in infections due to rare yeasts like *K. ohmeri*, *P. anomala* and *C. auris*. A nosocomial outbreak of *P. anomala* fungemia was reported in the paediatric wards of an Indian tertiary-care hospital over a period of 23 months (April 1996–February 1998), linked to poor hand hygiene practices amongst health-care personnel. A total of 379 neonates and children (4.2% admissions) were infected, and the outbreak could only be controlled after improvement of hand-washing practices and nystatin-fluconazole prophylaxis to all premature neonates and high-risk infants (Chakrabarti et al. 2001). The same hospital reported fungemia due to *K. ohmeri*, accounting for 9.5% of the cases in paediatric population. 25.7% of these isolates had been misidentified as *C. tropicalis* based on phenotypic characteristics, before molecular analysis was carried out. Compared with *C. tropicalis* fungemia, piperacillin-tazobactam use was found to be the only significant risk factor for acquiring *K. ohmeri* (Chakrabarti et al. 2014). Other studies from India have also reported *K. ohmeri* infection (Biswal et al. 2015; Capoor et al. 2015).

A new clonal strain of *C. auris* has recently emerged in India. In 2013, two North Indian hospitals reported case series of fungemia due to this strain, involving 12 inpatients (Chowdhary et al. 2013). Fifteen cases of infection due to the same strain were thereafter reported from a tertiary-care centre in South India, originating from fungemia (7 cases), diabetic gangrenous foot (3) and bronchopneumonia (1). These isolates were phenotypically and genotypically distinct from Korean and Japanese isolates. *C. auris* candidemia was reported from 19 of 27 ICUs in the multi-centric ICU study (Chakrabarti et al. 2015). The emergence of *C. auris* in India has important epidemiological and clinical implications, as the isolates are often multidrug resistant and are frequently misidentified as *Candida haemulonii* or

K. ohmeri by phenotypic methods. Amongst the 27 clonal *C. auris* isolates reported, all were resistant to fluconazole, 73% were resistant to voriconazole and 47% to flucytosine. 40% of these isolates also had high MIC (≥ 1 μg/ml) to caspofungin (Chowdhary et al. 2014). Chatterjee et al. (2015) have recently published the draft genome of *C. auris*, and reported that it is highly divergent compared with the genome of other common *Candida* species. This can now serve as a useful resource for developing accurate diagnostic markers for *C. auris*.

3.4 Antifungal Resistance

The resistance to azoles, including fluconazole, in the so-called susceptible species like *C. albicans* and *C. tropicalis* is increasing in South-Asian countries (Chakrabarti et al. 2015; Pfaller et al. 2010; Tan et al. 2016). This is a cause of concern, as fluconazole is the drug of choice when susceptible species like *C. albicans* and *C. tropicalis* are isolated, especially in the resource-limited set-ups. Even echinocandin resistance is emerging, especially in *C. glabrata* (Chakrabarti et al. 2015; Tan et al. 2016).

In the Indian multi-centric ICU study, resistance to amphotericin B, azoles and echinocandins was shown to be 2.1, 11.8 and 6.9%, respectively. Multidrug resistance was observed in 1.9% cases, with 3 cases (0.3%) being pan-resistant to all antifungal classes tested. *C. tropicalis*, *C. auris* and *C. krusei* were common species amongst multidrug-resistant isolates (Chakrabarti et al. 2015). Pan-resistance to azoles was noted in 10% of the *Candida* species in another report (Chakrabarti et al. 2009).

Differences in antifungal susceptibility have also been observed amongst various *Candida* species. NAC species show a decreased susceptibility compared with *C. albicans.* A recently published study involving 13 Asian centres from Brunei, Philippines, Singapore, South Korea, Taiwan, Thailand and Vietnam over a two year period showed lower fluconazole susceptibility in *C. tropicalis* (susceptible = 75.8%, susceptible dose-dependent = 6.1%), *C. glabrata* (susceptible dose-dependent = 94.9%) and *C. parapsilosis* (susceptible = 94.8%) than in *C. albicans* (susceptible = 99.7%). Majority of the isolates were susceptible to echinocandins (S > 99%), except *C. glabrata*, in which 6% of the tested isolates were susceptible dose-dependent for caspofungin (Tan et al. 2016). One isolate of *C. glabrata* from Pakistan was reportedly resistant to all the three available echinocandins (Farooqi et al. 2013).

Furthermore, the susceptibility pattern of *Candida* species in Asian region varies from that of the western world. In the ARTEMIS DISK Global Antifungal Surveillance Study carried out from 1997 to 2007 (Pfaller et al. 2010), higher rates of resistance to fluconazole and voriconazole were shown in *C. tropicalis* isolates from Asia-Pacific region (6.5 and 8.4%, respectively) compared to other areas (Latin America, 2.6 and 3.7%; Europe, 2.9 and 3.9%; North, America, 4.4 and 5.3%; Middle East and Africa, 2.6 and 2.4%, respectively).

3.5 Diagnostic Challenges

Presently available methods for diagnosis of invasive candidiasis include culturing of the fungus from blood or sterile body sites by conventional or automated culture systems, detection of *Candida* in blood by T2*Candida* panel; species identification using conventional phenotypic tests or different commercial phenotypic systems, matrix-assisted laser desorption ionization time-of-flight mass spectrometry (MALDI-TOF) and DNA sequencing; serological testing using surrogate markers such as β-D-glucan, mannan and anti-mannan antibodies; and PCR-based assays like SeptiFast or in-house methods (Chen 2013; Kullberg and Arendrup 2015).

Candida colonization index (ratio of the number of distinct body sites colonized with identical strains over the total number of distinct body sites tested daily) and *Candida* score (based on four significant predictors of proven infection—parenteral nutrition, surgery, multifocal colonization and severe sepsis) have been proposed as important indices to identify patients at high-risk of acquiring *Candida* infections (Pittet et al. 1994; León et al. 2006). In South Asia, the use of these scores is challenged due to high rate of *Candida* colonization in critically ill patients after frequent antibiotic use and misuse.

Blood culture is a sensitive technique for isolation of *Candida* species in candidemia cases. It is positive in nearly 50–60% of cases. The availability of automated systems has further improved the sensitivity. Blood culture, however, requires 40-ml blood at one time, and takes at least 2–3 days for isolation, which may cause significant delay in therapy. Further, patients with deep-seated infections sometimes yield negative blood cultures, as the fungus may have been cleared from the bloodstream by the time samples are taken. Prior use of antifungals and slower TTP (mean 26 h) are additional confounding factors (Lai et al. 2012; Chen 2013; Kullberg and Arendrup 2015; Pfeiffer et al. 2011). T2 magnetic resonance technique helps in detection of five predominant *Candida* species within 5–6 h. It is based on nanoparticle-dependent magnetic resonance and has been recently approved by FDA for diagnosis of *Candida* bloodstream infections (Chen 2013; Kullberg and Arendrup 2015). But the technique is still not available in South-Asian countries. Moreover, its application may not be so useful in these countries as spectrum of *Candida* species is broad (31 species in Indian study) and the technique cannot replace conventional blood culture technique due to its inability to perform susceptibility test.

Following culture from blood or other sterile body sites, species identification can be done either by conventional phenotypic tests or newer commercial automated assays. The commercially available VITEK 2 system for pathogenic yeasts has been employed in many settings for phenotypic identification of different species. However, studies have reported misidentification of several *Candida* species by VITEK 2, including *C. tropicalis, C. guilliermondii, C. auris, C. norvegensis, C. parapsilosis, C. famata* and *C. rugosa* (Valenza et al. 2008). In contrast, MALDI-TOF has shown superior results. In a recent study from an Indian tertiary-care centre, MALDI-TOF has been systematically standardized and database improved. After improvement of database, the assay could identify correctly 98.9% of the 354 PCR-sequenced yeast isolates, barring slight underperformance for *C. tropicalis* (1 isolate), *Pichia jadinii* (1 isolate) and *C. rugosa* (2 isolates) (Ghosh et al. 2015).

There are many commercial assays that can directly detect *Candida* in blood samples without prior culture, such as SeptiFast. SeptiFast assay is a multi-pathogen real-time PCR system capable of detecting several bacteria and fungi present in blood samples. It appears to have higher specificity than sensitivity, but its likely clinical utility in the setting of suspected sepsis remains unclear (Dark et al. 2015). Owing to the cost constraints, these techniques are also not easily accessible in all hospital settings of South Asia. Certain laboratories perform in-house PCR technique for diagnosis of invasive candidiasis, but the techniques require standardization in multi-centric studies.

Surrogate markers such as β-D-glucan, mannan and anti-mannan antibodies have also been employed for serological diagnosis of invasive candidiasis. The specificity and diagnostic accuracy of β-D-glucan assay is higher than *Candida* score and colonization index (Posteraro et al. 2011; Pappas et al. 2016). It particularly has a high negative predictive value, and is useful in identifying ICU patients at higher risk of IFIs as well as for monitoring the response to treatment (Jaijakul et al. 2012; Chen 2013; Kullberg and Arendrup, 2015). Parallel use of mannan antigen and anti-mannan antibodies improves sensitivity (Dark et al. 2015; Lunel et al. 2011; Rao et al. 2002), though the method is more suited for *C. albicans* but has low sensitivity for NAC species. These techniques are, however, costly and are used in very few laboratories of South Asia. The predicting values are also not clearly established (Chen 2013).

3.6 Management

Invasive candidiasis is associated with high mortality rates, approaching 35–75% across the globe (Chakrabarti et al. 2015), and early treatment has been related to better outcomes (Chen 2013). In the developed world, echinocandins have emerged as the favoured agents for most episodes of invasive candidiasis, except the ones involving central nervous system (CNS), ocular and urinary tract infections, where liposomal amphotericin B or fluconazole is preferred. Fluconazole is considered as an acceptable alternative to echinocandins in patients who are stable and not critically ill, unlikely to harbour fluconazole resistant isolates, and without prior azole exposure. It should be noted, however, that no single trial has demonstrated clear superiority of one therapeutic drug over another. Echinocandins are generally preferred due to their safety profile, early fungicidal effect, early clearance and reportedly better outcomes in azole resistant *Candida* species. During management of invasive candidiasis, central venous catheter should be removed as early as possible specially when it is the apparent source and can be removed safely (Pappas et al. 2016).

Owing to the resource-limited settings in South Asia, an echinocandin is used as first-line therapy only in cases where patients can afford this drug. Azoles are generally the preferred drugs (72%), followed by echinocandins (18.3%) and amphotericin B (including deoxycholate and lipid preparations (14.4%). While amphotericin B deoxycholate is used more commonly in public sector ICUs (11.5 vs. 5.3%), echinocandins appear to be the preferred agents in private/corporate settings (13.4 vs. 7.4%) (Chakrabarti et al. 2015). Although antifungals are not generally recommended in cases of primary *Candida* peritonitis, most of the patients present late to the hospital in our settings, and may therefore require antifungal treatment. Considering all the above issues including epidemiology, antifungal resistance, practical situation while handling critically ill patients in South Asia, a revised management guideline is required to treat invasive candidiasis.

3.7 Conclusion

Invasive candidiasis is a key public health problem worldwide, accounting for nearly 22% of IFIs in Asia with high morbidity and mortality rates. The disease is reported at a high prevalence in South-Asian countries (1–12 cases/1000 admissions), which is nearly 20–30 times the developed world. Candidemia remains the most common clinical manifestation of invasive candidiasis. *Candida* pancreatitis is an emerging problem. *C. tropicalis* is the leading *Candida* species implicated in invasive candidiasis in the tropical Asian countries, compared with the developed world where *C. glabrata* is common among NAC species. Differential species distribution has also been noted, dependent on risk factors, age groups and even between public versus private sector hospitals. Increasing cases of fungemia due to *K. ohmeri*, *P. anomala* and *C. auris* have been described in the past years. Automated blood culture and β-D-glucan assay are the cornerstones of diagnosing this disease. In the developed world, echinocandins are the favoured agents for most episodes of invasive candidiasis primarily because of their safety profile and fungicidal effect. However, in South-Asian countries, fluconazole and amphotericin B deoxycholate are commonly prescribed to manage invasive candidiasis, predominantly due to the cost constraints.

References

Almirante B, Rodríguez D, Park BJ et al (2005) Epidemiology and predictors of mortality in cases of *Candida* bloodstream infection: results from population-based surveillance, Barcelona, Spain, from 2002 to 2003. J Clin Microbiol 43(4):1829–1835

Apisarnthanarak A, Naknarongkij N, Kiratisin P et al (2009) Risk factors and outcomes of *Candida albicans* and non-*albicans Candida* species at a Thai tertiary care center. Am J Infect Control 37(9):781–782

Arendrup MC, Fuursted K, Gahrn-Hansen B et al (2005) Seminational surveillance of fungemia in Denmark: notably high rates of fungemia and numbers of isolates with reduced azole susceptibility. J Clin Microbiol 43(9):4434–4440

Biswal D, Sahu M, Mahajan A et al (2015) *Kodameae ohmeri* - an emerging yeast: Two cases and literature review. J Clin Diagn Res 9(3):DD01–3

Capoor MR, Gupta DK, Verma PK et al (2015) Rare yeasts causing fungemia in immunocompromised and haematology patients: case series from Delhi. Ind J Med Microbiol 33(4):576–579

Chakrabarti A (2013a) Introduction. In: Chakrabarti A (ed) Fungal infections in Asia—the eastern frontier of mycology, 1st edn (English). Elsevier, India, pp 3–7

Chakrabarti A (2013b) Opportunistic fungal infections in Asia. In: Chakrabarti A (ed) Fungal infections in Asia—the eastern frontier of mycology, 1st edn (English). Elsevier, India, pp 11–21

Chakrabarti A (2007) Epidemiology of central nervous system mycoses. Neurol India 55(3):191–197

Chakrabarti A, Padhye AA (2013) Endemic mycoses in Asia. In: Chakrabarti A (ed) Fungal infections in Asia—the eastern frontier of mycology, 1st edn (English). Elsevier, India, pp 22–33

Chakrabarti A, Singh R (2011) The emerging epidemiology of mould infections in developing countries. Curr Opin Infect Dis 24(6):521–526

Chakrabarti A, Sood P, Rudramurthy SM et al (2015) Incidence, characteristics and outcome of ICU-acquired candidemia in India. Intensive Care Med 41(2):285–295

Chakrabarti A, Rudramurthy SM, Kale P et al (2014) Epidemiological study of a large cluster of fungaemia cases due to *Kodamaea ohmeri* in an Indian tertiary care centre. Clin Microbiol Infect 20(2):083–089

Chakrabarti A, Chatterjee SS, Rao KL et al (2009) Recent experience with fungaemia: change in species distribution and azole resistance. Scand J Infect Dis 41(4):275–284

Chakrabarti A, Shivaprakash MR, Singh R et al (2008) Fungal endophthalmitis: fourteen years' experience from a center in India. Retina 28(10):1400–1407

Chakrabarti A, Rao P, Tarai B et al (2007) *Candida* in acute pancreatitis. Surg Today 37(3):207–211

Chakrabarti A, Singh K, Narang A et al (2001) Outbreak of *Pichia anomala* infection in the pediatric service of a tertiary-care center in Northern India. J Clin Microbiol 39(5):1702–1706

Chatterjee S, Alampalli SV, Nageshan RK et al (2015) Draft genome of a commonly misdiagnosed multidrug resistant pathogen *Candida auris*. BMC Genom 16:686

Chen Y-C (2013) Invasive candidiasis in Asia. In: Chakrabarti A (ed) Fungal infections in Asia—the eastern frontier of mycology, 1st edn (English). Elsevier, India, pp 91–105

Chen S, Slavin M, Nguyen Q et al (2006) Active surveillance for candidemia. Australia Emerg Infect Dis 12(10):1508–1516

Chowdhary A, Anil Kumar V, Sharma C et al (2014) Multidrug-resistant endemic clonal strain of *Candida auris* in India. Eur J Clin Microbiol Infect Dis 33(6):919–926

Chowdhary A, Sharma C, Duggal S et al (2013) New clonal strain of *Candida auris*, Delhi. India. Emerg Infect Dis 19(10):1670–1673

Chowdhary A, Becker K, Fegeler W et al (2003) An outbreak of candidemia due to *Candida tropicalis* in a neonatal intensive care unit. Mycoses 46(8):287–292

Dark P, Blackwood B, Gates S et al (2015) Accuracy of LightCycler (®) SeptiFast for the detection and identification of pathogens in the blood of patients with suspected sepsis: a systematic review and meta-analysis. Intensive Care Med 41(1):21–33

Faksri K, Kaewkes W, Chaicumpar K et al (2014) Epidemiology and identification of potential fungal pathogens causing invasive fungal infections in a tertiary care hospital in northeast Thailand. Med Mycol 52(8):810–818

Farooqi JQ, Jabeen K, Saeed N et al (2013) Invasive candidiasis in Pakistan: clinical characteristics, species distribution and antifungal susceptibility. J Med Microbiol 62(Pt 2):259–268

Ghosh AK, Paul S, Sood P et al (2015) Matrix-assisted laser desorption ionization time-of-flight mass spectrometry for the rapid identification of yeasts causing bloodstream infections. Clin Microbiol Infect 21(4):372–378

Hajjeh RA, Sofair AN, Harrison LH et al (2004) Incidence of bloodstream infections due to *Candida* species and *in vitro* susceptibilities of isolates collected from 1998 to 2000 in a population-based active surveillance program. J Clin Microbiol 42(4):1519–1527

Jaijakul S, Vazquez JA, Swanson RN et al (2012) (1,3)-β-D-glucan as a prognostic marker of treatment response in invasive candidiasis. Clin Infect Dis 55(4):521–526

Jain M, Dogra V, Mishra B et al (2011) Candiduria in catheterized intensive care unit patients: emerging microbiological trends. Indian J Pathol Microbiol 54(3):552–555

Juyal D, Adekhandi S, Negi V et al (2013) An outbreak of neonatal candidemia due to non-albicans *Candida* species in a resource constrained setting of Uttarakhand State. India J Clin Neonatol 2(4):183–186

Khan ZU, Randhawa HS, Lulla M (1982) Isolation of *Blastomyces dermatitidis* from the lungs of a bat, *Rhinopoma hardwickei* gray. Delhi. Sabouraudia 20(2):137–144

Kuhn DM, Chandra J, Mukherjee PK et al (2002) Comparison of biofilms formed by *Candida albicans* and *Candida parapsilosis* on bioprosthetic surfaces. Infect Immun 70(2):878–988

Kullberg BJ, Arendrup MC (2015) Invasive Candidiasis. N Engl J Med 373(15):1445–1456

Lai CC, Wang CY, Liu WL et al (2012) Time to positivity of blood cultures of different *Candida* species causing fungaemia. J Med Microbiol 61(Pt 5):701–704

León C, Ruiz-Santana S, Saavedra P et al (2006) A bedside scoring system ("Candida score") for early antifungal treatment in nonneutropenic critically ill patients with *Candida* colonization. Crit Care Med 34(3):730–737

Lunel FM, Donnelly JP, van der Lee HA et al (2011) Performance of the new platelia *Candida* plus assays for the diagnosis of invasive *Candida* infection in patients undergoing myeloablative therapy. Med Mycol 49(8):848–855

Malhotra P (2013) Opportunistic mycoses in hematological malignancies. In: Chakrabarti A (ed) Fungal infections in Asia—the eastern frontier of mycology, 1st edn (English). Elsevier, India, pp 71–87

Padhye AA, Pathak AA, Katkar VJ et al (1994) Oral histoplasmosis in India: a case report and an overview of cases reported during 1968–92. Med Vet Mycol 32(2):93–103

Pappas PG, Kauffman CA, Andes DR et al (2016) Clinical practice guideline for the management of candidiasis: 2016 update by the infectious diseases society of America. Clin Infect Dis 62 (4):15, e1–e50

Pappas PG, Kauffman CA, Andes D et al (2009) Clinical practice guidelines for the management of candidiasis: 2009 update by the Infectious Diseases Society of America. Clin Infect Dis 48 (5):503–535

Pfaller MA, Diekema DJ (2007) Epidemiology of invasive candidiasis: a persistent public health problem. Clin Microbiol Rev 20(1):133–163

Pfaller MA, Diekema DJ, Gibbs DL et al (2010) Results from the ARTEMIS DISK global antifungal surveillance study, 1997–2007: a 10.5-year analysis of susceptibilities of *Candida* species to fluconazole and voriconazole as determined by CLSI standardized disk diffusion. J Clin Microbiol 48(4):1366–1377

Pfeiffer CD, Samsa GP, Schell WA et al (2011) Quantitation of *Candida* CFU in initial positive blood cultures. J Clin Microbiol 49(8):2879–2883

Pittet D, Monod M, Suter PM et al (1994) *Candida* colonization and subsequent infections in critically ill surgical patients. Ann Surg 220(6):751–758

Polke M, Hube B, Jacobsen ID (2015) *Candida* survival strategies. Adv Appl Microbiol 91:139–235

Posteraro B, De Pascale G, Tumbarello M et al (2011) Early diagnosis of candidemia in intensive care unit patients with sepsis: a prospective comparison of (1 → 3)-β-D-glucan assay, candida score, and colonization index. Crit Care 15(5):R249

Randhawa HS, Chaturvedi VP, Kini S et al (1985) *Blastomyces dermatitidis* in bats: first report of its isolation from the liver of *Rhinopoma hardwickei* Gray. Sabouraudia 23(1):69–76

Rao DS, Ghosh A, Singhi S et al (2002) Mannan antigen detection in the diagnosis of patients with invasive candidiasis. Indian J Med Res 116:13–20

Roy A, Maiti PK, Adhya S et al (1993) Neonatal candidemia. Indian J Pediatr 60(6):799–801

Silva S, Negri M, Henriques M et al (2012) *Candida glabrata*, *Candida parapsilosis* and *Candida tropicalis*: biology, epidemiology, pathogenicity and antifungal resistance. FEMS Microbiol Rev 36(2):288–305

Singla N, Gulati N, Kaistha N et al (2012) *Candida* colonization in urine samples of ICU patients: determination of etiology, antifungal susceptibility testing and evaluation of associated risk factors. Mycopathologia 174(2):149–155

Sundaram C, Umabala P, Laxmi V et al (2006) Pathology of fungal infections of the central nervous system: 17 years' experience from Southern India. Histopathology 49(4):396–405

Tan TY, Hsu LY, Alejandria MM et al (2016) Antifungal susceptibility of invasive *Candida* bloodstream isolates from the Asia-Pacific region. Med Mycol. pii: myv114.

Tan BH, Chakrabarti A, Li RY et al (2015) Incidence and species distribution of candidaemia in Asia: a laboratory-based surveillance study. Clin Microbiol Infect 21(10):946–953

Taneja J, Mishra B, Thakur A et al (2009) *Candida albicans* meningitis with ventriculitis in a four-year-old child. Neurol India 57(5):682–684

Todi S (2013) Opportunistic mycoses in intensive care unit. In: Chakrabarti A (ed) Fungal infections in Asia—the eastern frontier of mycology, 1st edn (English). Elsevier, India, pp 37–44

Vaideeswar P (2015) Candidial endocarditis: a single-institute pathological analysis. Mycopathologia 180(1–2):81–87

Valenza G, Strasen J, Schäfer F et al (2008) Evaluation of new colorimetric vitek 2 yeast identification card by use of different source media. J Clin Microbiol 46(11):3784–3787

Vincent JL, Sakr Y, Sprung CL et al (2006) Sepsis in European intensive care units: results of the SOAP study. Crit Care Med 34(2):344–353

Yashavanth R, Shiju MP, Bhaskar UA et al (2013) Candiduria: prevalence and trends in antifungal susceptibility in a tertiary care hospital of Mangalore. J Clin Diagn Res 7(11):2459–2461

Chapter 4
Morphogenesis in *C. albicans*

David Kadosh

Abstract Many human fungal pathogens possess the ability to grow in a variety of different morphologies and *Candida albicans* is no exception. In dimorphic fungal pathogens, such as *Histoplasma capsulatum* and *Coccidioides immitis*, morphological transitions are critical for virulence (Klein and Tebbets, Curr Opin Microbiol, 10(4):314–319, 2007). However, other pathogens, such as *Candida glabrata*, rarely alter their morphology and the ability to change shape appears to play little, if any, role in their pathogenicity (Do Carmo-Sousa 1969; Fidel et al., Clin Microbiol Rev 12:80–96, 1999; Csank and Haynes, FEMS Microbiol Lett 189(1):115–120, 2000). The ability of *Candida albicans* to undergo a reversible morphological transition from yeast to filamentous form represents a fundamental aspect of this pathogen's biology. This transition is typically correlated with pathogenicity and important for a wide variety of virulence-related processes (Lo et al., Cell 90(5):939–949, 1997; Braun and Johnson, Science 277(5322):105–109, 1997; Braun et al., Genetics 156 (1):31–44, 2000; Saville et al., Eukaryot Cell 2(5):1053–1060, 2003; Carlisle et al., Proc Natl Acad Sci USA 106:599–604, 2009; Kumamoto and Vinces, Proc Natl Acad Sci USA 102(15):5576–5581, 2005; Korting et al., J Med Microbiol 52(Pt 8):623–632, 2003; Gow et al., Curr opin microbiol 5(4):366–371, 2002). As a consequence, a significant amount of research, mostly over the past 25 years, has focused on signaling pathways, regulators, and mechanisms that are involved in controlling the *C. albicans* morphological transition. In this chapter we will first describe the major *C. albicans* morphologies and the relationship between *C. albicans* morphology and virulence. Next, we will discuss the mechanics of hyphal growth as well as a variety of signaling pathways, regulators, and mechanisms important for regulating *C. albicans* morphogenesis in response to host environmental cues. Finally, we will discuss recent insights gained from genome-wide studies of the *C. albicans* morphological transition as well as the potential that this transition may hold to serve as a target for new therapeutic strategies.

D. Kadosh (✉)
Department of Microbiology, Immunology and Molecular Genetics,
University of Texas Health Science Center at San Antonio,
7703 Floyd Curl Dr., MC: 7758, San Antonio, TX 78229-3900, USA
e-mail: kadosh@uthscsa.edu

R. Prasad (ed.), *Candida albicans: Cellular and Molecular Biology*,
DOI 10.1007/978-3-319-50409-4_4

4.1 *C. albicans* Morphologies

C. albicans is most commonly found in either yeast or filamentous form (Fig. 4.1). Yeasts are single budding oval-shaped cells, typically several microns in diameter (Odds 1988; Sudbery et al. 2004). In contrast, filaments are elongated cells attached end-to-end. Hyphal filaments are typically about 2 μm in diameter, have parallel-sided walls, and lack constrictions at septal junctions. Pseudohyphal filaments, on the other hand, are usually somewhat wider in diameter (≥ 2.8 μm), lack parallel sides, and have constrictions at septal junctions as well as at the mother-bud neck (Odds 1988; Sudbery et al. 2004). Septum ring formation as well as nuclear division occurs at the mother-bud neck of pseudohyphae versus in the germ tube of hyphae during the first cell division (Gow et al. 1986; Sudbery 2001; Warenda and Konopka 2002). In addition, in hyphae initial formation of germ tubes occurs before the G_1/S transition, whereas growth of pseudohyphal and yeast cells is synchronized with the cell cycle (Barelle et al. 2003; Gow and Gooday 1984; Yokoyama and Takeo 1983). Hyphal cells also undergo a delay in G_1 phase following the first cell cycle and are consequently usually less branched than pseudohyphae (Sudbery et al. 2004). *C. albicans* cells are known to undergo a reversible morphological transition from yeast to pseudohyphal and hyphal filaments in response to a variety of environmental cues, many of which are found in the host. These cues include body temperature (37 °C), serum, high CO_2/O_2 ratio, pH greater than 6.5, carbon and/or nitrogen starvation, certain carbon sources (e.g., N-acetylglucosamine), alcohols, certain amino acids (e.g., proline), and a variety of human hormones (Odds 1988; Bramley et al. 1991; Caticha et al. 1992; Kinsman et al. 1988; Merson-Davies and Odds 1989; Brown 2002). Growth at 37 °C in the

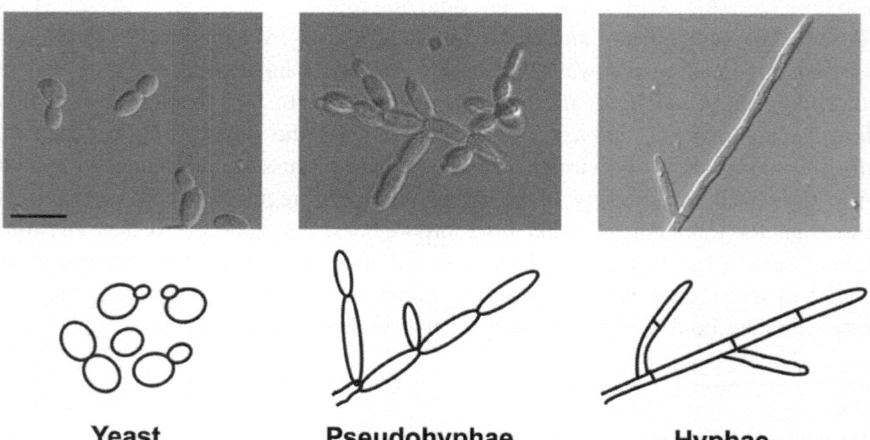

Fig. 4.1 Yeast and filamentous morphologies of *C. albicans*. (*Top*) Images depict the indicated morphologies, as visualized by differential interference contrast (DIC) microscopy (Bar = 10 μm). (*Bottom*) Indicated morphologies are represented schematically (Thompson et al. 2011)

presence of serum is generally considered one of the strongest filament-inducing conditions and can be optimized in vitro to generate a completely filamentous population (Kadosh and Johnson 2005).

Chlamydospores represent third *C. albicans* morphology. These cells are rounder and larger than yeast, have very thick walls, and usually form at the termini of hyphal filaments in response to reduced nutrient levels in the growth medium (Staib and Morschhauser 2007). While several *C. albicans* regulatory factors are known to be important for chlamydospore formation, the actual biological function of these cells has remained elusive. In addition, unlike yeast, pseudohyphae, and hyphae, chlamydospores have only rarely been found in infected tissues (Chabasse et al. 1988; Cole et al. 1991).

C. albicans also undergoes a phenotypic transition from white to opaque cells (Johnson 2003; Lohse and Johnson 2009; Soll et al. 1993; Morschhauser 2010). This epigenetic switch is heritable and can occur over multiple generations. Opaque cells are typically somewhat larger and more oblong than white cells and exhibit pimple-like structures at their surface. White cells grow as shiny domed colonies, whereas opaque cells grow as flatter, rougher, and darker colonies (Johnson 2003; Slutsky et al. 1987). White and opaque cells show differences in mating ability (opaque cells are the mating competent form), interaction with the host immune system, and metabolic preferences (Miller and Johnson 2002; Lan et al. 2002; Geiger et al. 2004; Lohse and Johnson 2008; Sasse et al. 2013). Switching frequency can be affected by a number of variables including carbon source, temperature, and carbon dioxide levels (Rikkerink et al. 1988; Huang et al. 2009; Huang et al. 2010; Lohse et al. 2013). *C. albicans* white-opaque switching has also been shown to be controlled by interlocking positive transcriptional feedback loops involving the master regulator Wor1 as well as Czf1, Efg1, and Wor2 (Zordan et al. 2007). Gray cells, which represent an intermediate morphology between white and opaque, have also been reported (Tao et al. 2014). These cells form smooth, dark colonies, and are similar in appearance to opaque cells but much smaller. Gray cells have a mating efficiency that is intermediate between white and opaque and also show differences in secreted aspartyl protease (Sap) activity, infection ability, and global gene expression when compared to white and opaque cells (Tao et al. 2014). A final *C. albicans* morphology, the GUT cell, has been observed upon overexpression of the *WOR1* master regulator of white-opaque switching in the mammalian gastrointestinal (GI) tract (Pande et al. 2013). While GUT cells are similar in appearance to opaque cells, they lack pimple structures, show a lower mating efficiency, and are stable at body temperature (37 °C). The white-GUT switch, which promotes *C. albicans* commensalism, is believed to be induced by environmental cues in the GI tract that result in increased *WOR1* expression. In the remainder of this chapter we will be focusing largely on the transition from yeast to pseudohyphal and hyphal morphologies as well as the role that this transition plays in *C. albicans* virulence and virulence-related processes.

4.2 Relationship Between *C. albicans* Morphology and Virulence

Initial studies to determine the relationship between *C. albicans* morphology and virulence involved the use of mutants (*efg1*Δ/Δ *cph1*Δ/Δ and *tup1*Δ/Δ) that are locked in either yeast or filamentous form (Lo et al. 1997; Braun and Johnson 1997; Braun et al. 2000). All of these mutants were highly attenuated for virulence in a mouse model of systemic candidiasis, suggesting that the ability of *C. albicans* to undergo a reversible morphological transition is important for virulence. A subsequent study used a strain that could be genetically manipulated to transition from yeast to filaments during the course of infection (Saville et al. 2003). When this strain was locked in the yeast form it was completely attenuated for virulence, although, interestingly, *C. albicans* cells still disseminated to various host organs. However, after the strain was allowed to form filaments at various post-infection time points, host survival was significantly reduced, strongly suggesting that the yeast-filament transition was important for this process. A subsequent complementary experiment showed that a strain which could be inoculated as yeast but genetically manipulated to strongly promote hyphal formation during infection caused enhanced virulence and tissue invasion in a mouse model of systemic candidiasis (Carlisle et al. 2009). One caveat to the experiments described above is that all the genes used to generate strains with altered morphology encoded transcriptional regulators, so the possibility that the observed effects on virulence could be due to misexpression of target genes not directly involved in morphology could not be excluded. However, deletion of *HGC1*, encoding a cyclin-like protein important for the physical process of filamentation per se but not transcription of filament-specific target genes, resulted in significantly reduced filamentous growth and highly attenuated virulence in the mouse systemic model, strongly suggesting a close link between the *C. albicans* morphological transition and virulence (Zheng et al. 2004). Consistent with this hypothesis, filamentation has also been shown to play an important role in a variety of virulence-related processes, including the invasion of epithelial cell layers, host immune evasion, breaching of endothelial cells, thigmotropism (contact sensing, which allows growth to be oriented toward sites of weakened integrity), and biofilm formation (Kumamoto and Vinces 2005; Jong et al. 2001; Zink et al. 1996; Korting et al. 2003; Lo et al. 1997; Gow et al. 1994; Gow 1997; Gow et al. 2002; Chandra et al. 2001; Gantner et al. 2005).

While multiple studies strongly suggest that filamentation is important for *C. albicans* virulence, a report by Noble et al., has challenged the dogma that there is a strict correlation between filamentation ability and pathogenicity (Noble et al. 2010). This study involved a screen of a large-scale homozygous knockout collection for specific defects in morphology and infectivity. A significant number of mutants were identified with defects in infectivity but not morphology (and vice versa). In this screen, infectivity was based on relative mutant strain abundance in mouse kidneys infected with a pool of bar-coded mutants, as determined by Q-PCR, and all morphology assays were performed in vitro. Although not practical on a large scale for the 674 mutants tested, it would be interesting to determine how many of these mutants show defects in virulence, invasion of other organs, and filamentation during infection in vivo. A more recent large-scale functional genomics study has shown a strong correlation between mutants that are defective for filamentation and those that are attenuated for virulence or avirulent (virulence data for 177 mutants was used in this study) (O'Meara et al. 2015). In addition, the large majority of strains that were competent for filamentation were either fully virulent or showed reduced virulence that was correlated with a severe growth defect. While the majority of evidence strongly suggests that filamentation is important for *C. albicans* virulence, the Noble et al., study (Noble et al. 2010) suggests that a more complex relationship exists between these two processes and that filamentation may not be absolutely required for certain aspects of pathogenicity.

4.3 Mechanics of *C. albicans* Hyphal Growth

C. albicans polarized growth is characterized by the movement of membrane-bound secretory vesicles toward the hyphal tip (Sudbery 2011). These vesicles most likely travel to the hyphal tip via actin cables and the myosin motor protein Myo2. The actin cables, in turn, are anchored to the tip by formins, such as Bni1. Bni1, along with vesicle-associated proteins (Sec2, Sec4, Mlc1) localizes to the Spitzenkörper, a vesicle-rich region just behind the hyphal tip (Crampin et al. 2005; Martin et al. 2005; Bishop et al. 2010; Jones and Sudbery 2010). The Spitzenkörper is believed to function as a vesicle supply center, coordinating the rate of hyphal growth by regulating both exocytosis and endocytosis of the vesicles at the hyphal tip. Both the shape of the hyphal tip as well as the direction of growth are determined by the position of the Spitzenkörper and the concentration gradient of vesicles that it produces (Bartnicki–Garcia et al. 1995; Lopez-Franco et al. 1994; Gale and Berman 2012; Sudbery 2011). The polarisome, a crescent-shaped structure at the hyphal tip, is comprised of the exocyst complex as well as the cell polarity markers Bud6 and Spa2 (Gale and Berman 2012; Sudbery 2011; Crampin et al. 2005). Bud6 and Spa2, in turn, help to anchor the Bni1 formin to the hyphal tip. Cdc42, a Rho-GTPase, plays a number of important roles in promoting *C. albicans* hyphal growth including exocyst complex formation, polarization of actin cables and patches to the hyphal tip, septin ring formation prior to bud emergence, and secretory vesicle docking (Sudbery 2011; Park and Bi 2007; Crampin et al. 2005). Both Cdc42 and Cdc24, the Cdc42 guanine exchange factor (GEF), are essential for viability and localize to the tip of hyphae (Sudbery 2011; Crampin et al. 2005; Park and Bi 2007; Bassilana et al. 2005; Bassilana et al. 2003; Ushinsky et al. 2002). Presence of both Cdc42 GTPase-activating proteins (GAPs), Rga2, and Bem3 (associated with bud emergence), is required for hyphal versus pseudohyphal growth (Court and Sudbery 2007). When Rga2 is phosphorylated by the Hgc1/Cdc28 cyclin/cyclin-dependent kinase (Cdk), its localization to the hyphal tip is blocked, leading to Cdc42 activation and sustained hyphal growth (Zheng et al. 2007). Hgc1/Cdc28 also phosphorylates septins, to promote septin ring formation, as well as the transcriptional regulator Efg1, to inhibit cell separation following cytokinesis (Sudbery 2011; Wang et al. 2009; Gonzalez-Novo et al. 2008). Two additional Rho-family GTPases, Rho3 and Rac1, also play important roles in actin polarization and hyphal growth (Dunkler and Wendland 2007; Bassilana and Arkowitz 2006; Hope et al. 2008).

4.4 Control of *C. albicans* Hyphal Growth by the Cell Cycle

C. albicans possesses three G_1 cyclins (Ccn1, Cln3, and Hgc1) as well as two G_2 cyclins (Clb2 and Clb4) that are thought to associate with the Cdc28 kinase (Gale and Berman 2012). Deletion of Cln3 results in large unbudded yeast cells which are still capable of forming hyphal filaments (Bachewich and Whiteway 2005; Chapa y Lazo et al. 2005). Interestingly, this finding suggests that Cln3 controls polarized growth of *C. albicans* yeast but not hyphae. Both Ccn1 and Hgc1are not required for initial formation of germ tubes but instead play a more important role in germ tube elongation and maintenance of hyphal growth. More specifically, sequential phosphorylation of the Cdc11 septin by Ccn1/Cdc28 and Hgc1/Cdc28 is associated with sustained hyphal development (Sinha et al. 2007). Hgc1/Cdc28 also plays an important role in maintaining hyphal growth through Cdc42 activation and inhibition of cell separation, as described above. The *HGC1* transcript has been shown to be induced by the key filamentous growth transcriptional regulator Ume6 (Carlisle and Kadosh 2010). *UME6*, in turn, is induced by a variety of filament-inducing conditions (Banerjee et al. 2008). Therefore, control of hyphal growth and maintenance by the Hgc1/Cdc28 complex is directly linked to environmental cues that promote filamentation.

The mitotic B cyclins Clb2 and Clb4 function to inhibit polarized growth of *C. albicans* yeast cells (Gale and Berman 2012). Strains depleted for Clb2 can still undergo hyphal growth, although there is a significant delay in mitosis and cells are very extended (Bensen et al. 2005). Unexpectedly, strains depleted for Clb4 are defective for hyphal growth and instead generate pseudohyphae when grown under hyphal-inducing conditions (Bensen et al. 2005). A similar phenotype is observed for strains bearing a mutation in *FKH2*, which encodes a transcriptional regulator important for *CLB4* induction (Bensen et al. 2002). Interestingly, these findings suggest that Clb4 controls polarized growth by different mechanisms in yeast versus hyphal cells (Gale and Berman 2012).

The cell cycle regulatory polo-like kinase, Cdc5, also plays an important role in *C. albicans* filamentous growth (Bachewich et al. 2003). Cells depleted for Cdc5 showed a block in nuclear division as well as defects in spindle elongation and chromatin separation. Surprisingly, these cells also generated hyphal-like filaments under nonfilament-inducing conditions. Although Cdc5 is not believed to directly repress hyphal growth, these results suggest an important link between *C. albicans* cell cycle regulation and morphogenesis. Consistent with this notion, *C. albicans* Cdc20 and Cdh1, anaphase-promoting complex co-activators, have been shown to play an important role in both mitotic progression and morphogenesis (Chou et al. 2011).

4.5 Regulation of the *C. albicans* Morphological Transition by Host Environmental Cues

Several key signal transduction pathways are important for the ability of *C. albicans* to sense host environmental cues and activate a transcriptional program important for morphogenesis (Fig. 4.2). A cAMP protein kinase A (cAMP-PKA) pathway responds to multiple host environmental conditions including serum, body temperature (37 °C), nitrogen starvation, high CO_2 levels, N-acetylglucosamine (GlcNAc, via the GlcNAc transporter Ngt1), and certain amino acids (e.g., methionine) as well as the quorum sensing molecules farnesol and dodecanol (Sudbery 2011; Shapiro et al. 2011; Alvarez and Konopka 2007). Several molecules are known to activate the cAMP-PKA pathway in response to these conditions (Fig. 4.2). The heat shock protein 90 (Hsp90) specifically responds to temperature (Shapiro et al. 2009). The cAMP-PKA pathway is also activated by Mep2, a transmembrane ammonium permease which senses nitrogen starvation conditions (Biswas and Morschhauser 2005). Another sensor is comprised of Gpr1 and the Gα protein Gpa2, which play an important role in activating the cAMP-PKA pathway in response to certain amino acids in the presence of glucose (Maidan et al. 2005). Ras1, a guanine nucleotide binding protein, functions to activate the Cyr1 adenylate cyclase (Rocha et al. 2001; Leberer et al. 2001). Cyr1 is also directly activated by CO_2/HCO_3 (Hall et al. 2010). Interestingly, the quorum sensing molecules farnesol and 3–oxo–C(12)–homoserine lactone (HSL), which is secreted by *Pseudomonas aeruginosa*, both directly inhibit *C. albicans* Cyr1 activity (Hall et al. 2011). Cyr1 generates cAMP which, in turn, functions to activate the PKA complex consisting of the two catalytic subunits Tpk1 and Tpk2 and the regulatory subunit Bcy1 (Rocha et al. 2001; Bockmuhl et al. 2001; Shapiro et al. 2011). Importantly, PKA has been shown to directly phosphorylate Efg1, a major transcriptional regulator of *C. albicans* filamentous growth and filament-specific target genes (Bockmuhl and Ernst 2001). Many components of the cAMP-PKA pathway, including Efg1, are known to be important for *C. albicans* virulence and pathogenicity (Lo et al. 1997; Shapiro et al. 2011).

A mitogen activated protein kinase (MAPK) pathway primarily responds to nutrient starvation conditions, particularly nitrogen starvation (Fig. 4.2) (Leberer et al. 1996; Liu et al. 1994). This pathway is homologous to the well-characterized MAPK pheromone signaling pathway of *S. cerevisiae*. The *C. albicans* MAPK pathway also shares several upstream components with the cAMP-PKA pathway.

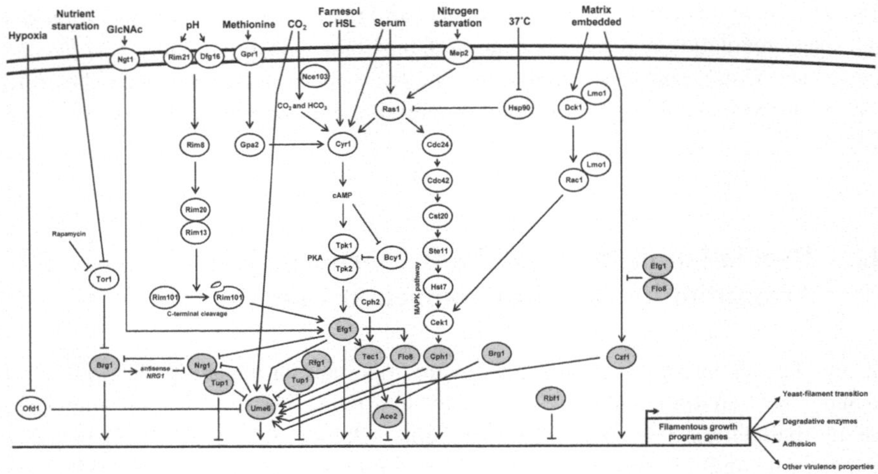

Fig. 4.2 Signal transduction pathways important for controlling *C. albicans* filamentous growth and expression of the filamentous growth program in response to environmental cues. Transcriptional regulators are indicated in *grey*. Please note that not all of the indicated interactions are direct and not all signaling components/transcriptional regulators are shown. In addition, positive and negative effects directed by certain signaling components/transcriptional regulators can vary depending on particular filament-inducing conditions. *PKA* Protein kinase A. *MAPK* Mitogen activated protein kinase. *HSL* 3–oxo–C(12)–homoserine lactone. Adapted from (Sudbery 2011)

More specifically, nitrogen starvation conditions can activate this pathway via the Mep2 sensor and Ras1 (Sudbery 2011; Biswas and Morschhauser 2005; Shapiro et al. 2011). In the context of the MAPK pathway, however, Ras1 specifically functions to activate the Cdc42 Rho–type GTPase via the Cdc24 GEF (Bassilana et al. 2003). Cdc42, in turn, promotes polarized growth, actin organization, and activation of the MAPK signal transduction cascade (Ushinsky et al. 2002; Sudbery 2011; Park and Bi 2007; Crampin et al. 2005). The terminal MAPK of this cascade, Cek1, is important for activation of Cph1, an ortholog of *S. cerevisiae* Ste20 and the main downstream transcription factor target of the *C. albicans* MAPK pathway (Liu et al. 1994). Cph1, in turn, is involved in controlling the level of activation of certain filament-specific target genes (Leng et al. 2001; Shapiro et al. 2011; Argimon et al. 2007). In general, filamentation and filament-specific gene expression defects resulting from mutations in Cph1 and MAPK pathway components are not as severe as those observed with mutants of Efg1 and components of the cAMP-PKA pathway (Shapiro et al. 2011).

In addition to the Cph1 MAPK pathway described above, other MAPK pathways have been shown to control *C. albicans* filamentous growth. These include the high osmolarity glycerol (HOG) pathway as well as the protein kinase C (PKC) cell wall integrity pathway. The HOG1 pathway is involved in mediating the response to a variety of stress conditions, including osmotic and heavy metal stresses (Enjalbert et al. 2006). Various components of this pathway also play roles in

C. albicans filamentation in response to multiple inducing conditions (Alonso-Monge et al. 1999; Shapiro et al. 2011). Mkc1, a MAPK component of the PKC pathway has been shown to be specifically important for the ability of *C. albicans* hyphae to invade solid media (Kumamoto 2005).

The highly conserved regulation of Ace2 and morphogenesis (RAM) pathway is also important for *C. albicans* filamentous growth in response to a variety of filament-inducing conditions including serum, Spider and Lee's medium at 37 °C (Saputo et al. 2012; Song et al. 2008). Mutations in multiple components of this pathway, including the critical Cbk1 kinase, lead to defects in cell wall integrity, cell separation, and filamentous growth (Song et al. 2008). Interestingly, mutation of *ACE2*, the downstream transcription factor target of the RAM pathway, results in constitutively pseudohyphal cells which are also defective for biofilm formation and adherence to plastic (Saputo et al. 2012; Kelly et al. 2004). Recent evidence suggests that the pseudohyphal phenotype of the *C. albicans ace2*Δ/Δ mutant may result from a compensatory increase in the activity of the PKA filamentous growth signaling pathway (Bharucha et al. 2011). In contrast to the findings described above, *C. albicans ace2*Δ/Δ mutants are defective for filamentous growth under hypoxic and embedded conditions, highlighting the complex regulation of this important transcription factor (Saputo et al. 2012; Mulhern et al. 2006).

Separate signaling pathways are known to direct *C. albicans* filamentous growth under embedded matrix conditions, as would be encountered in host tissues (Fig. 4.2). One pathway involves activation of Rac1, a G protein, by the GEF Dck1 (Bassilana and Arkowitz 2006; Hope et al. 2008). Both of these factors, in combination with Lmo1, appear to direct embedded filamentous growth via activation of the Cek1 MAPK (Hope et al. 2010). A separate embedded filamentous growth pathway is defined by the transcription factor Czf1 (Brown et al. 1999; Shapiro et al. 2011).

C. albicans also undergoes filamentation in response to alkaline pH conditions (Fig. 4.2) (Brown 2002). Both Rim21 and Dfg16 encode seven-transmembrane proteins that serve as sensors for the pH signaling pathway (Barwell et al. 2005; Davis 2003). These sensors are believed to function through Rim8 which, in turn, activates the Rim13 protease. Rim13, with assistance from Rim20, plays an important role in cleaving the C-terminal inhibitory domain of the Rim101 transcriptional regulator (Davis et al. 2000b; Li et al. 2004; Kullas et al. 2004). Following this proteolytic cleavage event under alkaline conditions, Rim101 is activated and will promote the expression of alkaline-induced genes such as *PHR1*, a glucosidase important for cell wall structure (Li et al. 2004; Davis et al. 2000b; Porta et al. 1999). Mutations in components of this pathway have been associated with a variety of filamentation and virulence defects. Rim101, in particular, plays an important role in oral epithelial cell damage and virulence in a mouse model of oropharyngeal candidiasis (Davis et al. 2000a; Mitchell et al. 2007; Nobile et al. 2008b; Yuan et al. 2010; Shapiro et al. 2011).

In large part, the signaling pathways described above target transcriptional regulators important for induction of both *C. albicans* filamentous growth and filament-specific transcripts. However, several key repressors of *C. albicans* filamentation have also been identified (Fig. 4.2). These repressors include the DNA-binding proteins Rfg1 and Nrg1, as well as the Tup1 co-repressor. Mutations in all three of these regulators, especially Nrg1 and Tup1, result in filamentation under non-filament-inducing conditions as well as highly attenuated virulence in a mouse model of systemic candidiasis (Braun et al. 2001; Kadosh and Johnson 2001; Braun et al. 2000; Braun and Johnson 1997; Murad et al. 2001; Khalaf and Zitomer 2001). Both Rfg1 and Nrg1 are believed to recruit Tup1 to the promoters of filament-specific genes to direct transcriptional repression. Importantly, the *NRG1* transcript is known to be down-regulated in response to serum at 37 °C, one of the strongest filament-inducing conditions (Braun et al. 2001; Murad et al. 2001). *NRG1* down-regulation is a key transcriptional event important not only for filamentation but also biofilm formation, tissue invasion, and virulence in a mouse model of systemic candidiasis (Saville et al. 2003; Uppuluri et al. 2010). This event requires *EFG1* and occurs via the cAMP-PKA pathway (Braun et al. 2001; Lu et al. 2011). Interestingly, induction of an antisense *NRG1* transcript by Brg1, a filament-induced transcriptional regulator, has been shown to reduce the stability of the *NRG1* transcript under strong filament-inducing conditions (Cleary et al. 2012). A quorum sensing pathway, mediated by farnesol, also inhibits filamentation by blocking degradation of the Nrg1 protein (Lu et al. 2014). As mentioned above, initiation of hyphal development requires a rapid and transient down-regulation of *NRG1* by the cAMP-PKA pathway. However, maintenance of hyphal-specific gene expression and hyphal growth requires recruitment of the Hda1 histone deacetylase to target promoters by Brg1 under conditions of reduced Tor signaling (Lu et al. 2011; Lu et al. 2012). Hda1, in turn, causes an alteration in promoter chromatin structure that prevents Nrg1 from binding. In addition, Ume6 then functions in a positive feedback loop to sustain hyphal-specific gene expression and promote hyphal elongation (Lu et al. 2011; 2012).

The Ume6 transcriptional regulator is induced in response to a variety of host filament-inducing conditions and represents a key downstream target for multiple *C. albicans* filamentous growth signaling pathways and transcription factors (Fig. 4.2) (Banerjee et al. 2008; Zeidler et al. 2009). *ume6*Δ/Δ mutants are specifically defective for hyphal extension and attenuated for virulence in a mouse model of systemic candidiasis. Ume6 also functions in a negative feedback loop with Nrg1 and plays an important role in promoting hyphal extension by controlling the level and duration of filament-specific gene expression (Banerjee et al. 2008). This is accomplished, in part, by induction of the Hgc1 cyclin-related protein and subsequent activities of the Hgc1/Cdc28 cyclin/Cdk complex, as discussed previously (Carlisle and Kadosh 2010; Sudbery 2011; Wang et al. 2009; Gonzalez-Novo et al. 2008; Zheng et al. 2007). Stabilization of Ume6 protein levels by the synergistic action of signaling pathways responding to hypoxia and CO_2 is important for maintaining hyphal elongation and promoting virulence (Lu et al. 2013). *UME6* expression has also recently been shown to be inhibited by a 5′ UTR-mediated translational efficiency mechanism (Childers et al. 2014). Interestingly, the level of translational inhibition appears to be modulated by a variety of different host filament-inducing conditions. Under non-filament-inducing conditions, expression levels of *UME6* are sufficient to promote a sequential morphological transition from yeast to pseudohyphae to hyphae (Carlisle et al. 2009). This result suggests that all three *C. albicans* morphologies are determined by a common dosage-dependent transcriptional mechanism. Given that *C. albicans* is a highly evolved opportunistic pathogen capable of survival in a wide variety of host environments, it is not surprising, as illustrated in this section, that a complex array of signaling mechanisms and regulators is involved in controlling morphogenesis in response to host environmental cues.

4.6 Genome-Wide Analyses of the *C. albicans* Morphological Transition

The availability of relatively new approaches, such as DNA microarray technology and RNA sequencing, has made it possible to determine the genome-wide transcriptional profile of *C. albicans* cells as they undergo the morphological transition from yeast to filaments. An initial DNA microarray experiment identified 28 genes showing >2 fold induction during the *C. albicans* yeast-hyphal transition in response to serum at 37 °C (Nantel et al. 2002). Stage-specific changes in gene expression were also identified during the serum and temperature induction time course. This study also found that a similar, but not identical, set of genes was induced when *C. albicans* cells were grown in Lee's medium, a different filament-inducing condition. A subsequent DNA microarray analysis using serum and temperature conditions optimized for complete filamentous growth, identified a total of 61 serum and temperature-induced genes, several of which had been previously identified in the prior study (Kadosh and Johnson 2005). A comparison of the *C. albicans* filamentous growth program with transcriptional profiles of the *rfg1*Δ/Δ, *nrg1*Δ/Δ, and *tup1*Δ/Δ mutants indicated that approximately half of all serum and temperature-induced genes were induced by relief of transcriptional repression. Genes involved in a wide variety of biological processes were found to be induced. A gene ontology (GO) analysis identified several gene classes that were overrepresented in the serum and temperature-induced gene set compared to their representation in the genome as a whole, including ER/Golgi/secretion, cell motility/budding/cell division, cell wall components, and secreted/degradative enzymes. Given the extensive cell wall remodeling and vesicle secretion that occurs during filamentous growth, overrepresentation of many of these gene classes appeared to make sense intuitively. Importantly, several of the induced cell wall components (e.g., Hwp1, Als3) had been previously identified as adhesins that are involved in adherence to host cells (Hoyer et al. 1998; Staab et al. 1999). Members of the secreted aspartyl protease (Sap) gene family, previously associated with tissue invasion and virulence, as well as Sod5, a superoxide dismutase important for the ability of *C. albicans* to tolerate oxidative stress in macrophages, were also induced (Hube 1996; Hube and Naglik 2002; Martchenko et al. 2004; Frohner et al.

2009). One of the most highly induced genes, *ECE1*, has recently been found to encode a cytolytic peptide toxin, candidalysin, important for epithelial cell damage and mucosal infection (Moyes et al. 2016). A subsequent RNA-Seq analysis has identified many of the same genes as being induced by growth in serum at 37 °C as well as several new and/or uncharacterized genes (Bruno et al. 2010). Overall, these studies demonstrated that the *C. albicans* filamentous growth program is not only comprised of genes important for the physical process of filamentation per se, but also genes associated with a variety of additional virulence-related processes.

One limitation of the studies described above is that because *C. albicans* filamentation is induced by external environmental conditions, it is difficult to distinguish between genes that are specifically associated with determination of pseudohyphal and hyphal morphologies and those that are simply expressed in response to the environmental condition. In order to determine sets of genes specifically associated with *C. albicans* morphology determination, a whole-genome DNA microarray analysis was performed using a *tetO-UME6* strain, which could be genetically manipulated to grow as yeast, pseudohyphae, or hyphae by altering *UME6* expression levels with doxycycline concentration (Carlisle and Kadosh 2013). Importantly, this experiment was carried out under standard non-filament-inducing conditions, without the added complications of external environmental cues. Interestingly, this study found that genes associated with the pseudohyphal morphology represented a subset of hyphal genes that were generally expressed at lower levels. Also, both the level and duration of filament-specific gene expression were sufficient to drive the *C. albicans* yeast-pseudohyphal-hyphal morphological transition. A GO term analysis indicated a strong correlation between gene classes that were induced by *UME6* during hyphal growth and those induced in response to serum and temperature (Carlisle and Kadosh 2013). Importantly, this study also identified a minimal core set of 15 genes that were consistently expressed in *C. albicans* hyphal filaments generated in response to *UME6* expression as well as growth in serum at 37 °C. Several of these genes, including *HGC1* as well as *CDC10* and *CDC12*, which encode septins, are specifically involved in the mechanics of hyphal growth (Zheng et al. 2004; Warenda and Konopka 2002; DiDomenico et al. 1994). Other genes in this set encode cell wall/cell surface proteins, including the adhesins Hwp1 and Als3, which are important for both host cell adhesion and biofilm formation (Hoyer et al. 1998; Staab et al. 1999; Nobile et al. 2006; Nobile et al. 2008a). The *tetO-UME6* system also allowed for transcriptional profiling of the reverse hyphal-pseudohyphal-yeast morphological transition in the absence of filament-inducing conditions (Carlisle and Kadosh 2013). Not surprisingly, many of the same gene classes induced during the forward yeast-pseudohyphal-hyphal transition were also down-regulated during the reverse hyphal-pseudohyphal-yeast transition. However, several additional gene classes, particularly those involved in protein synthesis, were down-regulated during the reverse transition as well. These findings are consistent with the notion that, overall, reduced protein production is associated with yeast phase growth. In addition to defining genes specifically associated with *C. albicans* pseudohyphal and hyphal morphologies, this study also highlighted several important differences in global gene expression patterns between the forward and reverse *C. albicans* morphological transitions (Carlisle and Kadosh 2013).

More recently, a functional genomics analysis of *C. albicans* filamentation was performed using the GRACE (gene replacement and conditional expression) collection of 2,356 *C. albicans* strains (O'Meara et al. 2015; Roemer et al. 2003). In this collection, one allele of each gene is deleted and the second allele placed under control of a regulatable promoter. 102 mutants were identified that showed enhanced filamentation under non-filament-inducing conditions (O'Meara et al. 2015). In addition to known regulators of *C. albicans* morphogenesis, approximately half of these genes appeared to be involved in cell cycle-related processes, such as regulation of DNA replication and structural maintenance of chromosomes. While certain cell cycle-related genes, such as *CDC28* and *CLB2*, had been previously implicated in *C. albicans* morphogenesis (Bensen et al. 2005; Umeyama et al. 2006), many others had not. Overall, 872 mutants were identified with filamentation defects (mild to severe) in response to growth in serum (O'Meara et al. 2015). A GO analysis indicated an enrichment for genes involved in vesicle and intracellular protein transport, which, as previously discussed, is known to play an important role in polarized growth. Interestingly, genes involved in ergosterol biosynthesis and N-linked glycosylation were also enriched in this gene set. Finally, the gene set contained many known regulators of *C. albicans* filamenatation, including *RAS1*, *FLO8*, *CDC24*, and *HSP90*, thus validating the functional genomics approach. One limitation of these experiments is that the GRACE collection is not complete and only contains mutants in about one-third of all *C. albicans* genes. Nevertheless, results from this functional genomics analysis have complemented those previously obtained using transcriptional profiling approaches and also provided new insights into processes that are required for *C. albicans* morphogenesis.

4.7 *C. albicans* Morphology as a Target for Novel Therapeutic Strategies

Given the importance of the *C. albicans* morphological transition for virulence, an obvious question is whether this transition can serve as a target for new antifungal therapies. Two recent studies have reported the identification of small molecule compounds that can inhibit *C. albicans* filamentation. The first study initially used a high-throughput screen to identify small molecules that inhibit *C. albicans* adhesion to polystyrene (Fazly et al. 2013). A novel compound, termed filastatin, was identified in this screen and found to also inhibit *C. albicans* adhesion to human epithelial cells, biofilm formation, the yeast-hyphal transition as well as pathogenicity in a nematode infection model. Importantly, filastatin was not toxic to human A549 cells, even at high concentrations. Filastatin also inhibited *C. albicans* filamentation driven by multiple signaling pathways, including the cAMP-PKA pathway. The second study involved a high-throughput screen of 20,000 small molecule compounds in the NOVACore™ library (Chembridge Corp.) (Pierce et al. 2015). Several novel diazaspiro-decane structural analogs were identified that significantly inhibited *C. albicans* biofilm formation. Importantly, the lead compound identified by this screen was able to inhibit both *C. albicans* biofilm formation and filamentation without eliciting resistance or affecting overall growth rate. Concentrations associated with cellular toxicity of the compound, as determined by exposure to a human hepatocyte cell line, were significantly higher than those required to inhibit biofilm formation. In vivo efficacy of the compound was demonstrated in both oral and systemic mouse models of candidiasis (Pierce et al. 2015). While the compounds described above are still in the very early stages of development, these studies have provided the best evidence so far that the *C. albicans* morphological transition can indeed serve as an effective target for antifungal compounds. Finally, by correlation, these studies suggest that many of the regulatory mechanisms, signaling pathways and target genes discussed in this chapter which are important for driving *C. albicans* filamentous growth may also serve as potential targets for the development of new and more effective antifungal strategies.

Acknowledgements This work was supported by National Institutes of Health grants 5R01AI083344 and 1R21AI117299 in addition to a Voelcker Young Investigator Award from the Max and Minnie Tomerlin Voelcker Fund. The content is solely the responsibility of the author and does not necessarily represent the official views of the National Institutes of Health.

References

Alonso-Monge R, Navarro-Garcia F, Molero G, Diez-Orejas R, Gustin M, Pla J, Sanchez M, Nombela C (1999) Role of the mitogen-activated protein kinase Hog1p in morphogenesis and virulence of *Candida albicans*. J Bacteriol 181(10):3058–3068

Alvarez FJ, Konopka JB (2007) Identification of an N-acetylglucosamine transporter that mediates hyphal induction in *Candida albicans*. Mol Biol Cell 18(3):965–975. doi:E06-10-0931 [pii]10.1091/mbc.E06-10-0931

Argimon S, Wishart JA, Leng R, Macaskill S, Mavor A, Alexandris T, Nicholls S, Knight AW, Enjalbert B, Walmsley R, Odds FC, Gow NA, Brown AJ (2007) Developmental regulation of an adhesin gene during cellular morphogenesis in the fungal pathogen *Candida albicans*. Eukaryot Cell 6(4):682–692. doi:EC.00340-06 [pii]10.1128/EC.00340-06

Bachewich C, Whiteway M (2005) Cyclin Cln3p links G1 progression to hyphal and pseudohyphal development in *Candida albicans*. Eukaryot Cell 4(1):95–102. doi:4/1/95 [pii] 10.1128/EC.4.1.95-102.2005

Bachewich C, Thomas DY, Whiteway M (2003) Depletion of a polo-like kinase in *Candida albicans* activates cyclase-dependent hyphal-like growth. Mol Biol Cell 14(5):2163–2180. doi:10.1091/mbc.02-05-0076

Banerjee M, Thompson DS, Lazzell A, Carlisle PL, Pierce C, Monteagudo C, Lopez-Ribot JL, Kadosh D (2008) *UME6*, a novel filament-specific regulator of *Candida albicans* hyphal extension and virulence. Mol Biol Cell 19(4):1354–1365

Barelle CJ, Bohula EA, Kron SJ, Wessels D, Soll DR, Schafer A, Brown AJ, Gow NA (2003) Asynchronous cell cycle and asymmetric vacuolar inheritance in true hyphae of *Candida albicans*. Eukaryot Cell 2(3):398–410

Bartnicki-Garcia S, Bartnicki DD, Gierz G, Lopez-Franco R, Bracker CE (1995) Evidence that Spitzenkorper behavior determines the shape of a fungal hypha: a test of the hyphoid model. Exp Mycol 19(2):153–159

Barwell KJ, Boysen JH, Xu W, Mitchell AP (2005) Relationship of *DFG16* to the Rim101p pH response pathway in *Saccharomyces cerevisiae* and *Candida albicans*. Eukaryot Cell 4 (5):890–899. doi:10.1128/EC.4.5.890-899.2005

Bassilana M, Arkowitz RA (2006) Rac1 and Cdc42 have different roles in *Candida albicans* development. Eukaryot Cell 5(2):321–329. doi:10.1128/EC.5.2.321-329.2006

Bassilana M, Blyth J, Arkowitz RA (2003) Cdc24, the GDP-GTP exchange factor for Cdc42, is required for invasive hyphal growth of *Candida albicans*. Eukaryot Cell 2(1):9–18

Bassilana M, Hopkins J, Arkowitz RA (2005) Regulation of the Cdc42/Cdc24 GTPase module during *Candida albicans* hyphal growth. Eukaryot Cell 4(3):588–603. doi:10.1128/EC.4.3.588-603.2005

Bensen ES, Filler SG, Berman J (2002) A forkhead transcription factor is important for true hyphal as well as yeast morphogenesis in *Candida albicans*. Eukaryot Cell 1(5):787–798

Bensen ES, Clemente-Blanco A, Finley KR, Correa-Bordes J, Berman J (2005) The mitotic cyclins Clb2p and Clb4p affect morphogenesis in *Candida albicans*. Mol Biol Cell 16 (7):3387–3400. doi:E04-12-1081 [pii]10.1091/mbc.E04-12-1081

Bharucha N, Chabrier-Rosello Y, Xu T, Johnson C, Sobczynski S, Song Q, Dobry CJ, Eckwahl MJ, Anderson CP, Benjamin AJ, Kumar A, Krysan DJ (2011) A large-scale complex haploinsufficiency-based genetic interaction screen in *Candida albicans*: analysis of the RAM network during morphogenesis. PLoS Genet 7(4):e1002058. doi:10.1371/journal.pgen.1002058

Bishop A, Lane R, Beniston R, Chapa-y-Lazo B, Smythe C, Sudbery P (2010) Hyphal growth in *Candida albicans* requires the phosphorylation of Sec2 by the Cdc28-Ccn1/Hgc1 kinase. EMBO J 29(17):2930–2942. doi:10.1038/emboj.2010.158

Biswas K, Morschhauser J (2005) The Mep2p ammonium permease controls nitrogen starvation-induced filamentous growth in *Candida albicans*. Mol Microbiol 56(3):649–669. doi:MMI4576 [pii]10.1111/j.1365-2958.2005.04576.x

Bockmuhl DP, Ernst JF (2001) A potential phosphorylation site for an A-type kinase in the Efg1 regulator protein contributes to hyphal morphogenesis of *Candida albicans*. Genetics 157 (4):1523–1530

Bockmuhl DP, Krishnamurthy S, Gerads M, Sonneborn A, Ernst JF (2001) Distinct and redundant roles of the two protein kinase A isoforms Tpk1p and Tpk2p in morphogenesis and growth of *Candida albicans*. Mol Microbiol 42(5):1243–1257. doi:2688 [pii]

Bramley TA, Menzies GS, Williams RJ, Kinsman OS, Adams DJ (1991) Binding sites for LH in *Candida albicans*: comparison with the mammalian corpus luteum LH receptor. J Endocrinol 130(2):177–190

Braun BR, Johnson AD (1997) Control of filament formation in *Candida albicans* by the transcriptional repressor *TUP1*. Science 277(5322):105–109

Braun BR, Head WS, Wang MX, Johnson AD (2000) Identification and characterization of *TUP1*-regulated genes in *Candida albicans*. Genetics 156(1):31–44

Braun BR, Kadosh D, Johnson AD (2001) *NRG1*, a repressor of filamentous growth in *C. albicans*, is down-regulated during filament induction. EMBO J 20:4753–4761

Brown AJ (2002) Expression of growth form-specific factors during morphogenesis in *Candida albicans*. In: Calderone RA (ed) *Candida* and Candidiasis. ASM Press, Washington, D.C., pp 87–93

Brown DH Jr, Giusani AD, Chen X, Kumamoto CA (1999) Filamentous growth of *Candida albicans* in response to physical environmental cues and its regulation by the unique *CZF1* gene. Mol Microbiol 34(4):651–662

Bruno VM, Wang Z, Marjani SL, Euskirchen GM, Martin J, Sherlock G, Snyder M (2010) Comprehensive annotation of the transcriptome of the human fungal pathogen *Candida albicans* using RNA-seq. Genome Res 20(10):1451–1458. doi:gr.109553.110 [pii]10.1101/gr. 109553.110

Carlisle PL, Kadosh D (2010) *Candida albicans* Ume6, a filament-specific transcriptional regulator, directs hyphal growth via a pathway involving Hgc1 cyclin-related protein. Eukaryot Cell 9(9):1320–1328. doi:EC.00046-10 [pii]10.1128/EC.00046-10

Carlisle PL, Kadosh D (2013) A genome-wide transcriptional analysis of morphology determination in *Candida albicans*. Mol Biol Cell 24(3):246–260. doi:10.1091/mbc.E12-01-0065

Carlisle PL, Banerjee M, Lazzell A, Monteagudo C, Lopez-Ribot JL, Kadosh D (2009) Expression levels of a filament-specific transcriptional regulator are sufficient to determine *Candida albicans* morphology and virulence. Proc Natl Acad Sci USA 106:599–604

Caticha O, Grover S, Winge D, Odell WD (1992) Stimulation of *Candida albicans* transition by human chorionic gonadotrophin and a bacterial protein. Endocr Res 18(2):133–143

Chabasse D, Bouchara JP, de Gentile L, Chennebault JM (1988) *Candida albicans* chlamydospores observed *in vivo* in a patient with AIDS. Ann Biol Clin (Paris) 46(10):817–818

Chandra J, Kuhn DM, Mukherjee PK, Hoyer LL, McCormick T, Ghannoum MA (2001) Biofilm formation by the fungal pathogen *Candida albicans*: development, architecture, and drug resistance. J Bacteriol 183(18):5385–5394

Chapa y Lazo B, Bates S, Sudbery P (2005) The G1 cyclin Cln3 regulates morphogenesis in *Candida albicans*. Eukaryot Cell 4(1):90–94. doi:10.1128/EC.4.1.90-94.2005

Childers DS, Mundodi V, Banerjee M, Kadosh D (2014) A 5' UTR-mediated translational efficiency mechanism inhibits the *Candida albicans* morphological transition. Mol Microbiol 92(3):570–585. doi:10.1111/mmi.12576

Chou H, Glory A, Bachewich C (2011) Orthologues of the anaphase-promoting complex/cyclosome coactivators Cdc20p and Cdh1p are important for mitotic progression and morphogenesis in *Candida albicans*. Eukaryot Cell 10(5):696–709. doi:10.1128/EC. 00263-10

Cleary IA, Lazzell AL, Monteagudo C, Thomas DP, Saville SP (2012) *BRG1* and *NRG1* form a novel feedback circuit regulating *Candida albicans* hypha formation and virulence. Mol Microbiol 85(3):557–573. doi:10.1111/j.1365-2958.2012.08127.x

Cole GT, Seshan KR, Phaneuf M, Lynn KT (1991) Chlamydospore-like cells of *Candida albicans* in the gastrointestinal tract of infected, immunocompromised mice. Can J Microbiol 37 (8):637–646

Court H, Sudbery P (2007) Regulation of Cdc42 GTPase activity in the formation of hyphae in *Candida albicans*. Mol Biol Cell 18(1):265–281

Crampin H, Finley K, Gerami-Nejad M, Court H, Gale C, Berman J, Sudbery P (2005) *Candida albicans* hyphae have a Spitzenkorper that is distinct from the polarisome found in yeast and pseudohyphae. J Cell Sci 118(Pt 13):2935–2947

Csank C, Haynes K (2000) *Candida glabrata* displays pseudohyphal growth. FEMS Microbiol Lett 189(1):115–120. doi:S0378-1097(00)00241-X [pii]

Davis D (2003) Adaptation to environmental pH in *Candida albicans* and its relation to pathogenesis. Curr Genet 44(1):1–7. doi:10.1007/s00294-003-0415-2

Davis D, Edwards JE Jr, Mitchell AP, Ibrahim AS (2000a) *Candida albicans RIM101* pH response pathway is required for host-pathogen interactions. Infect Immun 68(10):5953–5959

Davis D, Wilson RB, Mitchell AP (2000b) *RIM101*-dependent and -independent pathways govern pH responses in *Candida albicans*. Mol Cell Biol 20:971–978

DiDomenico BJ, Brown NH, Lupisella J, Greene JR, Yanko M, Koltin Y (1994) Homologs of the yeast neck filament associated genes.In: Isolation and sequence analysis of *Candida albicans CDC3* and *CDC10*. Mol Gen Genet 242(6):689–698

Do Carmo-Sousa L (1969) Distribution of yeasts in nature. In: Rose AH, Harrison JS (eds) The Yeasts, vol 1. Academic Press, pp 79–105

Dunkler A, Wendland J (2007) *Candida albicans* Rho-type GTPase-encoding genes required for polarized cell growth and cell separation. Eukaryot Cell 6(5):844–854. doi:10.1128/EC.00201-06

Enjalbert B, Smith DA, Cornell MJ, Alam I, Nicholls S, Brown AJ, Quinn J (2006) Role of the Hog1 stress-activated protein kinase in the global transcriptional response to stress in the fungal pathogen *Candida albicans*. Mol Biol Cell 17(2):1018–1032. doi:10.1091/mbc.E05-06-0501

Fazly A, Jain C, Dehner AC, Issi L, Lilly EA, Ali A, Cao H, Fidel PL Jr, Rao RP, Kaufman PD (2013) Chemical screening identifies filastatin, a small molecule inhibitor of *Candida albicans* adhesion, morphogenesis, and pathogenesis. Proc Natl Acad Sci USA 110(33):13594–13599. doi:10.1073/pnas.1305982110

Fidel PLJ, al. e (1999) *Candida glabrata*: a review of epidemiology, pathogenesis, and clinical disease. Clin Microbiol Rev 12:80–96

Frohner IE, Bourgeois C, Yatsyk K, Majer O, Kuchler K (2009) *Candida albicans* cell surface superoxide dismutases degrade host-derived reactive oxygen species to escape innate immune surveillance. Mol Microbiol 71(1):240–252. doi:MMI6528 [pii]10.1111/j.1365-2958.2008.06528.x

Gale CA, Berman J (2012) Cell cycle and growth control in *Candida* species. In: Calderone RA, Clancy CJ (eds) *Candida* and Candidiasis. ASM Press, Washington, D.C., pp 101–124

Gantner BN, Simmons RM, Underhill DM (2005) Dectin-1 mediates macrophage recognition of *Candida albicans* yeast but not filaments. EMBO J 24(6):1277–1286

Geiger J, Wessels D, Lockhart SR, Soll DR (2004) Release of a potent polymorphonuclear leukocyte chemoattractant is regulated by white-opaque switching in *Candida albicans*. Infect Immun 72(2):667–677

Gonzalez-Novo A, Correa-Bordes J, Labrador L, Sanchez M, Vazquez de Aldana CR, Jimenez J (2008) Sep7 is essential to modify septin ring dynamics and inhibit cell separation during *Candida albicans* hyphal growth. Mol Biol Cell 19(4):1509–1518. doi:E07-09-0876 [pii]10.1091/mbc.E07-09-0876

Gow NA (1997) Germ tube growth of *Candida albicans*. Curr Top Med Mycol 8(1–2):43–55

Gow NA, Gooday GW (1984) A model for the germ tube formation and mycelial growth form of *Candida albicans*. Sabouraudia 22(2):137–144

Gow NA, Henderson G, Gooday GW (1986) Cytological interrelationships between the cell cycle and duplication cycle of *Candida albicans*. Microbios 47(191):97–105

Gow NA, Perera TH, Sherwood-Higham J, Gooday GW, Gregory DW, Marshall D (1994) Investigation of touch-sensitive responses by hyphae of the human pathogenic fungus *Candida albicans*. Scan Microsc 8(3):705–710

Gow NA, Brown AJ, Odds FC (2002) Fungal morphogenesis and host invasion. Curr opin microbiol 5(4):366–371. doi:S1369527402003387 [pii]

Hall RA, De Sordi L, Maccallum DM, Topal H, Eaton R, Bloor JW, Robinson GK, Levin LR, Buck J, Wang Y, Gow NA, Steegborn C, Muhlschlegel FA (2010) CO(2) acts as a signalling molecule in populations of the fungal pathogen *Candida albicans*. PLoS Pathog 6(11):e1001193. doi:10.1371/journal.ppat.1001193

Hall RA, Turner KJ, Chaloupka J, Cottier F, De Sordi L, Sanglard D, Levin LR, Buck J, Muhlschlegel FA (2011) The quorum-sensing molecules farnesol/homoserine lactone and dodecanol operate via distinct modes of action in *Candida albicans*. Eukaryot Cell 10(8):1034–1042. doi:10.1128/EC.05060-11

Hope H, Bogliolo S, Arkowitz RA, Bassilana M (2008) Activation of Rac1 by the guanine nucleotide exchange factor Dck1 is required for invasive filamentous growth in the pathogen *Candida albicans*. Mol Biol Cell 19(9):3638–3651. doi:10.1091/mbc.E07-12-1272

Hope H, Schmauch C, Arkowitz RA, Bassilana M (2010) The *Candida albicans* ELMO homologue functions together with Rac1 and Dck1, upstream of the MAP Kinase Cek1, in invasive filamentous growth. Mol Microbiol 76(6):1572–1590. doi:10.1111/j.1365-2958.2010.07186.x

Hoyer LL, Payne TL, Bell M, Myers AM, Scherer S (1998) *Candida albicans ALS3* and insights into the nature of the *ALS* gene family. Curr Genet 33(6):451–459

Huang G, Srikantha T, Sahni N, Yi S, Soll DR (2009) CO(2) regulates white-to-opaque switching in *Candida albicans*. Curr Biol 19(4):330–334. doi:10.1016/j.cub.2009.01.018

Huang G, Yi S, Sahni N, Daniels KJ, Srikantha T, Soll DR (2010) N-acetylglucosamine induces white to opaque switching, a mating prerequisite in *Candida albicans*. PLoS Pathog 6(3): e1000806. doi:10.1371/journal.ppat.1000806

Hube B (1996) *Candida albicans* secreted aspartyl proteinases. Curr Top Med Mycol 7(1):55–69

Hube B, Naglik J (2002) Extracellular hydrolases. In: Calderone RA (ed) *Candida* and Candidiasis. ASM Press, Washington, D.C., pp 107–122

Johnson A (2003) The biology of mating in *Candida albicans*. Nat Rev Microbiol 1(2):106–116. doi:10.1038/nrmicro752

Jones LA, Sudbery PE (2010) Spitzenkorper, exocyst, and polarisome components in *Candida albicans* hyphae show different patterns of localization and have distinct dynamic properties. Eukaryot Cell 9(10):1455–1465. doi:10.1128/EC.00109-10

Jong AY, Stins MF, Huang SH, Chen SH, Kim KS (2001) Traversal of *Candida albicans* across human blood-brain barrier in vitro. Infect Immun 69(7):4536–4544. doi:10.1128/IAI.69.7.4536-4544.2001

Kadosh D, Johnson AD (2001) Rfg1, a protein related to the *Saccharomyces cerevisiae* hypoxic regulator Rox1, controls filamentous growth and virulence in *Candida albicans*. Mol Cell Biol 21:2496–2505

Kadosh D, Johnson AD (2005) Induction of the *Candida albicans* filamentous growth program by relief of transcriptional repression: a genome-wide analysis. Mol Biol Cell 16(6):2903–2912

Kelly MT, MacCallum DM, Clancy SD, Odds FC, Brown AJ, Butler G (2004) The *Candida albicans CaACE2* gene affects morphogenesis, adherence and virulence. Mol Microbiol 53 (3):969–983. doi:10.1111/j.1365-2958.2004.04185.xMMI4185 [pii]

Khalaf RA, Zitomer RS (2001) The DNA binding protein Rfg1 is a repressor of filamentation in *Candida albicans*. Genet 157(4):1503–1512

Kinsman OS, Pitblado K, Coulson CJ (1988) Effect of mammalian steroid hormones and luteinizing hormone on the germination of *Candida albicans* and implications for vaginal candidosis. Mycoses 31(12):617–626

Klein BS, Tebbets B (2007) Dimorphism and virulence in fungi. Curr Opin Microbiol 10(4):314–319. doi:S1369-5274(07)00080-X [pii]10.1016/j.mib.2007.04.002

Korting HC, Hube B, Oberbauer S, Januschke E, Hamm G, Albrecht A, Borelli C, Schaller M (2003) Reduced expression of the hyphal-independent *Candida albicans* proteinase genes *SAP1* and *SAP3* in the *efg1* mutant is associated with attenuated virulence during infection of oral epithelium. J Med Microbiol 52(Pt 8):623–632

Kullas AL, Li M, Davis DA (2004) Snf7p, a component of the ESCRT-III protein complex, is an upstream member of the RIM101 pathway in *Candida albicans*. Eukaryot Cell 3(6):1609–1618. doi:10.1128/EC.3.6.1609-1618.2004

Kumamoto CA (2005) A contact-activated kinase signals *Candida albicans* invasive growth and biofilm development. Proc Natl Acad Sci USA 102(15):5576–5581

Kumamoto CA, Vinces MD (2005) Contributions of hyphae and hypha-co-regulated genes to *Candida albicans* virulence. Cell Microbiol 7(11):1546–1554

Lan CY, Newport G, Murillo LA, Jones T, Scherer S, Davis RW, Agabian N (2002) Metabolic specialization associated with phenotypic switching in *Candida albicans*. Proc Natl Acad Sci USA 99(23):14907–14912. doi:10.1073/pnas.232566499

Leberer E, Harcus D, Broadbent ID, Clark KL, Dignard D, Ziegelbauer K, Schmidt A, Gow NAR, Brown AJP, Thomas DY (1996) Signal transduction through homologues of the Ste20p and Ste7p protein kinases can trigger hyphal formation in the pathogenic fungus *Candida albicans*. Proc Natl Acad Sci USA 93:13217–13222

Leberer E, Harcus D, Dignard D, Johnson L, Ushinsky S, Thomas DY, Schroppel K (2001) Ras links cellular morphogenesis to virulence by regulation of the MAP kinase and cAMP signalling pathways in the pathogenic fungus *Candida albicans*. Mol Microbiol 42(3):673–687

Leng P, Lee PR, Wu H, Brown AJ (2001) Efg1, a morphogenetic regulator in *Candida albicans*, is a sequence-specific DNA binding protein. J Bacteriol 183(13):4090–4093. doi:10.1128/JB.183.13.4090-4093.2001

Li M, Martin SJ, Bruno VM, Mitchell AP, Davis DA (2004) *Candida albicans* Rim13p, a protease required for Rim101p processing at acidic and alkaline pHs. Eukaryot Cell 3(3):741–751. doi:10.1128/EC.3.3.741-751.2004

Liu H, Kohler J, Fink GR (1994) Suppression of hyphal formation in *Candida albicans* by mutation of a *STE12* homolog. Sci 266(5191):1723–1726

Lo HJ, Kohler JR, DiDomenico B, Loebenberg D, Cacciapuoti A, Fink GR (1997) Nonfilamentous *C. albicans* mutants are avirulent. Cell 90(5):939–949

Lohse MB, Johnson AD (2008) Differential phagocytosis of white versus opaque *Candida albicans* by *Drosophila* and mouse phagocytes. PLoS ONE 3(1):e1473. doi:10.1371/journal.pone.0001473

Lohse MB, Johnson AD (2009) White-opaque switching in *Candida albicans*. Curr Opin Microbiol 12(6):650–654. doi:S1369-5274(09)00144-1 [pii]10.1016/j.mib.2009.09.010

Lohse MB, Hernday AD, Fordyce PM, Noiman L, Sorrells TR, Hanson-Smith V, Nobile CJ, DeRisi JL, Johnson AD (2013) Identification and characterization of a previously undescribed family of sequence-specific DNA-binding domains. Proc Natl Acad Sci USA 110(19):7660–7665. doi:10.1073/pnas.1221734110

Lopez-Franco R, Bartnicki-Garcia S, Bracker CE (1994) Pulsed growth of fungal hyphal tips. Proc Natl Acad Sci USA 91(25):12228–12232

Lu Y, Su C, Wang A, Liu H (2011) Hyphal development in *Candida albicans* requires two temporally linked changes in promoter chromatin for initiation and maintenance. PLoS Biol 9 (7):e1001105. doi:10.1371/journal.pbio.1001105 PBIOLOGY-D-10–01188 [pii]

Lu Y, Su C, Liu H (2012) A GATA transcription factor recruits Hda1 in response to reduced Tor1 signaling to establish a hyphal chromatin state in *Candida albicans*. PLoS Pathog 8(4):e1002663. doi:10.1371/journal.ppat.1002663 PPATHOGENS-D-11–02628 [pii]

Lu Y, Su C, Solis NV, Filler SG, Liu H (2013) Synergistic regulation of hyphal elongation by hypoxia, CO2, and nutrient conditions controls the virulence of *Candida albicans*. Cell Host Microbe 14(5):499–509. doi:10.1016/j.chom.2013.10.008

Lu Y, Su C, Unoje O, Liu H (2014) Quorum sensing controls hyphal initiation in *Candida albicans* through Ubr1-mediated protein degradation. Proc Natl Acad Sci USA 111(5):1975–1980. doi:10.1073/pnas.1318690111

Maidan MM, Thevelein JM, Van Dijck P (2005) Carbon source induced yeast-to-hypha transition in *Candida albicans* is dependent on the presence of amino acids and on the G-protein-coupled receptor Gpr1. Biochem Soc Trans 33(Pt 1):291–293. doi:10.1042/BST0330291

Martchenko M, Alarco AM, Harcus D, Whiteway M (2004) Superoxide dismutases in *Candida albicans*: transcriptional regulation and functional characterization of the hyphal-induced *SOD5* gene. Mol Biol Cell 15(2):456–467

Martin R, Walther A, Wendland J (2005) Ras1-induced hyphal development in *Candida albicans* requires the formin Bni1. Eukaryot Cell 4(10):1712–1724. doi:10.1128/EC.4.10.1712-1724.2005

Merson-Davies LA, Odds FC (1989) A morphology index for characterization of cell shape in *Candida albicans*. J Gen Microbiol 135(Pt 11)):3143–3152

Miller MG, Johnson AD (2002) White-opaque switching in *Candida albicans* is controlled by mating-type locus homeodomain proteins and allows efficient mating. Cell 110(3):293–302. doi:S0092867402008371 [pii]

Mitchell BM, Wu TG, Jackson BE, Wilhelmus KR (2007) *Candida albicans* strain-dependent virulence and Rim13p-mediated filamentation in experimental keratomycosis. Invest Ophthalmol Vis Sci 48(2):774–780. doi:10.1167/iovs.06-0793

Morschhauser J (2010) Regulation of white-opaque switching in *Candida albicans*. Med Microbiol Immunol 199(3):165–172. doi:10.1007/s00430-010-0147-0

Moyes DL, Wilson D, Richardson JP, Mogavero S, Tang SX, Wernecke J, Hofs S, Gratacap RL, Robbins J, Runglall M, Murciano C, Blagojevic M, Thavaraj S, Forster TM, Hebecker B,

Kasper L, Vizcay G, Iancu SI, Kichik N, Hader A, Kurzai O, Luo T, Kruger T, Kniemeyer O, Cota E, Bader O, Wheeler RT, Gutsmann T, Hube B, Naglik JR (2016) Candidalysin is a fungal peptide toxin critical for mucosal infection. Nat 532(7597):64–68. doi:10.1038/nature17625

Mulhern SM, Logue ME, Butler G (2006) *Candida albicans* transcription factor Ace2 regulates metabolism and is required for filamentation in hypoxic conditions. Eukaryot Cell 5(12):2001–2013. doi:10.1128/EC.00155-06

Murad AMA, Leng P, Straffon M, Wishart J, Macaskill S, MacCallum D, Schnell N, Talibi D, Marechal D, Tekaia F, d'Enfert C, Gaillardin C, Odds FC, Brown AJP (2001) *NRG1* represses yeast-hypha morphogenesis and hypha-specific gene expression in *Candida albicans*. EMBO J 20:4742–4752

Nantel A, Dignard D, Bachewich C, Harcus D, Marcil A, Bouin AP, Sensen CW, Hogues H, van het Hoog M, Gordon P, Rigby T, Benoit F, Tessier DC, Thomas DY, Whiteway M (2002) Transcription profiling of *Candida albicans* cells undergoing the yeast-to-hyphal transition. Mol Biol Cell 13(10):3452–3465

Nobile CJ, Nett JE, Andes DR, Mitchell AP (2006) Function of *Candida albicans* adhesin Hwp1 in biofilm formation. Eukaryot Cell 5(10):1604–1610

Nobile CJ, Schneider HA, Nett JE, Sheppard DC, Filler SG, Andes DR, Mitchell AP (2008a) Complementary adhesin function in *C. albicans* biofilm formation. Curr Biol 18(14):1017–1024. doi:10.1016/j.cub.2008.06.034

Nobile CJ, Solis N, Myers CL, Fay AJ, Deneault JS, Nantel A, Mitchell AP, Filler SG (2008b) *Candida albicans* transcription factor Rim101 mediates pathogenic interactions through cell wall functions. Cell Microbiol 10(11):2180–2196. doi:10.1111/j.1462-5822.2008.01198.x

Noble SM, French S, Kohn LA, Chen V, Johnson AD (2010) Systematic screens of a *Candida albicans* homozygous deletion library decouple morphogenetic switching and pathogenicity. Nat Genet 42(7):590–598. doi:ng.605 [pii]10.1038/ng.605

O'Meara TR, Veri AO, Ketela T, Jiang B, Roemer T, Cowen LE (2015) Global analysis of fungal morphology exposes mechanisms of host cell escape. Nat Commun 6:6741. doi:10.1038/ncomms7741

Odds FC (1988) *Candida* and Candidosis, 2nd edn. Baillière Tindall, London

Pande K, Chen C, Noble SM (2013) Passage through the mammalian gut triggers a phenotypic switch that promotes *Candida albicans* commensalism. Nat Genet 45(9):1088–1091. doi:10.1038/ng.2710

Park HO, Bi E (2007) Central roles of small GTPases in the development of cell polarity in yeast and beyond. Microbiol Mol Biol Rev 71(1):48–96. doi:10.1128/MMBR.00028-06

Pierce CG, Chaturvedi AK, Lazzell AL, Powell AT, Saville SP, McHardy SF, Lopez-Ribot JL (2015) A novel small molecule inhibitor of biofilm formation, filamentation and virulence with low potential for the development of resistance. NPJ Biofilms Microbiomes 1. doi:10.1038/npjbiofilms.2015.12

Porta A, Ramon AM, Fonzi WA (1999) *PRR1*, a homolog of *Aspergillus nidulans palF*, controls pH-dependent gene expression and filamentation in *Candida albicans*. J Bacteriol 181(24):7516–7523

Rikkerink EH, Magee BB, Magee PT (1988) Opaque-white phenotype transition: a programmed morphological transition in *Candida albicans*. J Bacteriol 170(2):895–899

Rocha CR, Schroppel K, Harcus D, Marcil A, Dignard D, Taylor BN, Thomas DY, Whiteway M, Leberer E (2001) Signaling through adenylyl cyclase is essential for hyphal growth and virulence in the pathogenic fungus *Candida albicans*. Mol Biol Cell 12(11):3631–3643

Roemer T, Jiang B, Davison J, Ketela T, Veillette K, Breton A, Tandia F, Linteau A, Sillaots S, Marta C, Martel N, Veronneau S, Lemieux S, Kauffman S, Becker J, Storms R, Boone C, Bussey H (2003) Large-scale essential gene identification in *Candida albicans* and applications to antifungal drug discovery. Mol Microbiol 50(1):167–181

Saputo S, Chabrier-Rosello Y, Luca FC, Kumar A, Krysan DJ (2012) The RAM network in pathogenic fungi. Eukaryot Cell 11(6):708–717. doi:10.1128/EC.00044-12

Sasse C, Hasenberg M, Weyler M, Gunzer M, Morschhauser J (2013) White-opaque switching of *Candida albicans* allows immune evasion in an environment-dependent fashion. Eukaryot Cell 12(1):50–58. doi:10.1128/EC.00266-12

Saville SP, Lazzell AL, Monteagudo C, Lopez-Ribot JL (2003) Engineered control of cell morphology *in vivo* reveals distinct roles for yeast and filamentous forms of *Candida albicans* during infection. Eukaryot Cell 2(5):1053–1060

Shapiro RS, Uppuluri P, Zaas AK, Collins C, Senn H, Perfect JR, Heitman J, Cowen LE (2009) Hsp90 orchestrates temperature-dependent *Candida albicans* morphogenesis via Ras1-PKA signaling. Curr Biol 19(8):621–629. doi:10.1016/j.cub.2009.03.017

Shapiro RS, Robbins N, Cowen LE (2011) Regulatory circuitry governing fungal development, drug resistance, and disease. Microbiol Mol Biol Rev 75(2):213–267. doi:10.1128/MMBR.00045-10

Sinha I, Wang YM, Philp R, Li CR, Yap WH, Wang Y (2007) Cyclin-dependent kinases control septin phosphorylation in *Candida albicans* hyphal development. Dev Cell 13(3):421–432

Slutsky B, Staebell M, Anderson J, Risen L, Pfaller M, Soll DR (1987) "White-opaque transition". In: A second high-frequency switching system in *Candida albicans*. J Bacteriol 169(1):189–197

Soll DR, Morrow B, Srikantha T (1993) High-frequency phenotypic switching in *Candida albicans*. Trends Genet 9(2):61–65

Song Y, Cheon SA, Lee KE, Lee SY, Lee BK, Oh DB, Kang HA, Kim JY (2008) Role of the RAM network in cell polarity and hyphal morphogenesis in *Candida albicans*. Mol Biol Cell 19(12):5456–5477. doi:10.1091/mbc.E08-03-0272

Staab JF, Bradway SD, Fidel PL, Sundstrom P (1999) Adhesive and mammalian transglutaminase substrate properties of *Candida albicans* Hwp1. Sci 283(5407):1535–1538

Staib P, Morschhauser J (2007) Chlamydospore formation in *Candida albicans* and *Candida dubliniensis*–an enigmatic developmental programme. Mycoses 50(1):1–12. doi:MYC1308 [pii]10.1111/j.1439-0507.2006.01308.x

Sudbery PE (2001) The germ tubes of *Candida albicans* hyphae and pseudohyphae show different patterns of septin ring localization. Mol Microbiol 41(1):19–31. doi:mmi2459 [pii]

Sudbery PE (2011) Growth of *Candida albicans* hyphae. Nat Rev Microbiol 9(10):737–748. doi:10.1038/nrmicro2636

Sudbery P, Gow N, Berman J (2004) The distinct morphogenic states of *Candida albicans*. Trends Microbiol 12(7):317–324

Tao L, Du H, Guan G, Dai Y, Nobile CJ, Liang W, Cao C, Zhang Q, Zhong J, Huang G (2014) Discovery of a "white-gray-opaque" tristable phenotypic switching system in *Candida albicans*: roles of non-genetic diversity in host adaptation. PLoS Biol 12(4):e1001830. doi:10.1371/journal.pbio.1001830

Thompson DS, Carlisle PL, Kadosh D (2011) Coevolution of morphology and virulence in *Candida* species. Eukaryot Cell 10(9):1173–1182. doi:EC.05085-11 [pii]10.1128/EC.05085-11

Umeyama T, Kaneko A, Niimi M, Uehara Y (2006) Repression of *CDC28* reduces the expression of the morphology-related transcription factors, Efg1p, Nrg1p, Rbf1p, Rim101p, Fkh2p and Tec1p and induces cell elongation in *Candida albicans*. Yeast 23(7):537–552. doi:10.1002/yea.1373

Uppuluri P, Pierce CG, Thomas DP, Bubeck SS, Saville SP, Lopez-Ribot JL (2010) The transcriptional regulator Nrg1p controls *Candida albicans* biofilm formation and dispersion. Eukaryot Cell 9(10):1531–1537. doi:EC.00111-10 [pii]10.1128/EC.00111-10

Ushinsky SC, Harcus D, Ash J, Dignard D, Marcil A, Morchhauser J, Thomas DY, Whiteway M, Leberer E (2002) *CDC42* is required for polarized growth in human pathogen *Candida albicans*. Eukaryot Cell 1(1):95–104

Wang A, Raniga PP, Lane S, Lu Y, Liu H (2009) Hyphal chain formation in *Candida albicans*: Cdc28-Hgc1 phosphorylation of Efg1 represses cell separation genes. Mol Cell Biol 29 (16):4406–4416. doi:10.1128/MCB.01502-08

Warenda AJ, Konopka JB (2002) Septin function in *Candida albicans* morphogenesis. Mol Biol Cell 13(8):2732–2746

Yokoyama K, Takeo K (1983) Differences of asymmetrical division between the pseudomycelial and yeast forms of *Candida albicans* and their effect on multiplication. Arch Microbiol 134 (3):251–253

Yuan X, Mitchell BM, Hua X, Davis DA, Wilhelmus KR (2010) The *RIM101* signal transduction pathway regulates *Candida albicans* virulence during experimental keratomycosis. Invest Ophthalmol Vis Sci 51(9):4668–4676. doi:10.1167/iovs.09-4726

Zeidler U, Lettner T, Lassnig C, Muller M, Lajko R, Hintner H, Breitenbach M, Bito A (2009) *UME6* is a crucial downstream target of other transcriptional regulators of true hyphal development in *Candida albicans*. FEMS Yeast Res 9(1):126–142. doi:FYR459 [pii]10.1111/j.1567-1364.2008.00459.x

Zheng X, Wang Y, Wang Y (2004) Hgc1, a novel hypha-specific G1 cyclin-related protein regulates *Candida albicans* hyphal morphogenesis. EMBO J 23(8):1845–1856

Zheng XD, Lee RT, Wang YM, Lin QS, Wang Y (2007) Phosphorylation of Rga2, a Cdc42 GAP, by CDK/Hgc1 is crucial for *Candida albicans* hyphal growth. EMBO J 26(16):3760–3769

Zink S, Nass T, Rosen P, Ernst JF (1996) Migration of the fungal pathogen *Candida albicans* across endothelial monolayers. Infect Immun 64(12):5085–5091

Zordan RE, Miller MG, Galgoczy DJ, Tuch BB, Johnson AD (2007) Interlocking transcriptional feedback loops control white-opaque switching in *Candida albicans*. PLoS Biol 5(10):e256. doi:10.1371/journal.pbio.0050256

Chapter 5
Candida *albicans* Biofilms

Priya Uppuluri and Jose Luis Lopez Ribot

Abstract Microorganisms have been conventionally studied in the laboratory as unicellular life forms. However, in their natural habitats, microbes are found in bionetworks attached to surfaces and not as free-floating (planktonic) organisms. *Candida* albicans is a part of the normal microbiota of humans, which enables contact with most medical devices and host surfaces, permitting formation of robust biofilms. Since the appearance of the first report on *C. albicans* biofilms two decades ago, there have been considerable advancements in the overall understanding of this complicated growth lifestyle. The areas of development have encompassed molecular understanding of the process, newer diagnostic modalities to identify biofilm-related cells, and identification of novel anti-biofilm drugs to curtail the process. In this chapter we have touched upon the current trends in the field of *C. albicans* biofilm research, including biofilm dispersal—the final stage of the biofilm lifecycle.

5.1 Introduction

Microorganisms have been conventionally studied in the laboratory as unicellular life forms. However, in their natural habitats, microbes are found in bionetworks attached to surfaces and not as free-floating (planktonic) organisms (Costerton et al. 1987, 1995; Donlan 2002). Thus, biofilms are classically defined as structured microbial communities, attached to a surface and encased in a matrix of self-produced exopolymeric material. A shift from free-living growth to biofilm is

P. Uppuluri (✉)
Los Angeles Biomedical Research Institute, Harbor-UCLA,
1124 W. Carson St, Torrance, CA 90502, USA
e-mail: puppuluri@labiomed.org

J.L. Lopez Ribot
Department of Biology, University of Texas San Antonio, San Antonio,
TX 98249, USA
e-mail: jose.lopezribot@utsa.edu

R. Prasad (ed.), *Candida albicans: Cellular and Molecular Biology*,
DOI 10.1007/978-3-319-50409-4_5

elicited by environmental variations and involves multiple regulatory networks that control gene expression changes, leading to spatial and temporal reorganization of the microbial cell (Desai et al. 2014; Monds and O'Toole 2009). Recent estimates by the National Institutes of Health indicate that pathogenic biofilms are responsible, directly or indirectly, for over 80% of all microbial infections (Gulati and Nobile 2016; Nobile and Johnson 2015).

Candida albicans is a part of the normal microbiota of humans, which enables contact with most medical devices and host surfaces. *Candida* biofilms have been found to develop readily on medical implants such as stents, shunts, pacemakers, endotracheal tubes, and various types of catheters (Kojic and Darouiche 2004). *C. albicans* are the third leading cause of catheter-related infections, representing the second highest colonization-to-infection rate and the overall highest crude mortality (Crump and Collignon 2000). Once a biofilm is formed, it becomes exceedingly difficult to get rid of the cells, because of their increased resistance to antifungal therapy and the ability of cells within biofilms to withstand host immune defenses. These cells thereby serve as a reservoir of infectious cells for future continuing infections, having direct access to the bloodstream, leading to invasive systemic diseases. The ability of *C. albicans* to form biofilms on medical devices has a profound effect on its capacity to cause human disease (Hawser and Douglas 1994). Studies of catheter-related *Candida* infections have unequivocally shown that retention of vascular catheters is associated with prolonged fungemia, high antifungal therapy failure rates, increased risk of metastatic complications and death. Catheter-related infections are the major cause of morbidity and mortality among hospitalized patients, and microbial biofilms are associated with 90% of these infections. In fact, catheter-associated *Candida* biofilms can lead to bloodstream infections with an approximate incidence of one episode per 100 hospital admissions (DiDone et al. 2011; Finkel and Mitchell 2011; Finkel et al. 2012), with device-associated mortality rates as high as 30% (Viudes et al. 2002). Removal of catheters remains the only choice to facilitate more rapid clearance of the bloodstream and better prognosis (Eppes et al. 1989; Viudes et al. 2002). Unfortunately, catheter removal may not always be an option in chronic catheter dependency because of a lack of alternative vascular access sites (Droste et al. 2003). The net effect is that *Candida* biofilms adversely impact the health of patients, with increasing frequency and severity, and with soaring economic sequelae (Beck-Sague and Jarvis 1993; Viudes et al. 2002; Wilson et al. 2002).

5.2 Characteristics of *C. albicans* Biofilm

Most evidence on the structural features connected with *C. albicans* biofilms comes from in vitro experiments. A wide variety of biofilm models have been established, and have utilized several different kinds of materials for biofilm development—catheter material, glass slides, tubing, a perfused biofilm fermenter, acrylic strips/disks, germanium substratum, polystyrene microtiter plates, and tissue culture flasks, among other systems. Biofilm formation has also been studied under both static and flow through conditions (Baillie and Douglas 1998a, b, 1999a, b, 2000; Chandra et al. 2001a, b; Hawser 1996; Hawser et al. 1998; Hawser and Douglas 1994; Lamfon et al. 2003; Raad et al. 2003; Ramage et al. 2002; Adams et al. 2002; Ramage et al. 2001a, b; Suci et al. 2001; Pierce et al. 2010; Uppuluri et al. 2009; Uppuluri and Lopez-Ribot 2010). Besides the in vitro approaches, several in vivo animal models of catheter-associated *Candida* infections have been described. Visualizations of the resulting in vivo-formed biofilms recapitulate structural features of biofilms formed in vitro (Andes et al. 2004; Schinabeck et al. 2004).

One of the most popular substrates to develop a *C. albicans* biofilms in the laboratory is silicone elastomer, a material used for intravascular catheters. Typically, a piece of the silicone elastomer is added to a culture of *C. albicans* (in any of a wide variety of growth media), and incubated for a couple of hours to initiate adhesion of the fungal cells to the surface of the substrate. After a series of wash steps to get rid of the non-adherent cells, biofilm is allowed to form in fresh medium. Throughout the process, the cells could either be incubated under static conditions (without shaking), or grown under a continuous flow of liquid across the biofilm, mimicking exposure of an implanted catheter to the sheer stress of blood flow. A mature biofilm typically forms within 24 h and can be visualized by naked eye as a layer of cells on top of the silicone square, and with a microscope as an organized collection of different cell types.

The biofilm growth is coordinated by events didactically divided into four stages (Fig. 5.1): (a) Early stage, where the founding yeasts adhere to the substrate forming the biofilm scaffold; (b) Intermediate stage of microcolony development and proliferation into filamentous forms; and (c) Mature stage, where the tridimensional structure enlarges to a thickness of tens to thousands of microns, and is encased in exopolymeric matrix secretions. Throughout the developmental cycle,

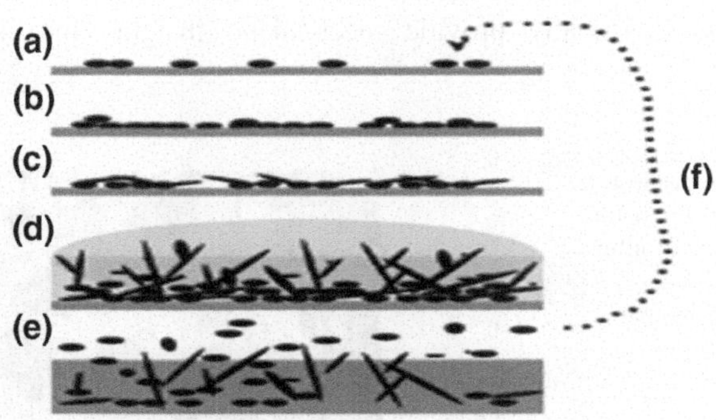

Fig. 5.1 Different stages of *C. albicans* biofilm formation. **a** Initial attachment of yeast cells. **b** Formation of a basal monolayer of cells. **c** Proliferation of microcolonies and filamentation. **d** Further filamentation during maturation and encasement within exopolymeric material; **e** Dispersion; **f** The arrow indicates how dispersed cells disseminate and go on to colonize new surfaces to complete the biofilm life cycle

yeasts cells from the topmost layers of the biofilm are released intermittently, and can disperse in the surrounding growth environment. Two steps—adherence of the original yeast cells to the bottom of the surface and differentiation into hyphae—are the backbone of a robust biofilm.

In groundbreaking work by Hawser and Douglas (Hawser et al. 1998), the structure of *C. albicans* biofilms formed on catheter disks was first examined by scanning electron microscopy (SEM). The initial attachment of yeast cells was followed by germ tube formation within 3–6 h. After 24–48 h of incubation, the fully mature *C. albicans* biofilms consisted of a dense network of yeasts, hyphae, and pseudohyphae, and extracellular polymeric material. Similar observations were made for *C. albicans* biofilms formed on acrylic material (Chandra et al. 2001a; Ramage et al. 2001b). Figure 5.2 shows a SEM image of a mature *C. albicans* biofilm.

Other method for visualizing biofilms in their natural growth habitat is confocal scanning laser microscopy which allows the visualization of fully hydrated living biofilms. The use of this technique has revealed the complex three-dimensional structure of mature C. *albicans* biofilms, in particular the spatial heterogeneity and architecture of microcolonies with ramifying water channels (Chandra et al. 2001a; Ramage et al. 2001b).

The well-established in vitro biofilm systems have for the most part, correlated well with in vivo and ex vivo biofilm models. For example, *Candida* biofilms recovered from patients with denture stomatitis or from patients with infected intravascular catheters confirm the presence of yeast, hyphae, and extracellular matrix (Nobile and Johnson 2015; Ramage et al. 2004). Biofilm architectures in several rat biofilm models, such as central venous, indwelling urinary catheter, and denture stomatitis models, also consist of numerous yeast cells in the basal region, and hyphae and extracellular matrix extending throughout the biofilm (Andes et al. 2004; Nett et al. 2014; Johnson et al. 2012; Schinabeck et al. 2004). In vivo models provide physiologically "natural" conditions to understand the biology of the biofilms, and can additionally provide mechanistic insights into host–pathogen

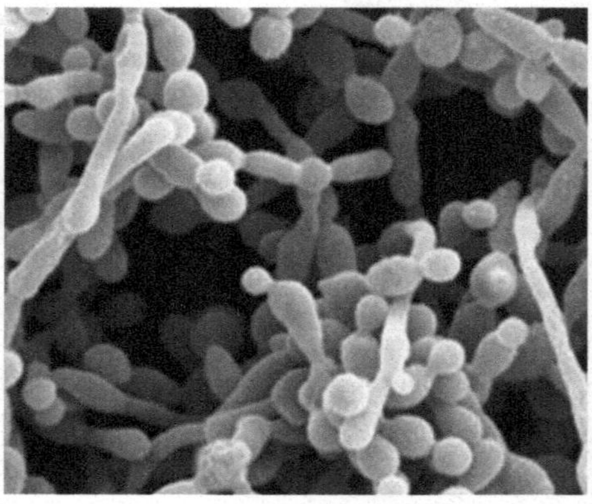

Fig. 5.2 Scanning electron microscopy image of a mature *C. albicans* biofilm. Biofilms are composed of yeast, hyphal, and pseudohyphal elements. Bar = 10 μm

interactions. Several such models will be discussed in subsequent chapters in this book authored by Dr. David Andes.

5.3 Genetic Regulation of *C. albicans* Biofilm Formation

Early studies to identify key molecular controllers of *C. albicans* biofilms were performed by screening mutant libraries of non-essential genes of the fungus; to date, nearly 1200 gene knockout mutants of this organism have been constructed (out of a total gene count of ~6000). Other pioneering approaches to identify biofilm-specific genes and proteins included genome-wide transcriptional profiling and proteomics techniques, respectively (Martinez-Gomariz et al. 2009; Zarnowski et al. 2014). These studies revealed a distinct biofilm-associated molecular signature, vastly different from those displayed under the planktonic growth condition. Substantial knowledge has been accumulated on the transcriptional regulation of biofilm formation—almost 50 transcriptional regulators and 101 non-regulatory genes that have functionally validated roles in biofilm formation have been identified.

C. albicans biofilm formation is regulated by a circuit of six transcription factors, *EFG1, TEC1, BCR1, NDT80, ROB1,* and *BRG1*, which have recently evolved to adapt precisely to host environment (Nobile et al. 2012). These transcription factors control the expression of each other and are activators and repressors of their target genes. Together the six regulators are responsible for the regulatory network of around 15% of the genes in the *C. albicans* genome [42]. These target genes also play distinct roles in different processes of biofilm development, such as adhesion, hyphal formation, drug resistance, and the production of extracellular matrix. Additionally, based on ontology mapping, the vast set of target genes is significantly enriched for "young" genes, suggesting that the ability of *C. albicans* to form biofilms evolved relatively recently with respect to evolutionary timescales. This could perhaps be the reason why *C. albicans* is one of the few fungal organisms capable of developing a biofilm in mammalian host.

In addition to the six master transcriptional regulators mentioned above, 44 additional transcriptional regulators have been identified, whose deletion has been shown to affect different aspects of *C. albicans* biofilm formation (Nobile et al. 2012; Nobile and Johnson 2015; Finkel and Mitchell 2011) [1, 7, 27]. Interestingly, the majority of these regulators are directly regulated by the core biofilm circuitry of the six genes.

5.3.1 *Adherence*

Every vital step of biofilm formation is strategically controlled by genes regulating those respective functions. To initiate a biofilm, yeast cells express adhesins on the cell surface to adhere to host tissues or abiotic surfaces. Two genes, a GPI-anchored cell wall adhesin, Eap1p and a cell surface adhesin, Csh1p confer attachment to inanimate surfaces, whereas the latter also mediates nonspecific adherence to host cells through hydrophobicity of cell surface (Ramage et al. 2009; de Souza et al. 2009; Klis et al. 2009). One of the biofilm master regulators, Bcr1, and some of its downstream targets, including the cell wall proteins Als1, Als3, and Hwp1, are all required for adherence during biofilm formation (Chandra et al. 2001a; Nobile et al. 2006a; Nobile and Mitchell 2005; Nobile et al. 2006b, 2008; Zhao et al. 2006). The adhesins interact with each other promoting cell-substrate adherence mediated by Als1p and Eap1p, and cell–cell interaction orchestrated by Als3p and Hwp1p (Finkel and Mitchell 2011; Nobile et al. 2008).

5.3.2 Hyphal Development

After adherence step, yeast cells form germ tubes that will grow to form true hyphae. Morphogenetic transition from yeast hyphae is an important characteristic of *C. albicans*, both under planktonic as well as biofilm conditions. Within 24 h of initiation, the biofilm displays a thick network of hyphal cells conferring a robust structure and stability to the entity. Genes and proteins involved in hyphal growth in suspension cultures are also required for proper biofilm formation. In biofilm phase, this process is mainly mediated by the *EFG1* gene involved in regulation of morphological transition and ability to form adherent structure on polystyrene, polyurethane, and glass. Yeast-to-hyphal transition is also mediated by other genes, such as *CPH1, TEC1, SUV3, NUP85, UME6, MDS3,* and *KEM3* (Nobile et al. 2006b; Ramage et al. 2009). Although dimorphism is necessary for the development of the spatially organized structure seen in mature biofilms, a few studies argue that it may not be an absolute prerequisite for biofilm formation, since substantial yeast-only biofilms have been described (Baillie and Douglas 1998b, 1999b).

5.3.3 Biofilm Dispersal

The active shedding of single cells from the body of the biofilm is referred to as dispersion. This phenomenon constitutes completion of a biofilm life cycle, as dispersed cells disseminate and go on to colonize new surfaces where a new biofilm is formed (Fig. 5.1). Biological processes associated with the early stages off *C. albicans* biofilm formation have become a major research focus over the past several years [reviewed in (Nobile and Mitchell 2006; Ramage et al. 2006)]. However, events associated with dispersion of cells from late stage of *C. albicans* biofilms (Fig. 5.1 steps E, F) have received little attention. Virtually nothing is known about the mechanisms of biofilm dispersal or the signals that trigger it. This lack of knowledge is particularly worrisome because biofilm dispersion is the pivotal mechanism associated with life-threatening clinical infections in susceptible patients: it is precisely detached cells that are responsible for septicemia and for dissemination to target organs prior to the establishment of invasive disease. Importantly, both of these clinical manifestations (candidemia and invasive candidiasis) are associated with the high morbidity and mortality rates (Gudlaugsson et al. 2003; Viudes et al. 2002). Rather than an end-stage process, release of cells from a biofilm occurs throughout its growth cycle (Uppuluri et al. 2010a). Apart from dispersal, *C. albicans* biofilms may also undergo a more dramatic massive detachment event, in which the entire biofilm can detach from the surface by hitherto unidentified mechanisms (Sellam et al. 2009).

Using genetically engineered *C. albicans* tetracycline-regulatable strains it was reported that the phenomenon of biofilm dispersion is under the control of the key transcriptional regulators *NRG1*, *UME6,* and *PES1* (Uppuluri et al. 2010a, b). The molecular chaperone Hsp90 has also been implicated in *C. albicans* biofilm dispersal, as depletion of Hsp90 leads to hyper-filamentation, and markedly reduces the number of dispersed cells from a biofilm (Robbins et al. 2011). The protein Ywp1, which is found present only on the cell wall of yeast cells, is also important for biofilm dispersal. Deletion of *YWP1* leads to decreased biofilm dispersal and increased biofilm adhesiveness (Granger et al. 2005). Current, limited knowledge on biofilm dispersal points out that any mutation that favors filamentous cells over yeast-form cells reduces biofilm dispersal. Consequently, hyphae to yeast transition increase the frequency of dispersal from a biofilm.

Whether dispersed cells display molecular characteristics similar to their source (the biofilm) or the population of cells they most appear like (planktonic cells) is not known. Compared to free-living cells, dispersed cells are reported to be infectious particles displaying ~40% increase in both adherence (to plastic) and biofilm-forming ability (Uppuluri et al. 2010a). Dispersed cells also display enhanced adhesion and damage to endothelial cells, which signify major hallmarks of the infectious process and have been found to be significantly more lethal compared to their planktonic counterparts in a hematogenously disseminated murine model of candidiasis. This later finding indicates that dispersed cells may be able to retain their virulence properties over several generations, perhaps indicating that heritable epigenetic modifications are responsible for enhanced adhesion, filamentation, and virulence of dispersed cells.

5.3.4 Extracellular Matrix (ECM)

Biofilm matrix is defined as an extracellular polymeric material that is maintained within a biofilm (Blankenship and Mitchell 2006; Branda et al. 2005; Sutherland 2001). It is secreted by cells in a biofilm, and its constituents may also come from the local site, such as an infected host. Thus the ECM provides support and protection of the microbial community embedded within it.

The *C. albicans* biofilm matrix is predominantly composed of proteins and glycoproteins (55%), carbohydrates (25%), lipids (15%), and nucleic acids (5%). Numerous proteins, mostly hydrolyzing enzymes, have been identified in the matrix, suggesting that the matrix may play an active role in breaking down biopolymers (Nobile and Johnson 2015; Zarnowski et al. 2014). Additionally, the *C. albicans* matrix is composed primarily of carbohydrate (probably β-1,3 glucan) and includes protein, hexosamine, phosphorus, and uronic acid (Al-Fattani and Douglas 2006). Nett et al. have shown that elevated β-1,3 glucan levels are characteristic of biofilm cells as compared to planktonic free-living *C. albicans* cells (Nett et al. 2007b). In fact, soluble β-1,3 glucan produced by *C. albicans* biofilms grown in an *in vivo* catheter infection model, can be used in diagnosis of catheter-based infection (Nett et al. 2007a). The gene that synthesizes β-1,3 glucan, *FKS1* and the predicted glucan transferases *BGL2* and *PHR1* and the exo-glucanase *XOG1* are responsible for the delivery and arrangement of -1,3 glucan in the matrix (Nett et al. 2010a; Taff et al. 2012).

The ECM production has many steps and regulatory networks. The zinc-response transcription factor Zap1 is one of the major negative regulators of the major matrix component, soluble glucan. It regulates *GCA1, GCA2*, and *ADH5* genes that are activators of matrix production and also binds to the promoters of *CSH1* and *IFD6* involved in inhibition of matrix production (Nobile et al. 2009). This pathway works independently from the *FKS1* matrix formation regulatory pathway (Taff et al. 2012).

The matrix also contributes to biofilm drug resistance, both acting as a physical barrier to drug penetration and support to the structural integrity of the biofilm (Al-Fattani and Douglas 2006; Nett et al. 2007b; Baillie and Douglas 2000; Nett et al. 2010b; Nobile and Johnson 2015). Overall, the biofilm matrix appears to perform as an extracellular, enzymatically active element of a *C. albicans* biofilm—one that both protects (from antifungal drugs and immune cells) and nurtures (acts as a nutrient source) the growing cells in a biofilm.

5.4 Future Perspectives

The ability to form biofilms is intimately associated with the ability to cause infection and as such should be considered an important virulence determinant during candidiasis. The biofilm lifestyle results in antifungal drug resistance and protection from host defenses, both of which carry important clinical repercussions. No biofilm-specific drugs exist today for the treatment of any biofilm-based microbial infection. A better understanding of the molecular mechanisms underlying biofilm formation is the key to development of new therapeutic agents that specifically target the biofilm state. Identification of the molecular mechanisms behind cell to surface adherence, filamentation or biofilm dispersal could lead to a drug-based strategy to prevent cells from attaching, proliferating or leaving a biofilm, respectively.

An alternative school of thought is targeting pathogenic mechanisms or virulence factors, rather than molecular processes for the development of new antibiotics (Vila et al. 2016). Discovery of novel drugs directed against certain virulence "phenotypes," such as filamentation or dispersal, and not their regulatory determinants could be a strategy fairly straightforward and simple. Because of its narrow spectrum of action, an anti-virulence approach should not alter the natural host's microbiota, and this could be of critical importance in the case of normal commensals such as *C. albicans* (Pierce et al. 2015b; Vila et al. 2016). In the last few years several groups have implemented more targeted efforts using high-content phenotypic screens for the identification of small molecule inhibitors of *C. albicans* filamentation. In the "phenotype-based" approach, large libraries of small molecules are screened in in vitro assays in a high-throughput manner to identify compounds that affect a biological process of interest (Berman and Sudbery 2002; Molero et al. 1998). Until recently, high-throughput screens were only feasible by pharmaceutical companies. Now, technological advances, including increased number of competitively priced compound collections available from companies, have made such screens a highly desirable strategy. Most recent studies elaborated on comprehensive screens of small and large molecule libraries consisting of off-patent drugs, many already approved by the Food and Drug Administration (FDA), to search for inhibitors of *C. albicans* biofilm formation. Several bioreactive molecules have been identified by such approaches. As an example, "Filastatin," named for its strong and long-lasting inhibition of filamentation, was discovered by screening 30,000 small molecules from a commercial library (DIVERset, Chembridge) (Fazly et al. 2013), to search for inhibitors of adhesion and filamentation. Similarly, screening of 20,000 small molecules from the research-intensive and medicinally relevant NOVACore™ chemical library (Chembridge) identified a novel hit series of diazaspiro-decane structural analogs, with activities against filamentation and biofilm formation (Pierce et al. 2015a). Identification of several inhibitory compounds without previously characterized antifungal activity indicates that chemi-genomic screening and/or repurposing orphan drugs may open up a valuable new avenue for identification and rapid development of antifungal agents (Siles et al. 2013).

References

Adams CC, Jakovljevic J, Roman J, Harnpicharnchai P, Woolford JL (2002) Saccharomyces cerevisiae nucleolar protein Nop7p is necessary for biogenesis of 60S ribosomal subunits. RNA 8:150–165

Al-Fattani MA, Douglas LJ (2006) Biofilm matrix of Candida albicans and Candida tropicalis: chemical composition and role in drug resistance. J Med Microbiol 55:999–1008

Andes D, Nett J, Oschel P, Albrecht R, Marchillo K, Pitula A (2004) Development and characterization of an in vivo central venous catheter Candida albicans biofilm model. Infect Immun 72:6023–6031

Baillie GS, Douglas LJ (1998a) Effect of growth rate on resistance of Candida albicans biofilms to antifungal agents. Antimicrob Agents Chemother 42:1900–1905
Baillie GS, Douglas LJ (1998b) Iron-limited biofilms of Candida albicans and their susceptibility to amphotericin B. Antimicrob Agents Chemother 42:2146–2149
Baillie GS, Douglas LJ (1999a) Candida biofilms and their susceptibility to antifungal agents. Methods Enzymol 310:644–656
Baillie GS, Douglas LJ (1999b) Role of dimorphism in the development of Candida albicans biofilms. J Med Microbiol 48:671–679
Baillie GS, Douglas LJ (2000) Matrix polymers of Candida biofilms and their possible role in biofilm resistance to antifungal agents. J Antimicrob Chemother 46:397–403
Beck-Sague C, Jarvis WR (1993) Secular trends in the epidemiology of nosocomial fungal infections in the United States, 1980–1990. National Nosocomial Infections Surveillance System. J Infect Dis 167:1247–1251
Berman J, Sudbery PE (2002) Candida Albicans: a molecular revolution built on lessons from budding yeast. Nat Rev Genet 3:918–930
Blankenship JR, Mitchell AP (2006) How to build a biofilm: a fungal perspective. Curr Opin Microbiol 9:588–594
Branda SS, Vik S, Friedman L, Kolter R (2005) Biofilms: the matrix revisited. Trends Microbiol 13:20–26
Chandra J, Kuhn DM, Mukherjee PK, Hoyer LL, Mccormick T, Ghannoum MA (2001a) Biofilm formation by the fungal pathogen Candida albicans: development, architecture, and drug resistance. J Bacteriol 183:5385–5394
Chandra J, Mukherjee PK, Leidich SD, Faddoul FF, Hoyer LL, Douglas LJ, Ghannoum MA (2001b) Antifungal resistance of candidal biofilms formed on denture acrylic in vitro. J Dent Res 80:903–908
Costerton JW, Cheng KJ, Geesey GG, Ladd TI, Nickel JC, Dasgupta M, Marrie TJ (1987) Bacterial biofilms in nature and disease. Annu Rev Microbiol 41:435–464
Costerton JW, Lewandowski Z, Caldwell DE, Korber DR, Lappin-Scott HM (1995) Microbial biofilms. Annu Rev Microbiol 49:711–745
Crump JA, Collignon PJ (2000) Intravascular catheter-associated infections. Eur J Clin Microbiol Infect Dis 19:1–8
De Souza RD, Mores AU, Cavalca L, Rosa RT, Samaranayake LP, Rosa EA (2009) Cell surface hydrophobicity of Candida albicans isolated from elder patients undergoing denture-related candidosis. Gerodontology 26:157–161
Desai JV, Mitchell AP, Andes DR (2014) Fungal biofilms, drug resistance, and recurrent infection. Cold Spring Harb Perspect Med 4
Didone L, Oga D, Krysan DJ (2011) A novel assay of biofilm antifungal activity reveals that amphotericin B and caspofungin lyse Candida albicans cells in biofilms. Yeast 28:561–568
Donlan RM (2002) Biofilms: microbial life on surfaces. Emerg Infect Dis 8:881–890
Droste JC, Jeraj HA, Macdonald A, Farrington K (2003) Stability and in vitro efficacy of antibiotic-heparin lock solutions potentially useful for treatment of central venous catheter-related sepsis. J Antimicrob Chemother 51:849–855
Eppes SC, Troutman JL, Gutman LT (1989) Outcome of treatment of candidemia in children whose central catheters were removed or retained. Pediatr Infect Dis J 8:99–104
Fazly A, Jain C, Dehner AC, Issi L, Lilly EA, Ali A, Cao H, Fidel PL, Rao RP, Kaufman PD (2013) Chemical screening identifies filastatin, a small molecule inhibitor of Candida albicans adhesion, morphogenesis, and pathogenesis. Proc Natl Acad Sci U S A 110:13594–13599
Finkel JS, Mitchell AP (2011) Genetic control of Candida albicans biofilm development. Nat Rev Microbiol 9:109–118
Finkel JS, Xu W, Huang D, Hill EM, Desai JV, Woolford CA, Nett JE, Taff H, Norice CT, Andes DR, Lanni F, Mitchell AP (2012) Portrait of Candida albicans adherence regulators. PLoS Pathog 8:e1002525
Granger BL, Flenniken ML, Davis DA, Mitchell AP, Cutler JE (2005) Yeast wall protein 1 of Candida albicans. Microbiology 151:1631–1644

Gudlaugsson O, Gillespie S, Lee K, Vande Berg J, Hu J, Messer S, Herwaldt L, Pfaller M, Diekema D (2003) Attributable mortality of nosocomial candidemia, revisited. Clin Infect Dis 37:1172–1177

Gulati M, Nobile CJ (2016) Candida albicans biofilms: development, regulation, and molecular mechanisms. Microbes Infect 18:310–321

Hawser S (1996) Adhesion of different Candida spp. to plastic: XTT formazan determinations. J Med Vet Mycol 34:407–410

Hawser SP, Baillie GS, Douglas LJ (1998) Production of extracellular matrix by Candida albicans biofilms. J Med Microbiol 47:253–256

Hawser SP, Douglas LJ (1994) Biofilm formation by Candida species on the surface of catheter materials in vitro. Infect Immun 62:915–921

Johnson CC, Yu A, Lee H, Fidel PL, Noverr MC (2012) Development of a contemporary animal model of Candida albicans-associated denture stomatitis using a novel intraoral denture system. Infect Immun 80:1736–1743

Klis FM, Sosinska GJ, De Groot PW, Brul S (2009) Covalently linked cell wall proteins of Candida albicans and their role in fitness and virulence. FEMS Yeast Res 9:1013–1028

Kojic EM, Darouiche RO (2004) Candida infections of medical devices. Clin Microbiol Rev 17:255–267

Lamfon H, Porter SR, Mccullough M, Pratten J (2003) Formation of Candida albicans biofilms on non-shedding oral surfaces. Eur J Oral Sci 111:465–471

Martinez-Gomariz M, Perumal P, Mekala S, Nombela C, Chaffin WL, Gil C (2009) Proteomic analysis of cytoplasmic and surface proteins from yeast cells, hyphae, and biofilms of Candida albicans. Proteomics 9:2230–2252

Molero G, Diez-Orejas R, Navarro-Garcia F, Monteoliva L, Pla J, Gil C, Sanchez-Perez M, Nombela C (1998) Candida albicans: genetics, dimorphism and pathogenicity. Int Microbiol 1:95–106

Monds RD, O'toole GA (2009) The developmental model of microbial biofilms: ten years of a paradigm up for review. Trends Microbiol 17:73–87

Nett J, Lincoln L, Marchillo K, Andes D (2007a) Beta-1,3 glucan as a test for central venous catheter biofilm infection. J Infect Dis 195:1705–1712

Nett J, Lincoln L, Marchillo K, Massey R, Holoyda K, Hoff B, Vanhandel M, Andes D (2007b) Putative role of beta-1,3 glucans in Candida albicans biofilm resistance. Antimicrob Agents Chemother 51:510–520

Nett JE, Brooks EG, Cabezas-Olcoz J, Sanchez H, Zarnowski R, Marchillo K, Andes DR (2014) Rat indwelling urinary catheter model of Candida albicans biofilm infection. Infect Immun 82:4931–4940

Nett JE, Crawford K, Marchillo K, Andes DR (2010a) Role of Fks1p and matrix glucan in Candida albicans biofilm resistance to an echinocandin, pyrimidine, and polyene. Antimicrob Agents Chemother 54:3505–3508

Nett JE, Sanchez H, Cain MT, Andes DR (2010b) Genetic basis of Candida biofilm resistance due to drug-sequestering matrix glucan. J Infect Dis 202:171–175

Nobile CJ, Andes DR, Nett JE, Smith FJ, Yue F, Phan QT, Edwards JE, Filler SG, Mitchell AP (2006a) Critical role of Bcr1-dependent adhesins in C. albicans biofilm formation in vitro and in vivo. PLoS Pathog 2:e63

Nobile CJ, Fox EP, Nett JE, Sorrells TR, Mitrovich QM, Hernday AD, Tuch BB, Andes DR, Johnson AD (2012) A recently evolved transcriptional network controls biofilm development in Candida albicans. Cell 148:126–138

Nobile CJ, Johnson AD (2015) Candida albicans Biofilms and Human Disease. Annu Rev Microbiol 69:71–92

Nobile CJ, Mitchell AP (2005) Regulation of cell-surface genes and biofilm formation by the C. albicans transcription factor Bcr1p. Curr Biol 15:1150–1155

Nobile CJ, Mitchell AP (2006) Genetics and genomics of Candida albicans biofilm formation. Cell Microbiol 8:1382–1391

Nobile CJ, Nett JE, Andes DR, Mitchell AP (2006b) Function of Candida albicans adhesin Hwp1 in biofilm formation. Eukaryot Cell 5:1604–1610

Nobile CJ, Nett JE, Hernday AD, Homann OR, Deneault JS, Nantel A, Andes DR, Johnson AD, Mitchell AP (2009) Biofilm matrix regulation by Candida albicans Zap1. PLoS Biol 7: e1000133

Nobile CJ, Schneider HA, Nett JE, Sheppard DC, Filler SG, Andes DR, Mitchell AP (2008) Complementary adhesin function in C. albicans biofilm formation. Curr Biol 18:1017–1024

Pierce CG, Chaturvedi AK, Lazzell AL, Powell AT, Saville SP, Mchardy SF, Lopez-Ribot JL (2015a) A novel small molecule inhibitor of biofilm formation, filamentation and virulence with low potential for the development of resistance. NPJ Biofilms Microbiomes 1

Pierce CG, Srinivasan A, Ramasubramanian AK, Lopez-Ribot JL (2015b) From biology to drug development: new approaches to combat the threat of fungal biofilms. Microbiol Spectr 3

Pierce CG, Uppuluri P, Tummala S, Lopez-Ribot JL (2010) A 96 well microtiter plate-based method for monitoring formation and antifungal susceptibility testing of Candida albicans biofilms. J Vis Exp

Raad I, Chatzinikolaou I, Chaiban G, Hanna H, Hachem R, Dvorak T, Cook G, Costerton W (2003) In vitro and ex vivo activities of minocycline and EDTA against microorganisms embedded in biofilm on catheter surfaces. Antimicrob Agents Chemother 47:3580–3585

Ramage G, Bachmann S, Patterson TF, Wickes BL, Lopez-Ribot JL (2002) Investigation of multidrug efflux pumps in relation to fluconazole resistance in Candida albicans biofilms. J Antimicrob Chemother 49:973–980

Ramage G, Martinez JP, Lopez-Ribot JL (2006) Candida biofilms on implanted biomaterials: a clinically significant problem. FEMS Yeast Res 6:979–986

Ramage G, Mowat E, Jones B, Williams C, Lopez-Ribot J (2009) Our current understanding of fungal biofilms. Crit Rev Microbiol 35:340–355

Ramage G, Tomsett K, Wickes BL, Lopez-Ribot JL, Redding SW (2004) Denture stomatitis: a role for Candida biofilms. Oral Surg Oral Med Oral Pathol Oral Radiol Endod 98:53–59

Ramage G, Vande Walle K, Wickes BL, Lopez-Ribot JL (2001a) Standardized method for in vitro antifungal susceptibility testing of Candida albicans biofilms. Antimicrob Agents Chemother 45:2475–2479

Ramage G, Vandewalle K, Wickes BL, Lopez-Ribot JL (2001b) Characteristics of biofilm formation by Candida albicans. Rev Iberoam Micol 18:163–170

Robbins N, Uppuluri P, Nett J, Rajendran R, Ramage G, Lopez-Ribot JL, Andes D, Cowen LE (2011) Hsp90 governs dispersion and drug resistance of fungal biofilms. PLoS Pathog 7: e1002257

Schinabeck MK, Long LA, Hossain MA, Chandra J, Mukherjee PK, Mohamed S, Ghannoum MA (2004) Rabbit model of Candida albicans biofilm infection: liposomal amphotericin B antifungal lock therapy. Antimicrob Agents Chemother 48:1727–1732

Sellam A, Al-Niemi T, Mcinnerney K, Brumfield S, Nantel A, Suci PA (2009) A Candida albicans early stage biofilm detachment event in rich medium. BMC Microbiol 9:25

Siles SA, Srinivasan A, Pierce CG, Lopez-Ribot JL, Ramasubramanian AK (2013) High-throughput screening of a collection of known pharmacologically active small compounds for identification of Candida albicans biofilm inhibitors. Antimicrob Agents Chemother 57:3681–3687

Suci PA, Geesey GG, Tyler BJ (2001) Integration of Raman microscopy, differential interference contrast microscopy, and attenuated total reflection Fourier transform infrared spectroscopy to investigate chlorhexidine spatial and temporal distribution in Candida albicans biofilms. J Microbiol Methods 46:193–208

Sutherland IW (2001) The biofilm matrix–an immobilized but dynamic microbial environment. Trends Microbiol 9:222–227

Taff HT, Nett JE, Zarnowski R, Ross KM, Sanchez H, Cain MT, Hamaker J, Mitchell AP, Andes DR (2012) A Candida biofilm-induced pathway for matrix glucan delivery: implications for drug resistance. PLoS Pathog 8:e1002848

Uppuluri P, Chaturvedi AK, Lopez-Ribot JL (2009) Design of a simple model of Candida albicans biofilms formed under conditions of flow: development, architecture, and drug resistance. Mycopathologia 168:101–109

Uppuluri P, Chaturvedi AK, Srinivasan A, Banerjee M, Ramasubramaniam AK, Kohler JR, Kadosh D, Lopez-Ribot JL (2010a) Dispersion as an important step in the Candida albicans biofilm developmental cycle. PLoS Pathog 6:e1000828

Uppuluri P, Lopez-Ribot JL (2010) An easy and economical in vitro method for the formation of Candida albicans biofilms under continuous conditions of flow. Virulence 1:483–487

Uppuluri P, Pierce CG, Thomas DP, Bubeck SS, Saville SP, Lopez-Ribot JL (2010b) The transcriptional regulator Nrg1p controls Candida albicans biofilm formation and dispersion. Eukaryot Cell 9:1531–1537

Vila T, Romo JA, Pierce, CG, Mchardy SF, Saville SP, Lopez-Ribot JL (2016) Targeting Candida albicans filamentation for antifungal drug development. Virulence 1–9

Viudes A, Peman J, Canton E, Ubeda P, Lopez-Ribot JL, Gobernado M (2002) Candidemia at a tertiary-care hospital: epidemiology, treatment, clinical outcome and risk factors for death. Eur J Clin Microbiol Infect Dis 21:767–774

Wilson LS, Reyes CM, Stolpman M, Speckman J, Allen K, Beney J (2002) The direct cost and incidence of systemic fungal infections. Value Health 5:26–34

Zarnowski R, Westler WM, Lacmbouh GA, Marita JM, Bothe JR, Bernhardt J, Lounes-Hadj Sahraoui A, Fontaine J, Sanchez H, Hatfield RD, Ntambi JM, Nett JE, Mitchell AP, Andes DR (2014) Novel entries in a fungal biofilm matrix encyclopedia. MBio 5:e01333–14

Zhao X, Daniels KJ, Oh SH, Green CB, Yeater KM, Soll DR, Hoyer LL (2006) Candida albicans Als3p is required for wild-type biofilm formation on silicone elastomer surfaces. Microbiology 152:2287–2299

Chapter 6
Candida Biofilm Tolerance: Comparison of Planktonic and Biofilm Resistance Mechanisms

Eddie G. Dominguez and David R. Andes

Abstract *Candida* species are opportunistic fungal pathogens residing as commensal organisms in approximately 70% of the human population. During times of decreased immune function, *Candida* spp. are able to transition from harmless members of the human microbiota into pathogens capable of causing life-threatening infections boasting mortality rates as high as 50%. Commonly adhering to implanted medical devices, *Candida* spp. grow as highly structured biofilms with inherent resistance to antifungal drug therapies and the host immune system. A multitude of investigations have found this resistance to be multifactorial involving mechanisms associated with planktonic antifungal resistance (efflux pump activity) along with biofilm-specific mechanisms. One biofilm-specific mechanism involves the complex extracellular matrix. Components of the matrix, specifically β-glucan, mannan, and extracellular DNA, have been found to promote resistance against multiple antifungal drug classes. Here we will review molecular mechanisms contributing to *Candida* biofilm drug resistance.

6.1 Introduction

Candida species is an opportunistic fungal pathogen which exists as a commensal organism in the elementary, gastrointestinal, and genitourinary tract of approximately 70% of the human population (Kabir et al. 2012; Meiller et al. 2009; Rosenbach et al. 2010; Ruhnke and Maschmeyer 2002; Schulze and Sonnenborn 2009; Sobel 1997). In a healthy human, the fungus typically exists in harmony with the normal microbiotic flora of the host. However, in an immunocompromised or immunologically weak host, such as patients receiving chemotherapy, transplant

E.G. Dominguez · D.R. Andes (✉)
Departments of Medicine and Medical Microbiology and Immunology,
University of Wisconsin, Madison, WI, USA
e-mail: dra@medicine.wisc.edu

R. Prasad (ed.), *Candida albicans: Cellular and Molecular Biology*,
DOI 10.1007/978-3-319-50409-4_6

recipients, and patients in the intensive care unit, *Candida* is among the most common pathogens and there is risk for spread beyond the mucosa which is associated with mortality in up to half of patients (Pfaller et al. 2005, 2010; Pfaller and Diekema 2007). These lethal cases of candidiasis are often a result of biofilm formation, often on implanted medical devices. In fact, some case series suggest that up to 70% of *Candida* bloodstream infection is linked to biofilm infection of vascular catheters. Considered to be the predominant microbial growth form found in nature, biofilms are an organized community of microbial cells adhered to a surface and enveloped in an extracellular matrix (ECM) with properties that are distinct from their planktonic counterparts (Kolter and Greenberg 2006; Nobile and Johnson 2015). Microbial biofilms exist within environments that are both biotic (aquatic, plant tissues, or mammalian tissues) and abiotic (indwelling medical devices). Microbial species that form biofilms on solid surfaces, such as *Candida* sp., *Staphylococcus* sp., *Streptococcus* sp., and *Escherichia coli,* each form a biofilm with structure, development, and unique properties that are distinct (Nobile and Johnson 2015). Whilst each of these organisms have the capability to form a biofilm on its own, it is becoming clear they often occur in multispecies biofilms due to different species of bacteria and bacterial and fungi that thrive as a result of shared virulence attributes (Roder et al. 2016).

Antimicrobial tolerance is an obstacle to the treatment of numerous biofilm infections (Mah 2012; Romling and Balsalobre 2012). *Candida* biofilms display innate resistance to all available drug classes and withstand antifungal concentrations up to 1000-fold higher than those which are effective toward non-biofilm planktonic cells (Chandra et al. 2001; Donlan and Costerton 2002; Douglas 2003; Ramage et al. 2005). Due to this inherent increased resistance to antifungal drugs, the recommended course of treatment for individuals afflicted with a *Candida* biofilm infection is extirpation of the afflicted device. *Candida* biofilm infections that are not successfully treated have a poor prognosis for the afflicted individual with an associated mortality rate as high as 50% (Al-Fattani and Douglas 2004; Kojic and Darouiche 2004; Mayer et al. 2013; Pfaller and Diekema 2007; Pfaller et al. 2005).

Antifungal resistance is an intrinsic biofilm characteristic and one of the many phenotypic changes that occurs upon transition to this mode of growth (Finkel and Mitchell 2011; Tobudic et al. 2012). During the later phases of development the resistant phenotype is most pronounced, however, drug resistance is able to be detected within minutes to hours of adhesion to a surface (Finkel and Mitchell 2011). Genetic mutations do not account for this observed resistance since biofilm cells re-cultured in planktonic conditions revert back to a susceptibility phenotype to antifungals. Furthermore, it is clear that multiple mechanisms contribute throughout the various stages of biofilm growth to the drug resistance phenotype (Marie and White 2009; Ramage et al. 2012; White et al. 1998a).

6.2 Overview of Antifungal Drug Classes and Planktonic *Candida* Resistance Mechanisms

6.2.1 Triazoles

The triazoles represent the most commonly used of all antifungal classes. Azole antifungal drugs inhibit *Candida* growth by targeting the enzyme lanosterol 14 α-demethylase (encoded by *ERG11*) which is necessary to convert lanosterol to ergosterol. This depletion of ergosterol in the fungal membrane causes the accumulation of toxic sterol intermediates which then lead to growth arrest (Kelly et al. 1995; Kelly et al. 1997; Lupetti et al. 2002; Shimokawa and Nakayama 1992). As with almost all types of antimicrobials, prolonged use has been linked to drug resistance, however, acquired *Candida* drug resistance is relatively uncommon (Cleveland et al. 2012; Ostrosky-Zeichner et al. 2010; White et al. 1998a). Long-term treatment of oral or esophageal candidiasis as reported in HIV/AIDS, especially in the pre-HAART era (Law et al. 1994; White et al. 1998a) was associated with high rates of resistance. The reported mechanisms of *C. albicans* azole resistance have included *ERG11* point mutations (S405F, Y132H, R467 K, and G464S), gene amplifications, and mitotic recombination events within *ERG11* or the drug efflux pumps (*Cdr1p, Cdr2p, and Mdr1p)* which result in their increased expression (Albertson et al. 1996; Coste et al. 2007; Lamb et al. 1997; Marichal et al. 1999; Sanglard et al. 1998; Sanglard et al. 1995; vanden Bossche et al. 1992; White 1997a, b). Acquired resistance in *C. glabrata* is even more common. Conversely, resistance in *C. parapsilosis* and *C. tropicalis* is uncommon (Bizerra et al. 2008; Cannon et al. 2009; Silva et al. 2011; vanden Bossche et al. 1992; Vandeputte et al. 2005). However, intrinsic resistance during biofilm growth is universal among all *Candida* species. Cells within a biofilm environment tolerate 1,000-fold higher azole concentrations than their planktonic counterparts, acquiring resistance as early as 4–6 h after initial adherence to a surface, resulting in ineffective triazole treatment (Lamfon et al. 2004; Ramage et al. 2001, 2012).

6.2.2 Polyenes

The polyene antifungals were the first available for systemic therapy. These Polyenes are amphiphilic molecules allowing for binding with sterols, primarily ergosterol, in the fungal cell membrane. This binding alters the transition temperature of the cell membrane, decreasing membrane fluidity and permeability. Release of monovalent ions (K^+, Na^+, H^+, and Cl^-) and small organic molecules ultimately leads to death of cell (Baginski and Czub 2009; Gray et al. 2012). Amphotericin B is a potent antifungal but is reserved for patients with severe systemic fungal infections due to the severe and potentially lethal side effects, the most important of which is renal toxicity. Resistance to amphotericin B is rare, but has been described in case reports from cancer patients undergoing chemotherapy and individuals undergoing prolonged prophylactic therapy (White et al. 1998a). The specific mechanisms of acquired polyene resistance are mechanistically poorly defined, but thought to involve alterations to cell membrane composition. Resistance has been linked to sterol changes in *C. glabrata* (Vandeputte et al. 2007) as well as in genetically altered strains that are defective in sterol C5,6—desaturase which produce little ergosterol (Kelly et al. 1996; Kelly et al. 1997). As with the triazoles, biofilms exhibit resistance to amphotericin B as well, but only 10–100 times as much as their planktonic counterparts in comparison to the 1,000 fold greater resistance associated with triazoles (Tobudic et al. 2010). Unfortunately, the concentrations needed for effective therapy are not achievable during systemic administration.

6.2.3 Echinocandins

The echinocandins represent the most recently available antifungal drug class. They act via inhibition of β-1,3 glucan synthase which is a key component of *Candida* cell walls (Denning 2003; Perlin 2015b; Shapiro et al. 2011). Inhibition of this enzyme at the cell wall results in osmotic instability within the cell ultimately leading to lysis of the cell (Chaffin et al. 1998). Clinical trial results suggest superiority over other antifungals for invasive candidiasis, likely due to its cidal activity and relative safety of this drug class. While, resistance remains relatively uncommon among all *Candida* species, treatment failures among *C. glabrata* have been emerging more rapidly. Point mutations along "hot spot" regions of the Fks1 subunits, located specifically among the range of amino acids from Phe641 to Pro649 and Arg1361, are the most commonly seen mechanisms of acquired resistance among this class of drugs and have been observed in *C. albicans* and homologous regions in *C. glabrata FKS2* gene (Balashov et al. 2006; Desnos-Ollivier et al. 2008; Garcia-Effron et al. 2009; Hernandez et al. 2004; Johnson et al. 2011; Katiyar and Edlind 2009; Laverdiere et al. 2006; Park et al. 2005; Perlin 2007, 2015a, b; Shapiro et al. 2011). As with the other antifungal drug classes, biofilms are intrinsically more resistant to echinocandins than their planktonic counterparts by approximately 2–20 fold (Nett et al. 2010a; Tobudic et al. 2010).

6.2.4 5-FC/Flucytosine

Flucytosine is a pyrimidine analogue that is metabolized in the pyrimidine salvage pathway by a cytosine deaminase into a toxic version of UTP. Upon incorporation, RNA synthesis is halted (Hope et al. 2004; White et al. 1998b). Flucytosine also decreases the availability of nucleotides for DNA synthesis via conversion into a metabolite that inhibits thymidylate synthetase (Hope et al. 2004). Emergence of resistance with flucytosine monotherapy is relatively rapid. This resistance is due to mutations in the cytosine permease gene *FCY2,* which is responsible for escorting flucytosine into the cell, or in the cytosine deaminase gene *FCY1* (White et al. 1998b). Due to this rapid rate of acquired resistance, flucytosine is almost generally administered to patients in conjunction with amphotericin B and/or azole antifungals (Pappas et al. 2009).

6.3 *Candida* Biofilm Resistance Mechanisms

Candida biofilms have been the subject of numerous investigations. Initial studies into the mechanisms of drug resistance primarily explored mechanisms previously linked to drug tolerance in planktonic cells. Mutations in genes that encode drug target enzymes, such as *ERG11* and *FKS1*and alterations in the composition of the plasma membrane have all been linked to planktonic cell resistance but have not been demonstrated to contribute to biofilm resistance (Balashov et al. 2006; Bizerra et al. 2011; Morschhauser 2002; Stevens et al. 2006).

6.3.1 Role of Efflux Pumps

As described above, the overexpression of efflux pumps, coupled with the reduction of antifungal accumulation within the cell, is a key mechanism of resistance for planktonic *Candida* (Morschhauser 2002). Ramage et al. examined if upregulation of efflux pumps may also contribute drug tolerance during biofilm growth. They found increased transcription of both *MDR1* and *CDR1* in 24 h *C. albicans* biofilms when compared to their planktonic counterparts of the same growth stage (Ramage et al. 2002a). Genetic manipulation, via deletion, of *MDR1, CDR1,* and/or *CDR2* was conducted in order to investigate the role of efflux pumps on triazole resistance during biofilm growth. Hypersensitivity to fluconazole was displayed by these mutants during both planktonic and biofilm growth, but not in biofilms grown for 24–48 h of the same mutant strain. This suggested that during the mature biofilm stage efflux pumps do not significantly contribute to drug resistance (Ramage et al. 2002a).

Mukherjee et al. explored the impact of efflux pumps at three phases of biofilm development including early (0–11 h), intermediate (12–30 h), and mature (31–72 h) time points in comparison to their planktonic counterparts (Mukherjee et al. 2003). Single, double, and triple mutants of these three main efflux pump genes showed no increase in susceptibility to fluconazole during the mature biofilm growth phase; however, in the early phase (6 h) the double and triple efflux pump mutants displayed a modest increase in azole susceptibility as compared to the parent strains (Mukherjee et al. 2003). Loss of a single efflux pump had little to no effect in regards to biofilm resistance, even at the earliest time points as seen in the double and triple mutants. This suggested that not only do efflux pumps function in a cooperative manner, but that they also contribute to resistance during the early biofilm developmental stages opposed to the later mature stages. To further investigate this observation, transcriptional analysis of efflux pump genes were conducted on 12 and 48 h biofilms which found with elevated expression levels of said genes during the earlier less mature phase in comparison to the older mature phase (Mukherjee et al. 2003). Additional studies directed at *C. glabrata* and *C. tropicalis* also suggested efflux pumps more than likely contribute to biofilm drug resistance during the early phases of growth (Bizerra et al. 2008; Ramage et al. 2012).

6.3.2 *Influence of Sterol Synthesis*

Mukherjee et al. explored the role of plasma membrane changes during biofilm formation, as changes in sterol synthesis have been linked to amphotericin B and ergosterol resistance during planktonic growth, as described above (Ghannoum and Rice 1999; Kontoyiannis 2000; Mukherjee et al. 2003). Initial studies examined the levels of sterols during different stages of biofilm development and found that early phase biofilms contained relatively similar levels of ergosterol as that of time matched planktonic cells. However, as biofilm development continued, the ergosterol levels reduced to 50% of the levels measured for planktonic conditions (Mukherjee et al. 2003). Furthermore, the sterol profile of intermediate and mature biofilms was different than planktonic cultures of the same age. Specifically, concentrations of ergosterol decreased as the biofilms aged and were replaced by intermediate sterols such as zymosterol, 4,14-dimethylzymosterol, and obtusifoliol (Mukherjee et al. 2003). This finding suggested that during the early stages of biofilm development ergosterol is an effective target for drug therapy, but as biofilms continue to grow and mature their dependency upon ergosterol decreases, potentially limiting the efficacy azole and polyene antifungals which target ergosterol.

Global transcriptional analysis also showed increased levels of *ERG11* transcription during the early phase growth stage of *C. albicans* biofilms in comparison to planktonic cells of the same age (Finkel and Mitchell 2011; Nett et al. 2009). A second gene that plays a role in ergosterol biosynthesis, *ERG25,* was found to be upregulated in intermediate and mature biofilms when compared to planktonic cultures of the same age (Nett et al. 2009). *ERG25* encodes a putative C-4 sterol methyl oxidase which is believed to play a role in the biosynthesis of ergosterol intermediates via C4-demethylation (Nett et al. 2009). The conversion of lanosterol to nonergosterol intermediates, such as eburicol and 14-methyl fecosterol, is a role that this enzyme is theorized to perform in biofilms. Additional studies directly testing the role of membrane changes in the biofilm drug-resistant phenotype have not been reported.

6.3.3 Impact of Cell Density and Quorum Sensing

Another trait contributing to the enhanced drug resistance observed during biofilm growth is the relatively large fungal burden (Perumal et al. 2007; Seneviratne et al. 2008). This relationship has also been described for planktonic cells, with higher inoculums producing higher MICs (Nguyen and Yu 1999; Riesselman et al. 2000). Based upon these studies, Perumal et al. examined the role of high cell density on antifungal resistance by comparing the susceptibility levels of planktonic yeast cultures with those of intact and disrupted biofilms (Perumal et al. 2007). Similar to findings in planktonic cells, high cell density cultures displayed higher levels of resistance to azoles when compared to cultures of lower density in the biofilm state.

Quorum sensing is the signaling process linked to control of cell density in the biofilm state. Two key quorum sensing molecules, tyrosol and farnesol, have opposing roles during biofilm development. Tyrosol promotes the hyphal state of biofilms whereas farnesol promotes the yeast state (Hornby 2001; Lindsay et al. 2012; Ramage et al. 2002b; Wongsuk et al. 2016). In addition to the roles of fungal morphogenesis and fungal development, quorum sensing molecules have also been implicated as having potential antifungal activity. Studies in which biofilm cells are co-treated with antifungal drugs (azoles and polyenes) and quorum sensing molecules have demonstrated a synergistic effect (Sharma and Prasad 2011; Wongsuk et al. 2016). Sharma et al. found that farnesol is able to reduce drug extrusion, of azoles, through the ABC transporters CaCdr1p and CaCdr2p, which may in part explain the synergistic effect observed with a triazole (Sharma and Prasad 2011).

6.3.4 Contribution of Biofilm Extracellular Matrix

The production of extracellular matrix is a distinctive feature of biofilms (Mitchell et al. 2015; Zarnowski et al. 2014). This matrix encompasses the cells within the biofilm and promotes cohesion among the cells within as well as adhesion to surfaces (Flemming and Wingender 2010; O'Toole 2003). Furthermore, the matrices of most microbes have been found capable of absorption numerous environmental components. For example, the surrounding matrix has been shown to retain water and nutrients (Flemming and Wingender 2010). However, the most studied aspect of the extracellular matrix is its ability to create a protective physical barrier between biofilm cells and its surrounding environment. This proves to be vital to the organism's survival when growing on the surface of an indwelling device by providing protection from pharmacological agents and the hosts innate immune system (Costerton et al. 1999; Donlan 2001).

The Douglas group performed the first studies on fungal biofilms using *C. albicans* to investigate the potential role of the extracellular matrix on drug resistance and the importance of environmental conditions on overall matrix production

(Douglas 2003). Since this initial study a number of labs have investigated the matrix composition and role in drug resistance in fungal biofilms. A recent detailed analysis of the *C. albicans* matrix identified each of the four macromolecules classes. Relative composition of the matrix based upon dry weight included 55% protein, 25% carbohydrate, 15% lipid, and 5% nucleic acid (Zarnowski et al. 2014). Proteomic analysis revealed 458 distinct entries which included protein classes involved in carbohydrate and amino acid metabolism (Al-Fattani and Douglas 2006; Faria-Oliveira et al. 2014; Thomas et al. 2006). Analysis of the carbohydrate fraction revealed the presence of three polysaccharides, β-1,3 glucan, β-1,6 glucan, and α-1,6 mannan with α-1,2 linked branches. Identified lipids included neutral and polar glycerolipids in addition to a small portion of sphingolipids (Zarnowski et al. 2014). The nucleic acids found consisted mainly of noncoding sequences of DNA (Zarnowski et al. 2014).

Nett et al. examined the relationship between biofilm resistance and matrix by taking purified matrix material from biofilms and adding it to planktonic cells prior to antifungal susceptibility testing (Nett et al. 2007). The addition of matrix rendered the planktonic cells resistant to antifungal drug to a degree that was similar to that seen in mature biofilms. This suggested that matrix material is interacting or sequestering antifungals preventing them from reaching their intended targets. Using radiolabeled fluconazole, they found that cultures containing matrix were able to sequester the azole drug from their intended targets, consistent with this theory. Genetic studies seeking the component responsible for this process implicated the carbohydrate β-1,3 glucan (Nett et al. 2007, 2010b). This drug sequestration phenomenon has been found to relatively nonspecific with regard to the antifungal drug as resistance linked to matrix protection has been shown important for triazoles, polyenes, flucytosine, and echinocandins (Nett et al. 2010a; Vediyappan et al. 2010). Studies conducted on biofilms formed by other *Candida* spp., namely *C. glabrata, C. parapsilosis,* and *C. tropicalis* also displayed this matrix antifungal mechanism as well as affinity toward β-1,3 glucans role in biofilm resistance (Fernandes et al. 2015; Kuhn et al. 2002; Mitchell et al. 2013; Nett et al. 2010a; Yi et al. 2011).

The first investigation into the genetic control of production of matrix β-1,3 glucan have shown that the β-1,3 glucan synthase Fks1p is required for production of this polysaccharide (Nett et al. 2010a). Subsequent delivery to the matrix has been found to be regulated in a complementary fashion by three glucan modifier proteins Bgl2p, Phr1p, and Xog1p (Nett et al. 2010a; Nett et al. 2010b; Taff et al. 2012). Additionally, Nobile et al. identified a transcription factor, Zap1p, that is critical for control of matrix production (Nobile et al. 2009). Zap1p was found to negatively regulate production of β-1,3 glucan by hydrolysis of matrix carbohydrates through the control of two glucoamylases *GCA1* and *GCA2*. It is also speculated to influence matrix production through quorum sensing pathways based upon its control over alcohol dehydrogenases ADH5, CSH1, and LFD6 enzymes (Nobile et al. 2009).

Recent work has also shown that the drug resistance phenotype is not solely due to β-1,3 glucan. In fact, each of the three polysaccharides identified in the carbohydrate fraction of the extracellular matrix were found to cooperate as a mannan-glucan complex to facilitate drug resistance in *C. albicans* biofilm. Mitchell et al. further investigated the genetic implications of the loss of genes which regulate production and modification of mannan and glucan within the extracellular matrix. They found that a subset of mutants had reduced levels of all three polysaccharides, lower levels of total matrix, and increased susceptibility toward antifungals. The β-1,3 glucan synthase gene (*FKS1),* two genes regulating matrix β-1,6 glucan (*BIG1* and *KRE5*), and seven genes regulating matrix mannan (*ALG11, MNN4-4, MNN9, PMR1, VAN1,* and *VRG4*) all prominently displayed the aforementioned phenotype. Surprisingly though when biofilms containing mutants from the various pathways were mixed and grown with one another, matrix structure and functionality was restored (Mitchell et al. 2015). This observation in addition to studies pharmacologically manipulating the matrix components demonstrated that the matrix constituents were assembled after export from the cell.

Extracellular DNA (eDNA) has also been found to impact the drug resistance phenotype seen in biofilms. This finding is based upon investigations that showing an increase in susceptibility of *C. albicans* biofilms to two classes of antifungals (echinocandins and polyenes) when co-treated with DNase (Martins et al. 2012; Martins et al. 2010). It is still unclear how the eDNA is mechanistically contributing to drug resistance (Zarnowski et al. 2014).

6.3.5 Presence of Persister Cells

Persister cells are a subpopulation of dormant cells found within biofilms, which exhibit a higher tolerance to multiple drug classes (LaFleur et al. 2006; Lewis 2008). They are defined as the population of microbes remaining after antimicrobial exposure and were first described for amphotericin B. The cells are capable of reconstituting new biofilms containing the same percentage of resistant persister cells (0.01–0.02%). These characteristics suggest they serve as the primary component of recurring fungal infections (Sun et al. 2016). Although the underlying mechanism for the production of persister cells is still unclear, it linked surface adhesion. Sun et al. showed they rapidly emerged upon surface adhesion, reaching a state of saturation by 2 h (Sun et al. 2016). The genetic basis underlying the persister cell lifestyle remains unclear.

6.4 Conclusion

The ability of *Candida* spp. to transition into a biofilm lifestyle allows this organism to thrive in even the healthy human host. This survival is attributed to the intrinsically high levels tolerance of conventional antifungal therapies and the host's innate immune response. The process appears multifactorial with resistance mechanisms varying by the phase of biofilm development.

Efflux pumps play an important role during the early phase of development. This mechanism of resistance is most prevalent during the stage at which the planktonic cells begin to adhere to a surface and transition into a biofilm state. As the biofilm matures the role played by these efflux pumps diminishes and instead biofilm-specific mechanisms are predominant. The earliest biofilm-specific mechanism is the production of persister cells, which are phenotypic variants of the parent cells that are resistant to antifungals and provide a mechanism for regrowth of the organism after high levels of drug exposure. As biofilms continue to mature an extracellular matrix is produced which contains a multitude of components that work together to provide multilayers of protection to the cells which now reside within this material. Carbohydrates of the matrix, specifically β-1,3 glucan, β-1,6 glucan, and α-1,6 mannan with α-1,2 linked branches, sequester a variety of antifungals providing drug resistance to mature biofilms. A secondary component,

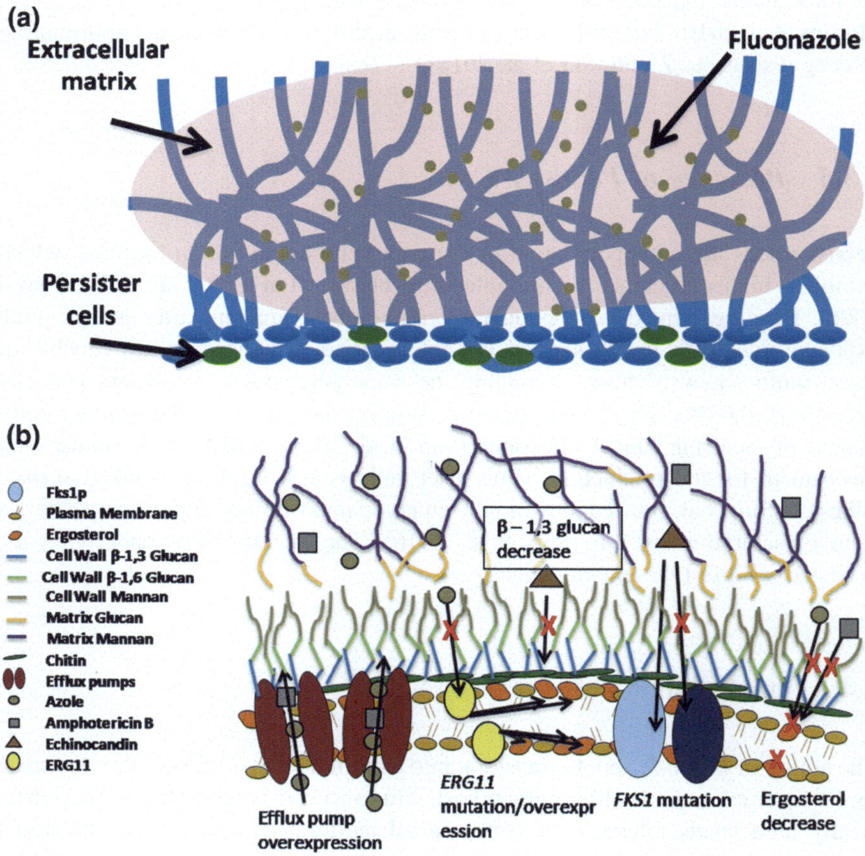

Fig. 6.1 *Candida* biofilm resistance mechanisms. **a** Resistance mechanisms at the biofilm community level. **b** Resistance mechanisms at the cellular level (Taff et al. 2013)

extracellular DNA, also promotes drug resistance to popular antifungals further contributing to this phenotype.

Biofilms contain many overlapping and redundant mechanisms which allow them to survive in hostile environments and evade drug treatments resulting in poor prognosis for patients (Fig. 6.1; Taff et al. 2013). Compositional, structural, and biochemical analysis of biofilms and their components have allowed us to better understand this complex organism and potentially develop innovative therapies to better combat infections. However, there are still many unknowns, such as the role of host factors and their interaction with matrix components during infection or do the components of duel species biofilms interact with one another to further enhance drug resistance? Additional investigations addressing questions such as these are still necessary in order to fully comprehend the nature and full potential of fungal biofilms.

References

Al-Fattani MA, Douglas LJ (2004) Penetration of *Candida* biofilms by antifungal agents. Antimicrob Agents Chemother 48:3291–3297

Al-Fattani MA, Douglas LJ (2006) Biofilm matrix of *Candida albicans* and *Candida tropicalis*: chemical composition and role in drug resistance. J Med Microbiol 55:999–1008

Albertson GD, Niimi M, Cannon RD, Jenkinson HF (1996) Multiple efflux mechanisms are involved in Candida albicans fluconazole resistance. Antimicrob Agents Chemother 40:2835–2841

Baginski M, Czub J (2009) Amphotericin B and its new derivatives—mode of action. Curr Drug Metab 10:459–469

Balashov SV, Park S, Perlin DS (2006) Assessing resistance to the echinocandin antifungal drug caspofungin in Candida albicans by profiling mutations in FKS1. Antimicrob Agents Chemother 50:2058–2063

Bizerra FC, Melo AS, Katchburian E, Freymuller E, Straus AH, Takahashi HK, Colombo AL (2011) Changes in cell wall synthesis and ultrastructure during paradoxical growth effect of caspofungin on four different Candida species. Antimicrob Agents Chemother 55:302–310

Bizerra FC, Nakamura CV, de Poersch C, Estivalet Svidzinski TI, Borsato Quesada RM, Goldenberg S, Krieger MA, Yamada-Ogatta SF (2008) Characteristics of biofilm formation by Candida tropicalis and antifungal resistance. FEMS Yeast Res 8:442–450

Cannon RD, Lamping E, Holmes AR, Niimi K, Baret PV, Keniya MV, Tanabe K, Niimi M, Goffeau A, Monk BC (2009) Efflux-mediated antifungal drug resistance. Clin Microbiol Rev 22:291–321, Table of Contents

Chaffin WL, Lopez-Ribot JL, Casanova M, Gozalbo D, Martinez JP (1998) Cell wall and secreted proteins of Candida albicans: identification, function, and expression. Microbiol Mol Biol Rev 62:130–180

Chandra J, Kuhn DM, Mukherjee PK, Hoyer LL, McCormick T, Ghannoum MA (2001) Biofilm formation by the fungal pathogen Candida albicans: development, architecture, and drug resistance. J Bacteriol 183:5385–5394

Cleveland AA, Farley MM, Harrison LH, Stein B, Hollick R, Lockhart SR, Magill SS, Derado G, Park BJ, Chiller TM (2012) Changes in incidence and antifungal drug resistance in candidemia: results from population-based laboratory surveillance in Atlanta and Baltimore, 2008–2011. Clin Infect Dis 55:1352–1361

Coste A, Selmecki A, Forche A, Diogo D, Bougnoux ME, d'Enfert C, Berman J, Sanglard D (2007) Genotypic evolution of azole resistance mechanisms in sequential Candida albicans isolates. Eukaryot Cell 6:1889–1904

Costerton JW, Stewart PS, Greenberg EP (1999) Bacterial biofilms: a common cause of persistent infections. Science (New York, NY) 284:1318–1322

Denning DW (2003) Echinocandin antifungal drugs. Lancet 362:1142–1151

Desnos-Ollivier M, Bretagne S, Raoux D, Hoinard D, Dromer F, Dannaoui E (2008) Mutations in the fks1 gene in Candida albicans, C-tropicalis, and C-krusei correlate with elevated caspofungin MICs uncovered in AM3 medium using the method of the European Committee on antibiotic susceptibility testing. Antimicrob Agents Chemother 52:3092–3098

Donlan RM (2001) Biofilm formation: a clinically relevant microbiological process. Clin Infect Dis 33:1387–1392

Donlan RM, Costerton JW (2002) Biofilms: survival mechanisms of clinically relevant microorganisms. Clin Microbiol Rev 15:167–193

Douglas LJ (2003) Candida biofilms and their role in infection. Trends Microbiol 11:30–36

Faria-Oliveira F, Carvalho J, Belmiro CL, Martinez-Gomariz M, Hernaez ML, Pavao M, Gil C, Lucas C, Ferreira C (2014) Methodologies to generate, extract, purify and fractionate yeast ECM for analytical use in proteomics and glycomics. BMC Microbiol 14:244

Fernandes T, Silva S, Henriques M (2015) Candida tropicalis biofilm's matrix–involvement on its resistance to amphotericin B. Diagn Microbiol Infect Dis 83:165–169

Finkel JS, Mitchell AP (2011) Genetic control of Candida albicans biofilm development. Nat Rev Microbiol 9:109–118

Flemming HC, Wingender J (2010) The biofilm matrix. Nat Rev Microbiol 8:623–633

Garcia-Effron G, Park S, Perlin DS (2009) Correlating echinocandin MIC and kinetic inhibition of fks1 mutant glucan synthases for Candida albicans: implications for interpretive breakpoints. Antimicrob Agents Chemother 53:112–122

Ghannoum MA, Rice LB (1999) Antifungal agents: mode of action, mechanisms of resistance, and correlation of these mechanisms with bacterial resistance. Clin Microbiol Rev 12:501–517

Gray KC, Palacios DS, Dailey I, Endo MM, Uno BE, Wilcock BC, Burke MD (2012) Amphotericin primarily kills yeast by simply binding ergosterol. Proc Natl Acad Sci USA 109:2234–2239

Hernandez S, Lopez-Ribot JL, Najvar LK, McCarthy DI, Bocanegra R, Graybill JR (2004) Caspofungin resistance in Candida albicans: correlating clinical outcome with laboratory susceptibility testing of three isogenic isolates serially obtained from a patient with progressive Candida esophagitis. Antimicrob Agents Chemother 48:1382–1383

Hope W, Tabernero L, Denning DW, Anderson MJ (2004) Molecular mechanisms of primary resistance to flucytosine in *Candida albicans*. Antimicrob Agents Chemother 48:4377–4386

Hornby JM, Jensen EC, Lisec AD, Tasto JJ, Jahnke B, Shoemaker R, et al (2001) Quorum sensing in the dimorphic fungus Candida albicans is mediated by farnesol. Appl Environ Microbiol 67 (7):2982–2992

Johnson ME, Katiyar SK, Edlind TD (2011) New Fks hot spot for acquired echinocandin resistance in Saccharomyces cerevisiae and its contribution to intrinsic resistance of Scedosporium species. Antimicrob Agents Chemother 55:3774–3781

Kabir MA, Hussain MA, Ahmad Z (2012) Candida albicans: a model organism for studying fungal pathogens. ISRN Microbiol 2012:538694

Katiyar SK, Edlind TD (2009) Role for Fks1 in the intrinsic echinocandin resistance of Fusarium solani as evidenced by hybrid expression in Saccharomyces cerevisiae. Antimicrob Agents Chemother 53:1772–1778

Kelly SL, Lamb DC, Corran AJ, Baldwin BC, Kelly DE (1995) Mode of action and resistance to azole antifungals associated with the formation of 14 alpha-methylergosta-8,24(28)-dien-3 beta,6 alpha-diol. Biochem Biophys Res Commun 207:910–915

Kelly SL, Lamb DC, Kelly DE, Loeffler J, Einsele H (1996) Resistance to fluconazole and amphotericin in Candida albicans from AIDS patients. Lancet 348:1523–1524

Kelly SL, Lamb DC, Kelly DE, Manning NJ, Loeffler J, Hebart H, Schumacher U, Einsele H (1997) Resistance to fluconazole and cross-resistance to amphotericin B in Candida albicans from AIDS patients caused by defective sterol delta5,6-desaturation. FEBS Lett 400:80–82

Kojic EM, Darouiche RO (2004) Candida infections of medical devices. Clin Microbiol Rev 17:255–267

Kolter R, Greenberg EP (2006) Microbial sciences: the superficial life of microbes. Nature 441:300–302

Kontoyiannis DP (2000) Efflux-mediated resistance to fluconazole could be modulated by sterol homeostasis in Saccharomyces cerevisiae. J Antimicrob Chemother 46:199–203

Kuhn DM, George T, Chandra J, Mukherjee PK, Ghannoum MA (2002) Antifungal susceptibility of Candida biofilms: unique efficacy of amphotericin B lipid formulations and echinocandins. Antimicrob Agents Chemother 46:1773–1780

LaFleur MD, Kumamoto CA, Lewis K (2006) Candida albicans biofilms produce antifungal-tolerant persister cells. Antimicrob Agents Chemother 50:3839–3846

Lamb DC, Kelly DE, Schunck WH, Shyadehi AZ, Akhtar M, Lowe DJ, Baldwin BC, Kelly SL (1997) The mutation T315A in Candida albicans sterol 14alpha-demethylase causes reduced enzyme activity and fluconazole resistance through reduced affinity. J Biol Chem 272:5682–5688

Lamfon H, Porter SR, McCullough M, Pratten J (2004) Susceptibility of Candida albicans biofilms grown in a constant depth film fermentor to chlorhexidine, fluconazole and miconazole: a longitudinal study. J Antimicrob Chemother 53:383–385

Laverdiere M, Lalonde RG, Baril JG, Sheppard DC, Park S, Perlin DS (2006) Progressive loss of echinocandin activity following prolonged use for treatment of Candida albicans oesophagitis. J Antimicrob Chemoth 57:705–708

Law D, Moore CB, Wardle HM, Ganguli LA, Keaney MG, Denning DW (1994) High prevalence of antifungal resistance in Candida spp. from patients with AIDS. The Journal of antimicrobial chemotherapy 34:659–668

Lewis K (2008) Multidrug tolerance of biofilms and persister cells. Curr Top Microbiol Immunol 322:107–131

Lindsay AK, Deveau A, Piispanen AE, Hogan DA (2012) Farnesol and cyclic AMP signaling effects on the hypha-to-yeast transition in Candida albicans. Eukaryot Cell 11:1219–1225

Lupetti A, Danesi R, Campa M, Del Tacca M, Kelly S (2002) Molecular basis of resistance to azole antifungals. Trends Mol Med 8:76–81

Mah TF (2012) Biofilm-specific antibiotic resistance. Future Microbiol 7:1061–1072

Marichal P, Koymans L, Willemsens S, Bellens D, Verhasselt P, Luyten W, Borgers M, Ramaekers FC, Odds FC, Bossche HV (1999) Contribution of mutations in the cytochrome P450 14alpha-demethylase (Erg11p, Cyp51p) to azole resistance in Candida albicans. Microbiology 145(Pt 10):2701–2713

Marie C, White TC (2009) Genetic basis of antifungal drug resistance. Curr Fungal Infect Rep 3:163–169

Martins M, Henriques M, Lopez-Ribot JL, Oliveira R (2012) Addition of DNase improves the in vitro activity of antifungal drugs against Candida albicans biofilms. Mycoses 55:80–85

Martins M, Uppuluri P, Thomas DP, Cleary IA, Henriques M, Lopez-Ribot JL, Oliveira R (2010) Presence of extracellular DNA in the Candida albicans biofilm matrix and its contribution to biofilms. Mycopathologia 169:323–331

Mayer FL, Wilson D, Hube B (2013) Candida albicans pathogenicity mechanisms. Virulence 4:119–128

Meiller TF, Hube B, Schild L, Shirtliff ME, Scheper MA, Winkler R, Ton A, Jabra-Rizk MA (2009) A novel immune evasion strategy of candida albicans: proteolytic cleavage of a salivary antimicrobial peptide. PLoS ONE 4:e5039

Mitchell KF, Taff HT, Cuevas MA, Reinicke EL, Sanchez H, Andes DR (2013) Role of matrix beta-1,3 glucan in antifungal resistance of non-albicans Candida biofilms. Antimicrob Agents Chemother 57:1918–1920

Mitchell KF, Zarnowski R, Sanchez H, Edward JA, Reinicke EL, Nett JE, Mitchell AP, Andes DR (2015) Community participation in biofilm matrix assembly and function. Proc Natl Acad Sci U S A 112:4092–4097

Morschhauser J (2002) The genetic basis of fluconazole resistance development in Candida albicans. Biochim Biophys Acta 1587:240–248

Mukherjee PK, Chandra J, Kuhn DM, Ghannoum MA (2003) Mechanism of fluconazole resistance in *Candida albicans* biofilms: phase-specific role of efflux pumps and membrane sterols. Infect Immun 71:4333–4340

Nett J, Lincoln L, Marchillo K, Massey R, Holoyda K, Hoff B, VanHandel M, Andes D (2007) Putative role of beta-1,3 glucans in Candida albicans biofilm resistance. Antimicrob Agents Chemother 51:510–520

Nett JE, Crawford K, Marchillo K, Andes DR (2010a) Role of Fks1p and matrix glucan in Candida albicans biofilm resistance to an echinocandin, pyrimidine, and polyene. Antimicrob Agents Chemother 54:3505–3508

Nett JE, Lepak AJ, Marchillo K, Andes DR (2009) Time course global gene expression analysis of an in vivo Candida biofilm. J Infect Dis 200:307–313

Nett JE, Sanchez H, Cain MT, Andes DR (2010b) Genetic basis of Candida biofilm resistance due to drug-sequestering matrix glucan. J Infect Dis 202:171–175

Nguyen MH, Yu CY (1999) Influence of incubation time, inoculum size, and glucose concentrations on spectrophotometric endpoint determinations for amphotericin B, fluconazole, and itraconazole. J Clin Microbiol 37:141–145

Nobile CJ, Johnson AD (2015) Candida albicans biofilms and human disease. Annu Rev Microbiol 69:71–92

Nobile CJ, Nett JE, Hernday AD, Homann OR, Deneault JS, Nantel A, Andes DR, Johnson AD, Mitchell AP (2009) Biofilm matrix regulation by Candida albicans Zap1. PLoS Biol 7: e1000133

O'Toole GA (2003) To build a biofilm. J Bacteriol 185:2687–2689

Ostrosky-Zeichner L, Casadevall A, Galgiani JN, Odds FC, Rex JH (2010) An insight into the antifungal pipeline: selected new molecules and beyond. Nat Rev Drug Discov 9:719–727

Pappas PG, Kauffman CA, Andes D, Benjamin DK Jr, Calandra TF, Edwards JE Jr, Filler SG, Fisher JF, Kullberg BJ, Ostrosky-Zeichner L et al (2009) Clinical practice guidelines for the management of candidiasis: 2009 update by the infectious diseases society of America. Clin Infect Dis 48:503–535

Park S, Kelly R, Kahn JN, Robles J, Hsu MJ, Register E, Li W, Vyas V, Fan H, Abruzzo G et al (2005) Specific substitutions in the echinocandin target Fks1p account for reduced susceptibility of rare laboratory and clinical Candida sp. isolates. Antimicrob Agents Chemother 49:3264–3273

Perlin DS (2007) Resistance to echinocandin-class antifungal drugs. Drug Resist Updat 10:121–130

Perlin DS (2015a) Echinocandin resistance in Candida. Clin Infect Dis 61(Suppl 6):S612–S617

Perlin DS (2015b) Mechanisms of echinocandin antifungal drug resistance. Ann N Y Acad Sci 1354:1–11

Perumal P, Mekala S, Chaffin WL (2007) Role for cell density in antifungal drug resistance in Candida albicans biofilms. Antimicrob Agents Chemother 51:2454–2463

Pfaller MA, Castanheira M, Messer SA, Moet GJ, Jones RN (2010) Variation in Candida spp. distribution and antifungal resistance rates among bloodstream infection isolates by patient age: report from the SENTRY antimicrobial surveillance program (2008–2009). Diagn Microbiol Infect Dis 68:278–283

Pfaller MA, Diekema DJ (2007) Epidemiology of invasive candidiasis: a persistent public health problem. Clin Microbiol Rev 20:133–163

Pfaller MA, Diekema DJ, Rinaldi MG, Barnes R, Hu B, Veselov AV, Tiraboschi N, Nagy E, Gibbs DL (2005) Results from the ARTEMIS DISK Global Antifungal Surveillance Study: a 6.5-year analysis of susceptibilities of Candida and other yeast species to fluconazole and voriconazole by standardized disk diffusion testing. J Clin Microbiol 43:5848–5859

Ramage G, Bachmann S, Patterson TF, Wickes BL, Lopez-Ribot JL (2002a) Investigation of multidrug efflux pumps in relation to fluconazole resistance in *Candida albicans* biofilms. J Antimicrob Chemother 49:973–980

Ramage G, Rajendran R, Sherry L, Williams C (2012) Fungal biofilm resistance. Int J Microbiol 2012:528521

Ramage G, Saville SP, Thomas DP, Lopez-Ribot JL (2005) Candida biofilms: an update. Eukaryot Cell 4:633–638

Ramage G, Saville SP, Wickes BL, Lopez-Ribot JL (2002b) Inhibition of Candida albicans biofilm formation by farnesol, a quorum-sensing molecule. Appl Environ Microbiol 68:5459–5463

Ramage G, Vande Walle K, Wickes BL, Lopez-Ribot JL (2001) Standardized method for in vitro antifungal susceptibility testing of Candida albicans biofilms. Antimicrob Agents Chemother 45:2475–2479

Riesselman MH, Hazen KC, Cutler JE (2000) Determination of antifungal MICs by a rapid susceptibility assay. J Clin Microbiol 38:333–340

Roder HL, Sorensen SJ, Burmolle M (2016) Studying bacterial multispecies biofilms: where to start? Trends Microbiol

Romling U, Balsalobre C (2012) Biofilm infections, their resilience to therapy and innovative treatment strategies. J Intern Med 272:541–561

Rosenbach A, Dignard D, Pierce JV, Whiteway M, Kumamoto CA (2010) Adaptations of Candida albicans for growth in the mammalian intestinal tract. Eukaryot Cell 9:1075–1086

Ruhnke M, Maschmeyer G (2002) Management of mycoses in patients with hematologic disease and cancer—review of the literature. Eur J Med Res 7:227–235

Sanglard D, Ischer F, Koymans L, Bille J (1998) Amino acid substitutions in the cytochrome P-450 lanosterol 14alpha-demethylase (CYP51A1) from azole-resistant Candida albicans clinical isolates contribute to resistance to azole antifungal agents. Antimicrob Agents Chemother 42:241–253

Sanglard D, Kuchler K, Ischer F, Pagani JL, Monod M, Bille J (1995) Mechanisms of resistance to azole antifungal agents in Candida albicans isolates from AIDS patients involve specific multidrug transporters. Antimicrob Agents Chemother 39:2378–2386

Schulze J, Sonnenborn U (2009) Yeasts in the gut: from commensals to infectious agents. Dtsch Arztebl Int 106:837–842

Seneviratne CJ, Jin LJ, Samaranayake H, Samaranayake LP (2008) Cell density and cell aging as factors modulating antifungal resistance of Candida albicans biofilms. Antimicrob Agents Chemother 52:3259–3266

Shapiro RS, Robbins N, Cowen LE (2011) Regulatory circuitry governing fungal development, drug resistance, and disease. Microbiol Mol Biol Rev 75:213–267

Sharma M, Prasad R (2011) The quorum-sensing molecule farnesol is a modulator of drug efflux mediated by ABC multidrug transporters and synergizes with drugs in Candida albicans. Antimicrob Agents Chemother 55:4834–4843

Shimokawa O, Nakayama H (1992) Increased sensitivity of Candida albicans cells accumulating 14 alpha-methylated sterols to active oxygen: possible relevance to in vivo efficacies of azole antifungal agents. Antimicrob Agents Chemother 36:1626–1629

Silva AP, Miranda IM, Guida A, Synnott J, Rocha R, Silva R, Amorim A, Pina-Vaz C, Butler G, Rodrigues AG (2011) Transcriptional profiling of azole-resistant Candida parapsilosis strains. Antimicrob Agents Chemother 55:3546–3556

Sobel JD (1997) Vaginitis. N Engl J Med 337:1896–1903

Stevens DA, Ichinomiya M, Koshi Y, Horiuchi H (2006) Escape of Candida from caspofungin inhibition at concentrations above the MIC (Paradoxical effect) accomplished by increased cell wall chitin; evidence for beta-1,6-glucan synthesis inhibition by caspofungin. Antimicrob Agents Chemother 50:3160–3161

Sun J, Li Z, Chu H, Guo J, Jiang G, Qi Q (2016) Candida albicans amphotericin B-Tolerant persister formation is closely related to surface adhesion. Mycopathologia 181:41–49

Taff HT, Nett JE, Zarnowski R, Ross KM, Sanchez H, Cain MT, Hamaker J, Mitchell AP, Andes DR (2012) A Candida biofilm-induced pathway for matrix glucan delivery: implications for drug resistance. PLoS Pathog 8:e1002848

Taff HT, Mitchell KF, Edward JA, Andes DR. (2013) Mechanisms of Candida biofilm drug resistance. Future Microbiol 8(10):1325–1337

Thomas DP, Bachmann SP, Lopez-Ribot JL (2006) Proteomics for the analysis of the Candida albicans biofilm lifestyle. Proteomics 6:5795–5804

Tobudic S, Kratzer C, Lassnigg A, Graninger W, Presterl E (2010) In vitro activity of antifungal combinations against Candida albicans biofilms. J Antimicrob Chemother 65:271–274

Tobudic S, Kratzer C, Lassnigg A, Presterl E (2012) Antifungal susceptibility of Candida albicans in biofilms. Mycoses 55:199–204

Vanden Bossche H, Marichal P, Odds FC, Le Jeune L, Coene MC (1992) Characterization of an azole-resistant Candida glabrata isolate. Antimicrob Agents Chemother 36:2602–2610

Vandeputte P, Larcher G, Berges T, Renier G, Chabasse D, Bouchara JP (2005) Mechanisms of azole resistance in a clinical isolate of Candida tropicalis. Antimicrob Agents Chemother 49:4608–4615

Vandeputte P, Tronchin G, Berges T, Hennequin C, Chabasse D, Bouchara JP (2007) Reduced susceptibility to polyenes associated with a missense mutation in the ERG6 gene in a clinical isolate of Candida glabrata with pseudohyphal growth. Antimicrob Agents Chemother 51:982–990

Vediyappan G, Rossignol T, d'Enfert C (2010) Interaction of Candida albicans biofilms with antifungals: transcriptional response and binding of antifungals to beta-glucans. Antimicrob Agents Chemother 54:2096–2111

White TC (1997a) Increased mRNA levels of ERG16, CDR, and MDR1 correlate with increases in azole resistance in Candida albicans isolates from a patient infected with human immunodeficiency virus. Antimicrob Agents Chemother 41:1482–1487

White TC (1997b) The presence of an R467 K amino acid substitution and loss of allelic variation correlate with an azole-resistant lanosterol 14alpha demethylase in Candida albicans. Antimicrob Agents Chemother 41:1488–1494

White TC, Marr KA, Bowden RA (1998a) Clinical, cellular, and molecular factors that contribute to antifungal drug resistance. Clin Microbiol Rev 11:382–402

White TC, Marr KA, Bowden RA (1998b) Clinical, cellular, and molecular factors that contribute to antifungal drug resistance. Clin Microbiol Rev 11, 382- +

Wongsuk T, Pumeesat P, Luplertlop N (2016) Fungal quorum sensing molecules: role in fungal morphogenesis and pathogenicity. J Basic Microbiol 56:440–447

Yi S, Sahni N, Daniels KJ, Lu KL, Srikantha T, Huang G, Garnaas AM, Soll DR (2011) Alternative mating type configurations (a/alpha versus a/a or alpha/alpha) of Candida albicans result in alternative biofilms regulated by different pathways. PLoS Biol 9:e1001117

Zarnowski R, Westler WM, Lacmbouh GA, Marita JM, Bothe JR, Bernhardt J, Lounes-Hadj Sahraoui A, Fontaine J, Sanchez H, Hatfield RD et al (2014) Novel entries in a fungal biofilm matrix encyclopedia. MBio 5:e01333–01314

Chapter 7
In Vivo Candida Device Biofilm Models

David R. Andes

Abstract During infection, fungi frequently transition to a biofilm lifestyle, proliferating as communities of surface-adherent aggregates of cells. Phenotypically, cells in a biofilm are distinct from free-floating or planktonic cells. Their high tolerance of antifungals and ability to withstand host defenses are two characteristics that foster infection persistence. Although many fungi have been observed to form biofilms, the most well-studied is *Candida albicans*. Animal models have been developed to mimic common *Candida* device-associated infections, including those involving vascular catheters, dentures, urinary catheters, and subcutaneous implants. Models have also reproduced the most common mucosal biofilm infections, oropharyngeal, and vaginal candidiasis. These models incorporate the anatomical site, immune components, and fluid dynamics of clinical niches and have been instrumental in the study of biofilm architecture and composition, drug resistance and investigation of novel therapies, and clinical exploration of the relevance of various gene products. This chapter describes the significance of *Candida* biofilm infections, the animal models developed for biofilm study, and how these models have contributed to our understanding of biofilm pathogenesis and for the development of new strategies for eradication of fungal biofilm infections.

7.1 Significance of Fungal Biofilms in Infection

Many fungal and bacterial pathogens build biofilms, establishing resilient communities on a variety of clinical surfaces (Donlan 2001b; Hoyle et al. 1990). Biofilm formation has become increasingly appreciated as one of the most common modes of growth. Medically, these adherent communities of cells pose a serious obstacle for successful treatment of infection. Compared to nonbiofilm, planktonic

D.R. Andes (✉)
Department of Medicine, Department of Medical Microbiology and Immunology,
Section of Infectious Diseases, University of Wisconsin-Madison,
1685 Highland Avenue, Madison, WI 53705, USA
e-mail: dra@medicine.wisc.edu

R. Prasad (ed.), *Candida albicans: Cellular and Molecular Biology*,
DOI 10.1007/978-3-319-50409-4_7

cells, they are extraordinarily tolerant to anti-infective therapies and resist killing by host defenses (Hawser et al. 1998). Biofilm formation has been well-described for *Candida albicans*, the most common fungal pathogen (Donlan 2001a; Douglas 2002). More recently, the majority of clinically encountered fungi have been shown to produce biofilms. This group includes filamentous fungi (*Aspergillus*, *Fusarium*, and zygomycetes), *Pneumocysitis*, and yeasts (*Blastoschizomyces*, *Saccharomyces*, *Malassezia*, *Trichosporon*, *Cryptococcus,* and numerous *Candida* spp.) (Cannizzo et al. 2007; D'Antonio et al. 2004; Davis et al. 2002; Di Bonaventura et al. 2006; Dyavaiah et al. 2007; Loussert et al. 2010; Ramage et al. 2012; Reynolds and Fink 2001; Seidler et al. 2008; Singh et al. 2011; Walsh et al. 1986).

One of the distinguishing traits of biofilm communities is their ability to adhere to a surface. In the medical setting, devices, such as catheters, provide an ideal niche for biofilm formation (Donlan 2001b; Passerini et al. 1992). As medical technology advances, the use of devices has continued to escalate. More than 35 million devices are implanted yearly in the United States alone and more than half of the 2 million hospital-acquired infections are device-associated (Kojic and Darouiche 2004). Many types of devices are at risk for biofilm infection, including catheters, dentures, implants, pacemakers, artificial heart valves, and central nervous system shunts (Donlan 2001b; Kojic and Darouiche 2004). Biofilm infections may be catastrophic, resulting in device malfunction or life-threatening, systemic infection (Donlan 2001b). Candidiasis in the hospital setting most frequently involves biofilm infection of a medical device. *Candida* spp. are the 4th most common nosocomial bloodstream pathogens and the 3rd most common cause of urinary tract pathogens (Edmond et al. 1999; Groeger et al. 1993; Pfaller and Diekema 2007; Richards et al. 1999). While biofilm infection was initially described for *C. albicans*, the majority of *Candida* spp., including *C. dubliniensis, C. glabrata*, *C. krusei*, *C. tropicalis*, and *C. parapsilosis*, have been shown to cause biofilm infections (Shin et al. 2002).

Mucosal candidiasis is widespread. Vaginal candidiasis effects approximately up to 70% of women and many suffer recurrent infections (Foxman et al. 2013). The clinical relevance of biofilms on biotic surfaces has become increasingly evident and mucosal biofilms have been described for *Candida*. In an animal model of oropharyngeal candidiasis, *Candida* forms a biofilm of yeast, hyphae, and commensal bacterial within the epithelial surface (Dongari–Bagtzoglou et al. 2009). *Candida* biofilms growing on the vaginal epithelial lining similarly demonstrate a typical biofilm architecture with adherence cells embedded in an extracellular matrix (Harriott et al. 2010). Mucosal biofilms appear to have many similarities to biofilms growing on abiotic surfaces, including sessile growth, protection from environmental factors, and variable access to nutrients (Dongari–Bagtzoglou 2008). However, biofilms on a mucosal surface participate in a dynamic interaction with the adjacent epithelial lining. The host epithelial lining may deliver immune component, nutrients, and antifungal components. Therefore, the basal aspect of a mucosal biofilm would be expected to be exposed to a vastly different environment when compared to a biofilm on an abiotic device surface.

7.2 Fungal Biofilm Traits

7.2.1 Structure and Composition

Fungal biofilms are comprised of adherent cells covered by an extracellular polymeric matrix. The process of fungal biofilm formation in vitro was initially described for *Candida albicans* in three main stages (Chandra et al. 2001). First, during early biofilm formation, *Candida* cells adhere to the biofilm substrate. For *C. albicans*, germ tube formation may be elicited. During intermediate biofilm formation stage, the extracellular matrix begins to appear and covers proliferating fungal cells. Release of cells, termed dispersion, is the final and key component of the dynamic nature of biofilms (Uppuluri et al. 2010). Spread of cells from biofilms is a regulated process by which organisms can disseminate throughout the host and establish new sites of infection. This developmental process appears to hold true for in vivo *Candida* biofilms in which biofilm cells have been identified in distant organs such as the kidney (Andes et al. 2004; Nett et al. 2010).

7.2.2 Tolerance to Antifungals

Biofilms are notoriously difficult to treat in the clinical setting and their physical removal, most often by removing a medical device, is often required for eradication of infection (Pappas et al. 2009). Available antifungal medications are seldom effective given the high tolerance of biofilms to these commonly used anti-infective therapies. *Candida* biofilms have been found to proliferate in the face of antifungal concentrations up to 1,000 fold higher than those needed to inhibit nonbiofilm, planktonic cells (Chandra et al. 2001; Hawser and Douglas 1994; Mah et al. 2003; O'Toole 2003).

The biofilm lifestyle of *C. albicans* is associated with resistance to available drug classes compared to activity against planktonic cells (Baillie and Douglas 1998; Chandra et al. 2001; Hawser and Douglas 1994; Kuhn et al. 2002; Lewis et al. 2002; Mukherjee et al. 2003; Ramage et al. 2002b). Resistance to the azoles (fluconazole, itraconazole, voriconazole, posaconazole) is particularly pronounced, while the echinocandins (caspofungin, micafungin, anidulafungin) and liposomal amphotericin are somewhat more effective. However, even these drugs fail to sterilize biofilm growth. Various mechanisms have been shown to contribute to this resistant phenotype, including the production of an extracellular matrix, an increase in efflux pump activity, alteration of sterols, production of resistant "persister cells," activation of stress responses, and an increase in cell density (Khot et al. 2006; Kumamoto 2005; LaFleur et al. 2006; Mukherjee et al. 2003, 2005; Perumal et al. 2007; Ramage et al. 2002a; Robbins et al. 2011; Uppuluri et al. 2008).

7.2.3 Immune Resistance

In addition to the well-described drug resistance, biofilm growth also appears to afford protection from the host immune response (Katragkou et al. 2010, 2011a, b; Xie et al. 2012). Compared to planktonic cells, both neutrophils and mononuclear cells are less effective in killing *Candida* biofilm cells. Mononuclear cells become entrapped in biofilms, but do not efficiently activate or phagocytize fungal cells (Chandra et al. 2007; Katragkou et al. 2010). Neutrophils have impaired function against both *C. albicans* and *C. parapsilosis* biofilms (Katragkou et al. 2011a; b; Xie et al. 2012). An intact immune response is not sufficient to clear *Candida* biofilms. Recent studies suggest that components of the *Candida* extracellular matrix impair the production and function of neutrophil extracellular traps (NETs) (Personal communication Jeniel Nett). When coupled with antifungal therapy, improved clinical outcomes are observed when *Candida*-infected medical devices are removed (Andes et al. 2012; Pappas et al. 2009).

7.3 Host Factors Influencing Fungal Biofilms

Fungal biofilms form in a variety of clinical niches. These sites of infection can vary quite significantly with regard to available nutrients, flow conditions, immune components, pH, and the substrate for cell adhesion and initiation of biofilm growth. Each of these factors is likely to influence the biofilm properties and structure, as discussed below. Models most closely mimicking the clinical niche are necessary to best reproduce the host environmental conditions and ultimately a clinical biofilm infection.

7.3.1 Flow Conditions

One of the greatest environmental differences among the sites of common fungal biofilms is the flow conditions. For example, *Candida* can form a biofilm in the face of a low rate of salivary flow (denture stomatitis), a rapid current of blood (endocarditis), or an intermittent flow (vascular and urinary catheter infection) (Kojic and Darouiche 2004). The Douglas group investigated the influence of flow conditions on *C. albicans* biofilm architecture using in vitro models (Al-Fattani and Douglas 2004, 2006). Compared to biofilms grown in static conditions, those propagated in a continuous-flow device were encased in a higher concentration of extracellular matrix. The continuous-flow biofilms also exhibited increased resistance to antifungals, including amphotericin B and fluconazole. Investigation of continuous flow by an independent laboratory confirmed the drug resistance phenotype, as well as the altered biofilm structure (Uppuluri et al. 2009). Biofilms formed in the flow environment were more dense and compact. One might expect that this architectural change to greatly impact other aspects of biofilm physiology, such the availability of oxygen and nutrients.

7.3.2 Substrates and Conditioning

One of the most influential factors for biofilm initiation is the substrate for adherence (Douglas 2002; Martinez and Casadevall 2007). Although many materials have been shown to support biofilm formation, the topography and hydrophobicity of the substrate may greatly impact fungal adherence. Indwelling medical devices are often designed to resist microbial adherence. Compared to other plastics, these materials, such as silicone, may even require a preconditioning, or protein coating, for robust biofilm formation in vitro (Nett et al. 2007b; Nobile and Mitchell 2005). However, in vivo, medical devices are rapidly conditioned with host factors from the surrounding fluids, such as blood, saliva, urine, or other fluids (Brash and Ten Hove 1993; Francois et al. 2000; Jenney and Anderson 2000; Proctor 2000; Yanagisawa et al. 2004). Several of the factors that may coat various devices, influencing adherence and biofilm formation, include fibrinogen, fibronectin, vitronectin, thrombospondin-1, albumin, and von Willebrand factor (VWF) (Brash and Ten Hove 1993; Francois et al. 2000; Jenney and Anderson 2000; Proctor 2000; Yanagisawa et al. 2004). As the concentrations of these substances differ among clinical niche sites, in vivo models best account for the conditioning of medical devices prior to biofilm initiation. In fact, recent study in three in vivo biofilm models found that a considerable portion of the biofilm extracellular matrix is of host origin (Nett et al. 2015). The host matrix proteome contained components that were unique to the infection site, such as amylase in the denture model. These findings underscore the relevance of the different host site in vivo environments. However, there were also a conserved set of more than a dozen host proteins across the three infection sites. Exploration of a functional role for a subset of these host components identified the importance of fibronectin for both early and late stages of biofilm formation (Nett et al. 2016). The identification of host factors in *Candida* biofilms provides the framework for future host-pathogen interaction studies. When considering mucosal biofilms, substrate representation becomes even more complex. Here, the epithelial layer of cells provides a surface for microbial adhesion which also involves receptor–ligand interactions (Dongari–Bagtzoglou 2008).

7.3.3 Nutrient Composition

Biofilm niche sites vary greatly with respect to availability of nutrients. For example, blood is a fairly nutrient-rich environment, while urine has a lower

Table 7.1 Animal models of *Candida* biofilm infection

Central venous catheter	Rat (Andes et al. 2004) Rabbit (Schinabeck et al. 2004) Mouse (Lazzell et al. 2009)
Urinary catheter	Mouse (Wang and Fries 2011) Rat (Nett et al. 2014a)
Subcutaneous implant	Mouse (Zumbuehl et al. 2007) Rat (Ricicova et al. 2010)
Denture stomatitis	Rat (Johnson et al. 2012; Nett et al. 2010)
Oral mucosal	Mouse (Dongari–Bagtzoglou et al. 2009)
Vaginal mucosal	Mouse (Harriott et al. 2010)

abundance of sugars and proteins. In addition, the conditions may be altered by both medical illness and diet. Untreated diabetes mellitus raises the glucose content throughout the host while a diet high in sugar primarily raises the glucose content in the oral cavity. The carbon source (galactose or glucose) and abundance has been shown to greatly influence biofilm integrity for *C. albicans* (Jin et al. 2004; Martinez and Casadevall 2007). Another factor shown to influence *Candida* biofilm formation is the concentration of metal ions (Ni^{2+}, Fe^{3+}, Cr^{3+}) (Ronsani et al. 2011). These variables can impact the rate of biofilm growth, production of extracellular matrix, and the strength of the biofilm. Animal models of biofilm infection which utilize an equivalent anatomical site provide the ideal composition of nutrients and minerals to mimic patient biofilm infections (Table 3.1).

7.3.4 Host Immune Components

Mounting evidence suggests a complex interaction between host immune cells and fungal biofilms. For example, leukocytes, important for controlling fungal infections, have also been shown to promote biofilm growth (Chandra et al. 2007). In vitro, *C. albicans* biofilms were observed to proliferate in response to a soluble factor released by mononuclear peripheral blood cells. Ultimately, mononuclear cells became entangled within the basal level of a *C. albicans* biofilm and were not able to phagocytize the biofilm cells. When examining oral mucosal *Candida* biofilms, Dongari–Bagtzoglou et al. observed the migration of neutrophils throughout the biofilm (Dongari–Bagtzoglou et al. 2009). Host immune cells appear to incorporate into biofilm, even augment biofilm growth, but are most often unable to contain the infection (Nett et al. 2015). To understand how the immune system impacts the biofilm lifestyle, models encompassing immune components at the site of infection are optimal.

7.4 In Vivo Models of Fungal Biofilms and Drug Discovery

Animal models best integrate the influence of host factors on the formation of biofilms and acquisition of their phenotypic traits. Utilization of animal models incorporates not only the influence of the immune system, but also niche-specific factors, such as the flow conditions, nutrients in the environment, pH, oxygen tension, and the substrate or surface of adherence. As *Candida* has served as a model organism for fungal biofilm infection in this arena, models involving this pathogen will be much of the focus of discussion in this chapter (Fig. 7.1).

7.4.1 Vascular Catheter Model

Perhaps the most commonly used animal models for in vivo biofilm study is the venous catheter model. This model has been adapted for use in a rat, a rabbit, and a

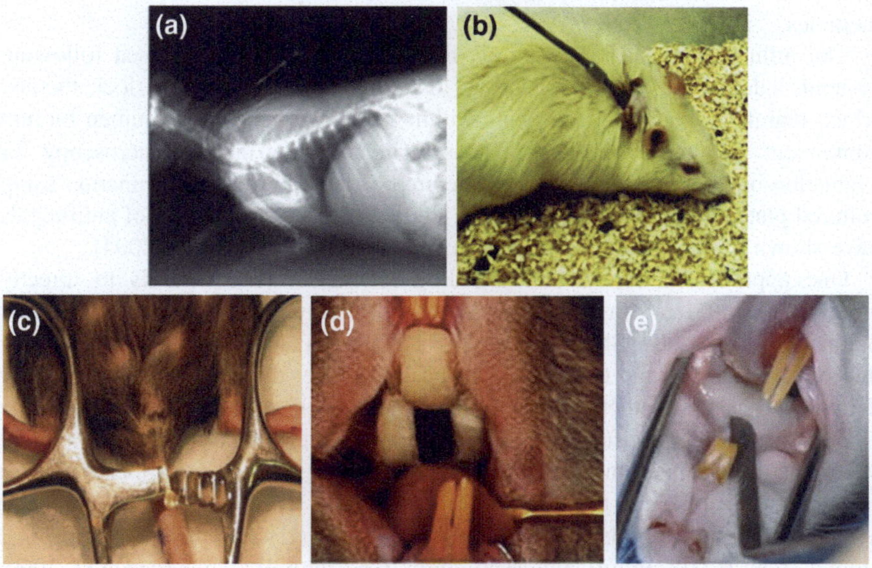

Fig. 7.1 Animal models of fungal biofilm infection. Animal models of fungal biofilm infection. Rabbit venous catheter-associated *Candida* biofilm infection (**a**) (Raad et al. 2007). Rat venous catheter-associated *Candida* biofilm infection (**b**) (Andes et al. 2004; Nett et al. 2007a). Mouse urinary catheter-associated *Candida* biofilm infection (**c**) (Wang and Fries 2011). Rat denture-associated *Candida* biofilm infection (removable intraoral device) (**d**) (Johnson et al. 2012). Rat denture-associated *Candida* biofilm infection (**e**) (Nett et al. 2010). Images adapted from prior publications (Nett et al. 2010; Raad et al. 2007; Johnson et al. 2012; Wang and Fries 2011; Nett et al. 2007a, 2014a, b)

mouse and has been instrumental for examining the efficacy of antifungals against biofilms formed in vivo (Andes et al. 2004; Lazzell et al. 2009; Schinabeck et al. 2004). As a close mimic one of the most common clinical biofilm infections, biofilms on the luminal catheter surface are exposed to host conditions, flow, serum proteins, and immune components. The model involves surgical vascular catheter insertion (jugular vein) followed by subcutaneous tunneling and securing with a protective device. Performing the procedure prior to luminal inoculation of organisms allows for a period for host protein conditioning of the device surface. Each of the animal species has subtle advantages and disadvantages. The smaller the animal such as the mouse are typically less expensive. However, the mouse model is technically the most challenging due to the surgical skill needed for the small vessels. For most laboratories use of this model requires purchase from a commercial vendor. Additionally, the small diameter of the catheter tubing needed for the mouse model can make it challenging to maintain patency due to blood clotting. The rabbit model is the most expensive thus there are often limitations in animal numbers for a given experiment. However, the larger animal and catheter are best for studies needing repeated sampling from the vascular device. In our opinion the rat offers a reasonable balance of cost and ability to maintain the catheter for prolonged periods of time.

Each of the models has been useful for assessing the efficacy of antifungal therapies.

The influence of anti-infectives on biofilm growth can be assessed following systemic administration of drug or by instilling in the lumen as a lock therapy which maintains high concentrations within the device next to the lumen biofilm component. Common techniques to assess the biofilms include microscopy for evaluation of biofilm extent and architecture or viable burden determination using cultured plate counts. Similar to in vitro studies, high concentrations of antifungals have shown minimal ability to clear the biofilm cells (Andes et al. 2004).

One approach to circumvent the drug tolerance of biofilms is to directly administer antifungal in the form of lock therapy (Mukherjee et al. 2009; Schinabeck et al. 2004; Shuford et al. 2006; Walraven and Lee 2013). By avoiding the majority of systemic toxicities, significantly higher drugs doses may be safely delivered. Vascular catheter animal models have been valuable for analyzing the efficacy of various antifungal lock therapies against *Candida* biofilms in vivo. When instilled in the lumen of *C. albicans* infected catheters, most lock therapies were found insufficient to clear biofilm infections. However, of the clinically available antifungals, the more efficacious solutions have included liposomal amphotericin B (10 mg/ml) and caspofungin (6.67 mg/ml) (Mukherjee et al. 2009; Schinabeck et al. 2004; Shuford et al. 2006; Walraven and Lee 2013). As might be predicted by in vitro biofilm susceptibility studies, the azole drugs and lower doses of echinocandin drugs have significantly less activity against vascular catheter biofilms and are not ideal for catheter lock therapy (Lazzell et al. 2009; Schinabeck et al. 2004). One concern regarding use of available antifungals for lock therapy is the potential for fostering a resistance-promoting environment. As a method to avoid this possibility, studies have also explored the use of alternative agents, such as biocides. Encouraging results have been observed in vitro for ethanol, ethylenediaminetetraacetic acid (EDTA), ethanol, and high dose minocycline (3 mg/ml) lock solutions (Raad et al. 2003, 2007; Sherertz et al. 2006).

A second tactic to overcome the profound antifungal tolerance of biofilms is delivery of combination drug therapy. The rat vascular catheter biofilm infection model has successfully been used to evaluate the in vivo efficacy of several combination therapy lock solutions (Robbins et al. 2011; Uppuluri et al. 2008). These studies have explored the impact of adding agents targeting cellular stress responses on azole drug resistance. Uppuluri et al. demonstrated the efficacy of combining calcineurin inhibitors and fluconazole for lock therapy treatment of *C. albicans* biofilms (Uppuluri et al. 2008). An agent inhibiting the calcineurin pathway (tacrolimus) was found to augment the activity of fluconazole against *C. albicans* catheter biofilms. Subsequent investigations suggest calcineurin inhibitors similarly potentiate the activity of agents in other drug classes, including the echinocandins and amphotericinB (Shinde et al. 2012). Robbins et al. also used a rat venous catheter model to test the efficacy of combination lock therapy (Robbins et al. 2011). Combining an inhibitor of the Hsp90 pathway (17-AAG) with fluconazole improved the activity against *C. albicans* biofilms. The mechanism of this action is thought to involve a decrease in biofilm extracellular matrix, limiting the capacity of the matrix to sequester antifungal.

The in vivo catheter models have also been important for exploring the clinical impact of *Candida* gene products on the biofilm lifestyle. The catheter model has been used in an exploratory fashion to explore gene product relevance using microarrays (Nett et al. 2009). Many of the differentially expressed transcripts have been shown relevant in follow on study of deletion mutants (Mitchell et al. 2015; Nobile et al. 2006; Taff et al. 2012). The catheter has been valuable for assessment of cell adhesion, mature biofilm formation, extracellular matrix formation, and cell dispersion (Andes et al. 2004). Most commonly a combination of viable cell counts and SEM imaging have been the most useful for assessing mutant phenotypes. Most frequently the in vitro biofilm phenotype has been consistent with the in vivo catheter biofilm phenotype. However, interesting differences have been identified in the relevance of genes in the adhesion pathway (Nobile et al. 2008). We speculate this could be in part due to important interactions with host protein components identified on these devices in the preconditioning period.

An additional component that can be varied in these models is the catheter material. The device substrate has been shown to impact adhesion of microbes. Numerous plastics are used in the manufacture of vascular catheters, although most are now silicone-based. Biomaterial investigations continue to test the protective effect of new materials as well as incorporation of antimicrobial compounds. Each of the animal models of biofilm infection should be useful is testing the biofilm prevention capabilities in vivo.

Both confocal and SEM imaging have been the primary tools used for assessment of extracellular matrix quantity. Most of the genes with reduced matrix component phenotypes have also demonstrated reduction in antifungal drug resistance (Mitchell et al. 2015). The dispersion phase of growth has also been explored by measuring viable cell counts in the kidney of rats in animals with biofilm infection of rat catheters (Andes et al. 2004). The impact of gene products in metabolic pathways have not thus far been explored but should be of interest in dissection of the relevance of nutrient limitations which appear common in the biofilm environment.

7.4.2 Denture Model

Advancing age and a number of pathologic conditions of the head and neck often result in a decline in oral health and a need for prosthetic dental devices. Denture stomatitis involves biofilm formation on a denture surface and inflammation of the adjacent oral mucosal surface (Webb et al. 1998a, b). These infections are common, occurring in up to 70% of denture-wearers, and are often painful, even impairing the ability to eat. Biofilms are frequently polymicrobial with *Candida* spp. playing a key role. In vitro models have attempted to incorporate host factors such as saliva into models in an attempt to better mimic this biofilm niche. Saliva plays several roles in oral health including the provision of a cleansing effect and innate immune defense proteins such as histatin and lactoferrin. However, some salivary proteins have been shown to facilitate *Candida* adhesion, such as mucins. Full incorporation of these and other niche factors is difficult to accomplish without use of an in vivo model.

Several models have been developed to explore the pathogenesis and treatment of denture stomatitis (Johnson et al. 2012; Nett et al. 2010; Samaranayake and Samaranayake 2001). Early models included primarily Macacairus monkeys with custom-fitted acrylic plates and Wistar rats fitted with prefabricated acrylic devices (Budtz–Jorgensen 1971; Lamb and Martin 1983; Norris et al. 1985). The focus of these investigations was examination of the mucosal inflammatory process associated with the infected device. *Candida*-infected animals with oral devices were observed to develop mucosal lesions similar to those seen in patients with denture stomatitis (Budtz-Jorgensen 1971; Lamb and Martin 1983; Norris et al. 1985). Although both models were useful for describing the host response to denture biofilms, the rat model was more suited for drug efficacy studies, primarily related to animal cost. In these investigations, the incorporation of either chlorhexidine or miconazole to the denture acrylic material prevented the development of mucosal lesions of palatal candidiasis (Lamb and Martin 1983; Norris et al. 1985). However, the chlorhexidine product was poorly tolerated with rat undergoing weight loss from poor dietary intake.

With the discovery of the role of biofilms in device-associated infections, there has been renewed interest in animal models to mimic denture stomatitis (Johnson et al. 2012; Lee et al. 2011; Nett et al. 2010). Two models have been developed to replicate this clinical scenario in rats. In the first model, a Sprague-Dawley rat undergoes placement of an acrylic dental device over the hard palate, which is secured in place by orthodontic wire (Nett et al. 2010). As the device is fitted to the individual rat, there is close approximation of the device with the oral mucosa and this space can be inoculated with *Candida* to produce a biofilm device infection and associated mucosal inflammation over the course of 24–72 h. This model represents an acute infection in the setting of immunosuppression, as rats are treated with a single dose of cortisone prior to infection. Unlike models with prolonged mucosal or device infection, epithelial atrophy was not observed, likely due to the shorter duration of the model. This model is the least expensive between the two rat models as there is no need for the molding process. The utility of this model is study of therapeutics and the impact of gene products. Thus far, *C. albicans* mutants with defects in the vascular catheter model have behaved similar in the denture model. However, given marked differences in the local environment from the standpoint of host cells, secreted host proteins, pH, and nutrient availability it is likely distinct *Candida* pathways are relevant for biofilm pathogenesis in this niche. Additionally, this animal model tool has been shown useful for study of *Candida*—oral bacterial interactions. Initial microbial characterization identified a variety of aerobic and anaerobic streptococci associated with the *Candida,* particularly the hyphal components of the biofilm.

In a second rat model, Wistar rats are custom fitted to palatal acrylic device (Johnson et al. 2012; Lee et al. 2011). However, a portion of the device is secured by embedded magnets and is easily removable throughout the experimental course. Following inoculation of *Candida*, biofilm develops on the device surface over weeks. In addition, mucosal biofilm infection and inflammation ensue, mimicking clinical infection occurs. As the devices can remain in place for an extended time (8 weeks), this model offers the opportunity for repeated sampling to longitudinally follow the course of an individual animal with a chronic infection. This model should be particularly useful for exploration of the host-pathogen interaction of chronic denture stomatitis as the preliminary histopathology is congruent with the process in patients.

Few studies have begun to investigate the antifungal treatment in the rodent denture models. As might be predicted from clinical scenarios and other biofilm infection models, the *C. albicans* communities on the denture surface were found to exhibit high tolerance of both fluconazole and micafungin upon either topical or systemic administration (Nett et al. 2010). The model has also been helpful for exploring the role of gene products on denture biofilm infection in vivo. Chen et al. described the importance of the calcineurin pathway for *C. dubliniensis* in the processes of both filamentation and biofilm formation in a rat denture biofilm (Chen et al. 2011). This suggests that calcineurin inhibitors may be a viable option for treatment of *C. dubliniensis*. As drugs in this class exert synergistic activity with azole and echinocandin drugs, combination therapy is an attractive possibility for treatment of *C. dubliniensis* biofilm infections (Chen et al. 2011).

7.4.3 Subcutaneous Implant Model

To study the activity of a novel antifungal formulation against biofilms, Zumbuehl et al. developed a murine model of subcutaneous *Candida* biofilm infection (Zumbuehl et al. 2007). Disks containing amphogel, a dextran-based hydrogel loaded with amphotericin B, were inoculated with *C. albicans* and surgically implanted in the subcuntaneous flank tissue of BALB/c mice. After 3 days, *Candida* had been cleared from the surface of the disks containing amphogel. In contrast, control disks with hydrogel only were coated with *C. albicans* biofilm and host cells. The amphogel was well-tolerated, eliciting only a minimal or mild inflammatory response. As this antifungal hydrogel maintains efficacy for over 50 days, it is ideally suited for prevention of device-associated infection.

As an alternative model for study of *Candida* biofilm infections in the subcutaneous space was adapted using catheter material. In this model, a polyurethane catheter segments are inoculated with *Candida* and implanted under the skin of a rat (Ricicova et al. 2010). Compared to other biofilm animal models, this procedure is technically easier to perform and multiple devices can be used in the same animal reducing cost and allowing for more experimental replicates. In terms of mimicking patient infection, the model has similarities to both vascular catheter infections and wound infections. The implanted catheter material is a close mimic of the vascular catheter material used in patients. The model is avascular, so biofilm cells are not subjected to blood flow conditions and not exposed to the same concentrations of serum protein and blood cells. However, the devices can be treated with serum prior to implantation to partially mimic this exposure. The anatomical location of the implantation is most similar to a biofilm wound infection. The model allows for the interaction between host immune components and *Candida* biofilms in the subcutaneous tissue. Numerous study assays have been successfully used as endpoints for study of biofilm architecture, biofilm burden, and gene expression. These include SEM and confocal microscopy, biofilm dry weight, quantitative viable plate counts, and RT-PCR. Similar to most in vitro and other in vivo models, time course studies identified mature biofilm formation after 48 h of incubation. Maintenance of biofilm structure was observed after more than a week of infection in rodents. Thus far common catheter materials have been utilized in this model. However, other materials could be easily investigated. For example, incorporation of orthopedic materials such as titanium and bone cements may be of interest given the relevance of the subcutaneous space for this infection in patients.

The subcutaneous implant model has been used most frequently to test for efficacy of novel anti-biofilm treatments. Bink et al. used this mode to test the efficacy of combining a nonsteroidal anti-inflammatory drug (NSAID) and an echinocandin for treatment of *C. albicans* biofilms in vivo (Bink et al. 2012). NSAIDs impair prostaglandin synthesis by targeting mammalian cycloogenases. Agents in this class are available for the treatment of pain and inflammation. However, the activity of this drug class is not limited to mammalian systems, as they have also been shown to disrupt filamentation and biofilm formation in

C. albicans, likely though inhibition of prostaglandin E2 synthesis (Alem and Douglas 2005; Ghalehnoo et al. 2010). To examine the impact of disrupting this pathway in vivo, rats received diclofenac treatment in the setting of subcutaneous catheter implant infection (Bink et al. 2012). In rats that had been treated with diclofenac prior to development of *C. albicans* biofilm infection, the anti-biofilm activity of caspofungin was enhanced.

Initial studies in the development and testing of the subcutaneous catheter model also demonstrated the utility of this tool to detect biofilm pathogenesis defects in *Candida* mutants in both adherence and filamentation pathways. Interestingly, however, the studies suggest a distinct role for certain adhesion proteins in this subcutaneous infection niche (Ricicova et al. 2010). Specifically, while a BCR1 mutant produced less biofilm growth than the wild type strain as has been found in the catheter and denture models. Conversely, ALS3 appeared dispensable for biofilm formation in this model. These findings underscore the importance of considering different host infection sites in discerning the role of various genetic pathways in the biofilm lifestyle.

7.4.4 Urinary Catheter Model

Biofilm infection of urinary catheters represents the most common (70%) infection acquired in the hospital setting with more than one million patients diagnosed in the United States alone. Some surveillance studies suggest up to 1 in 5 hospitalized patients will have a urinary catheter placed at some point during their stay. *Candida* species infection of these devices is the third most common microbe responsible for these infections. Two rodent models have been developed to mimic this infection process in patients. The niche components of hypothesized relevance for study of catheter urinary tract infection design includes the anatomic location, exposure to urine, flow, urinary epithelial cells, and commonly used device materials.

In vitro models have attempted to recapitulate a number of these factors by developing a synthetic urine media. The media includes adjusted electrolyte concentrations, pH, creatinine, and urea. However, it is difficult to account for the interaction with cells and proteins.

Two animal models have been developed to account for these factors in the study of biofilm pathogenesis. The first model published by Wang et al. utilized mice (Wang and Fries 2011). In this model, a guide wire is inserted through the urethra of a female mouse and a catheter segment is threaded over a guide wire and into the bladder. The segment is secured by suture through the bladder wall. After 5–7 days, the animal is infected with *C. albicans* by injection directly into the bladder. Candiduria is detectable quickly after infection and persists for 28 days. A dense biofilm of adherent yeast and hyphae forms on both the luminal and extraluminal surfaces. To increase the susceptibility to *Candida* infection, mice lacking lysozyme M, an important effector for mucosal innate immunity, can be utilized. The model closely mimics patient *Candida* biofilm formation with regard to the use of biofilm substrate (catheter) and anatomic location (bladder). As only a segment of catheter is in place and the catheter does not exit the urethra, the flow conditions are likely less than would be observed for a patient catheter functioning to drain the bladder. However, an advantage of the model includes the ability to incorporate the mammalian immune system with the option of using wildtype or specifically immunocompromised knockout animals.

A second model has been developed using larger rats (Nett et al. 2014b). The major difference between the mouse and rat model is the placement via the urethra which copies the process in patients and allows for urine flow through the catheter lumen as well as exposure to the external environment. A variety of biofilm and host response assessments have been undertaken in this model. SEM and quantitative culture of the catheter biofilm cells demonstrated mature biofilms 48 h after infection. Interestingly, we also observed biofilms overlying and invading urinary epithelia. Studies also demonstrated significant dispersion of cells either from the catheter or epithelial biofilms based upon quantitative assessment of viable *Candida* in the excreted urine which at times equaled the burden in the biofilms. The model was also used for transcriptomics and for assessing the phenotypic effect of select *Candida* mutants on biofilm formation. RT-PCR studies of select transcripts that were previously shown to be important for production and delivery of the extracellular matrix polysaccharide were found to be up-regulated during biofilm growth in this model. Additionally, study of a mutant with disruption of both ALS1 and ALS3 demonstrated a profound biofilm deficiency likely due to reduced adhesion in the early phase of biofilm formation.

7.4.5 Mucosal Candidiasis Models

Biofilms have frequently been described in association with medical devices and abiotic surfaces. However, there is mounting evidence that *Candida* spp. exhibit similar characteristics when growing on mucosal surfaces (Dongari–Bagtzoglou et al. 2009; Harriott et al. 2010). Murine models of both oropharyngeal and vaginal candidiasis demonstrate that *Candida* produces of conglomerates of yeast, hyphae, and extracellular material associated with mucosal surfaces. In an oropharyngeal candidiasis model, the biofilms appear to be complex, involving commensal bacteria, neutrophils, and keratin (Dongari–Bagtzoglou et al. 2009). A murine model of vaginal candidiasis shows *C. albicans* regulators of biofilm formation on abiotic surfaces are similar to those required for development of vaginal biofilms. Although mucosal biofilms share many characteristics with device-associated biofilms, it is not clear they exhibit the same degree of drug resistance. Clinically, mucosal biofilms are most often responsive to antifungal therapies, including azoles (Graybill et al. 1998; Sobel et al. 1998).

7.5 Future Opportunities

Recognition of the importance of animal models for the discovery of biofilm genetic pathways and new anti-biofilm drugs has only recently emerged. Most investigations have focused on *C. albicans* as a model pathogen and the vascular catheter models of biofilm infection have been the most popular. It will be interesting to see how study of biofilms formed under the conditions of other clinically relevant niches respond to antifungal therapies. The murine urinary catheter model, the rat subcutaneous model, and rat denture models should be of significant value for these investigations (Johnson et al. 2012; Nett et al. 2010; Ricicova et al. 2010; Wang and Fries 2011). The models allow for testing of anti-biofilm compounds under physiologic conditions very similar to those encountered clinically. In addition to these animal models of device-associated infections, models of mucosal *Candida* biofilms will surely be helpful for study of these common infections (Dongari–Bagtzoglou et al. 2009; Harriott et al. 2010).

Although *C. albicans* has been the model pathogen for many in vivo biofilm investigations, the in vivo biofilm models can likely be adapted to biofilm infections caused by a variety of non*albicans Candida* spp. Of note, the rat vascular catheter model has been successfully used for study of *C. parapsilosis* and *C. glabrata*, while a rat denture model has been employed for investigation of *C. dubliniensis* (Chen et al. 2011; Nett et al. 2007a).

There are many approaches to the discovery of new anti-infectives. One strategy is to screen large libraries of compounds. Using in vitro models, the mining of pharmaceutical and natural product libraries has identified novel compounds with anti-biofilm activity (Coleman et al. 2010; Lafleur et al. 2013; Sherry et al. 2012). An alternative approach is to determine a mechanism leading to drug resistance and identify or develop an anti-infective that disrupts the process. For *C. albicans*, the biofilm property most closely linked to resistance is the extracellular matrix. Enzymatic degradation of key matrix components, such as extracellular DNA and β-1,3 glucan, has been shown to enhance antifungal activity, suggesting these as potential drug targets (Martins et al. 2012; Nett et al. 2007b). In fact, a therapy directed at extracellular DNA degradation has shown to be a beneficial for patients with cystic fibrosis. It is thought dornase alfa (Pulmozyme), a clinically available inhaled enzymatic treatment, works by degrading extracellular DNA of bacterial biofilms (Frederiksen et al. 2006). Regardless of the path of drug discovery, animal models will be beneficial for testing the efficacy of compounds against clinical biofilms and establishing safety.

One of the unique aspects of exploring anti-biofilm activity of drugs in animal models is the opportunity to vary the mode of antifungal delivery. For example, compounds may be systemically administered, topically administered, coated on a device, or embedded in a device. Another interesting delivery method is direct administration of a gel with prolonged elution of high antifungal concentrations, such as was developed for amphotericin B (Hudson et al. 2010). Systemic administration is feasible to test in all models, while direct, topical administration of a compound is easily achievable in either the denture models via topical therapy or the venous catheter models via lock therapy. The subcutaneous tissue model may be ideal for exploring the utility of embedding or coating a device with an anti-biofilm compound, as numerous devices can be tested in a single animal. Another potential application is the investigation of vaccine efficacy, such as the NDV-3 vaccine in clinical trials, vaccines found to be efficacious in nonbiofilm models of infections, or future vaccines designed specifically to inhibit the biofilm mode of growth (Cassone 2013; Luo et al. 2010; Schmidt et al. 2012).

Acknowledgements D.A. is supported by NIH R01 AI073289.

References

Al-Fattani MA, Douglas LJ (2004) Penetration of *Candida* biofilms by antifungal agents. Antimicrob Agents Chemother 48:3291–3297

Al-Fattani MA, Douglas LJ (2006) Biofilm matrix of *Candida albicans* and *Candida tropicalis*: chemical composition and role in drug resistance. J Med Microbiol 55:999–1008

Alem MA, Douglas LJ (2005) Prostaglandin production during growth of Candida albicans biofilms. J Med Microbiol 54:1001–1005

Andes D, Nett J, Oschel P, Albrecht R, Marchillo K, Pitula A (2004) Development and characterization of an in vivo central venous catheter Candida albicans biofilm model. Infect Immun 72:6023–6031

Andes DR, Safdar N, Baddley JW, Playford G, Reboli AC, Rex JH, Sobel JD, Pappas PG, Kullberg BJ (2012) Impact of treatment strategy on outcomes in patients with candidemia and other forms of invasive candidiasis: a patient-level quantitative review of randomized trials. Clin Infect Dis 54:1110–1122

Baillie GS, Douglas LJ (1998) Effect of growth rate on resistance of *Candida albicans* biofilms to antifungal agents. Antimicrob Agents Chemother 42:1900–1905

Bink A, Kucharikova S, Neirinck B, Vleugels J, Van Dijck P, Cammue BP, Thevissen K (2012) The nonsteroidal antiinflammatory drug diclofenac potentiates the in vivo activity of caspofungin against Candida albicans biofilms. J Infect Dis 206:1790–1797

Brash JL, Ten Hove P (1993) Protein adsorption studies on 'standard' polymeric materials. J Biomater Sci Polym Ed 4:591–599

Budtz-Jorgensen E (1971) Denture stomatitis. IV. An experimental model in monkeys. Acta Odontol Scand 29:513–526

Cannizzo FT, Eraso E, Ezkurra PA, Villar-Vidal M, Bollo E, Castella G, Cabanes FJ, Vidotto V, Quindos G (2007) Biofilm development by clinical isolates of *Malassezia pachydermatis*. Med Mycol 45:357–361

Cassone A (2013) Development of vaccines for *Candida albicans*: fighting a skilled transformer. Nat Rev 11:884–891

Chandra J, Kuhn DM, Mukherjee PK, Hoyer LL, McCormick T, Ghannoum MA (2001) Biofilm formation by the fungal pathogen *Candida albicans*: development, architecture, and drug resistance. J Bacteriol 183:5385–5394

Chandra J, McCormick TS, Imamura Y, Mukherjee PK, Ghannoum MA (2007) Interaction of *Candida albicans* with adherent human peripheral blood mononuclear cells increases *C. albicans* biofilm formation and results in differential expression of pro- and anti-inflammatory cytokines. Infect Immun 75:2612–2620

Chen YL, Brand A, Morrison EL, Silao FG, Bigol UG, Malbas FF Jr, Nett JE, Andes DR, Solis NV, Filler SG et al (2011) Calcineurin controls drug tolerance, hyphal growth, and virulence in *Candida dubliniensis*. Eukaryot Cell 10:803–819

Coleman JJ, Okoli I, Tegos GP, Holson EB, Wagner FF, Hamblin MR, Mylonakis E (2010) Characterization of plant-derived saponin natural products against *Candida albicans*. ACS Chem Biol 5:321–332

D'Antonio D, Parruti G, Pontieri E, Di Bonaventura G, Manzoli L, Sferra R, Vetuschi A, Piccolomini R, Romano F, Staniscia T (2004) Slime production by clinical isolates of *Blastoschizomyces capitatus* from patients with hematological malignancies and catheter-related fungemia. Eur J Clin Microbiol Infect Dis 23:787–789

Davis LE, Cook G, Costerton JW (2002) Biofilm on ventriculo-peritoneal shunt tubing as a cause of treatment failure in coccidioidal meningitis. Emerg Infect Dis 8:376–379

Di Bonaventura G, Pompilio A, Picciani C, Iezzi M, D'Antonio D, Piccolomini R (2006) Biofilm formation by the emerging fungal pathogen *Trichosporon asahii:* development, architecture, and antifungal resistance. Antimicrob Agents Chemother 50:3269–3276

Dongari-Bagtzoglou A (2008) Mucosal biofilms: challenges and future directions. Expert review of anti-infective therapy 6:141–144

Dongari-Bagtzoglou A, Kashleva H, Dwivedi P, Diaz P, Vasilakos J (2009) Characterization of mucosal *Candida albicans* biofilms. PLoS ONE 4:e7967

Donlan RM (2001a) Biofilm formation: a clinically relevant microbiological process. Clin Infect Dis 33:1387–1392

Donlan RM (2001b) Biofilms and device-associated infections. Emerg Infect Dis 7:277–281

Douglas LJ (2002) Medical importance of biofilms in *Candida infections*. Rev Iberoam Micol 19:139–143

Dyavaiah M, Ramani R, Chu DS, Ritterband DC, Shah MK, Samsonoff WA, Chaturvedi S, Chaturvedi V (2007) Molecular characterization, biofilm analysis and experimental biofouling study of *Fusarium* isolates from recent cases of fungal keratitis in New York State. BMC ophthalmology 7:1

Edmond MB, Wallace SE, McClish DK, Pfaller MA, Jones RN, Wenzel RP (1999) Nosocomial bloodstream infections in United States hospitals: a three-year analysis. Clin Infect Dis 29:239–244

Foxman B, Muraglia R, Dietz JP, Sobel JD, Wagner J (2013) Prevalence of recurrent vulvovaginal candidiasis in 5 European countries and the United States: results from an internet panel survey. Journal of lower genital tract disease 17:340–345

Francois P, Schrenzel J, Stoerman-Chopard C, Favre H, Herrmann M, Foster TJ, Lew DP, Vaudaux P (2000) Identification of plasma proteins adsorbed on hemodialysis tubing that promote *Staphylococcus aureus* adhesion. J Lab Clin Med 135:32–42

Frederiksen B, Pressler T, Hansen A, Koch C, Hoiby N (2006) Effect of aerosolized rhDNase (Pulmozyme) on pulmonary colonization in patients with cystic fibrosis. Acta Paediatr 95:1070–1074

Ghalehnoo ZR, Rashki A, Najimi M, Dominguez A (2010) The role of diclofenac sodium in the dimorphic transition in Candida albicans. Microb Pathog 48:110–115

Graybill JR, Vazquez J, Darouiche RO, Morhart R, Greenspan D, Tuazon C, Wheat LJ, Carey J, Leviton I, Hewitt RG et al (1998) Randomized trial of itraconazole oral solution for oropharyngeal candidiasis in HIV/AIDS patients. Am J Med 104:33–39

Groeger JS, Lucas AB, Thaler HT, Friedlander-Klar H, Brown AE, Kiehn TE, Armstrong D (1993) Infectious morbidity associated with long-term use of venous access devices in patients with cancer. Ann Intern Med 119:1168–1174

Harriott MM, Lilly EA, Rodriguez TE, Fidel PL Jr, Noverr MC (2010) *Candida albicans* forms biofilms on the vaginal mucosa. Microbiology (Reading, England) 156:3635–3644

Hawser SP, Baillie GS, Douglas LJ (1998) Production of extracellular matrix by *Candida albicans* biofilms. J Med Microbiol 47:253–256

Hawser SP, Douglas LJ (1994) Biofilm formation by *Candida* species on the surface of catheter materials in vitro. Infect Immun 62:915–921

Hoyle BD, Jass J, Costerton JW (1990) The biofilm glycocalyx as a resistance factor. The Journal of antimicrobial chemotherapy 26:1–5

Hudson SP, Langer R, Fink GR, Kohane DS (2010) Injectable in situ cross-linking hydrogels for local antifungal therapy. Biomaterials 31:1444–1452

Jenney CR, Anderson JM (2000) Adsorbed serum proteins responsible for surface dependent human macrophage behavior. J Biomed Mater Res 49:435–447

Jin Y, Samaranayake LP, Samaranayake Y, Yip HK (2004) Biofilm formation of *Candida albicans* is variably affected by saliva and dietary sugars. Arch Oral Biol 49:789–798

Johnson CC, Yu A, Lee H, Fidel PL Jr, Noverr MC (2012) Development of a contemporary animal model of *Candida albicans*-associated denture stomatitis using a novel intraoral denture system. Infect Immun 80:1736–1743

Katragkou A, Chatzimoschou A, Simitsopoulou M, Georgiadou E, Roilides E (2011a) Additive antifungal activity of anidulafungin and human neutrophils against *Candida parapsilosis* biofilms. J Antimicrob Chemother 66:588–591

Katragkou A, Kruhlak MJ, Simitsopoulou M, Chatzimoschou A, Taparkou A, Cotten CJ, Paliogianni F, Diza-Mataftsi E, Tsantali C, Walsh TJ et al (2010) Interactions between human phagocytes and *Candida albicans* biofilms alone and in combination with antifungal agents. J Infect Dis 201:1941–1949

Katragkou A, Simitsopoulou M, Chatzimoschou A, Georgiadou E, Walsh TJ, Roilides E (2011b) Effects of interferon-gamma and granulocyte colony-stimulating factor on antifungal activity of human polymorphonuclear neutrophils against *Candida albicans* grown as biofilms or planktonic cells. Cytokine 55:330–334

Khot PD, Suci PA, Miller RL, Nelson RD, Tyler BJ (2006) A small subpopulation of blastospores in *Candida albicans* biofilms exhibit resistance to amphotericin B associated with differential regulation of ergosterol and beta-1,6-glucan pathway genes. Antimicrob Agents Chemother 50:3708–3716

Kojic EM, Darouiche RO (2004) *Candida* infections of medical devices. Clin Microbiol Rev 17:255–267

Kuhn DM, George T, Chandra J, Mukherjee PK, Ghannoum MA (2002) Antifungal susceptibility of *Candida* biofilms: unique efficacy of amphotericin B lipid formulations and echinocandins. Antimicrob Agents Chemother 46:1773–1780

Kumamoto CA (2005) A contact-activated kinase signals *Candida albicans* invasive growth and biofilm development. Proc Natl Acad Sci USA 102:5576–5581

LaFleur MD, Kumamoto CA, Lewis K (2006) Candida albicans biofilms produce antifungal-tolerant persister cells. Antimicrob Agents Chemother 50:3839–3846

Lafleur MD, Sun L, Lister I, Keating J, Nantel A, Long L, Ghannoum M, North J, Lee RE, Coleman K et al (2013) Potentiation of azole antifungals by 2-adamantanamine. Antimicrob Agents Chemother 57:3585–3592

Lamb DJ, Martin MV (1983) An in vitro and in vivo study of the effect of incorporation of chlorhexidine into autopolymerizing acrylic resin plates upon the growth of *Candida albicans*. Biomaterials 4:205–209

Lazzell AL, Chaturvedi AK, Pierce CG, Prasad D, Uppuluri P, Lopez-Ribot JL (2009) Treatment and prevention of *Candida albicans* biofilms with caspofungin in a novel central venous catheter murine model of candidiasis. J Antimicrob Chemother 64:567–570

Lee H, Yu A, Johnson CC, Lilly EA, Noverr MC, Fidel PL Jr (2011) Fabrication of a multi-applicable removable intraoral denture system for rodent research. J Oral Rehabil 38:686–690

Lewis RE, Kontoyiannis DP, Darouiche RO, Raad II, Prince RA (2002) Antifungal activity of amphotericin B, fluconazole, and voriconazole in an in vitro model of *Candida* catheter-related bloodstream infection. Antimicrob Agents Chemother 46:3499–3505

Loussert C, Schmitt C, Prevost MC, Balloy V, Fadel E, Philippe B, Kauffmann-Lacroix C, Latge JP, Beauvais A (2010) In vivo biofilm composition of *Aspergillus fumigatus*. Cell Microbiol 12:405–410

Luo G, Ibrahim AS, Spellberg B, Nobile CJ, Mitchell AP, Fu Y (2010) *Candida albicans* Hyr1p confers resistance to neutrophil killing and is a potential vaccine target. J Infect Dis 201:1718–1728

Mah TF, Pitts B, Pellock B, Walker GC, Stewart PS, O'Toole GA (2003) A genetic basis for *Pseudomonas aeruginosa* biofilm antibiotic resistance. Nature 426:306–310

Martinez LR, Casadevall A (2007) *Cryptococcus neoformans* biofilm formation depends on surface support and carbon source and reduces fungal cell susceptibility to heat, cold, and UV light. Appl Environ Microbiol 73:4592–4601

Martins M, Henriques M, Lopez-Ribot JL, Oliveira R (2012) Addition of DNase improves the in vitro activity of antifungal drugs against *Candida albicans* biofilms. Mycoses 55:80–85

Mitchell KF, Zarnowski R, Sanchez H, Edward JA, Reinicke EL, Nett JE, Mitchell AP, Andes DR (2015) Community participation in biofilm matrix assembly and function. Proc Natl Acad Sci USA 112:4092–4097

Mukherjee PK, Chandra J, Kuhn DM, Ghannoum MA (2003) Mechanism of fluconazole resistance in *Candida albicans* biofilms: phase-specific role of efflux pumps and membrane sterols. Infect Immun 71:4333–4340

Mukherjee PK, Long L, Kim HG, Ghannoum MA (2009) Amphotericin B lipid complex is efficacious in the treatment of *Candida albicans* biofilms using a model of catheter-associated *Candida* biofilms. Int J Antimicrob Agents 33:149–153

Mukherjee PK, Zhou G, Munyon R, Ghannoum MA (2005) *Candida* biofilm: a well-designed protected environment. Med Mycol 43:191–208

Nett J, Lincoln L, Marchillo K, Andes D (2007a) Beta -1,3 glucan as a test for central venous catheter biofilm infection. J Infect Dis 195:1705–1712

Nett J, Lincoln L, Marchillo K, Massey R, Holoyda K, Hoff B, VanHandel M, Andes D (2007b) Putative role of beta-1,3 glucans in *Candida albicans* biofilm resistance. Antimicrob Agents Chemother 51:510–520

Nett JE, Brooks E, Cabezas-Olcoz J, Sanchez H, Zarnowski R, Marchillo K, Andes DR (2014a) Rat Indwelling Urinary Catheter Model of Candida albicans Biofilm Infection. Infect Immun

Nett JE, Brooks EG, Cabezas-Olcoz J, Sanchez H, Zarnowski R, Marchillo K, Andes DR (2014b) Rat indwelling urinary catheter model of Candida albicans biofilm infection. Infect Immun 82:4931–4940

Nett JE, Cabezas-Olcoz J, Marchillo K, Mosher DF, Andes DR (2016) Targeting fibronectin to disrupt in vivo candida albicans biofilms. Antimicrob Agents Chemother 60:3152–3155

Nett JE, Lepak AJ, Marchillo K, Andes DR (2009) Time course global gene expression analysis of an in vivo Candida biofilm. J Infect Dis 200:307–313

Nett JE, Marchillo K, Spiegel CA, Andes DR (2010) Development and validation of an in vivo *Candida albicans* biofilm denture model. Infect Immun 78:3650–3659

Nett JE, Zarnowski R, Cabezas-Olcoz J, Brooks EG, Bernhardt J, Marchillo K, Mosher DF, Andes DR (2015) Host contributions to construction of three device-associated Candida albicans biofilms. Infect Immun 83:4630–4638

Nobile CJ, Andes DR, Nett JE, Smith FJ, Yue F, Phan QT, Edwards JE, Filler SG, Mitchell AP (2006) Critical role of Bcr1-dependent adhesins in C. albicans biofilm formation in vitro and in vivo. PLoS Pathog 2:e63

Nobile CJ, Mitchell AP (2005) Regulation of cell-surface genes and biofilm formation by the *C. albicans* transcription factor Bcr1p. Current biology: CB 15:1150–1155

Nobile CJ, Schneider HA, Nett JE, Sheppard DC, Filler SG, Andes DR, Mitchell AP (2008) Complementary adhesin function in C. albicans biofilm formation. Current biology: CB 18:1017–1024

Norris MM, Lamb DJ, Craig GT, Martin MV (1985) The effect of miconazole on palatal candidosis induced in the Wistar rat. J Dent 13:288–294

O'Toole GA (2003) To build a biofilm. J Bacteriol 185:2687–2689

Pappas PG, Kauffman CA, Andes D, Benjamin DK Jr, Calandra TF, Edwards JE Jr, Filler SG, Fisher JF, Kullberg BJ, Ostrosky-Zeichner L et al (2009) Clinical practice guidelines for the management of candidiasis: 2009 update by the Infectious Diseases Society of America. Clin Infect Dis 48:503–535

Passerini L, Lam K, Costerton JW, King EG (1992) Biofilms on indwelling vascular catheters. Crit Care Med 20:665–673

Perumal P, Mekala S, Chaffin WL (2007) Role for cell density in antifungal drug resistance in *Candida albicans* biofilms. Antimicrob Agents Chemother 51:2454–2463

Pfaller MA, Diekema DJ (2007) Epidemiology of invasive candidiasis: a persistent public health problem. Clin Microbiol Rev 20:133–163

Proctor RA (2000) Toward an understanding of biomaterial infections: a complex interplay between the host and bacteria. J Lab Clin Med 135:14–15

Raad I, Chatzinikolaou I, Chaiban G, Hanna H, Hachem R, Dvorak T, Cook G, Costerton W (2003) In vitro and ex vivo activities of minocycline and EDTA against microorganisms embedded in biofilm on catheter surfaces. Antimicrob Agents Chemother 47:3580–3585

Raad I, Hanna H, Dvorak T, Chaiban G, Hachem R (2007) Optimal antimicrobial catheter lock solution, using different combinations of minocycline, EDTA, and 25-percent ethanol, rapidly eradicates organisms embedded in biofilm. Antimicrob Agents Chemother 51:78–83

Ramage G, Bachmann S, Patterson TF, Wickes BL, Lopez-Ribot JL (2002a) Investigation of multidrug efflux pumps in relation to fluconazole resistance in *Candida albicans* biofilms. J Antimicrob Chemother 49:973–980

Ramage G, Rajendran R, Sherry L, Williams C (2012) Fungal biofilm resistance. Int J Microbiol 2012:528521

Ramage G, VandeWalle K, Bachmann SP, Wickes BL, Lopez-Ribot JL (2002b) In vitro pharmacodynamic properties of three antifungal agents against preformed *Candida albicans* biofilms determined by time-kill studies. Antimicrob Agents Chemother 46:3634–3636

Reynolds TB, Fink GR (2001) Bakers' yeast, a model for fungal biofilm formation. Science 291:878–881

Richards MJ, Edwards JR, Culver DH, Gaynes RP (1999) Nosocomial infections in medical intensive care units in the United States. National Nosocomial Infections Surveillance System. Crit Care Med 27:887–892

Ricicova M, Kucharikova S, Tournu H, Hendrix J, Bujdakova H, Van Eldere J, Lagrou K, Van Dijck P (2010) Candida albicans biofilm formation in a new in vivo rat model. Microbiology (Reading, England) 156:909–919

Robbins N, Uppuluri P, Nett J, Rajendran R, Ramage G, Lopez-Ribot JL, Andes D, Cowen LE (2011) Hsp90 governs dispersion and drug resistance of fungal biofilms. PLoS Pathog 7: e1002257

Ronsani MM, Mores Rymovicz AU, Meira TM, Trindade Gregio AM, Guariza Filho O, Tanaka OM, Ribeiro Rosa EA (2011) Virulence modulation of *Candida albicans* biofilms by metal ions commonly released from orthodontic devices. Microb Pathog 51:421–425

Samaranayake YH, Samaranayake LP (2001) Experimental oral candidiasis in animal models. Clin Microbiol Rev 14:398–429

Schinabeck MK, Long LA, Hossain MA, Chandra J, Mukherjee PK, Mohamed S, Ghannoum MA (2004) Rabbit model of *Candida albicans* biofilm infection: liposomal amphotericin B antifungal lock therapy. Antimicrob Agents Chemother 48:1727–1732

Schmidt CS, White CJ, Ibrahim AS, Filler SG, Fu Y, Yeaman MR, Edwards JE Jr, Hennessey JP Jr (2012) NDV-3, a recombinant alum-adjuvanted vaccine for *Candida* and *Staphylococcus aureus*, is safe and immunogenic in healthy adults. Vaccine 30:7594–7600

Seidler MJ, Salvenmoser S, Muller FM (2008) *Aspergillus fumigatus* forms biofilms with reduced antifungal drug susceptibility on bronchial epithelial cells. Antimicrob Agents Chemother 52:4130–4136

Sherertz RJ, Boger MS, Collins CA, Mason L, Raad II (2006) Comparative in vitro efficacies of various catheter lock solutions. Antimicrob Agents Chemother 50:1865–1868

Sherry L, Jose A, Murray C, Williams C, Jones B, Millington O, Bagg J, Ramage G (2012) Carbohydrate Derived Fulvic Acid: An in vitro Investigation of a Novel Membrane Active Antiseptic Agent Against *Candida albicans* Biofilms. Frontiers in microbiology 3:116

Shin JH, Kee SJ, Shin MG, Kim SH, Shin DH, Lee SK, Suh SP, Ryang DW (2002) Biofilm production by isolates of *Candida* species recovered from nonneutropenic patients: comparison of bloodstream isolates with isolates from other sources. J Clin Microbiol 40:1244–1248

Shinde RB, Chauhan NM, Raut JS, Karuppayil SM (2012) Sensitization of *Candida albicans* biofilms to various antifungal drugs by cyclosporine A. Annals of clinical microbiology and antimicrobials 11:27

Shuford JA, Rouse MS, Piper KE, Steckelberg JM, Patel R (2006) Evaluation of caspofungin and amphotericin B deoxycholate against *Candida albicans* biofilms in an experimental intravascular catheter infection model. J Infect Dis 194:710–713

Singh R, Shivaprakash MR, Chakrabarti A (2011) Biofilm formation by zygomycetes: quantification, structure and matrix composition. Microbiology (Reading, England) 157:2611–2618

Sobel JD, Faro S, Force RW, Foxman B, Ledger WJ, Nyirjesy PR, Reed BD, Summers PR (1998) Vulvovaginal candidiasis: epidemiologic, diagnostic, and therapeutic considerations. Am J Obstet Gynecol 178:203–211

Taff HT, Nett JE, Zarnowski R, Ross KM, Sanchez H, Cain MT, Hamaker J, Mitchell AP, Andes DR (2012) A Candida biofilm-induced pathway for matrix glucan delivery: implications for drug resistance. PLoS Pathog 8:e1002848

Uppuluri P, Chaturvedi AK, Lopez-Ribot JL (2009) Design of a simple model of *Candida albicans* biofilms formed under conditions of flow: development, architecture, and drug resistance. Mycopathologia 168:101–109

Uppuluri P, Chaturvedi AK, Srinivasan A, Banerjee M, Ramasubramaniam AK, Kohler JR, Kadosh D, Lopez-Ribot JL (2010) Dispersion as an important step in the *Candida albicans* biofilm developmental cycle. PLoS Pathog 6:e1000828

Uppuluri P, Nett J, Heitman J, Andes D (2008) Synergistic effect of calcineurin inhibitors and fluconazole against *Candida albicans* biofilms. Antimicrob Agents Chemother 52:1127–1132

Walraven CJ, Lee SA (2013) Antifungal lock therapy. Antimicrob Agents Chemother 57:1–8

Walsh TJ, Schlegel R, Moody MM, Costerton JW, Salcman M (1986) Ventriculoatrial shunt infection due to *Cryptococcus neoformans*: an ultrastructural and quantitative microbiological study. Neurosurgery 18:373–375

Wang X, Fries BC (2011) A murine model for catheter-associated candiduria. J Med Microbiol 60:1523–1529

Webb BC, Thomas CJ, Willcox MD, Harty DW, Knox KW (1998a) *Candida*-associated denture stomatitis. Aetiology and management: a review. Part 1. Factors influencing distribution of *Candida* species in the oral cavity. Aust Dent J 43:45–50

Webb BC, Thomas CJ, Willcox MD, Harty DW, Knox KW (1998b) *Candida*-associated denture stomatitis. Aetiology and management: a review. Part 2. Oral diseases caused by Candida species. Aust Dent J 43:160–166

Xie Z, Thompson A, Sobue T, Kashleva H, Xu H, Vasilakos J, Dongari-Bagtzoglou A (2012) *Candida albicans* biofilms do not trigger reactive oxygen species and evade neutrophil killing. J Infect Dis 206:1936–1945

Yanagisawa N, Li DQ, Ljungh A (2004) Protein adsorption on ex vivo catheters and polymers exposed to peritoneal dialysis effluent. Perit Dial Int 24:264–273

Zumbuehl A, Ferreira L, Kuhn D, Astashkina A, Long L, Yeo Y, Iaconis T, Ghannoum M, Fink GR, Langer R et al (2007) Antifungal hydrogels. Proc Natl Acad Sci USA 104:12994–12998

Chapter 8
Fungal–Bacterial Interactions: In Health and Disease

Patrick Van Dijck and Mary Ann Jabra-Rizk

Abstract Fungi and bacteria have long shared ecological niches and engaged in a spectrum of relationships ranging from synergistic to antagonistic to enhance survival or gain competitive advantages. Importantly, there is growing evidence that polymicrobial interactions involving *Candida* and bacteria may synergize their pathogenic potential by enhancing colonization in the host and contributing to recalcitrance to antimicrobial treatment. Increasing number of microbiome studies has enabled us to develop a greater understanding of how microbial interactions influence various aspects of host physiology and disease outcomes. However, studies exploring the influence of candidal populations on the composition of host bacterial microbiota have been lacking highlighting the importance of a dual approach to microbiome analysis. This chapter aims to appraise the clinical importance of the interactions between *C. albicans* and bacteria occupying the same sites in their human host, focusing on the most common infections from which *Candida* is co-isolated with various bacterial species. Specifically, emphasis is placed on quorum sensing, the complex system of cell-to-cell communication based on secreted molecules that govern behavior of cell populations. Understanding the physical and molecular interactions between *Candida* and bacteria will greatly aid in developing novel therapeutic strategies such as blocking of adhesion receptors or interrupting communications via targeting quorum sensing molecules. Therefore, in depth investigations are warranted to provide crucial insights into clinically relevant interactions between *Candida* and bacteria, which can be exploited for the benefit of mankind.

Keywords *Candida* · Bacteria · Mixed biofilms · Polymicrobial infections

P. Van Dijck (✉)
VIB Department of Molecular Microbiology, KU Leuven, Kasteelpark,
Arenberg, Louvain, Belgium
e-mail: Patrick.VanDijck@mmbio.vib-kuleuven.be

M.A. Jabra-Rizk
University of Maryland School of Dentistry, Baltimore, MD, USA
e-mail: MRizk@umaryland.edu

R. Prasad (ed.), *Candida albicans: Cellular and Molecular Biology*,
DOI 10.1007/978-3-319-50409-4_8

8.1 Introduction

Fungi and bacteria have long shared ecological niches and engaged in a spectrum of relationships ranging from synergistic to antagonistic to enhance survival or gain competitive advantages (Wargo and Hogan 2006; Kobayashi and Crouch 2009). Yet very little is known about the molecular mechanisms that govern interspecies communication and changes in microbial behaviors that occur as fungi and bacteria interact (Morales and Hogan 2010). This has led to a disparity of fundamental knowledge on the significance of candidal–bacterial interactions within the host environment. However, as our levels of knowledge increase, there is a growing awareness that *Candida* rarely exist within a mono-species environment, and that heterogeneous biofilm populations consisting of fungi and bacteria are highly prevalent and clinically important (Watnick and Kolter 2000; Stoodley et al. 2002; Moons et al. 2009; Tsui et al. 2016). Moreover, there is growing evidence that polymicrobial interactions may not only synergize the pathogenic potential of coexisting microorganisms, but also contribute to recalcitrance to antimicrobial treatment highlighting the importance of a dual approach to microbial analysis (Holmes et al. 1995a; Jenkinson and Lamont 2005; O'Connell et al. 2006; Bennett 2008).

Although the area of research exploring inter-kingdom interactions in biofilm is still in its infancy, there is increasing awareness of their clinical implications in the host between *Candida albicans* and various bacterial species (Joint et al. 2002; Hogan 2006; Shirtliff et al. 2009; Morales and Hogan 2010; Peters et al. 2012b). Therefore, it has become crucial to direct studies toward exploring the influence of candidal populations on the composition of microbial communities and importantly to understand how these interactions impact clinical outcomes. However, many hurdles stand in the way of understanding the nature of interspecies interactions due to the complexity of the microbial communities in the host. This is further complicated by the interplay between these microbial communities and their human host.

Candida bacteria mixed biofilms offer numerous benefits to the microorganisms involved; in addition to providing increased attachment sites, fungal–bacterial communities create environmental conditions that offer nutrients, stability, and importantly protection from the host immune defenses and antimicrobial drugs (Adam et al. 2002; Lynch and Robertson 2008; Bamford et al. 2009; Shirtliff et al. 2009; Harriott and Noverr 2011). Cross-feeding events have also been identified, in which metabolic end products of one species are used as carbon sources by another community member (Jenkinson et al. 1990, 2008; Holmes et al. 1995b). Similarly, within the host, the interactions between diverse microbial populations can be beneficial or detrimental to the host. In the case of *Candida* the normal bacterial flora maintains the fungus in its commensal state as evidenced by the rapid transition of *Candida* from a commensal to pathogen as the result of antibiotic treatment (Williams and Lewis 2011; Ursell et al. 2012; Mason et al. 2012a, b). However, numerous animal studies have shown increased persistence, severity, and mortality from *Candida* bacterial co-infections, as well as enhanced drug tolerance (Carlson 1983a, b; Lynch and Robertson 2008; Harriott and Noverr 2009; Schlecht et al. 2014; Kong et al. 2015; Diaz et al. 2014).

The most serious form of polymicrobial interactions are those involving bloodstream infections (BSI) as once microorganisms gain access to the bloodstream, they can disseminate and infect virtually any part of the human body. *Candida* species are currently ranked the third most commonly isolated blood stream pathogen and it is estimated that 27–56% of *C. albicans* BSIs are polymicrobial with *S. aureus* (20%), *P. aeruginosa* (8%), and *E. coli* (4%) the most commonly co-isolated bacterial species (Wisplinghoff et al. 2004; Klotz et al. 2007). Patients with bacteremia and candidemia have reduced survival rates compared to those with candidemia alone. Therefore, in depth investigations are warranted to contribute to our understanding of these interactions, which can be exploited for the benefit of mankind. Foremost are studies using clinically relevant models to mechanistically identify fungal–bacterial interactions that could serve as therapeutic targets for human diseases (Diaz et al. 2014). This chapter aims to appraise the clinical importance of the interactions between *C. albicans* and bacteria occupying the same sites in their human host, focusing on the most common infections from which *Candida* is co-isolated with various bacterial species to illustrate these points (O'Donnell et al. 2015) (Fig. 8.1).

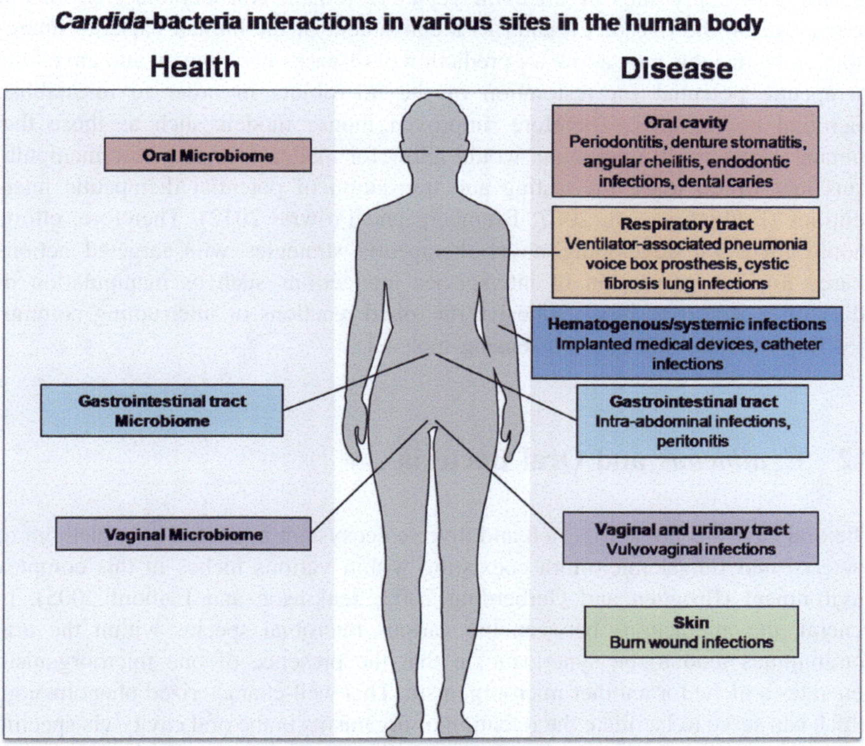

Fig. 8.1 *Candida*–bacteria interactions in health and disease. Importance of the interactions between *C. albicans* and bacteria occupying the same sites in their human host and the most common infections from which *Candida* and bacteria are co-isolated

In polymicrobial infections, the presence of one microorganism may predispose the host to colonization by others and in additive infections two or more non-pathogenic microorganisms together can cause disease (Brogden et al. 2005). For mixed biofilm-associated infections specifically, traditional therapies involve removal of devices, in addition to multi-drug administration that is generally targeted at individual causative agents without consideration for effect on a polymicrobial cause. However, the medical community is recognizing the significance of polymicrobial diseases and many therapies are now taking into account the cause of these conditions and the repercussions on treatment and prevention (Sancho et al. 2012). Therefore, understanding the physical and molecular interactions between diverse microorganisms will greatly aid in defining new strategies for disrupting these complex mixed infections (Klotz et al. 2007; Shirtliff et al. 2009; Harriott and Noverr 2011; Bouza et al. 2013).

Increasing number of microbiome studies has enabled us to develop a greater understanding of how microbial interactions influence various aspects of host physiology and disease outcomes (Ursell et al. 2012; Casadevall and Pirofski 2015). Studies, however, have primarily focused on either bacterial or fungal communities, but rarely with respect to each other. *C. albicans* is a highly evolved species with many sites of infection yet there remain considerable gaps in our knowledge on the role of *Candida* as a constituent of the human bacterial microbiota. Such insights may allow for prediction of disease development and unveil the therapeutic potential for restoration of the microbiota in order to re-establish microbial homeostasis. Therefore, improved mouse models such as those that contain a humanized microbiota would allow for malleable therapeutic manipulations that would assist the testing and translation of potential therapeutic interventions (Dethlefse et al. 2007; Blumberg and Prowrie 2012). Therefore, efforts should focus on developing novel therapeutic strategies with targeted actions geared towards prevention of interspecies interactions such as manipulation of adhesion receptors to block adhesive microbial reactions or interrupting communications via targeting quorum sensing molecules.

8.2 *C. albicans* and Oral Bacteria

The oral cavity is home to a rich and diverse ecosystem inhabited by a plethora of bacterial and fungal microflora coexisting within various niches in this complex environment (Brogden and Gerberding 2002; Jenkinson and Lamont 2005). In general, the interactions between the various microbial species within the oral communities tend to be synergistic in that the presence of one microorganism generates a niche for another microorganism. This well-characterized phenomenon, which can serve to facilitate the retention of organisms in the oral cavity via specific cell surface factors, is known as "co-aggregation" (Rickard et al. 2003; Bamford et al. 2009; Kuboniwa et al. 2012; O'Donnell et al. 2015). Similarly, for *C. albicans* the co-adhesion with oral bacteria is crucial for *C. albicans* colonization and persistence in the oral cavity (Jenkinson and Lamont 2005; Jenkinson et al. 2008). In addition to co-colonization, various types of metabolic communications among the microbial consortia also occur. For example, excretion of a metabolite by one organism can be used as a nutrient by other organisms and enzymatic breakdown of a substrate by one species creates available substrates for different organisms (Kleinberg 1999; Hojo et al. 2009; Metwalli et al. 2013; O'Donnell et al. 2015).

As there are limited surfaces for adherence in the oral cavity, *C. albicans* has to compete with other microbes (Kolenbrander et al. 2002, 2010). However, *C. albicans* can overcome this problem by binding directly to adhering bacteria (Jenkinson et al. 1990; Jenkinson and Lamont 2005). The streptococci are amongst the primary colonizers of the oral cavity comprising approximately 60–80% of the flora. Therefore, the best-studied fungal–bacterial interaction is that between *Candida* and *streptococci*, which is largely considered to be synergistic as streptococci augment the persistence of *Candida* through co-adherence (Jenkinson et al. 2008; Diaz et al. 2012; Xu et al. 2014). However, increasing studies indicate that this interaction is mutually beneficial as various streptococci species such as *Streptococcus oralis, Streptococcus mitis,* and *Streptococcus gordonii* were shown to avidly adhere to the hyphae of *C. albicans*. This increased surface area associated with hyphal networks enables the streptococci to form more robust oral biofilms (Bamford et al. 2009; Diaz et al. 2012; Xu et al. 2014).

The adhesive interactions between *C. albicans* and indigenous oral bacterial species can be mediated by protein–protein and lectin–carbohydrate interactions. For example, *S. gordonii* was shown to adsorb salivary proline-rich proteins that are recognized by *C. albicans* and act as receptors for the fungus (Holmes et al. 1995a, 1996; Bamford et al. 2009; Silverman et al. 2010). However, attachment of *S. gordonii* was also shown to involve direct binding via *C. albicans* adhesin proteins namely the hyphal cell wall proteins Als3p and Hwp1p, and the streptococcal cell surface adhesins SspA and SspB (Holmes et al. 1996; Klotz et al. 2007; Bamford et al. 2009; Silverman et al. 2010). Importantly, in addition to cell surface receptors, extracellular polysaccharide matrix (EPM) production by coexisting species supports mixed biofilm growth of dense communities particularly those formed on tooth surface (Xu et al. 2014). The EPM of streptococcal biofilms is composed of α-glucans, whereas *Candida* biofilm EPM is primarily composed of β-glucans (Al-Fattani and Douglas 2006; Taff et al. 2012). The cariogenic species *Streptococcus mutans* for example was shown to utilize its glucosyltransferase-derived EPM components to enhance adhesion to fungal cells. This process is mediated by deposition of streptococcal α-glucans on the surface of hyphae with the *Candida*-derived β1,3-glucans also contributing to the matrix structure (O'Sullivan et al. 2000; Dongari-Bagtzoglou et al. 2009; Falsetta et al. 2014). Therefore, the biofilm EPM contributes to this mutualistic behavior between *C. albicans* and streptococci favoring their coexistence in the oral environment. Importantly, in addition to providing adhesion sites, the streptococci excrete lactate that can act as a carbon source for yeast growth and through fermentation of carbohydrates, streptococci can create an acid environment (Douglas et al. 1981). Although at low pH *Candida* grows in its yeast form, streptococci can induce hyphal growth by inducing oxidative stress (Jenkinson et al. 1990). On the other hand, *C. albicans* can promote the survival of streptococci by reducing oxygen tension to levels preferred by streptococci and provide growth stimulatory factors for the bacteria (Jenkinson et al. 2008).

In the oral cavity, the onset and progression of the most common oral diseases is closely associated with disturbances in the ecosystem where population shifts lead to over-representation of pathogenic species (Kuboniwa et al. 2012; Casadevall and Pifroski 2015). In a recent clinical study, microbiome analysis of saliva showed an association between increased *Candida* load and a dysbiotic bacterial flora that favored the coexistence with oral streptococci to the exclusion of pathogenic anaerobic species (Kraneveld et al. 2011). In fact, although its contribution to oral disease is not fully elucidated, *C. albicans* has been isolated from periodontal pockets, root canals, orthodontic appliances, enamel, and dentures (Jabra-Rizk et al. 2001; Ramage et al. 2004; Dongari-Bagtzoglou et al. 2009; O'Donnell et al. 2015).

Recent attention has been directed toward characterizing the interaction between *Candida* and the cariogenic bacterium *S. mutans* within the context of dental caries. Dental caries or tooth decay is the most common oral disease characterized by irreversible destruction of the tooth (Isalm et al. 2007; Zero et al. 2009). Caries development is primarily mediated by the metabolic interactions between the microbial species embedded in the biofilm formed on tooth surface known as dental plaque (Kidd and Fejerskov 2004; Isalm et al. 2007; Rouabhia and Chmielewski 2012; Metwalli et al. 2013). These complex interactions cause fluctuations in pH ultimately resulting in dissolution of the dental hard tissues and formation of carious lesions (Isalm et al. 2007; Zero et al. 2009; Falsetta et al. 2012; Rouabhia and Chmielewski 2012; Lemos et al. 2013; Metwalli et al. 2013). Although a direct role has not yet been shown, there is growing evidence attributing a role for *C. albicans* in mediating cariogenic development via synergistic physical and metabolic interactions with *S. mutans* (Barbieri et al. 2007; Jarosz et al. 2009; Metwalli et al. 2013). Several in vitro studies have shown that *S. mutans* uses glucosyltransferases to attach to *C. albicans* exhibiting high affinity to the *C. albicans* hyphae, indicating a possible facilitation mechanism where *C. albicans* cells could be used by the bacteria as support for adherence (Jarosz et al. 2009; Raja et al. 2010; Metwalli et al. 2013) (Fig. 8.2). Supportive experimental evidence for this association was further provided by in vivo studies using rat models of caries where as a consequence of its pronounced ability to produce and tolerate acids, *C. albicans* was capable of causing advanced occlusal caries in rats at a high rate (Klinke et al. 2011). More importantly, clinical studies are increasingly reporting the isolation of *C. albicans* from patients with caries and the occurrence of caries in children was positively correlated with the frequency of oral candidal carriage (Raja et al. 2010). Collectively, the findings from these studies strongly indicate that the presence of *C. albicans* in the oral environment could be considered as an additional factor in evaluating risks of caries (Klinke et al. 2011; Metwalli et al. 2013).

Similar to dental caries, periodontitis is a prevalent and complex oral disease mediated by pathogens such as *Porphyromonas gingivalis* and *Fusobacterium nucleatum* (Khan et al. 2015). Although these pathogens are obligate anaerobes

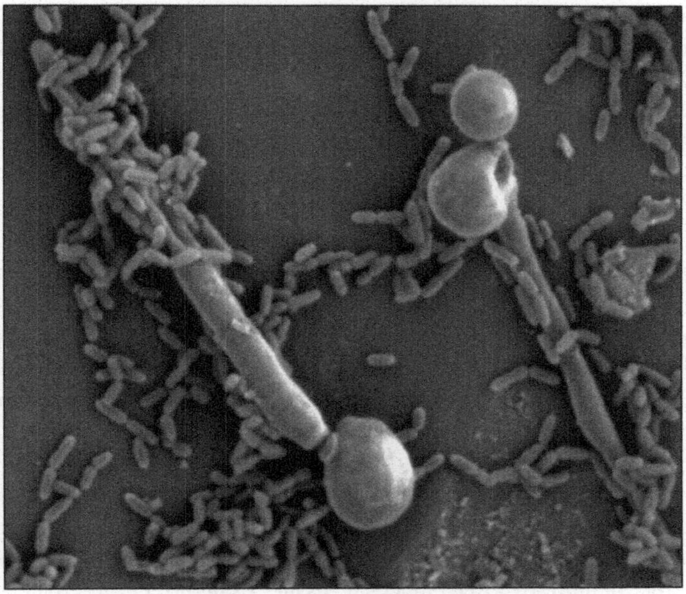

Fig. 8.2 Scanning electron micrograph demonstrating the affinity of *S. mutans* to *C. albicans* hyphae. Bacterial cells are seen attached in chains as they adhere to and wrap around the hyphae

and, therefore, reside in the highly anaerobic subgingival plaque, studies examining fungal colonization in periodontal pockets reported the co-isolation of *Candida* species from subgingival biofilms from patients with periodontitis. Further, high *Candida* levels were shown to correlate with chronic and aggressive forms of periodontitis (Canabarro et al. 2013; Haffajee and Socransky 2006). Similarly, other studies also reported the presence of *C. albicans* in the subgingival plaque microbiota of HIV+ patients (Jabra-Rizk et al. 2001). However, the relationship between *Candida* spp. and periodontitis remains undefined and a relatively neglected area of research. In fact, studies regarding *C. albicans* and *P. gingivalis* have produced conflicting results; where one study showed that *P. gingivalis* induces hyphae formation producing a more invasive phenotype thus increasing the risk of infection, others described *P. gingivalis* to exert an inhibitory effect on hyphae formation and a suppressive effect on candidal viability at high concentrations (Thein et al. 2006; Urzua et al. 2008). In addition to *P. gingivalis*, an association was also made between *Candida* and the periodontal pathogen *F. nucleatum* where studies demonstrated a strong ability for these species to coaggregate indicating that *C. albicans–F. nucleatum* interactions may be an important factor in oral yeast colonization (Grimaudo et al. 1996; Jabra-Rizk et al. 1999).

Endodontitis is another important oral disease characterized by an infection of the pulp within the dental root canal system. The bacterial species *Enterococcus faecalis* is considered the primary endodontic pathogen; however, *Candida* species and *E. faecalis* have become increasingly noted for their co-isolation within endodontic infections (Siqueira and Sen 2004). The association of *Candida* species with endodontic infections was based on findings from several studies demonstrating *C. albicans* penetration of dentine tubules with subsequent studies showing an association between *C. albicans* and *E. faecalis* and *C. albicans* presence in clinical root canal specimens (Peciuliene et al. 2001). In fact, *E. faecalis* was shown to adhere to *Candida* in both hyphal and yeast forms and is the third most predominant bacterial species found in mucosal fungal biofilms (Dongari-Bagtzoglou et al. 2009). Interestingly, studies demonstrated that *E. faecalis* inhibits hyphal morphogenesis, whereas *C. albicans* releases a surface protein Msb2, which binds to host antimicrobial peptides as well as antibiotics, thus conferring protection to both organisms (Swidergall and Ernst 2014). Collectively, these findings seem to suggest that both species influence one another's virulence and help maintain a commensal relationship. Yet, in spite of these observations there are no studies investigating candidal–bacterial interactions in the root canal environment. Overall, the adherence and co-aggregation of *C. albicans* and the various oral bacterial species are complex and species-specific, mediated by bacterial receptors that might be expressed under specific conditions. For example, denture usage highly predisposes to denture stomatitis, a prevalent *Candida*-associated inflammatory condition of the denture-exposed mucosal tissue (Salerno et al. 2011). Denture stomatitis stems from the ability of *Candida* adhere to and form biofilms on the denture acrylic and therefore, the physical interactions between *Candida* and various oral bacteria co-colonizing the denture surfaces likely contribute to the development and progression of denture stomatitis (Ramage et al. 2004).

Respiratory pathogens have been shown to colonize the oral cavity and oral pathogens to colonize the lungs indicating a potential link between oral and pulmonary colonization (Przybyłowska et al. 2015). Therefore, the high commensal presence of *Candida* in the oral cavity may also constitute a risk for pulmonary colonization. In fact *Candida* has been found to be one of the most predominant pathogens in the lungs, particularly in patients suffering from lung cancer and chronic pulmonary disease (Ader et al. 2011; Ibrahim et al. 2011). In examining the prevalence of microbial populations in the sputa of patients with cystic fibrosis, *C. albicans* was found to be present in 78% of the patients with *S. aureus* (63%), *P. aeruginosa* (50%), and *E. coli* (5%) co-isolated (Valenza et al. 2008). Further, patients with mechanical ventilation often suffer from ventilator-associated pneumonia (VAP) with *Pseudomonas*, *Staphylococcus,* and *Candida* species being the most common etiologic agents associated (Ader et al. 2011). There is clinical uncertainty regarding the significance of *Candida* colonization of the respiratory tract, yet *Candida* colonization of the lower respiratory tract affects 18–56% of patients receiving mechanical ventilation. However, although available data do not support a direct role for *Candida* as a causative agent in VAP, there is strong evidence that *Candida* colonization of the respiratory tract is associated with an increased risk of bacterial VAP and worse clinical outcomes (Azoulay et al. 2006; Delisle et al. 2008; Hamet et al. 2012).

8.3 *Candida–Pseudomonas aeruginosa*

Antagonistic interactions are typified by one organism's direct, deleterious impact on another. The interaction between *C. albicans* and *P. aeruginosa* remains the best explored with the vast majority of in vitro studies providing evidence for an antagonistic relationship (Kerr 1994; Hogan and Kolter 2002; Hogan et al. 2004; Morales and Hogan 2010; Mallick and Bennet 2013). In this scenario, *P. aeruginosa* gains the upper hand the majority of the time by this ability to attach to and form dense biofilms on *C. albicans* hyphae, ultimately killing the hyphae (Hogan and Kolter 2002; Ramage et al. 2005; McAlester et al. 2008; Kerr 1994; Hogan et al. 2004). In contrast, *P. aeruginosa* was unable to attach to or kill yeast cells (Hogan and Kolter 2002; Hogan et al. 2004). The presence of bacteria was required for the killing of hyphae, as bacterial culture media alone did not have this effect (Hogan and Kolter 2002; Brand et al. 2008). A number of factors were identified to be required for physical interaction between *P. aeruginosa* and hyphae such as type IV pili of *P. aeruginosa*, as type IV pili mutants are attenuated in their attachment to and killing of hyphae (Hogan and Kolter 2002; Hogan et al. 2004).

Interesting observations were reported from animal studies exploring the interactions between *C. albicans* and *P. aeruginosa* using murine models of lung infection. In one study, lung injury caused by *P. aeruginosa* infection was alleviated if preceded by a short-term *C. albicans* colonization (Ader et al. 2011). This phenomenon was attributed to *C. albicans* activation of innate lymphoid cells, which produced IL-22, providing protection against *P. aeruginosa*-induced injury (Mear et al. 2014). These observations were subsequently confirmed by a recent study demonstrating a similar beneficial effect of prior colonization with *C. albicans* on *P. aeruginosa*-mediated acute lung infection in a murine model of *C. albicans* airways colonization. This beneficial effect was also attributed to "priming" of the innate immune system through IL-22 by *C. albicans,* which resulted in increased bacterial clearance and reduced lung injury (Faure et al. 2016). In contrast, in patients on ventilators, colonization of *Candida* in the respiratory tract was reported to carry an increased risk of *Pseudomonas* VAP and in fact, treatment of *Candida* colonization with antifungals reduced the risk of *Pseudomonas*-associated VAP (Azoulay et al. 2006; Ader et al. 2011). Although the mechanism is not fully elucidated it is conceivable that *C. albicans* co-infections may impair the host immune system's ability to defend against *Pseudomonas*, particularly in patients with cystic fibrosis (Roux et al. 2013). Although *P. aeruginosa* suppressed the growth of *C. albicans* in CF patients, regrowth of *C. albicans* was seen when *P. aeruginosa* was eradicated (Kerr 1994). Another interesting clinical observation for these species was reported in the setting of burn wounds where *Pseudomonas* and *C. albicans* are almost never co-isolated indicating an antagonistic interaction, more in line with what has been demonstrated from in vitro studies (Gupta et al. 2005).

8.3.1 Candida–Staphylococcus aureus

In various niches in the host, *C. albicans* coexist with *S. aureus* (Rehm 2008; Shirtliff et al. 2009; Morales and Hogan 2010; Otto 2013; McGavin et al. 2012). With the emergence of methicillin-resistant *S. aureus* (MRSA), this ubiquitous pathogen is becoming an even greater therapeutic challenge (Sakoulas and Moellering 2008; Gordon and Lowy 2008). *S. aureus* is a poor former of biofilms; however, together with *C. albicans* these species form a resilient biofilm where the yeast creates a scaffold for the bacterium (Harriott and Noverr 2009; Peters et al. 2010, 2012b). Although *Candida* species and staphylococci exist as commensals colonizing human mucosal surfaces, they are considered important pathogens due to their increasing involvement in polymicrobial infections and escalating development of antimicrobial resistance (Fehrmann et al. 2013; Nair et al. 2014). In contrast to *P. aeruginosa*, *C. albicans* interaction with *S. aureus* is described as synergistic and in fact, it was shown that *S. aureus* binding to *C. albicans* hyphae was significantly stronger than all other bacteria tested (Peters et al. 2010; Fehrmann et al. 2013).

S. aureus forms poor biofilms but in a mixed biofilm with *C. albicans,* the bacteria can associate with the hyphae and form robust biofilms (Harriott and Noverr 2009). Using *C. albicans* mutant strains deficient in hyphal production, *C. albicans* morphogenesis was shown to be instrumental in formation of mixed biofilms (Harriott and Noverr 2009). Additionally, atomic force microscopy analysis indicated that *S. aureus* attachment to the hyphae is the strongest at the tip and the middle with weaker attachment to yeast cells (Ovchinnikova et al. 2012). Further, the *C. albicans* hyphal protein agglutinin-like sequence 3 (Als3p) was identified to be a cell wall receptor involved in the adherence of staphylococci to the hyphae as *S. aureus* inadequately attached to the *C. albicans als3* mutant strain and expression of Als3p in *Saccharomyces cerevisiae* allowed *S. aureus* to adhere to the yeast where otherwise it did not (Peters et al. 2012a; Beaussart et al. 2013). More importantly, in addition to demonstrating their ability to form mixed biofilms, characterization of this fungal–bacterial interaction at the molecular level demonstrated significant level of differential protein expression the result of mixed-species biofilm mode of growth (Peters et al. 2010). Interestingly, a number of these proteins were identified to be virulence factors in *S. aureus,* indicating a process whereby *C. albicans* may enhance *S. aureus* pathogenesis (Peters et al. 2012a, b).

The most clinically relevant findings, however, came from animal studies, where using various infection mouse models, several studies demonstrated that the interactions between these species are associated with enhanced pathogenicity, disease severity, and mortality. Earlier studies by Carlson (1983a, b) demonstrated that where administration of one species caused no mortality in a mouse model of systemic infection, co-infection led to 100% mortality. More interestingly, subsequent studies using a mouse model of oral co-infection demonstrated that tissue damage caused by hyphal invasion of oral tissue allowed *S. aureus* to penetrate and disseminate, causing high mortality (Schlecht et al. 2014; Kong and Jabra-Rizk 2015).

Interestingly, however, where hyphae was found to be crucial for the systemic bacterial disease to occur, the presence of hyphae was not found to be critical in a mouse peritoneal co-infection model, as demonstrated using a *C. albicans* strain genetically locked into the yeast state (Peters and Noverr 2013; Nash et al. 2014)

Clinically, *S. aureus* and *C. albicans* are often co-isolated from a multitude of diseases such as periodontitis, denture stomatitis, cystic fibrosis, keratitis, ventilator-associated pneumonia, urinary tract catheters, and burn wound infections among others (Tawara et al. 1996; Timsit et al. 2001; Baena-Monroy et al. 2005; Gupta et al. 2005; Pate et al. 2006). The growing use of implanted medical devices including voice prostheses, implants, and endotracheal is another reason why the incidence of *Candida* and staphylococci infections have steadily increased since the majority of these infections are emerging from biofilms formed on medical implants (Pfaller and Diekema 2007; Lazzell et al. 2009; Harriott and Noverr 2011; Tournu and Van Dijck 2012; Otto 2013; Nair et al. 2014). In fact, a recent analysis of cases of endocarditis associated with an implanted device found that ~25% of the infections to be polymicrobial and, in another study, 27% of nosocomial *C. albicans* bloodstream infections were estimated to be polymicrobial (Klotz et al. 2007; Chrissoheris et al. 2009). Importantly, *S. aureus* was found to be the third most commonly co-isolated species with *C. albicans* (Klotz et al. 2007). These polymicrobial infections represent a significant therapeutic challenge, and their co-isolation from blood is an indication of a dire prognosis.

8.4 *Candida–Lactobacilli*

It is well documented that lactobacilli antagonize candidal colonization which is why they are considered to play a key role in probiotics for reducing candidal levels at several sites in the human body (Orsi et al. 2014). This concept is based on studies indicating that the availability of nutrients for lactobacilli results in lactic acid production and low pH, which in turn inhibit candidal growth (Klinke et al. 2009). Another potential mechanism is the secretion of organic acids and production of hydrogen peroxide (H_2O_2) in some strains, which has been shown to exert anti-candidal activity as 96% of healthy women have H_2O_2-generating *Lactobacillus* species as part of their microflora, while these bacterial populations are lower in women with vaginitis (Morales and Hogan 2010). The inhibitory effects of *Lactobacillus* species on *Candida* growth and virulence were initially noted in light of the occurrence of vaginal candidiasis during antibiotic treatment, indicating a role for lactobacilli in maintaining vaginal homeostasis. However, immunomodulation mechanisms are also likely involved in disease, as lactobacilli cells were shown to up-regulate inflammatory cytokines when co-cultured with *C. albicans* (Martinez et al. 2009).

An estimated 75% of women will experience at least one vaginal infection, with about 5–10% of women having recurrent vaginal candidiasis (Sobel 1992). Similarly, VVC is a common side effect of antibiotic treatment, indicating that the vaginal bacterial microbiota might modulate colonization of yeast (Liu et al. 2013). However, the role of vaginal microbiota in VVC is controversial in the literature; where one study comparing the *Lactobacillus* species cultured from the vaginal secretions of women with or without VVC showed no significant differences, in another study, *Lactobacillus* colonization was associated with a $\geq$ fourfold increase in symptomatic VVC (Sobel and Chaim 1996; McClelland et al. 2009). A subsequent comprehensive study of the vaginal microbiota found no altered or unusual bacterial community in women with VVC, suggesting that commensal vaginal bacterial species might be incapable of preventing VVC (Zhou et al. 2009). In contrast to the vaginal milieu, certain oral *Lactobacillus* species namely *L. casei* was shown to have a stimulatory effect on *C. albicans* hyphal growth (Orsi et al. 2014). This is interesting as hyphae have the capacity to coaggregate and support lactobacilli levels in patients with higher levels of oral disease (Bilhan et al. 2009). Moreover, in the oral cavity, it was noted that *C. albicans* reduced *Lactobacillus* spp. while enhancing *E. faecalis* numbers. However, whether this effect was due to synergism with *E. faecalis* or an antagonistic interaction with lactobacilli remains to be investigated as the nature of interaction between *Candida* and lactobacilli may be dependent on the particular environment they cohabit (O'Donnell et al. 2015). Therefore, to understand the interactions of fungi and bacteria within the human GI and vaginal microbiota and their impact on health and disease states, more studies using high-throughput sequencing techniques with longitudinal samples are warranted (Liu et al. 2013).

8.5 *C. albicans* and Other Notable Bacterial Species

Similar to the association with *P. aeruginosa*, using a *Caenorhabditis elegans* infection model, *Acinetobacter baumannii* was shown to inhibit *C. albicans* filamentation by attaching to and killing the hyphae but not the yeast cells (Peleg et al. 2008). This attachment was found to be dependent on an outer membrane protein, ompA, of *A. baumannii,* which was also necessary for the killing of hyphae (Gaddy et al. 2009). Another study also used the *C. elegans* polymicrobial infection model to study the interaction between *C. albicans* and the intestinal pathogen *Salmonella typhimurium* (Ibrahim et al. 2009). The findings from the study identified an antagonistic interaction, as *S. typhimurium* inhibited *C. albicans* filamentation in the model. Interestingly, in vitro co-culture assays showed that although *S. typhimurium* reduced the viability of both yeast and filamentous forms of *C. albicans* and inhibited its ability to form biofilm, killing appeared more rapid for the filamentous cells. These findings are of interest as the interaction between *C. albicans* and intestinal bacteria might shape microbial virulence in the intestinal tract. *Escherichia coli* is another important intestinal bacterium shown to interact with *C. albicans*. Endotoxin (LPS) from *E. coli* is considered an important contributor to virulence in co-infection experiments, and *C. albicans* was shown to respond directly to LPS (Mallick and Bennet 2013). *E. coli* has a suppressive effect on

C. albicans growth as it was shown to inhibit germ tube formation; however, studies into their interactions have produced conflicting results. In one study, there was a decrease in overall *C. albicans* attachment when epithelial cells were pre-incubated with *E. coli*, whereas in a biofilm model, *E. coli* growth was increased when co-cultured with *C. albicans*, compared to the biofilm mass of *E. coli* alone (Nair and Samaranayake 1996). Another cooperative interaction was reported wherein *E. coli* enhanced adhesion of *C. albicans* to bladder mucosa and increased the likelihood of fungal urinary tract infections (Levison and Pitsakis 1987). In addition, *C. albicans* can also increase the virulence of *E. faecalis,* and *Serratia marcescens*, as co-infection with these species results in a more severe disease than infection with the bacterial species alone, presumed to involve unidentified molecules responsible for signaling between species resulting in increased virulence (Mallick and Bennet 2013). Nevertheless, whether these kinds of interactions are mediated by physical interactions, chemical interactions or a combination of the two has not been substantially examined.

8.6 Quorum Sensing and Inter-kingdom Communications

Microbial populations communicate and modulate their collective behavior using extracellular signals known as quorum sensing molecules (Miller and Bassler 2001). The mobilization of diffusible signal molecules among microbial populations facilitates coordination of cellular activities toward the benefit of the population as a whole. The ability to communicate is a particularly valuable asset in mixed microbial communities. Although this phenomenon of interspecies or inter-kingdom signaling is emerging as a key influence on the outcome of infectious diseases, currently a scarcity of knowledge exists regarding the signals involved in many of these interactions (Reen et al. 2011).

Processes like co-aggregation and biofilm formation promote the synthesis and secretion of molecules at levels that can induce a response by neighboring cells (Morales and Hogan 2010). In polymicrobial communities in particular, microbial species are highly interactive and have evolved a complex system of cell-to-cell communication termed quorum sensing (QS), which is based on the secretion and sensing of secreted molecules and increase in cell density of a population (Hogan 2006; Williams 2007; Jayaraman and Wood 2008). Quorum sensing requires production and release of chemical signal molecules called autoinducers that increase in concentration as a function of cell density but can also depend upon physiological conditions (Antunes and Ferreira 2009; Elias and Banin 2012). This phenomenon for promoting collective behavior within a population is important in ensuring survival and propagation by enhancing access to nutrients and niches, as well as providing protection (Nikolaev and Plankunov 2007; Moons et al. 2009; Nadell et al. 2009). This form of communication is a network that spans intraspecies, interspecies, and inter-kingdom interactions involving prokaryotic and eukaryotic microbes and therefore QS is an important factor in the relationship between *Candida* and bacteria. However, while QS has been studied extensively in single species systems, relatively little is known about the cross-kingdom communication in polymicrobial communities involving fungi and bacteria.

Several QS systems have been described to date in numerous bacterial species; in Gram-positive bacteria, small auto-inducing peptides (AIPs) are used as QS molecules and in Gram-negative bacteria, QS involves homoserine lactone molecules (HSL) (Miller and Bassler 2001). Quorum sensing involving the AI-2 system is not limited to interactions among different bacterial species, but also occurs between bacteria and fungi (Hartmann and Schikora 2012). Specifically, the AI-2 system is considered universal and can mediate interspecies communication and was shown to be required for the development of dental plaque. Moreover, cell wall-derived molecules such as bacterial muramyl dipeptides induce *C. albicans* hyphal growth, which may also promote fungal invasion of host tissues and virulence (Xu et al. 2008). In *C. albicans*, farnesol, a tetraprenoid alcohol and a key intermediate in the sterol biosynthetic pathway is a secreted auto-regulatory molecule identified as the primary QS molecule in *C. albicans*. Similar to other QS molecules, farnesol acts in a density-dependent manner as its accumulation in the environment increases with increase in cell density (Hornby et al. 2001; Sato et al. 2004). The key role farnesol plays in *C. albicans* physiology is in modulating morphology, as it induces a switch from hyphal to yeast growth by inhibiting the Ras1-controlled pathway involved in hyphal growth, thereby suppressing biofilm formation (Ramage et al. 2002; Davis-Hanna et al. 2008). Strikingly, this small molecule can also modulate bacterial behavior and virulence by altering the production of toxic phenazines (Cugini et al. 2010). Moreover, farnesol can also induce the generation of reactive oxygen species in a number of microorganisms, likely through effects on electron transport chain components and this process may play an important role in competition with bacteria (Machida et al. 1999).

During inter-kingdom signaling, no cell–cell contact is required. In *C. albicans* interactions with *S. gordonii*, the secreted bacterial AI-2 was shown to induce numerous changes in *C. albicans*, including promotion of hyphae formation via activation or repression of three mitogen-activated protein kinases involved in morphogenetic switching (Waters and Bassler 2005; Bamford et al. 2009; Federle 2009; Elias and Banin 2012; De Sordi and Muhschlegel 2009). However, although the mode of action has not been elucidated, *S. gordonii* was shown to enhance hyphal development, counteracting the inhibition of hyphal and biofilm formation induced by farnesol via its autoinducer 2 (AI-2) (Bamford et al. 2009). The streptococcal *luxS* gene is associated with AI-production and although *luxS* streptococcal mutants can form mono-species biofilms, when co-colonized with *C. albicans*, biofilm formation becomes abrogated, suggesting that this molecule is involved in cellular communication (Bamford et al. 2009; Jarosz et al. 2009). Another QS system in *S. gordonii* is the comCDE system, which may inhibit mono-species biofilm formation via modulation of production of extracellular DNA (eDNA), an important component of candidal EPM (Daniels et al. 2015; Jack et al. 2015). *Streptococcus mutans* also secretes a number of molecules that have effects on *C. albicans,* such as trans-2 decenoic acid, which shares similar structure to farnesol. This molecule, named Streptococcus diffusible factor (SDSF), was also shown to be secreted by *S. sanguinis, S. oralis, and S. mitis* (Vílchez et al. 2010). Interestingly, farnesol was shown to compromise *S. mutans* membrane permeability and inhibit biofilm accumulation and polysaccharide production, suggesting that it may be used by *C. albicans* to control its competitiveness in mixed-species biofilms (Koo et al. 2003; Jeon et al. 2011). In fact, when added to mixed oral bacterial biofilms, farnesol inhibited the competitiveness of *S. mutans* and reduced its dominance over other species in the polymicrobial biofilms (Jeon et al. 2011). Combined, these observations indicate that the interaction between streptococci and *Candida* is complex, involving diffusible molecules, as bacterial spent culture media could also induce morphological changes in *C. albicans* (Kolenbrander et al. 2010).

Chemically mediated signaling between *C. albicans* and *S. aureus* also likely plays an important role in orchestrating their interactions by potentiating both positive and negative interactions. Although *S. aureus* and *C. albicans* have intricate associations and are found together in numerous areas in the human body, relatively little is known about their chemical interactions. While *S. aureus* employs a peptide system of QS, it is not yet clear how these AIPs influence *C. albicans.* A recent study, however, showed that *S. aureus* conditioned media had a striking impact on *C. albicans* biofilm growth rate, indicating that *S. aureus* secretes a QS molecule that stimulates *C. albicans* growth (Lin et al. 2013). Although the interaction between *C. albicans* and *S. aureus* is considered to be mostly synergistic, competitive or antagonistic relationships have also been reported. Several studies have shown that at certain concentrations, farnesol decreased *S. aureus* and *S. epidermidis* biofilm formation and prevented further growth of preformed biofilms, and at high concentrations, disrupted cell membrane integrity and thereby cell viability (Akiyama et al. 2002; Jabra-Rizk et al. 2006). Moreover, farnesol was also shown to competitively inhibit *S. aureus* lipase activity, and importantly, to enhance susceptibility to a variety of clinically important antibiotics (Jabra-Rizk et al. 2006; Kuroda et al. 2007). However, whether *C. albicans* secretes farnesol under in vivo conditions and whether the secreted concentrations are sufficient to inhibit the growth of *S. aureus* in vivo is as yet unclear.

Perhaps the most studied fungal–bacterial chemical interaction is that between *C. albicans* and *P. aeruginosa,* which were shown to exhibit extensive crosstalk through secreted signaling molecules. *P. aeruginosa* possess two QS systems and an arsenal of secreted molecules and metabolites at its disposal several of which have been shown to affect *C. albicans* (Hogan and Kolter 2002; Hogan et al. 2004; Mallick and Bennet 2013). A key identified molecule secreted by *P. aeruginosa* is 3-oxo-C12 homoserine lactone (3OC12HSL) with a 12-carbon chain structure similar to that of farnesol (Hogan et al. 2004). Purified 3OC12HSL inhibited *C. albicans* filamentation, reverted preformed filaments to the yeast form, induced the expression of yeast genes, and decreased expression of hyphal genes (Hornby et al. 2001; Hogan et al. 2004; Mallick and Bennet 2013). HSL production was shown to be strain-specific and media from HSL-producing *P. aeruginosa* strains inhibited *C. albicans* morphological transition, indicating that in mixed-species biofilms, HSL reaches concentrations that repress *C. albicans* filamentation. In contrast, inhibition was not seen with reduced HSL-producing strains (Hogan et al. 2004; McAlester et al. 2008). Interestingly, however, the media supernatants from HSL producers and non-producers *P. aeruginosa* strains were able to inhibit biofilm formation at the maturation phase, but not at other biofilm stages (Holcombe et al. 2010). Further, gene expression of *C. albicans* biofilms exposed to *P. aeruginosa* supernatant from both HSL-producing and non-HSL-producing strains affected different genes but also regulated the expression of an overlapping 238 genes (Holcombe et al. 2010).

In addition to HSL, the pseudomonas quinolone signal (PQS) and its precursor 2-heptyl-4-quinolone (HHQ) which play a key role in coordinating *P. aeruginosa* virulence were also shown to inhibit *C. albicans* biofilm formation, although without affecting adhesion or morphology (Reen et al. 2011). Other important molecules produced by *P. aeruginosa* shown to act as modulators of *C. albicans* phenotype include phenazines, which reduce *C. albicans* viability and hyphae formation, impair biofilm formation, and alter *C. albicans* metabolism thereby decreasing virulence (Mallick and Bennet 2013). In addition, cis-2-decenoic acid induces *C. albicans* biofilm dispersion and pyocyanin and 1-hydroxyphenazine inhibits *C. albicans* growth (Kerr et al. 1999; Davies and Marques 2009; Reen et al. 2011). On the other hand, *C. albicans* uses farnesol as a chemical counter attack against *P. aeruginosa* as farnesol was shown to inhibit *P. aeruginosa* swarming motility and production of pyocyanin (McAlester et al. 2008). Cugini et al. (2010) also found that *P. aeruginosa* strains defective in the LasR QS system and QS production can be restored by the addition of farnesol, resulting in increased pyocyanin and phenazine production. In addition, the inhibitory activity of *P. aeruginosa* on *C. albicans* was also shown to be mediated by different phenazines produced by *P. aeruginosa*, such as pyocyanin and 1-hydroxyphenazine (Kerr 1994). Taken together, the numerous molecules shown to be involved in *Candida–Pseudomonas* interactions highlight the steps each of these species will take to gain an advantage.

Another interesting chemical interaction is between *C. albicans* and *Acinetobacter baumannii*, which implements a LuxI autoinducer system. Similar to the other bacterial species, *A. baumannii* spent media from stationary phase growth also inhibited *C. albicans* filamentation and biofilm formation, indicating that secreted molecules may play a role in this interaction (Peleg et al. 2008). Furthermore, synthetic farnesol added to growth media inhibited *A. baumannii* growth and *C. albicans* mutants defective in farnesol production were unable to inhibit *A. baumannii* (Peleg et al. 2008). Chemical interactions between *E. coli* and *C. albicans* remain ill defined. *E. coli* uses the QS molecule AI2; however, a study by Bandara et al. (2012) found no effect for AI2 on *C. albicans* on biofilm growth. Further, although bacterial lipopolysaccharide (LPS) can also serve as communication messengers between bacteria and *Candida, E. coli* LPS did not have an impact on *C. albicans* biofilm formation. Interestingly, however, *E. coli* LPS increased *C. albicans* virulence and tissue burden when injected into mice (Akagawa et al.1995).

The combined findings from the studies on *Candida*–bacteria interactions support the concept that both eukaryotic and prokaryotic microorganisms sense and respond to the diverse diffusible signaling molecules produced in the niches where they coexist. By responding to these signaling molecules, the fungus may disperse from sites where other co-inhabitants such as antifungal-producing bacteria are present, conferring a potential selective advantage. Importantly, the chemical warfare between bacteria and fungi may lead to increased toxin production and increased host damage and inflammation. Therefore, defining the influence of these molecules on microbial–eukaryotic–host interactions will facilitate future therapeutic strategies to combat microorganisms that are recalcitrant to conventional antimicrobial agents.

8.7 Therapeutic Implications and Host Immune Response

The cooperative effects that can occur in mixed fungal–bacterial infections particularly those that are biofilm associated can promote resistance to both host clearance pathways and antimicrobial agents. Although impact on drug resistance has attracted considerable interest, how the host immune response is perturbed when bacterial and fungal pathogens interact in a host is not fully explored. An important form of commensalism described in multispecies biofilms is indirect pathogenicity (IP), which was recognized during treatment failure of polymicrobial infections. This phenomenon describes an interactive association where one organism benefits, while the other is not affected. For example, in a mixed infection, an antibiotic-resistant microorganism of low intrinsic virulence protects an antibiotic-sensitive pathogen from eradication (Jenkinson and Lamont 2005; O'Connell et al. 2006). Although mixed fungal–bacterial infections correlate with increased frequency or severity of disease and are challenging to treat, the impact of these interactions on therapy remains largely understudied. In these situations, traditional therapies are generally targeted at individual causative agents without consideration for effect on a polymicrobial cause or on individual members of microbial communities. The standard treatment regimen employed for treatment of polymicrobial infections involves two or more antimicrobials, referred to as combination therapies, which tend to be complex when infections involve fungi and bacteria (Brogden and Gerberding 2002; Brook 2002).

One of the most alarming consequences of synergistic interactions between microorganisms is associated with resistance to antimicrobial agents (Bennett 2008). Within mixed biofilms, resistance relies on the ability of the community to cooperate in such a way that it can survive exposure to the antimicrobials. One mechanism of enhancing antibiotic resistance is largely due to the extracellular polysaccharide matrix (EPM) encasing the biofilm cells, which prevents drugs and other stresses from penetrating the biofilm. Although some EPM components are common to most microbial biofilms, the composition of the matrix varies greatly depending on the microbial species and environmental conditions (Flemming and

Wingender 2010). Accordingly, the composition of the EPM is different in mono- versus multispecies biofilms and the latter can provide a better defense against antimicrobial treatments. In the oral cavity, biofilms of *C. albicans* and oral streptococci were shown to be more resistant to antibiotics than their single species counterparts (Douglas 2003; Shirtliff et al. 2009). This phenomenon was primarily attributed to the physical interactions between *C. albicans* hyphae and oral streptococci, which increased tolerance of the polymicrobial biofilm to antimicrobial agents and enhanced resilience to physical disruption (Jenkinson and Lamont 2005).

In addition, matrix polymers produced by both *C. albicans* and bacteria result in a more viscous matrix that is more effective at restricting the penetration of drugs (Douglas 2003). This phenomenon was not limited to streptococci as *C. albicans* and *S. aureus* were also shown to form larger biofilms when co-cultured with increased *S. aureus* resistance to vancomycin (Harriott and Noverr 2009). In this scenario, matrix polysaccharides produced by both organisms similarly result in a more viscous matrix that is more effective at restricting the penetration of drugs. In fact, it has been shown that within the mixed biofilm environment, *C. albicans* EPM protected *S. aureus* against vancomycin treatment using concentrations as high as 1600 mg/mL (Adam et al. 2002; Harriott and Noverr 2009). Similarly, studies by Pammi et al. (2013) showed that *S. epidermidis* extracellular DNA (eDNA) released through autolysis is an important entity in supporting the integrity of mixed biofilm growth with *C. albicans*. In addition, the EPM produced by *S. epidermidis* was also shown to inhibit penetration of the antifungal drug fluconazole in the mixed *C. albicans*–*S. epidermidis* biofilm (Adam et al. 2002; Elias and Banin 2012). Therefore, it is not surprising that eDNA and the extracellular polysaccharides from both *C. albicans* and staphylococci biofilms are involved in affecting the action of antibacterial agents. There are, however, other adaptive resistance mechanisms that play a role in this resistance phenotype (Harriott et al. 2010). More recently, a study by De Brucker et al. (2015) reported a similar phenomenon involving *E. coli* tolerance to the antibiotic ofloxacin in mixed biofilm, and the *C. albicans* β-1,3-glucan matrix component was shown to be involved in the process. These findings are interesting as they indicate that the influence of *C. albicans* matrix on bacterial response to antibiotics is not drug or species-specific but impacts both Gram-positive and Gram-negative bacteria, as well as different classes of antibiotics.

In addition to impact on therapy, in mixed-species infections immunomodulatory interactions between species may also interfere with different lines of defense of the host immune system. In fact it was shown that during systemic infections, *C. albicans* and *S. aureus* might collaborate to evade phagocytic killing by polymorphonuclear leukocytes (PMNs). In one study, a *C. albicans*-secreted proteinase that degrades the Fc portion of immunoglobulin G (IgG) was shown to reduce the opsonizing activity of human PMNs against *S. aureus* (Nair et al. 2014). Further, *C. albicans* hyphae were shown to penetrate the cell membrane of murine macrophages causing cell lysis, in turn releasing co-ingested *S. aureus* cells (Kong and Jabra-Rizk 2015). Similarly, *S. aureus* secreted coagulase and extracellular fibrinogen-binding proteins (Efb), which protected *Candida* from PMN-mediated phagocytosis (Fehrmann et al. 2013). Combined these findings suggest that the synergism between *S. aureus* and *Candida* also includes strategies to circumvent host immune system, thereby facilitating persistence of infection.

The use of novel antibiotic combinations and antibiotic cycling may prolong the effectiveness of antibiotic therapies (Masterton 2005). However, careful attempt should be made to identify the causative microorganisms as appropriate management of mixed infections requires the administration of antimicrobials that are effective against both components of the infection (Brook 2002). Therefore, understanding the mechanisms of the interactions between bacteria and fungi coexisting within mixed microbial communities will provide crucial insights into the pathogenesis and enhanced tolerance of polymicrobial infections to antimicrobial therapy. Importantly, such insights will greatly aid our ability to effectively treat opportunistic polymicrobial infections and to modulate the behavior of potentially pathogenic bacteria and fungi in beneficial ways.

8.8 *C. albicans* an Important Member of the Host Microbiota

The human microbiome is a critical component of host health that plays a key role in the development of disease (Dewhirst et al. 2010; Rolph et al. 2001; Munson et al. 2002; Saito et al. 2006; Sakamoto et al. 2006). As in the oral cavity, it is well established that the GI tract microbiota plays a vital role in preventing fungal colonization as indicated by the enhanced susceptibility of germfree mice to *Candida* colonization (Ursell et al. 2012). During nonpathogenic colonization, *C. albicans* promoted the recovery of *Bacteroidetes* populations, antagonized *L. johnsonii* populations, and promoted the persistence of *E. faecalis* populations in the cecum. These data demonstrate that while the microbiota can prevent *C. albicans* overgrowth, *C. albicans* can also alter the microbiota indicating that there is a bidirectional interaction between normal members of the fungal and bacterial microbiota of the human gut. Surprisingly, however, little is known about the role of *C. albicans* in shaping the bacterial microbiota, particularly during antibiotic recovery (Mason et al. 2012a, b). Similarly, vulvovaginal candidiasis is a common side effect of antibiotic treatment, indicating that the vaginal microbiota might modulate colonization of yeast (Liu et al. 2013). Therefore, to understand the interactions of fungi and bacteria within the human oral, GI and vaginal microbiota and their impact on health and disease states, more studies using high-throughput sequencing techniques with longitudinal samples are warranted (Liu et al. 2013). Understanding the mechanisms by which host homeostasis is restored is critical for future therapies aimed at manipulating the microbiota. Such insights may allow for prediction of disease development and unveil the therapeutic potential for restoration of the microbiota in order to re-establish microbial homeostasis.

Specifically, a more complete understanding of the interactions occurring during antibiotic recovery could aid in our understanding of the relationship between the microbiota, *C. albicans,* and the host, leading to novel treatments for diseases (Mason et al. 2012a, b).

8.9 Perspectives

The medical community is recognizing the significance of polymicrobial diseases and the major types of microbial community interactions associated with human health and disease. The interaction between *Candida* and bacteria may be dependent on the nature of the interaction (chemical, physical, or both) and the particular environment they cohabit. It is clear that the interaction between *C. albicans* hyphae and different bacterial species is important in defining their interaction, whether mutualistic or antagonistic in nature. The secretion of bacterial and fungal signaling molecules is clearly important, with recent studies supporting the notion that the metabolome plays an integral part in defining the interaction between the host, *Candida,* and microbiota (Koo et al. 2003). Therefore, understanding the microbial communities and their interactions that drive sickness or health is a key to combating polymicrobial diseases (Jenkinson and Lamont 2005).

The key challenges now are to design strategies to prevent development of polymicrobial infections using emerging concepts such as targeting communication machinery through the use of QS inhibitors that block microbial communication (Jayaraman and Wood 2008; Høiby et al. 2010). Further, by manipulation of adhesion interactions it may be possible to develop new protocols to block adhesive reactions impeding development of polymicrobial diseases (Barbieri et al. 2007; Klinke et al. 2011; Bowen 2013; Lemos et al. 2013). However, to better understand the interaction between *C. albicans* and bacteria on disease development, it is crucial to determine mechanistically precise details of adhesion and signaling under conditions of coexistence, and to identify the molecular processes involved. To that end, future efforts should focus on clinical studies and on designing animal model systems to study in vivo-grown polymicrobial biofilms, with the goal of developing novel therapeutic strategies to prevent infections through targeted actions.

References

Adam B, Baillie GS, Douglas LJ (2002) Mixed species biofilms of *Candida albicans* and *Staphylococcus epidermidis*. J Med Microbiol 51:344–349

Ader F, Jawhara S., Nseir S, Kipnis E, Faure K, Vuotto F, Chemani C, Sendid B, Poulain D, Guery B. (2011) Short term *Candida albicans* colonization reduces *Pseudomonas aeruginosa*-related lung injury and bacterial burden in a murine model. Crit Care Med 15. doi:10.1186/cc10276

Akagawa G, Abe S, Yamaguchi H (1995) Mortality of *Candida albicans*-infected mice is facilitated by superinfection of *Escherichia coli* or administration of its lipopolysaccharide. J Infect Dis 171:1539–1544

Akiyama H, Oono T, Huh WK, Yamasaki O, Ogawa S, Katsuyama M, Ichikawa H, Iwatsuki K (2002) Actions of farnesol and xylitol against *Staphylococcus aureus*. Chemother 48:122–128

Al-Fattani MA, Douglas LJ (2006) Biofilm matrix of *Candida albicans* and *Candida tropicalis*: Chemical composition and role in drug resistance. J Med Microbiol 55:999–1008

Antunes LCM, Ferreira R (2009) Intercellular communication in bacteria. Crit Rev Microbiol 35:69–80

Azoulay E, Timsit JF, Tafflet M, de Lassence A, Darmon M, Zahar JR, Adrie C, Garrouste-Orgeas M, Cohen Y, Mourvillier B, Schlemmer B (2006) *Candida* colonization of the respiratory tract and subsequent pseudomonas ventilator-associated pneumonia. Chest 129:110–117

Baena-Monroy T, Moreno-Maldonado V, Franco-Martinez F, Aldape-Barrios B, Quindos G, Sanchez V, Octavio L (2005) *Candida albicans, Staphylococcus aureus* and *Streptococcus mutans* colonization in patients wearing dental prosthesis. Med Oral Patol Oral Cir Bucal 10: E27–E39

Bamford CV, d'Mello A, Nobbs AH, Dutton LC, Vickerman MM, Jenkinson HF (2009) *Streptococcus gordonii* modulates *Candida albicans* biofilm formation through intergeneric communication. Infect Immun 77:3696–3704

Bandara HM, Lam OL, Jin LJ, Samaranayake L (2012) Microbial chemical signaling: a current perspective. Crit Rev Microbiol 38:217–249

Barbieri DS, Vicente V, Fraiz FC, Lavoranti OJ, Svidzinski TI, Pinheiro RL (2007) Analysis of the in vitro adherence of *Streprococcus mutans* and *Candida albicans*. Brazilian J Microbiol 38:624–663

Beaussart A, Herman P, El-Kirat-Chatel S, Lipke PN, Kucharıkova S, Van Dijck P, Dufrene YF (2013) Single-cell force spectroscopy of the medically important *Staphylococcus epidermidis-Candida albicans* interaction. Nanoscale. doi:10.1039/c3nr03272h

Bennett P (2008) Plasmid encoded antibiotic resistance: acquisition and transfer of antibiotic resistance genes in bacteria. Br J Pharmacol 153:S347–S357

Bilhan H, Sulun T, Erkose G, Kurt H, Erturan Z, Kutay O, Bilgin T (2009) The role of *Candida albicans* hyphae and *Lactobacillus* in denture-related stomatitis. Clin Oral Investig 13:363–368

Blumberg R, Powrie F (2012) Microbiota, disease, and back to health: a metastable journey. Sci Transl Med 4(137)

Bouza E, Burillo A, Muñoz P, Guinea J, Marín M, Rodríguez-Créixems M (2013) Mixed bloodstream infections involving bacteria and *Candida* spp. J Antimicrob Chemother 68:1881–1888

Bowen W (2013) Rodent model in caries research. Odontology 101:9–14

Brand A, Barnes JD, Mackenzie KS, Odds FC, Gow NA (2008) Cell wall glycans and solule factors determine the interaction between the hyphae of *Candida albicans* and *Pseudomonas aeruginosa*. FEMS Microbiol Lett 287:48–55

Brogden KA, Gerberding JM (eds) (2002) Polymicrobial diseases. ASM Press, Washington

Brogden KA, Guthmiller JM, Taylor CE (2005) Human polymicrobial infections. Lancet 365:253–255

Brook I (2002) Microbiology of polymicrobial abscesses and implications for therapy. J Antimicrob Chemother 50:805–810

Canabarro A, Valle C, Farias M, Santos F, Lazera M, Wanke B (2013) Association of subgingival colonization of *Candida albicans* and other yeasts with severity of chronic periodontitis. J Periodontal Res 48:428–432

Carlson E (1983a) Effect of strain of *Staphylococcus aureus* on synergism with *Candida albicans* resulting in mouse mortality and morbidity. Infect Immun 42:285–292

Carlson E (1983b) Enhancement by *Candida albicans* of *Staphylococcus aureus*, *Serratia marcescens*, and *Streptococcus faecalis* in the establishment of infection in mice. Infect Immun 39:193–197

Casadevall A Pirofski L-A. (2015). What Is a Host? Incorporating the microbiota into the damage-response framework. Infect Immun 83(1)

Chrissoheris MP, Libertin C, Ali RG, Ghantous A, Bekui A, Donohue A (2009) Endocarditis complicating central venous catheter bloodstream infections: a unique form of health care associated endocarditis. Clin Cardiol 32:E48–E54

Cugini C, Morales D, Hogan DA (2010) *Candida albicans*-produced farnesol stimulates Pseudomonas quinolone signal production in LasR-defective *Pseudomonas aeruginosa* strains. Microbiology 156:3096–3107

Daniels DE, Jepson M, Vickerman MM, Lamont RJ, Jenkinson HJ, Nobbs AH (2015) *Streptococcus gordonii* comCDE (competence) operon modulates biofilm formation with *Candida albicans*. Microbiology 161:411–421

Davies DG, Marques CNH (2009) A fatty acid messenger is responsible for inducing dispersion in microbial biofilms. J Bacteriol 191:1393–1403

Davis-Hanna A, Piispanen AE, Stateva LI, Hogan DA (2008) Farnesol and dodecanol effects on the *Candida albicans* Ras1-cAMP signaling pathway and the regulation of morphogenesis. Mol Microbiol 67:47–62

De Brucker K, Tan Y, Vints K, De Cremer K, Braem A, Verstraeten N, Michiels J, Vleugels J, Cammue BPA, Thevissen K (2015) Fungal β-1,3-glucan increases ofloxacin tolerance of *Escherichia coli* in a polymicrobial *E. coli/Candida albicans* biofilm. Antimicrob Agents Chemother 59:3052–3058

De Sordi L, Mühlschlegel FA (2009) Quorum sensing and fungal–bacterial interactions in *Candida albicans*: A communicative network regulating microbial coexistence and virulence. FEMS Yeast Res 9:990–999

Delisle MS, Williamson DR, Perreault MM, Albert M, Jiang X, Heyland DK (2008) The clinical significance of *Candida* colonization of respiratory tract secretions in critically ill patients. J Crit Care 23:11–17

Dethlefse L, McFall-Ngai M, Relman DA (2007) An ecological and evolutionary perspective on human–microbe mutualism and disease. Nature 449:811–818

Dewhirst FE, Chen T, Izard J, Paster BJ, Tanner ACR, YU WH, Lakshmanan A, Wade WG (2010) The human oal microbiome. J Bacteriol: JB.00542-00510

Diaz PI, Xie Z, Sobue T, Thompson A, Biyikoglu B, Ricker A, Ikonomou L, Dongari-Bagtzoglou A (2012) Synergistic Interaction between *Candida albicans* and commensal oral *Streptococci* in a novel in nitro mucosal model. Infect Immun 80:620–632

Diaz PI, Strausbaugh LD, Dongari-Bagtzoglou A (2014) Fungal-bacterial interactions and their relevance to oral health: Linking the clinic and the bench. Front Cell Infect Microbiol 4. doi:10.3389/fcimb

Dongari-Bagtzoglou A, Kashleva H, Dwivedi P, Diaz P, Vasilakos J (2009) Characterization of mucosal *Candida albicans* biofilms. PLoS ONE 4:e7967

Douglas LJ (2003) *Candida* biofilms and their role in infection. Trends Microbiol 11:30–36

Douglas LJ, Houston JG, McCourtie J (1981) Adherence of *Candida albicans* to human buccal epithelial cells after growth on different carbon sources. FEMS Microbiol Lett 12:241–243

Elias S, Banin E (2012) Multi-species biofilms: living with friendly neighbors. FEMS Microbiol Rev 36:990–1004

Falsetta ML, Klein MI, Lemos JA, Silva BB, Agidi S, Scott-Anne KK, Koo H (2012) Novel antibiofilm chemotherapy targets exoplysaccharide synthesis ad stress tolerance in *Streptococcus mutans* to modulate virulence expression in vivo. Antimicrob Agents Chemother 56:6201–6211

Falsetta ML, Klein MI, Colone, PM, Scott-Anne K, Gregoire S, Pai C-H, Gonzalez M, Watson G, Krysan DJ, Bortolotti P, wen WH, Koo H (2014) Symbiotic relationship between *Streptococcus mutans* and *Candida albicans* synergizes the virulence of plaque-biofilms in vivo. Infect Immun IAI.00087-14

Faure E, Bortolotti P, Kipnis E, Faure K, Guery B (2016) Studying microbial communities in vivo: A model of host-mediated interaction between *Candida albicans* and *Pseudomonas aeruginosa* in the airways. J Vis Exp 107:e53218. doi:10.3791/53218

Federle M (2009) Autoinducer-2-based chemical communication in bacteria: Complexities of interspecies signaling. Contrib Microbiol 16:18–32
Fehrmann C, Jurk K, Bertling A, Seidel G, Fegeler W, Kehrel BE, Peters G, Becker K, Heilmann C (2013) "Role for the fibrinogen-binding proteins coagulase and Efb in the *Staphylococcus aureus-Candida* interaction. Int J Med Microbiol 303:230–238
Flemming HC, Wingender J (2010) The biofilm matrix. Nat Rev Microbiol 8:623–633
Gaddy JA, Tomaras A, Actis LA (2009) The *Acinetobacter baumannii* 19606 OmpA protein plays a role in biofilm formation on abiotic surfaces and in the interaction of this pathogen with eukaryotic cells. Infect Immun 77:3150–3160
Gordon RJ, Lowy FD (2008) Pathogenesis of methicillin-resistant *Staphylococcus aureus* infection. Clin Infect Dis 46:S350–S359
Grimaudo NJ, Nesbitt W, Clark W (1996) Coaggregation of *Candida albicans* with oral *Actinomyces* species. Oral Microbiol Immunol 11:59–61
Gupta N, Haque A, Mukhopadhyay G, Narayan RP, Prasad R (2005) Interactions between bacteria and *Candida* in the burn wound. Burns 31:375–378
Haffajee AD, Socransky S (2006) Introduction to microbial aspects of periodontal biofilm communities, development and treatment. Periodontology 42:7–12
Hamet M, Pavon A, Dalle F, Pechinot A, Prin S, Quenot JP, Charles PE (2012) *Candida* spp. airway colonization could promote antibiotic-resistant bacteria selection in patients with suspected ventilator-associated pneumonia. Intens Care Med 38:1272–1279
Harriott MM, Noverr MC (2009) *Candida albicans* and *Staphylococcus aureus* form polymicrobial biofilms: effects on antimicrobial resistance. Antimicrob Agents Chemother 53:3914–3922
Harriott, MM, Noverr MC (2011) Importance of *Candida*-bacterial polymicrobial biofilms in disease.Trends in Microbiol 19:557–563
Harriott MM, Lilly EA, Rodriguez TE, Fidel PL Jr, Noverr MC (2010) Ability of *Candida albicans* mutants to induce *Staphylococcus aureus* vancomycin resistance during polymicrobial biofilm formation. Antimicrob Agents Chemother 54:3746–3755
Hartmann A, Schikora A (2012) Quorum sensing of bacteria and trans-kingdom interactions of N-acyl homoserine lactones with eukaryotes. J Chem Ecol 38:704–713
Hogan DA (2006) Talking to themselves: autoregulation and quorum sensing in fungi. Eukaryot Cell 5:613–619
Hogan DA, Kolter R (2002) *Pseudomonas-Candida* interactions: an ecological role for virulence factors. Science 296:2229–2232
Hogan DA, Vik A, Kolter R (2004) A *Pseudomonas aeruginosa* quorum-sensing molecule influences *Candida albicans* morphology. Mol Microbiol 54:1212–1223
Høiby N, Bjarnsholt T, Givskov M, Molin S, Ciofu O (2010) Antibiotic resistance of bacterial biofilms. Intern J Antimicrob Agents 35:22–332
Hojo K, Nagaoka S, Ohshima T, Maeda N (2009) Bacterial interactions in dental biofilm development. J Dent Res 88:982–990
Holcombe LJ, McAlester G, Munro C, Enjalbert B, Brown AJP, Gow N, Ding C, Butler G, O'Gara F, Morrissey JP (2010) *Pseudomonas aeruginosa* secreted factors impair biofilm development in *Candida albicans*. Microbiol 156:1476–1486
Holmes AR, Cannon RD, Jenkinson HF (1995a) Interactions of *Candida albicans* with bacteria and salivary molecules in oral biofilms. J Ind Microbiol 15:208–213
Holmes AR, Gopal PK, Jenkinson HF (1995b) Adherence of *Candida albicans* to cell surface polysaccharide receptor on *Streptococcus gordonii*. Infect Immun 64:1827–1834
Holmes AR, McNab R, Jenkinson HF (1996) *Candida albicans* binding to the oral bacterium *Streptococcus gordonii* involves multiple adhesin-receptor interactions. Infect Immun 64:4680–4685
Hornby JM, Jensen EC, Lisec AD, Tasto JJ, Jahnke B, Shoemaker R, Dussault P, Nickerson KW (2001) Quorum sensing in the dimorphic fungus *Candida albicans* is mediated by farnesol. Appl Environ Microbiol 67:2982–2992

Ibrahim AS, Spellberg BJ, Avanesian V, Fu Y, Edwards JE, Tampakakis E, Peleg AY, Mylonakis E (2009) Interaction of *Candida albicans* with an intestinal pathogen, *Salmonella enterica Serovar typhimurium*. Eukaryot Cell 8:732–737

Ibrahim AS, Spellberg BJ, Avanesian V, Fu Y, Edwards JE, Laroumagne S, Salinas-Pineda A, Hermant C, Murris M, Gourraud PA, Do C, Segonds C, Didier A, Mazières J (2011) ncidence and characteristics of bronchial colonisation in patient with lung cancer: a retrospective study of 388 cases. Rev Mal Respir 28:328–335

Isalm B, Khan SN, Khan AU (2007) Dental caries: from infection to prevention. Med Sci Monit 13:196–203

Jabra-Rizk MA, Falkler WA Jr, Merz WG, Kelley JI, Baqui AAMA, Meiller TF (1999) Coaggregation of *Candida dubliniensis* with *Fusobacterium nucleatum*. J Clin Microbiol 37:1464–1468

Jabra-Rizk MA, Ferreira SMS, Sabet M, Falkler WA, Merz WG, Meiller TF (2001) Recovery of *Candida dubliniensis* and other yeast from human immunodeficiency virus-associated periodontal lesions. J Clin Microbiol 39:4520–4522

Jabra-Rizk MA, Meiller TF, James C, Shirtliff ME (2006) Effect of farnesol on *Staphylococcus aureus* biofilm formation and antimicrobial resistance. Antimicrob Agents Chemother 50:1463–1469

Jack AA, Daniels DE, Jepson MA, Vickerman MM, Lamont RJ, Jenkinson HF, Nobbs AH (2015) *Streptococcus gordonii* comCDE (competence) operon modulates biofilm formation with *Candida albicans*. Microbiol 161:411–421

Jarosz LM, Deng DM, van der Mei H, Crielaard W, Krom B (2009) *Streptococcus mutans* competence-stimulating peptide inhibits *Candida albicans* hypha formation. Eukaryot Cell 8:1658–1664

Jayaraman A, Wood TK (2008) Bacterial quorum sensing: Signals, circuits, and implications for biofilms and disease. Ann Rev Biomed Eng 10:145–167

Jenkinson HF, Lamont RJ (2005) Oral microbial communities in sickness and in health. Trends Microbiol 13:589–595

Jenkinson HF, Lala HC, Shepherd MG (1990) Coaggregation of *Streptococcus sanguis* and other streptococci with *Candida albicans*. Infect Immun 58:1429–1436

Jenkinson, HF, Barbour ME, Jagger DC, Miles M, M. Bamford C, Nobbs AH, Dutton LC, Silverman RJ, McNally L, Vickerman MM, Gill S (2008) *Candida albicans* - bacteria interactions in biofilms and disease. Univ Bristol Dent School 1–6

Jeon JG, Pandit S, Xiao J, Gregoire S, Falsetta ML, Klein MI, Koo H (2011) Influences of trans-trans farnesol, a membrane-targeting sesquiterpenoid, on *Streptococcus mutans* physiology and survival within mixed-species oral biofilms. Int J Oral Sci. 3:98–106

Joint I, Tait K, Callow ME, Callow JA, Milton D, Williams P, Camara M (2002) Cell-to-cell communication across the prokaryote-eukaryote boundary. Science 298:1207

Kerr JR (1994) Suppression of fungal growth exhibited by *Pseudomonas aeruginosa*. J Clin Microbiol 32:525–527

Kerr R, Taylor GW, Rutman A, Cole PJ, Wilson R (1999) *Pseudomonas aeruginosa* pyocyanin and 1-hydroxyphenazine inhibit fungal growth. J Clin Pathol 52:385–387

Khan SA, Kong EF, Meiller TF, Jabra-Rizk MA (2015) Periodontal diseases: bug induced, host promoted. PLoS Pathog 11:e1004952

Kidd EAM, Fejerskov O (2004) What constitutes dental caries? Histopathology of carious enamel and dentin related to the action of cariogenic biofilms. J Den Res 83:C35–C38

Kleinberg T (1999) Acids formed aerobically and anaerobically by pure and mixed cultures of oral bacteria from oxidizable sugar, organic acid and amino acid substrates. J Dent Res 78:211

Klinke T, Kneist S, Kuhlisch E, Mauersberger S, Forster A, Klimm W (2009) Acid production by oral strains of *Candida albicans* and lactobacilli. Caries Res 43:83–91

Klinke T, Guggenheim B, Klimm W, Thurnheer T (2011) Dental caries in rats associated with *Candida albicans*. Caries Res 45:100–106

Klotz SA, Chasin BS, Powell B, Gaur NK, Lipke PN (2007) Polymicrobial bloodstream infections involving *Candida* species: analysis of patients and review of the literature. Diagn Microbiol Infect Dis 59:401–406

Kobayashi DY, Crouch J (2009) Bacterial/fungal interactions: from pathogens to mutualistic endosymbionts. Ann Rev Phytopathol 47:63–82

Kolenbrander P, Anserson RX, Blehert DS, Egland PG, Foster JS, Palmer RJ Jr (2002) Communication among oral bacteria. Microbiol Mol Biol Rev 66:486–505

Kolenbrander PE, Palmer RJ, Periasamy S, Jakubovics NS (2010) Oral multispecies biofilm development and the key role of cell–cell distance. Nat Rev Microbiol 8:471–480

Kong EF, Jabra-Rizk MA (2015) The great escape: pathogen versus host. PLoS Path 11:e1004661

Kong E, Kucharikova S, Van Dijck P, Peters BM, Shirtliff ME, Jabra-Rizk MA (2015) Clinical implications of oral candidiasis: host tissue damage and disseminated bacterial disease. Infect Immun 83. doi:010.1128

Koo H, Hayacibara MF, Schobel BD, Cury JA, Rosalen PL, Park YK, Vacca-Smith AM, Bowen WH (2003) Inhibition of *Streptococcus mutans* biofilm accumulation and polysaccharide production by apigenin and *tt*-farnesol. J Antimicrob Chemother 52:782–789

Kraneveld EA, Buijs M, Bonder MJ, Visser M, Keijser BJF, Crielaard W, Zaura Egija (2011) The relation between oral *Candida* Load and bacterial microbiome profiles in dutch older adults. PLoS ONE 7(8):e4277

Kuboniwa M, Tribble GD, Hendrickson EL, Amano A, Lamont RJ, Hackett M (2012) Insights into the virulence of oral biofilms: discoveries from proteomics. Expert Rev Proteomics 9:311–323

Kuroda M, Nagasaki S, Ohta T (2007) Sesquiterpene farnesol inhibits recycling of the C_{55} lipid carrier of the murein monomer precursor contributing to increased susceptibility to beta-lactams in methicillin-resistant *Staphylococcus aureus*. J Antimicrob Chemother 59:425–432

Lazzell AL, Chaturvedi AK, Pierce CG, Prasad D, Uppuluri P, Lopez-Ribot JL (2009) Treatment and prevention of *Candida albicans* biofilms with caspofungin in a novel central venous catheter murine model of candidiasis. J Antimicrob Chemother 64:567–570

Lemos JA, Qquivey RG, Koo H, Abranches J (2013) *Streptococcus mutans:* a new Gram-positive paradigm? Microbiology 159:436–445

Levison ME, Pitsakis PG (1987) Susceptibility to experimental *Candida albicans* urinary tract infection in the urinary tract. J Infect Dis 155:841–846

Lin YJ, Alsad L, Vogel F, Koppar S, Nevarez L, Auguste F, Seymour J, Syed A, Christoph K, Loomis JS (2013) Interactions between *Candida albicans* and *Staphylococcus aureus* within mixed species biofilms. BIOS 84:30–39

Liu M-B, Xu S-R, He Y, Deng G-H, Sheng H-F, Huang X-M, Ouyang C-Y, Zhou H-W (2013) Diverse vaginal microbiomes in reproductive-age women with vulvovaginal candidiasis. PLoS ONE 8:e79812

Lynch AS, Robertson GT (2008) Bacterial and fungal biofilm infections. Ann Rev Med 59:415–428

Machida K, Tanaka T, Yano Y, Otani S, Taniguchi M (1999) Farnesol-induced growth inhibition in *Saccharomyces cerevisiae* by a cell cycle mechanism. Microbiology 145:293–299

Mallick EM, Bennett RJ (2013) Sensing of the microbial neighborhood by *Candida albicans*. PLoS Pathog 9:e1003661

Martinez RC, Franceschini S, Patta MC, Quintana SM, Candido RC, Ferreira JC, De Martinis EC, Reid G (2009) Improved treatment of vulvovaginal candidiasis with fluconazole plus probiotic Lactobacillus rhamnosus GR-1 and Lactobacillus reuteri RC-14. Lett Appl Microbiol 48:269–274

Mason KL, Downward JR, Mason KD, Falkowski NR, Eaton KA, Kao JY, Young VB, Huffnagle GB (2012a) *Candida albicans* and bacterial microbiota interactions in the cecum during recolonization following broad-spectrum antibiotic therapy. Infect Immun 80:3371–3380

Mason KL, Erb DJ, Mason KD, Falkowski NR, Eaton KA, Kao JY, Young VB, Huffnagle GB (2012b) *Candida albicans* and bacterial microbiota interactions in the cecum during recolonization following broad-spectrum antibiotic therapy. Infect Immun 80:3371–3380

Masterton RG (2005) Antibiotic cycling: more than it might seem? J Antimicrob Chemother 55:1–5

McAlester G, O'Gara F, Morrissey JP (2008) Signal-mediated interactions between *Pseudomonas aeruginosa* and *Candida albicans*. J Med Microbiol 57:563–569

McClelland RS, Richardson B, Hassan WM, Graham SM, Kiarie J, Baeten JM, Mandaliya K, Jaoko W, Ndinya-Achola JO, Holmes KK (2009) Prospective study of vaginal bacterial flora and other risk factors for vulvovaginal candidiasis. J Infect Dis 199:1883–1890

McGavin, Martin J, Heinrichs DE (2012) The staphylococci and staphylococcal pathogenesis. Front Cell Infect Microbiol. doi:10.3389/fcimb.2012.00066

Mear JB, Gosset P, Kipnis E, Faure E, Dessein R, Jawhara S, Fradin C, Faure K, Poulain D, Sendid B, Guery B (2014) *Candida albicans* airway exposure primes the lung innate immune response against *Pseudomonas aeruginosa* infection through innate lymphoid cell recruitment and interleukin-22-associated mucosal response. Infect Immun 82:306–315

Metwalli KH, Khan S, Krom BP, Jabra-Rizk MA (2013) *Streptococcus mutans*, *Candida albicans* and the human mouth: a sticky situation. PLoS Path 9:e1003616

Miller MB, Bassler B (2001) Quorum sensing in bacteria. Annu Rev Microbiol 55:165–199

Moons P, Michiels CW, Aertsen A (2009) Bacterial interactions in biofilms. Crit Rev Microbiol 35:57–168

Morales DK, Hogan DA (2010) *Candida albicans* interactions with bacteria in the context of human health and disease. PLoS Pathog 6:e1000886

Munson MA, Pitt-Ford T, Chong B, Weightman A, Wade WG (2002) Molecular and cultural analysis of the microflora associated with endodontic infections. J Dent Res 81:761–766

Nadell CD, Xavier JB, Foster KR (2009) The sociobiology of biofilms. FEMS Microbiol Rev 33:206–224

Nair RG, Samaranayake L (1996) The effect of oral commensal bacteria on candidal adhesion to human buccal epithelial cells in vitro. J Med Microbiol 45(3):179–185

Nair N, Biswas R, Götz F, Biswas L (2014) Impact of *Staphylococcus aureus* on pathogenesis in polymicrobial infections. Infect Immun 82:2162–2169

Nash E, Peters BM, Palmer GE, Fidel PL, Noverr MC (2014) Morphogenesis is not required for *Candida albicans-Staphylococcus aureus* intra-abdominal infection-mediated dissemination and lethal sepsis. Infect Immun 82:3426–3435

Nikolaev YA, Plankunov VK (2007) Biofilm-"city of microbes" or an analogue of multicellular organisms? Microbiology 76:125–138

O'Connell HA, Kottkamp GS, Eppelbaum JL, Stubblefield BA, Gilbert SE, Gilbert ES (2006) Influences of biofilm structure and antibiotic resistance mechanisms on indirect pathogenicity in a model polymicrobial biofilm. Appl Environ Microbiol 72:5013–5019

O'Donnell LE, Millhouse E, Sherry L, Kean R, Malcolm J, Nile CJ, Ramage G (2015) Polymicrobial *Candida* biofilms: friends and foe in the oral cavity. FEMS Yeast Res 15. doi:10.1093/femsyr/fov1077

O'Sullivan JM, Jenkinson HF, Cannon RD (2000) Adhesion of *Candida albicans* to oral streptococci is promoted by selective adsorption of salivary proteins to the streptococcal cell surface. Microbiology 146:41–48

Orsi CF, Sabia C, Ardizzoni A, Colombari B, Neglia RG, Peppoloni S, Morace G, Blasi E (2014) Inhibitory effects of different lactobacilli on *Candida* albicans hyphal formation and biofilm development. J Biol Regul Homeost Agents 28:743–752

Otto M (2013) Staphylococcal infections: Mechanisms of biofilm maturation and detachment as critical determinants of pathogenicity. Ann Rev Med 64:175–188

Ovchinnikova ES, Krom BP, Busscher HJ, van der Mei HC (2012) Evaluation of adhesion forces of *Staphylococcus aureus* along the length of *Candida albicans* hyphae. BMC Microbiol 12. doi:10.1186/1471

Pammi M, Liang R, Hicks J, Mistretta T-N, Versalovic M (2013) Biofilm extracellular DNA enhances mixed species biofilms of *Staphylococcus epidermidis* and *Candida albicans*. BMC Microbiol 13:257

Pate JC, Jones DB, Wilhelmus KR (2006) Prevalence and spectrum of bacterial co-infection during fungal keratitis. Brit J Ophthalmol 90:289–292

Peciuliene V, Reynaud A, Balciuniene I, Haapasalo M (2001) Isolation of yeasts and enteric bacteria in root-filled teeth with chronic apical periodontitis. Int Endod J 34:429–434

Peleg A, Tampakakis E, Fuchs BB, Eliopoulos GM, Moellering RC Jr, Mylonakis E (2008) Prokaryote-eukaryote interactions identified by using *Caenorhabditis elegans*. Proc Nat Acadf Sci USA 105:14585–14590

Peters BM, Noverr MC (2013) *Candida albicans-Staphylococcus aureus* polymicrobial peritonitis modulates host innate immunity. Infect Immun 81:2178–2189

Peters BM, Jabra-Rizk MA, Scheper MA, Leid JG, Costerton JW, Shirtliff ME (2010) Microbial interactions and differential protein expression in*Staphylococcus aureus* and *Candida albicans* dual-species biofilms. FEMS Imm Med Microbiol 59:493–503

Peters BM, Ovchinnikova E, Schlecht LM, Hoyer LL, Busscher HJ, van der Mei HC, Krom BP, Jabra-Rizk MA, Shirtliff ME (2012a) *Staphylococcus aureus* adherence to *Candida albicans* hyphae is mediated by the hyphal adhesin Als3p. Microbiology 158:2975–2986

Peters BM, Jabra-Rizk MA, O'May GA, Costerton JW, Shirtliff ME (2012b) Polymicrobial interactions in biofilms: impact on pathogenesis and human disease. Clin Microbiol Rev 25:193–213

Pfaller MA, Diekema DJ (2007) Epidemiology of invasive candidiasis: a persistent public health problem. Clin Microbiol Rev 20:133–163

Przybyłowska D, Rubinsztajn R, Chazan R, Swoboda-Kopeć E, Kostrzewa-Janicka J, Mierzwińska-Nastalska E (2015) The prevalence of oral inflammation among denture wearing patients with chronic obstructive pulmonary disease. Adv Exp Med Biol 858:87–91

Raja M, Hannan A, Ali K (2010) Association of oral candidal carriage with dental caries in children. Caries Res 44:272–276

Ramage G, Saville SP, Wickes BL, Lopez-Ribot JL (2002) Inhibition of *Candida albicans* biofilm formation by farnesol, a quorum-sensing molecule. App Environment Microbiol 68:5459–5463

Ramage G, Tomsett K, Wickes BL, Lopez-Ribot JL, Redding SW (2004) Denture stomatitis: a role for *Candida* biofilms. Oral Surg Oral Med Oral Pathol Oral Radiol Endodon 98:53–59

Ramage G, Saville SP, Thomas DP, Lopez-Ribot JL (2005) *Candida* biofilms: an update. Eukaryot Cell 4:633–638

Reen FJ, Mooij M, Holcombe LJ, McSweeney CM, McGlacken GP, Morrissey JP, O'Gara F (2011) The pseudomonas quinolone signal (PQS), and its precursor HHQ, modulate interspecies and interkingdom behaviour. FEMS Microbiol Ecol 77:413–428

Rehm SJ (2008) *Staphylococcus aureus:* the new adventures of a legendary pathogen. Clev Clin J Med 75:177–192

Rickard AH, Gilbert P, High NJ, Kolenbrander PE, Handley PS (2003) Bacterial coaggregation: an integral process in the development of multi-species biofilms. Trends in Microbiol 11:94–100

Rolph HJ, Lennon A, Riggio MP, Saunders WP, MacKenzie D, Coldero L, Bagg J (2001) Molecular identification of microorganisms from endodontic infections. J Clin Microbiol 39:3282–3289

Rouabhia M, Chmielewski W (2012) Diseases associated with oral polymicrobial biofilms. Open Mycol J 6:27–32

Roux D, Gaudry S, Khoy-Ear L, Aloulou M, Phillips-Houlbracq M, Bex J, Skurnik D, Denamur E, Monteiro RC, Dreyfuss D, Ricard JD (2013) Airway fungal colonization compromises the immune system allowing bacterial pneumonia to prevail. Crit Care Med 41:191–199

Saito D, Leonardo Rde T, Rodrigues JL, Tsai SM, Hofling JF, Goncalves RB (2006) Identification of bacteria in endodontic infections by sequence analysis of 16S rDNA clone libraries. J Med Microbiol 55:101–107

Sakamoto M, Rocas IN, Siqueira JF Jr, Benno Y (2006) Molecular analysis of bacteria in asymptomatic and symptomatic endodontic infections. Oral Microbiol Immunol 21:112–122

Sakoulas G, Moellering RC (2008) Increasing antibiotic resistance among methicillin-resistant *Staphylococcus aureus* strains. Clin Infect Dis 46:S360–S367

Salerno C, Pascale M, Contaldo M, Esposito V, Busciolano M, Milillo L, Petruzzi M, Serpico R (2011) *Candida*-associated denture stomatitis. Med Oral Patol Oral Cir Bucal 16:139–143

Sancho S, Artero A, Zaragoza R, Camarena JJ, González R, Nogueira JM (2012) Impact of nosocomial polymicrobial bloodstream infections on the outcome in critically ill patients. Europ J Clin Microbiol Infect Dis 31:1791–1796

Sato T, Watanabe T, Mikami T, Matsumoto T (2004) Farnesol, a morphogenetic autoregulatory substance in the dimorphic fungus *Candida albicans*, inhibits hyphae growth through suppression of a mitogen-activated protein kinase cascade. Biol Pharm Bull 27:751–752

Schlecht LS, Peters BM, Krom BP, Hänsch GM, Filler SG, Jabra-Rizk MA, Shirtliff ME (2014) Systemic *Staphylococcus aureus* infection mediated by *Candida albicans* hyphal invasion of mucosal tissue. Microbiology mic.0.083485-0

Shirtliff ME, Peters BM, Jabra-Rizk MA (2009) Cross-kingdom interactions: *Candida albicans* and bacteria. FEMS Microbiol Lett June:1-8

Silverman RJ, Nobbs AH, Vickerman MM, Barbour ME, Jenkinson HF (2010) Interaction of *Candida albicans* cell wall Als3 protein with *Streptococcus gordonii* AspB adhesin promotes development of mixed-species communities. Infect Immun 78:4644–4652

Siqueira JF Jr, Sen BH (2004) Fungi in endodontic infections. Oral Surg Oral Med Oral Pathol Oral Radiol Endod 97:632–641

Sobel J (1992) Pathogenesis and treatment of recurrent vulvovagnial candidiasis. Clin Infect Dis 14:S148–S153

Sobel JD, Chaim W (1996) Vaginal microbiology of women with acute recurrent vulvovaginal candidiasis. J Clin Microbiol 34:2497–2499

Stoodley P, Sauer K, Davies DG, Costerton JW (2002) Biofilms as complex differentiated communities. Ann Rev Microbiol 56:187–209

Swidergall M, Ernst J (2014) Interplay between *Candida albicans* and the antimicrobial peptide armory. Eukaryot Cell 13:950–957

Taff HT, Nett JE, Zarnowski R, Ross KM, Sanchez H, Cain MT, Hamaker J, Mitchell AP, Andes DR (2012) A *Candida* biofilm-induced pathway for matrix glucan delivery: implications for drug resistance. Plos Path 8:8

Tawara Y, Honma K, Naito Y (1996) Methicillin-resistant *Staphylococcus aureus* and *Candida albicans* on denture surfaces. Bulletin Tokyo Dent Coll 37:119–128

Thein ZM, Samaranayake YH, Samaranayake LP (2006) Effect of oral bacteria on growth and survival of *Candida albicans* biofilms. Arch Oral Biol 51:672–680

Timsit JF, Cheval C, Gachot B, Bruneel F, Wolff M, Carlet J, Regnier B (2001) Usefulness of a strategy based on bronchoscopy with direct examination of bronchoalveolar lavage fluid in the initial antibiotic therapy of suspected ventilator-associated pneumonia. Intens Care Med 27:640–647

Tournu H, Van Dijck P (2012) *Candida* biofilms and the host: Models and new concepts for eradication. Int J Microbiol ID 845352

Tsui C, Kong EF, Jabra-Rizk MA (2016) Pathogenesis of Candida albicans biofilm. Pathog Dis. doi:10.1093/femspd/ftw018

Ursell LK, Metcalf J, Parfrey LW, Knight R (2012) Defining the human microbiome. Nutr Rev 70: S38–S44

Urzua B, Hermosilla G, Gamonal J, Morales-Bozo I, Canals M, Barahona S, Cóccola C, Cifuentes (2008) Yeast diversity in the oral microbiota of subjects with periodontitis: *Candida albicans* and *Candida dubliniensis* colonize the periodontal pockets. Med Mycol 46:783–793

Valenza G, Tappe D, Turnwald D, Frosch M, König C, Hebestreit H, Abele-Horn M (2008) Prevalence and antimicrobial susceptibility of microorganisms isolated from sputa of patients with cystic fibrosis. J Cyst Fibros 7:123–127

Vílchez R, Lemme A, Ballhausen B, Thiel V, Schulz S, Jansen R, Sztajer H, Wagner-Döbler I (2010) *Streptococcus mutans* inhibits *Candida albicans* hyphal formation by the fatty acid signaling molecule trans-2-decenoic acid (SDSF). ChemBioChem 26:1552–1562

Wargo MJ, Hogan DA (2006) Fungal-bacterial interactions: a mixed bag of mingling microbes. Curr Opin Microbiol 9:359–364

Waters CM, Bassler B (2005) Quorum sensing: cell-to-cell communication in bacteria. Rev Cell Dev Biol 21:319–346

Watnick P, Kolter R (2000) Biofilm, city of microbes. J Bacteriol 182:2675–2679

Williams P (2007) Quorum sensing, communication and cross-kingdom signalling in the bacterial world. Microbiology 153:3923–3938

Williams D, Lewis M (2011) Pathogenesis and treatment of oral candidosis. Oral Microbiol 3 (5771)

Wisplinghoff H, Bischoff T, Tallent SM, Seifert H, Wenzel RP, Edmond MB (2004) Nosocomial bloodstream infections in U.S. hospitals: analysis of 24,179 cases from a prospective nationwide surveillance study. Clin Infect Dis 39:309–317

Xu X-L, Lee RTH, Fang H-M, Wang Y-M, Li R, Zou H, Zhu Y, Wang Y (2008) Bacterial peptidoglycan triggers *Candida albicans* hyphal growth by directly activating the adenylyl cyclase Cyr1p. Cell Host Microbe 4:28–39

Xu H, Sobue T, Thompson A, Xie Z, Poon K, Ricker A, Cervantes J, Diaz PI, Dongari-Bagtzoglou A (2014) Streptococcal co-infection augments *Candida* pathogenicity by amplifying the mucosal inflammatory response. Cell Microbiol 16:214–231

Zero DT, Fontana M, Martinez-Mier EA, Ferrera-Zandona A, Ando M, Gonzalez-Cabezas C, Bayne S (2009) The biology, prevention, diagnosis and treatment of dental caries. JADA 140:25S–34S

Zhou X, Westman R, Hickey R, Hansmann MA, Kennedy C, Osborn TW, Forney LJ (2009) Vaginal microbiota of women with frequent vulvovaginal candidiasis. Infect Immun 77: 4130–4135

Chapter 9
Host Immune Responses During Infections with *Candida albicans*

Mark H.T. Stappers and Gordon D. Brown

Abstract *Candida* species are commensals of the mucosal surfaces and the skin in the majority of healthy individuals. Under conditions in which the host defence is diminished, for example in patients who are immunocompromised or have endured major trauma or invasive clinical procedures, *Candida* species can initiate infections ranging from superficial mucosal to life-threatening invasive disease (Brown et al. Hidden killers: human fungal infections. Science translational medicine 4 (165):165rv113, 2012). In this chapter, we will explore the importance of the host's innate and adaptive immune defence against *Candida albicans*, the most abundant *Candida* species causing human infections. The majority of studies to date have focused on invasive *C. albicans* infections and considerably less is known about mucosal infections, such as oral and vulvovaginal candidiasis. We will discuss how *C. albicans* is recognized by pattern recognition receptors (PRRs) of the immune system and the role immune and non-immune cells play in the antifungal response to *C. albicans*. In addition, we will describe strategies developed by *C. albicans* to evade recognition and destruction by the host immune system.

Keywords *Candida albicans* · Candidiasis · Pattern recognition receptors · Fungal immunology

9.1 Pattern Recognition of *Candida albicans*

The crucial step in mounting an immune response to *C. albicans* is recognition of the fungus by the host. This is achieved by sensing of conserved molecular patterns of *C. albicans*, termed pathogen-associated molecular patterns (PAMPs), such as cell wall constituents and intracellular components (e.g. DNA). The cell wall of *C. albicans* is mainly composed of carbohydrate polymers and glycoproteins which

---

M.H.T. Stappers · G.D. Brown (✉)
Aberdeen Fungal Group, MRC Centre for Medical Mycology,
Institute of Medical Sciences, University of Aberdeen, Aberdeen, UK
e-mail: gordon.brown@abdn.ac.uk

R. Prasad (ed.), *Candida albicans: Cellular and Molecular Biology*,
DOI 10.1007/978-3-319-50409-4_9

give strength and shape to the fungus and are important for protection against the environment. The inner layer of the cell wall consists of a network of chitin, β-1,3-glucans and β-1,6-glucans, which is covered by an outer layer mainly comprised of O- and N-linked mannoproteins (Gow and Hube 2012). However, the cell wall is a dynamic structure and the composition and exposure of its components can differ significantly depending on morphological state (yeast, pseudohyphae or hyphae) and environment to which it is exposed (Gow and Hube 2012).

For direct recognition of *C. albicans* PAMPs, cells of the innate immune system are equipped with membrane-bound and cytoplasmic pattern recognition receptors (PRRs). Indirect recognition is also possible, in which soluble components, such as complement and antibodies, bind *C. albicans* and subsequently are detected by opsonizing receptors. PRRs can be subdivided in several families, including C type lectin receptors (CLRs), Toll-like receptors (TLRs), NOD-like receptors (NLRs) and RIG I-like receptors (RLRs) (Medzhitov 2007), and their roles in recognition of *C. albicans* are discussed below (Table 9.1 and Fig. 9.1).

9.1.1 C-Type Lectin Receptors

C-type lectin receptors are a large group of receptors with over 1000 members characterized by the expression of at least one C-type lectin-like domain (CTLD). Members of the "Dectin-1 and Dectin-2 family" of CLRs have been shown to be important for the development of an antifungal immune response, via the recognition of various cell wall components. CLR recognition and signalling, either direct (e.g. Dectin-1) or via association with the Fc receptor γ chain (e.g. Dectin-2, Mincle, Dectin-3), leads to activation of the Syk/PKCδ/CARD9/Bcl-10/MALT1 pathway and subsequently transcription factors, such as NF-κB, resulting in the development of an immune response. In addition, some CLRs can also signal via RAF1 kinase pathway (Hardison and Brown 2012). CARD9 is crucial for the immune defence against *C. albicans* infections, as defective signalling results in decreased killing by neutrophils and reduced induction of Th17 cells (Drewniak et al. 2013; LeibundGut-Landmann et al. 2007). Mice deficient for CARD9 are more susceptible to systemic *C. albicans* infections, exemplified by increased mortality and fungal burden (Gross et al. 2006) and in humans mutations in *CARD9* are associated with increased susceptibility to invasive *C. albicans* infections (Glocker et al. 2009; Lanternier et al. 2015).

9.1.1.1 CLR Recognition of Mannan and Mannoproteins

Mannans and mannoproteins are part of the outer layer of the *C. albicans* cell wall and can be directly recognized by CLRs. Mannoprotein fibril length is modulated by mannosyltransferases and deficiencies in these enzymes impact on the virulence and immune recognition of *C. albicans* (Hall et al. 2013; Zhang et al. 2016).

Table 9.1 Selected pattern recognition receptors, their *C. albicans* PAMPs, and the effect on *C. albicans* infection in PRR deficient mice or humans with PRR single nucleotide polymorphisms

Family	Receptor	Location	Ligand	Effect PRR −/− mice	Reference	Polymorphism	Effect SNPs in humans	Reference
CLR	Dectin-1	Surface	β-(1,3)-glucan	Increased susceptibility to systemic candidiasis	Taylor et al. (2007)	Y238X	Increased susceptibility to CMC	Ferwerda et al. (2009)
				No effect on susceptibility to systemic candidiasis	Saijo et al. (2007)	Y238X	Increased oral and gastrointestinal colonization in hematopoietic stem cell transplant patients	Plantinga et al. (2009)
						Y238X	No effect on susceptibility to candidemia	Rosentul et al. (2011)
						Y238X	No effect on susceptibility to RVVC	Rosentul et al. (2014)
CLR	Dectin-2	Surface	High-mannose structures	Increased susceptibility to systemic candidiasis	Ifrim et al. (2016) and Saijo et al. (2010)			
CLR	Dectin-3	Surface	α-mannan	Increased susceptibility to systemic candidiasis	Zhu et al. (2013)			
CLR	Mannose receptor	Surface, secreted	N-linked mannan	No effect on susceptibility to systemic candidiasis	Lee et al. (2003)			

(continued)

Table 9.1 (continued)

Family	Receptor	Location	Ligand	Effect PRR –/– mice	Reference	Polymorphism	Effect SNPs in humans	Reference
CLR	Mincle	Surface	α-mannan	Increased susceptibility to systemic candidiasis	Wells et al. (2008)			
CLR	DC-SIGN	Surface	N-linked mannan					
CLR	Galectin-3	Surface, secreted	β-1,2-linked oligomannans	Increased susceptibility to systemic candidiasis	Linden et al. (2013)			
CLR	Langerin	Surface	β-glucan, mannan					
CLR	SP-A	Secreted	Mannan					
CLR	SP-D	Secreted	Mannan					
CLR	MBL	Secreted	Mannan	Increased susceptibility to systemic candidiasis	Held et al. (2008)	Codon 54	Increased susceptibility to RVVC	Babula et al. (2003)
						Codons 52, 54, and 57	Increased susceptibility to abdominal *C. albicans* infection	van Till et al. (2008)
TLR	TLR1	Surface	Unknown	No effect on susceptibility to systemic candidiasis	Netea et al. (2008)	R80T, S248N, I602S	Increased susceptibility to candidemia	Plantinga et al. (2012)
						R80T	No effect on susceptibility to RVVC	Rosentul et al. (2014)

(continued)

Table 9.1 (continued)

Family	Receptor	Location	Ligand	Effect PRR −/− mice	Reference	Polymorphism	Effect SNPs in humans	Reference
TLR	TLR2	Surface	Phospholipomannan	Decreased susceptibility to systemic candidiasis	Netea et al. (2004b)	P631H	Increased susceptibility to RVVC	Rosentul et al. (2014)
				Increased susceptibility to systemic candidiasis	Villamon et al. (2004c)	P631H, R753Q	No effect on susceptibility to candidemia	Plantinga et al. (2012)
TLR	TLR3	Endosomal				L412F	Increased susceptibility to CMC	Nahum et al. (2011)
TLR	TLR4	Surface	O-mannan	No effect on susceptibility to systemic candidiasis	Gil and Gozalbo (2006)	D299G, T399I	Increased susceptibility to candidemia	Van der Graaf et al. (2006b)
				Increased susceptibility to systemic candidiasis	Netea et al. (2002)	D299G, T399I	No effect on susceptibility to candidemia	Plantinga et al. (2012)
						D299G, T399I	No effect on susceptibility to RVVC	Rosentul et al. (2014)
TLR	TLR6	Surface	Unknown	No effect on susceptibility to systemic candidiasis	Netea et al. (2008)	S249P	No effect on susceptibility to candidemia	Plantinga et al. (2012)

(continued)

Table 9.1 (continued)

Family	Receptor	Location	Ligand	Effect PRR –/– mice	Reference	Polymorphism	Effect SNPs in humans	Reference
TLR	TLR7	Endosomal	Single-stranded RNA	Increased susceptibility to systemic candidiasis	Biondo et al. (2012)			
TLR	TLR9	Endosomal	DNA	No effect on susceptibility to systemic candidiasis	van de Veerdonk et al. (2008)	Promotor (rs5743836)	No effect on susceptibility to candidemia	Plantinga et al. (2012)
				Increased susceptibility to systemic candidiasis	Biondo et al. (2012)			
NLR	NOD2	Cytoplasmic	Chitin			2104C/T, 2722G/C, and 3020insC	No effect on susceptibility to *C. albicans* infections	van der Graaf et al. (2006a)
NLR	NLRP3	Cytoplasmic	β-glucan and SAP2/6	Increased susceptibility to systemic candidiasis	Gross et al. (2009), Joly et al. (2009) and Hise et al. (2009)	Tandem repeat in intron 4	Increased susceptibility to RVVC	Lev-Sagie et al. (2009)
TLR	NLRP10	Cytoplasmic	Unknown	Increased susceptibility to systemic candidiasis	Joly et al. (2012)			

(continued)

Table 9.1 (continued)

Family	Receptor	Location	Ligand	Effect PRR −/− mice	Reference	Polymorphism	Effect SNPs in humans	Reference
TLR	NLRC4	Cytoplasmic	Unknown	Increased susceptibility to systemic candidiasis	Tomalka et al. (2011)			
RLR	MDA5	Cytoplasmic	Unknown			A946T, H843R	Increased susceptibility to candidemia	(Jaeger et al. 2015)
Other PRRs	CR3	Surface	β-(1,3)-glucan					
Other PRRs	CD14	Surface	Mannan					
Other PRRs	CD36	Surface	β-(1,3)-glucan					
Other PRRs	SCARF1	Surface	β-(1,3)-glucan					

CMC Chronic mucocutaneous candidiasis, *RVVC* Recurrent vulvovaginal candidiasis

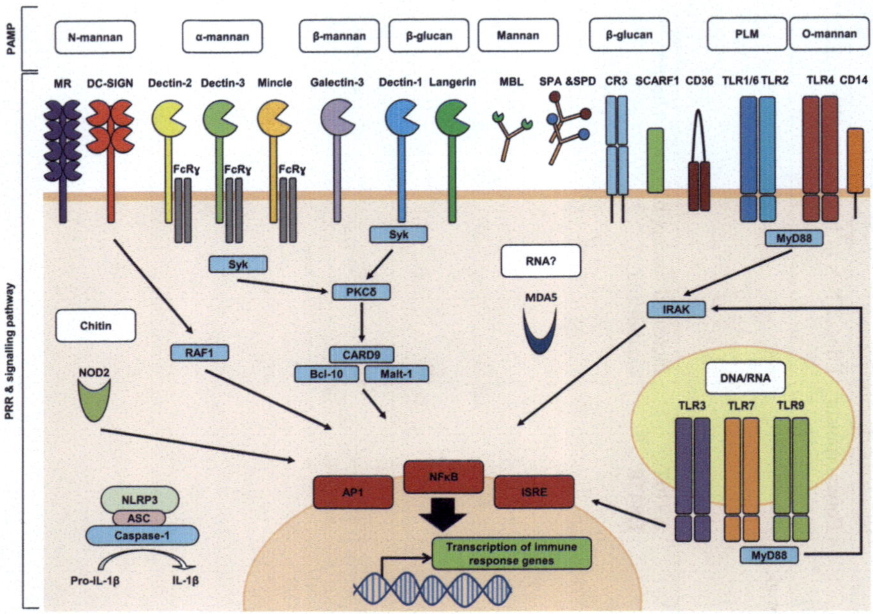

Fig. 9.1 Simplified overview of pattern recognition receptors (PRRs) involved in the immune response to *C. albicans*. Cell surface and soluble PRRs (including: Mannose receptor (MR), DC-SIGN, Dectin-2, Dectin-3, Mincle, Galectin-3, Dectin-1, Langerin, Mannose-binding lectin (MBL), Surfactant protein A and D (SPA and SPD), complement receptor 3 (CR3), Scarf1, CD36, TLR1, TLR6, TLR2, TLR4 and CD14) recognize various components of *C. albicans* (including: Mannans, β-glucans, chitin, RNA and DNA), resulting in downstream signalling via different pathways and activation of the immune response

Various CLRs have been implicated in the recognition of *C. albicans* mannans and mannoproteins including: Mannose receptor, Dectin-2, Dectin-3, Mincle, dendritic cell-specific ICAM3 grabbing non-integrin (DC-SIGN), Galectin-3, Langerin and collectins.

9.1.1.2 Mannose Receptor

The Mannose receptor (CD206) is predominantly expressed on macrophages, but its presence on other immune cells (DC's and microglia), non-immune cells (endothelial cells, kidney mesangial cells and retinal pigment epithelium) and a soluble form of MR have been described (Linehan et al. 1999; McKenzie et al. 2007; Takahashi et al. 1998; Shepherd et al. 1991). The receptor contains an extracellular region with a cysteine-rich domain, a fibronectin type II repeat domain and eight CTLDs, of which CTLD4-8 are involved in the recognition of *C. albicans* N-linked mannans (Netea et al. 2006), but lacks a known signalling domain. Recognition is associated with phagocytosis of *C. albicans* (Ezekowitz et al. 1990) and production of various pro-inflammatory cytokines (Yamamoto et al. 1997; Heinsbroek et al. 2008; van de Veerdonk et al. 2009). However, in a model of disseminated *C. albicans* infection, mice deficient of MR did not differ in survival compared with wild-type control mice (Lee et al. 2003).

9.1.1.3 Dectin-2

Dectin-2 (CLEC4n) is expressed on macrophages, neutrophils and DCs and recognizes Candida high-mannose structures and hyphae (McGreal et al. 2006; Sato et al. 2006). Via association with the Fc receptor γ (FcRγ) chain Dectin-2 induces pro-inflammatory cytokine production, recruitment of immune cells, phagocytosis and killing of *C. albicans* and Th17 cell responses (Ifrim et al. 2016; Saijo et al. 2010; Robinson et al. 2009). As a result, mice deficient of Dectin-2 are more susceptible to systemic *C. albicans* infection, exemplified by decreased survival and increased fungal burdens in the kidney (Ifrim et al. 2016; Saijo et al. 2010). In addition, Dectin-2 can form heterodimers with Dectin-3 (MCL, CLECSF8 or Clec4d) for the recognition of α-mannans and mice lacking Dectin-3 are highly susceptible to *C. albicans* infection (Zhu et al. 2013).

9.1.1.4 Mincle

Mincle (Clec4e or Clecsf9) is expressed on macrophages, neutrophils, DCs and B-cells and similar to Dectin-2 and Dectin-3 associates with FcRγ for intracellular signalling (Flornes et al. 2004). Mincle is not a phagocytic receptor for *C. albicans*, but absence of the receptor decreased TNFα production in response to the yeast. Mice deficient for Mincle showed increased kidney fungal burdens in a model of systemic candidiasis, whereas no differences in survival were reported (Wells et al. 2008).

9.1.1.5 DC-SIGN

DC-SIGN (CD209) recognizes Candida N-linked mannan and is expressed on DCs and macrophages (Cambi et al. 2008). Binding of the ligand leads to uptake of the fungus and production of the anti-inflammatory IL-10 and modulation of TLR signalling, via a Raf-1 dependent pathway (Gringhuis et al. 2007). The mouse has eight homologs of DC-SIGN, of which SIGNR1 and SIGNR3 have been shown to be involved in the recognition of *C. albicans* (Takahara et al. 2004). For SIGNR1, binding of *C. albicans* has been associated with the production of pro-inflammatory cytokines and reactive oxygen species (ROS) (Takahara et al. 2011; Taylor et al. 2004).

9.1.1.6 Galectin-3

Galectin-3 is expressed on macrophages, DCs and epithelial cells and binds Candida β-1,2-linked oligomannans (Kohatsu et al. 2006). Deletion of Galectin-3 is not important for binding and phagocytosis of *C. albicans* but reduces protective TNFα cytokine responses and in models of disseminated candidiasis increases susceptibility to infection, exemplified by faster mortality, increased fungal burden and abscess formation in the brain (Jouault et al. 2006; Linden et al. 2013).

9.1.1.7 Langerin

Langerin is expressed by Langerhans cells and Langerin-positive dermal DCs, subsets of DCs present in skin and mucosal surfaces. Langerin binds both mannans and β-glucans present on *C. albicans* cell wall (Takahara et al. 2004; de Jong et al. 2010), and antigen presentation by the Langerin expressing DC subsets is important in modulating adaptive immune responses (Igyarto et al. 2011).

9.1.1.8 Collectins

Collectins (collagen-containing C-type lectins) are a group of soluble CLRs and members mannose-binding lectin (MBL) and lung surfactant proteins A (SP-A) and D (SP-D) have been shown to be involved in the recognition of *C. albicans*. MBL can via complement-dependent (Brouwer et al. 2008) or independent (Li et al. 2012) methods mediate opsonophagocytosis of *C. albicans* by neutrophils. MBL-A and MBL-C double-deficient mice showed significantly increased mortality in intraperitoneal and intravenous induced disseminated candidiasis models (Held et al. 2008), and parental administration of MBL increased resistance to disseminated candidiasis (Lillegard et al. 2006). Furthermore, single nucleotide polymorphisms in *MBL* resulting in deficient or decreased levels of MBL protein, are associated with increased susceptibility to recurrent vulvovaginal candidiasis and abdominal *C. albicans* infection (Babula et al. 2003; van Till et al. 2008). Binding of SP-A to *C. albicans* inhibits phagocytosis by alveolar macrophages and monocytes and was found to suppress pro-inflammatory cytokine responses

(Rosseau et al. 1997, 1999). Binding of SP-D also inhibits phagocytosis of *C. albicans*, and directly influenced its ability to grow and form hyphae (van Rozendaal et al. 2000).

9.1.1.9 Recognition of β-Glucans by Dectin-1

In the *C. albicans* yeast, the outer layer of mannans and mannoproteins masks the exposure of β-glucans to immune cells, as discussed later. However, when the yeast starts the process of budding or hyphal formation, β-glucans become uncovered and can be recognized by Dectin-1. Dectin-1 (CLEC7A) recognizes β-1,3-glucans and is primarily expressed on macrophages, neutrophils and DCs, but expression in other cell types has been reported (Brown and Gordon 2001). The receptor mediates phagocytosis of *C. albicans*, induction of ROS, pro-inflammatory cytokines and Th17 cell differentiation (Taylor et al. 2007; LeibundGut-Landmann et al. 2007). Dectin-1 deficient mice were shown to be more susceptible to disseminated *C. albicans* infection than wildtype counterparts, exemplified by increased mortality, fungal burden and a decreased ability to recruit inflammatory cells to the site of infection (Taylor et al. 2007). In contrast, disseminated infection models in an independently generated Dectin-1 deficient mice did not show differences in susceptibility to *C. albicans* infection (Saijo et al. 2007), which was subsequently shown to be due to differential adaptation of *C. albicans* strains in vivo (Marakalala et al. 2013). Presence of a single nucleotide polymorphism in *CLEC7A* (Dectin-1) leads to the introduction of an early-stop-codon and deficient expression of the receptor, resulting in decreased production of pro-inflammatory cytokines in response to *C. albicans* (Ferwerda et al. 2009). This polymorphism was found to be associated with increased susceptibility to chronic mucocutaneous candidiasis (Ferwerda et al. 2009) and oral and gastrointestinal colonization in hematopoietic stem cell transplant patients (Plantinga et al. 2009), but not candidemia (Rosentul et al. 2011) and recurrent vulvovaginal candidiasis (Rosentul et al. 2014).

9.1.2 Toll-Like Receptors

After the discovery of a role for the *Drososphila* Toll-signalling pathway in the host defence against fungal infections, ten human homologues termed toll-like receptors have been characterized and extensively studied. TLRs can be broadly divided in two groups, those mainly expressed on the cellular surface involved in the recognition of lipids and proteins (TLR 1, 2, 4, 5, 6 and 10), and those mainly expressed in intracellular compartments involved in the recognition of nucleic acids (TLR3, 7, 8, 9). TLRs recognize PAMPs by an extracellular domain containing leucine-rich repeats, followed by a transmembrane region and finally an intracellular TIR homology domain involved in signal transduction. Upon recognition, TLRs dimerize and recruit the adaptor protein MyD88, which via IRAK, induces activation of transcription factors such as, NF-κB, AP1 and IFN-stimulated response elements, and development of the immune response. Although mice deficient for MyD88 have been shown to be more susceptible to systemic *C. albicans* infections, exemplified by increased mortality, fungal burden and decreased pro-inflammatory cytokine production (Villamon et al. 2004a), humans with mutations in *MyD88* or *IRAK* do not have increased incidence of fungal infections (Picard et al. 2003; von Bernuth et al. 2008). Most TLRs are expressed by various cells of the immune system including macrophages, monocytes, neutrophils, dendritic cells, T cells and epithelial cells (Medzhitov 2007).

9.1.2.1 Toll-Like Receptor 2

TLR2 recognizes both *C. albicans* yeast and hyphae forms via the glycolipid phospholipomannan (PLM) and induces pro-inflammatory cytokine induction (Gil and Gozalbo 2006; Jouault et al. 2003; Netea et al. 2002). However, conflicting results have been reported regarding the role for TLR2 in susceptibility to disseminated candidiasis in infection models. In one study, TLR2-deficient mice were more resistant to intravenous *C. albicans* infection, and this was associated with increased chemotaxis and enhanced candidacidal capacity due to decreased regulatory T cell responses (Netea et al. 2004b; Sutmuller et al. 2006). In studies by another group, TLR2-deficient mice have been shown to be more susceptible to *C. albicans* infection (Gil and Gozalbo 2006; Villamon et al. 2004b, c), and discrepancies can possibly be explained by differences in *C. albicans* strains. TLR2 can form heterodimers with both TLR1 and TLR6 for the recognition pathogens. Whereas, in vitro data suggests a role for TLR6 in the recognition of *C. albicans* (Jouault et al. 2003; Netea et al. 2008), mice deficient for either TLR1 or TLR6 showed a normal susceptibility to disseminated *C. albicans* infection (Netea et al. 2008). Additionally, in humans, single nucleotide polymorphisms in *TLR1*, but not *TLR6*, lead to decreased cytokine production by peripheral blood mononuclear cells in response to *C. albicans* and are associated with increased susceptibility to candidemia (Plantinga et al. 2012). A polymorphism in *TLR2* was not associated with candidemia (Plantinga et al. 2012) but was found to be linked to increased susceptibility to recurrent vulvovaginal candidiasis, due to decreased production of IFNγ and IL-17 in response to *C. albicans* (Rosentul et al. 2014).

9.1.2.2 Toll-Like Receptor 4

The role for TLR4 in *C. albicans* recognition is also a matter of debate. In vitro studies by one group suggest TLR4 recognizes Candida O-linked mannans which leads to pro-inflammatory cytokine production (Netea et al. 2006), whereas others observed minor differences of TLR4-deficient macrophages in response to *C. albicans* (Gil and Gozalbo 2006). Similarly, in in vivo models for disseminated candidiasis, TLR4-deficient mice were observed to be more susceptible to infection, due to impaired chemokine expression and neutrophil recruitment by one group (Netea et al. 2002), whereas no differences in survival were observed by the other (Gil and Gozalbo 2006). A reason opted for the discrepancies between these groups could be the variable dependency on TLR4 for the recognition of *C. albicans* strains (Netea et al. 2010). Furthermore, single nucleotide polymorphisms in *TLR4* have been associated with an increased susceptibility to Candida bloodstream infections (Van der Graaf et al. 2006b), however, a larger study could not confirm this association (Plantinga et al. 2012).

9.1.2.3 Nucleic-Acid Recognizing Toll-Like Receptors

Nucleic-acid recognizing TLRs 3, 7 and 9 are also involved in *C. albicans* recognition. In vitro stimulation of peripheral blood mononuclear cells carrying a single nucleotide polymorphism in *TLR3* showed decreased IFNγ responses to *C. albicans* (Nahum et al. 2012), and was found to be associated with susceptibility to cutaneous candidiasis (Nahum et al. 2011). TLR7 recognizes *C. albicans* RNA and results in decreased IL-12 responses, and mice deficient for the receptor showed increased mortality in disseminated *C. albicans* infection models (Biondo et al. 2012). In vitro, by inhibition or genetic manipulation of the receptor, TLR9 was shown to be involved in the recognition of *C. albicans*. In the same study, using TLR9-deficient mice, the receptor was found to be redundant in in vivo disseminated candidiasis model (van de Veerdonk et al. 2008), whereas another study found an increased mortality in mice deficient for the receptor (Biondo et al. 2012). In humans, a single nucleotide polymorphism in the promoter region of *TLR9*, failed to show an association with susceptibility to candidemia (Plantinga et al. 2012).

9.1.3 NOD-Like Receptors

NLRs are intracellular receptors and 3 subgroups can be defined: NODs, NLRPs and IPAF/NAIP. NOD1 and NOD2 belong to the NOD family, which are via association with RICK, result in the activation of NF-κB (Medzhitov 2007). *C. albicans* chitin has been shown to induce IL-10 cytokine responses via NOD2, in collaboration with MR and TLR9 (Wagener et al. 2014), and possibly dampen inflammatory responses to *C. albicans*. However, another group has concluded no differences between *C. albicans* stimulated-PBMCs carrying single nucleotide polymorphism in NOD2 versus wildtype PBMCs, and prevalence of *NOD2* polymorphisms was not increased in patients with Candida infections (van der Graaf et al. 2006a).

9.1.3.1 Inflammasome Activation by *C. albicans*

The NLRP and IPAF subfamilies take part in the formation of the inflammasome, which is a multiprotein complex involved in the processing of pro-IL-1β and pro-IL-18 into their biologically active forms (Medzhitov 2007). The most described NLRP3 inflammasome, in which NLPR3 interacts with ASC and caspase-1, can be activated by *C. albicans* hyphae, but not yeast cells, and mice with deficiencies in components of this inflammasome are more susceptible during disseminated *C. albicans* infection models (Gross et al. 2009; Joly et al. 2009; Hise et al. 2009). Possible activators of this response are β-glucans or secreted aspartic proteases 2 and 6 (Cheng et al. 2010; Pietrella et al. 2013; Gabrielli et al. 2015). In addition, *C. albicans* induced activation of the NLRP3 inflammasome has been shown to trigger programmed cell death pathway (pyroptosis), to escape from in macrophages (Wellington et al. 2014; Uwamahoro et al. 2014). Besides caspase-1, activation of a non-canonical caspase-8 inflammasome by *C. albicans* β-glucans has also been shown to induce processing of biologically active cytokines and programmed cell death (Gringhuis et al. 2012; Ganesan et al. 2014). Furthermore, mice deficient for NLRP10 also show increased susceptibility to disseminated candidiasis (Joly et al. 2012), whereas NLRC4 has a specific role in antimucosal defence against *C. albicans* (Tomalka et al. 2011). In humans, a single nucleotide polymorphism in *CIAS1*, the gene that encodes for NLRP3, has been associated with recurrent vulvovaginal candidiasis and decreased IL-1β production in response to *C. albicans* (Lev-Sagie et al. 2009).

9.1.4 Rig-I-Like Receptors

RLRs are a fairly newly discovered family of cytoplasmic PRRs that recognize nucleic acids. There are currently three known members: RIG-I, MDA5 and LGP2 and they are mainly involved in viral recognition (Medzhitov 2007). However, MDA5 (IFIH1) has recently been shown to play a role in recognition of *C. albicans*, as cells from MDA5-deficient mice and human PBMCs with different single nucleotide polymorphisms in *IFIH1* (MDA5) displayed an altered cytokine responses to *C. albicans* and genetic variation in *IFIH1* was strongly associated with susceptibility to systemic Candida infections (Jaeger et al. 2015).

9.1.5 Other PRR Receptors

CR3 (CD18/CD11b) belongs to the integrin family of cell surface receptors and is strongly expressed by neutrophils. CR3 recognizes *C. albicans* β-glucans and is important for uptake and killing of *C. albicans* conidia (Forsyth et al. 1998;

Gazendam et al. 2014). In the same study, killing of antibody-opsonized *C. albicans* was shown to be dependent on Fc-gamma receptors (Gazendam et al. 2014). CD14 is mainly expressed on monocytes and macrophages, and has been suggested to recognize mannans from *C. albicans* followed by signalling via TLR4 (Tada et al. 2002). Finally, SCARF1 and CD36, members of the scavenger receptor family, have also been shown to bind *C. albicans* in a β-glucan-dependent manner (Means et al. 2009).

9.2 Effector Cells in Defence Against *C. albicans*

Recognition of *C. albicans* PAMPS by PRRs leads to the activation of the innate and adaptive immune response, in which various effector cells contribute to clear the invading fungus. Innate immune responses are essential in the host defence against systemic and mucosal candidiasis. In contrast, the adaptive immune response is mainly involved in protection against mucosal candidiasis. Below, the role of these immune, and non-immune, cells in the antifungal response to *C. albicans* is described (Fig. 9.2).

9.2.1 Epithelial Cells

The epithelium acts as a physical barrier with the environment, and is the first line of defence against invading fungi. As a commensal, *C. albicans* is present at mucosal surfaces and discrimination between commensal and pathogenic states of the fungus are important for host defence. To accomplish this, epithelial cells respond to *C. albicans* with activation of NF-κB and a biphasic MAPK response. Independent of *C. albicans* morphology, the first pathway activates NF-κB and MAPK c-Jun. Activation of the second pathway, including MKP1 and c-FOS, is dependent on the formation of hyphae and fungal burden (Moyes et al. 2011, 2010). Activation of the second pathway was recently found to be dependent on secretion of Candidalysin (EceI-III_{62-92K}), a cytolytic protein capable of activation and damaging epithelial cells, by *C. albicans* hyphae (Moyes et al. 2016). Although epithelial cells are able to express various TLRs (Weindl et al. 2010), it remains to be determined which PRRs are important for the differential response to yeasts and hyphae. Activation of epithelial cells induces secretion of cytokines, chemokines and alarmins, important for the recruitment and activation of various immune cells (Moyes et al. 2010, 2011; Yano et al. 2010). In addition, epithelial cells produce antimicrobial peptides such as *β*-defensins and cathelicidin (LL-37) in response to *C. albicans* infection, which are able to directly kill the fungus (Tomalka et al. 2015; Vylkova et al. 2007a; Lopez-Garcia et al. 2005).

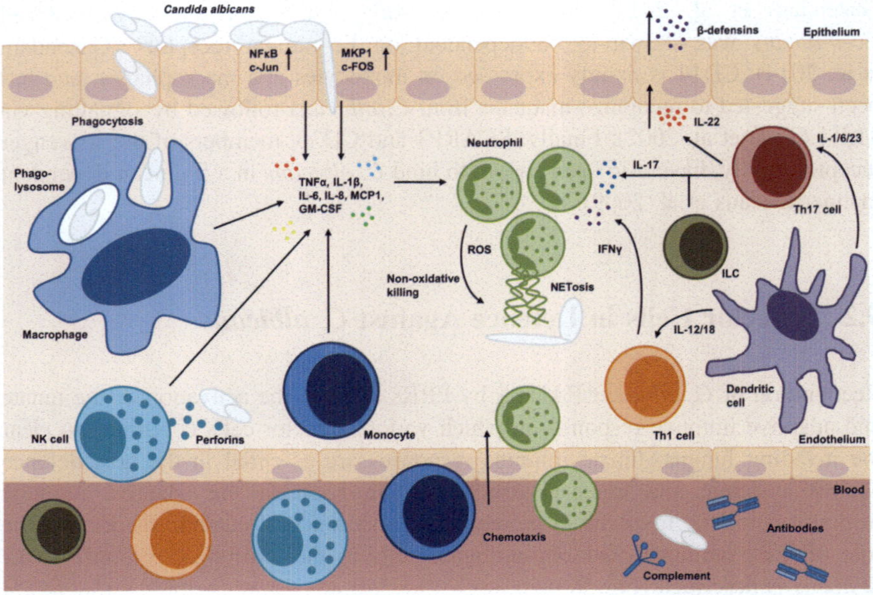

Fig. 9.2 Effector cells and their mechanisms in the host defence against *C. albicans*. The epithelium acts as a physical barrier against invading *C. albicans*. Formation of hyphae activates epithelial MAPK and FOS-dependent signalling and leads to production of cytokines and chemokines, for the recruitment of professional phagocytes, and secretion of antimicrobial peptides which can directly kill *C. albicans*. If the epithelium is breached, tissue macrophages, and neutrophils and monocytes attracted from the blood, will attempt to phagocytose and kill *C. albicans* using oxidative [e.g. reactive oxygen species (ROS)] and non-oxidative methods (e.g. NETosis for neutrophils). NK cells secrete perforins to kill *C. albicans*. Dendritic cells sample antigen and migrate to lymph nodes to stimulate differentiation of naïve T cells into T helper subsets, such as Th1 and Th17. Th1 cells produce IFNγ, which activates phagocytes. Th17 cells, and innate lymphoid cells (ILCs), produce IL-17, which recruits and activates neutrophils, and IL-22 which induces the secretion of β-defensins by epithelial cells

9.2.2 Phagocytes

Invasion of the epithelial barrier by *C. albicans* induces the production of chemokines, which attract professional phagocytes of the innate immune system, such as polymorphonuclear neutrophils, monocytes/macrophages and dendritic cells, to sites of infection. These, as their name suggests, have been shown to be crucial for the phagocytosis and killing of *C. albicans*. In vitro, the presence of neutrophils was required for control of *C. albicans* growth and hyphal formation (Fradin et al. 2005). Neutropenia is a risk factor in patients with invasive infections and depletion of neutrophils in mouse models of disseminated infection resulted in increased susceptibility to *C. albicans* (Horn et al. 2009; van't Wout et al. 1988; Kullberg et al. 1990). Depletion of macrophages in mice systemically challenged with *C. albicans* resulted in increased mortality due to slower clearance and higher fungal burden in the kidneys (Qian et al. 1994). Similar for monocytes, one study reported that depletion resulted in increased mortality due to uncontrolled fungal growth in the kidneys and brain (Ngo et al. 2014), however, another study found no effect of monocytopenia in a disseminated *C. albicans* infection model (van't Wout et al. 1988). Dendritic cells are also able to phagocytose and kill *C. albicans*, but less effectively than other professional phagocytes (Netea et al. 2004a).

Phagocytosis requires recognition of PAMPs by PRRs and uptake of non-opsonized *C. albicans* is dependent on multiple receptors including, Dectin-1, MR and DC-SIGN, whereas CR3 and Fc receptors are important for uptake of opsonized *C. albicans* (Heinsbroek et al. 2008; Cambi et al. 2003; Gazendam et al. 2014). After uptake of *C. albicans*, the phagosome undergoes a process of maturation, which includes multiple rounds of fusion events with lysosomes, containing various components required for oxidative and non-oxidative mechanisms to kill *C. albicans*.

9.2.2.1 Oxidative Killing of *C. albicans*

Oxidative methods include the production of ROS and reactive nitrogen species are crucial for phagocytic killing of *C. albicans*. Activation of phagocytes leads to the assembly of the NADPH oxidase complex which results in production of superoxide, which in turn leads to the formation of hydroxyl radicals and hydrogen peroxide. This hydrogen peroxide can subsequently be converted to hypochlorous acid by myeloperoxidase, an enzyme produced by phagocytes. Phagocytes also express inducible nitric oxide synthase, or iNOS, which forms nitric oxide from arginine and oxygen, and in turn can react with superoxide leading to the formation of peroxynitrite. Mice deficient in components of the NADPH oxidase complex, myeloperoxidase and iNOS have all been shown to be susceptible to systemic candidiasis, exemplified by increased fungal burden and mortality (Aratani et al. 2002; Balish et al. 2005), whereas no difference in susceptibility to oral candidiasis was observed in mice deficient for iNOS (Farah et al. 2009). Interestingly, in vitro, reactive oxygen and nitrogen species were not required for the killing of *C. albicans*, indicating an important role for non-oxidative mechanisms (Balish et al. 2005).

9.2.2.2 Non-oxidative Killing of *C. albicans*

Non-oxidative mechanisms comprise the production of antimicrobial peptides, hydrolases, nutrient restriction and the formation of neutrophil extracellular traps. Several phagocyte-derived antimicrobial peptides have been shown to be able to kill *C. albicans*, including α-defensins, histatin 5 and cathelicidin, by influencing cell wall integrity and release of ATP from the fungus (Vylkova et al. 2007b;

Lehrer et al. 1988; Lopez-Garcia et al. 2005). In addition, several hydrolases have potential anti-Candida activity, including lysozyme (Wu et al. 1999), seprocidins [proteinase-3, cathepsin G and elastase (Sorensen et al. 2001)] and chitinases (Chen et al. 2009). Besides these more active approaches for the killing of *C. albicans*, the host also attempts to restrict access to essential nutrients required for growth and pathogenicity of *C. albicans*, a process called nutritional immunity. In response to *C. albicans*, phagocytes produce lactoferrin and calprotectin (Voganatsi et al. 2001), which respectively are important for limiting availability of free iron and zinc metals and inhibition of *C. albicans* growth (Sohnle et al. 1996).

Both proteins also have shown to be involved in a process termed neutrophil extracellular traps (NETs) formation (Urban et al. 2006, 2009). In response to *C. albicans* hyphae, which are too large to phagocytose, neutrophils swarm around hyphae and undergo a process of programmed cell death, leading to decondensation of DNA and formation of large, extracellular, web-like fibrils. These NETs capture *C. albicans* and are covered with factors such as calprotectin, elastase, lactoferrin, myeloperoxidase and cathelicidin, which subsequently are involved in killing of the fungus (Urban et al. 2006, 2009). Interestingly, Dectin-1 was found to be an important mediator of NETosis, as Dectin-1-induced phagocytosis inhibited nuclear trafficking of neutrophil elastase and release of NETs (Branzk et al. 2014).

9.2.3 Natural Killer Cells

Natural killer (NK) cells are innate lymphocytes mainly known for their role in killing of tumour and virus-infected cells, but are also involved in the clearance of fungi. NK cells kill target cells via the release of granules containing perforin and granzymes, or Fas-ligand-mediated apoptosis, but also influence other cells of the immune system through the production of cytokines and chemokines. Depletion of NK cells had variable responses in *C. albicans* infection models. In one study, depletion in NK cells only resulted in increased susceptibility to systemic *C. albicans* infections if mice were also deficient of T and B cell responses (Quintin et al. 2014). In another, depletion of NK cells alone was sufficient to increase susceptibility to systemic candidiasis. NK cells induced secretion of GM-CSF, promoting the fungicidal activity of neutrophils (Bar et al. 2014). The NK cell receptor NKp30 has been shown to be responsible for recognition and killing of *C. albicans*, via mediation of perforin release (Li et al. 2013). Human NK cells also recognize and kill *C. albicans* via release of perforin, but acquire assistance of neutrophils via the production of cytokines to kill filamentous form of *C. albicans* (Voigt et al. 2014).

9.2.4 Innate Lymphoid Cells

Over the past couple of years new lineages of innate lymphocytes have been characterized, which have been termed innate lymphoid cells (ILCs) and play an important role in mucosal immunity. Grouping of ILCs resembles that of the T helper cell subsets, but differs due to the lack of a T cell receptor. Three different ILC populations have been described: ILC1 express transcription factor T-bet and produce IFNγ and TNF, ILC2 express transcription factor GATA3 and produce IL-5 and IL-13, and ILC3 express transcription factor RORγt and produce IL-17 and IL-22. These last two groups, ILC2s and ILC3s, have recently been implicated in protection against *C. albicans* infection. Production of IL-17 was shown to be crucial for protection in an oropharyngeal *C. albicans* infection model, and ILCs were found to be the main producers of this cytokine. Antibody-mediated depletion of all ILCs led to increased susceptibility to *C. albicans* infection, and Rorc-deficient mice which lack ILC3s, failed to control *C. albicans* infection (Gladiator et al. 2013). A subset of ILC2s, termed inflammatory ILC2s, are able to develop into ILC3-like cells and acquire the potential to produce IL-17 and provide partial protection in an *C. albicans* oral infection model (Huang et al. 2015). Other innate-like T-lymphocytes, γδT cells and natural Th17 cells, also have been implicated in the mucosal host defence against *C. albicans* (Conti et al. 2014).

9.2.5 T-Lymphocytes

Besides their role in phagocytosis, DCs are of major importance in the presentation of antigen and priming of naive T cells subsets. Depending on the stimulus and subsequent signalling via PRRs and production of cytokines, DCs can instruct naïve CD4 T cells to polarize into various T helper (Th) cells: Th1, Th2, Th17 or regulatory T cells (Treg). In response to *C. albicans*, signalling via Dectin-1, Dectin-2 and MR resulted in the induction of Th1 and Th17 cell responses in mouse models (Gringhuis et al. 2009; Robinson et al. 2009; van de Veerdonk et al. 2009). Differentiation is dependent on the antigen presenting DC subset and morphological form of the fungus. Skin-resident DCs promote opposite Th cell responses directed against *C. albicans*, with Langerhans cells driving Th17 cells and Langerin-positive dermal DCs inducing Th1 cell generation (Igyarto et al. 2011). Similarly, opposite responses are induced based on *C. albicans* morphology, with *C. albicans* yeast inducing Th17 cell formation, whereas *C. albicans* hyphae drives Th1 cell formation (Kashem et al. 2015), indicating the host response can provide tailor-made protection according to tissue-specific needs.

9.2.5.1 T Helper 1 Cells

Th1 cells are characterized by the production of IFNγ, important for phagocyte activation and killing of *C. albicans* (Nathan et al. 1983), and are mainly implicated in defence against systemic candidiasis (Kashem et al. 2015). Mice deficient for IL-18, which helps drive Th1 responses, were more susceptible to disseminated *C. albicans* infection, due to the absence of IFNγ and influx of phagocytes (Netea et al. 2003). Supplementation of either IL-18 or IFNγ in these models improves outcome of infection (Stuyt et al. 2004; Kullberg et al. 1990), an immunotherapy which has shown to be promising in a recent clinical trial in patients with invasive fungal infections (Delsing et al. 2014).

9.2.5.2 T Helper 17 Cells

Th17 cells are characterized by the production of cytokines IL-17 and IL-22 and are crucial in the immune response against *C. albicans*. Mice deficient for various components important in the generation of Th17 cells were more susceptible to mucosal (Conti et al. 2009; Whibley et al. 2016), skin (Kagami et al. 2010) and systemic candidiasis (Huang et al. 2004). Production of IL-17 is required for recruitment and activation neutrophils (Huppler et al. 2014; Huang et al. 2004), whereas IL-22 is important for activation and barrier integrity of the epithelium (De Luca et al. 2010). In humans, *C. albicans*-specific Th cells were found to produce both IL-17 and IFNγ (Zielinski et al. 2012). Over the recent years, various mutations in genes involved in the production and signalling of IL-17, including *IL-17RA*, *IL-17F*, *CARD9*, *RORC*, *STAT1*, *STAT3* and *ACT1* (Puel et al. 2011; Milner et al. 2008; Glocker et al. 2009; Okada et al. 2015; van de Veerdonk et al. 2011; Boisson et al. 2013), have been linked to the development of chronic mucocutaneous candidiasis, enforcing the importance of Th17 cells in human infections.

9.2.5.3 T Helper 2 and Regulatory T Cells

In contrast to Th1 and Th17 subsets, less in clear about the role for Th2 and Treg cells in *C. albicans* infections, but they are generally considered detrimental in fungal infections. Hallmark cytokines, IL-4 and IL-10, were shown to be to inhibit killing of *C. albicans* (Cenci et al. 1993; Romani et al. 1994), and mice treated with anti-IL-4R or IL-10R antibodies or deficient for IL-10 were more resistant to systemic candidiasis (Mencacci et al. 2001; Tavares et al. 2000). Overexpression of GATA3, which drives Th2 responses, is associated with increased susceptibility to *C. albicans* infection (Haraguchi et al. 2010). Furthermore, Tregs were shown to enhance Th17 cell induction, however this has been shown to be both detrimental (Whibley et al. 2014) and beneficial (Pandiyan et al. 2011) for clearance of *C. albicans*.

9.2.6 *B-Lymphocytes and the Humoral Antifungal Response*

B-lymphocytes, or B-cells, are part of the adaptive immune system and function in the humoral immune response, by the secretion of antibodies. In addition, they are able to present antigens and produce cytokines and chemokines. The role for B-lymphocytes and their antibodies in the protection against *C. albicans* infections has been suggested to be modest in comparison to the cellular responses. In various candidiasis models with mice lacking B-lymphocytes no differences in susceptibility to *C. albicans* infection were observed (Carrow et al. 1984; Jensen et al. 1993; Wagner et al. 1996).

No vaccine is available for fungal infections, however, antibodies directed against various components of *C. albicans* have been shown to be protective in the immune response to the fungus. An antibody against *C. albicans* invasion ALS3 was found to inhibit the adherence to human cells, inhibition of germination of yeast, iron acquisition and possesses direct candidacidal activity (Brena et al. 2011; Moragues et al. 2003). This antibody (NDV-3) is the only vaccine candidate against Candida infections currently in clinical trials, and has been successfully tested in a phase-1 clinical trial (Schmidt et al. 2012). Injection of monoclonal antibodies directed against the mannan layer of the cell wall conferred protection in murine models of disseminated candidiasis (Viudes et al. 2004; Zhang et al. 2006; Han and Cutler 1995) and vaginal candidiasis (Han et al. 1998). Antibodies directed against cell wall β-glucan inhibit fungal growth and adherence to human epithelial cells and are protective in systemic and mucosal murine models of infection (Torosantucci et al. 2009, 2005). Besides cell wall components, antibodies mediated against *C. albicans* secreted aspartyl proteinases (Sandini et al. 2011; De Bernardis et al. 1997) and heat-shock protein 90 (Matthews et al. 1991), have also been shown to confer protection in *C. albicans* infection models.

9.3 Evasion and Modulation of the Host Immune Response by *C. albicans*

As the host response has been established to protect itself against fungal invasion, *C. albicans* has also developed mechanisms to evade recognition and clearance by the immune system. Here, we will briefly highlight these strategies employed by *C. albicans* to escape and modulate host defences.

9.3.1 Morphological Transition from Yeast to Hyphae

The ability of *C. albicans* to change between yeast and hyphal growth is considered as one of its main virulence factors. *C. albicans* strains, defective in transcription factors required for hyphal growth, are avirulent in a mouse model for invasive candidiasis (Lo et al. 1997). Upon engulfment, hyphal growth is associated with increased phagocyte lysis and escape of *C. albicans* (Marcil et al. 2002; McKenzie et al. 2010). One obvious explanation is that hyphal growth of the fungus leads to rupture and lysis of the phagocyte. However, recent studies suggest that formation of hyphae induces pyroptosis, a programmed cell death pathway, which leads to death of the phagocyte and escape of the fungus (Wellington et al. 2014; Uwamahoro et al. 2014). In addition, non-lytic expulsion of *C. albicans* from phagocytes, in which both phagocyte and pathogen remain viable, has also been described (Bain et al. 2012).

9.3.2 Masking of PAMPs

As previously mentioned, the cell wall of *C. albicans* is a dynamic structure and the composition can differ significantly depending on environment and morphological state. Mannans and mannoproteins of the outer layer, although PAMPs in their own right, mask the recognition of the more immunogenic inner layer containing β-glucans. *C. albicans* strains deficient for O- and N-linked mannans were more rapidly engulfed and killed by phagocytes (McKenzie et al. 2010). In addition, heat-killing of *C. albicans*, which due to disruption of the cell wall increases the exposure of β-glucans, resulted in higher cytokine responses in human PBMCs (Gow et al. 2007). *C. albicans* hyphae, unlike their yeast form, do not expose the β-glucans on their surface, resulting in a weaker inflammatory response to the hyphal form (Gantner et al. 2005; van der Graaf et al. 2005). The immune system counteracts masking by a recently discovered mechanism in which neutrophil NET-mediated attack induces changes in *C. albicans* cell wall architecture. This remodelling of the cell wall is an active fungal response to immune-mediated stress, and resulted in enhanced recognition of β-glucan by Dectin-1 on macrophages (Hopke et al. 2016). And finally, antifungal drugs of the echinocandin family also impact on cell wall composition by inhibition of β-glucan synthesis resulting in lysis of the fungus (Douglas et al. 1997). Sub-inhibitory concentrations of the echinocandin caspofungin have been shown to be involved in remodelling of *C. albicans* cell wall, with increased exposure of β-glucan (Wheeler et al. 2008) and deposition of chitin (Walker et al. 2008).

9.3.3 Inhibition of Phagosome Maturation

After engulfment of the *C. albicans*, the phagosome undergoes various fusion events with lysosomes, resulting in the acquisition of antimicrobial agents, enzymes and ROS, and killing of the fungus. To prevent this, *C. albicans* modulates intracellular membrane trafficking by preventing the formation of phagolysosomes (Fernandez-Arenas et al. 2009). Inhibition of phagosome maturation has been shown to be dependent on fungal viability, hyphal morphogenesis and cell wall composition. Viable *C. albicans*, but not killed *C. albicans*, was able to delay phagosome maturation, suggesting an active process of inhibition. In engulfed *C. albicans* cells deficient in O-mannosylation, phagosome maturation occurred more rapidly, which, in part was due to a diminished intraphagosomal hyphal formation and enhanced recognition of Dectin-1 via unmasking of β-glucan (Bain et al. 2014).

9.3.4 Inhibition of Reactive Oxygen Species

Production of ROS is a major defence mechanism utilized by phagocytes. In return, *C. albicans* has developed several strategies to counteract killing via ROS. *C. albicans* produces superoxide dismutase on its cell surface, an enzyme which can convert the toxic superoxide into oxygen and less toxic hydrogen peroxide. *C. albicans* cells lacking the superoxide dismutase show accumulation of ROS and severe loss of viability in the presence of macrophages (Frohner et al. 2009). Besides superoxide dismutase, *C. albicans* also produces the enzyme catalase, which converts hydrogen peroxide into oxygen and water. Catalase-deficient *C. albicans* cells are less virulent and are associated with decreased survival in mouse infection models (Nakagawa et al. 2003). Furthermore, glutathione and thioredoxin pathways are also major anti-oxidant systems in *C. albicans* for the conversion of hydrogen peroxide and evasion of killing by ROS (Komalapriya et al. 2015). And finally, farnesol, originally identified as a quorum-sensing molecule involved in the transition from yeast to hyphae, has a dual role in evasion of ROS-induced killing. From the host's perspective, farnesol has been suggested to induce ROS, leading to decreased viability of phagocytes and inhibited killing of *C. albicans* (Abe et al. 2009). From the pathogen's perspective, farnesol protects *C. albicans* from oxidative stress via upregulating of catalase and superoxide dismutase enzymes (Westwater et al. 2005).

9.3.5 Inhibition of Complement

C. albicans has developed efficient strategies to evade attack by the complement system of the host. *C. albicans* expresses Pra1 on the cell surface which is able to bind the human complement regulatory proteins, C4b-binding protein, Factor H and FHL-1, in order to inhibit the activation of the complement system (Luo et al. 2011, 2009). In addition, secreted Pra1 binds complement C3, blocks its conversion to active forms C3a and C3b, and thereby decreases complement-mediated recognition and uptake of *C. albicans* (Luo et al. 2010). Besides inhibition of complement activation, *C. albicans* is also able to degrade complement proteins. *C. albicans* secreted aspartic proteases Sap1, Sap2 and Sap3, can degrade host complement proteins C3b, C4b and C5, inhibiting opsonisation and suppressing the host immune response (Gropp et al. 2009).

9.3.6 Modulation of Immune Cell Responses

Recognition by the PRRs is required for the induction of an immune response, however *C. albicans* has also developed strategies to exploit and modulate recognition by the host immune response to assist in evasion. *C. albicans* activation of TLR2 promotes induction of tolerogenic antigen presenting cells and regulatory T cells (Dillon et al. 2006; Sutmuller et al. 2006), which may interfere with optimal clearance of the fungus (Netea et al. 2004b). In addition, *C. albicans* can influence polarization of macrophages from an inflammatory M1, to a more anti-inflammatory M2 phenotype, which may enhance survival of the fungus (Reales-Calderon et al. 2014). Furthermore, *C. albicans* modulates cytokine secretion by the production of soluble factors, as demonstrated by selective inhibition of IL-12 secretion (Xiong et al. 2000) and inhibition of IL-17 responses by modulation of the tryptophan metabolism (Cheng et al. 2010).

9.3.7 Epithelial and Endothelial Invasion

Invasion of the epithelial and endothelial barriers can occur via active tissue penetration or induced endocytosis. *C. albicans* can induce its own endocytosis to into epithelial and endothelial cells, via binding of invasin ALS3 to surface proteins E-cadherin and N-cadherin respectively (Phan et al. 2007). Induced endocytosis, is dependent on hyphal formation as yeast-locked variants were not taken up, and it is a passive process as killed *C. albicans* hyphae are endocytosed similarly to live hyphae. Uptake of *C. albicans* into epithelial and endothelial cells may prevent recognition and destruction by phagocytes (Phan et al. 2000, 2005).

9.4 Conclusion

We have made great advances in our understanding of the underlying mechanisms that lead to the induction of an efficient antifungal immune response. *C. albicans* has played a very important role as model pathogen in many of these discoveries. Studies have shown the role of PRRs in recognition and uptake of *C. albicans*, how effector cells discriminate between commensal and pathogenic states and exploit various mechanisms to clear *C. albicans*, and what the effect of deficiencies in recognition and effector mechanisms is on the susceptibility to infection. On the side of the fungus, studies have shown how *C. albicans* has developed strategies to evade these host defence mechanisms. Increased understanding of both mechanisms employed by the host and the fungus will hopefully allow the development of new therapies that will have a significant impact on the treatment of fungal infections.

Acknowledgements We thank the Wellcome Trust and the MRC Centre for Medical Mycology for funding.

References

Abe S, Tsunashima R, Iijima R, Yamada T, Maruyama N, Hisajima T, Abe Y, Oshima H, Yamazaki M (2009) Suppression of anti-Candida activity of macrophages by a quorum-sensing molecule, farnesol, through induction of oxidative stress. Microbiol Immunol 53(6):323–330. doi:10.1111/j.1348-0421.2009.00128.x

Aratani Y, Kura F, Watanabe H, Akagawa H, Takano Y, Suzuki K, Dinauer MC, Maeda N, Koyama H (2002) Critical role of myeloperoxidase and nicotinamide adenine dinucleotide phosphate-oxidase in high-burden systemic infection of mice with Candida albicans. J Infect Dis 185(12):1833–1837. doi:10.1086/340635

Babula O, Lazdane G, Kroica J, Ledger WJ, Witkin SS (2003) Relation between recurrent vulvovaginal candidiasis, vaginal concentrations of mannose-binding lectin, and a mannose-binding lectin gene polymorphism in Latvian women. Clin Infect Dis 37(5):733–737. doi:10.1086/377234 (an official publication of the Infectious Diseases Society of America)

Bain JM, Lewis LE, Okai B, Quinn J, Gow NA, Erwig LP (2012) Non-lytic expulsion/exocytosis of Candida albicans from macrophages. Fungal Genet Biol: FG & B 49(9):677–678. doi:10.1016/j.fgb.2012.01.008

Bain JM, Louw J, Lewis LE, Okai B, Walls CA, Ballou ER, Walker LA, Reid D, Munro CA, Brown AJ, Brown GD, Gow NA, Erwig LP (2014) Candida albicans hypha formation and mannan masking of beta-glucan inhibit macrophage phagosome maturation. mBio 5(6): e01874. doi:10.1128/mBio.01874-14

Balish E, Warner TF, Nicholas PJ, Paulling EE, Westwater C, Schofield DA (2005) Susceptibility of germfree phagocyte oxidase- and nitric oxide synthase 2-deficient mice, defective in the production of reactive metabolites of both oxygen and nitrogen, to mucosal and systemic candidiasis of endogenous origin. Infect Immun 73(3):1313–1320. doi:10.1128/IAI.73.3.1313-1320.2005

Bar E, Whitney PG, Moor K, Reis e Sousa C, LeibundGut-Landmann S (2014) IL-17 regulates systemic fungal immunity by controlling the functional competence of NK cells. Immunity 40 (1):117–127. doi:10.1016/j.immuni.2013.12.002

Biondo C, Malara A, Costa A, Signorino G, Cardile F, Midiri A, Galbo R, Papasergi S, Domina M, Pugliese M, Teti G, Mancuso G, Beninati C (2012) Recognition of fungal RNA by TLR7 has a nonredundant role in host defense against experimental candidiasis. Eur J Immunol 42(10):2632–2643. doi:10.1002/eji.201242532

Boisson B, Wang C, Pedergnana V, Wu L, Cypowyj S, Rybojad M, Belkadi A, Picard C, Abel L, Fieschi C, Puel A, Li X, Casanova JL (2013) An ACT1 mutation selectively abolishes interleukin-17 responses in humans with chronic mucocutaneous candidiasis. Immunity 39 (4):676–686. doi:10.1016/j.immuni.2013.09.002

Branzk N, Lubojemska A, Hardison SE, Wang Q, Gutierrez MG, Brown GD, Papayannopoulos V (2014) Neutrophils sense microbe size and selectively release neutrophil extracellular traps in response to large pathogens. Nat Immunol 15(11):1017–1025. doi:10.1038/ni.2987

Brena S, Cabezas-Olcoz J, Moragues MD, Fernandez de Larrinoa I, Dominguez A, Quindos G, Ponton J (2011) Fungicidal monoclonal antibody C7 interferes with iron acquisition in Candida albicans. Antimicrob Agents Chemother 55(7):3156–3163. doi:10.1128/AAC.00892-10

Brouwer N, Dolman KM, van Houdt M, Sta M, Roos D, Kuijpers TW (2008) Mannose-binding lectin (MBL) facilitates opsonophagocytosis of yeasts but not of bacteria despite MBL binding. J immunol 180(6):4124–4132

Brown GD, Gordon S (2001) Immune recognition. A new receptor for beta-glucans. Nature 413 (6851):36–37. doi:10.1038/35092620

Brown GD, Denning DW, Gow NA, Levitz SM, Netea MG, White TC (2012) Hidden killers: human fungal infections. Science translational medicine 4 (165):165rv113. doi:10.1126/scitranslmed.3004404

Cambi A, Gijzen K, de Vries lJ, Torensma R, Joosten B, Adema GJ, Netea MG, Kullberg BJ, Romani L, Figdor CG (2003) The C-type lectin DC-SIGN (CD209) is an antigen-uptake receptor for Candida albicans on dendritic cells. Eur J Immunol 33(2):532–538. doi:10.1002/immu.200310029

Cambi A, Netea MG, Mora-Montes HM, Gow NA, Hato SV, Lowman DW, Kullberg BJ, Torensma R, Williams DL, Figdor CG (2008) Dendritic cell interaction with Candida albicans critically depends on N-linked mannan. J Biol Chem 283(29):20590–20599. doi:10.1074/jbc.M709334200

Carrow EW, Hector RF, Domer JE (1984) Immunodeficient CBA/N mice respond effectively to Candida albicans. Clin Immunol Immunopathol 33(3):371–380

Cenci E, Romani L, Mencacci A, Spaccapelo R, Schiaffella E, Puccetti P, Bistoni F (1993) Interleukin-4 and interleukin-10 inhibit nitric oxide-dependent macrophage killing of Candida albicans. Eur J Immunol 23(5):1034–1038. doi:10.1002/eji.1830230508

Chen L, Shen Z, Wu J (2009) Expression, purification and in vitro antifungal activity of acidic mammalian chitinase against Candida albicans, Aspergillus fumigatus and Trichophyton rubrum strains. Clin Exp Dermatol 34(1):55–60. doi:10.1111/j.1365-2230.2008.03092.x

Cheng SC, van de Veerdonk F, Smeekens S, Joosten LA, van der Meer JW, Kullberg BJ, Netea MG (2010) Candida albicans dampens host defense by downregulating IL-17 production. J Immunol 185(4):2450–2457. doi:10.4049/jimmunol.1000756

Conti HR, Shen F, Nayyar N, Stocum E, Sun JN, Lindemann MJ, Ho AW, Hai JH, Yu JJ, Jung JW, Filler SG, Masso-Welch P, Edgerton M, Gaffen SL (2009) Th17 cells and IL-17 receptor signaling are essential for mucosal host defense against oral candidiasis. J Exp Med 206(2):299–311. doi:10.1084/jem.20081463

Conti HR, Peterson AC, Brane L, Huppler AR, Hernandez-Santos N, Whibley N, Garg AV, Simpson-Abelson MR, Gibson GA, Mamo AJ, Osborne LC, Bishu S, Ghilardi N, Siebenlist U, Watkins SC, Artis D, McGeachy MJ, Gaffen SL (2014) Oral-resident natural Th17 cells and gammadelta T cells control opportunistic Candida albicans infections. J Exp Med 211 (10):2075–2084. doi:10.1084/jem.20130877

De Bernardis F, Boccanera M, Adriani D, Spreghini E, Santoni G, Cassone A (1997) Protective role of antimannan and anti-aspartyl proteinase antibodies in an experimental model of Candida albicans vaginitis in rats. Infect Immunol 65(8):3399–3405

de Jong MA, Vriend LE, Theelen B, Taylor ME, Fluitsma D, Boekhout T, Geijtenbeek TB (2010) C-type lectin Langerin is a beta-glucan receptor on human Langerhans cells that recognizes opportunistic and pathogenic fungi. Mol Immunol 47(6):1216–1225. doi:10.1016/j.molimm.2009.12.016

De Luca A, Zelante T, D'Angelo C, Zagarella S, Fallarino F, Spreca A, Iannitti RG, Bonifazi P, Renauld JC, Bistoni F, Puccetti P, Romani L (2010) IL-22 defines a novel immune pathway of antifungal resistance. Mucosal Immunol 3(4):361–373. doi:10.1038/mi.2010.22

Delsing CE, Gresnigt MS, Leentjens J, Preijers F, Frager FA, Kox M, Monneret G, Venet F, Bleeker-Rovers CP, van de Veerdonk FL, Pickkers P, Pachot A, Kullberg BJ, Netea MG (2014) Interferon-gamma as adjunctive immunotherapy for invasive fungal infections: a case series. BMC Infect Dis 14:166. doi:10.1186/1471-2334-14-166

Dillon S, Agrawal S, Banerjee K, Letterio J, Denning TL, Oswald-Richter K, Kasprowicz DJ, Kellar K, Pare J, van Dyke T, Ziegler S, Unutmaz D, Pulendran B (2006) Yeast zymosan, a stimulus for TLR2 and Dectin-1, induces regulatory antigen-presenting cells and immunological tolerance. J Clin Invest 116(4):916–928. doi:10.1172/JCI27203

Douglas CM, D'Ippolito JA, Shei GJ, Meinz M, Onishi J, Marrinan JA, Li W, Abruzzo GK, Flattery A, Bartizal K, Mitchell A, Kurtz MB (1997) Identification of the FKS1 gene of Candida albicans as the essential target of 1,3-beta-D-glucan synthase inhibitors. Antimicrob Agents Chemother 41(11):2471–2479

Drewniak A, Gazendam RP, Tool AT, van Houdt M, Jansen MH, van Hamme JL, van Leeuwen EM, Roos D, Scalais E, de Beaufort C, Janssen H, van den Berg TK, Kuijpers TW (2013) Invasive fungal infection and impaired neutrophil killing in human CARD9 deficiency. Blood 121(13):2385–2392. doi:10.1182/blood-2012-08-450551

Ezekowitz RA, Sastry K, Bailly P, Warner A (1990) Molecular characterization of the human macrophage mannose receptor: demonstration of multiple carbohydrate recognition-like domains and phagocytosis of yeasts in Cos-1 cells. J Exp Med 172(6):1785–1794

Farah CS, Saunus JM, Hu Y, Kazoullis A, Ashman RB (2009) Gene targeting demonstrates that inducible nitric oxide synthase is not essential for resistance to oral candidiasis in mice, or for killing of Candida albicans by macrophages in vitro. Oral Microbiol Immunol 24(1):83–88. doi:10.1111/j.1399-302X.2008.00462.x

Fernandez-Arenas E, Bleck CK, Nombela C, Gil C, Griffiths G, Diez-Orejas R (2009) Candida albicans actively modulates intracellular membrane trafficking in mouse macrophage phagosomes. Cell Microbiol 11(4):560–589. doi:10.1111/j.1462-5822.2008.01274.x

Ferwerda B, Ferwerda G, Plantinga TS, Willment JA, van Spriel AB, Venselaar H, Elbers CC, Johnson MD, Cambi A, Huysamen C, Jacobs L, Jansen T, Verheijen K, Masthoff L, Morre SA, Vriend G, Williams DL, Perfect JR, Joosten LA, Wijmenga C, van der Meer JW, Adema GJ, Kullberg BJ, Brown GD, Netea MG (2009) Human Dectin-1 deficiency and mucocutaneous fungal infections. New Engl J Med 361(18):1760–1767. doi:10.1056/NEJMoa0901053

Flornes LM, Bryceson YT, Spurkland A, Lorentzen JC, Dissen E, Fossum S (2004) Identification of lectin-like receptors expressed by antigen presenting cells and neutrophils and their mapping to a novel gene complex. Immunogenetics 56(7):506–517. doi:10.1007/s00251-004-0714-x

Forsyth CB, Plow EF, Zhang L (1998) Interaction of the fungal pathogen Candida albicans with integrin CD11b/CD18: recognition by the I domain is modulated by the lectin-like domain and the CD18 subunit. J Immunol 161(11):6198–6205

Fradin C, De Groot P, MacCallum D, Schaller M, Klis F, Odds FC, Hube B (2005) Granulocytes govern the transcriptional response, morphology and proliferation of Candida albicans in human blood. Mol Microbiol 56(2):397–415. doi:10.1111/j.1365-2958.2005.04557.x

Frohner IE, Bourgeois C, Yatsyk K, Majer O, Kuchler K (2009) Candida albicans cell surface superoxide dismutases degrade host-derived reactive oxygen species to escape innate immune surveillance. Mol Microbiol 71(1):240–252. doi:10.1111/j.1365-2958.2008.06528.x

Gabrielli E, Pericolini E, Luciano E, Sabbatini S, Roselletti E, Perito S, Kasper L, Hube B, Vecchiarelli A (2015) Induction of caspase-11 by aspartyl proteinases of Candida albicans and implication in promoting inflammatory response. Infect Immunol 83(5):1940–1948. doi:10.1128/IAI.02895-14

Ganesan S, Rathinam VA, Bossaller L, Army K, Kaiser WJ, Mocarski ES, Dillon CP, Green DR, Mayadas TN, Levitz SM, Hise AG, Silverman N, Fitzgerald KA (2014) Caspase-8 modulates Dectin-1 and complement receptor 3-driven IL-1beta production in response to beta-glucans and the fungal pathogen, Candida albicans. J Immunol 193(5):2519–2530. doi:10.4049/jimmunol.1400276

Gantner BN, Simmons RM, Underhill DM (2005) Dectin-1 mediates macrophage recognition of Candida albicans yeast but not filaments. EMBO J 24(6):1277–1286. doi:10.1038/sj.emboj.7600594

Gazendam RP, van Hamme JL, Tool AT, van Houdt M, Verkuijlen PJ, Herbst M, Liese JG, van de Veerdonk FL, Roos D, van den Berg TK, Kuijpers TW (2014) Two independent killing mechanisms of Candida albicans by human neutrophils: evidence from innate immunity defects. Blood 124(4):590–597. doi:10.1182/blood-2014-01-551473

Gil ML, Gozalbo D (2006) TLR2, but not TLR4, triggers cytokine production by murine cells in response to Candida albicans yeasts and hyphae. Microbes Infect/Institut Pasteur 8(8):2299–2304. doi:10.1016/j.micinf.2006.03.014

Gladiator A, Wangler N, Trautwein-Weidner K, LeibundGut-Landmann S (2013) Cutting edge: IL-17-secreting innate lymphoid cells are essential for host defense against fungal infection. J Immunol 190(2):521–525. doi:10.4049/jimmunol.1202924

Glocker EO, Hennigs A, Nabavi M, Schaffer AA, Woellner C, Salzer U, Pfeifer D, Veelken H, Warnatz K, Tahami F, Jamal S, Manguiat A, Rezaei N, Amirzargar AA, Plebani A, Hannesschlager N, Gross O, Ruland J, Grimbacher B (2009) A homozygous CARD9 mutation in a family with susceptibility to fungal infections. New Engl J Med 361(18):1727–1735. doi:10.1056/NEJMoa0810719

Gow NA, Hube B (2012) Importance of the Candida albicans cell wall during commensalism and infection. Curr Opin Microbiol 15(4):406–412. doi:10.1016/j.mib.2012.04.005

Gow NA, Netea MG, Munro CA, Ferwerda G, Bates S, Mora-Montes HM, Walker L, Jansen T, Jacobs L, Tsoni V, Brown GD, Odds FC, Van der Meer JW, Brown AJ, Kullberg BJ (2007) Immune recognition of Candida albicans beta-glucan by Dectin-1. J Infect Dis 196(10):1565–1571. doi:10.1086/523110

Gringhuis SI, den Dunnen J, Litjens M, van Het Hof B, van Kooyk Y, Geijtenbeek TB (2007) C-type lectin DC-SIGN modulates toll-like receptor signaling via Raf-1 kinase-dependent acetylation of transcription factor NF-kappaB. Immunity 26(5):605–616. doi:10.1016/j.immuni.2007.03.012

Gringhuis SI, den Dunnen J, Litjens M, van der Vlist M, Wevers B, Bruijns SC, Geijtenbeek TB (2009) Dectin-1 directs T helper cell differentiation by controlling noncanonical NF-kappaB activation through Raf-1 and Syk. Nat Immunol 10(2):203–213. doi:10.1038/ni.1692

Gringhuis SI, Kaptein TM, Wevers BA, Theelen B, van der Vlist M, Boekhout T, Geijtenbeek TB (2012) Dectin-1 is an extracellular pathogen sensor for the induction and processing of IL-1beta via a noncanonical caspase-8 inflammasome. Nat Immunol 13(3):246–254. doi:10.1038/ni.2222

Gropp K, Schild L, Schindler S, Hube B, Zipfel PF, Skerka C (2009) The yeast Candida albicans evades human complement attack by secretion of aspartic proteases. Mol Immunol 47(2–3):465–475. doi:10.1016/j.molimm.2009.08.019

Gross O, Gewies A, Finger K, Schafer M, Sparwasser T, Peschel C, Forster I, Ruland J (2006) Card9 controls a non-TLR signalling pathway for innate anti-fungal immunity. Nature 442 (7103):651–656. doi:10.1038/nature04926

Gross O, Poeck H, Bscheider M, Dostert C, Hannesschlager N, Endres S, Hartmann G, Tardivel A, Schweighoffer E, Tybulewicz V, Mocsai A, Tschopp J, Ruland J (2009) Syk kinase signalling couples to the Nlrp3 inflammasome for anti-fungal host defence. Nature 459(7245):433–436. doi:10.1038/nature07965

Hall RA, Bates S, Lenardon MD, Maccallum DM, Wagener J, Lowman DW, Kruppa MD, Williams DL, Odds FC, Brown AJ, Gow NA (2013) The Mnn2 mannosyltransferase family modulates mannoprotein fibril length, immune recognition and virulence of Candida albicans. PLoS Pathog 9(4):e1003276. doi:10.1371/journal.ppat.1003276

Han Y, Cutler JE (1995) Antibody response that protects against disseminated candidiasis. Infect Immunol 63(7):2714–2719

Han Y, Morrison RP, Cutler JE (1998) A vaccine and monoclonal antibodies that enhance mouse resistance to Candida albicans vaginal infection. Infect Immunol 66(12):5771–5776

Haraguchi N, Ishii Y, Morishima Y, Yoh K, Matsuno Y, Kikuchi N, Sakamoto T, Takahashi S, Hizawa N (2010) Impairment of host defense against disseminated candidiasis in mice overexpressing GATA-3. Infect Immunol 78(5):2302–2311. doi:10.1128/IAI.01398-09

Hardison SE, Brown GD (2012) C-type lectin receptors orchestrate antifungal immunity. Nat Immunol 13(9):817–822. doi:10.1038/ni.2369

Heinsbroek SE, Taylor PR, Martinez FO, Martinez-Pomares L, Brown GD, Gordon S (2008) Stage-specific sampling by pattern recognition receptors during Candida albicans phagocytosis. PLoS Pathog 4(11):e1000218. doi:10.1371/journal.ppat.1000218

Held K, Thiel S, Loos M, Petry F (2008) Increased susceptibility of complement factor B/C2 double knockout mice and mannan-binding lectin knockout mice to systemic infection with Candida albicans. Mol Immunol 45(15):3934–3941. doi:10.1016/j.molimm.2008.06.021

Hise AG, Tomalka J, Ganesan S, Patel K, Hall BA, Brown GD, Fitzgerald KA (2009) An essential role for the NLRP3 inflammasome in host defense against the human fungal pathogen Candida albicans. Cell Host Microbe 5(5):487–497. doi:10.1016/j.chom.2009.05.002

Hopke A, Nicke N, Hidu EE, Degani G, Popolo L, Wheeler RT (2016) Neutrophil attack triggers extracellular trap-dependent candida cell wall remodeling and altered immune recognition. PLoS Pathog 12(5):e1005644. doi:10.1371/journal.ppat.1005644

Horn DL, Neofytos D, Anaissie EJ, Fishman JA, Steinbach WJ, Olyaei AJ, Marr KA, Pfaller MA, Chang CH, Webster KM (2009) Epidemiology and outcomes of candidemia in 2019 patients: data from the prospective antifungal therapy alliance registry. Clin Infect Dis Off Publ Infect Dis Soc Am 48(12):1695–1703. doi:10.1086/599039

Huang W, Na L, Fidel PL, Schwarzenberger P (2004) Requirement of interleukin-17A for systemic anti-Candida albicans host defense in mice. J Infect Dis 190(3):624–631. doi:10.1086/422329

Huang Y, Guo L, Qiu J, Chen X, Hu-Li J, Siebenlist U, Williamson PR, Urban JF Jr, Paul WE (2015) IL-25-responsive, lineage-negative KLRG1(hi) cells are multipotential 'inflammatory' type 2 innate lymphoid cells. Nat Immunol 16(2):161–169. doi:10.1038/ni.3078

Huppler AR, Conti HR, Hernandez-Santos N, Darville T, Biswas PS, Gaffen SL (2014) Role of neutrophils in IL-17-dependent immunity to mucosal candidiasis. J Immunol 192(4):1745–1752. doi:10.4049/jimmunol.1302265

Ifrim DC, Quintin J, Courjol F, Verschueren I, van Krieken JH, Koentgen F, Fradin C, Gow NA, Joosten LA, van der Meer JW, van de Veerdonk F, Netea MG (2016) The role of Dectin-2 for host defense against disseminated Candidiasis. J Interferon Cytokine Res Off J Int Soc Interferon Cytokine Res 36(4):267–276. doi:10.1089/jir.2015.0040

Igyarto BZ, Haley K, Ortner D, Bobr A, Gerami-Nejad M, Edelson BT, Zurawski SM, Malissen B, Zurawski G, Berman J, Kaplan DH (2011) Skin-resident murine dendritic cell subsets promote distinct and opposing antigen-specific T helper cell responses. Immunity 35(2):260–272. doi:10.1016/j.immuni.2011.06.005

Jaeger M, van der Lee R, Cheng SC, Johnson MD, Kumar V, Ng A, Plantinga TS, Smeekens SP, Oosting M, Wang X, Barchet W, Fitzgerald K, Joosten LA, Perfect JR, Wijmenga C, van de Veerdonk FL, Huynen MA, Xavier RJ, Kullberg BJ, Netea MG (2015) The RIG-I-like helicase receptor MDA5 (IFIH1) is involved in the host defense against Candida infections. Eur J Clin Microbiol Infect Dis: off Publ Eur Soc Clin Microbiol 34(5):963–974. doi:10.1007/s10096-014-2309-2

Jensen J, Warner T, Balish E (1993) Resistance of SCID mice to Candida albicans administered intravenously or colonizing the gut: role of polymorphonuclear leukocytes and macrophages. J Infect Dis 167(4):912–919

Joly S, Ma N, Sadler JJ, Soll DR, Cassel SL, Sutterwala FS (2009) Cutting edge: Candida albicans hyphae formation triggers activation of the Nlrp3 inflammasome. J Immunol 183(6):3578–3581. doi:10.4049/jimmunol.0901323

Joly S, Eisenbarth SC, Olivier AK, Williams A, Kaplan DH, Cassel SL, Flavell RA, Sutterwala FS (2012) Cutting edge: Nlrp10 is essential for protective antifungal adaptive immunity against Candida albicans. J Immunol 189(10):4713–4717. doi:10.4049/jimmunol.1201715

Jouault T, Ibata-Ombetta S, Takeuchi O, Trinel PA, Sacchetti P, Lefebvre P, Akira S, Poulain D (2003) Candida albicans phospholipomannan is sensed through toll-like receptors. J Infect Dis 188(1):165–172. doi:10.1086/375784

Jouault T, El Abed-El Behi M, Martinez-Esparza M, Breuilh L, Trinel PA, Chamaillard M, Trottein F, Poulain D (2006) Specific recognition of Candida albicans by macrophages requires galectin-3 to discriminate Saccharomyces cerevisiae and needs association with TLR2 for signaling. J Immunol 177(7):4679–4687

Kagami S, Rizzo HL, Kurtz SE, Miller LS, Blauvelt A (2010) IL-23 and IL-17A, but not IL-12 and IL-22, are required for optimal skin host defense against Candida albicans. J Immunol 185 (9):5453–5462. doi:10.4049/jimmunol.1001153

Kashem SW, Igyarto BZ, Gerami-Nejad M, Kumamoto Y, Mohammed J, Jarrett E, Drummond RA, Zurawski SM, Zurawski G, Berman J, Iwasaki A, Brown GD, Kaplan DH (2015) Candida albicans morphology and dendritic cell subsets determine T helper cell differentiation. Immunity 42(2):356–366. doi:10.1016/j.immuni.2015.01.008

Kohatsu L, Hsu DK, Jegalian AG, Liu FT, Baum LG (2006) Galectin-3 induces death of Candida species expressing specific beta-1,2-linked mannans. J Immunol 177(7):4718–4726

Komalapriya C, Kaloriti D, Tillmann AT, Yin Z, Herrero-de-Dios C, Jacobsen MD, Belmonte RC, Cameron G, Haynes K, Grebogi C, de Moura AP, Gow NA, Thiel M, Quinn J, Brown AJ, Romano MC (2015) integrative model of oxidative stress adaptation in the fungal pathogen Candida albicans. PLoS ONE 10(9):e0137750. doi:10.1371/journal.pone.0137750

Kullberg BJ, van't Wout JW, van Furth R (1990) Role of granulocytes in increased host resistance to Candida albicans induced by recombinant interleukin-1. Infection and immunity 58 (10):3319–3324

Lanternier F, Mahdaviani SA, Barbati E, Chaussade H, Koumar Y, Levy R, Denis B, Brunel AS, Martin S, Loop M, Peeters J, de Selys A, Vanclaire J, Vermylen C, Nassogne MC, Chatzis O, Liu L, Migaud M, Pedergnana V, Desoubeaux G, Jouvion G, Chretien F, Darazam IA, Schaffer AA, Netea MG, De Bruycker JJ, Bernard L, Reynes J, Amazrine N, Abel L, Van der Linden D, Harrison T, Picard C, Lortholary O, Mansouri D, Casanova JL, Puel A (2015) Inherited CARD9 deficiency in otherwise healthy children and adults with Candida species-induced meningoencephalitis, colitis, or both. J Allergy Clin Immunol 135(6):1558–1568 e1552. doi:10.1016/j.jaci.2014.12.1930

Lee SJ, Zheng NY, Clavijo M, Nussenzweig MC (2003) Normal host defense during systemic candidiasis in mannose receptor-deficient mice. Infect Immunol 71(1):437–445

Lehrer RI, Ganz T, Szklarek D, Selsted ME (1988) Modulation of the in vitro candidacidal activity of human neutrophil defensins by target cell metabolism and divalent cations. J Clin Invest 81 (6):1829–1835. doi:10.1172/JCI113527

LeibundGut-Landmann S, Gross O, Robinson MJ, Osorio F, Slack EC, Tsoni SV, Schweighoffer E, Tybulewicz V, Brown GD, Ruland J, Reis e Sousa C (2007) Syk- and CARD9-dependent coupling of innate immunity to the induction of T helper cells that produce interleukin 17. Nat Immunol 8(6):630–638. doi:10.1038/ni1460

Lev-Sagie A, Prus D, Linhares IM, Lavy Y, Ledger WJ, Witkin SS (2009) Polymorphism in a gene coding for the inflammasome component NALP3 and recurrent vulvovaginal candidiasis in women with vulvar vestibulitis syndrome. Am J Obstet Gynecol 200(3):303 e301–306. doi:10.1016/j.ajog.2008.10.039

Li D, Dong B, Tong Z, Wang Q, Liu W, Wang Y, Liu W, Chen J, Xu L, Chen L, Duan Y (2012) MBL-mediated opsonophagocytosis of Candida albicans by human neutrophils is coupled with intracellular Dectin-1-triggered ROS production. PLoS ONE 7(12):e50589. doi:10.1371/journal.pone.0050589

Li SS, Kyei SK, Timm-McCann M, Ogbomo H, Jones GJ, Shi M, Xiang RF, Oykhman P, Huston SM, Islam A, Gill MJ, Robbins SM, Mody CH (2013) The NK receptor NKp30 mediates direct fungal recognition and killing and is diminished in NK cells from HIV-infected patients. Cell Host Microbe 14(4):387–397. doi:10.1016/j.chom.2013.09.007

Lillegard JB, Sim RB, Thorkildson P, Gates MA, Kozel TR (2006) Recognition of Candida albicans by mannan-binding lectin in vitro and in vivo. J Infect Dis 193(11):1589–1597. doi:10.1086/503804

Linden JR, De Paepe ME, Laforce-Nesbitt SS, Bliss JM (2013) Galectin-3 plays an important role in protection against disseminated candidiasis. Med Mycol 51(6):641–651. doi:10.3109/13693786.2013.770607

Linehan SA, Martinez-Pomares L, Stahl PD, Gordon S (1999) Mannose receptor and its putative ligands in normal murine lymphoid and nonlymphoid organs: In situ expression of mannose receptor by selected macrophages, endothelial cells, perivascular microglia, and mesangial cells, but not dendritic cells. J Exp Med 189(12):1961–1972

Lo HJ, Kohler JR, DiDomenico B, Loebenberg D, Cacciapuoti A, Fink GR (1997) Nonfilamentous C. albicans mutants are avirulent. Cell 90(5):939–949

Lopez-Garcia B, Lee PH, Yamasaki K, Gallo RL (2005) Anti-fungal activity of cathelicidins and their potential role in Candida albicans skin infection. J Invest Dermatol 125(1):108–115. doi:10.1111/j.0022-202X.2005.23713.x

Luo S, Poltermann S, Kunert A, Rupp S, Zipfel PF (2009) Immune evasion of the human pathogenic yeast Candida albicans: Pra1 is a Factor H, FHL-1 and plasminogen binding surface protein. Mol Immunol 47(2–3):541–550. doi:10.1016/j.molimm.2009.07.017

Luo S, Hartmann A, Dahse HM, Skerka C, Zipfel PF (2010) Secreted pH-regulated antigen 1 of Candida albicans blocks activation and conversion of complement C3. J Immunol 185 (4):2164–2173. doi:10.4049/jimmunol.1001011

Luo S, Blom AM, Rupp S, Hipler UC, Hube B, Skerka C, Zipfel PF (2011) The pH-regulated antigen 1 of Candida albicans binds the human complement inhibitor C4b-binding protein and mediates fungal complement evasion. J Biol Chem 286(10):8021–8029. doi:10.1074/jbc.M110.130138

Marakalala MJ, Vautier S, Potrykus J, Walker LA, Shepardson KM, Hopke A, Mora-Montes HM, Kerrigan A, Netea MG, Murray GI, Maccallum DM, Wheeler R, Munro CA, Gow NA, Cramer RA, Brown AJ, Brown GD (2013) Differential adaptation of Candida albicans in vivo modulates immune recognition by Dectin-1. PLoS Pathog 9(4):e1003315. doi:10.1371/journal.ppat.1003315

Marcil A, Harcus D, Thomas DY, Whiteway M (2002) Candida albicans killing by RAW 264.7 mouse macrophage cells: effects of Candida genotype, infection ratios, and gamma interferon treatment. Infect Immunol 70(11):6319–6329

Matthews RC, Burnie JP, Howat D, Rowland T, Walton F (1991) Autoantibody to heat-shock protein 90 can mediate protection against systemic candidosis. Immunology 74(1):20–24

McGreal EP, Rosas M, Brown GD, Zamze S, Wong SY, Gordon S, Martinez-Pomares L, Taylor PR (2006) The carbohydrate-recognition domain of Dectin-2 is a C-type lectin with specificity for high mannose. Glycobiology 16(5):422–430. doi:10.1093/glycob/cwj077

McKenzie EJ, Taylor PR, Stillion RJ, Lucas AD, Harris J, Gordon S, Martinez-Pomares L (2007) Mannose receptor expression and function define a new population of murine dendritic cells. J Immunol 178(8):4975–4983

McKenzie CG, Koser U, Lewis LE, Bain JM, Mora-Montes HM, Barker RN, Gow NA, Erwig LP (2010) Contribution of Candida albicans cell wall components to recognition by and escape from murine macrophages. Infect Immunol 78(4):1650–1658. doi:10.1128/IAI.00001-10

Means TK, Mylonakis E, Tampakakis E, Colvin RA, Seung E, Puckett L, Tai MF, Stewart CR, Pukkila-Worley R, Hickman SE, Moore KJ, Calderwood SB, Hacohen N, Luster AD, El Khoury J (2009) Evolutionarily conserved recognition and innate immunity to fungal pathogens by the scavenger receptors SCARF1 and CD36. J Exp Med 206(3):637–653. doi:10.1084/jem.20082109

Medzhitov R (2007) Recognition of microorganisms and activation of the immune response. Nature 449(7164):819–826. doi:10.1038/nature06246

Mencacci A, Perruccio K, Bacci A, Cenci E, Benedetti R, Martelli MF, Bistoni F, Coffman R, Velardi A, Romani L (2001) Defective antifungal T-helper 1 (TH1) immunity in a murine model of allogeneic T-cell-depleted bone marrow transplantation and its restoration by treatment with TH2 cytokine antagonists. Blood 97(5):1483–1490

Milner JD, Brenchley JM, Laurence A, Freeman AF, Hill BJ, Elias KM, Kanno Y, Spalding C, Elloumi HZ, Paulson ML, Davis J, Hsu A, Asher AI, O'Shea J, Holland SM, Paul WE, Douek DC (2008) Impaired T(H)17 cell differentiation in subjects with autosomal dominant hyper-IgE syndrome. Nature 452(7188):773–776. doi:10.1038/nature06764

Moragues MD, Omaetxebarria MJ, Elguezabal N, Sevilla MJ, Conti S, Polonelli L, Ponton J (2003) A monoclonal antibody directed against a Candida albicans cell wall mannoprotein exerts three anti-C. albicans activities. Infect Immunol 71(9):5273–5279

Moyes DL, Runglall M, Murciano C, Shen C, Nayar D, Thavaraj S, Kohli A, Islam A, Mora-Montes H, Challacombe SJ, Naglik JR (2010) A biphasic innate immune MAPK response discriminates between the yeast and hyphal forms of Candida albicans in epithelial cells. Cell Host Microbe 8(3):225–235. doi:10.1016/j.chom.2010.08.002

Moyes DL, Murciano C, Runglall M, Islam A, Thavaraj S, Naglik JR (2011) Candida albicans yeast and hyphae are discriminated by MAPK signaling in vaginal epithelial cells. PLoS ONE 6(11):e26580. doi:10.1371/journal.pone.0026580

Moyes DL, Wilson D, Richardson JP, Mogavero S, Tang SX, Wernecke J, Hofs S, Gratacap RL, Robbins J, Runglall M, Murciano C, Blagojevic M, Thavaraj S, Forster TM, Hebecker B, Kasper L, Vizcay G, Iancu SI, Kichik N, Hader A, Kurzai O, Luo T, Kruger T, Kniemeyer O, Cota E, Bader O, Wheeler RT, Gutsmann T, Hube B, Naglik JR (2016) Candidalysin is a fungal peptide toxin critical for mucosal infection. Nature 532(7597):64–68. doi:10.1038/nature17625

Nahum A, Dadi H, Bates A, Roifman CM (2011) The L412F variant of toll-like receptor 3 (TLR3) is associated with cutaneous candidiasis, increased susceptibility to cytomegalovirus, and autoimmunity. J Allergy Clin Immunol 127(2):528–531. doi:10.1016/j.jaci.2010.09.031

Nahum A, Dadi H, Bates A, Roifman CM (2012) The biological significance of TLR3 variant, L412F, in conferring susceptibility to cutaneous candidiasis, CMV and autoimmunity. Autoimmun Rev 11(5):341–347. doi:10.1016/j.autrev.2011.10.007

Nakagawa Y, Kanbe T, Mizuguchi I (2003) Disruption of the human pathogenic yeast Candida albicans catalase gene decreases survival in mouse-model infection and elevates susceptibility to higher temperature and to detergents. Microbiol Immunol 47(6):395–403

Nathan CF, Murray HW, Wiebe ME, Rubin BY (1983) Identification of interferon-gamma as the lymphokine that activates human macrophage oxidative metabolism and antimicrobial activity. J Exp Med 158(3):670–689

Netea MG, Van Der Graaf CA, Vonk AG, Verschueren I, Van Der Meer JW, Kullberg BJ (2002) The role of toll-like receptor (TLR) 2 and TLR4 in the host defense against disseminated candidiasis. J Infect Dis 185(10):1483–1489. doi:10.1086/340511

Netea MG, Vonk AG, van den Hoven M, Verschueren I, Joosten LA, van Krieken JH, van den Berg WB, Van der Meer JW, Kullberg BJ (2003) Differential role of IL-18 and IL-12 in the host defense against disseminated Candida albicans infection. Eur J Immunol 33(12):3409–3417. doi:10.1002/eji.200323737

Netea MG, Gijzen K, Coolen N, Verschueren I, Figdor C, Van der Meer JW, Torensma R, Kullberg BJ (2004a) Human dendritic cells are less potent at killing Candida albicans than both monocytes and macrophages. Microbes infect/Institut Pasteur 6(11):985–989. doi:10.1016/j.micinf.2004.05.013

Netea MG, Sutmuller R, Hermann C, Van der Graaf CA, Van der Meer JW, van Krieken JH, Hartung T, Adema G, Kullberg BJ (2004b) Toll-like receptor 2 suppresses immunity against Candida albicans through induction of IL-10 and regulatory T cells. J Immunol 172(6):3712–3718

Netea MG, Gow NA, Munro CA, Bates S, Collins C, Ferwerda G, Hobson RP, Bertram G, Hughes HB, Jansen T, Jacobs L, Buurman ET, Gijzen K, Williams DL, Torensma R, McKinnon A, MacCallum DM, Odds FC, Van der Meer JW, Brown AJ, Kullberg BJ (2006) Immune sensing of Candida albicans requires cooperative recognition of mannans and glucans by lectin and toll-like receptors. J Clin Investig 116(6):1642–1650. doi:10.1172/JCI27114

Netea MG, van de Veerdonk F, Verschueren I, van der Meer JW, Kullberg BJ (2008) Role of TLR1 and TLR6 in the host defense against disseminated candidiasis. FEMS Immunol Med Microbiol 52(1):118–123. doi:10.1111/j.1574-695X.2007.00353.x

Netea MG, Gow NA, Joosten LA, Verschueren I, van der Meer JW, Kullberg BJ (2010) Variable recognition of Candida albicans strains by TLR4 and lectin recognition receptors. Med Mycol 48(7):897–903. doi:10.3109/13693781003621575

Ngo LY, Kasahara S, Kumasaka DK, Knoblaugh SE, Jhingran A, Hohl TM (2014) Inflammatory monocytes mediate early and organ-specific innate defense during systemic candidiasis. J Infect Dis 209(1):109–119. doi:10.1093/infdis/jit413

Okada S, Markle JG, Deenick EK, Mele F, Averbuch D, Lagos M, Alzahrani M, Al-Muhsen S, Halwani R, Ma CS, Wong N, Soudais C, Henderson LA, Marzouqa H, Shamma J, Gonzalez M, Martinez-Barricarte R, Okada C, Avery DT, Latorre D, Deswarte C, Jabot-Hanin F, Torrado E, Fountain J, Belkadi A, Itan Y, Boisson B, Migaud M, Arlehamn CS, Sette A, Breton S, McCluskey J, Rossjohn J, de Villartay JP, Moshous D, Hambleton S, Latour S, Arkwright PD, Picard C, Lantz O, Engelhard D, Kobayashi M, Abel L, Cooper AM, Notarangelo LD, Boisson-Dupuis S, Puel A, Sallusto F, Bustamante J, Tangye SG, Casanova JL (2015) IMMUNODEFICIENCIES. Impairment of immunity to Candida and mycobacterium in humans with bi-allelic RORC mutations. Science 349(6248):606–613. doi:10.1126/science.aaa4282

Pandiyan P, Conti HR, Zheng L, Peterson AC, Mathern DR, Hernandez-Santos N, Edgerton M, Gaffen SL, Lenardo MJ (2011) CD4(+)CD25(+)Foxp3(+) regulatory T cells promote Th17 cells in vitro and enhance host resistance in mouse Candida albicans Th17 cell infection model. Immunity 34(3):422–434. doi:10.1016/j.immuni.2011.03.002

Phan QT, Belanger PH, Filler SG (2000) Role of hyphal formation in interactions of Candida albicans with endothelial cells. Infect Immunol 68(6):3485–3490

Phan QT, Fratti RA, Prasadarao NV, Edwards JE Jr, Filler SG (2005) N-cadherin mediates endocytosis of Candida albicans by endothelial cells. J Biol Chem 280(11):10455–10461. doi:10.1074/jbc.M412592200

Phan QT, Myers CL, Fu Y, Sheppard DC, Yeaman MR, Welch WH, Ibrahim AS, Edwards JE Jr, Filler SG (2007) Als3 is a Candida albicans invasin that binds to cadherins and induces endocytosis by host cells. PLoS Biol 5(3):e64. doi:10.1371/journal.pbio.0050064

Picard C, Puel A, Bonnet M, Ku CL, Bustamante J, Yang K, Soudais C, Dupuis S, Feinberg J, Fieschi C, Elbim C, Hitchcock R, Lammas D, Davies G, Al-Ghonaium A, Al-Rayes H, Al-Jumaah S, Al-Hajjar S, Al-Mohsen IZ, Frayha HH, Rucker R, Hawn TR, Aderem A, Tufenkeji H, Haraguchi S, Day NK, Good RA, Gougerot-Pocidalo MA, Ozinsky A, Casanova JL (2003) Pyogenic bacterial infections in humans with IRAK-4 deficiency. Science 299(5615):2076–2079. doi:10.1126/science.1081902

Pietrella D, Pandey N, Gabrielli E, Pericolini E, Perito S, Kasper L, Bistoni F, Cassone A, Hube B, Vecchiarelli A (2013) Secreted aspartic proteases of Candida albicans activate the NLRP3 inflammasome. Eur J Immunol 43(3):679–692. doi:10.1002/eji.201242691

Plantinga TS, van der Velden WJ, Ferwerda B, van Spriel AB, Adema G, Feuth T, Donnelly JP, Brown GD, Kullberg BJ, Blijlevens NM, Netea MG (2009) Early stop polymorphism in human Dectin-1 is associated with increased candida colonization in hematopoietic stem cell transplant recipients. Clin Infect Dis Off Publ Infect Dis Soc Am 49(5):724–732. doi:10.1086/604714

Plantinga TS, Johnson MD, Scott WK, van de Vosse E, Velez Edwards DR, Smith PB, Alexander BD, Yang JC, Kremer D, Laird GM, Oosting M, Joosten LA, van der Meer JW, van Dissel JT, Walsh TJ, Perfect JR, Kullberg BJ, Netea MG (2012) Toll-like receptor 1 polymorphisms increase susceptibility to candidemia. J Infect Dis 205(6):934–943. doi:10.1093/infdis/jir867

Puel A, Cypowyj S, Bustamante J, Wright JF, Liu L, Lim HK, Migaud M, Israel L, Chrabieh M, Audry M, Gumbleton M, Toulon A, Bodemer C, El-Baghdadi J, Whitters M, Paradis T, Brooks J, Collins M, Wolfman NM, Al-Muhsen S, Galicchio M, Abel L, Picard C, Casanova JL (2011) Chronic mucocutaneous candidiasis in humans with inborn errors of interleukin-17 immunity. Science 332(6025):65–68. doi:10.1126/science.1200439

Qian Q, Jutila MA, Van Rooijen N, Cutler JE (1994) Elimination of mouse splenic macrophages correlates with increased susceptibility to experimental disseminated candidiasis. J Immunol 152(10):5000–5008

Quintin J, Voigt J, van der Voort R, Jacobsen ID, Verschueren I, Hube B, Giamarellos-Bourboulis EJ, van der Meer JW, Joosten LA, Kurzai O, Netea MG (2014) Differential role of NK cells against Candida albicans infection in immunocompetent or immunocompromised mice. Eur J Immunol 44(8):2405–2414. doi:10.1002/eji.201343828

Reales-Calderon JA, Aguilera-Montilla N, Corbi AL, Molero G, Gil C (2014) Proteomic characterization of human proinflammatory M1 and anti-inflammatory M2 macrophages and their response to Candida albicans. Proteomics 14(12):1503–1518. doi:10.1002/pmic.201300508

Robinson MJ, Osorio F, Rosas M, Freitas RP, Schweighoffer E, Gross O, Verbeek JS, Ruland J, Tybulewicz V, Brown GD, Moita LF, Taylor PR, Reis e Sousa C (2009) Dectin-2 is a Syk-coupled pattern recognition receptor crucial for Th17 responses to fungal infection. J Exp Med 206(9):2037–2051. doi:10.1084/jem.20082818

Romani L, Puccetti P, Mencacci A, Cenci E, Spaccapelo R, Tonnetti L, Grohmann U, Bistoni F (1994) Neutralization of IL-10 up-regulates nitric oxide production and protects susceptible mice from challenge with Candida albicans. J Immunol 152(7):3514–3521

Rosentul DC, Plantinga TS, Oosting M, Scott WK, Velez Edwards DR, Smith PB, Alexander BD, Yang JC, Laird GM, Joosten LA, van der Meer JW, Perfect JR, Kullberg BJ, Netea MG, Johnson MD (2011) Genetic variation in the Dectin-1/CARD9 recognition pathway and susceptibility to candidemia. J Infect Dis 204(7):1138–1145. doi:10.1093/infdis/jir458

Rosentul DC, Delsing CE, Jaeger M, Plantinga TS, Oosting M, Costantini I, Venselaar H, Joosten LA, van der Meer JW, Dupont B, Kullberg BJ, Sobel JD, Netea MG (2014) Gene polymorphisms in pattern recognition receptors and susceptibility to idiopathic recurrent vulvovaginal candidiasis. Front Microbiol 5:483. doi:10.3389/fmicb.2014.00483

Rosseau S, Guenther A, Seeger W, Lohmeyer J (1997) Phagocytosis of viable Candida albicans by alveolar macrophages: lack of opsonin function of surfactant protein A. J Infect Dis 175 (2):421–428

Rosseau S, Hammerl P, Maus U, Gunther A, Seeger W, Grimminger F, Lohmeyer J (1999) Surfactant protein A down-regulates proinflammatory cytokine production evoked by Candida albicans in human alveolar macrophages and monocytes. J Immunol 163(8):4495–4502

Saijo S, Fujikado N, Furuta T, Chung SH, Kotaki H, Seki K, Sudo K, Akira S, Adachi Y, Ohno N, Kinjo T, Nakamura K, Kawakami K, Iwakura Y (2007) Dectin-1 is required for host defense against Pneumoc carinii but not against Candida albicans. Nat Immunol 8(1):39–46. doi:10.1038/ni1425

Saijo S, Ikeda S, Yamabe K, Kakuta S, Ishigame H, Akitsu A, Fujikado N, Kusaka T, Kubo S, Chung SH, Komatsu R, Miura N, Adachi Y, Ohno N, Shibuya K, Yamamoto N, Kawakami K, Yamasaki S, Saito T, Akira S, Iwakura Y (2010) Dectin-2 recognition of alpha-mannans and induction of Th17 cell differentiation is essential for host defense against Candida albicans. Immunity 32(5):681–691. doi:10.1016/j.immuni.2010.05.001

Sandini S, La Valle R, Deaglio S, Malavasi F, Cassone A, De Bernardis F (2011) A highly immunogenic recombinant and truncated protein of the secreted aspartic proteases family (rSap2t) of Candida albicans as a mucosal anticandidal vaccine. FEMS Immunol Med Microbiol 62(2):215–224. doi:10.1111/j.1574-695X.2011.00802.x

Sato K, Yang XL, Yudate T, Chung JS, Wu J, Luby-Phelps K, Kimberly RP, Underhill D, Cruz PD Jr, Ariizumi K (2006) Dectin-2 is a pattern recognition receptor for fungi that couples with the Fc receptor gamma chain to induce innate immune responses. J Biol Chem 281 (50):38854–38866. doi:10.1074/jbc.M606542200

Schmidt CS, White CJ, Ibrahim AS, Filler SG, Fu Y, Yeaman MR, Edwards JE Jr, Hennessey JP Jr (2012) NDV-3, a recombinant alum-adjuvanted vaccine for Candida and Staphylococcus aureus, is safe and immunogenic in healthy adults. Vaccine 30(52):7594–7600. doi:10.1016/j.vaccine.2012.10.038

Shepherd VL, Tarnowski BI, McLaughlin BJ (1991) Isolation and characterization of a mannose receptor from human pigment epithelium. Invest Ophthalmol Vis Sci 32(6):1779–1784

Sohnle PG, Hahn BL, Santhanagopalan V (1996) Inhibition of Candida albicans growth by calprotectin in the absence of direct contact with the organisms. J Infect Dis 174(6):1369–1372

Sorensen OE, Follin P, Johnsen AH, Calafat J, Tjabringa GS, Hiemstra PS, Borregaard N (2001) Human cathelicidin, hCAP-18, is processed to the antimicrobial peptide LL-37 by extracellular cleavage with proteinase 3. Blood 97(12):3951–3959

Stuyt RJ, Netea MG, van Krieken JH, van der Meer JW, Kullberg BJ (2004) Recombinant interleukin-18 protects against disseminated Candida albicans infection in mice. J Infect Dis 189(8):1524–1527. doi:10.1086/382955

Sutmuller RP, den Brok MH, Kramer M, Bennink EJ, Toonen LW, Kullberg BJ, Joosten LA, Akira S, Netea MG, Adema GJ (2006) Toll-like receptor 2 controls expansion and function of regulatory T cells. J Clin Invest 116(2):485–494. doi:10.1172/JCI25439

Tada H, Nemoto E, Shimauchi H, Watanabe T, Mikami T, Matsumoto T, Ohno N, Tamura H, Shibata K, Akashi S, Miyake K, Sugawara S, Takada H (2002) Saccharomyces cerevisiae- and Candida albicans-derived mannan induced production of tumor necrosis factor alpha by human monocytes in a CD14- and Toll-like receptor 4-dependent manner. Microbiol Immunol 46 (7):503–512

Takahara K, Yashima Y, Omatsu Y, Yoshida H, Kimura Y, Kang YS, Steinman RM, Park CG, Inaba K (2004) Functional comparison of the mouse DC-SIGN, SIGNR1, SIGNR3 and Langerin, C-type lectins. Int Immunol 16(6):819–829. doi:10.1093/intimm/dxh084

Takahara K, Tokieda S, Nagaoka K, Takeda T, Kimura Y, Inaba K (2011) C-type lectin SIGNR1 enhances cellular oxidative burst response against C. albicans in cooperation with Dectin-1. Eur J Immunol 41(5):1435–1444. doi:10.1002/eji.200940188

Takahashi K, Donovan MJ, Rogers RA, Ezekowitz RA (1998) Distribution of murine mannose receptor expression from early embryogenesis through to adulthood. Cell Tissue Res 292 (2):311–323

Tavares D, Ferreira P, Arala-Chaves M (2000) Increased resistance to systemic candidiasis in athymic or interleukin-10-depleted mice. J Infect Dis 182(1):266–273. doi:10.1086/315674

Taylor PR, Brown GD, Herre J, Williams DL, Willment JA, Gordon S (2004) The role of SIGNR1 and the beta-glucan receptor (Dectin-1) in the nonopsonic recognition of yeast by specific macrophages. J Immunol 172(2):1157–1162

Taylor PR, Tsoni SV, Willment JA, Dennehy KM, Rosas M, Findon H, Haynes K, Steele C, Botto M, Gordon S, Brown GD (2007) Dectin-1 is required for beta-glucan recognition and control of fungal infection. Nat Immunol 8(1):31–38. doi:10.1038/ni1408

Tomalka J, Ganesan S, Azodi E, Patel K, Majmudar P, Hall BA, Fitzgerald KA, Hise AG (2011) A novel role for the NLRC4 inflammasome in mucosal defenses against the fungal pathogen Candida albicans. PLoS Pathog 7(12):e1002379. doi:10.1371/journal.ppat.1002379

Tomalka J, Azodi E, Narra HP, Patel K, O'Neill S, Cardwell C, Hall BA, Wilson JM, Hise AG (2015) beta-Defensin 1 plays a role in acute mucosal defense against Candida albicans. J Immunol 194(4):1788–1795. doi:10.4049/jimmunol.1203239

Torosantucci A, Bromuro C, Chiani P, De Bernardis F, Berti F, Galli C, Norelli F, Bellucci C, Polonelli L, Costantino P, Rappuoli R, Cassone A (2005) A novel glyco-conjugate vaccine against fungal pathogens. J Exp Med 202(5):597–606. doi:10.1084/jem.20050749

Torosantucci A, Chiani P, Bromuro C, De Bernardis F, Palma AS, Liu Y, Mignogna G, Maras B, Colone M, Stringaro A, Zamboni S, Feizi T, Cassone A (2009) Protection by anti-beta-glucan antibodies is associated with restricted beta-1,3 glucan binding specificity and inhibition of fungal growth and adherence. PLoS ONE 4(4):e5392. doi:10.1371/journal.pone.0005392

Urban CF, Reichard U, Brinkmann V, Zychlinsky A (2006) Neutrophil extracellular traps capture and kill Candida albicans yeast and hyphal forms. Cell Microbiol 8(4):668–676. doi:10.1111/j.1462-5822.2005.00659.x

Urban CF, Ermert D, Schmid M, Abu-Abed U, Goosmann C, Nacken W, Brinkmann V, Jungblut PR, Zychlinsky A (2009) Neutrophil extracellular traps contain calprotectin, a cytosolic protein complex involved in host defense against Candida albicans. PLoS Pathog 5 (10):e1000639. doi:10.1371/journal.ppat.1000639

Uwamahoro N, Verma-Gaur J, Shen HH, Qu Y, Lewis R, Lu J, Bambery K, Masters SL, Vince JE, Naderer T, Traven A (2014) The pathogen Candida albicans hijacks pyroptosis for escape from macrophages. mBio 5(2):e00003–00014. doi:10.1128/mBio.00003-14

van de Veerdonk FL, Netea MG, Jansen TJ, Jacobs L, Verschueren I, van der Meer JW, Kullberg BJ (2008) Redundant role of TLR9 for anti-Candida host defense. Immunobiology 213(8):613–620. doi:10.1016/j.imbio.2008.05.002

van de Veerdonk FL, Marijnissen RJ, Kullberg BJ, Koenen HJ, Cheng SC, Joosten I, van den Berg WB, Williams DL, van der Meer JW, Joosten LA, Netea MG (2009) The macrophage mannose receptor induces IL-17 in response to Candida albicans. Cell Host Microbe 5(4):329–340. doi:10.1016/j.chom.2009.02.006

van de Veerdonk FL, Plantinga TS, Hoischen A, Smeekens SP, Joosten LA, Gilissen C, Arts P, Rosentul DC, Carmichael AJ, Smits-van der Graaf CA, Kullberg BJ, van der Meer JW, Lilic D, Veltman JA, Netea MG (2011) STAT1 mutations in autosomal dominant chronic mucocutaneous candidiasis. New Engl J Med 365(1):54–61. doi:10.1056/NEJMoa1100102

van der Graaf CA, Netea MG, Verschueren I, van der Meer JW, Kullberg BJ (2005) Differential cytokine production and toll-like receptor signaling pathways by Candida albicans blastoconidia and hyphae. Infect Immunol 73(11):7458–7464. doi:10.1128/IAI.73.11.7458-7464.2005

van der Graaf CA, Netea MG, Franke B, Girardin SE, van der Meer JW, Kullberg BJ (2006a) Nucleotide oligomerization domain 2 (Nod2) is not involved in the pattern recognition of Candida albicans. Clin Vaccine Immunol: CVI 13(3):423–425. doi:10.1128/CVI.13.3.423-425.2006

Van der Graaf CA, Netea MG, Morre SA, Den Heijer M, Verweij PE, Van der Meer JW, Kullberg BJ (2006b) Toll-like receptor 4 Asp299Gly/Thr399Ile polymorphisms are a risk factor for Candida bloodstream infection. Eur Cytokine Netw 17(1):29–34

van Rozendaal BA, van Spriel AB, van De Winkel JG, Haagsman HP (2000) Role of pulmonary surfactant protein D in innate defense against Candida albicans. J Infect Dis 182(3):917–922. doi:10.1086/315799

van Till JW, Modderman PW, de Boer M, Hart MH, Beld MG, Boermeester MA (2008) Mannose-binding lectin deficiency facilitates abdominal Candida infections in patients with secondary peritonitis. Clin Vaccine Immunol: CVI 15(1):65–70. doi:10.1128/CVI.00297-07

van't Wout JW, Linde I, Leijh PC, van Furth R (1988) Contribution of granulocytes and monocytes to resistance against experimental disseminated Candida albicans infection. Eur J Clin Microbiol Infect Dis: Off Publ Eur Soc Clin Microbiol 7(6):736–741

Villamon E, Gozalbo D, Roig P, Murciano C, O'Connor JE, Fradelizi D, Gil ML (2004a) Myeloid differentiation factor 88 (MyD88) is required for murine resistance to Candida albicans and is critically involved in Candida -induced production of cytokines. Eur Cytokine Netw 15 (3):263–271

Villamon E, Gozalbo D, Roig P, O'Connor JE, Ferrandiz ML, Fradelizi D, Gil ML (2004b) Toll-like receptor 2 is dispensable for acquired host immune resistance to Candida albicans in a murine model of disseminated candidiasis. Microbes infect/Institut Pasteur 6(6):542–548. doi:10.1016/j.micinf.2004.02.015

Villamon E, Gozalbo D, Roig P, O'Connor JE, Fradelizi D, Gil ML (2004c) Toll-like receptor-2 is essential in murine defenses against Candida albicans infections. Microbes infect/Institut Pasteur 6(1):1–7

Viudes A, Lazzell A, Perea S, Kirkpatrick WR, Peman J, Patterson TF, Martinez JP, Lopez-Ribot JL (2004) The C-terminal antibody binding domain of Candida albicans mp58 represents a protective epitope during candidiasis. FEMS Microbiol Lett 232(2):133–138. doi:10.1016/S0378-1097(04)00042-4

Voganatsi A, Panyutich A, Miyasaki KT, Murthy RK (2001) Mechanism of extracellular release of human neutrophil calprotectin complex. J Leukoc Biol 70(1):130–134

Voigt J, Hunniger K, Bouzani M, Jacobsen ID, Barz D, Hube B, Loffler J, Kurzai O (2014) Human natural killer cells acting as phagocytes against Candida albicans and mounting an inflammatory response that modulates neutrophil antifungal activity. J Infect Dis 209 (4):616–626. doi:10.1093/infdis/jit574

von Bernuth H, Picard C, Jin Z, Pankla R, Xiao H, Ku CL, Chrabieh M, Mustapha IB, Ghandil P, Camcioglu Y, Vasconcelos J, Sirvent N, Guedes M, Vitor AB, Herrero-Mata MJ, Arostegui JI, Rodrigo C, Alsina L, Ruiz-Ortiz E, Juan M, Fortuny C, Yague J, Anton J, Pascal M, Chang HH, Janniere L, Rose Y, Garty BZ, Chapel H, Issekutz A, Marodi L, Rodriguez-Gallego C, Banchereau J, Abel L, Li X, Chaussabel D, Puel A, Casanova JL (2008) Pyogenic bacterial infections in humans with MyD88 deficiency. Science 321 (5889):691–696. doi:10.1126/science.1158298

Vylkova S, Nayyar N, Li W, Edgerton M (2007a) Human beta-defensins kill Candida albicans in an energy-dependent and salt-sensitive manner without causing membrane disruption. Antimicrob Agents Chemother 51(1):154–161. doi:10.1128/AAC.00478-06

Vylkova S, Sun JN, Edgerton M (2007b) The role of released ATP in killing Candida albicans and other extracellular microbial pathogens by cationic peptides. Purinergic Signal 3(1–2):91–97. doi:10.1007/s11302-006-9040-0

Wagener J, Malireddi RK, Lenardon MD, Koberle M, Vautier S, MacCallum DM, Biedermann T, Schaller M, Netea MG, Kanneganti TD, Brown GD, Brown AJ, Gow NA (2014) Fungal chitin dampens inflammation through IL-10 induction mediated by NOD2 and TLR9 activation. PLoS Pathog 10(4):e1004050. doi:10.1371/journal.ppat.1004050

Wagner RD, Vazquez-Torres A, Jones-Carson J, Warner T, Balish E (1996) B cell knockout mice are resistant to mucosal and systemic candidiasis of endogenous origin but susceptible to experimental systemic candidiasis. J Infect Dis 174(3):589–597

Walker LA, Munro CA, de Bruijn I, Lenardon MD, McKinnon A, Gow NA (2008) Stimulation of chitin synthesis rescues Candida albicans from echinocandins. PLoS Pathog 4(4):e1000040. doi:10.1371/journal.ppat.1000040

Weindl G, Wagener J, Schaller M (2010) Epithelial cells and innate antifungal defense. J Dent Res 89(7):666–675. doi:10.1177/0022034510368784

Wellington M, Koselny K, Sutterwala FS, Krysan DJ (2014) Candida albicans triggers NLRP3-mediated pyroptosis in macrophages. Eukaryot Cell 13(2):329–340. doi:10.1128/EC.00336-13

Wells CA, Salvage-Jones JA, Li X, Hitchens K, Butcher S, Murray RZ, Beckhouse AG, Lo YL, Manzanero S, Cobbold C, Schroder K, Ma B, Orr S, Stewart L, Lebus D, Sobieszczuk P, Hume DA, Stow J, Blanchard H, Ashman RB (2008) The macrophage-inducible C-type lectin, mincle, is an essential component of the innate immune response to Candida albicans. J Immunol 180(11):7404–7413

Westwater C, Balish E, Schofield DA (2005) Candida albicans-conditioned medium protects yeast cells from oxidative stress: a possible link between quorum sensing and oxidative stress resistance. Eukaryot Cell 4(10):1654–1661. doi:10.1128/EC.4.10.1654-1661.2005

Wheeler RT, Kombe D, Agarwala SD, Fink GR (2008) Dynamic, morphotype-specific Candida albicans beta-glucan exposure during infection and drug treatment. PLoS Pathog 4(12): e1000227. doi:10.1371/journal.ppat.1000227

Whibley N, Maccallum DM, Vickers MA, Zafreen S, Waldmann H, Hori S, Gaffen SL, Gow NA, Barker RN, Hall AM (2014) Expansion of Foxp3(+) T-cell populations by Candida albicans enhances both Th17-cell responses and fungal dissemination after intravenous challenge. Eur J Immunol 44(4):1069–1083. doi:10.1002/eji.201343604

Whibley N, Tritto E, Traggiai E, Kolbinger F, Moulin P, Brees D, Coleman BM, Mamo AJ, Garg AV, Jaycox JR, Siebenlist U, Kammuller M, Gaffen SL (2016) Antibody blockade of IL-17 family cytokines in immunity to acute murine oral mucosal candidiasis. J Leukoc Biol 99 (6):1153–1164. doi:10.1189/jlb.4A0915-428R

Wu T, Samaranayake LP, Leung WK, Sullivan PA (1999) Inhibition of growth and secreted aspartyl proteinase production in Candida albicans by lysozyme. J Med Microbiol 48(8):721–730. doi:10.1099/00222615-48-8-721

Xiong J, Kang K, Liu L, Yoshida Y, Cooper KD, Ghannoum MA (2000) Candida albicans and Candida krusei differentially induce human blood mononuclear cell interleukin-12 and gamma interferon production. Infect Immunol 68(5):2464–2469

Yamamoto Y, Klein TW, Friedman H (1997) Involvement of mannose receptor in cytokine interleukin-1beta (IL-1beta), IL-6, and granulocyte-macrophage colony-stimulating factor responses, but not in chemokine macrophage inflammatory protein 1beta (MIP-1beta), MIP-2, and KC responses, caused by attachment of Candida albicans to macrophages. Infect Immunol 65(3):1077–1082

Yano J, Lilly E, Barousse M, Fidel PL Jr (2010) Epithelial cell-derived S100 calcium-binding proteins as key mediators in the hallmark acute neutrophil response during Candida vaginitis. Infect Immunol 78(12):5126–5137. doi:10.1128/IAI.00388-10

Zhang MX, Bohlman MC, Itatani C, Burton DR, Parren PW, St Jeor SC, Kozel TR (2006) Human recombinant antimannan immunoglobulin G1 antibody confers resistance to hematogenously disseminated candidiasis in mice. Infect Immunol 74(1):362–369. doi:10.1128/IAI.74.1.362-369.2006

Zhang SQ, Zou Z, Shen H, Shen SS, Miao Q, Huang X, Liu W, Li LP, Chen SM, Yan L, Zhang JD, Zhao JJ, Xu GT, An MM, Jiang YY (2016) Mnn10 Maintains pathogenicity in Candida albicans by extending alpha-1,6-Mannose backbone to evade host Dectin-1 mediated antifungal immunity. PLoS Pathog 12(5):e1005617. doi:10.1371/journal.ppat.1005617

Zhu LL, Zhao XQ, Jiang C, You Y, Chen XP, Jiang YY, Jia XM, Lin X (2013) C-type lectin receptors Dectin-3 and Dectin-2 form a heterodimeric pattern-recognition receptor for host defense against fungal infection. Immunity 39(2):324–334. doi:10.1016/j.immuni.2013.05.017
Zielinski CE, Mele F, Aschenbrenner D, Jarrossay D, Ronchi F, Gattorno M, Monticelli S, Lanzavecchia A, Sallusto F (2012) Pathogen-induced human TH17 cells produce IFN-gamma or IL-10 and are regulated by IL-1beta. Nature 484(7395):514–518. doi:10.1038/nature10957

Chapter 10
AMPlified Defense: Antimicrobial Peptides During *Candida albicans* Infection

Joachim F. Ernst and Marc Swidergall

Abstract The fungal pathogen *Candida albicans* is a successful colonizer of the human host as part of the normal mycobiota. Under certain circumstances the fungus can cause superficial, as well as life-threatening infections. A complex balance of host immune defense mechanisms and fungal responses determine the level of fungal colonization or infection. Humans express a large number of various antimicrobial peptides (AMPs) with antifungal action that are key effectors of innate immunity. In this chapter, we review the host AMP response to fungal infection including AMP production mechanisms and the immunomodulatory and antifungal actions of AMPs. Furthermore, *C. albicans* counterattack strategies leading to AMP resistance are presented.

10.1 Introduction

Antimicrobial peptides (AMPs) represent an evolutionary old humoral system that establishes the first defense line against invading microbes (Ostaff et al. 2013). Human AMPs exert main influences on the composition of microbial communities. The ability to evade killing by AMPs provided pathogenic bacterial and fungal species with a selective evolutionary advantage (Koprivnjak and Peschel 2011; Swidergall and Ernst 2014).

Candida species are the fourth most frequent cause of nosocomial bloodstream infections in the United States and systemic candidiasis is still associated with a high mortality rate (up to 50%) (Lewis 2009; Perlroth et al. 2007). Systemic candidemia is mostly related to infections with one of five species: *C. albicans*, *C.*

J.F. Ernst
Department Biologie, Molekulare Mykologie, Heinrich-Heine-Universität Düsseldorf, Düsseldorf, Germany

M. Swidergall (✉)
Division of Infectious Diseases, Los Angeles Biomedical Research Institute, Harbor-UCLA Medical Center, Torrance, CA 90502, USA
e-mail: mswidergall@labiomed.org

R. Prasad (ed.), *Candida albicans: Cellular and Molecular Biology*,
DOI 10.1007/978-3-319-50409-4_10

glabrata, *C. parapsilosis*, *C. tropicalis*, and *C. krusei* (Brown et al. 2012; Polke et al. 2015).

C. albicans is a polymorphic fungus belonging to the endogenous human microbiota that colonizes the digestive tract and other mucosal surfaces of the body (e.g., oral cavity and vagina) of 30–70% of healthy individuals (Ghannoum et al. 2010; Gow et al. 2012; Pfaller and Diekema 2007). The fungus causes superficial mucosal disease including oral and vaginal thrush and is also capable to trigger life-threatening illness in immunocompromised patients (Odds et al. 1998). The decision between harmless fungal commensalism and the inception of a life-threatening mycosis is influenced by the strength of the host response and the ability of the fungus to evade the respective mechanisms (Brown and Netea 2012; Cheng et al. 2012; Richardson and Moyes 2015). The ability of *C. albicans* to colonize different niches of the human body implies that even a 'healthy' interaction between the commensal fungus and its host is based on a highly complex equilibrium (Hube 2004), which is a result of long coevolution of the opportunistic pathogen and the host (Cottier and Pavelka 2012).

10.2 Antimicrobial Peptides: Structure and Function

The first antimicrobial peptides were described in the 1920s as a component of nasal secretions from a patient suffering from acute coryza (Fleming 1922). In the following decades, over 1000 peptide antimicrobials were described from different species (Nakatsuji and Gallo 2012). AMPs are effector molecules of the innate immune system with direct antimicrobial function (Huttner and Bevins 1999; Lehrer and Ganz 1999). AMPs interact with microbial membranes or intracellular targets to kill or inhibit growth of pathogens (Peschel and Sahl 2006; Swidergall and Ernst 2014). Peptide antimicrobials are released by macrophages and granulocytes, as well as by several epithelial cells including cells of the intestine, the vaginal epithelium, the oral cavity epithelium, and airway epithelium (Aarbiou et al. 2002; Beisswenger and Bals 2005; Wah et al. 2006; Zhang and Falla 2010).

Antimicrobial peptides are classified according to size, structure, and amino acid composition and can be divided into two large groups: linear and cyclic AMPs (Andreu and Rivas 1998; Hancock 1997). Active peptides consist of 10–50 amino acids and are mostly positively charged as a result of several lysine and arginine residues, but they also contain a substantial proportion ($\sim$30%) of hydrophobic residues (Peschel and Sahl 2006). AMPs that can form an amphipathic, α-helical structure are currently categorized on the basis of their structural characteristics, while others, such as the defensins, are classified by their number of disulfide bonds (van 't Hof et al. 2001). However, there are numerous additional possibilities for AMP classification, e.g., based on their biosynthesis, biological source, biological functions, peptide properties, covalent bonding patterns or on their molecular targets (Wang 2015; Wang et al. 2009).

One of the best investigated examples for a linear, α-helical human AMP, which contains no cysteine residues, is the cathelicidin LL-37, a cationic peptide derived from a 18 kDa precursor protein by proteolytic cleavage (Sorensen et al. 2001). It is found in epithelial cells, the skin, the gastrointestinal and respiratory tracts, as well as in neutrophils (Zanetti 2004). In the oral cavity, a large variety of antimicrobial peptides including human defensins and histatins are found (Edgerton and Koshlukova 2000). Defensins contain six conserved cysteine residues forming three disulfide bridges. Depending on the topology of the disulfide bonds defensins are classified in α-, β-, or θ- subgroups (Auvynet and Rosenstein 2009). Defensins are translated as an inactive precursor, which is cleaved into an active form. The primary sequence is an 87–94 amino acid peptide including a hydrophobic leader sequence, a short acidic pro-segment that neutralizes the peptide and a highly cationic mature sequence (Kaiser and Diamond 2000). Histatins (Hsts) were initially described as basic components of the secretions of human salivary glands (Oppenheim et al. 1988). Among all secreted Hsts, histatin 5 presents the major AMP of this class (Johnson et al. 2000). The Hst family comprises a group of 12 peptides derived from the gene products of *HTN1* and *HTN2* genes. Hst 1 and Hst 3 are full-length peptides (lacking the secretion signal peptide), while the peptides Hst 2 and Hsts 4–12 are generated by further proteolytic cleavage (Puri and Edgerton 2014).

10.3 Expression of Mammalian AMP Genes

Epithelial cells produce AMPs either constitutively or upon the first contact with *C. albicans* or other invading pathogens (Selsted and Ouellette 2005; Swidergall and Ernst 2014; Vandamme et al. 2012). While human cathelicidin LL-37 and β-defensins 2–4 (hBD2–hBD4) are inducible AMPs, human α-defensin 5 (HD-5), HD-6, and β-defensin 1 (hBD1) are constitutively expressed by epithelial cells (Zhao et al. 1996). hBD1, hBD2, and hBD3 are expressed in the oral cavity (Abiko et al. 2007). hBD1 and hBD2 are enriched in the supra-basal layer of epithelium and differentiated epithelial cells, while hBD3 is produced in undifferentiated epithelial cells in the basal layer of epithelia (Dale et al. 2001; Lu et al. 2005). Apart from hBD1, all β-defensins can also be found in the small and large intestine (Kopp et al. 2015), where their genes are induced upon immune stimulation (Krisanaprakornkit et al. 1998). The total amount of Hsts concentrations secreted from parotids reaches 53 μg/ml including Hst 5 as a major component (Johnson et al. 2000). An important pathway of pathogen-induced AMP expression is through the recognition of pathogen-associated molecular patterns (PAMPs) via Toll-like receptors (TLRs), which ultimately leads to activation and translocation of NF-κB to genes encoding AMPs and other pro-inflammatory molecules cytokines (Gilmore and Wolenski 2012; Lai and Gallo 2009).

In the *C. albicans* commensal life-style, continuous but relatively rare PAMP-PRR (Pattern-Recognition Receptors) interactions trigger low levels of NF-κB activation that drives basal transcription of AMP-encoding genes. Fungal proliferation favored by a weak immune system or by disturbed bacterial flora increases fungal PAMP–PRR interactions resulting in a 'boosted' AMP response. Epithelial cells have developed a mechanism to discriminate between the commensal yeast form and the invasive hyphal form of *C. albicans* (Moyes et al. 2015). Hyphae trigger the production of proinflammatory cytokines including IL-1β, IL-6 and IL-8 that activate co-habitating immune cells including Th17 cells producing IL-17, IL-22, and TNF-α, which are especially relevant for AMP production (Eyerich et al. 2011; Moyes et al. 2010; Wolk et al. 2004). A *C. albicans* peptide of Ece1 protein, shed by the hyphal growth form, is critical for epithelial damage and innate recognition by inducing c-Fos and MKP1 phosphorylation (Moyes et al. 2016). This fungal peptide, Candidalysin, may act similarly to cationic AMPs or peptide toxins by its cytolytic activity, which in turn activates an epithelial 'danger response'. *C. albicans* hyphae stimulate an initial release of proinflammatory cytokines through activation of the inflammasome complex (Kao et al. 2004) but prolonged hyphal exposure results in decreased production of β-defensins (Rehaume et al. 2010). In keratinocytes, AMP upregulation is caused mainly by a strong increase of ERK/JNK MAP kinase activities and activation of their dedicated transcription factors (e.g., AP-1) immediately after binding of *C. albicans* phospholipomannan to TLR2 (Li et al. 2009). Co-habitating neutrophils activate further inflammatory signaling of epithelial cells by upregulation of TLR4 (Lu et al. 2006), which binds to *O*-mannans in the *C. albicans* cell wall (Bourgeois and Kuchler 2012). Interestingly, TLR4 not only binds 'professional' PAMPs but also the defensin hBD2, which generates an autostimulatory feedback loop to increase antifungal transcriptional responses in epithelial cells (Kawai et al. 2002). Fungal foci of proliferation become increasingly inflamed because of the chemoattractive activities of AMPs that recruit more lymphocytes (Pahl et al. 2011). Although it has been well established that transcriptional activation is mediated by transcription factors NF-κB and AP-1 in epithelial cells (Moyes et al. 2015), interestingly, increased transcription of the hBD3-encoding gene is independent of NF-κB and requires an EGFR/MAPK/AP-1-dependent pathway by initial activation of TGF-α through ADAM17-processing (Pahl et al. 2011; Steubesand et al. 2009). By inducing c-Fos and AP-1 activity, antifungal drugs are able to increase defensin production in keratinocytes, conceivably contributing to the therapeutic success of such antifungals (Kanda et al. 2011).

Transcriptional activation is the major mechanism of *C. albicans*-induced AMP induction in host cells, but peptide release by processing of a precursor protein is additionally required. Neutrophils, keratinocytes, and Paneth intestinal cells store precursors of mature AMPs in intracellular vesicles. Paneth cells shed α-defensins into intestinal crypts and, for unknown reasons, this secretion is enhanced by bacterial co-habitants and not by fungi (Rumio et al. 2004). In neutrophils, ingested

Table 10.1 Antimicrobial peptide expression

AMPs	Expression pattern	Source
LL-37	Constitutively/induced	Epithelial cells, neutrophils
Histatin5	Constitutively	Saliva (parotid and submandibular)
hBD1	Induced	Oral cavity
hBD2–4	Constitutively	Oral cavity, intestine
hNP1–4	Constitutively/induced	Neutrophils
Lactoferrin	Constitutively	Epithelial cells, neutrophils

microbes are attacked intracellularly by F061-defensins, which fuse with phagosomes (Sorensen et al. 1997).

Vitamin D has been linked to susceptibility to infections and may also have a role in candidemia (Lim et al. 2015). It was shown that vitamin D is critical for the regulation of both LL-37 and DEFB4 genes encoding β-defensin 2 in epithelial cells (Gombart 2009). TLR signaling leads to 1-α-hydroxylation of the 25(OH)D vitamin D precursor and the resulting 1.25(OH)2D molecule forms a complex with the D-binding protein that binds to the vitamin D receptor (VDR), a transcription factor. The cathelicidin gene is thereby activated and the protein (hCAP18/LL-37) is synthesized. Interestingly, candidemic patients have significantly lower 25(OH)D concentrations (Lim et al. 2015). *Candida*-infected mice treated with low-dose 1.25 (OH)2D3 showed reduced fungal burden and increased survival relative to untreated mice resulting in proinflammatory benefits during *Candida* infection.

At sites of fungal infections, influx of immune cells and tissue necrosis caused by the invading pathogen generate hypoxic microenvironments to which both the pathogen and host cells must adapt (Ernst and Tielker 2009; Grahl and Cramer 2010). During exposure to environmental low-oxygen concentrations, mammalian cells trigger a hypoxic response pathway by the regulated expression of the hypoxia-inducible transcription factor-alpha (HIF-1α), which upregulates cathelicidin expression in keratinocytes (Peyssonnaux et al. 2008). The reduction of hBD1 by thioredoxin expressed in the epidermis enhances the antimicrobial activity of this AMP (Schroeder et al. 2011). This suggests that redox regulation and microenvironments in general are crucial for the innate immune protection by hBD1. In conclusion, host cells boost AMP expression during fungal colonization, when the constitutive production of AMPs as the primary line of defense fails to inhibit or eliminate the invading pathogen (Table 10.1).

10.4 AMPs as Immunomodulatory Molecules

Similarities between chemokines and host defense peptides have been previously observed. Indeed, many chemokines have modest antimicrobial activity (Hieshima et al. 2003; Yang et al. 2003). Beside their broad antimicrobial function (see Sect. 10.5), cathelicidins such as LL-37 and CRAMP (Cathelicidin-related

Antimicrobial Peptide; mouse homolog for human LL-37) are able to signal cell damage resulting in chemoattraction of immune cells, or stimulation and modulation of cytokine release (Vandamme et al. 2012). Thus, cathelicidins recruit and activate cells of the innate immune system including dendritic cells and as a result promote adaptive immune responses. This LL-37-mediated immunomodulatory mechanism functions through activation of multiple receptors and depends on cell type and environmental setting (Vandamme et al. 2012). LL-37 and the HNPs are neutrophil-derived peptides released upon neutrophil degranulation, which activate the transcription of genes encoding chemokines, such as IL-8 (CXCL8) that attract neutrophils (Bowdish et al. 2006). Besides their chemoattractive action, AMPs have anti-inflammatory functions by neutralizing LPS (Rosenfeld and Shai 2006; Scott et al. 2011; Scott et al. 2000); furthermore, they activate the adaptive immune response by (i) chemotaxis of immature dendritic cells (iDC), monocytes/Pre-DCs, PMN and T cells, (ii) modulation of lymphocyte activity and/or proliferation, (iii) alteration of the local cytokine environment, (iv) direct iDC activation via TLR4, or (v) generation of primed iDCs with enhanced antigen uptake and presentation capacity (Bowdish et al. 2006).

10.5 AMP Antifungal Activities

It is known that at high AMP concentrations, pores are formed in microbial membranes that lead to cytoplasmic membrane dysfunction and depolarization by release of ATP and ions; these events effect osmotic dysregulation and finally lead to cell death (Brogden 2005). In *C. albicans*, LL-37 associates first with the cell wall and/or the cytoplasmic membrane of the pathogen (den Hertog et al. 2005). Treatment of fungal cells with high amounts of LL-37 resulted in vesicle formation of the membrane, rapid efflux of small molecules such as ATP and larger molecules with molecular masses up to 40 kDa. The cell wall-linked β1,3-exoglucanase Xog1 of *C. albicans* has been identified as a LL-37 receptor (Tsai et al. 2011a). This peptide–enzyme interaction leads to cell wall remodeling and decreased adhesive properties of *C. albicans* (Tsai et al. 2011b). LL-37 treated *C. albicans* cell walls contain reduced amounts of glucan (by ~25%) and mannan (by ~30%) suggesting that besides cell wall disruption, LL-37 alters the cell wall architecture to expose β-1,3-glucan, which triggers immune responses (Tsai et al. 2014). Interestingly, peptides derived from processing of LL-37 by host proteases were found on human skin (Murakami et al. 2004). These truncated peptides induce leakage of nucleotides and proteins in host cells (Den Hertog et al. 2006). Another described antifungal mechanism triggered by LL-37 is the formation of neutrophil extracellular traps (NETs). NETosis is a process of generation of NETs, whose main components are DNA, granular antimicrobial peptides, and nuclear and cytoplasmic proteins. NETs become associated with AMPs including defensins and

LL-37 (Vorobjeva and Pinegin 2014). Their high local concentrations on chromatin fibrils seem to determine the antimicrobial activity of NETs. Interestingly, LL-37 is able to stimulate NET formation via interaction with the CD32 surface receptor protein (Brinkmann and Zychlinsky 2007).

The candidacidal action of Hst 5 has been well studied and was described as a multistep mechanism (Puri and Edgerton 2014). Initially, this defense peptide binds cell wall β-glucans, as well as the heat-shock proteins Ssa1 and Ssa2, on the cell surface of *C. albicans* (Jang et al. 2010a; Sun et al. 2008), which is followed by its rapid uptake via the polyamine influx transporters Dur3 and Dur31 (Kumar et al. 2011; Mayer et al. 2012; Ordonez et al. 2014). Intracellularly, Hst 5 induces formation of reactive oxygen species (ROS) and efflux of ions and ATP, resulting in cell death (Helmerhorst et al. 1999; Helmerhorst et al. 2001; Komatsu et al. 2011; Koshlukova et al. 1999). Hst 5 downregulates mitochondrial proteins involved in *C. albicans* energy metabolism leading to a drastic decrease in mitochondrial ATP synthesis (Komatsu et al. 2011; Koshlukova et al. 1999); additionally Hst 5 is able to reduce fungal colonization on the saliva-derived protein pellicle of the mucosal surface by inhibiting fungal adhesion (Moffa et al. 2015). Furthermore, the Hst 5 binds iron to reduce growth of *C. albicans* (Puri et al. 2015b).

Physiological concentrations of α-defensins 1–3 (hNP1–3) amount to 6 mg/ml in neutrophils. While similar to the action of histatins, hNP1 kills *C. albicans* by acting on its energy metabolism by causing non-lytic depletion of intracellular ATP, β-defensins hBD1–3 cause membrane permeabilization leading to cell death (Edgerton et al. 2000; Krishnakumari et al. 2009); however, unlike hBD1 and hBD2, hBD3 kills *C. albicans* by energy-independent mechanisms. Furthermore, hBD3 compromises the cell wall integrity of *C. albicans* by inducing an elevated activity of Xog1 exoglucanase, while the binding of it to Xog1 results in reduced fungal adherence (Chang et al. 2012).

Besides 'specialized' host defense peptides humans possess protein fragments with antifungal activity, which are derived from proteins with different functions including RNase 7, expressed by human keratinocytes, which exhibits candidicial activity independent of the RNase activity (Harder and Schroder 2002). Lactoferrin, a 77 kDa iron-binding glycoprotein, exerts is antifungal activity by a highly basic N-terminal region containing a 25-amino acid-domain termed lactoferricin (LF) (Lupetti et al. 2000). LF interacts, similar to LL-37, with the microbial membrane resulting in disintegration of the membrane bilayer, as well as efflux of ATP and proteins in *C. albicans* cells (Bolscher et al. 2012). A peptide derived from human glyceraldehyde-3-phosphate dehydrogenase (hGAPDH) killed the fungal pathogen by initiating apoptosis and induced secretion of IL-8 and GM-CSF to attract immune cells to the site of infection (Wagener et al. 2013). Furthermore, the hGAPDH-peptide (2–32) counteracts an important *C. albicans* virulence trait by inhibiting Sap1 and 2 protease activity, which influence induced endocytosis (Dalle et al. 2010; Zhu and Filler 2010). Peptides derived from the N-terminal portion of human mucin MUC7 associate with the fungal plasma membrane, become internalized and exert fungicidal activity without affecting cellular metabolic activity (Situ et al. 2003). On the other hand, it was shown that mucins are able to bind AMPs (Antoni et al. 2013; Swidergall et al. 2013); binding may possibly serve to concentrate AMPs on epithelial surfaces, although AMP activity is lowered (Bucki et al. 2008).

10.6 Antifungal AMP Activities in Vivo

Host epithelial surfaces are colonized by complex bacterial and fungal communities (McFall-Ngai et al. 2013; Underhill and Iliev 2014). This commensal scenario advantages physiological functions of the host including immune system maturation and resistance to pathogens (Mazmanian et al. 2005; Stecher and Hardt 2008; Underhill and Pearlman 2015).

Commensal fungi, frequently *Candida* spp., have been detected in the gastrointestinal (GI) tracts of numerous mammals (Iliev et al. 2012). It is reported that the bacterial phyla *Firmicutes* and *Bacteroidetes* account for >95% of the bacteria in the distal guts of healthy adult mice and humans (Ley et al. 2005); interestingly these commensal anaerobic bacteria are critical for inhibiting *C. albicans* colonization in mice (Fan et al. 2015) by increasing expression of transcription factor HIF-1α, a known regulator of innate immunity (Nizet and Johnson 2009; Peyssonnaux et al. 2008) and of the defense peptide LL-37 (CRAMP in mice). Furthermore, murine β-defensin 1 (mBD1)-deficient mice (homolog to human β-defensin 1) exhibited increased mucosal and systemic fungal burdens during early infection after oral inoculation with *C. albicans* (Tomalka et al. 2015). mBD1-deficient mice show reduced neutrophil infiltration to sites of mucosal fungal invasion and defective expression of other AMPs including mBD2 and CRAMP, while production of antifungal inflammatory mediators IL-1β, IL-6, KC, and IL-17 is increased. Patients with a dominant-negative mutation in signal transducer and activator of transcription 3 (STAT3), a downstream molecule of the T_H-17-inductive cytokines IL-6 and IL-23, are highly susceptible to mucosal *C. albicans* infection. These patients show significant impairment in salivary AMPs β-defensin 2 and histatins (Conti et al. 2011). Additionally, using a mouse model of oropharyngeal candidiasis (OPC) it was shown that mice deficient for IL-17 receptor A (IL-17RA) or IL-17RC fail to induce defense peptides in the epithelium, resulting in fungal persistence (Trautwein-Weidner et al. 2015). During the early stage of OPC the host highly increases expression of AMPs (Conti et al. 2009) including DEFB3 (homolog to human β-defensin 2) and the alarmins S100A8 and S100A9, neutrophil-chemoattractants with direct antimicrobial activity (Batycka-Baran et al. 2014), highlighting the importance of peptide antimicrobials for fungal innate immune defense.

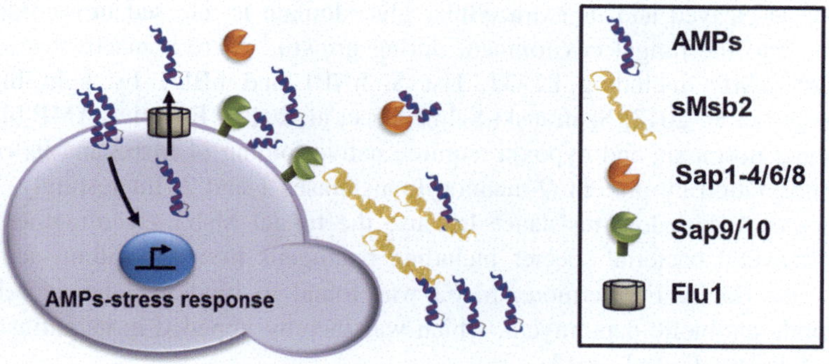

Fig. 10.1 *C. albicans* three-phase AMP evasion strategy. The shed exodomain of the *C. albicans* Msb2 membrane sensor (sMsb2) binds several AMPs extracellularly to provide broad-range protection against AMPs. In addition, several shed proteases, as well as the cell wall-anchored proteases Sap9 and Sap10 cleave and inactivate AMPs on the outside of fungal cells. The AMP Histatin 5 (Hst 5) is taken up and induces the formation of reactive oxygen species. Fungal cells decrease toxicity of Hst 5 by its extrusion via the polyamine efflux transporter Flu1. Different response pathways, including the Hog1 MAPK-, Cek1 MAPK-, and the RAM pathway, are activated during AMP stress to overcome antifungal AMP activity

10.7 The *C. albicans* Three-Phase AMP Evasion Strategy

To conquer antifungal activity of host defense peptides *C. albicans* has developed a three-phase AMP evasion strategy for secretion of AMP effectors, for peptide efflux pumps and for regulation of signaling pathways (Fig. 10.1). The first fungal defense line to combat AMPs includes secreted proteins, which inhibit peptide activity by degradation or binding. *C. albicans* uses the GPI-anchored proteases Sap9 and Sap10 (highly expressed during oral infection), as well as the secreted protease Sap6 (most abundant secreted protease during hyphal growth (Sorgo et al. 2010)), to degrade the salivary peptide Hst 5 (Meiller et al. 2009; Puri et al. 2015a). Besides Sap6, Sap9, and Sap10, *C. albicans* expresses seven more members of the aspartyl protease family (Sap1–10) that exert multiple functions during infection (Naglik et al. 2003) including cleavage of various AMPs. The AMP LL-37 is cleaved into multiple products by Sap1–4, Sap8, and also by Sap9 (Rapala-Kozik et al. 2015). Interestingly, one major intermediate peptide fragment named LL-25 still exerts antifungal activity but lacks the immunomodulatory properties of LL-37. The intermediate LL-25 has a lower ability to recruit neutrophils and to inhibit neutrophil apoptosis; therefore, LL-37 cleavage decreases the lifespan of these important defense cells. In addition to AMP-degrading enzymes, *C. albicans* secretes a fragment of the Msb2 glycoprotein, which acts as a broad-range protectant against AMPs (Swidergall et al. 2013). Msb2 is the upstream sensor of the mitogen-activated protein kinase (MAPK) Cek1 (Román et al. 2009). The precursor of Msb2 is cleaved and its extracellular glycodomain is released in considerable amounts into the fungal environment during growth, where it inactivates a wide range of AMPs including LL-37, Hst 5, hNP1 and hBD1 by tight binding (Swidergall et al. 2013; Szafranski-Schneider et al. 2012). Effective AMP binding both under normoxia and hypoxia required native folding of Msb2 and its correct *O*-mannosylation by protein-*O*-mannosyltransferases 1 and 2. Interestingly, Msb2 mediates cross-kingdom resistance because the fungal Msb2 glycofragment also rescued several bacterial species including pathogens from the killing action of LL-37 and Hst 5. Furthermore, Msb2 was found to bind the clinical relevant lipopeptide antibiotic daptomycin, which was thereby impeded in its antibacterial action (Swidergall et al. 2013).

AMPs can be actively exported from the cytoplasm of fungal cells to reduce peptide toxicity. It was observed that the amount of internalized Hst 5 was reduced from live *C. albicans* cells by using the polyamine efflux transporter Flu1 (Jang et al. 2010b; Li et al. 2013). Efflux of Hst 5 was reduced in *flu1*Δ/Δ cells but did not stop suggesting that additional mechanisms or transporters are involved in the export of Hst 5.

A well-established response of eukaryotic microbes to different stressors is the rapid phosphorylation of signaling pathways to induce mRNAs that encode rescue and repair proteins. *C. albicans* stress response pathways have been shown to be crucial for basal resistance to Hst 5. Physiological levels of Hst 5 activated the MAPK Hog1 in wild-type cells, while *hog1* mutants were hypersusceptible to this AMP (Vylkova et al. 2007). Human β-defensins trigger Hog1 activity by interaction with the upstream MAPK kinase Pbs2 to induce a cell rescue response (Argimon et al. 2011). Accordingly, mutants of the HOG (High-Osmolarity Glycerol) pathway were shown to be supersensitive to β-defensins. Besides its protective role in hyperosmotic environments, the HOG pathway is involved in regulating ROS production and ATP efflux by mitochondria (Alonso-Monge et al. 2009); interestingly, ROS production and ATP release are also often induced by AMPs (see Sect. 16.5). Thus, by this mechanism the HOG pathway may function as a key component in basal resistance of *C. albicans* to different AMPs. Physiological concentrations of some AMPs such as LL-37 are low but increase drastically at sites of pathogen infection and inflammation (Nijnik and Hancock 2009). In a commensal scenario with a low fungal burden, *C. albicans* may cope with relatively low concentrations of AMPs and continue to grow. Under these circumstances, target genes of the Msb2/Cek1 signaling pathway, e.g., genes encoding mannosyltransferases, are required to restore cell wall integrity (Swidergall et al. 2015). Some of the AMP-response pathways, which mediate basal resistance to host defense peptides, are distinct from general stress adaptation pathways. The *C. albicans* RNA-binding protein Ssd1, a component of the RAM pathway (Regulation of Ace2 and Morphogenesis), and its downstream transcription factor Bcr1 regulate basal resistance to different AMPs including the helical cationic polypeptide protamine, hBD2 and platelet microbicidal proteins found in the bloodstream by maintaining mitochondrial integrity and by reducing membrane permeabilization (Gank et al. 2008; Jung et al. 2013). The RAM pathway orchestrates multiple processes including cell wall integrity (Bharucha et al. 2011); thus, Ssd1 and Bcr1 proteins may regulate cell surface associated AMP targets and membrane composition to adapt to AMP-derived antifungal mechanism (Gank et al. 2008). Furthermore, a peptide derived from processing of the human salivary mucin MUC7 activated the calcineurin pathway as well as the activity of the 20S and 26S proteasome in *C. albicans* (Lis et al. 2010). The *C. albicans* phosphatase calcineurin regulates membrane integrity resulting in tolerance to antifungals (Cruz et al. 2002). In addition, calcineurin coordinately regulates, in combination with Hog1 and PKC (protein kinase C), cell wall integrity pathways the biosynthesis of chitin biosynthesis in response to cell wall stress (Lenardon et al. 2007). Some AMPs (Hst 5, hNP1 and LF) suppress the synthesis of chitin, while stimulation of chitin synthesis was shown to rescue *C. albicans* from the action of echinocandin (Walker et al. 2008). Conceivably, enhanced chitin levels effected by calcineurin, Hog1 and Mkc1 pathways increase protection against different AMPs by fortifying the fungal cell wall. In conclusion, *C. albicans* has evolved various AMP evasion mechanisms to overcome the antifungal activity of host defense peptides.

10.8 AMPs of the Human Microbiome

AMPs are produced by organisms of all types including microorganisms (Zasloff 2002). Therefore, it is not surprising that peptide antimicrobials, which contribute to the human innate immune defense, are not only of human origin. For example, unique peptides namely phenol-soluble modulin (PSM)γ and PSMδ produced by *Staphylococcus epidermidis* could be beneficial to the host and serve as additional defense peptide on the skin surface (Cogen et al. 2010). PSM peptides functionally cooperate with LL-37 to enhance antimicrobial activity and to stimulate the innate immune killing capacity of neutrophils by inducing NETs. The AMP nisin produced by *Lactococcus lactis* also contributes to the huge AMP diversity in the human body (Akerey et al. 2009). Nisin reduces *C. albicans* adhesion and blocks its yeast-to-hyphae transition. *C. albicans* colonizes different niches in the human body, where the yeast may network with many other microbial species including bacterial pathogens. These fungal–bacterial interactions may affect colonization, survival, and pathogenesis of both organisms. A myriad of microbes co-habitate the human host, some of which may produce AMPs that inhibit *C. albicans* growth. Considering the diversity of the human microbiome it is likely that many other microbial peptide antimicrobials which benefit our immune defense will be discovered. These discoveries are only at the very early stages.

References

Aarbiou J, Rabe KF, Hiemstra PS (2002) Role of defensins in inflammatory lung disease. Ann Med 34(2):96–101

Abiko Y, Saitoh M, Nishimura M, Yamazaki M, Sawamura D, Kaku T (2007) Role of beta-defensins in oral epithelial health and disease. Med Mol Morphol 40(4):179–184. doi:10.1007/s00795-007-0381-8

Akerey B, Le-Lay C, Fliss I, Subirade M, Rouabhia M (2009) In vitro efficacy of nisin Z against *Candida albicans* adhesion and transition following contact with normal human gingival cells. J Appl Microbiol 107(4):1298–1307

Alonso-Monge R, Carvaihlo S, Nombela C, Rial E, Pla J (2009) The Hog1 MAP kinase controls respiratory metabolism in the fungal pathogen *Candida albicans*. Microbiology 155(Pt 2):413–423. doi:10.1099/mic.0.023309-0

Andreu D, Rivas L (1998) Animal antimicrobial peptides: an overview. Biopolymers 47(6):415–433. doi:10.1002/(SICI)1097-0282(1998)47:6<415:AID-BIP2>3.0.CO;2-D

Antoni L, Nuding S, Weller D, Gersemann M, Ott G, Wehkamp J, Stange EF (2013) Human colonic mucus is a reservoir for antimicrobial peptides. J Crohns Colitis 7(12):e652–664. doi:10.1016/j.crohns.2013.05.006

Argimon S, Fanning S, Blankenship JR, Mitchell AP (2011) Interaction between the *Candida albicans* high-osmolarity glycerol (HOG) pathway and the response to human beta-defensins 2 and 3. Eukaryot Cell 10(2):272–275. doi:10.1128/ec.00133-10

Auvynet C, Rosenstein Y (2009) Multifunctional host defense peptides: antimicrobial peptides, the small yet big players in innate and adaptive immunity. FEBS J 276(22):6497–6508. doi:10.1111/j.1742-4658.2009.07360.x

Batycka-Baran A, Maj J, Wolf R, Szepietowski JC (2014) The new insight into the role of antimicrobial proteins-alarmins in the immunopathogenesis of psoriasis. J Immunol Res 2014:628289. doi:10.1155/2014/628289

Beisswenger C, Bals R (2005) Functions of antimicrobial peptides in host defense and immunity. Curr Protein Pept Sci 6(3):255–264

Bharucha N, Chabrier-Rosello Y, Xu T, Johnson C, Sobczynski S, Song Q, Dobry CJ, Eckwahl MJ, Anderson CP, Benjamin AJ, Kumar A, Krysan DJ (2011) A large-scale complex haploinsufficiency-based genetic interaction screen in *Candida albicans*: analysis of the RAM network during morphogenesis. PLoS Genet 7(4):e1002058. doi:10.1371/journal.pgen.1002058

Bolscher J, Nazmi K, van Marle J, van 't Hof W, Veerman E (2012) Chimerization of lactoferricin and lactoferrampin peptides strongly potentiates the killing activity against *Candida albicans*. Biochem Cell Biol 90 (3):378–388. doi:10.1139/o11-085

Bourgeois C, Kuchler K (2012) Fungal pathogens-a sweet and sour treat for toll-like receptors. Front Cell Infect Microbiol 2:142. doi:10.3389/fcimb.2012.00142

Bowdish DM, Davidson DJ, Hancock RE (2006) Immunomodulatory properties of defensins and cathelicidins. Curr Top Microbiol Immunol 306:27–66

Brinkmann V, Zychlinsky A (2007) Beneficial suicide: why neutrophils die to make NETs. Nat Rev Microbiol 5(8):577–582. doi:10.1038/nrmicro1710

Brogden KA (2005) Antimicrobial peptides: pore formers or metabolic inhibitors in bacteria? Nat Rev Microbiol 3(3):238–250. doi:10.1038/nrmicro1098

Brown GD, Netea MG (2012) Exciting developments in the immunology of fungal infections. Cell Host Microbe 11(5):422–424. doi:10.1016/j.chom.2012.04.010

Brown GD, Denning DW, Gow NA, Levitz SM, Netea MG, White TC (2012) Hidden killers: human fungal infections. Sci Transl Med 4(165):165rv113. doi:10.1126/scitranslmed.3004404

Bucki R, Namiot DB, Namiot Z, Savage PB, Janmey PA (2008) Salivary mucins inhibit antibacterial activity of the cathelicidin-derived LL-37 peptide but not the cationic steroid CSA-13. J Antimicrob Chemother 62(2):329–335. doi:10.1093/jac/dkn176

Chang HT, Tsai PW, Huang HH, Liu YS, Chien TS, Lan CY (2012) LL37 and hBD-3 elevate the beta-1,3-exoglucanase activity of *Candida albicans* Xog1p, resulting in reduced fungal adhesion to plastic. Biochem J 441(3):963–970. doi:10.1042/bj20111454

Cheng SC, Joosten LA, Kullberg BJ, Netea MG (2012) Interplay between *Candida albicans* and the mammalian innate host defense. Infect Immun 80(4):1304–1313. doi:10.1128/iai.06146-11

Cogen AL, Yamasaki K, Sanchez KM, Dorschner RA, Lai Y, MacLeod DT, Torpey JW, Otto M, Nizet V, Kim JE, Gallo RL (2010) Selective antimicrobial action is provided by phenol-soluble modulins derived from Staphylococcus epidermidis, a normal resident of the skin. J Invest Dermatol 130(1):192–200. doi:10.1038/jid.2009.243

Conti HR, Shen F, Nayyar N, Stocum E, Sun JN, Lindemann MJ, Ho AW, Hai JH, Yu JJ, Jung JW, Filler SG, Masso-Welch P, Edgerton M, Gaffen SL (2009) Th17 cells and IL-17 receptor signaling are essential for mucosal host defense against oral candidiasis. J Exp Med 206(2):299–311. doi:10.1084/jem.20081463

Conti HR, Baker O, Freeman AF, Jang WS, Holland SM, Li RA, Edgerton M, Gaffen SL (2011) New mechanism of oral immunity to mucosal candidiasis in hyper-IgE syndrome. Mucosal Immunol 4(4):448–455. doi:10.1038/mi.2011.5

Cottier F, Pavelka N (2012) Complexity and dynamics of host-fungal interactions. Immunol Res 53(1–3):127–135. doi:10.1007/s12026-012-8265-y

Cruz MC, Goldstein AL, Blankenship JR, Del Poeta M, Davis D, Cardenas ME, Perfect JR, McCusker JH, Heitman J (2002) Calcineurin is essential for survival during membrane stress in *Candida albicans*. EMBO J 21(4):546–559

Dale BA, Kimball JR, Krisanaprakornkit S, Roberts F, Robinovitch M, O'Neal R, Valore EV, Ganz T, Anderson GM, Weinberg A (2001) Localized antimicrobial peptide expression in human gingiva. J Periodontal Res 36(5):285–294

Dalle F, Wachtler B, L'Ollivier C, Holland G, Bannert N, Wilson D, Labruere C, Bonnin A, Hube B (2010) Cellular interactions of *Candida albicans* with human oral epithelial cells and enterocytes. Cell Microbiol 12(2):248–271. doi:10.1111/j.1462-5822.2009.01394.x

Den Hertog AL, van Marle J, van Veen HA, Van't Hof W, Bolscher JG, Veerman EC, Nieuw Amerongen AV (2005) Candidacidal effects of two antimicrobial peptides: histatin 5 causes small membrane defects, but LL-37 causes massive disruption of the cell membrane. Biochem J 388(Pt 2):689–695. doi:10.1042/BJ20042099

Den Hertog AL, Van Marle J, Veerman ECI, Valentijn-Benz M, Nazmi K, Kalay H, Grün CH, Van't Hof W, Bolscher JGM, Amerongen AVN (2006) The human cathelicidin peptide LL-37 and truncated variants induce segregation of lipids and proteins in the plasma membrane of *Candida albicans*. Biol Chem 387(10–11):1495–1502

Edgerton M, Koshlukova SE (2000) Salivary histatin 5 and its similarities to the other antimicrobial proteins in human saliva. Adv Dent Res 14:16–21

Edgerton M, Koshlukova SE, Araujo MW, Patel RC, Dong J, Bruenn JA (2000) Salivary histatin 5 and human neutrophil defensin 1 kill *Candida albicans* via shared pathways. Antimicrob Agents Chemother 44(12):3310–3316

Ernst JF, Tielker D (2009) Responses to hypoxia in fungal pathogens. Cell Microbiol 11(2):183–190. doi:10.1111/j.1462-5822.2008.01259.x

Eyerich S, Wagener J, Wenzel V, Scarponi C, Pennino D, Albanesi C, Schaller M, Behrendt H, Ring J, Schmidt-Weber CB, Cavani A, Mempel M, Traidl-Hoffmann C, Eyerich K (2011) IL-22 and TNF-alpha represent a key cytokine combination for epidermal integrity during infection with *Candida albicans*. Eur J Immunol 41(7):1894–1901. doi:10.1002/eji.201041197

Fan D, Coughlin LA, Neubauer MM, Kim J, Kim MS, Zhan X, Simms-Waldrip TR, Xie Y, Hooper LV, Koh AY (2015) Activation of HIF-1alpha and LL-37 by commensal bacteria inhibits *Candida albicans* colonization. Nat Med 21(7):808–814. doi:10.1038/nm.3871

Fleming A (1922) On a remarkable bacteriolytic element found in tissues and secretions, vol 93, 653. doi:10.1098/rspb.1922.0023

Gank KD, Yeaman MR, Kojima S, Yount NY, Park H, Edwards JE Jr, Filler SG, Fu Y (2008) SSD1 is integral to host defense peptide resistance in *Candida albicans*. Eukaryot Cell 7 (8):1318–1327. doi:10.1128/ec.00402-07

Ghannoum MA, Jurevic RJ, Mukherjee PK, Cui F, Sikaroodi M, Naqvi A, Gillevet PM (2010) Characterization of the oral fungal microbiome (mycobiome) in healthy individuals. PLoS Pathog 6(1):e1000713. doi:10.1371/journal.ppat.1000713

Gilmore TD, Wolenski FS (2012) NF-kappaB: Where did it come from and why? Immunol Rev 246(1):14–35. doi:10.1111/j.1600-065X.2012.01096.x

Gombart AF (2009) The vitamin D-antimicrobial peptide pathway and its role in protection against infection. Future Microbiol 4(9):1151–1165. doi:10.2217/fmb.09.87

Gow NA, van de Veerdonk FL, Brown AJ, Netea MG (2012) *Candida albicans* morphogenesis and host defence: discriminating invasion from colonization. Nat Rev Microbiol 10(2):112–122. doi:10.1038/nrmicro2711

Grahl N, Cramer RA Jr (2010) Regulation of hypoxia adaptation: an overlooked virulence attribute of pathogenic fungi? Med Mycol 48(1):1–15. doi:10.3109/13693780902947342

Hancock RE (1997) Peptide antibiotics. Lancet 349(9049):418–422. doi:10.1016/S0140-6736(97)80051-7

Harder J, Schroder JM (2002) RNase 7, a novel innate immune defense antimicrobial protein of healthy human skin. J Biol Chem 277(48):46779–46784. doi:10.1074/jbc.M207587200

Helmerhorst EJ, Breeuwer P, Van 't Hof W, Walgreen-Weterings E, Oomen LCJM, Veerman ECI, Amerongen AVN, Abee T (1999) The cellular target of histatin 5 on *Candida albicans* is the energized mitochondrion. J Biol Chem 274 (11):7286-7291

Helmerhorst EJ, Troxler RF, Oppenheim FG (2001) The human salivary peptide histatin 5 exerts its antifungal activity through the formation of reactive oxygen species. Proc Natl Acad Sci U S A 98(25):14637–14642. doi:10.1073/pnas.141366998

Hieshima K, Ohtani H, Shibano M, Izawa D, Nakayama T, Kawasaki Y, Shiba F, Shiota M, Katou F, Saito T, Yoshie O (2003) CCL28 has dual roles in mucosal immunity as a chemokine with broad-spectrum antimicrobial activity. J Immunol 170(3):1452–1461

Hube B (2004) From commensal to pathogen: stage—and tissue-specific gene expression of *Candida albicans*. Curr Opin Microbiol 7(4):336–341. doi:10.1016/j.mib.2004.06.003

Huttner KM, Bevins CL (1999) Antimicrobial peptides as mediators of epithelial host defense. Pediatr Res 45(6):785–794

Iliev ID, Funari VA, Taylor KD, Nguyen Q, Reyes CN, Strom SP, Brown J, Becker CA, Fleshner PR, Dubinsky M, Rotter JI, Wang HL, McGovern DP, Brown GD, Underhill DM (2012) Interactions between commensal fungi and the C-type lectin receptor Dectin-1 influence colitis. Science 336(6086):1314–1317. doi:10.1126/science.1221789

Jang WS, Bajwa JS, Sun JN, Edgerton M (2010a) Salivary histatin 5 internalization by translocation, but not endocytosis, is required for fungicidal activity in *Candida albicans*. Mol Microbiol 77(2):354–370. doi:10.1111/j.1365-2958.2010.07210.x

Jang WS, Bajwa JS, Sun JN, Edgerton M (2010b) Salivary histatin 5 internalization by translocation, but not endocytosis, is required for fungicidal activity in *Candida albicans*. Mol Microbiol 77(2):354–370

Johnson DA, Yeh CK, Dodds MW (2000) Effect of donor age on the concentrations of histatins in human parotid and submandibular/sublingual saliva. Arch Oral Biol 45(9):731–740

Jung SI, Finkel JS, Solis NV, Chaili S, Mitchell AP, Yeaman MR, Filler SG (2013) Bcr1 functions downstream of Ssd1 to mediate antimicrobial peptide resistance in *Candida albicans*. Eukaryot Cell 12(3):411–419. doi:10.1128/ec.00285-12

Kaiser V, Diamond G (2000) Expression of mammalian defensin genes. J Leukoc Biol 68(6):779–784

Kanda N, Kano R, Ishikawa T, Watanabe S (2011) The antimycotic drugs itraconazole and terbinafine hydrochloride induce the production of human beta-defensin-3 in human keratinocytes. Immunobiology 216(4):497–504. doi:10.1016/j.imbio.2010.08.008

Kao CY, Chen Y, Thai P, Wachi S, Huang F, Kim C, Harper RW, Wu R (2004) IL-17 markedly up-regulates beta-defensin-2 expression in human airway epithelium via JAK and NF-kappaB signaling pathways. J Immunol 173(5):3482–3491

Kawai K, Shimura H, Minagawa M, Ito A, Tomiyama K, Ito M (2002) Expression of functional Toll-like receptor 2 on human epidermal keratinocytes. J Dermatol Sci 30(3):185–194

Komatsu T, Salih E, Helmerhorst EJ, Offner GD, Oppenheim FG (2011) Influence of histatin 5 on *Candida albicans* mitochondrial protein expression assessed by quantitative mass spectrometry. J Proteome Res 10(2):646–655. doi:10.1021/pr100861k

Kopp ZA, Jain U, Van Limbergen J, Stadnyk AW (2015) Do antimicrobial peptides and complement collaborate in the intestinal mucosa? Front Immunol 6:17. doi:10.3389/fimmu.2015.00017

Koprivnjak T, Peschel A (2011) Bacterial resistance mechanisms against host defense peptides. Cell Mol Life Sci 68(13):2243–2254. doi:10.1007/s00018-011-0716-4

Koshlukova SE, Lloyd TL, Araujo MW, Edgerton M (1999) Salivary histatin 5 induces non-lytic release of ATP from *Candida albicans* leading to cell death. J Biol Chem 274(27):18872–18879

Krisanaprakornkit S, Weinberg A, Perez CN, Dale BA (1998) Expression of the peptide antibiotic human beta-defensin 1 in cultured gingival epithelial cells and gingival tissue. Infect Immun 66 (9):4222–4228

Krishnakumari V, Rangaraj N, Nagaraj R (2009) Antifungal activities of human beta-defensins HBD-1 to HBD-3 and their C-terminal analogs Phd1 to Phd3. Antimicrob Agents Chemother 53(1):256–260. doi:10.1128/aac.00470-08

Kumar R, Chadha S, Saraswat D, Bajwa JS, Li RA, Conti HR, Edgerton M (2011) Histatin 5 uptake by *Candida albicans* utilizes polyamine transporters Dur3 and Dur31 proteins. J Biol Chem 286(51):43748–43758. doi:10.1074/jbc.M111.311175

Lai Y, Gallo RL (2009) AMPed up immunity: how antimicrobial peptides have multiple roles in immune defense. Trends Immunol 30(3):131–141. doi:10.1016/j.it.2008.12.003

Lehrer RI, Ganz T (1999) Antimicrobial peptides in mammalian and insect host defence. Curr Opin Immunol 11(1):23–27

Lenardon MD, Whitton RK, Munro CA, Marshall D, Gow NAR (2007) Individual chitin synthase enzymes synthesize microfibrils of differing structure at specific locations in the *Candida albicans* cell wall. Mol Microbiol 66(5):1164–1173

Lewis RE (2009) Overview of the changing epidemiology of candidemia. Curr Med Res Opin 25 (7):1732–1740. doi:10.1185/03007990902990817

Ley RE, Backhed F, Turnbaugh P, Lozupone CA, Knight RD, Gordon JI (2005) Obesity alters gut microbial ecology. Proc Natl Acad Sci U S A 102(31):11070–11075. doi:10.1073/pnas.0504978102

Li M, Chen Q, Shen Y, Liu W (2009) *Candida albicans* phospholipomannan triggers inflammatory responses of human keratinocytes through Toll-like receptor 2. Exp Dermatol 18(7):603–610. doi:10.1111/j.1600-0625.2008.00832.x

Li R, Kumar R, Tati S, Puri S, Edgerton M (2013) *Candida albicans* flu1-mediated efflux of salivary histatin 5 reduces its cytosolic concentration and fungicidal activity. Antimicrob Agents Chemother 57(4):1832–1839. doi:10.1128/aac.02295-12

Lim JH, Ravikumar S, Wang YM, Thamboo TP, Ong L, Chen J, Goh JG, Tay SH, Chengchen L, Win MS, Leong W, Lau T, Foo R, Mirza H, Tan KS, Sethi S, Khoo AL, Chng WJ, Osato M, Netea MG, Wang Y, Chai LY (2015) Bimodal influence of vitamin D in host response to systemic candida infection-vitamin D dose matters. J Infect Dis 212(4):635–644. doi:10.1093/infdis/jiv033

Lis M, Liu TT, Barker KS, Rogers PD, Bobek LA (2010) Antimicrobial peptide MUC7 12-mer activates the calcium/calcineurin pathway in *Candida albicans*. FEMS Yeast Res 10(5):579–586. doi:10.1111/j.1567-1364.2010.00638.x

Lu Q, Samaranayake LP, Darveau RP, Jin L (2005) Expression of human beta-defensin-3 in gingival epithelia. J Periodontal Res 40(6):474–481. doi:10.1111/j.1600-0765.2005.00827.x

Lu Q, Jayatilake JA, Samaranayake LP, Jin L (2006) Hyphal invasion of *Candida albicans* inhibits the expression of human beta-defensins in experimental oral candidiasis. J Invest Dermatol 126 (9):2049–2056. doi:10.1038/sj.jid.5700346

Lupetti A, Paulusma-Annema A, Welling MM, Senesi S, van Dissel JT, Nibbering PH (2000) Candidacidal activities of human lactoferrin peptides derived from the N terminus. Antimicrob Agents Chemother 44(12):3257–3263

Mayer FL, Wilson D, Jacobsen ID, Miramon P, Grosse K, Hube B (2012) The novel *Candida albicans* transporter Dur31 Is a multi-stage pathogenicity factor. PLoS Pathog 8(3):e1002592. doi:10.1371/journal.ppat.1002592

Mazmanian SK, Liu CH, Tzianabos AO, Kasper DL (2005) An immunomodulatory molecule of symbiotic bacteria directs maturation of the host immune system. Cell 122(1):107–118. doi:10.1016/j.cell.2005.05.007

McFall-Ngai M, Hadfield MG, Bosch TC, Carey HV, Domazet-Loso T, Douglas AE, Dubilier N, Eberl G, Fukami T, Gilbert SF, Hentschel U, King N, Kjelleberg S, Knoll AH, Kremer N, Mazmanian SK, Metcalf JL, Nealson K, Pierce NE, Rawls JF, Reid A, Ruby EG, Rumpho M, Sanders JG, Tautz D, Wernegreen JJ (2013) Animals in a bacterial world, a new imperative for the life sciences. Proc Natl Acad Sci U S A 110(9):3229–3236. doi:10.1073/pnas.1218525110

Meiller TF, Hube B, Schild L, Shirtliff ME, Scheper MA, Winkler R, Ton A, Jabra-Rizk MA (2009) A novel immune evasion strategy of *Candida albicans*: Proteolytic cleavage of a salivary antimicrobial peptide. PLoS ONE 4 (4)

Moffa EB, Mussi MC, Xiao Y, Garrido SS, Machado MA, Giampaolo ET, Siqueira WL (2015) Histatin 5 inhibits adhesion of C. albicans to reconstructed human oral epithelium. Front Microbiol 6:885. doi:10.3389/fmicb.2015.00885

Moyes DL, Runglall M, Murciano C, Shen C, Nayar D, Thavaraj S, Kohli A, Islam A, Mora-Montes H, Challacombe SJ, Naglik JR (2010) A biphasic innate immune MAPK response discriminates between the yeast and hyphal forms of *Candida albicans* in epithelial cells. Cell Host Microbe 8(3):225–235. doi:10.1016/j.chom.2010.08.002

Moyes DL, Richardson JP, Naglik JR (2015) *Candida albicans*-epithelial interactions and pathogenicity mechanisms: scratching the surface. Virulence 6(4):338–346. doi:10.1080/21505594.2015.1012981

Moyes DL, Wilson D, Richardson JP, Mogavero S, Tang SX, Wernecke J, Hofs S, Gratacap RL, Robbins J, Runglall M, Murciano C, Blagojevic M, Thavaraj S, Forster TM, Hebecker B, Kasper L, Vizcay G, Iancu SI, Kichik N, Hader A, Kurzai O, Luo T, Kruger T, Kniemeyer O, Cota E, Bader O, Wheeler RT, Gutsmann T, Hube B, Naglik JR (2016) Candidalysin is a fungal peptide toxin critical for mucosal infection. Nature. doi:10.1038/nature17625

Murakami M, Lopez-Garcia B, Braff M, Dorschner RA, Gallo RL (2004) Postsecretory processing generates multiple cathelicidins for enhanced topical antimicrobial defense. J Immunol 172 (5):3070–3077

Naglik JR, Challacombe SJ, Hube B (2003) *Candida albicans* secreted aspartyl proteinases in virulence and pathogenesis. Microbiol Mol Biol Rev 67 (3):400–428, table of contents

Nakatsuji T, Gallo RL (2012) Antimicrobial peptides: old molecules with new ideas. J Invest Dermatol 132(3):887–895

Nijnik A, Hancock REW (2009) The roles of cathelicidin LL-37 in immune defences and novel clinical applications. Curr Opin Hematol 16(1):41–47

Nizet V, Johnson RS (2009) Interdependence of hypoxic and innate immune responses. Nat Rev Immunol 9(9):609–617. doi:10.1038/nri2607

Odds FC, Van Gerven F, Espinel-Ingroff A, Bartlett MS, Ghannoum MA, Lancaster MV, Pfaller MA, Rex JH, Rinaldi MG, Walsh TJ (1998) Evaluation of possible correlations between antifungal susceptibilities of filamentous fungi in vitro and antifungal treatment outcomes in animal infection models. Antimicrob Agents Chemother 42(2):282–288

Oppenheim FG, Xu T, McMillian FM, Levitz SM, Diamond RD, Offner GD, Troxler RF (1988) Histatins, a novel family of histidine-rich proteins in human parotid secretion. Isolation, characterization, primary structure, and fungistatic effects on *Candida albicans*. J Biol Chem 263(16):7472–7477

Ordonez SR, Amarullah IH, Wubbolts RW, Veldhuizen EJ, Haagsman HP (2014) Fungicidal mechanisms of cathelicidins LL-37 and CATH-2 revealed by live-cell imaging. Antimicrob Agents Chemother 58(4):2240–2248. doi:10.1128/aac.01670-13

Ostaff MJ, Stange EF, Wehkamp J (2013) Antimicrobial peptides and gut microbiota in homeostasis and pathology. EMBO Mol Med 5(10):1465–1483. doi:10.1002/emmm.201201773

Pahl R, Brunke G, Steubesand N, Schubert S, Bottner M, Wedel T, Jurgensen C, Hampe J, Schafer H, Zeissig S, Schreiber S, Rosenstiel P, Reiss K, Arlt A (2011) IL-1beta and ADAM17 are central regulators of beta-defensin expression in Candida esophagitis. Am J Physiol Gastrointest Liver Physiol 300(4):G547–553. doi:10.1152/ajpgi.00251.2010

Perlroth J, Choi B, Spellberg B (2007) Nosocomial fungal infections: epidemiology, diagnosis, and treatment. Med Mycol 45(4):321–346. doi:10.1080/13693780701218689

Peschel A, Sahl HG (2006) The co-evolution of host cationic antimicrobial peptides and microbial resistance. Nat Rev Microbiol 4(7):529–536. doi:10.1038/nrmicro1441

Peyssonnaux C, Boutin AT, Zinkernagel AS, Datta V, Nizet V, Johnson RS (2008) Critical role of HIF-1alpha in keratinocyte defense against bacterial infection. J Invest Dermatol 128(8):1964–1968. doi:10.1038/jid.2008.27

Pfaller MA, Diekema DJ (2007) Epidemiology of invasive candidiasis: a persistent public health problem. Clin Microbiol Rev 20(1):133–163. doi:10.1128/CMR.00029-06

Polke M, Hube B, Jacobsen ID (2015) Candida survival strategies. Adv Appl Microbiol 91:139–235. doi:10.1016/bs.aambs.2014.12.002

Puri S, Edgerton M (2014) How does it kill?: understanding the candidacidal mechanism of salivary histatin 5. Eukaryot Cell 13(8):958–964. doi:10.1128/ec.00095-14

Puri S, Friedman J, Saraswat D, Kumar R, Li R, Ruszaj D, Edgerton M (2015a) *Candida albicans* shed Msb2 and host mucins affect the candidacidal activity of salivary Hst 5. Pathogens 4 (4):752–763. doi:10.3390/pathogens4040752

Puri S, Li R, Ruszaj D, Tati S, Edgerton M (2015b) Iron binding modulates candidacidal properties of salivary histatin 5. J Dent Res 94(1):201–208. doi:10.1177/0022034514556709

Rapala-Kozik M, Bochenska O, Zawrotniak M, Wolak N, Trebacz G, Gogol M, Ostrowska D, Aoki W, Ueda M, Kozik A (2015) Inactivation of the antifungal and immunomodulatory properties of human cathelicidin LL-37 by aspartic proteases produced by the pathogenic yeast *Candida albicans*. Infect Immun 83(6):2518–2530. doi:10.1128/iai.00023-15

Rehaume LM, Jouault T, Chamaillard M (2010) Lessons from the inflammasome: a molecular sentry linking Candida and Crohn's disease. Trends Immunol 31(5):171–175. doi:10.1016/j.it.2010.01.007

Richardson JP, Moyes DL (2015) Adaptive immune responses to *Candida albicans* infection. Virulence 6(4):327–337. doi:10.1080/21505594.2015.1004977

Román E, Cottier F, Ernst JF, Pla J (2009) Msb2 signaling mucin controls activation of Cek1 mitogen-activated protein kinase in *Candida albicans*. Eukaryot Cell 8(8):1235–1249

Rosenfeld Y, Shai Y (2006) Lipopolysaccharide (Endotoxin)-host defense antibacterial peptides interactions: role in bacterial resistance and prevention of sepsis. Biochim Biophys Acta 1758 (9):1513–1522. doi:10.1016/j.bbamem.2006.05.017

Rumio C, Besusso D, Palazzo M, Selleri S, Sfondrini L, Dubini F, Menard S, Balsari A (2004) Degranulation of paneth cells via toll-like receptor 9. Am J Pathol 165(2):373–381. doi:10.1016/S0002-9440(10)63304-4

Schroeder BO, Wu Z, Nuding S, Groscurth S, Marcinowski M, Beisner J, Buchner J, Schaller M, Stange EF, Wehkamp J (2011) Reduction of disulphide bonds unmasks potent antimicrobial activity of human β 2-defensin 1. Nature 469(7330):419–423

Scott MG, Vreugdenhil AC, Buurman WA, Hancock RE, Gold MR (2000) Cutting edge: cationic antimicrobial peptides block the binding of lipopolysaccharide (LPS) to LPS binding protein. J Immunol 164(2):549–553

Scott A, Weldon S, Buchanan PJ, Schock B, Ernst RK, McAuley DF, Tunney MM, Irwin CR, Elborn JS, Taggart CC (2011) Evaluation of the ability of LL-37 to neutralise LPS in vitro and ex vivo. PLoS ONE 6(10):e26525. doi:10.1371/journal.pone.0026525

Selsted ME, Ouellette AJ (2005) Mammalian defensins in the antimicrobial immune response. Nat Immunol 6(6):551–557. doi:10.1038/ni1206

Situ H, Wei G, Smith CJ, Mashhoon S, Bobek LA (2003) Human salivary MUC7 mucin peptides: effect of size, charge and cysteine residues on antifungal activity. Biochem J 375(Pt 1):175–182. doi:10.1042/bj20030779

Sorensen O, Arnljots K, Cowland JB, Bainton DF, Borregaard N (1997) The human antibacterial cathelicidin, hCAP-18, is synthesized in myelocytes and metamyelocytes and localized to specific granules in neutrophils. Blood 90(7):2796–2803

Sorensen OE, Follin P, Johnsen AH, Calafat J, Tjabringa GS, Hiemstra PS, Borregaard N (2001) Human cathelicidin, hCAP-18, is processed to the antimicrobial peptide LL-37 by extracellular cleavage with proteinase 3. Blood 97(12):3951–3959

Sorgo AG, Heilmann CJ, Dekker HL, Brul S, De Koster CG, Klis FM (2010) Mass spectrometric analysis of the secretome of *Candida albicans*. Yeast 27(8):661–672

Stecher B, Hardt WD (2008) The role of microbiota in infectious disease. Trends Microbiol 16 (3):107–114. doi:10.1016/j.tim.2007.12.008

Steubesand N, Kiehne K, Brunke G, Pahl R, Reiss K, Herzig KH, Schubert S, Schreiber S, Folsch UR, Rosenstiel P, Arlt A (2009) The expression of the beta-defensins hBD-2 and hBD-3 is differentially regulated by NF-kappaB and MAPK/AP-1 pathways in an in vitro model of Candida esophagitis. BMC Immunol 10:36. doi:10.1186/1471-2172-10-36

Sun JN, Li W, Jang WS, Nayyar N, Sutton MD, Edgerton M (2008) Uptake of the antifungal cationic peptide Histatin 5 by *Candida albicans* Ssa2p requires binding to non-conventional sites within the ATPase domain. Mol Microbiol 70(5):1246–1260

Swidergall M, Ernst JF (2014) Interplay between *Candida albicans* and the antimicrobial peptide armory. Eukaryot Cell 13(8):950–957. doi:10.1128/ec.00093-14

Swidergall M, Ernst AM, Ernst JF (2013) *Candida albicans* mucin Msb2 is a broad-range protectant against antimicrobial peptides. Antimicrob Agents Chemother 57(8):3917–3922. doi:10.1128/AAC.00862-13

Swidergall M, van Wijlick L, Ernst JF (2015) Signaling domains of mucin Msb2 in *Candida albicans*. Eukaryot Cell 14(4):359–370. doi:10.1128/ec.00264-14

Szafranski-Schneider E, Swidergall M, Cottier F, Tielker D, Roman E, Pla J, Ernst JF (2012) Msb2 shedding protects *Candida albicans* against antimicrobial peptides. PLoS Pathog 8(2): e1002501. doi:10.1371/journal.ppat.1002501

Tomalka J, Azodi E, Narra HP, Patel K, O'Neill S, Cardwell C, Hall BA, Wilson JM, Hise AG (2015) Beta-Defensin 1 plays a role in acute mucosal defense against *Candida albicans*. 194 (4):1788–1795. doi:10.4049/jimmunol.1203239

Trautwein-Weidner K, Gladiator A, Nur S, Diethelm P, LeibundGut-Landmann S (2015) IL-17-mediated antifungal defense in the oral mucosa is independent of neutrophils. Mucosal Immunol 8 (2):221–231. doi:10.1038/mi.2014.57

Tsai PW, Yang CY, Chang HT, Lan CY (2011a) Characterizing the role of cell-wall beta-1,3-exoglucanase Xog1p in *Candida albicans* adhesion by the human antimicrobial peptide LL-37. PLoS ONE 6(6):e21394. doi:10.1371/journal.pone.0021394

Tsai PW, Yang CY, Chang HT, Lan CY (2011b) Human antimicrobial peptide LL-37 inhibits adhesion of *Candida albicans* by interacting with yeast cell-wall carbohydrates. PLoS ONE 6 (3)

Tsai PW, Cheng YL, Hsieh WP, Lan CY (2014) Responses of *Candida albicans* to the human antimicrobial peptide LL-37. J Microbiol 52 (7):581–589. doi:10.1007/s12275-014-3630-2

Underhill DM, Iliev ID (2014) The mycobiota: interactions between commensal fungi and the host immune system. Nat Rev Immunol 14(6):405–416. doi:10.1038/nri3684

Underhill DM, Pearlman E (2015) Immune interactions with pathogenic and commensal fungi: a two-way street. Immunity 43(5):845–858. doi:10.1016/j.immuni.2015.10.023

van 't Hof W, Veerman EC, Helmerhorst EJ, Amerongen AV (2001) Antimicrobial peptides: properties and applicability. Biol Chem 382 (4):597–619. doi:10.1515/BC.2001.072

Vandamme D, Landuyt B, Luyten W, Schoofs L (2012) A comprehensive summary of LL-37, the factotum human cathelicidin peptide. Cell Immunol 280(1):22–35. doi:10.1016/j.cellimm.2012.11.009

Vorobjeva NV, Pinegin BV (2014) Neutrophil extracellular traps: mechanisms of formation and role in health and disease. Biochemistry (Mosc) 79(12):1286–1296. doi:10.1134/s0006297914120025

Vylkova S, Jang WS, Li W, Nayyar N, Edgerton M (2007) Histatin 5 initiates osmotic stress response in *Candida albicans* via activation of the Hog1 mitogen-activated protein kinase pathway. Eukaryot Cell 6(10):1876–1888. doi:10.1128/ec.00039-07

Wagener J, Schneider JJ, Baxmann S, Kalbacher H, Borelli C, Nuding S, Kuchler R, Wehkamp J, Kaeser MD, Mailander-Sanchez D, Braunsdorf C, Hube B, Schild L, Forssmann WG, Korting HC, Liepke C, Schaller M (2013) A peptide derived from the highly conserved protein GAPDH is involved in tissue protection by different antifungal strategies and epithelial immunomodulation. J Invest Dermatol 133(1):144–153. doi:10.1038/jid.2012.254

Wah J, Wellek A, Frankenberger M, Unterberger P, Welsch U, Bals R (2006) Antimicrobial peptides are present in immune and host defense cells of the human respiratory and gastrointestinal tracts. Cell Tissue Res 324(3):449–456. doi:10.1007/s00441-005-0127-7

Walker LA, Munro CA, de Bruijn I, Lenardon MD, McKinnon A, Gow NA (2008) Stimulation of chitin synthesis rescues *Candida albicans* from echinocandins. PLoS Pathog 4(4):e1000040. doi:10.1371/journal.ppat.1000040

Wang G (2015) Improved methods for classification, prediction, and design of antimicrobial peptides. Methods Mol Biol 1268:43–66. doi:10.1007/978-1-4939-2285-7_3

Wang G, Li X, Wang Z (2009) APD2: the updated antimicrobial peptide database and its application in peptide design. Nucleic acids Res 37 (Database issue):D933–937. doi:10.1093/nar/gkn823

Wolk K, Kunz S, Witte E, Friedrich M, Asadullah K, Sabat R (2004) IL-22 increases the innate immunity of tissues. Immunity 21(2):241–254. doi:10.1016/j.immuni.2004.07.007

Yang D, Chen Q, Hoover DM, Staley P, Tucker KD, Lubkowski J, Oppenheim JJ (2003) Many chemokines including CCL20/MIP-3alpha display antimicrobial activity. J Leukoc Biol 74 (3):448–455

Zanetti M (2004) Cathelicidins, multifunctional peptides of the innate immunity. J Leukoc Biol 75 (1):39–48. doi:10.1189/jlb.0403147

Zasloff M (2002) Antimicrobial peptides of multicellular organisms. Nature 415(6870):389–395. doi:10.1038/415389a

Zhang L, Falla TJ (2010) Potential therapeutic application of host defense peptides. Methods Mol Biol 618:303–327. doi:10.1007/978-1-60761-594-1_19

Zhao C, Wang I, Lehrer RI (1996) Widespread expression of beta-defensin hBD-1 in human secretory glands and epithelial cells. FEBS Lett 396(2–3):319–322

Zhu W, Filler SG (2010) Interactions of *Candida albicans* with epithelial cells. Cell Microbiol 12 (3):273–282

Chapter 11
Genome Diversity and Dynamics in *Candida albicans*

Christophe d'Enfert, Marie-Elisabeth Bougnoux, Adeline Feri, Mélanie Legrand, Raphaël Loll-Krippleber, Timea Marton, Corinne Maufrais, Jeanne Ropars, Natacha Sertour and Emilie Sitterlé

Abstract The fungal pathogen *Candida albicans* shows significant diversity at the genetic and phenotypic levels. In this Chapter, we review our current knowledge of the *C. albicans* diploid genome and its variability, the genetic structure of the *C. albicans* population and the mechanisms that are involved in *C. albicans* genome dynamics, with a focus on the parasexual cycle and loss-of-heterozygosity events. We further explore the impact of genetic diversity and genome dynamics on *C. albicans* phenotypic diversity. Finally, we discuss how our current knowledge of *C. albicans* genetic diversity could be leveraged in the future in order to get insights in the mechanisms underlying important biological attributes that are subject to variations across *C. albicans* isolates.

C. d'Enfert (✉) · M.-E. Bougnoux · A. Feri · M. Legrand · R. Loll-Krippleber · T. Marton · J. Ropars · N. Sertour · E. Sitterlé
Unité Biologie et Pathogénicité Fongiques, Institut Pasteur, INRA, Paris, France
e-mail: christophe.denfert@pasteur.fr

M.-E. Bougnoux
Unité de Parasitologie-Mycologie, Service de Microbiologie Clinique, Hôpital Necker-Enfants-Malades, Assistance Publique Des Hôpitaux de Paris (APHP), Paris, France

A. Feri · R. Loll-Krippleber · E. Sitterlé
Cellule Pasteur, Univ. Paris Diderot, Sorbonne Paris Cité, rue du Docteur Roux, Paris, France

C. Maufrais
Institut Pasteur, Centre d'Informatique pour la Biologie, Paris, France

R. Loll-Krippleber (*Present address*)
Donnelly Centre for Cellular and Biomolecular Research, University of Toronto, Toronto, Canada

R. Prasad (ed.), *Candida albicans: Cellular and Molecular Biology*,
DOI 10.1007/978-3-319-50409-4_11

11.1 Introduction

Because of the centrality of *Candida albicans* as a fungal pathogen of humans (Brown et al. 2012; Azie et al. 2012), diversity at the phenotypic and genotypic levels within this species has long attracted interest. Indeed, variations in phenotypes such as virulence, biofilm formation, sensitivity to antifungals, etc., are often seen across clinical isolates (Maccallum et al. 2009; Wu et al. 2007; Li et al. 2003; Pujol et al. 2004). Moreover, it has been years since genotypic diversity has been described across *C. albicans* isolates at the karyotype and nucleotide levels (Chibana et al. 2000; Thrash-Bingham and Gorman 1993; Navarro-Garcia et al. 1995; Chu et al. 1993; Reagan et al. 1990; Schmid et al. 1990; Robert et al. 1995). The last two decades have seen major progresses in our description of *C. albicans* genotypic diversity through the application of highly discriminant typing methods to large cohorts of isolates (Bougnoux et al. 2007; Odds 2010; McManus and Coleman 2014) as well as in our knowledge of the genomic rearrangements that *C. albicans* isolates undergo as part of their commensal lifestyle or in response to stresses such as antifungal exposure (Selmecki et al. 2010; Loll-Krippleber et al. 2015a).

These progresses have been in part supported by the availability of the genome sequence of the *C. albicans* SC5314 reference strain (Jones et al. 2004) that has helped pioneer methods such as comparative genome hybridization (CGH) (Selmecki et al. 2005) and genome-scale single-nucleotide polymorphism (SNP) typing (Forche et al. 2004), the combination of which allows defining the nature and extent of genome changes. Nowadays, high-throughput sequencing is changing the scale and resolution at which analyses of the *C. albicans* genome can be conducted, providing the opportunity to develop population genomic studies and to track the genomic changes that occur in the course of commensalism, infection, and in response to antifungal treatments. Because access to genotypic variation at the genome-wide level can be obtained for large panels of *C. albicans* isolates, we can start to envision applying statistical genetics approaches such as genome-wide association studies (GWAS) in order to identify the genetic polymorphisms that underlie the phenotypic variations observed across *C. albicans* isolates. Hence, our aim in this Chapter is to review our knowledge of the genomics of *C. albicans*, its population structure, the mechanisms that are involved in *C. albicans* genome dynamics, and to discuss how this knowledge could be leveraged in the future in order to get insights in the mechanisms underlying important biological attributes that are subject to variations across *C. albicans* isolates.

11.2 The *Candida albicans* Genome Through the Lens of Whole Genome Sequencing

C. albicans is a predominantly diploid species with no reported sexual cycle. As will be discussed below, haploid, tetraploid, and aneuploid isolates have been described and yet a majority of the commensal and clinical isolates are found in the diploid state. Genome sequencing of strain SC5314, a clinical isolate obtained from a patient with disseminated infection (Gillum et al. 1984) and now used as a reference strain by those studying *C. albicans* at the molecular level, has landed strong support for diploidy (Jones et al. 2004). Indeed, heterozygosity was observed throughout much of the eight chromosomes (noted Chr1-7 and ChrR for the chromosome that harbors the rDNA repeats) that form the 2 × 15.845 Mb (2 × 14.28 Mb if rDNA repeats are excluded) *C. albicans* genome (Jones et al. 2004; van het Hoog et al. 2007). More than 60,000 polymorphisms—SNPs, insertions, deletions—have been identified in strain SC5314 and further characterization of derivatives of SC5314 that have undergone large loss-of-heterozygosity (LOH) events by genome sequencing has enabled to define the two haplotypes across each chromosome (Jones et al. 2004; Muzzey et al. 2013). Annotation and curation of the *C. albicans* genome has now revealed 6198 ORFs (Braun et al. 2005; Binkley et al. 2014; van het Hoog et al. 2007). Notably, 404 ORFs differ in size according to the haplotype (Genome version: A22-s06-m01-r03). These differences are often some multiples of three, consistent with deletions or expansions of repeated aminoacids. However, this is not the case for 89 instances, suggesting that at least in these cases one of the two haplotypes encodes a frame-shift induced shorter and possibly defective allele. Interestingly also, allele-specific expression (ASE) at the transcription and translation levels has been documented (Muzzey et al. 2013, 2014). These studies showed that ASE arising from transcription and translation are of a similar magnitude, both in terms of affected genes number and magnitude of the bias. Moreover, transcriptional and translational ASE often acted in common to favor the same allele, consistent with selective pressure tuning ASE at multiple regulatory steps. Overall, characterization of the *C. albicans* strain SC5314 indicates that diploidy contributes an important source of diversity at the isolate level with variations in transcription, translation, and protein composition that may impact strain behavior.

Genome sequences have now been established for other, nonepidemiologically-linked isolates of *C. albicans*. Butler et al. (2009) first reported the genome sequence of *C. albicans* strain WO-1 that is characterized by its ability to undergo white-opaque switching (see Sect. 11.3). More recently, Hirakawa et al. (2014) have reported the genome sequence of 21 isolates originating from invasive or superficial infections of oral or vaginal tissues. Again, heterozygosity was observed throughout much of the eight chromosomes for all these strains, with on average, a heterozygous SNP every 267 bases (Hirakawa et al. 2014). Inter-strain variation ranged from 1 variant every 1404 bases to 1 variant every 235 bases, with an average genome-wide nucleotide diversity between any two *C. albicans* isolates estimated at 0.37%. Higher diversity was observed at neutrally evolving codon positions, suggesting that many sites in the genome are under selective constraint. Importantly, comparison of the distribution of heterozygous SNPs along chromosomes has revealed that genome heterozygosity varies from 48 to 89% (Jones et al. 2004; Butler et al. 2009; Hirakawa et al. 2014). These variations in heterozygosity are predominantly due to large LOH events encompassing whole chromosomes or extending from varying sites within chromosomes to the telomere, consistent with observations made previously for other isolates using SNP typing or targeted sequencing (Diogo et al. 2009; Coste et al. 2007; Forche et al. 2004). Sites at which these LOH arise differ between strains, suggesting that they are isolate-specific events. This is except for LOH events on the right arm of ChrR that tend to originate from the rDNA region in all sequenced isolates. A similar phenomenon has been observed in *Saccharomyces cerevisiae* diploid isolates where LOH events extending from the rDNA to the right arm of ChrXII occur (Magwene et al. 2011). This possibly reflects conflicts between replication origin firing and transcription in the vicinity of the highly transcribed rDNA locus, such perturbations yielding DNA double strand breaks (DSB) whose repair by mitotic crossover or break-induced replication leads to LOH from the DNA DSB site to the telomere (see Sect. 11.4 for further details on the mechanisms involved in LOH).

In addition to LOH events, the characterization of the genomes of *C. albicans* isolates by Hirakawa et al. (2014) has revealed a high frequency of aneuploidies in these isolates, landing support to the idea that *C. albicans* is tolerant to aneuploidy (Selmecki et al. 2010). Indeed, 6 of the 21 sequenced isolates showed trisomy for one or several of the smallest chromosomes (Chr4–7). In addition, segmental regions of Chr1 and ChrR were found at trisomic and tetrasomic levels in two isolates. Another isolate was hemizygous for the final 200 kb on the right arm of Chr3. Such hemizygosity of the Chr3 right arm has also been observed in *C. albicans* strain WO-1 but encompassing a larger part of this chromosome (Butler et al. 2009). However, in a recent study where we have characterized the genomes of 145 *C. albicans* isolates of commensal and clinical origin, we only observed 7 strains with whole chromosome aneuploidies and one strain with segmental increase in ploidy (MEB, CD, CM, JR, NS and G. Sherlock, manuscript in preparation), suggesting that aneuploidies are not widespread across *C. albicans* isolates. The reason for the discrepancy between our study and that of Hirakawa et al. (2014) may possibly lie in previous exposure to antifungals, especially azoles, that has driven aneuploidies in the isolates sequenced by these latter authors. Strikingly, the isolates characterized by Hirakawa et al. (2014) present frequent homozygosity at the mating type like (*MTL*) locus: while nine isolates are *MTL*-heterozygous, six are ***MTLa***-homozygous and five, *MTL*α-homozygous. This contrasts with the frequency that has been observed in the *C. albicans* population for *MTL*-heterozygosity and -homozygosity. Indeed, studies by Lockhart et al. (2002) and Odds et al. (2007) using large collections of commensal and clinical isolates have shown only 3–8% of the isolates to be homozygous at the *MTL* locus. As will be discussed in Sects. 11.3 and 11.4, azole antifungal exposure is associated with the formation of LOH as well as aneuploidies of Chr5, on which the *MTL* locus is located. Thus, aneuploid/*MTL*-homozygous isolates sequenced by Hirakawa et al. (2014) may have been previously exposed to azole antifungals and this may somehow have biased the estimate of the frequency of aneuploidies in the *C. albicans* species.

Translocations, inversions, and chromosome size variations have also been reported in *C. albicans* (Selmecki et al. 2010). These events often involve the Major Repeat Sequence (MRS) that is composed of the length-variable RPS region flanked by the so-called HOK and RB2 regions (Chindamporn et al. 1998, 1995). RPS size variability is due to copy number variation of the $\sim$172 bp alts repeats (Chibana et al. 1994; Iwaguchi et al. 1992a). All chromosomes, except Chr3 contain a MRS (Chibana et al. 1998; Lephart et al. 2005). The MRS affects karyotypic variation by serving as a hot spot for chromosome translocation and by expanding and contracting internal repeats, thereby changing chromosome length (Chibana and Magee 2009). Additional chromosome size variations result from changes in the number of rDNA repeats on ChrR (Iwaguchi et al. 1992b). Strikingly, genome sequencing of *C. albicans* isolates could not reveal any example of translocation except for strain WO-1 that is characterized by reciprocal translocations between Chr7 and Chr4, Chr6 and Chr5, and Chr5 and Chr1, all involving the MRS (Chu et al. 1993; Butler et al. 2009; Hirakawa et al. 2014). In contrast, inversions were observed in almost all sequenced isolates, which contained between 1 and 12 total inverted regions ranging in size from 705 bp to 144 kb (Butler et al. 2009; Hirakawa et al. 2014). It is interesting to note that some inversions were shared by several of the 21 isolates sequenced by Hirakawa et al. (2014) suggesting an ancestral origin.

The study of Hirakawa et al. (2014) also provides insights in the conservation of ORFs across *C. albicans* isolates, suggesting little variation in ORF content in this species. Indeed, each of the 21 sequenced strains shares at least 6046 out of the 6189 ORFs identified in the SC5314 strain genome. Inversely, gene calling in these strains has identified an average of 6268 genes, above the number of identified genes in strain SC5314 (Hirakawa et al. 2014). These relatively minor differences in ORF number suggest that the *C. albicans* pan genome (*i.e.,* all genes identified in the *C. albicans* species) does not differ markedly from the *C. albicans* core genome (i.e., the genes shared by all *C. albicans* isolates). Changes in copy number of ORFs, independent of aneuploidies, have also been observed, such as in ORFs of retrotransposons, in the *CYP5* gene encoding a cyclophilin, and in members of the subtelomeric family of *TLO* genes. *TLO* genes encode Med2-like subunits of the Mediator complex with a proposed role in host adaptation and virulence (Anderson et al. 2012; Zhang et al. 2012; Haran et al. 2014; van het Hoog et al. 2007). Notably, besides changes in copy number, the *TLO* genes present variations in genome locations and the biological consequences for such variability remain to be explored.

Further differences between ORFs in sequenced isolates are reflected at the aminoacid level, due to heterozygous and homozygous SNPs relative to the reference genome. Although a precise count of all SNPs identified in the 21 sequenced genomes was not provided by Hirakawa et al. (2014), our unpublished data for 145 *C. albicans* isolates have identified 588,273 polymorphic sites of which 48.7% are located in coding regions. Notably, a majority of these SNPs are found in very few isolates (e.g., 20.0% are unique to one strain) suggesting recent emergence. By analyzing nucleotide substitution rates in ORFs, Hirakawa et al. (2014) have identified rapidly evolving proteins among which cell wall glycosylphosphatidylinositol (GPI)-anchored proteins and cell wall biogenesis enzymes. This is consistent with the central role played by the cell wall in host-pathogen interactions and adaptation to changing environments and in line with previous studies that have explored genes under positive selection when comparing pathogenic and nonpathogenic yeasts (Butler et al. 2009).

In summary, the genomic comparison of several *C. albicans* isolates has revealed significant variation in terms of heterozygosity and—debatably—ploidy but to a lesser extent at the gene content level. Changes in gene expression, protein composition, and protein structure due to SNP genotypes may therefore represent the major source of phenotypic variation in this species.

11.3 *Candida albicans* Genetic Diversity Through the Lens of Molecular Typing Studies

Although genome sequencing of independent isolates could provide a comprehensive view of the genetic diversity within the *C. albicans* species, this is still partial, as only a limited number of isolates have been sequenced. Our current understanding of *C. albicans* genetic diversity at the population level is the consequence of molecular typing of isolates collected worldwide. Three major typing methods with high discriminatory power and good inter-laboratory portability have been predominantly used over the years: DNA fingerprinting by Southern hybridization with the probe Ca3, Multi-Locus Sequence Typing (MLST), and Multi-Locus Microsatellite Typing (MLMT). The Ca3 probe contains sequences of repeated elements dispersed throughout the *C. albicans* genome. Ca3 typing allows high discrimination and quantification of genetic distances between unrelated strains and has emerged as being especially helpful for understanding the epidemiology of *C. albicans* infections, assessing the genetic structure of *C. albicans* and identifying microevolutions (indels of the repeated element) within infecting strains over time (Pujol et al. 2002; Soll 2000; Lockhart et al. 1995; Pujol et al. 1999). MLST is a highly discriminatory method based on the analysis of nucleotide polymorphisms within the sequences of six–eight PCR-generated 400–500 bp internal fragments of housekeeping genes (loci) (Bougnoux et al. 2002; Tavanti et al. 2003). A consensus *C. albicans* MLST scheme with seven loci is now commonly used (Bougnoux et al. 2003) and data obtained by laboratories worldwide are collectively available through a web-based database now at http://pubmlst.org/calbicans/ (Bougnoux et al. 2004). Because *C. albicans* is a diploid species, nucleotide sequences obtained through direct sequencing of PCR products show heterozygous positions and strains are described by a Diploid Sequence Type (DST) (Bougnoux et al. 2002). As of May 25, 2016, the *C. albicans* database contained 3080 unique DSTs and information for 4142 isolates. Finally, Multi-Locus Microsatellite Typing (MLMT) involves sizing of amplified microsatellites through capillary electrophoresis and shows high discriminatory power when a well-defined combination of microsatellites is used (Botterel et al. 2001; Sampaio et al. 2005; L'Ollivier et al. 2012). However, in contrast to Ca3 typing and MLST, MLMT may have less value for population studies as the homology of alleles may not always reflect common inheritance but independent mutations (Orti et al. 1997).

These different typing methods have revealed groups of genetically related isolates, referred to as clades. Importantly, typing of the same isolates by these different methods reveals the same grouping (Bougnoux et al. 2008; Odds et al. 2007). Because MLST has been applied to the largest set of *C. albicans* isolates, it currently provides the most accurate representation of the population structure and diversity in this species. A landmark study by Odds et al. (2007), analyzing a panel of 1391 *C. albicans* isolates has revealed 17 clades of which clades 1, 2, 3, 4, and 11 (equivalent to clades I, II, III, SA, and E defined by Ca3 typing) are the most populous. While a majority of the isolates (96.7%) could be assigned to these clades, some remained singletons. These singletons could reflect the occurrence of undersampled clades in the *C. albicans* population or possible recombinants between isolates from different clades. In this respect, a more recent study investigating isolates originating from South Korea has revealed an eighteenth clade (Shin et al. 2011).

The study by Odds et al. (2007) and the discovery of an additional clade with isolates mostly originating from Eastern Asia (Shin et al. 2011) has landed support to the geographic origin of clades although isolates from any given clade are found worldwide. For instance, North American isolates were predominantly assigned to clades 1 and 3, South American isolates tended to cluster in clade 8, African isolates were often found in clade 4 (Odds et al. 2007). Clade 13 is distantly related to the other clades and clusters isolates that have been referred to as *C. africana* (Tietz et al. 2001). Yet, their genetic distance to other *C. albicans* isolates does not support that they form an independent species. It is also notable that the clade distribution of isolates from animal origin seems to differ from what we observed with human isolates (Jacobsen et al. 2008; Bougnoux et al. 2004; Wrobel et al. 2008). Yet, these studies should be taken with caution because of the paucity of animal isolates and the specific context in which these are obtained. Therefore, *C. albicans* clades may have their origins in ancestral geographic isolation of human and (possibly) animal populations and it is thus questionable whether their divergence at the genetic level simply corresponds to drift or reflects adaptation to specific niches. This question will probably be best resolved through large population genomics studies. In this respect, Hirakawa et al. (2014) could recapitulate the population structure inferred from MLST when using data for 112,223 informative SNPs across 21 genome-sequenced isolates. Yet, this isolate sample is likely too small to interrogate *C. albicans* long-term evolutionary history. Similarly, we could recapitulate the MLST-deduced population structure of 145 isolates when using 266,468 informative SNPs deduced from genome sequencing of these isolates (MEB, CD, CM, JR, NS, and G. Sherlock, manuscript in preparation). This confirms the power of MLST in establishing genuine genetic relationships between isolates, although isolates with identical DST often show differences at the genomic level, especially with respect to LOH events (MEB, CD, NS, and ES, manuscript in preparation) or microsatellites (Odds 2010).

One of the key questions that may arise from strain typing studies is whether there is any correlation between clade assignment and phenotypes, among which the ability of strains to cause different forms of infection. This has been explored by

Odds et al. (2007) who did not reveal any enrichment for invasive isolates in any of the 17 clades encompassing 1391 *C. albicans* isolates. However, these authors did find that MLST clade 1 (equivalent to clade I and group A defined by Ca3 typing; (Bougnoux et al. 2008; Odds et al. 2007; Schmid et al. 1999)) was enriched with isolates from superficial infection or commensal carriage, an observation that needs to be firmly confirmed through investigation of additional cohorts of *C. albicans* isolates. Importantly, clade 1 is characterized by a significantly higher prevalence than other clades in most geographical areas and in most categories of patients or types of infection and has been coined a general-purpose genotype (GPG) (Schmid et al. 1999; Odds 2010). Zhang et al. (2003) have identified the *ALS7* gene as one of the genes that could be responsible for the success of the GPG/clade 1 strains as pathogens. *ALS7* encodes a member of the Als family of agglutinin-like cell surface adhesins that have roles in adherence of *C. albicans* to various surfaces and interaction with epithelial and endothelial cells (Hoyer and Cota 2016). Als proteins are characterized by tandem-repeat (108-bp repeats) and VASES domain (15-bp repeats) domains. Isolates within the GPG cluster tend to have more tandem repeats than other isolates and to differ by the nature of their VASES repeats (Zhang et al. 2003). Variations across clades in the number of tandem-repeats in members of the Als family, namely Als2–7 and Als9, and in members of the Hyr/Iff family of cell wall proteins, namely Hyr1, Iff1/Rbr3, and Iff2/Hyr3 have been observed in other studies (Oh et al. 2005; Maccallum et al. 2009; Boisrame et al. 2011). Yet, it is unclear whether these structural differences reflect variations in phenotypes. On the one hand, clade differentiation for these repeats and the fact that they are under selection suggest a genuine contribution (Zhang et al. 2003). Moreover, variation in the number of tandem-repeats in the Als3 protein impacts the interaction of *C. albicans* with host cells (Oh et al. 2005). On the other hand, there was no correlation between alleles in the *ALS* and *HYR* genes and virulence in a mouse model of systemic infection (Maccallum et al. 2009).

Another distinctive feature of isolates in MLST clade 1 lies in their frequent resistance or lower susceptibility to flucytosine (5-FC), one of the antifungal agents used for treatment of invasive fungal diseases (Pujol et al. 2004). This has been ascribed to a unique polymorphism (C301T resulting in a change of arginine to cysteine) in the *FUR1* gene encoding uracil phosphoribosyltransferase (Dodgson et al. 2004; Hope et al. 2004). Isolates with the CT and TT genotypes are unique to clade 1 and show reduced susceptibility and resistance to 5-FC, respectively, (Dodgson et al. 2004). Consistently, 5-FC-resistant isolates that are not from clade 1 have the CC genotype and use other resistance mechanisms (Tavanti et al. 2005; Hope et al. 2004). It has been reported that clade 1 isolates, and to a lesser extent clade 3 isolates, show frequent resistance to terbinafine, an antifungal agent that is never used for the treatment of *C. albicans* infections (Odds 2009), and that MLST clade 4 strains (equivalent to clade SA defined by Ca3 typing) could have increased amphotericin B resistance (Blignaut et al. 2005). However, further studies will be needed to confirm these associations. In contrast, no association between azole resistance and specific clades has been observed (Tavanti et al. 2005).

Maccallum et al. (2009) have thoroughly investigated phenotypic variations across 10 isolates in each of clades 1–4: growth rate at different temperatures, biofilm formation, ability to use proteins as nitrogen source, growth in the presence of high salt, acid phosphatase production, alcian blue staining, adherence to biliary epithelial cells, adherence to catheter, gene expression at the genome-wide level in media that support growth in the yeast or hyphal forms, and virulence in a mouse model of disseminated infection were tested. Overall, few significant differences could be observed, with clade 1 showing a higher proportion of salt-tolerant isolates and clade 2 isolates showing low acid phosphatase activity (Maccallum et al. 2009). A limited number of genes showed clade-specific expression, most of them distinguishing clade 3 isolates from those in clades 1, 2, and 4. Yet, a link between these gene expression differences and a particular phenotype has not been established. In a more recent study, Jung et al. (2016) have shown that clade 18 bloodstream isolates secrete increased levels of the quorum-sensing molecule farnesol in biofilms than isolates from other clades. These isolates are also associated to lower clinical severity scores but it is unknown whether there is a link between these two observations (Jung et al. 2016).

The work summarized above illustrates the search for phenotypes (and possibly their underlying genotypes) that distinguish isolates in one clade from isolates in the others. This might be limited to a few cases and it is also possible that some phenotypes differentiate isolates from a group of clades to isolates from a complementary group of clades. In a recent study, Calderon-Norena et al. (2015) have shown that alternative translation initiation results in the production of long and short forms of the Ace2 transcription factor, namely Ace2^{L} and Ace2^{S}, and that Ace2^{L} is necessary for septum integrity in *C. albicans* hyphae. Interestingly, a polymorphism was observed in some *C. albicans* isolates that introduced a stop codon between the start codons for Ace2^{L} and Ace2^{S}, resulting in the sole production of Ace2^{S} and defects in hyphal septation. When 144 isolates distributed across the major clades were scanned for their genotypes at this polymorphic site, almost all clade 1 and clade 3 isolates had a heterozygous genotype allowing for the production of the Ace2^{L} and Ace2^{S}, whereas isolates from clades 2, 4, and 11 and other minor clades had a homozygous genotype only allowing the production of Ace2^{S}. At this stage, it is still unclear why a new, *C. albicans*-specific function for Ace2 during hyphal growth has been selected in some clades, possibly fine-tuning the persistence of multicellular structures in the host.

In summary, molecular typing methods have revealed a highly structured population in *C. albicans*, with at least 18 well-separated clades that may have emerged due to geographic or host isolation. Isolates of clade 1 are the most frequently encountered but the basis for such predominance remains to be understood. Some clade-specific phenotypes and genotypes have been identified such as *FUR1*-dependent 5-FC resistance in clade 1 isolates. However, the search for phenotypes that distinguish clades or groups of clades and the underlying genotypes is still an ongoing challenge that may inform us on the evolutionary pressures that have led to the emergence of clades and the extant population structure of *C. albicans*. While the search has so far focused on the phenotypes, it is likely that population genomics will reverse the approach as it allows accessing the full repertoire of the genetic diversity across a large number of isolates and possibly revealing clade(s)-specific, positively selected polymorphisms, and their associated phenotypes.

11.4 *Candida albicans*: A Clonally Reproducing Species Equipped with a Parasexual Cycle

As mentioned above, *C. albicans* is a predominantly diploid species and has long been thought to be devoid of any form of sexuality, even though isolates with altered ploidy had been observed. Genome sequencing of *C. albicans* SC5314 (Jones et al. 2004) has dramatically changed this view. Indeed, Hull et al. (1999), by mining the *C. albicans* SC5314 genome, could identify a mating-type like (*MTL*) locus with similarity to the mating type (*MAT*) locus of *S. cerevisiae* and other sexually-reproducing yeasts (Lee et al. 2010) (Fig. 11.1a). In *C. albicans* SC5314, one homolog of Chr5 harbors the *MTL**a*** idiotype that encodes the Mtl**a**1 homeodomain protein with orthology to *S. cerevisiae* Mat**a** and the Mtl**a**2 HMG domain protein, whereas the other Chr5 homolog harbors the *MTLα* idiotype encoding Mtlα1 and Mtlα2 that are orthologues of *S. cerevisiae* Matα1 and Matα2, respectively. Notably, the *C. albicans MTL**a*** and *MTLα* loci differ from the *S. cerevisiae MAT**a*** and *MATα* loci by the occurrence of additional genes besides *MTL**a**1*, *MTL**a**2*, *MTLα1*, and *MTLα2*, namely *PAP**a**/PAPα*, *OBP**a**/OBPα*, and *PIK**a**/PIKα* encoding proteins with similarities to poly(A) polymerases, oxysterol binding proteins, and phosphatidylinositol kinases. Further works by the Johnson's and Magee's groups demonstrated that strains that had become homozygous at the *MTL* locus through genetic engineering (*MTL**a**/Δ* or *Δ/MTLα*) or induced LOH (*MTL**a**/MTL**a*** or *MTLα/MTLα*) and that had complementary *MTL* genotypes could undergo mating and form tetraploids (Hull et al. 2000; Magee and Magee 2000). It was then observed that *MTL* homozygotes were prone to undergo white-to-opaque phenotypic switching, white cells being the commonly-encountered yeast form of *C. albicans* and opaque cells having an elongated shape with their surface decorated by pimples and being exquisitely efficient at mating (Lockhart et al. 2002; Miller and Johnson 2002). Finally, it was shown that tetraploids could return to the diploid stage as a consequence of concerted random chromosome losses, establishing the concept of a parasexual cycle that allows alternating between diploid and tetraploid forms in *C. albicans* (Fig. 11.1b; Bennett and Johnson 2003). Indeed, despite the occurrence in the *C. albicans* genome of genes encoding known components of the meiotic process (Tzung et al. 2001), no evidence for meiosis has been obtained to date in this species and this diploid-tetraploid cycle cannot be coined a sexual cycle. In addition, no active mechanism allowing alternation between *MTL**a**/α*, MTL***a**/**a***, and MTL*α/α*, such as the action of an endonuclease enzyme within the *MTL* locus, has been found. Nevertheless, the parasexual cycle results in chromosome shuffling as well as rare recombination events, mainly gene conversions, between

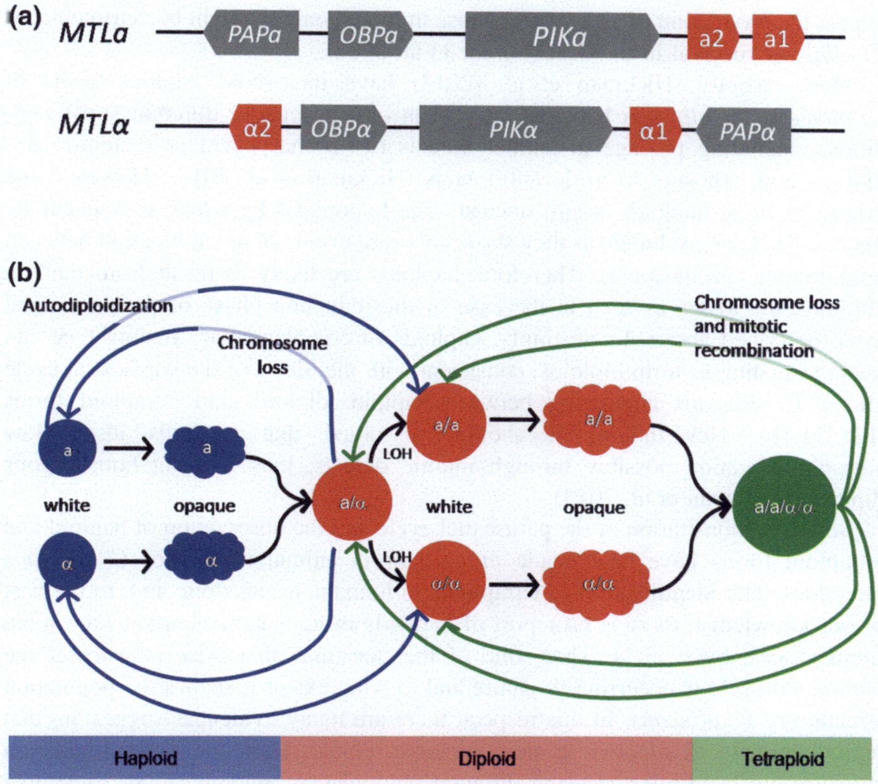

Fig. 11.1 Parasexuality and the haploid, diploid and tetraploid states of *Candida albicans*. **a** Schematic representation of the *MTLa* and *MTLα* idiotypes. The genes encoding transcription factors responsible for cellular identity are shown in *red* while the *OBP, PAP* and *PIK* genes that are not involved in cellular identity but differ between the two idiotypes are shown in *grey*. **b** Schematic representation of the haploid-diploid-tetraploid life cycle of *C. albicans. C. albicans* is predominantly existing in the diploid state with heterozygosity at the *MTL* locus. Homozygosis at the *MTL* locus allows white-opaque phenotypic switching and mating. Transition from tetraploidy to diploidy or diploidy to haploidy is independent of meiosis and involves random concerted chromosome loss with the intermediate production of aneuploids. Haploids are shown in *blue*, diploids in *red* and tetraploids in *green*

homologous chromosomes indicating that it can be a source of genetic (and phenotypic) diversity in *C. albicans* (Bennett and Johnson 2003; Forche et al. 2008). Interestingly, recombination during the parasexual cycle is dependent on a functional Spo11 protein, whose homolog in *S. cerevisiae* is involved in initiating cross-overs during meiosis (Keeney et al. 1997). It has been proposed that parasexuality might be beneficial to *C. albicans* over sexuality as it can generate aneuploid strains with increased fitness under certain growth conditions (see below) or bypass the production of ascospores whose immunogenicity might be detrimental to *C. albicans* survival in the host (Forche et al. 2008).

More recently, Hickman et al. (2013) have uncovered haploid forms of *C. albicans*. *C. albicans* haploids have been obtained under different stress conditions, including passage in animal models of oropharyngeal or systemic candidiasis and exposure to azole antifungals (Hickman et al. 2013). However, the origin for these haploids is still unclear. The hypothesis by which they occur by meiosis has been excluded, as they show only rare events of recombination between homologous chromosomes. Therefore, haploids are likely to result from random chromosome losses as seen in the case of the reduction phase of the parasexual cycle described above. Interestingly, haploids of complementary mating type can undergo mating to form diploids, consistent with the idea that the parasexual cycle allows *C. albicans* alternating between haploid, diploid, and tetraploid forms (Fig. 11.1b). Nevertheless, it should be noted that haploids also show auto-diploidization possibly through mitotic defects, thus forming homozygous diploids (Hickman et al. 2013).

So far, demonstration of the parasexual cycle and the observation of haploid and tetraploid forms have been made *in vitro* or in animal models of *C. albicans* infections. The identification of tetraploids in humans is anecdotic and, to the best of our knowledge, there is no report of haploids in humans. Aneuploids are more frequent (see Sect. 11.1). Thus, one of the questions that arise is whether the parasexual cycle is occurring in nature and to what extent it shapes the population structure of *C. albicans*. In this respect, there are many evidences suggesting that reproduction in *C. albicans* is predominantly clonal. First, the same population structure has been inferred using different phylogenetic markers, SNPs, repetitive sequences or microsatellites (Odds et al. 2007; Bougnoux et al. 2008) (MEB, CD, CM, JR, NS, and G. Sherlock, manuscript in preparation). Second, highly significant linkage disequilibrium was found upon analysis of MLST data (Bougnoux et al. 2008). In addition, an excess of homozygosity has been observed at polymorphic sites population-wide, a phenomenon referred to as the Wahlund effect and indicative of the absence of gene flow between clades. In contrast, departure from the Hardy–Weinberg equilibrium and an excess of heterozygosity was observed within each clade at the clade-specific polymorphic sites, indicating low recombination within clades. Nevertheless, these observations do not fully negate the possibility of recombination events within the *C. albicans* population that could arise from parasexuality. For instance, it has been proposed that the diversity in the haplotypes at MLST loci could result from the combination of mitotic recombination events and parasexual chromosome exchange (Odds et al. 2007). The presence of shared polymorphisms across clades could also reflect genetic exchange, even though it is probably best explained by their presence in highly heterozygous and sexual ancestors of extant *C. albicans* (Anderson et al. 2001). Further indications that parasexuality is relevant to *C. albicans* diversity were recently provided by Zhang et al. (2015). Indeed, these authors observed that the *MTL**a**1*, *MTL**a**2*, *MTLα1*, and *MTLα2* mating genes are under natural selection based on the analysis of nucleotide substitution rates across *Candida* species with parasexual or sexual reproduction. The low frequency of mutations abolishing mating in these genes was also indicative that they were under selection (Zhang et al. 2015). Further work by these authors showed that mating could generate "fusants" with increased fitness relative to their *MTL*-homozygous parents as well as *MTL*-heterozygous isolates (Zhang et al. 2015). Fusants with augmented fitness were more frequent in laboratory conditions (a new environment) than in an animal model of *C. albicans* commensalism (a natural environment), consistent with the notion that mating could lead to new allele combinations that are more likely to be beneficial in new environments. Moreover, fusants derived from strains with the lowest fitness benefited most from mating thus providing a probable explanation as why mating remains under selection in *C. albicans,* i.e., to provide a salvage pathway to isolates that have lost fitness (Zhang et al. 2015). As will be discussed in the next section, LOH is one of the frequent genomic rearrangements that occurs in *C. albicans,* either spontaneously or in response to environmental stresses, and can lead to reduced fitness, in particular due to the homozygosis of recessive deleterious alleles. LOH also allows for *MTL*-homozygosity and mating capability. Hence, mating of individuals with different forms of LOH might provide *C. albicans* the restoration of heterozygosity and fitness (Zhang et al. 2015; Schmid et al. 2016). Yet, it is likely that such rare parasexuality events barely influence the *C. albicans* population structure, thus explaining why this structure is predominantly clonal even though parasexuality is still active.

In summary, population genetics, molecular, and experimental evolution data point to a model whereby the *C. albicans* species has lost sexual reproduction and adopted a predominantly clonal mode of reproduction that should ineluctably result in the accumulation of deleterious mutations and species extinction, although diploidy would contribute to limit this fate. In this context, infrequent parasexuality appears to have been retained in order to allow isolates of reduced fitness emerging by loss-of-heterozygosity to regain fitness through mating and chromosome reassortments (Schmid et al. 2016). Further support for this model is needed including data in the human host, which are still lacking to illustrate the genuine functionality of the parasexual cycle. Tracing the events that have led from sexual to clonal reproduction and parasexualilty in *C. albicans* also represents a challenge for future research.

11.5 Genomic Plasticity in *Candida albicans*: The Case of Loss-of-Heterozygosity

Previous sections have alluded to the high plasticity of the *C. albicans* genome with mentions of chromosome translocations, aneuploidies, and LOH events. In particular, LOH events are especially frequent as they are observed in all isolates that have been characterized extensively by SNP typing or sequencing, possibly affecting all chromosomes (Diogo et al. 2009; Jones et al. 2004; Butler et al. 2009;

Angebault et al. 2013; Hirakawa et al. 2014; Ford et al. 2015). In this respect, it is noteworthy that more than 50% of the events that distinguish isolates within clonal clusters as defined by the analysis of MLST data are LOH events (Odds et al. 2007; Bougnoux et al. 2008). Importantly, LOH are a source of microevolution during commensalism (Bougnoux et al. 2006; Diogo et al. 2009). As discussed in the previous section, heterozygosity is probably beneficial to *C. albicans* adaptation to new environments. Consistently, there is evidence that *C. albicans* fitness in vitro seems negatively correlated with homozygosity and *C. albicans* autodiploids, i.e., fully homozygous diploids, show dramatically reduced fitness (Hickman et al. 2013; Hirakawa et al. 2014; Abbey et al. 2011). Hence, this section will focus on our understanding of the mechanisms that lead to LOH, the impact of the environment on LOH and the beneficial or detrimental effects of LOH to *C. albicans* fitness.

LOH events can reflect the mechanisms used by *C. albicans* to handle DNA double strand breaks (DSBs) or be the consequence of chromosome nondisjunction events during mitosis. As shown in Fig. 11.2a, repair of DNA DSBs by gene conversion without crossover (GC) explains short-range LOH, while repair of DNA DSBs by either break-induced replication (BIR) or mitotic crossover (MCO) leads to LOH events that extend from the DNA DSB site to the telomere. In the absence of DNA DSB repair or upon chromosome nondisjunction, segmental or whole chromosome losses (SCL/WCL) are observed and the loss of a chromosome is often followed by a reduplication event (Symington et al. 2014). As mentioned above, examples of BIR/MCO or WCL followed by reduplication are easily detectable in the genome of sequenced isolates with BIR/MCO events being more frequent than WCL (Jones et al. 2004; Butler et al. 2009; Hirakawa et al. 2014). For instance, in their analysis of 21 genomes, Hirakawa et al. (2014) could detect BIR/MCO at an average frequency of 7.86 events/strain and WCL at an average frequency of 0.33 events/strain. SCL have also been observed although at a much lower frequency (one example in the set of 21 genomes analyzed by Hirakawa et al. (2014)). GC events, as they lead to short tracts of homozygosis within heterozygous regions, are more difficult to detect. Yet, our experience comparing closely related isolates shows that these strains differ by many short-range LOH events that can only be explained by GC events (ES, MEB, and CD, unpublished results).

Molecular tools have been developed in order to analyze the frequency at which LOH events occur in strain SC5314 and the extent to which they are influenced by changes in environmental conditions, including passage in the host, and mutations. These tools rely on the use of counter-selectable marker genes such as *URA3* and *GAL1* that confer distinctive phenotypes when in a heterozygous or a homozygous null state (Boeke et al. 1984; Forche et al. 2003; Gorman et al. 1992). *URA3* encodes the orotidine-5-phosphate decarboxylase; while *URA3/URA3* and *URA3/ura3* strains are prototrophic for uridine and sensitive to 5-fluoroorotic acid (5-FOA), *ura3/ura3* strains are auxotrophic for uridine and resistant to 5-FOA (Fig. 11.2b). *GAL1* encodes a galactokinase; while *GAL1/GAL1* and *GAL1/gal1* strains are sensitive to 2-deoxygalactose (2-DG), *gal1/gal1* strains can form colonies on 2-DG containing medium (Fig. 11.2c). Hence, measuring the number of

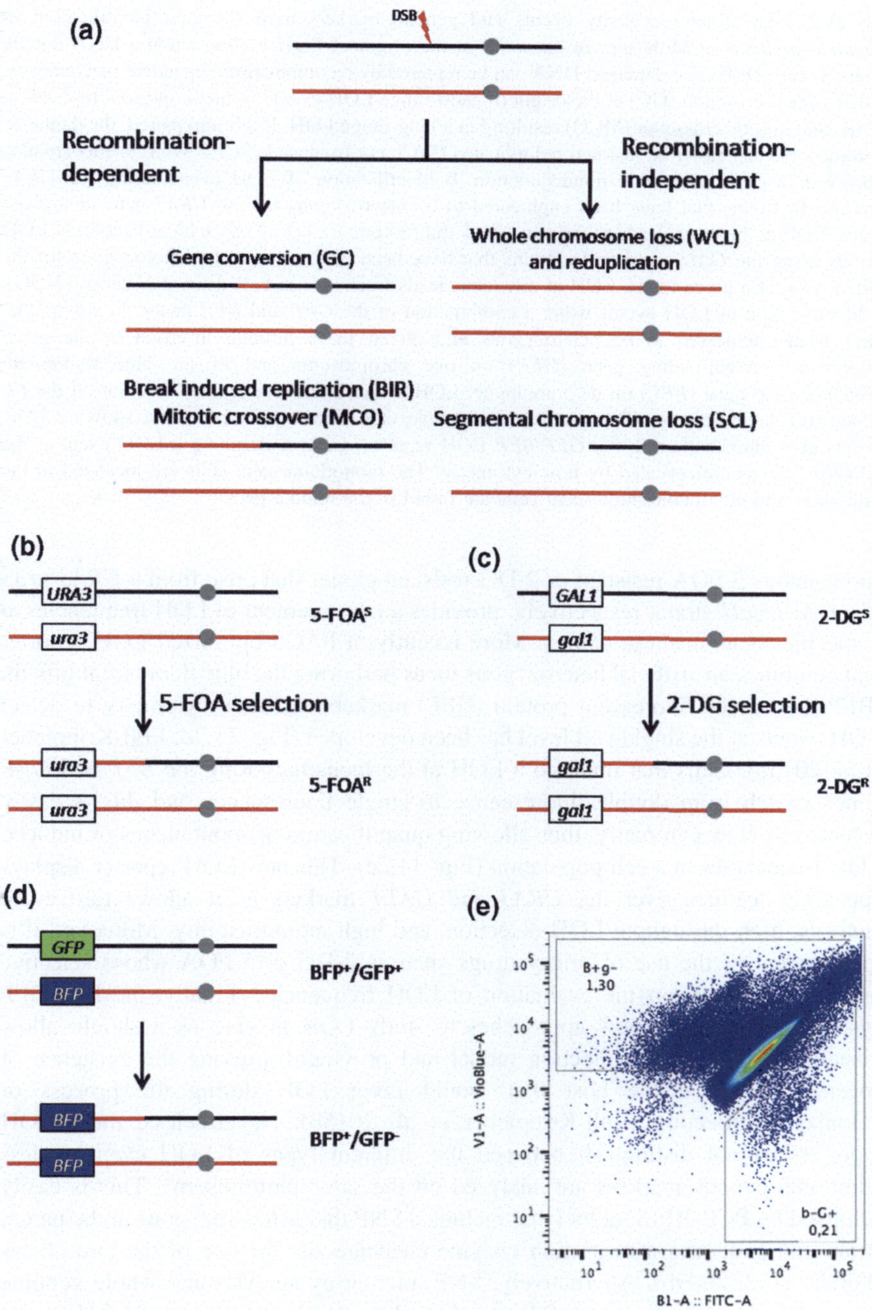
(a)
DSB
Recombination-dependent
Recombination-independent
Gene conversion (GC)
Whole chromosome loss (WCL) and reduplication
Break induced replication (BIR)
Mitotic crossover (MCO)
Segmental chromosome loss (SCL)
(b)
URA3
ura3
5-FOA^S
5-FOA selection
5-FOA^R
(c)
GAL1
gal1
2-DG^S
2-DG selection
2-DG^R
(d)
GFP
BFP
BFP+/GFP+
BFP+/GFP-
(e)
B+g-
1,30
b-G+
0,21
V1-A :: VioBlue-A
B1-A :: FITC-A

◂**Fig. 11.2** Loss-of-heterozygosity events and genetic markers used for their identification in *Candida albicans*. **a** Molecular mechanisms at the origin of LOH events. Upon a DNA double strand break (DSB), the damaged DNA can be repaired by recombination-dependent mechanisms, either gene conversion (GC) at the origin of short-range LOH events or break-induced replication (BIR) and mitotic crossover (MCO) resulting in a long-range LOH. If left unrepaired, the damaged chromosome can either be lost and reduplicated (WCL) or truncated (SCL). WCL events are also observed in cases of mitotic nondisjunction. **b** Identification of LOH events using the *URA3* marker. In strains that have been engineered to be heterozygous for the *URA3* gene at a given locus, LOH at this locus leads to resistance to 5-fluoroorotic acid (5-FOA). **c** Identification of LOH events using the *GAL1* marker. In strains that have been engineered to be heterozygous for the *GAL1* gene at a given locus, LOH at this locus leads to resistance to 2-deoxygalacatose (2-DG). **d** Identification of LOH events using a combination of the *GFP* and *BFP* genes. In strains that have been engineered to be heterozygous at a given locus through insertion of the green fluorescent protein-coding gene (*GFP*) on one chromosome and of the blue fluorescent protein-coding gene (*BFP*) on its homologue, LOH at this locus results in loss of one of the *FP* genes and therefore mono-fluorescence. **e** Example of a flow cytometry output showing LOH events in a strain harboring the *GFP/BFP* LOH reporter. Cells undergoing a LOH event at the *GFP/BFP* locus are revealed by flow cytometry. The monofluorescent cells are localized in the side gates and the double fluorescent cells are found in the middle gate

spontaneous 5-FOA-resistant or 2-DG-resistant clones that arise from a *URA3/ura3* or a *GAL1/gal1* strain, respectively, provides a measurement of LOH frequencies at a specific locus in these strains. More recently, a FACS-optimized LOH reporter that combines an artificial heterozygous locus harboring the blue fluorescent protein (BFP) and green fluorescent protein (GFP) markers and flow cytometry to detect LOH events at the single-cell level has been developed (Fig. 11.2d; Loll-Krippleber et al. 2015b). Cells that undergo a LOH at the locus harboring the *BFP* and *GFP* genes switch from double fluorescence to single fluorescence and this is easily detected by flow cytometry, thus allowing quantification of spontaneous or induced LOH frequencies in a cell population (Fig. 11.2e). This new LOH reporter displays appealing features over the *URA3* and *GAL1* markers as it allows rare event analysis, high-throughput LOH detection, and high reproducibility. Moreover, it is independent of the use of pricey drugs such as 2-DG or 5-FOA whose selective pressure might distort the evaluation of LOH frequencies. Finally, the BFP/GFP system may open up new approaches to study LOH in vivo as it should allow visualizing LOH in an infection model and proving/disproving the existence of specific niches in the host that would favor LOH during the process of colonization/infection (Loll-Krippleber et al. 2015b). Nevertheless, these LOH reporters cannot distinguish between the different types of LOH events unless additional genetic markers are analyzed on the same chromosome. This is easily achieved by PCR-RFLP at loci that include a SNP that is heterozygous in the parent strain and generates a restriction enzyme cleavage site for one of the two alleles (Forche et al. 2009b). Alternatively, SNP microarray analysis and whole genome sequencing have been successfully used to characterize the extent of LOH events (Forche et al. 2004, 2005; Loll-Krippleber et al. 2014, 2015b).

The impact of different stresses on LOH has been studied and it has been shown that LOH rates are elevated during in vitro exposure to oxidative stress, elevated temperature, and antifungal drugs (Forche et al. 2011). DNA transformation is also associated with LOH and other genomic rearrangements (Abbey et al. 2011; Arbour et al. 2009; Bouchonville et al. 2009). Importantly, it was also shown that these stresses differently influence the frequency at which the different forms of LOH occur. While oxidative stress (H_2O_2) led to increased proportion of GC events, high temperature (39 °C), and fluconazole led to increased proportion of WCL (Forche et al. 2011). This certainly reflects the mechanisms by which these stresses act on the *C. albicans* genome. H_2O_2-mediated oxidative stress generates DNA DSBs that should be predominantly repaired by GC or BIR/MCO. In this respect, a recent study in our laboratory has demonstrated that DNA DSBs are predominantly repaired by GC in *C. albicans* (Feri et al. 2016). Here, a DNA DSB-inducing system was developed through conditional expression of the *S. cerevisiae* I-*Sce*I meganuclease (Jacquier and Dujon 1985; Monteilhet et al. 1990) in a *C. albicans* strain engineered to harbor a unique I-*Sce*I cleavage site. Characterization of repair events at this I-*Sce*I site showed that they almost always correspond to GC events but some instances of BIR/MCO or WCL were also observed, as well as combination of independent events. Temperature has been proposed to act by limiting chaperones and cochaperones that have roles in the assembly of the kinetochore and spindle pole body and the function of the mitotic checkpoints, thus favoring events of mitotic nondisjunction and WCL events (Forche et al. 2011). Fluconazole, by acting on ergosterol biosynthesis, is likely to impact nuclear membrane fluidity, leading to impairment of chromosome segregation and WCL events. In this respect, Harrison et al. (2014) have shown that fluconazole exposure of *C. albicans* leads to cell cycle deregulation with nuclear and spindle cycles initiating prior to bud emergence. This results in cells where two nuclei eventually collapse, forming a tetraploid, which upon subsequent cell cycles generates aneuploids due to abnormal number of spindles and unequal DNA segregation. It is notable that the tetraploid state, that is also formed in the context of the parasexual cycle, appears to favor ploidy variations, LOH and population heterogeneity, with random aneuploids tending to rapidly resolve to the euploid state (Hickman et al. 2015). Consistently, Ford et al. (2015) have observed that aneuploidies appearing in the course of the acquisition of azole resistance in strains collected from patients with oropharyngeal candidiasis are often transient.

Passage through an animal model of *C. albicans* infection also appears to trigger genomic rearrangements, among which LOH. Indeed, Forche et al. (2009a) have observed that long-range LOH, especially those due to WCL, are more frequent in isolates recovered from a model of invasive candidiasis than in isolates that have been grown in vitro. This contrasts with short-range LOH events (i.e., those due to GC) that occur at similar frequency in in vitro or in vivo-passaged isolates. Interestingly, these authors have observed that in vivo-passaged strains that had undergone a LOH at the *GAL1* LOH reporter locus had a higher chance to present additional LOH at other loci and chromosomes (Forche et al. 2009a). Moreover, they observed that the frequency of appearance of colony morphology variants was dramatically higher upon passage in the animal and that these variants showed increased frequency of LOH events. As aforementioned, haploids have been shown to arise at high frequency during passage of *C. albicans* in animal models of candidaemia and oropharyngeal candidiasis (Hickman et al. 2013). Hence, the host environment appears to promote chromosome nondisjunction events, genomic rearrangements and the emergence of variants. Similar events have also been observed in commensal isolates sampled in humans, suggesting that the observation made in animal models reflects the genome dynamics occurring in the actual *C. albicans* host (Diogo et al. 2009). Yet, only three pairs of isolates have been analyzed in this study and it is notable that our sequencing of 145 isolates recovered from healthy individuals as well as patients with superficial or invasive infections did not reveal frequent WCL (MEB, CD, CM, JR, NS, and G. Sherlock, manuscript in preparation; see Sect. 11.1). This discrepancy might suggest that passaging *C. albicans* through a nonnatural host results in increased frequency of genomic rearrangements or that sampling methods used in the clinics tend to select for subsequent analysis of the fittest isolates and therefore those that are devoid of WCL and/or aneuploidies.

In previous sections we have emphasized the contribution of heterozygosity to diversity in *C. albicans.* Indeed, SNPs may contribute to changes in gene expression, protein composition, and structure and represent a major source of phenotypic variation. In this respect, LOH may unravel SNPs that have detrimental or beneficial effects. Because of clonal reproduction, it is expected that deleterious mutations will accumulate in the *C. albicans* genome with their impact being revealed upon LOH. Indeed, such mutations have now been uncovered in the genome of *C. albicans* strain SC5314. For instance, Gomez-Raja et al. (2008) could identify a nonfunctional allele of the *HIS4* gene located on Chr4. As *HIS4* is necessary for histidine biosynthesis, homozygosis of this nonfunctional allele leads to histidine auxotrophy in *C. albicans* laboratory strains. In another study, Ciudad et al. (2016) have uncovered an allele of the Chr3-located *MBP1* gene that, upon homozygosis or hemizygosis, confers sensitivity to the DNA alkylating agent methyl methane sulfonate to the *C. albicans* SC5314 strain. Besides these mutations that have consequences only under specific conditions, it has been hypothesized that some chromosomes in *C. albicans* strain SC5314 harbor recessive lethal mutations. Indeed, long-range LOH that maintained one of the two haplotypes could not be observed for several chromosomes. For instance, Forche et al. (2008) have observed that LOH in progeny originating from the parasexual cycle had a strong bias towards one of the two haplotypes for chromosomes R, 2, 4, 6, and 7. A similar bias was observed in a *C. albicans rad52Δ/rad52Δ* mutant (Andaluz et al. 2011). *C. albicans* haploids never harbor one of the two haplotypes for chromosomes 3, 4, 6, and 7 and most of Chr1 is never observed in the homozygous state (Hickman et al. 2013). Finally, a detailed investigation of the events associated to LOH at a specific locus on *C. albicans* strain SC5314 Chr4 revealed that chromosome loss events maintaining the sole haplotype B were never observed (Loll-Krippleber et al. 2015b). In a recent study, we have successfully identified the recessive lethal allele located on *C. albicans* SC5314 Chr4B. This corresponds to a nonsense mutation in the *GPI16* gene that encodes an essential component of the glycosylphosphatidylinositol anchor biosynthesis machinery (Feri et al. 2016). Interestingly, introduction of an ectopic copy of *GPI16* allowed recovering progenies with homozygosity of Chr4B and revealed additional recessive deleterious alleles on this chromosome, in particular a nonsense mutation in the *MRF2* gene that encodes a mitochondrial translation release factor. Homozygosis of this allele leads to respiratory defects in *C. albicans* (Feri et al. 2016). Surprisingly, the nonsense mutations in the *GPI16* and *MRF2* genes appear unique to strain SC5314, i.e., they are absent from 144 commensal and clinical genome-sequenced isolates. Preliminary results indicate that other recessive lethal alleles are present on other chromosomes (e.g., Chr5) in these genome-sequenced isolates (Feri et al. 2016). Taken together, these data indicate that recessive deleterious and lethal alleles are frequently found in the heterozygous state in *C. albicans* isolates and will directly influence the outcome of LOH events that these isolates may undergo. Evidence for sign epistasis whereby one LOH may affect the phenotype associated to another LOH have also been obtained and may further complicate the picture (Ciudad et al. 2016).

The above examples focus on LOH events that have a detrimental effect on *C. albicans* fitness in general or in specific conditions. Remarkably, other LOH events have been shown to allow *C. albicans* adaptation to changing environments. Janbon et al. (1998) have observed that the ability of *C. albicans* to use L-sorbose as a carbon source is associated to monosomy of Chr5. This is explained by the occurrence on Chr5 of negative regulators of the *SOU1* gene that encodes a sorbose reductase required for L-sorbose assimilation (Greenberg et al. 2005; Kabir et al. 2005). Further examples of beneficial LOH events have arisen through the study of antifungal resistant isolates. Briefly, LOH events allowing homozygosity of mutant alleles for genes involved in the sensitivity to azole, polyene, and echinocandin antifungals have been shown to lead to enhanced antifungal resistance. This is true for mutations in *ERG11*, encoding the target of azole drugs; *UPC2* encoding a transcriptional activator of ergosterol biosynthesis genes; *TAC1* and *MRR1* encoding transcription factors that regulate the expression of the *CDR1/2* and *MDR1* genes for azole-efflux pumps, respectively; *ERG3*, encoding an enzyme of the ergosterol biosynthesis pathway whose reduced activity leads to amphotericin B resistance; *FKS1* encoding the target of echinocandins (Coste et al. 2006, 2007; Morschhauser et al. 2007; Dunkel et al. 2008; Niimi et al. 2010; Morio et al. 2012). The case of *ERG11* and *TAC1* is especially interesting as both genes lie on the left arm of Chr5, next to the *MTL* locus. Hence, according to the type of LOH events experienced by Chr5—GC, BIR/MCO or WCL, preexisting mutations that render Erg11 insensitive to azole inhibition or result in an hyperactive form of Tac1 will be independently or simultaneously brought to homozygosity, leading to enhanced resistance to azoles, and strains will become *MTL*-homozygous (Coste et al. 2007). Importantly, these LOH events can be complemented by the formation of an isochromosome that joins two left arms of Chr5, therefore further amplifying the mutant *ERG11* and *TAC1* genes and increasing azole resistance (Coste et al. 2007; Selmecki et al. 2006, 2008). Other examples of genomic rearrangements (e.g., triplications and chromosome fusions) that lead to amplification of azole-resistance mutations have been observed (reviewed in Loll-Krippleber et al. 2015a).

In summary, LOH events are frequent in *C. albicans* resulting in homozygosis of recessive or dominant mutations that can have detrimental or beneficial effects on *C. albicans* fitness. In particular, LOH have been shown to contribute to expand antifungal-resistance mutations and consequently enhance antifungal resistance. It could be expected, given *C. albicans* clonality, that LOH would unmask recessive lethal and deleterious mutations that are often unique to a particular *C. albicans* isolate. Importantly, these mutations have been so far identified in vitro and it is likely that additional recessive deleterious mutations might impact *C. albicans* fitness in vivo. Indeed, viable autodiploids show reduced fitness in vivo compared to heterozygous diploids (Hickman et al. 2013). Identifying recessive mutations that show their deleterious effect in vivo will represent a significant challenge. Finally, it has been proposed that haploids may represent a suitable alternative to study the biology of *C. albicans* as their genetic engineering is facilitated compared to that in diploids (Hickman et al. 2013; Seneviratne et al. 2015). This should be regarded with caution as recessive mutations in haploids will be unmasked and may exert synthetic effects with engineered mutations. Thus, according to the regions of the *C. albicans* genome that have been retained in the engineered haploid strain, phenotypes associated to an engineered mutation may vary.

11.6 Outlook

In this chapter, we have reviewed some of the current knowledge of *C. albicans* genetic diversity and genome dynamics. While a diploid species, *C. albicans* appears to be prone to the formation of aneuploidies and genomic rearrangements, in particular LOH, that can impact its ability to survive in challenging environments. Notably, the host environment appears to favor *C. albicans* genome dynamics (Bougnoux et al. 2006; Diogo et al. 2009; Forche et al. 2009a; Hickman et al. 2013) and it will be of major interest deciphering which host and, possibly, microbiota factors act as triggers for genome rearrangements. It will also be of interest to investigate the mechanisms by which *C. albicans* adapts to genomic challenges. In this respect, several genes have already been identified that impact the frequency at which LOH arise in *C. albicans* (reviewed in Loll-Krippleber et al. 2015a) and tools are becoming available to perform larger investigations (Loll-Krippleber et al. 2015b).

Despite frequent genome rearrangements and a parasexual cycle, the *C. albicans* population appears predominantly clonal. Clonality should ineluctably lead to the accumulation of deleterious mutations and species extinction. Evidence has been obtained in vitro suggesting that the parasexual cycle might, to some extent, provide a solution to this destiny. The challenge is now to obtain evidence that this is indeed the case in the natural host of *C. albicans*. Another challenge certainly lies in understanding the evolutionary history of *C. albicans*. What led to the loss of meiosis and maintenance of a parasexual cycle? When have clades diverged and what genomic changes are involved in the origin and processes of population divergence? Does this support an emergence through geographic isolation of human populations or suggest another origin? Has clade divergence been associated with the emergence of specific phenotypes and what were the evolutionary pressures for the acquisition of these phenotypes?

C. albicans isolates show considerable phenotypic diversity, in clade-dependent or -independent manners. With the advent of high-throughput genome sequencing, we can envision the development of genome-wide association studies as a mean to uncover the polymorphisms that underlie phenotypic differences within the *C. albicans* population. We can also envision revealing clade-specific polymorphisms and, possibly, new clade-specific phenotypes. Moreover, we can envision revealing whether *C. albicans* association with humans leads, under conditions other than commensalism, to some form of adaptation. Recent results by Kim et al. (2015) have for instance shown that *C. albicans* isolates collected from sputum of cystic fibrosis patients often present loss-of-function mutations in the *NRG1* gene, leading to uncontrolled hyphal formation and possibly to an increased persistence in the lung. Similarly, Ford et al. (2015) have observed that long-term carriage of *C. albicans* during oropharyngeal candidiasis is associated with the fixation of mutations that could be related to host adaptation. Further understanding of such adaptive processes in these contexts of *C. albicans*-host interaction or other contexts will deserve attention.

Acknowledgements Research in the Fungal Biology and Pathogenicty Unit on the topic of genetic diversity and genome dynamics in *C. albicans* has been funded by Institut Pasteur and Institut National de la Recherche Agronomique. AF is the recipient of a PhD fellowship from INRA and Institut Pasteur. RLK was the recipient of a PhD fellowship from INRA and Institut Pasteur. JR is the recipient of Roux post-doctoral fellowship from Institut Pasteur. ES is the recipient of a PhD fellowship from the DIM-MalInf from Région Ile-de-France. We acknowledge support from the French Government's Investissement d'Avenir program (Laboratoire d'Excellence Integrative Biology of Emerging Infectious Diseases, ANR-10-LABX-62-IBEID).

References

Abbey D, Hickman M, Gresham D, Berman J (2011) High-resolution SNP/CGH microarrays reveal the accumulation of loss of heterozygosity in commonly used Candida albicans strains. G3 (Bethesda) 1(7):523–530. doi:10.1534/g3.111.000885

Andaluz E, Bellido A, Gomez-Raja J, Selmecki A, Bouchonville K, Calderone R, Berman J, Larriba G (2011) Rad52 function prevents chromosome loss and truncation in Candida albicans. Mol Microbiol 79(6):1462–1482. doi:10.1111/j.1365-2958.2011.07532.x

Anderson JB, Wickens C, Khan M, Cowen LE, Federspiel N, Jones T, Kohn LM (2001) Infrequent genetic exchange and recombination in the mitochondrial genome of *Candida albicans*. J Bacteriol 183:865–872

Anderson MZ, Baller JA, Dulmage K, Wigen L, Berman J (2012) The three clades of the telomere-associated TLO gene family of Candida albicans have different splicing, localization, and expression features. Eukaryot Cell 11(10):1268–1275. doi:10.1128/EC.00230-12

Angebault C, Djossou F, Abelanet S, Permal E, Ben Soltana M, Diancourt L, Bouchier C, Woerther PL, Catzeflis F, Andremont A, d'Enfert C, Bougnoux ME (2013) Candida albicans is not always the preferential yeast colonizing humans: a study in Wayampi Amerindians. J Infect Dis 208(10):1705–1716. doi:10.1093/infdis/jit389

Arbour M, Epp E, Hogues H, Sellam A, Lacroix C, Rauceo J, Mitchell A, Whiteway M, Nantel A (2009) Widespread occurrence of chromosomal aneuploidy following the routine production of Candida albicans mutants. FEMS Yeast Res 9(7):1070–1077. doi:10.1111/j.1567-1364.2009.00563.x

Azie N, Neofytos D, Pfaller M, Meier-Kriesche HU, Quan SP, Horn D (2012) The PATH (Prospective Antifungal Therapy) Alliance(R) registry and invasive fungal infections: update 2012. Diag Microbiol Infec Dis 73(4):293–300. doi:10.1016/j.diagmicrobio.2012.06.012

Bennett RJ, Johnson AD (2003) Completion of a parasexual cycle in Candida albicans by induced chromosome loss in tetraploid strains. EMBO J 22(10):2505–2515

Binkley J, Arnaud MB, Inglis DO, Skrzypek MS, Shah P, Wymore F, Binkley G, Miyasato SR, Simison M, Sherlock G (2014) The Candida genome database: the new homology information page highlights protein similarity and phylogeny. Nucleic Acids Res 42 (Database issue): D711–716. doi:10.1093/nar/gkt1046

Blignaut E, Molepo J, Pujol C, Soll DR, Pfaller MA (2005) Clade-related amphotericin B resistance among South African Candida albicans isolates. Diagn Microbiol Infect Dis 53 (1):29–31

Boeke JD, Lacroute F, Fink GR (1984) A positive selection for mutants lacking orotidine-5'-phosphate decarboxylase activity in yeast: 5-fluoro-orotic acid resistance. Mol Gen Genet 197:345–346

Boisrame A, Cornu A, Da Costa G, Richard ML (2011) Unexpected role for a serine/threonine-rich domain in the Candida albicans Iff protein family. Eukaryot Cell 10 (10):1317–1330. doi:10.1128/EC.05044-11

Botterel F, Desterke C, Costa C, Bretagne S (2001) Analysis of microsatellite markers of Candida albicans used for rapid typing. J Clin Microbiol 39(11):4076–4081

Bouchonville K, Forche A, Tang KE, Selmecki A, Berman J (2009) Aneuploid chromosomes are highly unstable during DNA transformation of Candida albicans. Eukaryot Cell 8(10):1554–1566. doi:10.1128/EC.00209-09

Bougnoux ME, Aanensen DM, Morand S, Theraud M, Spratt BG, d'Enfert C (2004) Multilocus sequence typing of Candida albicans: strategies, data exchange and applications. Infect Genet Evol 4(3):243–252

Bougnoux ME, Diogo D, Francois N, Sendid B, Veirmeire S, Colombel JF, Bouchier C, Van Kruiningen H, d'Enfert C, Poulain D (2006) Multilocus sequence typing reveals intrafamilial transmission and microevolutions of Candida albicans isolates from the human digestive tract. J Clin Microbiol 44(5):1810–1820

Bougnoux ME, Diogo D, Pujol C, Soll DR, d'Enfert C (2007) Molecular epidemiology and population dynamics in Candida albicans. In: d'Enfert C, Hube B (eds) Candida: comparative and fucntional genomics. Caister Academic Press, Wymondham, United Kingdom, pp 51–70

Bougnoux ME, Morand S, d'Enfert C (2002) Usefulness of multilocus sequence typing for characterization of clinical isolates of Candida albicans. J Clin Microbiol 40(4):1290–1297

Bougnoux ME, Pujol C, Diogo D, Bouchier C, Soll DR, d'Enfert C (2008) Mating is rare within as well as between clades of the human pathogen Candida albicans. Fungal Genet Biol 45 (3):221–231

Bougnoux ME, Tavanti A, Bouchier C, Gow NA, Magnier A, Davidson AD, Maiden MC, D'Enfert C, Odds FC (2003) Collaborative consensus for optimized multilocus sequence typing of Candida albicans. J Clin Microbiol 41(11):5265–5266

Braun BR, van Het Hoog M, d'Enfert C, Martchenko M, Dungan J, Kuo A, Inglis DO, Uhl MA, Hogues H, Berriman M, Lorenz M, Levitin A, Oberholzer U, Bachewich C, Harcus D, Marcil A, Dignard D, Iouk T, Zito R, Frangeul L, Tekaia F, Rutherford K, Wang E, Munro CA, Bates S, Gow NA, Hoyer LL, Kohler G, Morschhauser J, Newport G, Znaidi S, Raymond M, Turcotte B, Sherlock G, Costanzo M, Ihmels J, Berman J, Sanglard D, Agabian N, Mitchell AP, Johnson AD, Whiteway M, Nantel A (2005) A Human-Curated Annotation of the Candida albicans Genome. PLoS Genet 1(1):36–57

Brown GD, Denning DW, Gow NA, Levitz SM, Netea MG, White TC (2012) Hidden killers: human fungal infections. Science Transl Med 4(165):165rv113. doi:10.1126/scitranslmed.3004404

Butler G, Rasmussen MD, Lin MF, Santos MA, Sakthikumar S, Munro CA, Rheinbay E, Grabherr M, Forche A, Reedy JL, Agrafioti I, Arnaud MB, Bates S, Brown AJ, Brunke S, Costanzo MC, Fitzpatrick DA, de Groot PW, Harris D, Hoyer LL, Hube B, Klis FM, Kodira C, Lennard N, Logue ME, Martin R, Neiman AM, Nikolaou E, Quail MA, Quinn J, Santos MC, Schmitzberger FF, Sherlock G, Shah P, Silverstein KA, Skrzypek MS, Soll D, Staggs R, Stansfield I, Stumpf MP, Sudbery PE, Srikantha T, Zeng Q, Berman J, Berriman M, Heitman J, Gow NA, Lorenz MC, Birren BW, Kellis M, Cuomo CA (2009) Evolution of pathogenicity and sexual reproduction in eight Candida genomes. Nature 459(7247):657–662. doi:10.1038/nature08064

Calderon-Norena DM, Gonzalez-Novo A, Orellana-Munoz S, Gutierrez-Escribano P, Arnaiz-Pita Y, Duenas-Santero E, Suarez MB, Bougnoux ME, Del Rey F, Sherlock G, d'Enfert C, Correa-Bordes J, de Aldana CR (2015) A single nucleotide polymorphism uncovers a novel function for the transcription factor Ace2 during Candida albicans hyphal development. PLoS Genet 11(4):e1005152. doi:10.1371/journal.pgen.1005152

Chibana H, Beckerman JL, Magee PT (2000) Fine-resolution physical mapping of genomic diversity in Candida albicans. Genome Res 10(12):1865–1877

Chibana H, Iwaguchi S, Homma M, Chindamporn A, Nakagawa Y, Tanaka K (1994) Diversity of tandemly repetitive sequences due to short periodic repetitions in the chromosomes of Candida albicans. J Bacteriol 176(13):3851–3858

Chibana H, Magee BB, Grindle S, Ran Y, Scherer S, Magee PT (1998) A physical map of chromosome 7 of Candida albicans. Genetics 149(4):1739–1752

Chibana H, Magee PT (2009) The enigma of the major repeat sequence of Candida albicans. Future Microbiol 4(2):171–179. doi:10.2217/17460913.4.2.171

Chindamporn A, Nakagawa Y, Homma M, Chibana H, Doi M, Tanaka K (1995) Analysis of the chromosomal localization of the repetitive sequences (RPSs) in Candida albicans. Microbiology 141 (Pt 2):469–476. doi:10.1099/13500872-141-2-469

Chindamporn A, Nakagawa Y, Mizuguchi I, Chibana H, Doi M, Tanaka K (1998) Repetitive sequences (RPSs) in the chromosomes of Candida albicans are sandwiched between two novel stretches, HOK and RB2, common to each chromosome. Microbiology 144(Pt 4):849–857. doi:10.1099/00221287-144-4-849

Chu WS, Magee BB, Magee PT (1993) Construction of an SfiI macrorestriction map of the Candida albicans genome. J Bacteriol 175(20):6637–6651

Ciudad T, Hickman M, Bellido A, Berman J, Larriba G (2016) The phenotypic consequences of a spontaneous loss of heterozygosity in a common laboratory strain of Candida albicans. Genetics. doi:10.1534/genetics.116.189274

Coste A, Selmecki A, Forche A, Diogo D, Bougnoux ME, d'Enfert C, Berman J, Sanglard D (2007) Genotypic evolution of azole resistance mechanisms in sequential Candida albicans isolates. Eukaryot Cell 6(10):1889–1904

Coste A, Turner V, Ischer F, Morschhauser J, Forche A, Selmecki A, Berman J, Bille J, Sanglard D (2006) A mutation in Tac1p, a transcription factor regulating CDR1 and CDR2, is coupled with loss of heterozygosity at chromosome 5 to mediate antifungal resistance in Candida albicans. Genetics 172(4):2139–2156

Diogo D, Bouchier C, d'Enfert C, Bougnoux ME (2009) Loss of heterozygosity in commensal isolates of the asexual diploid yeast Candida albicans. Fungal Genet Biol 46 (2):159–168. doi:10.1016/j.fgb.2008.11.005

Dodgson AR, Dodgson KJ, Pujol C, Pfaller MA, Soll DR (2004) Clade-specific flucytosine resistance is due to a single nucleotide change in the FUR1 gene of Candida albicans. Antimicrob Agents Chemother 48(6):2223–2227

Dunkel N, Liu TT, Barker KS, Homayouni R, Morschhauser J, Rogers PD (2008) A gain-of-function mutation in the transcription factor Upc2p causes upregulation of ergosterol biosynthesis genes and increased fluconazole resistance in a clinical *Candida albicans* isolate. Eukaryot Cell 7(7):1180–1190. doi:10.1128/EC.00103-08

Feri A, Loll-Krippleber, Commere PH, Maufrais C, Sertour N, Schwartz K, Sherlock G, Bougnoux ME, d'Enfert C, Legrand M (2016) Analysis of repair mechanisms following an induced double-strand break uncovers recessive deleterious alleles in the *Candida albicans* diploid genome. MBio 7(5):e01109–16. doi:10.1128/mBio.01109-16

Forche A, Abbey D, Pisithkul T, Weinzierl MA, Ringstrom T, Bruck D, Petersen K, Berman J (2011) Stress alters rates and types of loss of heterozygosity in Candida albicans. MBio 2(4): e00129–00111. doi:10.1128/mBio.00129-11

Forche A, Alby K, Schaefer D, Johnson AD, Berman J, Bennett RJ (2008) The parasexual cycle in Candida albicans provides an alternative pathway to meiosis for the formation of recombinant strains. PLoS Biol 6(5):e110

Forche A, Magee PT, Magee BB, May G (2004) Genome-wide single-nucleotide polymorphism map for Candida albicans. Eukaryot Cell 3(3):705–714

Forche A, Magee PT, Selmecki A, Berman J, May G (2009a) Evolution in Candida albicans populations during a single passage through a mouse host. Genetics 182(3):799–811. doi:10.1534/genetics.109.103325

Forche A, May G, Beckerman J, Kauffman S, Becker J, Magee PT (2003) A system for studying genetic changes in *Candida albicans* during infection. Fungal Genet Biol 39(1):38–50

Forche A, May G, Magee PT (2005) Demonstration of loss of heterozygosity by single-nucleotide polymorphism microarray analysis and alterations in strain morphology in Candida albicans strains during infection. Eukaryot Cell 4(1):156–165

Forche A, Steinbach M, Berman J (2009b) Efficient and rapid identification of Candida albicans allelic status using SNP-RFLP. FEMS Yeast Res 9(7):1061–1069. doi:10.1111/j.1567-1364.2009.00542.x

Ford CB, Funt JM, Abbey D, Issi L, Guiducci C, Martinez DA, Delorey T, Li BY, White TC, Cuomo C, Rao RP, Berman J, Thompson DA, Regev A (2015) The evolution of drug resistance in clinical isolates of Candida albicans. Elife 4:e00662. doi:10.7554/eLife.00662

Gillum AM, Tsay EY, Kirsch DR (1984) Isolation of the *Candida albicans* gene for orotidine-5'-phosphate decarboxylase by complementation of *S. cerevisiae ura3* and *E. coli pyrF* mutations. Mol Gen Genet 198(1):179–182

Gomez-Raja J, Andaluz E, Magee B, Calderone R, Larriba G (2008) A single SNP, G929T (Gly310Val), determines the presence of a functional and a non-functional allele of HIS4 in Candida albicans SC5314: detection of the non-functional allele in laboratory strains. Fungal Genet Biol 45(4):527–541. doi:10.1016/j.fgb.2007.08.008

Gorman JA, Gorman JW, Koltin Y (1992) Direct selection of galactokinase-negative mutants of Candida albicans using 2-deoxy-galactose. Curr Genet 21(3):203–206

Greenberg JR, Price NP, Oliver RP, Sherman F, Rustchenko E (2005) Candida albicans SOU1 encodes a sorbose reductase required for L-sorbose utilization. Yeast 22(12):957–969. doi:10.1002/yea.1282

Haran J, Boyle H, Hokamp K, Yeomans T, Liu Z, Church M, Fleming AB, Anderson MZ, Berman J, Myers LC, Sullivan DJ, Moran GP (2014) Telomeric ORFs (TLOs) in Candida spp. Encode mediator subunits that regulate distinct virulence traits. PLoS Genet 10(10): e1004658. doi:10.1371/journal.pgen.1004658

Harrison BD, Hashemi J, Bibi M, Pulver R, Bavli D, Nahmias Y, Wellington M, Sapiro G, Berman J (2014) A tetraploid intermediate precedes aneuploid formation in yeasts exposed to fluconazole. PLoS Biol 12(3):e1001815. doi:10.1371/journal.pbio.1001815

Hickman MA, Paulson C, Dudley A, Berman J (2015) Parasexual ploidy reduction drives population heterogeneity through random and transient aneuploidy in Candida albicans. Genetics 200(3):781–794. doi:10.1534/genetics.115.178020

Hickman MA, Zeng G, Forche A, Hirakawa MP, Abbey D, Harrison BD, Wang YM, Su CH, Bennett RJ, Wang Y, Berman J (2013) The 'obligate diploid' Candida albicans forms mating-competent haploids. Nature 494(7435):55–59. doi:10.1038/nature11865

Hirakawa MP, Martinez DA, Sakthikumar S, Anderson MZ, Berlin A, Gujja S, Zeng Q, Zisson E, Wang JM, Greenberg JM, Berman J, Bennett RJ, Cuomo CA (2014) Genetic and phenotypic intra-species variation in Candida albicans. Genome Res in press. doi:10.1101/gr.174623.114

Hope WW, Tabernero L, Denning DW, Anderson MJ (2004) Molecular mechanisms of primary resistance to flucytosine in Candida albicans. Antimicrob Agents Chemother 48(11):4377–4386. doi:10.1128/AAC.48.11.4377-4386.2004

Hoyer LL, Cota E (2016) Candida albicans Agglutinin-Like Sequence (Als) family vignettes: a review of Als protein structure and function. Front Microbiol 7:280. doi:10.3389/fmicb.2016.00280

Hull CM, Johnson AD (1999) Identification of a mating type-like locus in the asexual pathogenic yeast Candida albicans. Science 285(5431):1271–1275

Hull CM, Raisner RM, Johnson AD (2000) Evidence for mating of the "asexual" yeast Candida albicans in a mammalian host. Science 289(5477):307–310

Iwaguchi S, Homma M, Chibana H, Tanaka K (1992a) Isolation and characterization of a repeated sequence (RPS1) of Candida albicans. J Gen Microbiol 138(9):1893–1900. doi:10.1099/00221287-138-9-1893

Iwaguchi S, Homma M, Tanaka K (1992b) Clonal variation of chromosome size derived from the rDNA cluster region in Candida albicans. J Gen Microbiol 138(6):1177–1184. doi:10.1099/00221287-138-6-1177

Jacobsen MD, Bougnoux ME, d'Enfert C, Odds FC (2008) Multilocus sequence typing of Candida albicans isolates from animals. Res Microbiol 159(6):436–440

Jacquier A, Dujon B (1985) An intron-encoded protein is active in a gene conversion process that spreads an intron into a mitochondrial gene. Cell 41(2):383–394

Janbon G, Sherman F, Rustchenko E (1998) Monosomy of a specific chromosome determines L-sorbose utilization: a novel regulatory mechanism in Candida albicans. Proc Natl Acad Sci U S A 95(9):5150–5155

Jones T, Federspiel NA, Chibana H, Dungan J, Kalman S, Magee BB, Newport G, Thorstenson YR, Agabian N, Magee PT, Davis RW, Scherer S (2004) The diploid genome sequence of Candida albicans. Proc Natl Acad Sci U S A 101(19):7329–7334

Jung SI, Shin JH, Kim SH, Kim J, Kim JH, Choi MJ, Chung EK, Lee K, Koo SH, Chang HH, Bougnoux ME, d'Enfert C (2016) Comparison of E, E-Farnesol secretion and the clinical characteristics of Candida albicans bloodstream isolates from different multilocus sequence typing clades. PLoS ONE 11(2):e0148400. doi:10.1371/journal.pone.0148400

Kabir MA, Ahmad A, Greenberg JR, Wang YK, Rustchenko E (2005) Loss and gain of chromosome 5 controls growth of Candida albicans on sorbose due to dispersed redundant negative regulators. Proc Natl Acad Sci U S A 102(34):12147–12152. doi:10.1073/pnas.0505625102

Keeney S, Giroux CN, Kleckner N (1997) Meiosis-specific DNA double-strand breaks are catalyzed by Spo11, a member of a widely conserved protein family. Cell 88(3):375–384

Kim SH, Clark ST, Surendra A, Copeland JK, Wang PW, Ammar R, Collins C, Tullis DE, Nislow C, Hwang DM, Guttman DS, Cowen LE (2015) Global analysis of the fungal microbiome in cystic fibrosis patients reveals loss of function of the transcriptional repressor Nrg1 as a mechanism of pathogen adaptation. PLoS Pathog 11(11):e1005308. doi:10.1371/journal.ppat.1005308

L'Ollivier C, Labruere C, Jebrane A, Bougnoux ME, d'Enfert C, Bonnin A, Dalle F (2012) Using a multi-locus microsatellite typing method improved phylogenetic distribution of Candida albicans isolates but failed to demonstrate association of some genotype with the commensal or clinical origin of the isolates. Infection, Genetics Evolution: J Molecular Epidemiol Evolutionary Genetics Infectious Diseases 12(8):1949–1957. doi:10.1016/j.meegid.2012.07.025

Lee SC, Ni M, Li W, Shertz C, Heitman J (2010) The evolution of sex: a perspective from the fungal kingdom. Microbiol Mol Biol Rev 74(2):298–340. doi:10.1128/MMBR.00005-10

Lephart PR, Chibana H, Magee PT (2005) Effect of the major repeat sequence on chromosome loss in Candida albicans. Eukaryot Cell 4(4):733–741

Li X, Yan Z, Xu J (2003) Quantitative variation of biofilms among strains in natural populations of Candida albicans. Microbiology 149(Pt 2):353–362. doi:10.1099/mic.0.25932-0

Lockhart SR, Fritch JJ, Meier AS, Schroppel K, Srikantha T, Galask R, Soll DR (1995) Colonizing populations of Candida albicans are clonal in origin but undergo microevolution through C1 fragment reorganization as demonstrated by DNA fingerprinting and C1 sequencing. J Clin Microbiol 33(6):1501–1509

Lockhart SR, Pujol C, Daniels KJ, Miller MG, Johnson AD, Pfaller MA, Soll DR (2002) In Candida albicans, white-opaque switchers are homozygous for mating type. Genetics 162 (2):737–745

Loll-Krippleber R, d'Enfert C, Feri A, Diogo D, Perin A, Marcet-Houben M, Bougnoux ME, Legrand M (2014) A study of the DNA damage checkpoint in Candida albicans: uncoupling of the functions of Rad53 in DNA repair, cell cycle regulation and genotoxic stress-induced polarized growth. Mol Microbiol 91(3):452–471. doi:10.1111/mmi.12471

Loll-Krippleber R, Feri A, d'Enfert C, M. L (2015a) Genome integrity: mechanisms and contribution to antifungal resistance. In: Coste A, Vandeputte P (eds) Antifungals: from genomics to resistance and the development of novel agents. Caister Academic Press, Norfolk, UK, pp 211–236

Loll-Krippleber R, Feri A, Nguyen M, Maufrais C, Yansouni J, d'Enfert C, Legrand M (2015) A FACS-optimized screen identifies regulators of genome stability in Candida albicans. Eukaryot Cell 14(3):311–322. doi:10.1128/EC.00286-14

Maccallum DM, Castillo L, Nather K, Munro CA, Brown AJ, Gow NA, Odds FC (2009) Property differences among the four major Candida albicans strain clades. Eukaryot Cell 8:373–387

Magee BB, Magee PT (2000) Induction of mating in Candida albicans by construction of MTLa and MTLalpha strains. Science 289(5477):310–313

Magwene PM, Kayikci O, Granek JA, Reininga JM, Scholl Z, Murray D (2011) Outcrossing, mitotic recombination, and life-history trade-offs shape genome evolution in Saccharomyces cerevisiae. Proc Natl Acad Sci U S A 108(5):1987–1992. doi:10.1073/pnas.1012544108

McManus BA, Coleman DC (2014) Molecular epidemiology, phylogeny and evolution of Candida albicans. Infect Genet Evol 21:166–178. doi:10.1016/j.meegid.2013.11.008

Miller MG, Johnson AD (2002) White-opaque switching in Candida albicans is controlled by mating-type locus homeodomain proteins and allows efficient mating. Cell 110(3):293–302

Monteilhet C, Perrin A, Thierry A, Colleaux L, Dujon B (1990) Purification and characterization of the in vitro activity of I-Sce I, a novel and highly specific endonuclease encoded by a group I intron. Nucleic Acids Res 18(6):1407–1413

Morio F, Pagniez F, Lacroix C, Miegeville M, Le Pape P (2012) Amino acid substitutions in the Candida albicans sterol Delta 5,6-desaturase (Erg3p) confer azole resistance: characterization of two novel mutants with impaired virulence. J Antimicrob Chemother 67(9):2131–2138. doi:10.1093/jac/dks186

Morschhauser J, Barker KS, Liu TT, Bla BWJ, Homayouni R, Rogers PD (2007) The transcription factor Mrr1p controls expression of the MDR1 efflux pump and mediates multidrug resistance in Candida albicans. PLoS Pathog 3(11):e164. doi:10.1371/journal.ppat.0030164

Muzzey D, Schwartz K, Weissman JS, Sherlock G (2013) Assembly of a phased diploid Candida albicans genome facilitates allele-specific measurements and provides a simple model for repeat and indel structure. Genome Biol 14(9):R97. doi:10.1186/gb-2013-14-9-r97

Muzzey D, Sherlock G, Weissman JS (2014) Extensive and coordinated control of allele-specific expression by both transcription and translation in Candida albicans. Genome Res 24(6):963–973. doi:10.1101/gr.166322.113

Navarro-Garcia F, Perez-Diaz RM, Magee BB, Pla J, Nombela C, Magee P (1995) Chromosome reorganization in Candida albicans 1001 strain. J Med Vet Mycol 33(6):361–366

Niimi K, Monk BC, Hirai A, Hatakenaka K, Umeyama T, Lamping E, Maki K, Tanabe K, Kamimura T, Ikeda F, Uehara Y, Kano R, Hasegawa A, Cannon RD, Niimi M (2010) Clinically significant micafungin resistance in Candida albicans involves modification of a glucan synthase catalytic subunit GSC1 (FKS1) allele followed by loss of heterozygosity. J Antimicrob Chemother 65(5):842–852. doi:10.1093/jac/dkq073

Odds FC (2009) In Candida albicans, resistance to flucytosine and terbinafine is linked to MAT locus homozygosity and multilocus sequence typing clade 1. FEMS Yeast Res 9(7):1091–1101. doi:10.1111/j.1567-1364.2009.00577.x

Odds FC (2010) Molecular phylogenetics and epidemiology of Candida albicans. Future Microbiol 5:67–79. doi:10.2217/fmb.09.113

Odds FC, Bougnoux ME, Shaw DJ, Bain JM, Davidson AD, Diogo D, Jacobsen MD, Lecomte M, Li SY, Tavanti A, Maiden MC, Gow NA, d'Enfert C (2007) Molecular phylogenetics of Candida albicans. Eukaryot Cell 6:1041–1052

Oh SH, Cheng G, Nuessen JA, Jajko R, Yeater KM, Zhao X, Pujol C, Soll DR, Hoyer LL (2005) Functional specificity of Candida albicans Als3p proteins and clade specificity of ALS3 alleles discriminated by the number of copies of the tandem repeat sequence in the central domain. Microbiology 151(Pt 3):673–681

Orti G, Pearse DE, Avise JC (1997) Phylogenetic assessment of length variation at a microsatellite locus. Proc Natl Acad Sci U S A 94(20):10745–10749

Pujol C, Joly S, Nolan B, Srikantha R, Lockhart S, Soll D (1999) Microevolutionary changes in Candida albicans identified par le complex Ca3 probe involve insertions and deletions of RPS units at specific genomic sites. Microbiology 145:2635–2644

Pujol C, Pfaller M, Soll DR (2002) Ca3 fingerprinting of Candida albicans bloodstream isolates from the United States, Canada, South America, and Europe reveals a European clade. J Clin Microbiol 40(8):2729–2740

Pujol C, Pfaller MA, Soll DR (2004) Flucytosine resistance is restricted to a single genetic clade of Candida albicans. Antimicrob Agents Chemother 48(1):262–266

Reagan DR, Pfaller MA, Hollis RJ, Wenzel RP (1990) Characterization of the sequence of colonization and nosocomial candidemia using DNA fingerprinting and a DNA probe. J Clin Microbiol 28(12):2733–2738

Robert F, Lebreton F, Bougnoux ME, Paugam A, Wassermann D, Schlotterer M, Tourte-Schaefer C, Dupouy-Camet J (1995) Use of random amplified polymorphic DNA as a typing method for Candida albicans in epidemiological surveillance of a burn unit. J Clin Microbiol 33(9):2366–2371

Sampaio P, Gusmao L, Correia A, Alves C, Rodrigues AG, Pina-Vaz C, Amorim A, Pais C (2005) New microsatellite multiplex PCR for Candida albicans strain typing reveals microevolutionary changes. J Clin Microbiol 43(8):3869–3876. doi:10.1128/JCM.43.8.3869-3876.2005

Schmid J, Herd S, Hunter PR, Cannon RD, Yasin MS, Samad S, Carr M, Parr D, McKinney W, Schousboe M, Harris B, Ikram R, Harris M, Restrepo A, Hoyos G, Singh KP (1999) Evidence for a general-purpose genotype in Candida albicans, highly prevalent in multiple geographical regions, patient types and types of infection. Microbiology 145(Pt 9):2405–2413

Schmid J, Magee PT, Holland BR, Zhang N, Cannon RD, Magee BB (2016) Last hope for the doomed? Thoughts on the importance of a parasexual cycle for the yeast Candida albicans. Curr Genet 62(1):81–85. doi:10.1007/s00294-015-0516-8

Schmid J, Voss E, Soll DR (1990) Computer-assisted methods for assessing strain relatedness in Candida albicans by fingerprinting with the moderately repetitive sequence Ca3. J Clin Microbiol 28(6):1236–1243

Selmecki A, Bergmann S, Berman J (2005) Comparative genome hybridization reveals widespread aneuploidy in Candida albicans laboratory strains. Mol Microbiol 55(5):1553–1565

Selmecki A, Forche A, Berman J (2006) Aneuploidy and isochromosome formation in drug-resistant Candida albicans. Science 313(5785):367–370. doi:10.1126/science.1128242

Selmecki A, Forche A, Berman J (2010) Genomic plasticity of the human fungal pathogen Candida albicans. Eukaryot Cell 9 (7):991–1008. doi:10.1128/EC.00060-10

Selmecki A, Gerami-Nejad M, Paulson C, Forche A, Berman J (2008) An isochromosome confers drug resistance in vivo by amplification of two genes, ERG11 and TAC1. Mol Microbiol 68 (3):624–641. doi:10.1111/j.1365-2958.2008.06176.x

Seneviratne CJ, Zeng G, Truong T, Sze S, Wong W, Samaranayake L, Chan FY, Wang YM, Wang H, Gao J, Wang Y (2015) New "haploid biofilm model" unravels IRA2 as a novel regulator of Candida albicans biofilm formation. Sci Rep 5:12433. doi:10.1038/srep12433

Shin JH, Bougnoux ME, d'Enfert C, Kim SH, Moon CJ, Joo MY, Lee K, Kim MN, Lee HS, Shin MG, Suh SP, Ryang DW (2011) Genetic diversity among Korean Candida albicans bloodstream isolates: assessment by multilocus sequence typing and restriction endonuclease analysis of genomic DNA by use of BssHII. J Clin Microbiol 49(7):2572–2577. doi:10.1128/JCM.02153-10

Soll DR (2000) The ins and outs of DNA fingerprinting the infectious fungi. Clin Microbiol Rev 13(2):332–370

Symington LS, Rothstein R, Lisby M (2014) Mechanisms and regulation of mitotic recombination in Saccharomyces cerevisiae. Genetics 198(3):795–835. doi:10.1534/genetics.114.166140

Tavanti A, Davidson AD, Fordyce MJ, Gow NA, Maiden MC, Odds FC (2005) Population structure and properties of Candida albicans, as determined by multilocus sequence typing. J Clin Microbiol 43(11):5601–5613

Tavanti A, Gow NA, Senesi S, Maiden MC, Odds FC (2003) Optimization and validation of multilocus sequence typing for Candida albicans. J Clin Microbiol 41(8):3765–3776

Thrash-Bingham C, Gorman JA (1993) Identification, characterization and sequence of Candida albicans repetitive DNAs Rel-1 and Rel-2. Curr Genet 23(5–6):455–462

Tietz HJ, Hopp M, Schmalreck A, Sterry W, Czaika V (2001) Candida africana sp. nov., a new human pathogen or a variant of Candida albicans? Mycoses 44(11–12):437–445

Tzung KW, Williams RM, Scherer S, Federspiel N, Jones T, Hansen N, Bivolarevic V, Huizar L, Komp C, Surzycki R, Tamse R, Davis RW, Agabian N (2001) Genomic evidence for a complete sexual cycle in Candida albicans. Proc Natl Acad Sci U S A 98(6):3249–3253

van het Hoog M, Rast TJ, Martchenko M, Grindle S, Dignard D, Hogues H, Cuomo C, Berriman M, Scherer S, Magee BB, Whiteway M, Chibana H, Nantel A, Magee PT (2007) Assembly of the Candida albicans genome into sixteen supercontigs aligned on the eight chromosomes. Genome Biol 8(4):R52

Wrobel L, Whittington JK, Pujol C, Oh SH, Ruiz MO, Pfaller MA, Diekema DJ, Soll DR, Hoyer LL (2008) Molecular phylogenetic analysis of a geographically and temporally matched set of Candida albicans isolates from humans and nonmigratory wildlife in central Illinois. Eukaryot Cell 7 (9):1475–1486. doi:10.1128/EC.00162-08

Wu W, Lockhart SR, Pujol C, Srikantha T, Soll DR (2007) Heterozygosity of genes on the sex chromosome regulates Candida albicans virulence. Mol Microbiol 64(6):1587–1604

Zhang A, Petrov KO, Hyun ER, Liu Z, Gerber SA, Myers LC (2012) The Tlo proteins are stoichiometric components of Candida albicans mediator anchored via the Med3 subunit. Eukaryot Cell 11(7):874–884. doi:10.1128/EC.00095-12

Zhang N, Harrex AL, Holland BR, Fenton LE, Cannon RD, Schmid J (2003) Sixty alleles of the ALS7 open reading frame in Candida albicans: ALS7 is a hypermutable contingency locus. Genome Res 13(9):2005–2017

Zhang N, Magee BB, Magee PT, Holland BR, Rodrigues E, Holmes AR, Cannon RD, Schmid J (2015) Selective advantages of a parasexual cycle for the yeast Candida albicans. Genetics 200 (4):1117–1132. doi:10.1534/genetics.115.1771

Chapter 12
Chromosome Components Important for Genome Stability in *Candida albicans* and Related Species

Lakshmi Sreekumar, Neha Varshney and Kaustuv Sanyal

Abstract Pathogenic microorganisms have been constantly evolving to battle the various responses elicited by the host upon their invasion. *Candida albicans* bears no exception to this trend. It is not only a fitter pathogen but also a successful gut commensal in humans. The extraordinary genome plasticity of *C. albicans* makes its survival possible in such widely diverse host niches. This chapter focuses on various chromosomal elements that are involved in maintaining the genome dynamics and stability in *C. albicans*. Here, we discuss molecular players of the basic cellular processes that lead to duplication of chromosomes, their faithful segregation in progeny, and chromosome maintenance in *C. albicans* and its related species.

12.1 Insights into DNA Replication

DNA is a self-replicating molecule. The faithful duplication of the genetic material is achieved at the synthetic or S-phase of the cell cycle. "Origins" are the sites on the chromosome where DNA replication initiates. These initiator sequences act as a platform for multi-protein sub-complexes to assemble and facilitate the opening up of the double-stranded DNA to form "replication bubbles". The rate limiting step in this pool is the assembly of the pre-replicative complex, which primarily comprises of evolutionarily conserved proteins like the Origin Recognition Complex (ORC), Cdc6, Cdt1, and the MiniChromosome Maintenance (MCM) helicase (see Table 12.1). Sequential activation of MCM helicase by loading of accessory factors like Cdc6 and Cdt1 makes it a tightly regulated and highly coordinated process. Upon loading of the helicase machinery, DNA polymerase is ready to start the addition of nucleotides to the template DNA (Masai et al. 2010). Termination of replication is governed by termination sites. Any damage to DNA during this

L. Sreekumar · N. Varshney · K. Sanyal (✉)
Molecular Mycology Laboratory, Molecular Biology and Genetics Unit, Jawaharlal Nehru Centre for Advanced Scientific Research, Jakkur, Bangalore 560064, India
e-mail: sanyal@jncasr.ac.in

R. Prasad (ed.), *Candida albicans: Cellular and Molecular Biology*,
DOI 10.1007/978-3-319-50409-4_12

process due to genomic insults like gamma rays or chemical mutagens or even hostile endogenous conditions in the cell is well taken care by the "sensors" or DNA damage check point proteins. All of these mechanisms are conserved across eukaryotes.

A replication origin or "*ori*" is a sequence where the replication bubble is first formed to initiate the templated synthesis of new DNA. These sequences may or may not have strict sequence dependence (Leonard and Mechali 2013; Rui 1999). Initially, when origins were examined in *Saccharomyces cerevisiae*, they were cloned in episomal vectors containing metabolic markers to yield high frequency transformants in auxotrophic strains. These Autonomously Replicating Sequences (*ARS*s) give rise up to a thousand colonies per microgram of the transforming DNA. *ARS* plasmids have proved to be efficient tools to clone yeast DNA sequences. Moreover, several genomic library preparations, in the past decades, have provided information about sequences that can act as putative origins in the genome.

Candida genus includes a diverse group of organisms possessing both pathogenic and non-pathogenic species where several aspects of DNA metabolism have been studied. Additionally, the pathogenic members are known to be genetically more resistant to DNA-damaging agents than the non-pathogenic ones (Rodrigues et al. 2014). Among *Candida* species, *ARSs* have been reported in *C. albicans, Candida guilliermondii, Candida utilis, and Candida glabrata*. In *C. albicans*, a genomic DNA library was constructed in an *ADE2* containing plasmid of *S. cerevisiae* (Kurtz et al. 1987). Upon transformation in an adenine auxotroph of *C. albicans*, a thousand transformants were obtained of which a minor fraction were genomic integrants as well. However, after eight generations of growth in non-selective media, only 1–2% of the progeny were still Ade$^+$, indicating that *ARS* plasmids, if not integrated into the genome, have very low mitotic stability in this organism. In one of the earlier studies, an 8.6 kb fragment was isolated from *C. albicans* genome and a part of it was subcloned in a bacterial plasmid containing *CaLEU2* and *CaURA3* genes (Cannon et al. 1990). These plasmids could yield a transformation efficiency of 2.15×10^3 CFU/μg and 1.91×10^3 CFU/μg of plasmid DNA in *S. cerevisiae* and *C. albicans,* respectively. Upon further sequence analysis, these *ARS* components were shown to be similar to *S. cerevisiae ARS* elements. Subsequently, a 15.3 kb fragment was cloned in a bacterial plasmid and it showed properties of an *ARS* plasmid. However, these *ARS* plasmids are subject to random integration and multimer formation in the genome (Herreros et al. 1992).

Centromeres (*CENs*) are DNA loci that act as platforms for chromosome segregation. *CENs* in *C. albicans* were found to be the earliest replicating regions in the S-phase of every cell cycle (Koren et al. 2010). The replication profile of chromosome 1 showed that origins flanking the *CENs* are the first to fire during every S-phase. Genomic origins were identified by the binding of ORC proteins and validated using the approach of 2-dimensional gel electrophoresis. A more detailed analysis of the chromosome 5 and chromosome 7 in *C. albicans* revealed a strong correlation between the effects of the *CEN* proximal origin on *CEN* function (Mitra et al. 2014). *CEN* proximal origins contributed to replication fork stalling at the *CEN*. This fork stalling or random termination was shown to be a kinetochore-mediated phenomenon. Upon

Table 12.1 Protein components associated with major chromosomal elements in *C. albicans*

Name	Allele name	Feature type	Description	Orf no.
Replication origin-associated proteins				
MCM2	CR_03830C_B	Uncharacterized	Phosphorylated protein of unknown function; transcription is periodic with a peak at M/G1 phase	orf19.4354
MCM3	C2_07350W_B	Uncharacterized	Putative DNA replication protein; periodic mRNA expression that peaks at the M/G1 phase	orf19.1901
CDC54	C1_12550C_B	Uncharacterized	Putative pre-replication complex helicase subunit; periodic mRNA expression, peak at cell cycle M/G1 phase	orf19.3761
CDC46	C2_06250C_B	Uncharacterized	Putative hexameric MCM complex subunit; predicted role in control of cell division; periodic mRNA expression. peak at cell cycle M/G1 phase	orf19.5487
MCM6	CR_02110W_B	Uncharacterized	Putative MCM DNA replication initiation complex component; mRNA expression peaks at cell cycle M/G1 phase	orf19.2611
CDC47	C2_09020W_B	Uncharacterized	Phosphorylated protein described as having role in control of cell division; RNA abundance regulated by tyrosol and cell density	orf19.202
ORC1	C1_03070C_B	Uncharacterized	Putative origin recognition complex (ORC) large subunit; essential for viability; similar to *S. cerevisiae* Orc1p subunit	orf19.3000
ORC2	C2_10760C_B	Uncharacterized	Phosphorylated protein of unknown function	orf19.5358
ORC3	C3_03750C_B	Verified	Protein similar to *S. cerevisiae* Orc3p, ORC component	orf19.6942
ORC4	C5_02150C_B	Uncharacterized	Phosphorylated protein similar to *S. cerevisiae* Orc4	orf19.4221
ORC5	CR_06960W_B	Uncharacterized	Ortholog(s) have ATP binding, DNA replication origin binding activity	orf19.2369
ORC6	C1_01000C_B	Uncharacterized	Phosphorylated protein of unknown function	orf19.3289

(continued)

Table 12.1 (continued)

Name	Allele name	Feature type	Description	Orf no.
POL30	C4_01770W_B	Verified	Proliferating Cell Nuclear Antigen (PCNA), forms homotrimeric sliding clamp for DNA polymerases; cell density; stationary phase enriched protein	orf19.4616
CDC13	C1_00360W_B	Verified	Essential protein with similarity to *S. cerevisiae* Cdc13p, involved in telomere maintenance	orf19.6072
Centromere-associated proteins				
CSE4	C3_00860W_B	Verified	Centromeric histone H3 variant; role in structural changes of centromeric nucleosomes during cell cycle, centromeric DNA binding	orf19.6163
MIF2	C6_02780C_B	Verified	Centromere-associated protein; similar to CENP-C proteins; Cse4p and Mif2p colocalize at *C. albicans* centromeres	orf19.5551
MTW1	C2_09840W_B	Verified	Kinetochore component; the amount of Nuf2p and Mtw1p protein detected at each centromere is consistent with a single kinetochore-microtubule attachment site	orf19.1367
DAM1	C1_09730W_B	Verified	Component of DASH complex, microtubule plus end binding, chromosome segregation by coupling kinetochores to spindle microtubules	orf19.4837
DAD1	C1_13710C_B	Verified	Component of DASH complex, microtubule plus end binding, chromosome segregation by coupling kinetochores to spindle microtubules	orf19.5008.1
DAD2	C2_05210W_B	Verified	Component of DASH complex, microtubule plus end binding, chromosome segregation by coupling kinetochores to spindle microtubules	orf19.3551
ASK1	C4_01150W_B	Verified	Component of DASH complex, microtubule binding, chromosome segregation by coupling kinetochores to spindle microtubules	orf19.4675

(continued)

Table 12.1 (continued)

Name	Allele name	Feature type	Description	Orf no.
SPC19	C1_03980W_B	Verified	Component of DASH complex, microtubule plus end binding, chromosome segregation by coupling kinetochores to spindle microtubules	orf19.4473
RAD51	CR_02200C_B	Verified	Homologous recombination and DNA repair; slow growth and increased white-to-opaque switching frequency in null mutant	orf19.3752
RAD52	C6_00510C_B	Verified	Required for homologous DNA recombination, repair of UV- or MMS-damaged DNA, telomere length, UV-induced LOH; constitutive expression; slow growth, increased white-to-opaque switch	orf19.4208
MRE11	C7_01340W_B	Verified	Putative DNA double-strand break repair factor, 3′–5′ exonuclease activity; involved in response to oxidative stress and drug resistance	orf19.6915
SCM3	C3_01770C_B	Not verified	Unknown	orf19.1668
Telomere-associated proteins				
STN1	C2_08470C_B	Verified	Protein involved in telomere maintenance; forms a complex with Ten1p	orf19.3631
TEN1	CR_01010W_B	Verified	Protein involved in telomere maintenance; forms a complex with Stn1; transcription is regulated upon yeast-hypha switch	orf19.3255
RAP1	C2_10080W_B	Verified	Transcription factor; binds telomeres and regulatory sequences in DNA; involved in telomere maintenance; represses hyphal growth under yeast-favoring conditions	orf19.1773
EST1	C5_05470W_B	Verified	Telomerase subunit; allosteric activator of catalytic activity, but not required for catalytic activity	orf19.4045

(continued)

Table 12.1 (continued)

Name	Allele name	Feature type	Description	Orf no.
TERT	C1_08130C_B	Verified	Telomerase reverse transcriptase; catalytic protein subunit of telomere synthesis; essential for telomerase activity; has telomerase-specific motif T and other conserved reverse transcriptase motifs	orf19.5089
EST3	C3_00430W_B	Verified		orf19.5423

Source: *Candida* genome database

deletion of these *CEN* proximal origins, the *CEN* function was compromised and occupancy of CENP-A over the central core of *CEN7* was seen to be ablated.

The dual nature of origins in the genome of *C. albicans* was shown in a study where the nature of origins was claimed to differ on the basis of the chromosomal location context (Tsai et al. 2014). A mini-*ARS* screen was used to identify bona fide *ARSs*. *CEN* proximal origins were termed "epigenetic," whereas arm origins were called "hard wired" sequence-dependent origins. Genome-wide origins were mapped based on ORC binding, nucleosome depletion patterns, and the aforementioned *ARS* screen. All the eight chromosomes of *C. albicans* showed the characteristic GC-skew pattern that is reminiscent of bacterial chromosomes.

Several *ARSs* have been reported in other *Candida* species. In the industrially important yeast *C. utilis,* a 6.6 kb *ARS* has been reported which was obtained after a library preparation and it had a mitotic stability of less than 1% (Iwakiri et al. 2005). Unlike other yeasts where the size of a functional *ARS* is limited to 100–200 bp, in *C. utilis* the smallest *ARS* is 2.8 kb, the rest being 5 kb and above. *C. guilliermondii* is an important yeast of biotechnological interest (Papon et al. 2013). In an attempt to standardize the electroporation method of transformation in this system, scientists stumbled upon an *ARS* located upstream of the *URA5* gene (Foureau et al. 2013). In *silico* analysis of the sequence predicted *ARS*-like elements (ALS) which apparently increased the transformation efficiency by a thousand fold. The most recent addition of a *Candida* species in the collectible of *ARS*s is in *C. glabrata* (Descorps-Declere et al. 2015). Though it is a member of the *Saccharomycota*, it underwent a whole-genome duplication (WGD) event followed by extensive loss of genes (Fitzpatrick et al. 2006). Its genome has several long tandem repeats or megasatellites. Using deep sequencing and chromosome conformation capture (3C) experiments, the replication landscape of *C. glabrata* was determined (Descorps-Declere et al. 2015). This study identified 253 replication origins and 275 *ARS*s in this pathogenic yeast. Using a time course S-phase study, centromeres, *MAT* loci, and most histone genes were found to be located in early replicating domains of the chromosome, whereas fragile sites and chromosome breakpoints were shown to be late-replicating. The chromosome conformation capture (3C) experiment showed clustering of early replicating origins which were not observed in non-subtelomeric megasatellites. In spite of their significant divergence, the replication program of *C. glabrata* bore striking similarities to that of *S. cerevisiae*, once again proving that binding specificities of the MCM/ORC complexes are highly conserved on diverse DNA sequences.

Members of the pre-replicative complex (pre-RC) have not been characterized fully in the *Candida* species. Homologs of the MCM/ORC complex are yet to be characterized. The *C. albicans* Proliferating Cell Nuclear Antigen (PCNA) was shown to be an essential protein that functions as a sliding clamp for the DNA polymerase (Manohar and Acharya 2015). It is loaded by the clamp loader RFC. It is a homotrimeric ring encircling the double-stranded DNA. PCNA interacts physically with DNA polymerase ε. Cell cycle regulation is mediated tightly by the concerted activity of cyclin-CDK complexes. *C. albicans* has two B-type cyclins—Clb2 and Clb4. Clb2 is the functional homolog of Clb5/Clb6 (Ofir and Kornitzer 2010).

12.2 Segregating the Nuclear Material: When and How?

Aneuploidy is a major feature for a wide range of health disorders including cancers. Even in certain fungal species, like *C. albicans*, aneuploidy provides a unique advantage to the organism to maintain genome plasticity (reviewed in Sanyal 2012). The altered ploidy state seen in terms of genomic rearrangements and segmental aneuploidy of chromosomes stems from various exposures to stress conditions, one of them being a prolonged treatment of this fungal pathogen to antifungals thereby providing it a survival strategy in hostile environment niches.

Chromosomes are required to be segregated with utmost fidelity in a cell before it divides. Centromeres (*CEN*s) are DNA loci responsible for precise chromosome segregation. The proper attachment of the protein machinery residing on the *CEN* DNA, the kinetochore with the microtubules emanating from opposite spindle poles, is the determinant of high fidelity chromosome segregation. *CEN*s are typically located in regions of a genome with less gene density and a reduced rate of recombination. While in most organisms they occupy discrete sites on monocentric chromosomes, they may extend up to the entire length of a holocentric chromosome. A centromere can be physically located either at the middle of a chromosome (metacentric) or at the end of a chromosome (acrocentric or telocentric). Several centromeric proteins (CENPs) were first identified serendipitously by isolating anti-centromere antibodies from individuals with the CREST disease (Palmer et al. 1987). Subsequently, other members of CENP family were identified in many organisms. Centromere protein A (CENP-A)/Cse4 is a variant of histone H3 present exclusively at functional *CEN*s.

Centromeres are broadly classified as point, small regional, and large regional centromeres (reviewed in Roy and Sanyal 2011). The small regional centromeres of 3–5 kb non-repetitive, unique sequences were identified in three *Candida* species—*C. albicans*, *Candida dubliniensis,* and *Candida lusitaniae* (Sanyal et al. 2004; Padmanabhan et al. 2008; Kapoor et al. 2015). The *CEN* DNA of all eight chromosomes of *C. albicans* were identified and characterized by chromatin immunoprecipitation using anti-CaCse4 antibodies (Sanyal et al. 2004). The Cse4-binding region was mapped to a 3–4.5 kb stretch, deletion of which resulted in a high frequency loss of only the altered chromosome. Similarly, *CEN* sequences on all eight chromosomes were identified in the closely related yeast, *C. dubliniensis* (Padmanabhan et al. 2008). Interestingly, all the eight *CEN*s in these two organisms lack any conserved sequence including any motifs or repeats (Mishra et al. 2007). However, *CEN1, CEN4, CEN5,* and *CENR* of both *C. albicans* and *C. dubliniensis* possess unique chromosome-specific pericentric-inverted repeats (IRs). In addition to IRs, LTRs were found within *CEN2, CEN3, CEN5,* and *CEN6* of *C. albicans*. The conservation of the relative position of the *CEN* in orthologous regions with respect to the adjacent *ORF*s (synteny) in these two *Candida* species hinted toward the possibility that the *CEN* DNA sequence alone may not determine its identity. This was further validated by an elegant experiment where an 85-kb chromosome fragment (CF) exhibiting *CEN* function in vivo failed to show de novo *CEN* activity when shuttled as a naked DNA from *C. albicans* back into *C. albicans* (Baum et al. 2006). Even the naked *CEN* DNA introduced 6.7 kb away from the native locus on chromosome 7 could not recruit CENP-A in *C. albicans*. Inability of a circular *ARS* plasmid carrying the CENP-A-rich *CEN* region to produce a stable minichromosome in *Candida* cells also bolstered the fact that DNA sequence is not the only determinant of the *CEN* identity. In budding yeast, the CENP-A occupying *CEN* chromatin has a distinct nucleosome occupancy pattern from the bulk chromatin in having a ladder of DNA fragments after partial digestion with micrococcal nuclease (MNase) which is a characteristic of regularly spaced nucleosomes. However, *C. albicans* gives a more distinct pattern in this assay, where mononucleosomes, dinucleosomes, and a smeared pattern were detected around *CEN7* with an absence of the nucleosome ladders. In the three species of *Candida* clade, *C. lusitaniae*, *Pichia stipitis,* and *Yarrowia lipolytica*, one obvious GC-poor region was found to be located on each chromosome speculated to be the *CEN*, even though the *CEN*s in *C. albicans* and *C. dubliniensis* are not seen to be GC-poor (Lynch et al. 2010). These *CEN* locations were experimentally validated in *C. lusitaniae* by ChIP-sequencing of two key centromeric proteins, Cse4 and Mif2 (Kapoor et al. 2015). The centromere DNA sequence in *C. lusitaniae* was unique for each chromosome and spanned 4–4.5 kb, similar to regional centromeres of *C. albicans*. A distinct pattern of histone modifications, methylated H3K79 and H3R2 but lack of methylation of H3K4, which is otherwise seen in regional centromeres, was enriched at centromeric chromatin in *C. lusitaniae*. However, unlike other regional centromeres, there was no evidence for the presence of pericentromeric heterochromatin in *C. lusitaniae*. The pericentromeric regions in *C. albicans* are assembled into an intermediate chromatin state harboring features of both heterochromatin and euchromatin (Freire-Beneitez et al. 2016). The regions flanking the central core are associated with nucleosomes that are hyperacetylated-like euchromatic regions and hypomethylated on H3K4-like heterochromatin. However, the pericentromeric regions in *C. lusitaniae* are not associated with hypoacetylated histones or sirtuin deacetylases that generates heterochromatin in other yeasts. In other words, the lack of the *URA3* reporter gene repression when inserted adjacent to the centromere supports the fact that there is no pericentromeric heterochromatin in *C. lusitaniae* (Kapoor et al. 2015).

Strikingly, the structure of the *CEN* has been rewired in the closely related *Candida* species, *Candida tropicalis* where each of the seven chromosomes comprises a 2–5 kb non-repetitive mid-core region that forms the binding sites for CENP-A (Cse4) and CENP-C (Mif2) flanked by 2–5 kb inverted repeats (IRs) (Chatterjee et al. 2016). The repeat associated centromeres of *C. tropicalis* also share a high degree of sequence conservation with each other. In spite of the observed rapid change in the sequence and organization of *CEN*s in these closely related species, the AT-content of the CENP-A bound *CEN* DNA sequence was found to be similar in *C. albicans* and *C. tropicalis*. All *C. albicans CEN*s are free of transposons while one *CEN* in *C. tropicalis* is associated with retrotransposons (Mishra et al. 2007; Chatterjee et al. 2016). Interestingly, these small regional *CEN*s maintain their uniform *CEN* sizes (~3 kb) despite the absence of any obvious boundary elements such as the tRNA genes.

*C. glabrat*a exhibits similarities to *S. cerevisiae* in its *CEN* organization. A 451 bp fragment exhibiting *CEN* activity was isolated and sequenced (Kitada et al. 1997). It was shown to possess three elements similar to *S. cerevisiae*, Cg*CDEI*, Cg*CDEII,* and Cg*CDEIII*. Substitution mutation analysis revealed the requirement of at least Cg*CDEI* and Cg*CDEIII* for *CEN* function in this organism. Despite the presence of functional similarities between *CEN*s of *C. glabrata* and *S. cerevisiae,* there exists species specificity in *CEN* function. Similarly, in *C. maltosa* a 325 bp fragment was shown to harbor *CEN* activity with *CDEI* and *CDEII*-like regions of *S. cerevisiae* (Ohkuma et al. 1995).

One of the most astonishing examples of epigenetic changes within a genome is the formation of neocentromeres. Neocentromeres are *CEN*s arising at atypical chromosomal loci. *C. albicans* provides an excellent model system to study this process. Upon deletion of a native centromere (*CEN*5), neocentromeres could form efficiently in *C. albicans* in various locations on the same chromosome. They could be proximal neocentromeres, formed close to the location of the native *CEN* and distal neocentromeres which are formed at other locations on the chromosomes (Thakur and Sanyal 2013; Ketel et al. 2009). A more comprehensive study in multiple chromosomes (chromosome 1, 5 and 7) of *C. albicans* and additionally involving the closely related *C. dubliniensis* suggested that neocentromere formation is a conserved mechanism in these organisms and occurs in *CEN* proximal regions (Thakur and Sanyal 2013). The distal neocentromeres show low CENP-A enrichment as compared to the native *CEN*s leading to chromosome loss, asserting the fact that *CEN* proximal sites are the preferred sites for neocentromere formation (Thakur and Sanyal 2013). Since *CEN* chromatin can have a negative effect on the gene expression in *C. albicans*, all neocentromeres are formed in the intergenic regions of the chromosomes. Although deletion of the endogenous *CEN7* leads to neocentromere formation in *C. albicans*, in a fraction of strains the *CEN* was repositioned to the endogenous locus by gene conversion events through copying *CEN7* of the unaltered homolog (Thakur and Sanyal 2013). Additionally, there is no correlation between the neocentromere formed and length of the deleted *CEN* region. Deletion of *CEN1* and *CEN5* led to neocentromere formation which was two to four times the size of native *CEN*s suggesting that length of the neocentromere is variable across chromosomes (Scott and Sullivan 2014). *C. albicans* has a combination of neocentromere properties observed in various plants, animals, and fungal species asserting the conservation of such mechanisms across species (Marshall and Choo 2009).

The early replicating timing of *CEN*s could be related to loading of CENP-A, that is also seen to take place in the early S-phase of other yeast species (Aravamudhan et al. 2013; Pearson et al. 2004; Takahashi et al. 2005). However, the formation of a neocentromere at a late-replicating domain in *C. albicans* has created a paradigm shift in the view that replication timing is not the sole determinant of de novo centromere assembly (Koren et al. 2010). The neocentromeres occurring in the late-replicating domain were accompanied by shifts in replication timing. These neocentromeres became the first to replicate and became associated with the origin recognition complex (Koren et al. 2010).

Kinetochores are large macromolecular complexes assembled on *CEN* DNA. The kinetochores are tri-layered structures revealed in the early microscopic images of mitotic chromosomes in human cells. It comprises an inner layer interacting directly with the *CEN* DNA, the outer layer forming the chromosomal attachment site for the microtubule plus end and middle layer, bridging the two layers (see Table 12.1). However, due to the small cellular size of unicellular organisms like yeasts, the ultra-structure of a kinetochore cannot be ascertained. Immuno-localization of kinetochore proteins in these organisms appears as puncta of clustered kinetochores at the nuclear peripheral regions located close to spindle pole bodies.

CaCse4, the yeast homolog of CENP-A, was identified and localized as an intense dot-like signal (a cluster of 16 kinetochores), co-localizing with the nucleus in *C. albicans* (Sanyal and Carbon 2002). Like in other organisms, CaCse4 is an essential protein and is involved in kinetochore formation in *C. albicans*. CaCse4 is required for proper chromosome segregation as depletion of this protein results in the accumulation of large buds in the population. Similarly, *CaMIF2* (CENP-C homolog) was shown to be an essential gene in *C. albicans*. CaMif2 colocalizes with CaCse4 and is enriched at all *CEN*s in *C. albicans* (Sanyal et al. 2004). Subsequently, CaMtw1, a homolog of human Mis12/Mtw1 in *C. albicans,* was characterized for its function in the process of kinetochore-microtubule mediated chromosome segregation in *C. albicans* (Roy et al. 2011). CaMtw1 is an essential protein required for G2/M progression and proper chromosome segregation during mitosis. It is required for spindle positioning and morphogenesis as well. A fungus-specific outer kinetochore protein complex, the Dam1 complex mediates attachment of the chromosomes to the mitotic spindle. It comprises of ten different subunits which oligomerize in various ways to form a ring that interacts with the microtubules. Dam1, Dad1, Dad2, Ask1, and Spc19 subunits of the Dam1 complex are shown to be essential for viability and are indispensable for proper chromosome segregation in *C. albicans* (Thakur and Sanyal 2011; Burrack et al. 2011). Dad2 shares functional similarity with Dam1 and was shown to be localized at the mid-zone in addition to its kinetochore localization. Although the recruitment of the Dam1 complex was shown to be independent of the kinetochore-microtubule interactions, the function of this complex was shown to be monitored by the spindle assembly checkpoint (SAC). Also, the Dam1 complex is required to prevent spindle elongation in early mitosis. The first biochemical study on a kinetochore protein in *C. albicans* was done on the Dam1 complex. Dad1 was shown to be an intrinsically disordered protein with a structure similar to its *S. cerevisiae* counterpart (Waldo et al. 2010).

The correlation of "one microtubule/kinetochore" was established in *S. cerevisiae* having a point *CEN* in contrast to multiple microtubules/kinetochore in *S. pombe* which contains a large regional *CEN* (Winey et al. 1995; Ding et al. 1993). Interestingly, the essentiality of the Dam1 complex can be correlated with one microtubule-one kinetochore type of interaction. Dam1 is essential in the budding yeasts, *S. cerevisiae* and *C. albicans,* while it is non-essential in the fission yeast, *S. pombe. C. albicans,* harboring a regional *CEN*, supports one microtubule/kinetochore-like interaction (Joglekar et al. 2008). Strikingly, CENP-A over-expression could rescue the depletion of Dam1 by increasing the level of other kinetochore proteins and hence microtubules to form a functional kinetochore (Burrack et al. 2011).

Recruitment of most of kinetochore proteins studied so far is regulated by CENP-A. However, a few master regulators such as Ndc10, Scm3 (*S. cerevisiae*), Mis6, the Mis16–Mis18 complex, and Ams2 (*S. pombe*) and Rad51–Rad52/CENP-C (*C. albicans*) were shown to influence CENP-A localization in various yeast species (Camahort et al. 2007; Hayashi et al. 2004; Takahashi et al. 2005; Mitra et al. 2014; Roy et al. 2011). Recruitment of CENP-A at the *CEN* in *C. albicans* is mediated by homologous recombination (HR) proteins, where the replication forks coming from *CEN* proximal origins stall at the *CEN* in a kinetochore-dependent manner (Mitra et al. 2014). Fork stalling at the *CEN* is reduced in the absence of HR proteins Rad51 or Rad52. Null *rad51* or *rad52* mutants exhibit an increased kinetochore declustering and degradation of CENP-A. The physical association of CENP-A and Rad51/Rad52 in a complex is an indicator of an HR-mediated CENP-A recruitment mechanism in this organism. CENP-A levels in HR mutants, such as *mre11, rad51,* and *rad52* null mutants, have been observed to be low in *C. albicans* (Mitra et al., 2014). This emerging role of HR proteins at the centromere raises an interesting possibility of the involvement of a Holliday Junction Recognition Protein, HJURP/Scm3 at the *CEN*. The timing of loading of CENP-A with respect to cell cycle varies in different species. Experiments suggest that CENP-A is deposited during S-phase in *S. cerevisiae* in contrast to its biphasic loading observed in *S. pombe*. Paradoxically, anaphase-specific loading of CENP-A was also indicated in contrast to the previous report of CENP-A deposition during S-phase in *S. cerevisiae*. Strikingly, *C. albicans* was also shown to have an anaphase coupled loading of CENP-A (Shivaraju et al. 2012).

While the kinetochore assembly occurs in a step-wise manner on a point *CEN* in *S. cerevisiae*, the kinetochore architecture is stabilized in a coordinated interdependent manner by its individual components in *C. albicans*. Intriguingly in *C. albicans*, the kinetochore proteins from the outer and middle layers influence the localization of the inner kinetochore protein, CENP-A. The *CEN* localization of CENP-A was shown to be remarkably reduced in the absence of the inner (Mif2/CENP-C), middle (Mis12/Mtw1), and outer (Dam1 complex and Nuf2) kinetochore proteins (Thakur and Sanyal 2012). Even Mif2 exhibited dependency on the Mtw1 for its recruitment (Roy et al. 2011). The unprecedented observation of outer kinetochore proteins influencing the localization of CENP-A in *C. albicans* suggests that the kinetochore sub-complex assembles in a unique interdependent concerted manner to form a stable kinetochore. The kinetochore collapses in the absence of its essential components indicating that the kinetochore may not be a layered structure in *C. albicans*. Most strikingly, the kinetochore protects CENP-A from proteasomal degradation in *C. albicans*. Even the newly synthesized CENP-A molecules fail to rescue the kinetochore integrity defects strengthening the fact that individual kinetochore components are absolutely essential for protecting the *CEN*-bound CENP-A molecules in *C. albicans*.

Several *Candida* species reside as a harmless commensal in the human gastrointestinal tract and genitourinary tract. However, they can be opportunistic human pathogens causing superficial to fatal systemic infections in immuno-compromised patients. Candidemia is the fourth most common cause of hospital-acquired infections with huge annual costs for medicare of patients. Hence, it is important to identify potential drug targets in these organisms. The outer kinetochore protein, the Dam1 complex, is essential and localized at the kinetochore throughout the cell cycle in *C. albicans*. Interestingly, no homologue of Dam1 has been found in metazoan system. The fungal specificity of the Dam1 complex and its crucial role in kinetochore-microtubule attachments makes it a potent drug target in this pathogenic fungus and its related species.

12.3 Cell Longevity Measures Taken by *Candida albicans*

To prevent the breakage and fusion of linear chromosomes in the nucleus, the cell employs a machinery to sequester the ends or "telomeres" such that it is masked from the exposure of DNA damage response. Telomeres act as caps at chromosomal ends and also maintain the replicative lifespan of chromosomes. Telomeres are G-rich repetitive sequences present at the end of every chromosome and are maintained by the reverse transcriptase enzyme, telomerase, that uses the 3'G-rich overhang as a template to synthesize these repetitive sequences (O'Sullivan and Karlseder 2010).

C. albicans is an interesting model to study telomere biology because of various reasons. Unlike other species, it contains unusually long (23 bp-long) distinct and regular telomere repeat units. Also, the overall length of telomeres can be varied depending on the growth conditions provided to the organism, making this a novel yet interesting mechanism of telomere length regulation (McEachern and Hicks 1993). There exists a definite interplay between telomerase activity and telomere recombination for telomere maintenance. In wild-type *Candida* cells, it has been difficult to detect a senescent phenotype owing to heterogeneity in the sizes of Terminal Restriction Fragments (TRFs) of a single telomere studied over a period of time (Singh et al. 2002). Using computational and experimental data, homologs of telomerase proteins have been identified in *Candida* genomes. A rapid evolutionary divergence of telomere-associated proteins has been evident. Most of the studies on telomeres and telomerase components have been carried out in *C. albicans* so far with a very little documentation available in other *Candida* species. Telomere binding factors can be categorized as single-stranded and double-stranded nucleotide-binding proteins (see Table 12.1).

The Cdc13-Stn1-Ten1 (CST) complex forms a major heterotrimeric complex involved in telomere protection, via its oligosaccharide/oligonucleotide-binding (OB)-fold domain that recognizes single-stranded DNA (Lue et al. 2013). The *Candida* Cdc13 homologs are smaller than the *S. cerevisiae* counterpart, as they lack the OB1 domain at the N-terminus required for dimerization of Cdc13 (Lue and Chan 2013). However, the DNA-binding domain and the C-terminal OB4 domains are shared between *C. albicans, C. tropicalis,* and *S. cerevisiae.* Unlike Cdc13, the *Candida* Stn1 and Ten1 share considerable homology with their *S. cerevisiae* counterpart. Crystal structure analysis of the *C. tropicalis* Stn1–Ten1 complex reveals that these proteins do contain a single OB-fold domain. Additionally, the Stn1–Ten1 complex is structurally similar to the RPA complex (Rpa2–Rpa3) (Sun et al. 2009). Both these proteins and their interactions regulate telomere length as their null mutants exhibit long and heterogeneous telomeres. In the *C. albicans* strain BWP17, telomeres normally range from 1 to 5 kb in length. If the telomeres are "de-protected," it leads to the formation of extra-chromosomal telomeric circles resulting from recombination mediated telomere maintenance and Telomere Rapid Deletion (TRD), called t-loops (Tomaska et al. 2009).

Rap1 is a double-stranded DNA-binding protein which is required for mating type and transcriptional silencing. It is a multifunctional protein owing to its complex domain architecture. Rap1 homologs have three functional domains—the BRCT domain required for protein–protein interaction, the MYB domain for DNA binding, and the C-terminal RCT (Rap1 C-terminus) which is thought to interact with four other proteins, namely Rif1, Rif2, Sir3, and Sir4. The Rap1 homolog in *C. albicans* lacks the RCT domain (Uemura et al. 2004). Rap1 is not essential for cell viability, but null mutants show aberrant telomeric recombination leading to longer telomere lengths and an increased formation of t-circles (Yu et al. 2010; Biswas et al. 2003). Rap1 seems to have overlapping functions with Stn1–Ten1, but they are not redundant. Due to the lack of a C-terminal domain, the Sir-mediated subtelomeric silencing by Rap1 may not be mediated the same way in *C. albicans* as it occurs in other species studies so far (Feeser and Wolberger 2008). Apart from these proteins, Rif1/Rif2 and the Ku complex have been studied in other species, but their *Candida* homologs are yet to be characterized.

The telomerase complex is known to consist of a catalytic protein subunit (Est2p/TERT) and two non-catalytic subunits (Est1p, Est3p). The TERT protein is well conserved across species owing to its catalytic function. On the other hand, Est1 exhibits a low level of sequence conservation. The C-terminus of CaEst1, being involved in its binding to the telomerase RNA, Cdc13 and the telomerase DNA, is quite similar to ScEst1. CaEst3 shows the least level of conservation with its *S. cerevisiae* counterpart (Singh et al. 2002).

CaTER1 and CaEst2 (TERT) have been found to be essential for the catalytic activity of the telomerase complex (Singh et al. 2002). In the same report, the authors have proposed a potential telomere "capping" role of Est1p and TERT. In *C. albicans*, deletion of *EST2* results in progressive telomere attrition, thereby increasing the level of G-strand overhangs, whereas *EST1* or *EST3* deletion did not yield the same phenotype indicating that telomerase can physically protect telomeres in a catalytically independent manner. It has been speculated that the catalytic core complex physically blocks the access of degrading enzymes to telomere ends, thereby rendering its protection. Such an accumulation of G-strand overhangs was seen in null mutants of telomere protecting proteins like Cdc13 and the Ku complex. A similar phenotype was observed in a null mutant of the RNA component of telomerase, *TER1*. The notable feature of the *CaTER1* gene is that it possesses the largest *TER* gene identified so far (1,554 bp); however, the template for *CaTER1* gene is smaller (250 bp). The significance of this finding is still unknown (Hsu et al. 2007).

Telomeric *ORFs* (*TLOs*) are a family of telomere associated *ORFs* in *C. albicans* and *C. dubliniensis* that encode a subunit of the Mediator complex which is used for the recruitment of RNA pol II during transcription initiation. *C. albicans* has 15 TLOs, whereas *C. dubliniensis* has two (Haran et al. 2014; Anderson et al. 2012; Sullivan et al. 2015). This expansion of the *TLO* gene family could suggest a flexible transcriptional network in this organism that gives it the plasticity to thrive in varying environmental niches. These 15 *TLO* genes can be classified into three clades—*TLO*α (6 members), *TLO*β (one member), and *TLO*γ (7 members) and a pseudogene. They primarily differ by the presence of a Long Terminal Repeat (LTR) that gives rise to splice variants. Members of *TLO*γ clade produce both spliced and unspliced transcripts. *TLO*α genes are expressed at the highest levels, whereas *TLO*γ genes are expressed at very low levels. A plethora of Mediator subunit composition, caused by a broad range of expression levels of *TLO* genes, provides a greater flexibility to *C. albicans* allowing it to acclimatize to a broad range of host niches.

Transient silencing by transcription is mediated by chromatin complexes. In several organisms, genes on telomeres are subject to Telomere Positioning Effect (TPE), where a biphasic open or closed chromatin state at a given telomere, switches the gene expression status as ON or OFF (Gottschling et al. 1990). TPE is dependent on Sir2p in *S. cerevisiae*. Cell to cell variation in transcription/translation levels, termed as "gene expression noise" gives rise to phenotypic variation in an isogenic population. Noise can be extrinsic from different environment cues or intrinsic, arising from allele to allele variation (Raser and O'Shea 2005). Telomere

Table 12.2 Unique genomic features of *C. albicans*

1. It shows enormous genome plasticity—segmental duplication, trisomy, monosomy for most chromosomes. These features provide *C. albicans* a survival advantage in the host
2. An epigenetic mechanism for centromere formation is evident in *C. albicans*. All eight chromosomes possess unique *CEN* DNA sequences
3. Efficient neocentromere formation has been observed when a native centromere is deleted in *C. albicans*. Genomic mechanisms such as gene conversion prevent centromere repositioning
4. A unique interdependent concerted manner of kinetochore formation takes place in *C. albicans*. The kinetochore integrity protects CENP-A from proteasomal-mediated degradation
5. *C. albicans* contains unusually long telomeric repeats, the length of which can be varied depending upon the environmental/culture conditions
6. *C. albicans* TLOs provide a wide range of transcriptional flexibility to adapt to diverse host niches

proximal genes exhibit higher noise levels largely due to intrinsic noise that is dependent on genome position, or Telomere-Adjacent Gene Expression Noise (TAGEN) (Anderson et al. 2014). TAGEN generates expression variability due to local chromatin-mediated gene silencing. Likewise, *TLO* genes, when placed in the internal locations of a chromosome, were shown to exhibit reduced noise levels. In *C. albicans,* TAGEN is regulated in a Sir2-dependent manner, largely promoter independent and is tightly associated with telomere position effect dynamics. A similar effect of colony to colony variation due to noise was seen for one gene, *EPA1* in *C. glabrata* (Juarez-Reyes et al. 2012).

Hence, *C. albicans* is an ideal system to study telomere biology, owing to its highly plastic genome and adaptability to diverse environment niches. Components of the telomere machinery that have diverged evolutionarily in terms of domain structure and loss or emergence of key factors suggests that this organism has a wealth of genomic tools that is different from other budding yeasts (see Table 12.2).

References

Anderson MZ, Baller JA, Dulmage K, Wigen L, Berman J (2012) The three clades of the telomere-associated TLO gene family of Candida albicans have different splicing, localization, and expression features. Eukaryot Cell 11:1268–1275

Anderson MZ, Gerstein AC, Wigen L, Baller JA, Berman J (2014) Silencing is noisy: population and cell level noise in telomere-adjacent genes is dependent on telomere position and sir2. PLoS Genet 10:e1004436

Aravamudhan P, Felzer-Kim I, Joglekar AP (2013) The budding yeast point centromere associates with two Cse4 molecules during mitosis. Curr Biol 23:770–774

Baum M, Sanyal K, Mishra PK, Thaler N, Carbon J (2006) Formation of functional centromeric chromatin is specified epigenetically in Candida albicans. Proc Natl Acad Sci U S A 103:14877–14882

Biswas K, Rieger KJ, Morschhauser J (2003) Functional analysis of CaRAP1, encoding the Repressor/activator protein 1 of Candida albicans. Gene 307:151–158

Burrack LS, Applen SE, Berman J (2011) The requirement for the Dam1 complex is dependent upon the number of kinetochore proteins and microtubules. Curr Biol 21:889–896

Camahort R, Li B, Florens L, Swanson SK, Washburn MP, Gerton JL (2007) Scm3 is essential to recruit the histone h3 variant cse4 to centromeres and to maintain a functional kinetochore. Mol Cell 26:853–865

Cannon RD, Jenkinson HF, Shepherd MG (1990) Isolation and nucleotide sequence of an autonomously replicating sequence (ARS) element functional in Candida albicans and Saccharomyces cerevisiae. Mol Gen Genet 221:210–218

Chatterjee G, Sankaranarayanan SR, Guin K, Thattikota Y, Padmanabhan S, Siddharthan R, Sanyal K (2016) Repeat-associated fission yeast-like regional centromeres in the ascomycetous budding yeast candida tropicalis. PLoS Genet 12:e1005839

Descorps-Declere S, Saguez C, Cournac A, Marbouty M, Rolland T, Ma L, Bouchier C, Moszer I, Dujon B, Koszul R, Richard GF (2015) Genome-wide replication landscape of Candida glabrata. BMC Biol 13:69

Ding R, McDonald KL, McIntosh JR (1993) Three-dimensional reconstruction and analysis of mitotic spindles from the yeast, Schizosaccharomyces pombe. J Cell Biol 120:141–151

Feeser EA, Wolberger C (2008) Structural and functional studies of the Rap1 C-terminus reveal novel separation-of-function mutants. J Mol Biol 380:520–531

Fitzpatrick DA, Logue ME, Stajich JE, Butler G (2006) A fungal phylogeny based on 42 complete genomes derived from supertree and combined gene analysis. BMC Evol Biol 6:1–15

Foureau E, Courdavault V, Navarro Gallon SM, Besseau S, Simkin AJ, Creche J, Atehortua L, Giglioli-Guivarc'h N, Clastre M, Papon N (2013) Characterization of an autonomously replicating sequence in candida guilliermondii. *microbiol res,* 168, 580–8

Freire-Beneitez V, Price RJ, Buscaino A (2016) The Chromatin of Candida albicans Pericentromeres Bears Features of Both Euchromatin and Heterochromatin. Front Microbiol 7:759

Gottschling DE, Aparicio OM, Billington BL, Zakian VA (1990) Position effect at S. cerevisiae telomeres: reversible repression of Pol II transcription. Cell 63:751–762

Haran J, Boyle H, Hokamp K, Yeomans T, Liu Z, Church M, Fleming AB, Anderson MZ, Berman J, Myers LC, Sullivan DJ, Moran GP (2014) Telomeric ORFs (TLOs) in Candida spp. Encode mediator subunits that regulate distinct virulence traits. PLoS Genet 10:e1004658

Hayashi T, Fujita Y, Iwasaki O, Adachi Y, Takahashi K, Yanagida M (2004) Mis16 and Mis18 are required for CENP—A loading and histone deacetylation at centromeres. Cell 118:715–729

Herreros E, Garcia-Saez MI, Nombela C, Sanchez M (1992) A reorganized Candida albicans DNA sequence promoting homologous non-integrative genetic transformation. Mol Microbiol 6:3567–3574

Hsu M, McEachern MJ, Dandjinou AT, Tzfati Y, Orr E, Blackburn EH, Lue NF (2007) Telomerase core components protect Candida telomeres from aberrant overhang accumulation. Proc Natl Acad Sci U S A 104:11682–11687

Iwakiri R, Eguchi S, Noda Y, Adachi H, Yoda K (2005) Isolation and structural analysis of efficient autonomously replicating sequences (ARSs) of the yeast Candida utilis. Yeast 22:1049–1060

Joglekar AP, Bouck D, Finley K, Liu X, Wan Y, Berman J, He X, Salmon ED, Bloom KS (2008) Molecular architecture of the kinetochore-microtubule attachment site is conserved between point and regional centromeres. J Cell Biol 181:587–594

Juarez-Reyes A, Ramirez-Zavaleta CY, Medina-Sanchez L, De Las Penas A, Castano I (2012) A protosilencer of subtelomeric gene expression in Candida glabrata with unique properties. *Genet,* 190, 101–111

Kapoor S, Zhu L, Froyd C, Liu T, Rusche LN (2015) Regional centromeres in the yeast Candida lusitaniae lack pericentromeric heterochromatin. Proc Natl Acad Sci U S A 112:12139–12144

Ketel C, Wang HS, McClellan M, Bouchonville K, Selmecki A, Lahav T, Gerami-Nejad M, Berman J (2009) Neocentromeres form efficiently at multiple possible loci in Candida albicans. PLoS Genet 5:e1000400

Kitada K, Yamaguchi E, Hamada K, Arisawa M (1997) Structural analysis of a Candida glabrata centromere and its functional homology to the Saccharomyces cerevisiae centromere. Curr Genet 31:122–127

Koren A, Tsai HJ, Tirosh I, Burrack LS, Barkai N, Berman J (2010) Epigenetically-inherited centromere and neocentromere DNA replicates earliest in S-phase. PLoS Genet 6:e1001068

Kurtz MB, Cortelyou MW, Miller SM, Lai M, Kirsch DR (1987) Development of autonomously replicating plasmids for Candida albicans. Mol Cell Biol 7:209–217

Leonard AC, Mechali M (2013) DNA replication origins. Cold Spring Harb Perspect Biol 5: a010116

Lue NF, Chan J (2013) Duplication and functional specialization of the telomere-capping protein Cdc13 in Candida species. J Biol Chem 288:29115–29123

Lue NF, Zhou R, Chico L, Mao N, Steinberg-Neifach O, Ha T (2013) The telomere capping complex CST has an unusual stoichiometry, makes multipartite interaction with G-Tails, and unfolds higher-order G-tail structures. PLoS Genet 9:e1003145

Lynch DB, Logue ME, Butler G, Wolfe KH (2010) Chromosomal G + C content evolution in yeasts: Systematic interspecies differences, and GC-poor troughs at centromeres. Genome Biol Evol 2:572–583

Manohar K, Acharya N (2015) Characterization of proliferating cell nuclear antigen (PCNA) from pathogenic yeast Candida albicans and its functional analyses in S. Cerevisiae. BMC Microbiol 15:257

Marshall OJ, Choo KH (2009) Neocentromeres come of age. PLoS Genet 5:e1000370

Masai H, Matsumoto S, You Z, Yoshizawa-Sugata N, Oda M (2010) Eukaryotic chromosome DNA replication: Where, when, and how? Annu Rev Biochem 79:89–130

McEachern MJ, Hicks JB (1993) Unusually large telomeric repeats in the yeast Candida albicans. Mol Cell Biol 13:551–560

Mishra PK, Baum M, Carbon J (2007) Centromere size and position in Candida albicans are evolutionarily conserved independent of DNA sequence heterogeneity. Mol Genet Genomics 278:455–465

Mitra S, Gomez-Raja J, Larriba G, Dubey DD, Sanyal K (2014) Rad51–Rad52 mediated maintenance of centromeric chromatin in Candida albicans. PLoS Genet 10:e1004344

O'Sullivan RJ, Karlseder J (2010) Telomeres: Protecting chromosomes against genome instability. Nat Rev Mol Cell Biol 11:171–181

Ofir A, Kornitzer D (2010) Candida albicans cyclin Clb4 carries S-phase cyclin activity. Eukaryot Cell 9:1311–1319

Ohkuma M, Kobayashi K, Kawai S, Hwang CW, Ohta A, Takagi M (1995) Identification of a centromeric activity in the autonomously replicating TRA region allows improvement of the host-vector system for Candida maltosa. Mol Gen Genet 249:447–455

Padmanabhan S, Thakur J, Siddharthan R, Sanyal K (2008) Rapid evolution of Cse4p-rich centromeric DNA sequences in closely related pathogenic yeasts, Candida albicans and Candida dubliniensis. Proc Natl Acad Sci U S A 105:19797–19802

Palmer DK, O'Day K, Wener MH, Andrews BS, Margolis RL (1987) A 17-kD centromere protein (CENP-A) copurifies with nucleosome core particles and with histones. J Cell Biol 104:805–815

Papon N, Savini V, Lanoue A, Simkin AJ, Creche J, Giglioli-Guivarc'h N, Clastre M, Courdavault V, Sibirny AA (2013) Candida guilliermondii: Biotechnological applications, perspectives for biological control, emerging clinical importance and recent advances in genetics. Curr Genet 59:73–90

Pearson CG, Yeh E, Gardner M, Odde D, Salmon ED, Bloom K (2004) Stable kinetochore-microtubule attachment constrains centromere positioning in metaphase. Curr Biol 14:1962–1967

Raser JM, O'Shea EK (2005) Noise in gene expression: Origins, consequences, and control. Science 309:2010–2013

Rodrigues CF, Silva S, Henriques M (2014) Candida glabrata: A review of its features and resistance. Eur J Clin Microbiol Infect Dis 33:673–688

Roy B, Burrack LS, Lone MA, Berman J, Sanyal K (2011) CaMtw1, a member of the evolutionarily conserved Mis12 kinetochore protein family, is required for efficient inner kinetochore assembly in the pathogenic yeast Candida albicans. Mol Microbiol 80:14–32

Roy B, Sanyal K (2011) Diversity in requirement of genetic and epigenetic factors for centromere function in fungi. Eukaryot Cell 10:1384–1395

Rui WJ (1999) Regulation of eukaryotic DNA replication and nuclear structure. Cell Res 9:163–170

Sanyal K (2012) How do microbial pathogens make CENs? PLoS Pathog 8:e1002463

Sanyal K, Baum M, Carbon J (2004) Centromeric DNA sequences in the pathogenic yeast Candida albicans are all different and unique. Proc Natl Acad Sci U S A 101:11374–11379

Sanyal K, Carbon J (2002) The CENP-A homolog CaCse4p in the pathogenic yeast Candida albicans is a centromere protein essential for chromosome transmission. Proc Natl Acad Sci U S A 99:12969–12974

Scott KC, Sullivan BA (2014) Neocentromeres: a place for everything and everything in its place. Trends Genet 30:66–74

Shivaraju M, Unruh JR, Slaughter BD, Mattingly M, Berman J, Gerton JL (2012) Cell-cycle-coupled structural oscillation of centromeric nucleosomes in yeast. Cell 150:304–316

Singh SM, Steinberg-Neifach O, Mian IS, Lue NF (2002) Analysis of telomerase in Candida albicans: Potential role in telomere end protection. Eukaryot Cell 1:967–977

Sullivan DJ, Berman J, Myers LC, Moran GP (2015) Telomeric ORFS in Candida albicans: does mediator tail wag the yeast? PLoS Pathog 11:e1004614

Sun J, Yu EY, Yang Y, Confer LA, Sun SH, Wan K, Lue NF, Lei M (2009) Stn1-Ten1 is an Rpa2–Rpa3-like complex at telomeres. Genes Dev 23:2900–2914

Takahashi K, Takayama Y, Masuda F, Kobayashi y, Saitoh S (2005) Two distinct pathways responsible for the loading of CENP-A to centromeres in the fission yeast cell cycle. *Philos Trans R Soc Lond B Biol Sci,* 360, 595–606; Discussion 606–607

Thakur J, Sanyal K (2011) The essentiality of the fungus-specific Dam1 complex is correlated with a one-kinetochore-one-microtubule interaction present throughout the cell cycle, independent of the nature of a centromere. Eukaryot Cell 10:1295–1305

Thakur J, Sanyal K (2012) A coordinated interdependent protein circuitry stabilizes the kinetochore ensemble to protect CENP-A in the human pathogenic yeast Candida albicans. PLoS Genet 8:e1002661

Thakur J, Sanyal K (2013) Efficient neocentromere formation is suppressed by gene conversion to maintain centromere function at native physical chromosomal loci in Candida albicans. Genome Res 23:638–652

Tomaska L, Nosek J, Kramara J, Griffith JD (2009) Telomeric circles: Universal players in telomere maintenance? Nat Struct Mol Biol 16:1010–1015

Tsai HJ, Baller JA, Liachko I, Koren A, Burrack LS, Hickman MA, Thevandavakkam MA, Rusche LN, Berman J (2014) Noise in gene expression: Origins, consequences, and control. MBio 5:e01703–e01714

Uemura H, Watanabe-Yoshida M, Ishii N, Shinzato T, Haw R, Aoki Y (2004) Isolation and characterization of Candida albicans homologue of RAP1, a repressor and activator protein gene in Saccharomyces cerevisiae. Yeast 21:1–10

Waldo JT, Greagor SA, Iqbal AJ, Gittens AS, Grant KK (2010) The Dad1 subunit of the yeast kinetochore Dam1 complex is an intrinsically disordered protein. Biochem Biophys Res Commun 400:313–317

Winey M, Mamay CL, O'toole ET, Mastronarde DN, Giddings TH, Jr Mcdonald KL, Mcintosh JR (1995) Three–dimensional ultrastructural analysis of the saccharomyces cerevisiae mitotic spindle. *J Cell Biol,* 129, 1601–1615

Yu EY, Yen WF, Steinberg-Neifach O, Lue NF (2010) Rap1 in Candida albicans: an unusual structural organization and a critical function in suppressing telomere recombination. Mol Cell Biol 30:1254–1268

Chapter 13
Infection-Associated Gene Expression—The Pathogen Perspective

Saranna Fanning and Aaron P. Mitchell

Abstract It is exciting to think that we now have data that link *C. albicans* gene expression responses to diverse environmental and genetic regulatory signals. We also have several technologies that allow precise measurement of RNA levels during infection. Together these capabilities allow us to infer many major signals that drive pathogen gene expression during infection. These signals and responses in turn yield insight into therapeutic targets for which inhibition may block infection. However, our deepening understanding of infection biology also reveals gaps in our understanding of how transcriptional networks may be revised during infection. That issue emphasizes the urgency of defining regulatory relationships in the infection environment.

13.1 Introduction

Candida species are major fungal pathogens (Brown et al. 2012). They cause bloodstream and deep tissue infections in patients who are immunocompromised, have had invasive clinical procedures, or have an implanted medical device. There are an estimated 400,000 life-threatening *Candida* infections per year worldwide, with mortality rates that exceed 40% (Brown et al. 2012). How does *Candida* achieve this level of notoriety? Most pathogenic *Candida* species are commensal inhabitants of the gastrointestinal and genitourinary tracts, and thus are poised to cause opportunistic infection. Yet among the many fungi associated with humans—the components of the mycobiome—a handful of *Candida* species are most adept at

S. Fanning (✉)
Whitehead Institute for Biomedical Research, 9 Cambridge Center,
Cambridge, MA 02142, USA
e-mail: sfanning@wi.mit.edu

A.P. Mitchell
Department of Biological Sciences, Carnegie Mellon University,
Pittsburgh, PA 15213, USA
e-mail: apm1@cmu.edu

R. Prasad (ed.), *Candida albicans: Cellular and Molecular Biology*,
DOI 10.1007/978-3-319-50409-4_13

causing infection (Huffnagle and Noverr 2013). It has long been thought that the ability to adapt to diverse environments is the key pathogenicity trait that supports *Candida* species' remarkable infection capability. In this chapter, we explore the adaptation ability of the most frequently isolated *Candida* pathogen, *Candida albicans* (Pfaller and Diekema 2007), as revealed through assays of gene expression during infection.

To date, most *C. albicans* gene expression studies have focused on effects of genotype or environment during growth in vitro. In fact, roughly 200 such datasets are publicly available (Xu et al. 2015). These datasets comprise a valuable resource that can guide inferences from gene expression data derived from infection environments. The key virtue of gene expression data from in vitro-grown cells is the simplicity of interpretation: typically, samples that differ by a single variable are compared, and the impact of that specific variable can be tested and validated through precise controls. Gene expression data from infected tissue samples might be considered a mess by comparison, because infection samples are heterogeneous, the course of infection is dynamic, and the infection environment presents a convergence of multiple stresses (Brown et al. 2014). One might despair at the prospect of learning even one useful fact from infection-associated gene expression. However, *C. albicans* is not a model organism; it is generally not studied with the goal of learning fundamental biological principles. We believe that the major goal of studying *C. albicans*, or any pathogen for that matter, is to understand mechanisms of pathogenesis and drug resistance. Those processes occur in the infection environment. Therefore, gene expression and its regulation in the infection environment provide the ultimate validation of the precise analyses that can be carried out during growth in vitro.

What do gene expression and its regulation tells us? One of the articles of faith that drives the study of gene expression in any organism is that gene expression predicts gene function: genes that are up-regulated in a specific situation will be required in that situation. This idea was called into question by the elegant "functional profiling" studies of the model yeast *Saccharomyces cerevisiae*, in which a comprehensive set of gene deletion mutants was used in competitive growth assays to identify genes that functioned in specific environments (Giaever et al. 2002). The authors reported the unexpected finding that only a small fraction of genes that are up-regulated under a specific growth condition make a measurable contribution to growth in that condition. Nonetheless, gene expression data can be exceptionally useful in several ways. First, regardless of the functions of specific genes, the similarities in gene expression changes under two conditions can reveal common response pathways that are operative under both conditions. One of the most elegant demonstrations of this principle also came from early genome-wide studies of *S. cerevisiae*, in which a compendium of microarray datasets was used to identify genes that function together in regulatory pathways (Hughes et al. 2000). This rationale has proven successful time and again (for example, see Singh et al. 2010; Gower et al. 2011; Reimand et al. 2010; Finkel et al. 2012). Second, while the up- or down-regulation of specific genes may not yield precise functional information, the overall expression level can do so (Xu et al. 2015). Third, in vivo expression data allow a test of whether the detailed regulatory relationships deduced from in vitro experiments provide valid descriptions of regulatory relationships in the complex infection environment. As we will see, regulatory relationships are often distinct in vivo, and lead to an understanding of novel functional relationships that are not apparent during growth in vitro (Cheng et al. 2013; Fanning et al. 2012; Xu et al. 2015).

13.2 Technologies That Enable In Vivo Gene Expression Analysis

Several kinds of gene expression assays allow measurements of small numbers of genes in infecting or colonizing *C. albicans* cells. These approaches include QRTPCR (Fairman et al. 1999; Nailis et al. 2010), elegant recombinase-based IVET assays (Staib et al. 1999; Kretschmar et al. 2002; Pande et al. 2013), fluorescent reporter gene fusion assays (Barelle et al. 2006), and bioluminescent luciferase imaging in whole animals (Mosci et al. 2013; Brock 2012). These approaches have shed light on the regulation of specific genes: for example, the work of Staib et al. provided the first example of a niche-specific gene, *SAP2* (Staib et al. 1999). However, these tools are not sufficiently scalable to enable examination of large numbers of genes.

Microarrays have been used to attempt genome-wide analysis of colonizing and infecting organisms (Walker et al. 2009; Rosenbach et al. 2010). This tool is most reliable for detection of highly expressed genes with large expression changes. The challenge of microarray insensitivity is magnified in infection samples by contaminating host RNA, which can be more than 99% of the total infected tissue RNA. RNA Seq has greater sensitivity and dynamic range than microarray analysis and enables sequencing of multiple species concurrently without experimenter bias. However, amplification or enrichment is often associated with this method, which can affect sampling (Westermann et al. 2012). Further development of direct RNASeq and the more targeted approach Capture RNASeq (Mercer et al. 2014; Ozsolak et al. 2009) will no doubt overcome these drawbacks.

The nanoString RNA-based approach commands exquisite sensitivity to analyze host and pathogen gene expression at the same time, even when pathogen RNA is <1% of total RNA. Moreover, to this date, nanoString profiling is the only method that has been reported in publications for profiling of mutants with extremely low colonization levels due to attenuated virulence. No amplification or enrichment is required for this rapid method. However, probe selection can be limiting (Geiss et al. 2008; Malkov et al. 2009).

Ultimately, there is much room for improvement! For example, RNA levels do not indicate which RNAs are translated, and whether proteins are post-translationally modified. Proteomic analysis (Nett et al. 2015) and, potentially, ribosome profiling (Ingolia et al. 2012) will augment the depth of understanding from transcriptomic data. Although transcriptomic data is subject to criticism from both conceptual and technical standpoints, our goal in this chapter is to capture and summarize the insights that it has provided into *C. albicans* gene expression responses during infection.

13.3 Pathogen RNA Levels as a Reflection of the Infection Environment

Transcriptional profiling gives an indication of the environment that *C. albicans* senses. It can help define the cues that trigger infection establishment through impact on adherence and morphogenesis, and its responses to host defense factors that help maintain and enhance its status in the host. Profiles give insight into common hallmarks of infection (e.g., hyphal and adhesin gene expression) as well as niche-specific environmental responses such as general amino acid control. Profiling of the mouse kidney at early time points (Xu et al. 2015; Amorim–Vaz et al. 2015), mouse liver at early time points (Thewes et al. 2007), as well as other infection types (Fanning et al. 2012; Cheng et al. 2013; Thewes et al. 2007), indicate that environmental sensing drives a major portion of pathogen gene expression. We focus below primarily on the features of *C. albicans* gene expression, primarily as seen during invasive infection in mouse infection models.

13.3.1 Iron-Limitation Response

Free iron is scarce in the kidney (Potrykus et al. 2013) and other tissues (Hood and Skaar 2012), and this situation is a driver of pathogen gene expression. There is a vast regulation of iron-related genes at early time points in the kidney (Xu et al. 2015; Amorim-Vaz et al. 2015; Walker et al. 2009). *IRO1*, an ortholog of *S. cerevisiae* iron regulatory gene *AFT1*, is up-regulated in the kidney early in infection. *BUD2*, known to be induced during low-iron conditions and *CCP1*, known to be up-regulated in response to iron starvation, as well as iron sensors and responders *CFL1*, *CFL2*, *CFL5* (Amorim-Vaz et al. 2015) are activated. Aconitase-related genes induced in high-iron environments are down-regulated at early time points. Many ferric reductase and iron permease genes as well as the well-defined indicator of low-iron conditions, *HAP43*, are up during the early kidney time points (Xu et al. 2015). Iron regulation genes *HMX1* and *MRS4*, important for iron homeostasis, are up-regulated in the liver also (Thewes et al. 2007). Finally, *SIT1*, a ferrichrome siderophore transporter observed to be regulated in the kidney throughout infection, is one of the most highly expressed genes in the intra-abdominal candidiasis model (Cheng et al. 2013). Iron limitation is not evident in the gastrointestinal environment, as up-regulation of iron acquisition genes is not as prominent a feature of the cecum and ileum profiles (Rosenbach et al. 2010; Pierce et al. 2013). Suzanne Noble has argued convincingly that adaptation from a high-iron to low-iron environment is pivotal for dissemination from the gastrointestinal tract to the bloodstream and organs (Noble 2013), and mutant analysis fits well with gene expression data.

13.3.2 Zinc-Limitation Response

Several observations indicate that zinc is limiting for *C. albicans* during infection, as is the case for many bacterial pathogens (Hood and Skaar 2012). Up-regulation of the zinc transporter *ZRT1* suggests that zinc may be limiting for *C. albicans* in the liver (Thewes et al. 2007) and kidney (Xu et al. 2015; Amorim-Vaz et al. 2015). This inference is strengthened for the kidney by the accompanying up-regulation of a broader set of zinc-limitation response genes, including of *ZAP1/CSR1, SUT1, PRA1*, orf19.1534, and *ZRT2* (Xu et al. 2015; Amorim-Vaz et al. 2015). In fact, two positive transcriptional regulators of zinc-limitation response genes, *SUT1* and *ZAP1*, are required for proliferation in the kidney (Xu et al. 2015). A response to zinc limitation is also apparent in intra-abdominal candidiasis, as reflected by high-level expression of *SIT1, ZRT1, ZRT2*, and *PRA1* (Cheng et al. 2013). Zinc limitation is thus a feature of many infection sites.

13.3.3 Copper and Manganese Limitation Responses

Metal ions other than iron and zinc are known to impact pathogenesis significantly (Diaz-Ochoa et al. 2014; Hood and Skaar 2012). For *C. albicans*, down-regulation in the liver of manganese transporter gene *CCC1* (Thewes et al. 2007) suggests that manganese levels are not limiting for growth in that tissue. On the other hand, up-regulation of copper acquisition genes *CUP1* and *CTR1* (Thewes et al. 2007) may represent a copper-limitation response or part of the related iron-limitation response. The transporter gene *CTR1* is required for efficient iron acquisition, as is its transcriptional activator specified by *MAC1* (Marvin et al. 2004). In fact, *MAC1* expression is also up-regulated in the liver (Thewes et al. 2007). Kidney infection is also accompanied by a copper-limitation response, as reflected in up-regulation of *MAC1* and the copper transporter gene *CRP1* (Amorim-Vaz et al. 2015; Xu et al. 2015). Because metal ions can have complex concentration-dependent effects—limiting proliferation, competing with one another, and even toxicity at high concentrations (Diaz-Ochoa et al. 2014; Hood and Skaar 2012)—the pathogen's "perception" of concentration is not always predictable from chemical concentration data alone.

13.3.4 Neutral/Alkaline pH Adaptation Response

The neutral/alkaline environment in the kidney is reflected in down-regulation of alkaline repressed genes *CCC1, CCP1, CYC1, CYT1,* and *PHR2*, and by up-regulation of orf19.7539.1, *CSA1*, *IHD1*, *RBT1*, *RBT5*, and *PHR1* (Amorim-Vaz et al. 2015; Xu et al. 2015). The pH response regulator *RIM101* and alkaline pH induced gene *PHR1* are among the most highly expressed genes in the intra-abdominal candidiasis model, and a large number of neutral/alkaline pH induced genes are highly up-regulated during intra-abdominal candidiasis (Cheng et al. 2013). A neutral/alkaline pH response extends to the liver, as indicated by up-regulation of *PTP3* and *IHD1* (Thewes et al. 2007). A neutral/alkaline pH environment can induce hyphal morphogenesis, and the high-level expression of hypha-associated genes at many infection sites may be in part a reflection of this response. In that context, intra-abdominal candidiasis is noteworthy in that expression of many neutral/alkaline pH response genes is uncoupled from hyphal morphogenesis: the infecting population comprises predominantly yeast cells (Cheng et al. 2013).

For many fungi, one of the major regulators of pH-dependent responses is the Rim101/PacC pathway (Cornet and Gaillardin 2014). Indeed, it is critical for virulence of numerous pathogens in diverse infection models (Kronstad et al. 2013; Davis 2009; Cornet and Gaillardin 2014). For *C. albicans*, Rim101 is required for disseminated infection, intra-abdominal candidiasis, and oropharyngeal candidiasis (Cheng et al. 2013; Nobile et al. 2008; Xu et al. 2015). Surprisingly, though, the specific set of genes that depend upon Rim101 for expression is quite different in the two niches examined, kidney (representing disseminated infection) and peritoneal fluid (representing intra-abdominal candidiasis) (Cheng et al. 2013; Xu et al. 2015). Moreover, the Rim101 network in both infection niches is different from that defined under in vitro growth conditions (Cheng et al. 2013; Xu et al. 2015)! We return to the environmental contingency of regulatory networks later in the chapter.

13.3.5 General Amino Acid Control

The transcriptional activator gene *GCN4,* which governs general amino acid control response, is up-regulated during early infection in the kidney (Xu et al. 2015) and during oropharyngeal candidiasis (Fanning et al. 2012). By contrast *GCN4* is down-regulated early in the liver (Thewes et al. 2007). Inferences about Gcn4 function are generally not based on RNA levels alone, because the major mechanism that regulates Gcn4 protein levels acts at the translational level (Hinnebusch 2005). In *C. albicans*, though, the predominant level of Gcn4 regulation appears to be transcriptional (Tournu et al. 2005). A recent study has demonstrated that *C. albicans* lysine biosynthetic genes are direct Gcn4 targets (Priyadarshini and Natarajan 2016), and several of these genes are up-regulated during kidney infection (Amorim-Vaz et al. 2015). Among other amino acid biosynthetic genes, several arginine and ornithine biosynthesis and transporter genes have increased RNA levels in the kidney early in infection (Xu et al. 2015). *ARG1* is regulated in the mouse cecum but direction of regulation differs according to two studies (Rosenbach et al. 2010; Pierce et al. 2013). These differences may stem from differing in vitro comparison datasets or because of expression level and amplification for microarray. Arginine biosynthesis genes are not substantially regulated during liver infection, though leucine, methionine, and lysine genes are up-regulated. Methionine biosynthesis genes are generally down-regulated in the kidney (Amorim-Vaz et al. 2015; Xu et al. 2015). While most genes in the aspartate pathway are not regulated during kidney infection, there is some regulation of aspartate-related genes such as up-regulation of *APR1* and *AAT1* at later time points (Amorim-Vaz et al. 2015). This may indicate a stage at which the pathogen is capable of generating precursors for other amino acids that would be advantageous to the pathogen to maintain its position in the host. Interestingly, *AAT1* is also up-regulated in the cecum (Rosenbach et al. 2010). The many examples of amino acid biosynthesis up-regulation during infection illustrate that the protein-rich environment in host tissue does not make excess amino acids available.

13.3.6 Carbon Utilization Control

The ability to utilize a range of carbon sources is central to metabolic adaptation and, perhaps surprisingly, impacts stress adaptation as well (Brown et al. 2014). Carbon source utilization, based on gene expression, seems to be dynamic during kidney invasion, as indicated by up-regulation of several genes that respond to acetate as a carbon source (Xu et al. 2015). In addition, the carbon utilization regulatory gene *TYE7* (Askew et al. 2009) is one of the most highly expressed transcription factor genes (Xu et al. 2015). There is up-regulation of galactose genes early during kidney invasion (Amorim-Vaz et al. 2015) and the glucose transcriptional regulator (*RGT1*) and glucose transporters are also highly regulated in both directions in the kidney (Amorim-Vaz et al. 2015; Xu et al. 2015). Galactose genes are also up-regulated in mouse cecum (Pierce et al. 2013; Rosenbach et al. 2010) and in the ileum (Pierce et al. 2013) and to a lesser degree in the liver (Thewes et al. 2007). Maltose and mannose sugar utilization genes are highly regulated during kidney infection (Amorim-Vaz et al. 2015; Xu et al. 2015). *C. albicans* may sample many carbon sources during infection, perhaps reflecting the dynamic environment it encounters.

13.3.7 Oxidative Stress Response

Oxidative stress is an infection hurdle for a pathogen (Briones-Martin-Del-Campo et al. 2014; Dantas Ada et al. 2015; Cuellar-Cruz et al. 2014; de Dios et al. 2010). The oxidative stress-response genes *TRX1* and *TRR1* are highly expressed in the intra-abdominal candidiasis model (Cheng et al. 2013) and are up-regulated early during kidney infection (Xu et al. 2015). *YFH1,* which promotes oxidative stress resistance, is up-regulated late in kidney infection (Xu et al. 2015). Up-regulation of *PBS2* early during liver infection may be an indication of osmotic or oxidative stress (Thewes et al. 2007). Significant regulation of superoxide dismutase (*SOD*) genes may be indicative of a response to stress, with *SOD2* and *SOD3* down-regulated, and *SOD4* and *SOD5* up-regulated in the mouse kidney and liver (Thewes et al. 2007; Amorim-Vaz et al. 2015; Xu et al. 2015), during oropharyngeal candidiasis (Fanning et al. 2012), and in the cecum and ileum (Pierce et al. 2013). *SOD* genes are differentially regulated at later times during kidney invasion, down-regulation, or no regulation of many genes and up-regulation of *SOD6* (Amorim-Vaz et al. 2015; Xu et al. 2015). *CAT1*, another indicator of oxidative stress, is down-regulated in early and late kidney infection (Xu et al. 2015) but is up-regulated in mouse cecum (Pierce et al. 2013) and during oropharyngeal candidiasis. *CAT1* up-regulation is accompanied by high expression levels of oxidative stress regulatory gene *CAP1* in intra-abdominal candidiasis (Cheng et al. 2013). *CAT1* and the *SOD* genes are governed by multiple regulatory pathways, so it can be rationalized that they are not co-regulated during infection. It is not simple to deduce which of the many possible inducing or repressing signals may have overriding influence on their expression during infection.

13.4 Pathogen RNA Levels as a Reflection of Biological Processes

13.4.1 Adherence

Simultaneously, with sensing and responding to the niche environment the pathogen must adhere to cells and surfaces. In keeping with this need, many cell surface genes are up-regulated during infection. These genes specify GPI-anchored proteins such as adhesins and their relatives, including *ALS1*, *ALS3*, *MSB2*, *FAV1,* and *FAV2*, all of which are up-regulated at early time points in the kidney (Amorim-Vaz et al. 2015; Walker et al. 2009; Xu et al. 2015). *BCR1*, which specifies a transcription factor known to regulate *ALS1* and *ALS3* (Nobile and Mitchell 2005), is up-regulated at this time as well (Amorim-Vaz et al. 2015; Xu et al. 2015). The adhesin regulatory gene *AHR1* is also up-regulated in the kidney both at early and late stages (Amorim-Vaz et al. 2015; Xu et al. 2015). Interestingly, nanoString analysis of early time points during kidney infection detected an increase in *ALS3* with a concurrent decrease in *ALS2* and *ALS4* RNA levels (Xu et al. 2015). The changes in *ALS2* may not have been identified with other methods due to its low level expression, particularly early in infection. Interestingly, *ALS2* appears to be up-regulated later in infection (Xu et al. 2015) and is up-regulated during oropharyngeal candidiasis (Fanning et al. 2012). *ALS1* is one of the most highly expressed genes in the intra-abdominal candidiasis model (Cheng et al. 2013) and is up-regulated in the cecum (Rosenbach et al. 2010), while *ALS3* is significantly up-regulated in the ileum, in the cecum (Rosenbach et al. 2010; Pierce et al. 2013), during oropharyngeal candidiasis infection (Fanning et al. 2012) and in the *Galleria* model (Amorim-Vaz et al. 2015). We infer that Als3 and the closely related Als1 may be the major adhesins operative during infection, in keeping with the broad success of an anti-Als1/3 vaccine (Spellberg et al. 2006). Other Als gene expression changes are more variable with *ALS5* being down-regulated during oropharyngeal candidiasis (Fanning et al. 2012) and down-regulated along with *ALS6* in late stages of kidney infection (Amorim-Vaz et al. 2015). The fact that *ALS3* defects have modest or variable impact on pathogenicity, depending on the infection model, may reflect compensatory activity of Als1 and other adhesins (Liu and Filler 2011).

13.4.2 Hyphal Morphogenesis

Diverse environmental cues induce *C. albicans* to undergo morphogenesis to form hyphae. Hypha-associated genes that are up-regulated early during liver infection include *ECE1, FGR44, ZCF3, DEF1, IHD1*, and *MUC1* (Thewes et al. 2007). Up-regulation of hypha-associated genes *TEC1, BRG1, CEK1, CEK2, DEF1, ECE1, HYR1, HWP1, IHD1, IHD2,* and *UME6* as well as a contributor to invasion, *LMO1,* occur in the kidney early in infection (Amorim-Vaz et al. 2015). NanoString analysis reveals, in addition, up-regulation of *HYR1* and *RBT1*. Hypha-associated genes are also significantly up-regulated in the cecum (Rosenbach et al. 2010) and in oropharyngeal candidiasis (Fanning et al. 2012) but to a much lesser extent in the ileum (Pierce et al. 2013).

13.4.3 Host Tissue/Cell Invasion

The up-regulation of several secreted aspartyl protease (*SAP*) genes is observed in many infection models, revealing a potential for host tissue destruction and invasion. *SAP5* appears to be the most notable of the *SAP* genes in this regard; it is up-regulated in mouse kidney and during *Galleria* infection (Amorim-Vaz et al. 2015) and has been shown to be up-regulated along with *SAP6* in the kidney by nanoString (Xu et al. 2015), in the liver at early time points (Thewes et al. 2007), during oropharyngeal candidiasis infection (Fanning et al. 2012), in oral pseudomembranous *Candida* infection in HIV+ patients (Zakikhany et al. 2007), in the cecum (accompanied by up-regulation of *SAP4*) (Pierce et al. 2013) and in the ileum (accompanied by up-regulation of *SAP4* and *SAP6*) (Pierce et al. 2013). *SAP4* is up-regulated in the ileum and cecum and down-regulated in the early infection stages in the kidney (Xu et al. 2015). *CCD11* (Amorim-Vaz et al. 2015) and *CDC10* (Xu et al. 2015) are also up-regulated during early kidney infection and have been associated with hyphae and kidney invasion.

13.4.4 Cell Wall and Secretion

The elegant experiments of Wheeler and Fink showed that the *C. albicans* cell wall is remodeled during invasive infection of the kidney (Wheeler et al. 2008), and cell wall-related genes are expressed prominently during infection. The transcriptional profile shows up-regulation of the cell wall regeneration gene *EXG2* and cell wall assembly regulator *SMI1B*, a cell wall assembly regulator (Amorim-Vaz et al. 2015). Other cell wall-related genes such as *ECM15*, *CSP1*, and the hyphal cell wall genes *CSP37* and *CSA1* have altered expression (Amorim-Vaz et al. 2015; Xu et al. 2015). Regulators of cell wall integrity *BCK1*, *IRS4*, and *IRE1* are up-regulated as well. There is significant regulation of the cell wall chitin distributor *BNI4* as well as other chitin-related proteins, *CHS2, CHS3*, and *CHT1* (Amorim-Vaz et al. 2015). There are conflicting data regarding *RLM1,* a transcriptional regulator of cell wall integrity: it was reported to be up-regulated based on Capture RNASeq (Amorim-Vaz et al. 2015) but not based on nanoString analysis (Xu et al. 2015). Overall, the data point to numerous candidate genes that may participate in cell wall remodeling and, ultimately, immune evasion (Wheeler et al. 2008; Wheeler and Fink 2006).

Secreted proteins, like the *SAP* gene products, and cell wall functions like *ECM15* require suitable cellular machinery for their biogenesis and function. Many genes that specify secretory pathway components are up-regulated during infection. They include, during liver infection, *KAR2* and the secretory pathway genes *SEC10, SEC24, SEC26, SEC31* (Thewes et al. 2007). Many trafficking genes are highly expressed in the kidney environment throughout infection including many *VPS* genes (Amorim-Vaz et al. 2015). Another indicator of transcriptional adaptations to accommodate augmented protein secretion is increased expression of *HAC1* in the kidney (Amorim-Vaz et al. 2015) and accompanying increase in *IRE1* indicate up-regulation of the unfolded protein response continuing throughout early and late infection (Amorim-Vaz et al. 2015). In contrast, *HAC1* is down-regulated in the cecum (Pierce et al. 2013). This distinction raises the interesting possibility that hyphal morphogenesis, which is accompanied by high-level expression of numerous cell wall and secreted protein genes, and is prominent in infected tissue but not in the cecum, may be a driver of induction of the unfolded protein response.

13.5 Regulatory Relationships Defined In Vitro and Observed In Vivo

Much time and effort have been devoted to defining the relationships between transcription factors and their target genes—the genes that respond to alterations of the respective transcription factor. Our current understanding of these relationships has been summarized and elegantly applied recently by Nantel and colleagues (Sellam et al. 2014), and an extensive gene expression dataset collection is available (Xu et al. 2015). Experiments to define these relationships as gene regulatory networks have of necessity been carried out under in vitro culture conditions. However, the studies detailed above suggest that the infection environment may be distinct at various sites, and in addition that it is a unique intersection of multiple stresses and signals that impact gene expression. Is there a single "gene regulatory network" that can predict *C. albicans* gene expression responses under many different growth conditions? There are numerous examples in which predictions that derived from in vitro experiments are validated by animal infection assay outcomes. However, our thinking on this subject was affected by a case in which we apparently got lucky.

The *C. albicans* transcription factor Bcr1 is required in vitro for expression of the cell surface protein gene *HWP1.* This relationship has been validated functionally in a murine oropharyngeal candidiasis model: a *bcr1Δ/Δ* mutant is attenuated in this model, and increased *HWP1* expression in the mutant (from a promoter replacement construct) restores virulence (Dwivedi et al. 2011; Fanning et al. 2012). This result supports the idea that the *bcr1Δ/Δ* mutant strain is attenuated because it fails to express *HWP1* during oropharyngeal candidiasis. Surprisingly, though, direct RNA measurements indicate that the *bcr1Δ/Δ* mutant is not defective in *HWP1* expression in vivo in the oropharyngeal candidiasis model. Therefore, overexpression of *HWP1* seems to restore *bcr1Δ/Δ* mutant virulence through a bypass mechanism. However, without measurement of RNA levels, one would reasonably conclude that *HWP1* is a critical Bcr1-dependent gene in vivo (Fanning et al. 2012). Our overall finding was that Bcr1 targets are affected by the *bcr1Δ/Δ* mutation both in vitro and in vivo (in the oropharyngeal candidiasis model), such as *ALS1* and *ALS3.* However, many other targets that are affected by the *bcr1Δ/Δ* mutation in vitro, such as *HWP1* and *ECE1,* are expressed independently of the *bcr1Δ/Δ* mutation in vivo in the oropharyngeal candidiasis model.

More recent studies support the idea that many regulatory relationships are niche specific. There are many genes that are only regulated by Rim101 in vivo in the intra-abdominal candidiasis model and not during kidney infection, including those that specify adherence functions (*ALS1, ALS3, CSH1*), biofilm regulators (*TYE7, ACE2, NRG1*), chaperones (*HSP104, HSP70*), metabolic functions (*PFK2, HGT7, MAE1, SGA1*), and many others. Genes that are regulated by Rim101 only in the kidney and not in intra-abdominal candidiasis include pH responsive genes *PHR2* and *SCW4*, and others such as *GAP2* and orf19.1862. In fact, many genes are oppositely regulated by Rim101 in the kidney and intra-abdominal candidiasis models, including orf19.1691, *RTA3*, and *DDR48*. Interestingly, several genes depend upon Rim101 in vivo in both the kidney and intra-abdominal candidiasis locations, but have not shown Rim101 dependence in vitro, including zinc responsive genes (*ZRT1, PRA1*) as well as *CSA1, FDH1, PIR1,* and *RBT1*. Our initial bias was that only more indirect target genes—those whose promoters are not bound by Rim101—would manifest niche-specific dependence, our analysis below suggests that the situation is more complex.

The transcription factor gene *EFG1* is perhaps the most extensively studied gene in the *C. albicans* genome. It is a positive regulator of hyphal morphogenesis (Lo et al. 1997) and many other *C. albicans* biological processes (see Huang (2012) for a recent review). Its target genes are substantially different in vivo, during kidney invasion (Xu et al. 2015), compared to those identified through in vitro studies (Nobile et al. 2012). A significant number of genes are regulated by *EFG1* in vivo but not in vitro (59/63 genes in these datasets). In addition, many genes that are direct Efg1 targets in vitro—their promoters are bound by Efg1 in chromatin immunoprecipitation experiments, and their RNA levels are affected by an *efg1Δ/Δ* mutation (Nobile et al. 2012)—do not respond to an *efg1Δ/Δ* mutation, or respond differently, during kidney invasion (Xu et al. 2015). The net result is that it is difficult to predict the in vivo expression properties of both direct and indirect Efg1 targets based on their properties defined in vitro.

We can make a broader appraisal of the similarity between in vitro and in vivo gene expression alterations that arise from defects in individual transcription factors. In Table 13.1 we have compared gene expression changes associated with three *C. albicans* transcription factor mutants during invasive kidney infection (Xu et al. 2015) and during in vitro culture (Nobile et al. 2009, 2012). All three transcription factors have been analyzed by high-quality chromatin immunoprecipitation analysis (Nobile et al. 2009, 2012), and these data allow us to distinguish behavior of direct and indirect target genes. For the zinc-acquisition regulator Zap1, the in vitro data predict in vivo responses well for direct target genes, but not for

Table 13.1 Expression responses among transcription factor target genes assayed under in vitro growth conditions or during invasive kidney infection

	Transcription factor		
	Zap1	Rob1	Efg1
Correlation coefficient (R) for all genes[a]	0.256	0.153	0.003
Correlation coefficient (R) for direct target genes[b]	0.536	0.269	0.097

[a]Expression ratios (mutant vs. wild type) were compared for datasets generated under in vitro growth conditions (Nobile et al. 2009, 2012) and from invasion of the kidney (Xu et al. 2015) were compared. The comparison was limited to the 148 RNAs measured by nanoString in vivo (Xu et al. 2015). Correlation coefficients are listed

[b]Expression ratios (mutant vs. wild type) were compared for the datasets listed above, but restricted to direct target genes as defined by RNA levels and chromatin immunoprecipitation experiments (Nobile et al. 2009, 2012), and limited to the 148 RNAs measured by nanoString in vivo (Xu et al. 2015). Correlation coefficients are listed

indirect target genes. This outcome can be rationalized with the assumption that zinc limitation provokes different stress-response pathways in vitro and in vivo. Two regulators of hyphal morphogenesis, Rob1 and Efg1, provide sobering counterexamples. For these two transcription factor mutants, the in vitro data correlate very poorly with in vivo data. The correlation is weak for both indirect and direct targets. These two transcription factors function in an interconnected network that governs formation of hyphae and biofilm (Nobile et al. 2012). Many of the target genes are shared among transcription factors in this network, so the poor in vitro-in vivo agreement may represent compensatory interactions.

Is it surprising that *C. albicans* regulatory relationships are environmentally contingent? Extensively studied regulatory relationships have been defined for the model yeast *S. cerevisiae*, reflecting the culmination of decades of analysis and diverse experimental approaches. In a masterful review of this topic, Hughes and de Boer note that there is no single gene regulatory network for *S. cerevisiae*; rather, transcription factor activity can vary widely in response to environmental signals (Hughes and de Boer 2013). We believe that the in vitro-in vivo disparity that we have presented for *C. albicans* is a manifestation of this same biological principle. We are fortunate to now have the technology to define and dissect this significant phenomenon.

References

Amorim-Vaz S, Tran Vdu T, Pradervand S, Pagni M, Coste AT, Sanglard D (2015) RNA enrichment method for quantitative transcriptional analysis of pathogens in vivo applied to the fungus Candida albicans. MBio 6(5):e00942–00915. doi:10.1128/mBio.00942-15

Askew C, Sellam A, Epp E, Hogues H, Mullick A, Nantel A, Whiteway M (2009) Transcriptional regulation of carbohydrate metabolism in the human pathogen Candida albicans. PLoS Pathog 5(10):e1000612. doi:10.1371/journal.ppat.1000612

Barelle CJ, Priest CL, Maccallum DM, Gow NA, Odds FC, Brown AJ (2006) Niche-specific regulation of central metabolic pathways in a fungal pathogen. Cell Microbiol 8(6):961–971. doi:CMI676 [pii]10.1111/j.1462-5822.2005.00676.x

Briones-Martin-Del-Campo M, Orta-Zavalza E, Juarez-Cepeda J, Gutierrez-Escobedo G, Canas-Villamar I, Castano I, De Las Penas A (2014) The oxidative stress response of the opportunistic fungal pathogen Candida glabrata. Rev Iberoam Micol 31(1):67–71. doi:10.1016/j.riam.2013.09.012

Brock M (2012) Application of bioluminescence imaging for in vivo monitoring of fungal infections. Int J Microbiol 2012:956794. doi:10.1155/2012/956794

Brown AJ, Budge S, Kaloriti D, Tillmann A, Jacobsen MD, Yin Z, Ene IV, Bohovych I, Sandai D, Kastora S, Potrykus J, Ballou ER, Childers DS, Shahana S, Leach MD (2014) Stress adaptation in a pathogenic fungus. J Exp Biol 217(Pt 1):144–155. doi:10.1242/jeb.088930

Brown GD, Denning DW, Gow NA, Levitz SM, Netea MG, White TC (2012) Hidden killers: human fungal infections. Sci Transl Med 4(165):165rv113. doi:10.1126/scitranslmed.3004404

Cheng S, Clancy CJ, Xu W, Schneider F, Hao B, Mitchell AP, Nguyen MH (2013) Profiling of Candida albicans gene expression during intra-abdominal candidiasis identifies biologic processes involved in pathogenesis. J Infect Dis 208(9):1529–1537. doi:10.1093/infdis/jit335

Cornet M, Gaillardin C (2014) pH signaling in human fungal pathogens: a new target for antifungal strategies. Eukaryot Cell 13(3):342–352. doi:10.1128/EC.00313-13

Cuellar-Cruz M, Lopez-Romero E, Ruiz-Baca E, Zazueta-Sandoval R (2014) Differential response of Candida albicans and Candida glabrata to oxidative and nitrosative stresses. Curr Microbiol 69(5):733–739. doi:10.1007/s00284-014-0651-3

Dantas Ada S, Day A, Ikeh M, Kos I, Achan B, Quinn J (2015) Oxidative stress responses in the human fungal pathogen. Candida albicans Biomolecules 5(1):142–165. doi:10.3390/biom5010142

Davis DA (2009) How human pathogenic fungi sense and adapt to pH: the link to virulence. Curr Opin Microbiol 12(4):365–370. doi:S1369-5274(09)00055-1 [pii]10.1016/j.mib.2009.05.006

de Dios CH, Roman E, Monge RA, Pla J (2010) The role of MAPK signal transduction pathways in the response to oxidative stress in the fungal pathogen Candida albicans: implications in virulence. Curr Protein Pept Sci 11(8):693–703

Diaz-Ochoa VE, Jellbauer S, Klaus S, Raffatellu M (2014) Transition metal ions at the crossroads of mucosal immunity and microbial pathogenesis. Front Cell Infect Microbiol 4:2. doi:10.3389/fcimb.2014.00002

Dwivedi P, Thompson A, Xie Z, Kashleva H, Ganguly S, Mitchell AP, Dongari-Bagtzoglou A (2011) Role of Bcr1-activated genes Hwp1 and Hyr1 in Candida albicans oral mucosal biofilms and neutrophil evasion. PLoS ONE 6(1):e16218. doi:10.1371/journal.pone.0016218

Fairman J, Roche L, Pieslak I, Lay M, Corson S, Fox E, Luong C, Koe G, Lemos B, Grove R, Fradkin L, Vernachio J (1999) Quantitative RT-PCR to evaluate in vivo expression of multiple transgenes using a common intron. Biotechniques 27(3):566–570, 572–564

Fanning S, Xu W, Solis N, Woolford CA, Filler SG, Mitchell AP (2012) Divergent targets of Candida albicans biofilm regulator Bcr1 in vitro and in vivo. Eukaryot Cell 11(7):896–904. doi:10.1128/EC.00103-12

Finkel JS, Xu W, Huang D, Hill EM, Desai JV, Woolford CA, Nett JE, Taff H, Norice CT, Andes DR, Lanni F, Mitchell AP (2012) Portrait of Candida albicans adherence regulators. PLoS Pathog 8(2):e1002525. doi:10.1371/journal.ppat.1002525PPATHOGENS-D-11-02358 [pii]

Geiss GK, Bumgarner RE, Birditt B, Dahl T, Dowidar N, Dunaway DL, Fell HP, Ferree S, George RD, Grogan T, James JJ, Maysuria M, Mitton JD, Oliveri P, Osborn JL, Peng T, Ratcliffe AL, Webster PJ, Davidson EH, Hood L, Dimitrov K (2008) Direct multiplexed measurement of gene expression with color-coded probe pairs. Nat Biotechnol 26(3):317–325. doi:10.1038/nbt1385

Giaever G, Chu AM, Ni L, Connelly C, Riles L, Veronneau S, Dow S, Lucau-Danila A, Anderson K, Andre B, Arkin AP, Astromoff A, El-Bakkoury M, Bangham R, Benito R, Brachat S, Campanaro S, Curtiss M, Davis K, Deutschbauer A, Entian KD, Flaherty P, Foury F, Garfinkel DJ, Gerstein M, Gotte D, Guldener U, Hegemann JH, Hempel S, Herman Z, Jaramillo DF, Kelly DE, Kelly SL, Kotter P, LaBonte D, Lamb DC, Lan N, Liang H, Liao H, Liu L, Luo C, Lussier M, Mao R, Menard P, Ooi SL, Revuelta JL, Roberts CJ, Rose M, Ross-Macdonald P, Scherens B, Schimmack G, Shafer B, Shoemaker DD, Sookhai-Mahadeo S, Storms RK, Strathern JN, Valle G, Voet M, Volckaert G, Wang CY, Ward TR, Wilhelmy J, Winzeler EA, Yang Y, Yen G, Youngman E, Yu K, Bussey H, Boeke JD, Snyder M, Philippsen P, Davis RW, Johnston M (2002) Functional profiling of the Saccharomyces cerevisiae genome. Nature 418(6896):387–391. doi:10.1038/nature00935nature00935 [pii]

Gower AC, Spira A, Lenburg ME (2011) Discovering biological connections between experimental conditions based on common patterns of differential gene expression. BMC Bioinformatics 12:381. doi:1471-2105-12-381 [pii]10.1186/1471-2105-12-381

Hinnebusch AG (2005) Translational regulation of GCN4 and the general amino acid control of yeast. Annu Rev Microbiol 59:407–450. doi:10.1146/annurev.micro.59.031805.133833

Hood MI, Skaar EP (2012) Nutritional immunity: transition metals at the pathogen-host interface. Nat Rev Microbiol 10(8):525–537. doi:nrmicro2836 [pii]10.1038/nrmicro2836

Huang G (2012) Regulation of phenotypic transitions in the fungal pathogen Candida albicans. Virulence 3(3):251–261. doi:20010 [pii]10.4161/viru.20010

Huffnagle GB, Noverr MC (2013) The emerging world of the fungal microbiome. Trends Microbiol 21(7):334–341. doi:S0966-842X(13)00073-5 [pii]10.1016/j.tim.2013.04.002

Hughes TR, de Boer CG (2013) Mapping yeast transcriptional networks. Genetics 195(1):9–36. doi:195/1/9 [pii]10.1534/genetics.113.153262

Hughes TR, Marton MJ, Jones AR, Roberts CJ, Stoughton R, Armour CD, Bennett HA, Coffey E, Dai H, He YD, Kidd MJ, King AM, Meyer MR, Slade D, Lum PY, Stepaniants SB, Shoemaker DD, Gachotte D, Chakraburtty K, Simon J, Bard M, Friend SH (2000) Functional discovery via a compendium of expression profiles. Cell 102(1):109–126. doi:S0092-8674(00) 00015-5 [pii]

Ingolia NT, Brar GA, Rouskin S, McGeachy AM, Weissman JS (2012) The ribosome profiling strategy for monitoring translation in vivo by deep sequencing of ribosome-protected mRNA fragments. Nat Protoc 7(8):1534–1550. doi:nprot.2012.086 [pii]10.1038/nprot.2012.086

Kretschmar M, Felk A, Staib P, Schaller M, Hess D, Callapina M, Morschhauser J, Schafer W, Korting HC, Hof H, Hube B, Nichterlein T (2002) Individual acid aspartic proteinases (Saps) 1–6 of Candida albicans are not essential for invasion and colonization of the gastrointestinal tract in mice. Microb Pathog 32(2):61–70. doi:10.1006/mpat.2001.0478

Kronstad JW, Hu G, Jung WH (2013) An encapsulation of iron homeostasis and virulence in Cryptococcus neoformans. Trends Microbiol 21(9):457–465. doi:S0966-842X(13)00093-0 [pii]10.1016/j.tim.2013.05.007

Liu Y, Filler SG (2011) Candida albicans Als3, a multifunctional adhesin and invasin. Eukaryot Cell 10(2):168–173. doi:EC.00279-10 [pii]10.1128/EC.00279-10

Lo HJ, Kohler JR, DiDomenico B, Loebenberg D, Cacciapuoti A, Fink GR (1997) Nonfilamentous C. albicans mutants are avirulent. Cell 90(5):939–949. doi:S0092-8674(00) 80358-X [pii]

Malkov VA, Serikawa KA, Balantac N, Watters J, Geiss G, Mashadi-Hossein A, Fare T (2009) Multiplexed measurements of gene signatures in different analytes using the Nanostring nCounter Assay System. BMC Res Notes 2:80. doi:10.1186/1756-0500-2-80

Marvin ME, Mason RP, Cashmore AM (2004) The CaCTR1 gene is required for high-affinity iron uptake and is transcriptionally controlled by a copper-sensing transactivator encoded by CaMAC1. Microbiology 150(Pt 7):2197–2208. doi:10.1099/mic.0.27004-0150/7/2197 [pii]

Mercer TR, Clark MB, Crawford J, Brunck ME, Gerhardt DJ, Taft RJ, Nielsen LK, Dinger ME, Mattick JS (2014) Targeted sequencing for gene discovery and quantification using RNA CaptureSeq. Nat Protoc 9(5):989–1009. doi:10.1038/nprot.2014.058

Mosci P, Pericolini E, Gabrielli E, Kenno S, Perito S, Bistoni F, d'Enfert C, Vecchiarelli A (2013) A novel bioluminescence mouse model for monitoring oropharyngeal candidiasis in mice. Virulence 4(3):250–254. doi:10.4161/viru.23529

Nailis H, Kucharikova S, Ricicova M, Van Dijck P, Deforce D, Nelis H, Coenye T (2010) Real-time PCR expression profiling of genes encoding potential virulence factors in Candida albicans biofilms: identification of model-dependent and -independent gene expression. BMC Microbiol 10:114. doi:10.1186/1471-2180-10-114

Nett JE, Zarnowski R, Cabezas-Olcoz J, Brooks EG, Bernhardt J, Marchillo K, Mosher DF, Andes DR (2015) Host contributions to construction of three device-associated Candida albicans biofilms. Infect Immun 83(12):4630–4638. doi:10.1128/IAI.00931-15

Nobile CJ, Fox EP, Nett JE, Sorrells TR, Mitrovich QM, Hernday AD, Tuch BB, Andes DR, Johnson AD (2012) A recently evolved transcriptional network controls biofilm development in Candida albicans. Cell 148(1–2):126–138. doi:S0092-8674(11)01361-4 [pii]10.1016/j.cell. 2011.10.048

Nobile CJ, Mitchell AP (2005) Regulation of cell-surface genes and biofilm formation by the C. albicans transcription factor Bcr1p. Curr Biol 15(12):1150–1155. doi:S0960-9822(05)00562-2 [pii]10.1016/j.cub.2005.05.047

Nobile CJ, Nett JE, Hernday AD, Homann OR, Deneault JS, Nantel A, Andes DR, Johnson AD, Mitchell AP (2009) Biofilm matrix regulation by Candida albicans Zap1. PLoS Biol 7(6): e1000133. doi:10.1371/journal.pbio.1000133

Nobile CJ, Solis N, Myers CL, Fay AJ, Deneault JS, Nantel A, Mitchell AP, Filler SG (2008) Candida albicans transcription factor Rim101 mediates pathogenic interactions through cell

wall functions. Cell Microbiol 10(11):2180–2196. doi:CMI1198 [pii] 10.1111/j.1462-5822.2008.01198.x
Noble SM (2013) Candida albicans specializations for iron homeostasis: from commensalism to virulence. Curr Opin Microbiol 16(6):708–715. doi:S1369-5274(13)00157-4 [pii]10.1016/j.mib.2013.09.006
Ozsolak F, Platt AR, Jones DR, Reifenberger JG, Sass LE, McInerney P, Thompson JF, Bowers J, Jarosz M, Milos PM (2009) Direct RNA sequencing. Nature 461(7265):814–818. doi:10.1038/nature08390
Pande K, Chen C, Noble SM (2013) Passage through the mammalian gut triggers a phenotypic switch that promotes Candida albicans commensalism. Nat Genet 45(9):1088–1091. doi:ng.2710 [pii]10.1038/ng.2710
Pfaller MA, Diekema DJ (2007) Epidemiology of invasive candidiasis: a persistent public health problem. Clin Microbiol Rev 20(1):133–163. doi:20/1/133 [pii]10.1128/CMR.00029-06
Pierce JV, Dignard D, Whiteway M, Kumamoto CA (2013) Normal adaptation of Candida albicans to the murine gastrointestinal tract requires Efg1p-dependent regulation of metabolic and host defense genes. Eukaryot Cell 12(1):37–49. doi:10.1128/EC.00236-12
Potrykus J, Stead D, Maccallum DM, Urgast DS, Raab A, van Rooijen N, Feldmann J, Brown AJ (2013) Fungal iron availability during deep seated candidiasis is defined by a complex interplay involving systemic and local events. PLoS Pathog 9(10):e1003676. doi:10.1371/journal.ppat.1003676PPATHOGENS-D-13-01432 [pii]
Priyadarshini Y, Natarajan K (2016) Reconfiguration of transcriptional control of lysine biosynthesis in Candida albicans involves a central role for the Gcn4 transcriptional activator. mSphere 1(1). doi:10.1128/mSphere.00016-15
Reimand J, Vaquerizas JM, Todd AE, Vilo J, Luscombe NM (2010) Comprehensive reanalysis of transcription factor knockout expression data in Saccharomyces cerevisiae reveals many new targets. Nucleic Acids Res 38(14):4768–4777. doi:gkq232 [pii]10.1093/nar/gkq232
Rosenbach A, Dignard D, Pierce JV, Whiteway M, Kumamoto CA (2010) Adaptations of Candida albicans for growth in the mammalian intestinal tract. Eukaryot Cell 9(7):1075–1086. doi:10.1128/EC.00034-10
Sellam A, van het Hoog M, Tebbji F, Beaurepaire C, Whiteway M, Nantel A (2014) Modeling the transcriptional regulatory network that controls the early hypoxic response in Candida albicans. Eukaryot Cell 13(5):675–690. doi:EC.00292-13 [pii]10.1128/EC.00292-13
Singh AK, Elvitigala T, Cameron JC, Ghosh BK, Bhattacharyya-Pakrasi M, Pakrasi HB (2010) Integrative analysis of large scale expression profiles reveals core transcriptional response and coordination between multiple cellular processes in a cyanobacterium. BMC Syst Biol 4:105. doi:1752-0509-4-105 [pii]10.1186/1752-0509-4-105
Spellberg BJ, Ibrahim AS, Avanesian V, Fu Y, Myers C, Phan QT, Filler SG, Yeaman MR, Edwards JE Jr (2006) Efficacy of the anti-Candida rAls3p-N or rAls1p-N vaccines against disseminated and mucosal candidiasis. J Infect Dis 194(2):256–260. doi:JID36169 [pii]10.1086/504691
Staib P, Kretschmar M, Nichterlein T, Kohler G, Michel S, Hof H, Hacker J, Morschhauser J (1999) Host-induced, stage-specific virulence gene activation in Candida albicans during infection. Mol Microbiol 32(3):533–546
Thewes S, Kretschmar M, Park H, Schaller M, Filler SG, Hube B (2007) In vivo and ex vivo comparative transcriptional profiling of invasive and non-invasive Candida albicans isolates identifies genes associated with tissue invasion. Mol Microbiol 63(6):1606–1628. doi:10.1111/j.1365-2958.2007.05614.x
Tournu H, Tripathi G, Bertram G, Macaskill S, Mavor A, Walker L, Odds FC, Gow NA, Brown AJ (2005) Global role of the protein kinase Gcn2 in the human pathogen Candida albicans. Eukaryot Cell 4(10):1687–1696. doi:4/10/1687 [pii]10.1128/EC.4.10.1687-1696.2005
Walker LA, Maccallum DM, Bertram G, Gow NA, Odds FC, Brown AJ (2009) Genome-wide analysis of Candida albicans gene expression patterns during infection of the mammalian kidney. Fungal Genet Biol 46(2):210–219. doi:10.1016/j.fgb.2008.10.012

Westermann AJ, Gorski SA, Vogel J (2012) Dual RNA-seq of pathogen and host. Nat Rev Microbiol 10(9):618–630. doi:10.1038/nrmicro2852

Wheeler RT, Fink GR (2006) A drug-sensitive genetic network masks fungi from the immune system. PLoS Pathog 2(4):e35. doi:10.1371/journal.ppat.0020035

Wheeler RT, Kombe D, Agarwala SD, Fink GR (2008) Dynamic, morphotype-specific Candida albicans beta-glucan exposure during infection and drug treatment. PLoS Pathog 4(12): e1000227. doi:10.1371/journal.ppat.1000227

Xu W, Solis NV, Ehrlich RL, Woolford CA, Filler SG, Mitchell AP (2015) Activation and alliance of regulatory pathways in C. albicans during mammalian infection. PLoS Biol 13(2):e1002076. doi:10.1371/journal.pbio.1002076

Zakikhany K, Naglik JR, Schmidt-Westhausen A, Holland G, Schaller M, Hube B (2007) In vivo transcript profiling of Candida albicans identifies a gene essential for intraepithelial dissemination. Cell Microbiol 9(12):2938–2954. doi:10.1111/j.1462-5822.2007.01009.x

Chapter 14
Signaling Mechanisms in Pathogenesis and Virulence of *Candida albicans*

Neeraj Chauhan

Abstract Invasive fungal infections kill over 1.5 million people a year worldwide, and this number is on the rise due to increasing numbers of people living with compromised immunity, including the elderly, premature infants, transplant recipients, and cancer patients. Fungal pathogens, such as, *Candida albicans* rely on complex network of signal transduction pathways that allow fungus to not only survive in the human host but also contribute to its pathogenesis. This chapter will focus on the up-to-date information on the signaling pathways and downstream target proteins that contribute to *C. albicans* virulence, which has been obtained primarily through the analysis of null mutants or inference from genome annotation. However, before addressing these issues in detail, a brief introduction of the magnitude and the economic impact of the healthcare problems caused by fungal pathogens and the currently available treatment options are discussed.

14.1 The Impact of Fungal Diseases

More people die from the invasive fungal diseases than from tuberculosis or malaria, with *Candida* species alone being responsible for ~400,000 bloodstream infections annually worldwide with an associated mortality of 46–75% (Brown et al. 2012). Treatment of fungal diseases, such as invasive candidiasis, remains problematic for clinicians, and new antifungal drugs are urgently sought because the number of currently available antifungal is highly limited, while those that are

N. Chauhan (✉)
Public Health Research Institute, New Jersey Medical School,
Rutgers, The State University of New Jersey, 225 Warren Street, Newark,
NJ 07103, USA
e-mail: chauhan1@njms.rutgers.edu

N. Chauhan
Department of Microbiology, Biochemistry and Molecular Genetics,
New Jersey Medical School, Rutgers, The State University of New Jersey,
225 Warren Street, Newark, NJ 07103, USA

R. Prasad (ed.), *Candida albicans: Cellular and Molecular Biology*,
DOI 10.1007/978-3-319-50409-4_14

available have serious limitations, including toxicity and increasing loss of effectiveness due to emerging drug resistance. According to a recent epidemiological survey, the risk of death due to candidiasis in hospitals today is no less than it was in the 1980s and 1990s, underscoring an urgent need to identify new drug targets in fungal pathogens (Pfaller and Diekema 2007). However, discovery of new drugs acting on fungal targets is complicated by the fact that fungi are eukaryotes, greatly limiting the number of fungal-specific drug targets.

14.2 Currently Available Antifungal Treatments

Currently available standard antifungal therapies are limited because of toxicity, increasingly prevalent drug resistance, and drug–drug interactions. The former "gold standard" antifungal, amphotericin B, which binds to fungal membrane component ergosterol and causes changes in membrane permeability, invariably causes toxicity in patients, compromising its effectiveness as a fungicidal agent. Currently, triazole drugs, which target ergosterol synthesis, are a preferred agent to treat fungal disease because of their excellent toxicity profiles and ease of administration (oral bioavailability). However, the majority of triazoles are fungistatic and not fungicidal, allowing for the emergence of resistant mutants. Furthermore, some non-*albicans Candida* species, most notably *C. glabrata*, display elevated intrinsic resistance to triazoles, and these species are increasingly recovered among bloodstream clinical isolates (Pfaller and Diekema 2007). The third and last antifungal drug class currently in clinical use against invasive fungal infections are echinocandins (e.g., caspofungin and micafungin), which target fungal cell wall enzyme β-1,3-glucan synthase. Echinocandins are fungicidal; however, these drugs are ineffective against *Cryptococcus neoformans* and of questionable value in invasive aspergillosis patients (Vehreschild and Cornely 2006). Furthermore, even though echinocandins were only introduced on the market 15 years ago, recent reports indicate increasing prevalence of echinocandin-resistant clinical *Candida* isolates (Chandrasekar and Sobel 2006; Deresinski and Stevens 2003; Pfaller 2004). Indeed, recent studies have shown that the incidence of echinocandin-resistant *C. glabrata* at the examined medical centers grew from 2 to 3% in 2001 to >13% in 2010 (Alexander et al. 2013). Furthermore, the identification of multi-drug (azole and echinocandin) resistant *C. glabrata* clinical isolates (Pfaller et al. 2012) is a cause for major concern, as treatment options for patients infected with such strains are very limited and associated with significant toxicity. Thus, the limited number of available antifungal drug classes together with increasing occurrence of bloodstream fungal infections and emerging antifungal drug resistance underscore the critical need for discovering new types of antifungal drugs.

14.3 Signal Transduction Pathways Important for Virulence of *C. albicans*

The yeast *Candida albicans* does not have a natural environmental reservoir instead it is part of normal microflora of humans (Kumamoto 2011). As a result of its close association with human host, *C. albicans* encounter significant challenges in its natural habitat (human host) in the form of fluctuating temperature, pH, availability of nutrients, and interactions with phagocytes and other microorganisms. Indeed, interactions with phagocytes (neutrophils and macrophages) occupy a large portion of the disease cycle, be it in the blood, tissues, or at mucosal surfaces.

C. albicans is also an opportunistic fungal pathogens of humans causing both mucosal (e.g., oral, vaginal) and disseminated infections. Candidiasis caused by *Candida* spp. is the leading cause of nosocomial infectious disease in the US, and there is a similar trend world wide (Ascioglu et al. 2002; Stover et al. 2001; Wenzel 1995; Wilson et al. 2002; Wisplinghoff et al. 2003, 2004). Annual death toll due to candidiasis is approximately 10,000 in the United States alone (Brown et al. 2012). Virulence mechanisms in *C. albicans* have been extensively studied and several virulence factors have been identified, including ability to form hyphae and secretion of enzymes that contribute to invasiveness (Calderone and Fonzi 2001). In this section, we will discuss fungal signaling networks that facilitate *C. albicans* evade host defense and cause disease.

A class of proteins known as "sensor histidine kinases" usually recognizes environmental signals received by *C. albicans*. The signal is then subsequently transferred to a downstream protein via a series of phosphorylation steps. The final outcome of these phosphorylation events often results in the activation of a downstream MAP kinase cascade, which, in turn, activates the transcription factors whose target genes typically participate in the cellular response to environmental change. One such signaling pathway that *C. albicans* utilizes in response to a variety of environmental stimuli is a so-called two-component signal transduction (TCST) pathway. The TCST has important function(s) in the regulation of morphogenesis (yeast–hypha transition), adhesion, stress response, drug resistance and the virulence factor expression (Chauhan et al. 2006; Kruppa and Calderone 2006). A short overview of fungal two-component signaling systems is presented below.

14.4 A Brief History of Two-Component Signal Transduction Pathways

In this section we will provide a brief synopsis of the evolution and architecture of two-component signal transduction pathway in *C. albicans*, specifically highlighting the differences between prokaryotes and eukaryotes. A comprehensive list of all known *C. albicans* two-component signal transduction proteins can be found in Table 14.1.

Table 14.1 Function(s) of two-component signaling proteins in *C. albicans*

Fungal pathogen	HK	Function	HPt	Function	RR	Function	References
C. albicans	Chk1	Quorum sensing, Cell wall biogenesis, virulence, morphogenesis, stress response	Ypd1	Not essential for viability, localizes to both nucleus and cytosol. Regulator of stress and cell membrane integrity	Ssk1	Stress response, adhesion, morphogenesis, virulence	Du et al. (2006), Fassler and West (2013), Gabaldon and Huynen (2004), Grant (2001), Gray (1993), Hagiwara et al. (2013), Hancock and Perego (2004), Koretke et al. (2000), Kruppa and Calderone (2006), Kruppa et al. (2003), Kruppa et al. (2004a, b), Mascher et al. (2006), Mavrianos et al. (2013), Mavrianos et al. (2014), Menon et al. (2006), Menon et al. (2008), Nagahashi et al. (1998), Nemecek et al. (2006), Nierman et al. (2005), Posas et al. (1996), Stover et al. (2001)
	Sln1	Osmosensor, virulence			Skn7	Oxidative stress, morphogenesis	
	Nik1	Morphogenesis, virulence			Srr1	Stress response, morphogenesis, virulence, apoptosis	
					Rim15	uncharacterized	

The two-component signaling systems were first identified in bacteria where the phosphorelay generally involves two proteins: a histidine kinase (HK) and a response regulator (RR) protein (Fig. 14.1). In response to an environmental signal, the HK, which is frequently localized in the bacterial outer membrane, is autophosphorylated on a conserved histidine residue, followed by a transfer of the

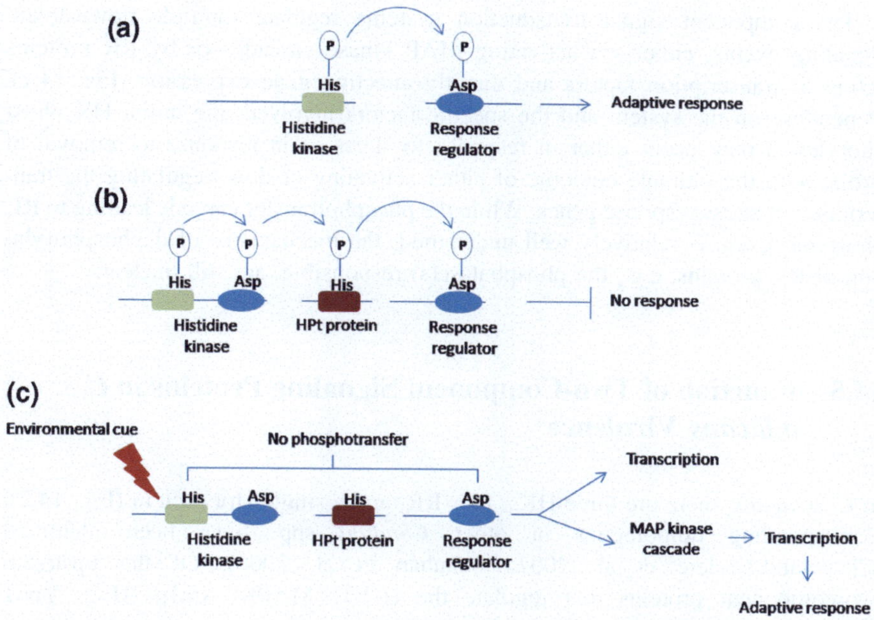

Fig. 14.1 Comparative model of phosphotransfer reactions from HK to RR in bacteria (**a**) and fungi (**b**, **c**). In eukaryotes a more intricate phosphotransfer network exists. In the absence of stress normal phosphotransfer occurs from HK to RR via Hpt (**b**). However, under conditions of stress, there is no phosphotransfer (**c**) which typically results in activation of downstream MAPK Hog1

phosphoryl group to a cognate response regulator protein (RR) on a conserved aspartate residue. The phosphorylated RR then usually acts directly as a transcription factor to activate genes associated with chemotaxis, stress response, quorum sensing, sporulation, virulence factor expression, and antibiotic resistance (Mascher et al. 2006). However, fungal two-component phosphorelays are more intricate in two respects (Fig. 14.1). First, the signaling cascade involves three proteins: HK, RR, and a histidine phosphotransferase (HPt) whose function is to shuttle the phosphate moiety from HK to RR (Fassler and West 2013). Second, in fungi the phosphorelay typically comprises four phosphorylation events: (1) the HK is autophosphorylated on a histidine residue within its histidine kinase domain, (2) the phosphate is transferred intramolecularly to an aspartate (His → Asp) in the HK receiver domain, (3) a third, intermolecular phosphotransfer occurs to the histidine residue present in the HPt domain on the transferase (His → Asp → His), and (4) the phosphoryl group is relayed to an aspartate on the RR protein (His → Asp → His → Asp). Thus, two-component like phosphorelay systems are unusual in terms of mechanism: the amino acids that accept phosphoryl groups are either aspartate or histidine residues. These unique features may be exploited in designing specific inhibitors that would not affect the activity of conventional Ser/Thr/Tyr kinases more prevalent in mammalian systems.

Two-component signal transduction systems regulate multiple downstream signaling events, either via activating MAP kinase cascades or by RR proteins acting as transcription factors and directly affecting gene expression (Fig. 14.1). Depending on the system and the specific factors involved, the initial HK phosphorylation may occur either in response to stress or in response to removal of stress, with the ultimate outcome of either activating or downregulating the transcription of stress response genes. While the phosphotransfer cascade leading to RR phosphorylation is relatively well understood, the mechanisms of dephosphorylation of RR proteins, e.g., the phosphatase(s) responsible, are still unclear.

14.5 Function of Two-Component Signaling Proteins in *C. albicans* Virulence

In *C. albicans*, there are three HKs, four RR, and a single Hpt protein (Fig. 14.2). Corresponding homologues in other *Candida* spp. have been identified (Chapeland-Leclerc et al. 2007; Chauhan et al. 2006). Of the upstream two-component proteins that regulate the HOG1 MAPK, Sln1p (HK), Ypd1 (Hpt) and Ssk1p (RR) have been extensively characterized phenotypically (Calera et al. 2000; Chauhan et al. 2003; Chen et al. 2005; Li et al. 2002; Mavrianos et al. 2014; Nagahashi et al. 1998; Srikantha et al. 1998; Yamada-Okabe et al. 1999).

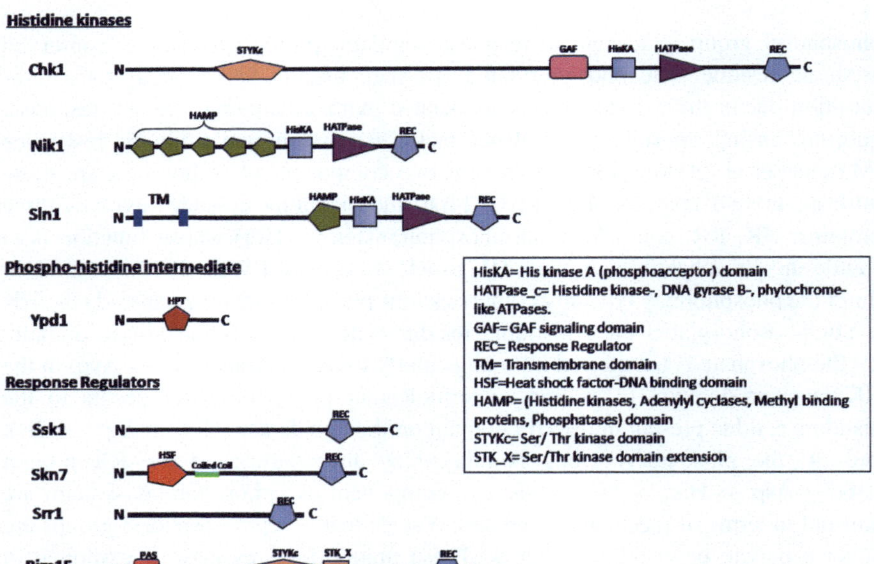

Fig. 14.2 Two-component signal transduction proteins in *C. albicans* including domains present in each protein. The *black lines* represent the relative sizes of proteins

With the exception of Skn7 response regulator, all other HK and RR mutants (*sln1, chk1, nik1, ssk1*, and *srr1*) have been shown to be avirulent in a murine model of hematogenously disseminated candidiasis (Calera et al. 1999; Calera et al. 2000; Desai et al. 2011; Singh et al. 2004; Yamada-Okabe et al. 1999). The biological changes in each mutant associated with avirulence vary according to the specific gene deletion. All HK and RR mutants, tested thus far, have impaired patterns of morphogenesis, which is a response that is typical of many *C. albicans* mutants.

The role of two-component signal proteins and MAPK pathways in antifungal drug sensitivity is also studied (Chauhan et al. 2007). The scientific premise for these studies was based on the previous published reports indicating several bacterial pathogens that use two-component signaling in adaptation to antibacterial drugs (Hancock and Perego 2004; Mascher et al. 2006; Nishino and Yamaguchi 2002). To investigate the activity of two-component proteins in antifungal drug sensitivity/resistance in *C. albicans*, standardized MIC assays were performed with a number of antifungal drugs and general cell inhibitors. Interestingly, only the *chk1*Δ/Δ and *ssk1*Δ/Δ null mutants were found to be sensitive to triazoles (fluconazole and voriconzaole). However, quite strikingly, no changes in sensitivity to amphotericin B, 5-FC, or imidizoles such as ketoconazole and miconazole were observed. This observation was further corroborated by increased uptake of [^{3}H]-fluconazole in the *ssk1*Δ/Δ *and chk1*Δ/Δ mutants compared to parental and a gene-reconstituted strain (Chauhan et al. 2007). Collectively, these observations suggest that in addition to their prominent roles in virulence, histidine kinases and response regulator proteins regulate antifungal drug resistance in *C. albicans*. Therefore, it is perhaps not surprising that similar reports are beginning to emerge from other fungal pathogens such as *Cryptococcus neoformans* (Bahn et al. 2006) and *Aspergillus fumigatus* (Hagiwara et al. 2013). The following section discusses individual constituent of fungal two-component signaling pathways, with especial emphasis on the virulence related output responses.

14.5.1 Histidine Kinases (HK)

Most fungal HKs are hybrid histidine kinases which contain both HK and RR domains in a single polypeptide. Although present in all major fungal pathogens, HKs have been most extensively studied in *C. albicans* (Table 14.1) (Kruppa and Calderone 2006). The data generated from previous published studies have shown that *C. albicans* HK mutants have impaired patterns of morphogenesis (yeast to hyphae transition) and attenuated virulence (Calera and Calderone 1999). As stated above, in *C. albicans*, there are three HK proteins, Chk1, Sln1, and Nik1 (Table 14.1 and Fig. 14.2). Sln1 is the homologue of the *S. cerevisiae* Sln1. While Chk1 is not found in *S. cerevisiae*, two homologues have been identified in *Schizosaccharomyces pombe* (Aoyma et al. 2001; Buck et al. 2001), but a functional pathway has not been assigned. The Nik1 is a homologue of the *Nik1* of *Neurospora crassa* and is also not found in *S. cerevisiae* (Yamada-Okabe et al.

1999). The Chk1 histidine kinase is most extensively studied due to its role in cell wall biosynthesis, quorum sensing and virulence (Calera et al. 1999; Kruppa et al. 2003, 2004a, b; Li et al. 2009). The deletion of *CHK1* also impairs the ability of *C. albicans* to adhere to human esophageal tissue in vitro, renders it refractory to quorum sensing, and significantly sensitizes it to killing by human neutrophils (Calera et al. 1999; Li et al. 2002; Torosantucci et al. 2002).

In summary, the *chk1*Δ/Δ mutant appears to be in an apparent communication failure in its ability to respond to human cells as well as its ability to communicate with self. The changes in the mutant at a biochemical level strongly suggest defects in its cell wall (Kruppa et al. 2003, 2004a, b; Li et al. 2009). This conclusion is based upon the following observations indicating: (1) extensive flocculation of mutant cells in tissue culture media such as M199 compared to the wildtype and gene-reconstituted strains; (2) altered ratio of β-1,3 to β-1,6 glucans determined via cell wall mass spectrometry, culminating in an increase in sensitivity to the cell wall disrupting agent Congo red; and (3) acid stable mannan side chains appear to be truncated, suggesting lower molecular mass of mannoproteins in the *chk1*Δ/Δ mutant.

The *sln1*Δ/Δ and *nik1*Δ/Δ mutants are attenuated for virulence in a mouse model of candidiasis (Yamada-Okabe et al. 1999). In addition, the *C. albicans* cells lacking Nik1 are also impaired in its ability to undergo the opaque-white switch phenotype that is required for mating and virulence (Srikantha et al. 1998).

Due to their importance in regulating fungal virulence, there is considerable interest in histidine kinases in other medically important human fungal pathogens. The *A. fumigatus* genome contains ~15 HKs (Nierman et al. 2005), of which only three (Fos1, TcsB and NikA/TcsC) have thus far been functionally characterized (Clemons et al. 2002; Du et al. 2006; Hagiwara et al. 2013). While TcsB has been reported to be dispensable for most cellular functions (Du et al. 2006), Fos1 has been shown to regulate cell wall assembly and virulence (Clemons et al. 2002). More recently, NikA/TcsC was reported to be involved in conidia and hyphae production, resistance to osmotic stress, and resistance to cell wall perturbing reagents and fungicides (Hagiwara et al. 2013).

C. neoformans has seven HKs (Tco1-7) (Bahn et al. 2006), all of which lack a transmembrane domain suggesting that they are cytosolic proteins. The Tco1 HK regulates melanin biosynthesis and is required for *C. neoformans* pathogenesis in a mouse model of cryptococcal meningitis (Bahn et al. 2006). One of the *C. neoformans* HKs, Tco2, is unique in the sense that it has two HK domains and two RR domains—a feature that has not been reported for any other known fungal hybrid HK.

Among the fungi that cause endemic mycoses, only one HK (Drk1) has been studied in *B. dermatitidis* and *H. capsulatum* (Nemecek et al. 2006). The phenotypes of *drk1* mutants in both of these fungi include altered growth at 37 °C, sensitivity to cell wall inhibitors, and significantly attenuated virulence in murine model of pulmonary infection (Nemecek et al. 2006). Orthologs of Drk1 are also reported in *C. immits* but their functions have not yet been examined (Nemecek et al. 2006).

14.5.2 Histidine Phosphotransferases (HPt)

The major function of phosphohistidine intermediate protein Ypd1 is to shuttle phosphate from HK to RR proteins in all known fungal two-component systems (Fassler and West 2013). Most of the information about this class of proteins has been obtained from studies of model yeast *Saccharomyces cerevisiae* (Fassler and West 2013; Posas et al. 1996). Ypd1 is essential for viability in *S. cerevisiae* and *C. neoformans* (Fassler and West 2013; Lee et al. 2011) but not in *C. albicans* (Mavrianos et al. 2014). This apparent difference could be due to mechanisms unique to *C. albicans*, whereby the phosphotransfer can proceed from the upstream HK to the RR while bypassing Ypd1. Deletion of *YPD1* results in increased flocculation and constitutive filamentation in *C. albicans* (Mavrianos et al. 2014). Furthermore, deletion of Ypd1 also results in constitutive phosphorylation of Hog1 MAPK in *C. albicans*. This could be due to disruption of Ypd1 mediated phosphorelay, resulting in unphosphorylated Ssk1 response regulator, which leads to activation of downstream Hog1 MAPK pathway. It was also reported that *C. albicans* Ypd1 is localized to both the nucleus and the cytoplasm (Fig. 14.3). The subcellular segregation of Ypd1 hints at an important role(s) of Ypd1 in regulation of Ssk1 (cytosolic) and Skn7 (nuclear) response regulator proteins via phosphorylation in *C. albicans*. This discovery has profound implications for a

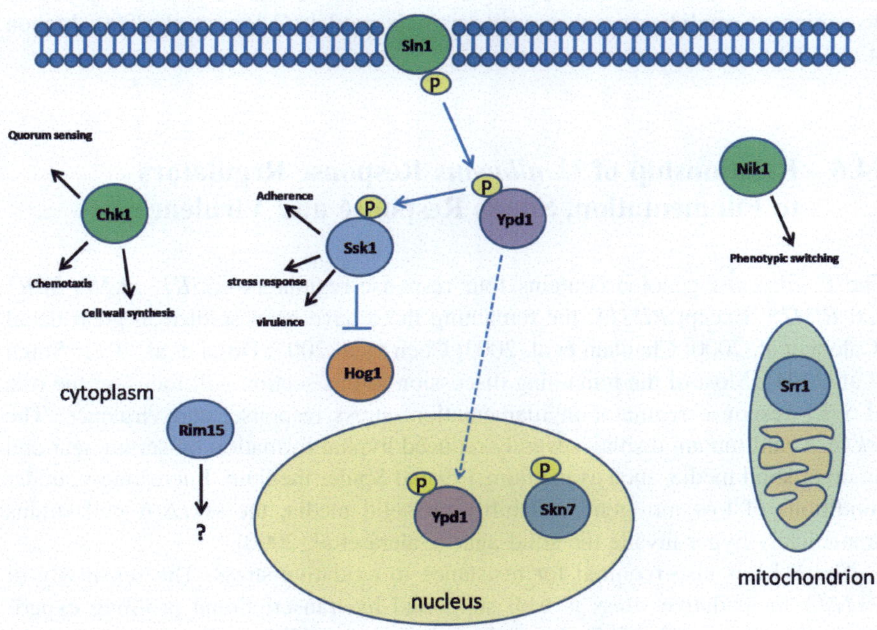

Fig. 14.3 Two-component signal pathways that includes each of *C. albicans* HK, Hpt and RR proteins and their apparent function. Subcellular localization of each protein is also depicted below

mechanistic understanding of two-component signaling pathways in *C. albicans*, and perhaps in other pathogenic fungi.

14.5.3 Response Regulators (RR)

Fungal pathogens usually contain significantly fewer RRs than HKs, suggesting that diverse environmental stimuli are detected by a variety of HKs and are then transduced via Ypd1 to a smaller number of RR proteins. With the exception of *C. albicans*, most fungal pathogens contain either 2 or 3 RR proteins. In *C. albicans*, the genes that encode putative RRs include *SSK1, SKN7, SRR1, and RIM15* (Fig. 14.3). While homologues of these proteins are found in *S. cerevisiae*, it is important to realize that *C. albicans* homologues have functions that are different from their counterparts in *S. cerevisiae*. For example, the Ssk1 of *C. albicans* is not essential for the phosphorylation of Hog1 when cells are osmostressed (Chauhan et al. 2003), while it is for *S. cerevisiae*. The Ssk1 of *C. albicans* is required for adaptation to oxidative stress and regulates some aspects of cell wall biosynthesis (Chauhan et al. 2003), functions not reported for the *S. cerevisiae SSK1*. The Ssk1 is required for morphogenesis (yeast to hyphae transition), resistance to oxidative stress, normal cell wall biosynthesis, and adherence to human esophageal tissues in vitro (Calera et al. 2000; Chauhan et al. 2003; Li et al. 2002). Importantly, Ssk1 is also required for *C. albicans* pathogenesis in a murine model of hematogenously disseminated candidiasis and in a rat vaginitis model (Calera et al. 2000; Menon et al. 2008).

14.6 Relationship of *C. albicans* Response Regulators to Filamentation, Stress Response and Virulence

The *C. albicans* genome contains four response regulators—*SSK1*, *SKN7*, *SRR1* and *RIM15*. Except *RIM15*, the remaining three have been studied in great detail (Calera et al. 2000; Chauhan et al. 2003; Chen et al. 2005; Desai et al. 2011; Singh et al. 2004). Most of the remaining discussion in this section will focus on the role of Ssk1 response regulator in filamentation, stress response, and virulence. The *ssk1*Δ/Δ null mutant display severely reduced hyphal formation on serum agar and on other solid media, such as medium 199 and Spider medium. Interestingly, under conditions of low nitrogen availability on solid media, the *ssk1*Δ/Δ null strains dramatically hyper-invade the solid agar (Calera et al. 2000).

The Ssk1 is also required for resistance to oxidative stress. The sensitivity of *ssk1*Δ/Δ to oxidative stress is well supported by transcriptional profiling experiments (Chauhan et al. 2003). Several genes that are differentially expressed in the *ssk1*Δ/Δ mutant are among those whose function is associated partially or totally with adaptation to stress conditions, including oxidative stress. Some of the upregulated genes such as, *AHP1* encode one of the several oxidoreductases whose function is to maintain cells in a reduced state (redox potential) during oxidative metabolism (Grant 2001; Park et al. 2000). The oxidoreductases are small, heat-stable proteins that contain two conserved cysteine residues in their active sites and are of two types, the thioredoxins and glutathione–glutaredoxins (Grant 2001; Park et al. 2000). While they are highly conserved with overlapping functions, differences exist between the two types of oxidoreductases. For example, the thioredoxins utilize NADPH as an H donor, while the glutaredoxins use reduced glutathione (Grant 2001). Importantly, the expression of thioredoxin-related genes, such as *AHP1*, is regulated by the transcription factors YAP1 and Skn7 (Grant 2001). The latter transcription factor is a response regulator two-component signal protein in *C. albicans*. Thus, a consequence of cell growth and oxidative metabolism is the generation of ROS that are highly toxic to cells unless proteins such as Ahp1 are produced. While the oxidoreductases of *C. albicans* have not been characterized in detail, it appears that *C. albicans* processes oxidative stress in a manner that is somewhat different from the *S. cerevisiae* process (Chauhan et al. 2003). Furthermore, *SSK1* appears to regulate the expression of several cell wall proteins (Mnn4, Als1, and Flo1) and a two-component histidine kinase, *CHK1* that regulates cell wall synthesis (Chauhan et al. 2003). Up regulation of *MNN4* in is often associated with stress responses (Odami et al. 1997). Importantly, expression of the adhesin *ALS1* is down regulated in *ssk1*Δ/Δ mutant. This observation correlates well with a reduced adherence to human esophageal cells phenotype by the *ssk1*Δ/Δ mutant of *C. albicans* (Li et al. 2002).

The biological relationship between the Ssk1 and downstream MAP kinase Hog1 is well established in *C. albicans*. The Hog1 MAPK is not phosphorylated in the *ssk1*Δ/Δ mutant when the latter is grown in the presence of hydrogen peroxide. These data indicate that *C. albicans* exploits the Ssk1 response regulator protein to adapt cells to oxidative stress, while its role in the adaptation to osmotic stress, through Hog1 MAPK is less certain.

C. albicans Skn7 acts as a transcription factor, functions in resistance to oxidative stress, and regulates morphogenesis (Singh et al. 2004). Recently, another *C. albicans* RR protein, Srr1, was shown to be located in the mitochondria to and play an important role in hyphal development (morphogenesis), resistance to stress, virulence, and apoptosis (Mavrianos et al. 2013). A detailed discussion about function of Srr1 is given below in a separate section.

In summary, response regulators control multiple downstream signaling events either via activating MAP kinase cascades or by RR proteins acting as transcription factors and directly affecting gene expression. The *C. albicans* Ssk1 response regulator has functions (direct or indirect) in morphogenesis, cell wall biosynthesis, adaptation to stress conditions, survival in human PMNs, and virulence in an invasive model of candidiasis. Thus, at least in the case of the Ssk1 two-component protein of *C. albicans*, the functional circuitry has assumed a more expansive role than in *S. cerevisiae*.

14.7 Mitochondrial Two-Component Signal Transduction Pathways: A Plausible Link Between Metabolism and Virulence

Mitochondria are cytoplasmic organelles that are essential for cellular metabolism. These organelles are thought to have originated as a result of endosymbiotic events over the course of evolution of eukaryotic cells from prokaryotic ancestors. (Andersson et al. 2003; Gabaldon and Huynen 2004; Gray 1993). The acquisition of two-component signal transduction pathways by eukaryotes is believed to have occurred through horizontal gene transfer from chloroplasts and/or mitochondria (Koretke et al. 2000). This hypothesis is supported by several recent observations indicating that both mitochondria and chloroplasts retain components of two-component signaling cascades from their bacterial ancestors (Mavrianos et al. 2013; Puthiyaveetil et al. 2008). Recently it was shown that mitochondrion of *C. albicans* contains a two-component response regulator protein Srr1 (Mavrianos et al. 2013). The *C. albicans SRR1* appears to be unique to the *Candida* clade, which is known to uniquely translate the CUG codon as serine, rather than leucine (Butler et al. 2009). Srr1 contributes to pathogenesis of *C. albicans* by regulating hyphal formation, a major virulence factor of *C. albicans*, and causing resistance to oxidative stress and apoptosis (Desai et al. 2011; Mavrianos et al. 2013). For example, the *srr1*Δ/Δ mutant is defective in hyphal formation, sensitive to hydrogen peroxide and is also avirulent in a mouse model of disseminated candidiasis (Desai et al. 2011). Transcriptional analysis of the *srr1*Δ/Δ mutant has shown that two gene families were significantly affected in the mutant, including those associated with stress adaptation and genes whose putative functions are related to mitochondrial function (Mavrianos et al. 2013). Furthermore, deletion of *SRR1* also resulted in increased apoptosis of *C. albicans* (Mavrianos et al. 2013). Remarkably, apoptosis was observed in the absence of any external stimuli, suggesting regulation of the mitochondrial apoptotic pathway by Srr1. Thus far, Srr1 remains the only two-component protein identified in fungal mitochondria. Orthologs of Srr1 are present in other pathogenic *Candida* species. These observations raise an interesting possibility of similar mitochondrial two-component systems in other fungi.

A critical step in eukaryotic two-component phosphorelay is faithful transfer of phosphate from histidine kinase to its cognate response regulator. In most cases it is expected that a single HK will phosphorylate a single RR. However, in some cases, it has been proposed that a single RR can receive signal from multiple HKs. This is one of the apparent advantages of multistep phosphorelay because it allows integration of signal at intermediate steps. This has been demonstrated by elegant in vitro studies in model yeast *S. cerevisiae* and pathogenic fungus *C. albicans* (Menon et al. 2006). In both organisms in vitro phosphorelay assays were used to show phosphotransfer from Sln1-Ypd1-Ssk1 (Menon et al. 2006). Because Srr1 is a response regulator and localized to the mitochondrion there is an interest to determine its upstream histidine kinase protein. However, no reports exist of a fungal histidine kinase in the mitochondrion. Future research efforts aimed at identification of similar mitochondrial two-component systems in fungal pathogens and characterization of their biological role in fungal pathogenesis may not only provide new avenues to tackle fungal disease but may also lay foundation for antifungal drug discovery.

14.8 Two-Component Signaling Systems as Antifungal Drug Targets

An ideal drug target must fulfill several requirements: the target must be (1) present in most if not all pathogenic fungi, (2) absent in humans, and (3) required for the disease process. One set of proteins that fit all of these criteria functions in two-component signal transduction pathways. As discussed above, these signaling pathways are based on the transfer of phosphoryl groups (phosphorelay) among their components (Fig. 14.1). The proteins that participate in this pathway are unique in regard to the amino acids that accept phosphoryl groups, which include either aspartate or histidine residues. While the genes encoding two-component systems are frequently not essential for viability, multiple studies have demonstrated their importance in virulence of several leading fungal pathogens, suggesting that their exploitation could lead to the development of drugs that may have broad activity spectra. Finally, two-component systems are only found in bacteria, plants, and fungi, but not in humans (Koretke et al. 2000), suggesting that their pharmacological targeting should not be associated with significant off-target effects in the host.

References

Alexander BD et al (2013) Increasing echinocandin resistance in Candida glabrata: clinical failure correlates with presence of FKS mutations and elevated minimum inhibitory concentrations. Clin Infect Dis 56(12):1724–1732

Andersson SG et al (2003) On the origin of mitochondria: a genomics perspective. Philos Trans R Soc Lond B Biol Sci 358:165–177, discussion 177–179

Aoyama K et al (2001) Genetic analysis of the His-to-Asp phosphorelay implicated in mitotic cell cycle control: involvement of histidine-kinase genes of Schizosaccharomyces pombe. Biosci Biotechnol Biochem 65(10):2347–2352

Ascioglu S et al (2002) Defining opportunistic fungal infections in immunocompromised patients with cancer and hematopoietic stem cell transplants: an international consensus. Clin Infect Dis 34(1):7–14

Bahn YS et al (2006) A unique fungal two-component system regulates stress responses, drug sensitivity, sexual development, and virulence of Cryptococcus neoformans. Mol Biol Cell 17:3122–3135

Brown GD et al (2012) Hidden killers: human fungal infections. Sci Transl Med 4(165):165rv13

Buck V et al (2001) Peroxide sensors for the fission yeast stress-activated mitogen-activated protein kinase pathway. Mol Biol Cell 12(2):407–419

Butler G et al (2009) Evolution of pathogenicity and sexual reproduction in eight Candida genomes. Nature 459:657–662

Calderone RA, Fonzi WA (2001) Virulence factors of *Candida albicans*. Trends Microbiol 9 (7):327–335

Calera JA, Calderone RA (1999) Flocculation of hyphae is associated with a deletion in the putative CaHK1 two-component histidine kinase gene from *Candida albicans*. Microbiology 145:1431–1442

Calera JA et al (1999) Avirulence of *Candida albicans* CaHK1 mutants in a murine model of hematogenously disseminated candidiasis. Infect Immun 67:4280–4284

Calera JA et al (2000) Defective hyphal formation and avirulence caused by a deletion of the CSSK1 response regulator gene in *Candida albicans*. Infect Immun 68:518–525

Chandrasekar PH, Sobel JD (2006) Micafungin: a new echinocandin. Clin Infect Dis 42:1171–1178

Chapeland-Leclerc F et al (2007) Differential involvement of histidine kinase receptors in pseudohyphal development, stress adaptation, and drug sensitivity of the opportunistic yeast Candida lusitaniae. Eukaryot Cell 6(10):1782–1794

Chauhan N et al (2003) The SSK1 of *Candida albicans* is associated with oxidative stress adaptation and cell wall biosynthesis. Eucaryot Cell 2:1018–1024

Chauhan N et al (2006) Signalling and oxidant adaptation in *Candida albicans* and Aspergillus fumigatus. Nat Rev Microbiol 4(6):435–444

Chauhan N et al (2007) The Ssk1p response regulator and Chk1p histidine kinase mutants of *Candida albicans* are hypersensitive to fluconazole and voriconazole. Antimicrob Agents Chemother 51(10):3747–3751

Chen D et al (2005) Deletion of the SSK1 response regulator gene in *Candida albicans* contributes to enhanced killing by human polymorphonuclear neutrophils. Infect Immun 73:865–871

Clemons KV et al (2002) fos-1, a putative histidine kinase as a virulence factor for systemic aspergillosis. Med Mycol 40:259–262. doi:10.1080/mmy.40.3.259.262

Deresinski CC, Stevens DA (2003) Caspofungin. Clin Infect Dis 36:1445–1457

Desai C et al (2011) *Candida albicans* SRR1, a putative two component response regulator gene, is required for stress adaptation, morphogenesis and virulence. Eukaryot Cell 10(10):1370–1374

Du C et al (2006) The role of the sakA (Hog1) and tcsB (Sln1) genes in the oxidant adaptation of Aspergillus fumigatus. Med Mycol 44:211–218

Fassler JS and West AH (2013) Histidine phosphotransfer proteins in fungal two-component signal transduction pathways. Eukaryot Cell 12(8):1052–1060

Gabaldon T, Huynen MA (2004) Shaping the mitochondrial proteome. Biochim Biophys Acta 1659:212–220

Grant CM (2001) Role of the glutathione/glutaredoxin and thioredoxin systems in yeast growth and response to stress conditions. Mol Microbiol 39:533–541

Gray MW (1993) Origin and evolution of organelle genomes. Curr Opin Genet Dev 3:884–890

Hagiwara D et al (2013) NikA/TcsC histidine kinase is involved in conidiation, hyphal morphology, and responses to osmotic stress and antifungal chemicals in Aspergillus fumigatus. PLoS One 8(12):e80881. doi:10.1371/journal.pone.0080881. eCollection 2013

Hancock LE, Perego M (2004) Systematic inactivation and phenotypic characterization of two-component signal transduction systems of Enterococcus faecalis V583. J Bacteriol 186 (23):7951–7958

Koretke KK et al (2000) Evolution of two-component signal transduction. Mol Biol Evol 17 (12):1956–1970

Kruppa M, Calderone R (2006) Two-component signal transduction in human fungal pathogens. FEMS Yeast Res 6(2):149–159

Kruppa M et al (2003) The Chk1p of *C. albicans* and its role in the regulation of cell wall synthesis. FEMS Yeast Res 3:289–299

Kruppa M et al (2004a) The histidine kinases of *Candida albicans*: regulation of cell wall mannan biosynthesis. FEMS Yeast Res 4(4–5):409–416

Kruppa M et al (2004b) The two-component signal transduction protein Chk1p regulates quorum sensing in *Candida albicans*. Eukaryot Cell 3(4):1062–1065

Kumamoto CA (2011) Inflammation and gastrointestinal Candida colonization. Curr Opin Microbiol 14(4):386–391

Lee JW et al (2011) Multiple roles of Ypd1 phosphotransfer protein in viability, stress response, and virulence factor regulation in Cryptococcus neoformans. Eukaryot Cell 10(7):998–1002. doi:10.1128/EC.05124-11

Li D et al (2002) Temporal expression of the *Candida albicans* genes CHK1 and CSSK1, adherence and morphogenesis in a model of reconstituted human esophageal epithelial candidiasis. Infect Immun 70:1558–1565

Li D et al (2009) The *Candida albicans* histidine kinase Chk1p: signaling and cell wall mannan. Fungal Genet Biol 46(10):731–741

Mascher T et al (2006) Stimulus perception in bacterial signal-transducing histidine kinases. Microbiol Mol Biol Rev 70(4):910–938

Mavrianos J et al (2013) Mitochondrial two-component signaling systems in *Candida albicans*. Eukaryot Cell 12(6):913–922

Mavrianos J et al (2014) Two-component histidine phosphotransfer protein Ypd1 is not essential for viability in *Candida albicans*. Eukaryot Cell 13(4):452–460

Menon V et al (2006) Functional studies of the Ssk1p response regulator protein of *Candida albicans* as determined by phenotypic analysis of receiver domain point mutants. Mol Microbiol 62(4):997–1013

Menon V et al (2008) Transcriptional profiling of the *Candida albicans* Ssk1p receiver domain point mutants and their virulence. FEMS Yeast Res 8(5):756–763

Nagahashi S et al (1998) Isolation of CaSLN1 and CaNIK1, the genes for osmosensing histidine kinase homologues, from the pathogenic fungus, *Candida albicans*. Microbiology 144:425–432

Nemecek J et al (2006) Global control of dimorphism and virulence in fungi. Science 312:583–588

Nierman WC et al (2005) Genome sequencing of the pathogenic/allergenic filamentous fungus Aspergillus fumigatus. Nature 438:1151–1156

Nishino K, Yamaguchi A (2002) EvgA of the two-component signal transduction system modulates production of the yhiUV multidrug transporter in Escherichia coli. J Bacteriol 184 (8):2319–2323

Odami T et al (1997) Mannosyl phosphate transfer to cell wall mannan is regulated by the transcriptional level of the MNN4 gene in Saccharomyces cerevisiae. FEBS Lett 420:186–190

Park SG et al (2000) Distinct physiological functions of thiol peroxidase isoenzymes in Saccharomyces cerevisiae. J Biol Chem 275:5723–5732

Pfaller MA (2004) Anidulafungin: an echinocandin antifungal. Expert Opin Investig Drugs 13:1183–1197

Pfaller MA, Diekema DJ (2007) Epidemiology of invasive candidiasis: a persistent public health problem. Clin Microbiol Rev 20(1):133–163

Pfaller MA et al (2012) Frequency of decreased susceptibility and resistance to echinocandins among fluconazole-resistant bloodstream isolates of Candida glabrata. J Clin Microbiol 50 (4):1199–1203

Posas F et al (1996) Yeast HOG1 MAP kinase cascade is regulated by a multistep phosphorelay mechanism in the SLN1-YPD1-SSK1 "two-component" osmosensor. Cell 86(6):865–875

Puthiyaveetil S et al (2008) The ancestral symbiont sensor kinase CSK links photosynthesis with gene expression in chloroplasts. Proc Natl Acad Sci USA 105:10061–10066

Singh P et al (2004) SKN7 of *Candida albicans*: mutant construction and phenotype analysis. Infect Immun 72:2390–2394

Srikantha T et al (1998) The two-component hybrid kinase regulator CaNIK1 of *Candida albicans*. Microbiology 144:2715–2729

Stover BH et al (2001) Nosocomial infection rates in US children's hospital's neonatal and pediatric intensive care units. Am J Infect Control 29(3):152–157

Torosantucci AP et al (2002) Deletion of the two-component histidine kinase gene (CHK1) of *Candida albicans* contributes to enhanced growth inhibition and killing by human neutrophils in vitro. Infect Immun 70:985–987

Vehreschild J, Cornely O (2006) Micafungin sodium, the second of the echinocandin class of antifungals: theory and practice. Future Microbiol. 1:161–170

Wenzel RP (1995) Nosocomial candidemia: risk factors and attributable mortality. Clin Infect Dis 20:153–154

Wilson LS et al (2002) The direct cost and incidence of systemic fungal infections. Value Health 5 (1):26–34

Wisplinghoff H et al (2003) Nosocomial blood stream infections in pediatric patients in US hospitals: epidemiology, clinical features, and susceptibilities. Pediatric Infect Dis J 22 (8):686–691

Wisplinghoff H et al (2004) Nosocomial bloodstream infections in US hospitals: analysis of 24,179 cases from a prospective nationwide surveillance study. Clin Infect Dis 39:309–317

Yamada-Okabe T et al (1999) Roles of three histidine kinase genes in hyphal development and virulence of the pathogenic fungus, *Candida albicans*. J Bacteriol 181:7243–7247

Chapter 15
Mechanisms of Drug Resistance in *Candida albicans*

Dominique Sanglard

Abstract Antifungal drug resistance in *Candida albicans* started to emerge with the introduction of antifungal agents in the treatment of diseases caused in human by this fungal pathogen. The decreased activity of these drugs in *C. albicans* as a result of resistance has been observed in different scales for every currently used drug classes. The occurrence of resistance, generally acquired by genetic alterations, can be considered as a natural defense mechanism of *C. albicans* against drugs introduced by humans. Several other mechanisms contribute to drug efficacy in *C. albicans*, one of which is being attributed to tolerance. Antifungal tolerance is rather an adaptive mechanism not necessarily requiring genome changes. Even though the incidence of antifungal resistance in *C. albicans* is still generally low, clinical samples have given the opportunity to dissect resistance mechanisms. Resistance mechanisms have resulted in a significant gain of knowledge in diverse cellular functions and also genome organization. This knowledge can now be implemented in other fields including clinical diagnostics and design of novel drugs or novel therapies. In this chapter, I will not only review the general principles mediating antifungal drug resistance and latest developments, but also update on the mechanisms mediating antifungal tolerance. Moreover, I will illustrate how the understanding of resistance mechanisms can help to establish novel approaches in detecting drug resistance and designing alternative treatment strategies.

15.1 Introduction

Candida albicans is among the most diagnosed fungal species in clinical samples. This fungal species lives as a commensal in the microbiome of healthy individuals, but can propagate as a pathogen in immunocompromised patients (Pfaller and

D. Sanglard (✉)
Institute of Microbiology, University of Lausanne and University Hospital Center, Rue du Bugnon 48, CH-1011 Lausanne, Switzerland
e-mail: Dominique.Sanglard@chuv.ch

R. Prasad (ed.), *Candida albicans: Cellular and Molecular Biology*,
DOI 10.1007/978-3-319-50409-4_15

Diekema 2007). *C. albicans* infections are treated with a few antifungal agents, including principally polyenes, azoles, and echinocandins. Each of these agents possesses specific cellular targets and is declined in several formulations with different derivatives (Sanglard and Odds 2002). In general, all these different compounds are active against *C. albicans*, however later in this book chapter exceptions will be extensively discussed.

Amphotericin B binds to ergosterol in the fungal cell membrane and acts as a sterol "sponge". By depriving rapidly ergosterol from membrane, this leads to loss of membrane integrity and ultimately cell death (Anderson et al. 2014). Amphotericin B is formulated as a deoxycholate salt but also as liposomal and lipid complexes. While the first formulation is associated with toxicity issues, the liposomal and lipid complex formulations exhibit more favorable outcomes (Lemke et al. 2005).

Azoles belong to the class of sterol biosynthesis inhibitors and target specifically a lanosterol 14α-demethylase in fungi. This class of antifungal agent contains a large spectrum of derivatives. In human disease, triazoles are mostly used including fluconazole, itraconazole, voriconazole, posaconazole, and most recently, isavuconazole. Fluconazole, due its versatility in formulation designs and low price, still remains a major option for the treatment of *C. albicans* diseases. The other different azole derivatives see their use mostly implemented in fungal pathogens in which fluconazole has no or limited activity (Mukherjee et al. 2005).

Echinocandins belong to a class of semi-synthetic compounds targeting cell wall biogenesis and more specifically β-1,3 glucan synthase in *C. albicans* and other fungal pathogens. This class of antifungal agents comprises three major derivatives with related chemical structures including caspofungin, micafungin, and anidulafungin. These echinocandins are generally active against *C. albicans* isolates but at different drug concentrations they are used in intravenous formulations (Perlin 2011).

C. albicans infections treated with the above-mentioned agents respond with average outcomes in patients. Since 40–60% of patient with invasive candidiasis do not survive, this suggests that antifungal treatments are still far from being optimal. There are several reasons behind these mortality rates, including non-adapted treatments, the complexity of patient status, and the existence of several microbial diseases in the same patient (Lortholary et al. 2014). One other factor is the lack of drug response, which itself is not only dependent on clinical factors such as drug pharmacokinetics and pharmacodynamics (PK/PD) but also on the development of antifungal resistance in *C. albicans* (Lepak and Andes 2014).

Below I will give an overview of the different antifungal drug resistance mechanisms prevailing in *C. albicans* and also summarize current trends in this research topic.

15.2 Antifungal Activity and Antifungal Resistance

Before describing resistance mechanisms, it is useful to first define what is understood about the activity of antifungals in fungal pathogens. Antifungal activity is defined by a measure, which is the reduction of growth in vitro as compared to drug-untreated cells. Liquid media in microtiter formats are usually used, although solid surface agars with drug gradients (E-tests) can also be used (Pfaller and Diekema 2012). Antifungal susceptibility is measured by two major protocols from antifungal susceptibility testing subcommittees (CLSI: Clinical Laboratory Standards Institute; EUCAST: European Committee on Antimicrobial Susceptibility Testing). When running these protocols, a so-called minimum inhibition concentration (MIC) values (given in µg/ml) is obtained that reflects antifungal activity. The agreement between the two methods for antifungal activities is now generally high (Table 15.1) (Pfaller et al. 2014).

When testing *C. albicans* isolates collections for susceptibility to a single agent (for example fluconazole), the resulting MICs are distributed along a Gaussian bell-shaped curve. These analyses are helpful in order to discriminate between wild-type and non-wild-type isolates by epidemiological cutoff (ECOFF) values (Delarze and Sanglard 2015). The ECOFF value is defined as the upper limit of the wild-type population and will, in general, encompass about 95–99% of a given population for a specific agent. The ECOFF is useful in the sense that it can identify non-wild-type isolates that may exhibit specific antifungal resistance mechanisms (Arendrup et al. 2013). The ECOFF value therefore sets a threshold above which in vitro resistance is likely to occur.

Ideally, in vitro resistance (or microbiological resistance) may predict in vivo resistance (or clinical resistance), however in reality this assumption is far from being a straight relationship. Several clinical studies enrolling patient records have established clinical breakpoints (CBP) for specific agents using clinical parameters including in vivo drug pharmacokinetics, resistance mechanisms, and clinical response (Alastruey-Izquierdo and Cuenca-Estrella 2012). With MICs above CBPs, the success of therapy with a given agent is limited or not achievable. The CBPs for fluconazole and *C. albicans* are 2 and 4 µg/ml by EUCAST and CLSI, respectively. Table 15.1 gives an overview of current CBPs for *C. albicans* and currently available antifungal agents.

Antifungal resistance can be acquired in vitro by drug exposure or during therapy. Antifungal resistance can be measured by elevated MICs as compared to those of a wild-type population and is generally measured above the ECOFF values. Acquired antifungal resistance has been reported virtually for all existing antifungal agents and major fungal pathogens (Perlin et al. 2015).

Table 15.1 ECOFF and CBP of different antifungal agents and fungal species[a]

		ECOFF (μg/ml)[b]			CBP (μg/ml)[c]					
Species	Method	Fluconazole	Anidulafungin	Micafungin	Fluconazole		Anidulafungin		Micafungin	
					S[d]	R[d]	S	R	S	R
C. albicans	CLSI	0.5	≤0.12	≤0.03	2	4	0.25	0.5	0.25	0.5
	EUCAST	1	0.03	0.015	2	4	0.03	0.03	0.016	0.016

[a]Data obtained from published studies (EUCAST-AFST 2008; Pfaller et al. 2010; Arendrup et al. 2013, 2014; Pfaller et al. 2014; Maubon et al. 2014)
[b]ECOFF: epidemiological cutoff
[c]CBP: clinical breakpoint
[d]Categorical discrimination between resistant (R) and susceptible (S)

15.3 Epidemiology of Antifungal Resistance

The frequency at which antifungal resistance occurs in hospitalized patients varies from one study to the other. In order to assess the incidence of resistance in a given fungal population, one needs data collected by sentinel- and population-based surveillance programs that are launched on nationwide scale.

One important issue to consider in the estimation of rates of resistance is with which CPBs these rates are calculated as recently discussed (Fothergill et al. 2014). For example, the CBP for fluconazole in *C. albicans* recommended by the CLSI and EUCAST committees used to be very divergent, but now tends to be more harmonized (Table 15.1). Given these issues, the comparisons between old and more recent epidemiological studies are not an easy task (Fothergill et al. 2014). As far as *C. albicans* is concerned, antifungal resistance rates in *C. albicans* are generally low. In a study from two different US areas between 2008 and 2011, resistance to fluconazole (CBP: $\geq$ 64 µg/ml) or echinocandins (CBP: $\geq$ 4 µg/ml) ranged between 1 and 2% in bloodstream isolates (Cleveland et al. 2012). These values are different for *Candida* spp. such as *Candida glabrata*. According to the ARTEMIS Antifungal Surveillance Program, *C. glabrata* increased as a cause of invasive candidiasis from 18% of all BSI isolates in 1992–2001 to 25% in 2001–2007. Fluconazole resistance rates in *C. glabrata* increased over the same period from 9 to 14% (CBP: $\geq$ 64 µg/ml) (Pfaller et al. 2003, 2009). Resistance of *C. glabrata* to the class of echinocandins is now reaching significant proportions. It was reported that, within a 10-year survey (2001–2010) in an US hospital (Duke University Hospital), echinocandin resistance rate increased from 4.9 to 12.3% (Alexander et al. 2013). Similar trends are reported in Europe, although resistance rates range between 1 and 4% (Arendrup and Perlin 2014).

15.4 Mechanisms of Antifungal Resistance

The occurrence of resistant isolates from different origins has resulted in a vast number of studies with the aim to understand antifungal resistance at the molecular level. Based on these studies, general principles were deduced and can be split into three major categories including (1) decrease of effective drug concentration within the cell, (2) drug–target alterations, and (3) metabolic bypasses. In the following, the major features for each of these principles will be summarized according to the most current knowledge (Fig. 15.1).

15.4.1 Decreasing the Effective Drug Concentrations Within Cells Is a Classical Way to Achieve Resistance to Antifungal Drugs

There are several distinct mechanisms resulting in this effect.

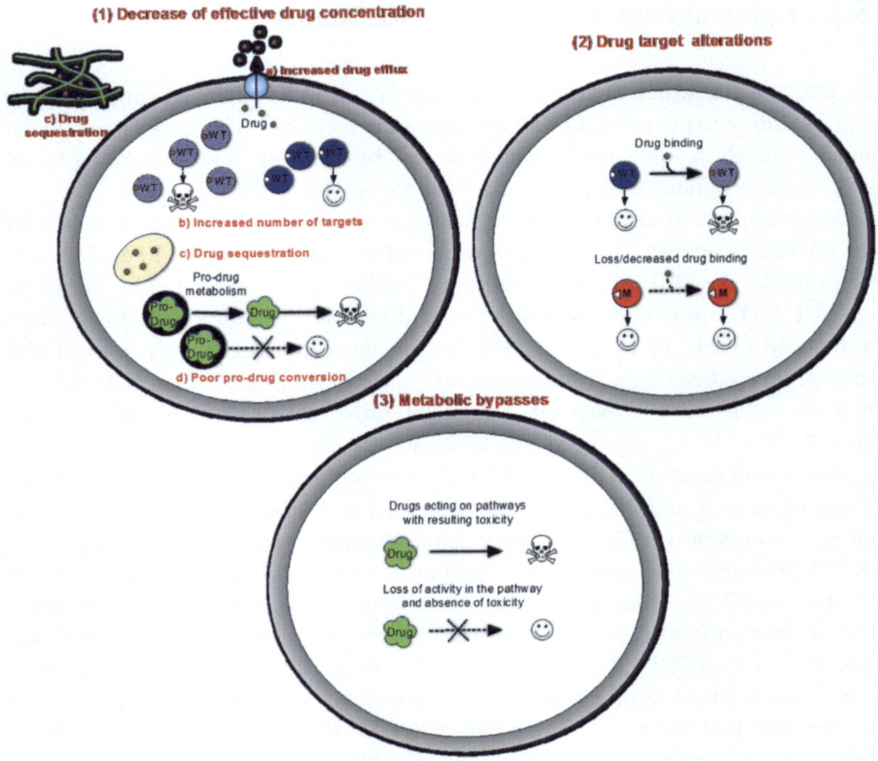

Fig. 15.1 Three basic resistance mechanisms to antifungal drugs. These mechanisms include: *1* Decrease of effective drug concentration with specific mechanisms including increased drug efflux, increased number of targets, drug sequestration of extracellular and intracellular origins and poor pro-drug conversion; *2* Drug–target alterations, *3* metabolic bypasses. Wild-type proteins are represented by blue circles catalyzing cellular functions; Blue-shaded circles represent proteins blocked by drugs in which cellular functions are inhibited, thus causing poor growth or death. Mutant proteins are represented by red circles. Drugs are represented with different symbols. Symbols: *WT*: wild type; *M*: mutant

15.4.1.1 The Drug Intracellular Concentrations Can Be Decreased by Active Efflux

Drug efflux is intrinsically existing in many different types of microbial cells. In wild-type cells, efflux systems are functioning with a low residual activity. Drug resistance can result from their enhanced activity. Several efflux transport systems exist in *C. albicans* including ATP-binding cassette (ABC) transporters and transporters of the Major Facilitator Superfamily (MFS). *C. albicans* is predicted to contain 28 ABC proteins and 96 potential MFS transporters (Paulsen et al. 1998; Gaur et al. 2008; Prasad and Goffeau 2012). ABC transporters are arranged in different subfamilies, however, they all contain membrane spanning domains and use ATP hydrolysis for drug transport. MFS transporters are transmembrane proteins which use the electrochemical proton-motive force to mediate drug efflux. MFS are involved in multidrug resistance (MDR; MFS-MDR transporters) function as proton antiporters and are classified into two groups: the drug: H+ antiporter-1 DHA1 family, and the drug: H+ antiporter-2 DHA2 family (Gaur et al. 2008; Costa et al. 2014). Fungal ABC transporters have been arranged into several classes, however only ABC transporters of the PDR (Pleiotropic Drug Resistance) class contained those relevant for antifungal drug resistance. In *C. albicans*, the PDR class comprises the major ABC transporters involved in azole resistance such as *CDR1* (for Candida Drug Resistance) and *CDR2*, but other transporters have not yet been involved in antifungal resistance (*CDR3*, *CDR4*, *CDR11*, *SNQ2*). The upregulation of both *CDR1* and *CDR2* mediates azole resistance by enhanced drug efflux and reduces azole accumulation in some *C. albicans* clinical strains (Sanglard et al. 1995, 2009).

MFS involved in the development of azole resistance in clinical isolates is restricted to *MDR1* from *C. albicans*. *MDR1* is upregulated in specific strains which results in enhanced azole efflux (Sanglard et al. 1995; Lamping et al. 2007). *FLU1* (for Fluconazole resistance) from *C. albicans* is another MFS, and its heterologous expression in *S. cerevisiae* revealed that it served as a fluconazole efflux transporter (Calabrese et al. 2000). Until now, however, no studies have shown the participation of *FLU1* in azole resistance in clinical isolates. *FLU1* was identified as a transporter for histatin 5, which is a human antimicrobial peptide (Li et al. 2013).

Upregulation of ABC- and MFS transporters is mediated by specific regulators in resistant fungal pathogens. In *C. albicans, CDR1* and *CDR2* are known to be regulated by a zinc cluster finger transcriptional regulator called *TAC1* and *MDR1* by another regulator called *MRR1* (Coste et al. 2004; Morschhäuser et al. 2007). Mutations (gain-of-function or GOF mutations) in these regulators have been described and they confer a hyperactivation state which does not require additional stimulation, thus explaining the inherent high expression levels of the transporters in drug-resistant isolates (Coste et al. 2006; Dunkel et al. 2008a). Other transcriptional regulators of drug transporters relevant to azole resistance such as *PDR1* have been described in *C. glabrata* (Vermitsky et al. 2006; Ferrari et al. 2009).

While drug efflux may contribute to antifungal resistance, drug import may also result in decreased drug concentrations. It has been suggested that azoles could be imported by facilitated diffusion and the putative protein(s) involved in this process have not been yet identified (Mansfield et al. 2010). Membrane permeability may also be altered in *C. albicans*, which can be obtained, for example, in mutants of the sterol biosynthesis pathway. It has been reported that deficiencies in *ERG2* to *ERG6* may contribute to membrane fluidity alterations followed by changes in drug uptake (Mukhopadhyay et al. 2002).

Less-efficient drug import may also contribute to resistance to fluorocytosine (5-FC) that is currently not a drug of choice for the treatment of *C. albicans* infections. Loss of functions of cytosine permeases from clinical isolate is not yet reported in *C. albicans*, but more frequent in other *Candida* spp. (Edlind and Katiyar 2010).

15.4.1.2 The Drug Target Can Be Overexpressed

When the number of drug targets is increased within cells, the effective drug concentration needs to be also increased to bind all target molecules. Ideally, at an equimolar ratio between target and drug, microbial growth will be inhibited. When this ratio is imbalanced at the expense of the drug, drug-free targets are still functional and thus higher drug concentrations are required for growth inhibition, all of which results in drug resistance. For example, ERG11 upregulation has been associated with azole resistance in *C. albicans*. This transcriptional regulation is mediated by a zinc cluster finger transcription factor called *UPC2*. As in the case of other drug resistance transcriptional regulators, GOF mutations in *UPC2* have been described and result in upregulation of various genes, among which is ERG11 (Dunkel et al. 2008b).

15.4.1.3 The Drug Is Sequestered in Extra- or Intracellular Compartments

Fungal pathogens have the ability to sequester drugs within extracellular compartments. Several fungal pathogens including *C. albicans* are able to form biofilms in specific growth conditions (Ramage et al. 2005). Biofilms are multicellular structures in which cells form a dense network that is covered by a so-called matrix. The matrix is composed of different elements in *C. albicans* biofilms, including several cell wall polymers (Mitchell et al. 2015). Biofilm formation is known to be associated to resistance to several drugs such as polyenes and pyrimidine analogues (Desai et al. 2014). Interestingly, recent data showed that the matrix participates to this process by its capacity to sequester antifungal agents. This process has been clearly documented for fluconazole (Nett et al. 2007; Bonhomme and d'Enfert 2013; Mitchell et al. 2015) and was suggested for amphotericin B in *C. albicans* (Vediyappan et al. 2010).

Much less is known in drug sequestration in intracellular compartments. A single report documents the accumulation of fluconazole in *C. albicans* within organelles that were described as vesicular vacuoles. Whether or not this type of mechanism could occur in other isolates remains unknown (Maebashi et al. 2002).

15.4.1.4 A Pro-drug Is Poorly Converted to an Active Drug

Poor drug metabolization as a principle of antifungal resistance is also observed when 5-fluorocytosine (5-FC) resistance occurs. 5-FC is a pro-drug which is metabolized by cells into fluorinated pyrimidine analogues, thus inhibiting nucleic acid and protein biosynthesis. After import into cells, cytosine deaminase converts 5-FC into 5-FU. 5-FU is a toxic metabolite and therefore the deficiency of this step diminishes further processing and toxicity of the drug. Mutations in cytosine deaminase in *C. albicans* (*FCA1*) (Hope et al. 2004) and *C. glabrata* (*FCY1*) have been reported to result in 5-FC resistance (Edlind and Katiyar 2010; Vandeputte et al. 2011).

15.4.2 Altering Drug Targets

Drug–target alterations have been reported for at least two classes of antifungal agents, i.e., azoles and echinocandins. The targets of these two drugs are a 14α-lanosterol demethylase and a β-1,3 glucan synthase, respectively. Lanosterol demethylase is encoded by ERG11 in *C. albicans*. Mutations in ERG11 resulting in non-synonymous amino acid substitutions that are present in azole-resistant *C. albicans* isolates are numerous and were shown to decrease the affinity of the target to azoles (Lamb et al. 2000). The effects of ERG11 mutations have different outcomes on azole MICs that depend on structural features of azole drugs. While most known mutations decrease affinity to fluconazole, they have only a moderate effect on posaconazole affinity (Sanglard and Coste 2015). In many cases, simultaneous ERG11 mutations can be present on the same ERG11 allele and be accompanied by drug transport modifications, thus resulting in azole-resistant isolates with high MIC values against azoles (for example, fluconazole MIC > 128 μg/ml) (Morio et al. 2010). The 3D structure of Erg11 has been resolved recently in its native membrane-bound state (Monk et al. 2014). The structure was obtained from the *Saccharomyces cerevisiae* Erg11, a close homolog of the *C. albicans* Erg11. This study constitutes a milestone in the further understanding of drug–target interactions. The authors could map with high-resolution substrate and azole bound to Erg11 and thus identify critical amino acid residues participating to efficient azole-Erg11 complexes. Figure 15.2 illustrates a slice cut in the 3D structure of Erg11 in the presence of itraconazole obtained from published crystal structures (Monk et al. 2014). The position of itraconazole along the substrate channel can be recognized. The imidazole ring of itraconazole is placed directly on the heme planar structure in order to block monooxygenation reactions. Such detailed structural elements may predict the binding of alternative novel azole drugs.

Decreased affinity to the target is also known for echinocandins. Beta-1,3 glucan synthases are encoded by *FKS* genes in different fungal species. Up to now, echinocandin resistance has been attributed to specific mutations leading to amino acid substitutions in two different regions of these genes (Hot spot 1 and 2, or HS1 and HS2). *FKS1* mutations have been reported in these two regions (HS1: region 640–650; HS2: 1345–1365) in clinical isolates of *C. albicans* isolated from refractory therapies (Perlin 2015). Equivalent mutations in the HS1 of *FKS2* (an homologue to *FKS1*) of *C. glabrata* and *FKS1* of *C. lusitaniae* (Desnos-Ollivier et al. 2011; Asner et al. 2015), C. *tropicalis*, and *C. krusei* (Desnos-Ollivier et al. 2008) have been reported. The effect of *FKS1* mutations was shown to decrease by 500–1000-fold, the echinocandin concentrations need to inhibit by 50% of the

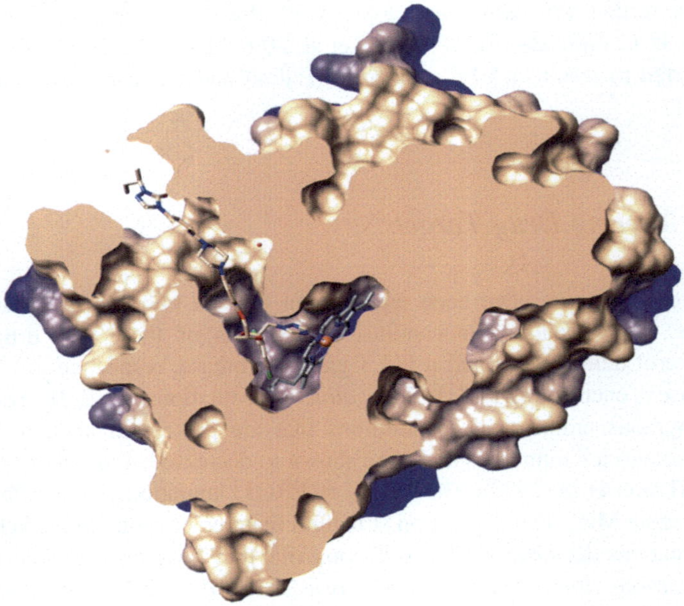

Fig. 15.2 Slice view of the *S. cerevisiae* Erg11 structure bound to itraconazole. The view was obtained with the structure coordinates obtained from 4KF0.pdb (Monk et al. 2014). The slice view was imaged with the UCSF Chimera software

activity of the enzyme (IC_{50}) (Perlin 2011). Depending on the site of the mutation and the type of substitution, the IC_{50} values can be quite different. For example, the substitution S645Y results generally in 4–5 higher IC_{50} values as compared to S645P. This trend is shared by the three major echinocandins (Perlin 2011).

15.4.3 Occurrence of Metabolic Bypasses

Metabolic bypasses occur in fungal cells when given metabolic pathways are perturbed by loss or strong decrease of specific functions. Metabolic bypasses can be compared to compensatory mechanisms in which cells circumvent drug toxic effects exerted by some antifungal agents. For example, it is known that resistance to azoles can be mediated by loss of function mutations in the gene *ERG3* that encodes a sterol $\Delta^{5,6}$ desaturase. If this enzyme remains active, it converts 14α-methylated sterols arising from azole exposure into a toxic 3,6-diol derivative (Kelly et al. 1995). Therefore, fungi that are unable to produce this metabolite acquire azole resistance. Several studies have reported *ERG3* loss of function mutants with corresponding acquisition of azole resistance (Chau et al. 2005; Martel et al. 2010b; Morio et al. 2012; Eddouzi et al. 2013). These mutants exhibit deficiencies in ergosterol biosynthesis and are less competitive than wild-type isolates in conditions encountered in the host. Recently, it was suggested that loss of ergosterol biosynthesis will have a negative impact on the switch between the yeast and hyphal morphology and thus could potentially contribute to decreased fitness in vivo (O'Meara et al. 2015). As a result of loss of function of *ERG3*, ergosterol is absent from cell membranes. This way, the mutants also escape the toxic effect of amphotericin B, which normally acts as a "sponge" for ergosterol to rapidly destabilize membrane functions (Anderson et al. 2014). Several other mutations in the ergosterol biosynthetic pathway (*ERG6*, *ERG24*, *ERG2*) lead to the same effect and have also a compensatory effect (Jensen-Pergakes et al. 1998; Jia et al. 2002; Vincent et al. 2013).

Another illustration of metabolic bypass as mediator of drug resistance is a mutation in the gene *FUR1* encoding uracil phosphoribosyl-transferase. The mutation (amino acid substitution R101C) results in a decrease in the conversion from 5-FC deamination of 5-fluorouridine (5-FU) into a toxic metabolite (5-fluorouridine monophosphate). Due to decreased 5-fluorouridine monophosphate production, the toxicity of 5-FC is decreased, thus resulting in 5-FC resistance (Dodgson et al. 2004).

15.5 Multidrug Resistance (MDR)

Antifungal resistance has been observed in most occasions as a process involving resistance to single class of drugs. There are cases when a single resistance mechanism results in cross-resistance to several agents simultaneously. MDR is the simultaneous resistance to at least two different classes of antifungal agents. It is known that the expression of ABC transporters (i.e., *CDR1* or *CDR2*) mediate cross-resistance not only to all azoles used in medicine (Sanglard and Coste 2015), but also can support resistance to other less common agents from other antifungal classes such as terbinafine (Kohli et al. 2002). In the recent years, several other reports documenting cases of MDR in fungal pathogens have been published. In clinical settings, it is not unusual to treat fungal infections with different antifungal agent classes including azoles, polyenes, and echinocandins. MDR between azoles and polyenes has been reported in *C. albicans* by the loss of function mutations in *ERG3* (Sanglard et al. 2003b; Martel et al. 2010a; Vale-Silva et al. 2012; Morio et al. 2012). Other *ERG* gene defects may also confer MDR to both drug classes, such as the loss of function mutation in *ERG2* observed in *C. albicans* (Vincent et al. 2013). Some specific isolates may show simultaneous mutations in several genes as in *C. albicans* by ERG11/*ERG5* mutations (Martel et al. 2010a).

Echinocandins are being increasingly used for the therapy of fungal infections, especially those caused by *Candida* spp. Resistance to echinocandins logically appeared soon after its introduction in medicine in 2005 (Park et al. 2005). A first report of MDR to caspofungin and azoles in *C. glabrata* isolated from blood cultures was made in 2010 after caspofungin therapy (Chapeland-Leclerc et al. 2010). Resistance mechanisms were combining mutations in the β-1,3 glucan synthase *FKS2* (S663P) and ABC transporters upregulation. Closely related isolates became resistant to 5-FC after therapy with this drug, however it was still susceptible to the two other drugs. These isolates exhibited a non-synonymous mutation (G190D) in *FUR1*, which probably accounted for decreased 5-FC toxicity.

A recent case illustrated the evolution of MDR in *C. albicans* sequential isolates taken from a patient at different sites (oropharynx, esophagus, feces, colon) treated over time (Jensen et al. 2015). The isolates were related to each other as confirmed by genotyping methods. The evolution of drug resistance followed the course of drug treatments. Fluconazole treatment induced first a GOF mutation in *TAC1* with corresponding azole resistance (MIC fluconazole > 16 μg/ml). Caspo- and anidulafungin treatment resulted in resistance (MIC caspofungin > 32 μg/ml) with a corresponding *FKS1* mutation (S645P). Lastly, amphotericin B treatment established polyene resistance (MIC > 32 μg/ml) with a loss of function mutation in *ERG2*. All three mutations were conserved in the final MDR strain (Jensen et al. 2015). MDR evolution took place within a time lapse of 5 years.

15.6 Genome Adaptation and Antifungal Resistance

C. albicans naturally occurs as a diploid fungal species. Mutations that are associated with antifungal resistance in *C. albicans* first occur in the heterozygous state but are mostly identified as homozygous mutations in clinical isolates with high antifungal MIC values. The transition from the heretozygous state to homozygocity is important, since it contributes to increase resistance. In *C. albicans*, several studies have investigated this mechanism at the level of single chromosomes or of whole genomes. For example, Coste et al. (2006) investigated loss of heterozygocity (LOH) in the *TAC1* locus (a gene critical for the regulation of *CDR1* and *CDR2*, see above) in the development of azole resistance in *C. albicans*. *TAC1* is situated in chromosome 5 left arm and in close proximity to the mating locus (*MTL*). *TAC1* GOF mutations that are acquired upon azole exposure are converted to the homozygous state. These events were the result of loss of one sister chromosome and reduplication, thus resulting in whole chromosome homozygocity. However, *TAC1* LOH could cover only partial chromosome lengths up to the end of chromosome 5 left arm. It is likely that these LOH events were resulting from break-induced replications or single crossover recombination events. Interestingly, ERG11 is also situated on the extremity of chromosome 5 left arm and thus ERG11 mutations can be rendered homozygotes by similar LOH events than for *TAC1* (Coste et al. 2006). Recently, genome-wide analysis of seventeen *C. albicans* sequential isolates having acquired resistance to azoles upon a long time lapse (approximatively 2 years) was carried out (Ford et al. 2015). The authors were able to identify specific regions undergoing LOH and aneupoidies, some of which contain genes relevant for azole resistance development. Upon single nucleotide polymorphisms (SNP) analysis, the latest isolates accumulated about 4600 SNPs as compared to the initial isolate, thus highlighting that a strong selective pressure was taking place under host conditions.

A remarkable genome alteration that may occur in azole-resistant *C. albicans* isolate is the formation of small isochromosomes by segmental aneuploidy. These isochromosomes were found as duplication of two chromosome 5 left arms (5iL) connected by a centromere. Interestingly, 5iL contains two genes (ERG11 and *TAC1*) important for azole resistance. Thus, 5iL formation contributes to increase azole resistance by increasing the copy number of these genes (Selmecki et al. 2006). These studies highlighted that antifungal drugs have profound effect in the maintenance of genome integrity. In addition to these long-term effects, it has been reported that short time drug exposure can also perturb chromosome stability and distribution of nuclei within cells during DNA replication. These additional events may contribute as well to establish the diverse chromosomal rearrangements and aneuploidies of azole-resistant clinical isolates (Harrison et al. 2014).

15.7 Antifungal Drug Tolerance

There are different parameters for characterizing the activity of a given drug. While antifungal activity can be defined with MICs as a possible parameter, one often neglected other parameter is the ability of fungal pathogens to tolerate the presence of a drug. In the field of antimicrobial agents, drug tolerance is used to describe the ability, whether inherited or not, of microorganisms to survive transient exposure to high concentrations of an antibiotic without a change in the MIC (Brauner et al. 2016). Antibiotic tolerance is reversible and thus is rather the result of epigenetics as opposed to drug resistance which depends on genes and their mutations. Drug tolerance can result in the presence of so-called persisters, which are a subpopulation of clonal cells able to survive exposure to an antibiotic at supra MIC concentrations. Persistence is observed when most of the bacterial population is killed, while a subpopulation persists for longer period of time, despite the cell population being clonal (Gefen and Balaban 2009).

Drug tolerance is a concept that was proposed with antibacterial drugs possessing a cidal effect. Tolerant persisters can therefore be simply distinguished by their ability to withstand antibiotic killing above the MIC. The presence of persisters in bacterial biofilms and indeed the presence of persister cells in antibiotic treatments might be important in many recalcitrant infectious diseases (Lewis 2006). Recent data show that persister bacterial cells are favored by nutrient starvation and mechanisms that induce decrease in membrane potential, thus resulting in a dormant stage (Verstraeten et al. 2015).

In agreement with bacterial definitions, antifungal tolerance could be also defined as the ability of cells to survive at drug concentrations exceeding the MIC. Accordingly, antifungal tolerance is different from antifungal resistance, which itself reflects an increase of MIC value independent of the ability to survive at drug concentrations higher than this value. One difficulty is that antifungal drug classes contain either fungistatic (azoles) or fungicidal drugs (echinocandins, amphotericin B) and the notion of survival as defined for antibacterial agents needs to be adjusted for antifungal agents. For fungicidal agents such as echinocandins and polyenes, tolerance defined for antibacterials can be used. For example, it is known that echinocandins fragilize the fungal cell wall and induce stresses against which fungal cells respond with adaptive protective mechanisms (Walker et al. 2010). These mechanisms create a subpopulation of drug-tolerant persister cells. With regards to echinocandins, the reversible adaptive mechanisms are mediated by signaling cascades of the cell wall integrity pathway involving protein kinase C (PKC), HOG (High Osmolarity Glycerol) signaling and calcium-dependent calcineurin (Walker et al. 2010).

The concept of drug tolerance in the sense of resisting antibacterials killing for is less well adapted for fungistatic agents such as azoles. Basically, fungal pathogens such as *C. albicans* can be considered as generally tolerant to azoles, since these cells are not killed by drug concentrations above individual MICs. Rather, cells can grow at different degrees at drug concentrations higher than the MIC value. Azole tolerance reflects a phenomenon that is well known as the "trailing growth" in MIC susceptibility tests with azoles (Marr et al. 1999; Marchetti et al. 2000). Trailing growth can be more or less pronounced and mirror the degree of azole tolerance. Figure 15.3 illustrates the different degrees of tolerance to fluconazole in several *C. albicans* isolates. Isolate 3 shows the highest degree of azole tolerance with a MIC at 0.5 μg/ml. Isolate 1 and 2 exhibit lower degrees of azole tolerance as compared to isolate 3 and have both fluconazole MIC values between 0.0625 and 0.125 μg/ml. Isolate 4 is fluconazole-resistant (MIC > 128 μg/ml).

C. albicans biofilms are known to be resistant to several antifungal agents. Contrary to resistance in planktonic cells, which can acquire resistance by genetic modifications, resistance of *C. albicans* biofilm cells is only a transient state. The question is whether or not biofilms can be considered as resistant or tolerant to antifungal agents. When taking amphotericin B as an example, MICs of biofilm cells are 16-fold higher than planktonic cells, thus implying that biofilms are

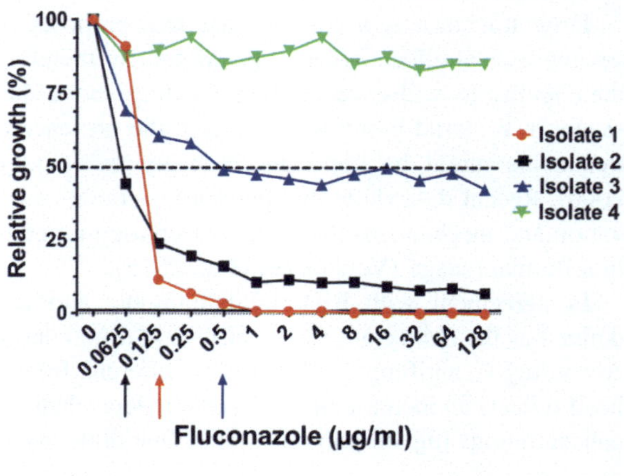

Fig. 15.3 MIC susceptibility tests with fluconazole illustrating growth profiles of several *C. albicans* isolates and exhibiting resistance or different levels of tolerance. *Colored arrows* indicate the MIC corresponding to each of the growth profiles. The relative growth threshold for MIC assignment was set at 50%

resistant to amphotericin B (Tobudic et al. 2012). Above this MIC value, biofilm cells can survive at higher amphotericin B concentrations (Li et al. 2015). Cells surviving in these conditions are tolerant to amphotericin B. Among tolerant cells, some subpopulations may be persister cells, thus indicating that persisters originate from drug tolerance mechanisms.

15.8 Drug Tolerance Mechanisms

Tolerance mechanisms to antifungal drugs have still received little attention up to now. One first approach to tackle tolerance mechanisms was to search for substances inhibiting tolerance. For example, it was reported that cyclosporine A (CsA), an inhibitor of calcineurin activity, converts fluconazole into a fungicidal drug in *C. albicans* (Marchetti et al. 2003; Sanglard et al. 2003a). Later, it was found that *C. albicans* cells lacking the gene *CNA/CMP1* gene encoding calcineurin subunit A were not surviving in the presence of fluconazole (Sanglard et al. 2003a). Importantly, calcineurin was critical for drug tolerance only and not drug resistance (Sanglard et al. 2003a). More recently, calcineurin has been declared as client protein of the chaperone Hsp90. Thus, inhibition of Hsp90 functions phenocopies a loss of calcineurin activity with respect to drug tolerance (Cowen 2013). Hsp90 has also multiple other chaperoning functions when fungal cells are exposed to various stresses. Therefore, Hsp90 is critical in defining trajectories not only of antifungal resistance but also of antifungal tolerance (Cowen and Lindquist 2005).

Another example of chemo-dependent azole tolerance is the combination between tetracycline and fluconazole in *C. albicans*. Tetracycline (and other derivatives, for example doxyclycine or minocycline) inhibits protein synthesis. Synergism of doxycycline (DOX) with fluconazole has also been reported when high DOX concentrations were used (>200 μg/ml), which also resulted in decreased tolerance of *C. albicans* to fluconazole (Fiori and van Dijck 2012). Iron availability was critical in this process and was associated to the iron-chelating ability of DOX. Thus, iron homeostasis is able to determine the level of antifungal tolerance (Fiori and van Dijck 2012).

Forward genetic approaches were also used to understand antifungal tolerance mechanisms by screening of yeast mutants in which azoles have a fungicidal effect. Such screens highlighted that the transcriptional activator *UPC2* regulating sterol homeostasis was important for maintaining viability of *C. albicans* in the presence of fluconazole (Vasicek et al. 2014). A more recent study reported that a *C. albicans* *vps21*Δ/Δ mutant blocked in membrane trafficking through the late endosomal prevacuolar compartment was able to behave like isolates with high azole trailing characteristics. *VPS21* might therefore mediate azole tolerance. The basic mechanisms behind this observation remain largely unexplained, however the authors speculated that sterol homeostasis may be altered upon azole exposure, thus creating a less toxic cellular environment for the fungus (Luna-Tapia et al. 2015).

In *C. glabrata*, the calcium channels genes *CCH1* and *MID1* were determining fluconazole tolerance, since absence of these genes converted this azole into a fungicidal agent (Kaur et al. 2004). These data suggest a connection between calcium homeostasis and azole tolerance, which is also consistent with the involvement of the calcium-dependent calcineurin activity in drug tolerance (see above).

Tolerance mechanisms to other antifungal classes different than azoles have still not been fully addressed. As above mentioned, the calcineurin/HSP90 complex is critical for tolerance of *C. albicans* to echinocandins. Compromising calcineurin and/or HSP90 activity increases to variable extent this fungicidal activity (Singh et al. 2009). One important issue in identifying novel mediators of tolerance is that many studies report antifungal susceptibility assays which cannot predict and measure drug tolerance. The undertaken assays are in many cases not measuring cell survival above the point of growth inhibition, which is mandatory to assess the degree of drug tolerance. Antifungal tolerance can be however correctly addressed by time-kill curves as illustrated in a published study (Miyazaki et al. 2010) in which *SLT2,* a protein kinase involved in cell wall integrity, was adjusting tolerance of *C. glabrata* to micafungin. Interestingly, *SLT2* is also involved in the response of *C. glabrata* to fluconazole, however it is not yet clear if this gene participates in fluconazole tolerance (Schwarzmüller et al. 2014). Echinocandins have been shown to stimulate this pathway in *C. albicans* (Reinoso-Martín et al. 2003).

15.9 Combatting Antifungal Resistance

Acquiring detailed knowledge in antifungal resistance mechanisms is useful for different purposes, one of them being to directly inhibit the mediators of resistance by specific molecules. As above mentioned, the major mechanism by which fungal pathogen escape inhibition by antifungal agents is by active efflux. Drug efflux in fungal pathogens is carried out by two basic transport systems including ATP-Binding Cassette (ABC) transporters and Major Facilitators (Sanglard et al. 2009). Targeting these efflux systems by inhibitors may enhance antifungal activity and/or decrease the development of resistance. Most of the studies aimed to identify such inhibitors used a combination of drugs with fluconazole and measured the decrease of drug resistance (Hayama et al. 2012; Holmes et al. 2012; Silva et al. 2013; Maurya et al. 2013). One of these studies exploited the Prestwick chemical library and identified compounds inhibiting drug efflux via Cdr1 and Cdr2. One efflux inhibitor, clorgyline, was a monoamine oxidase A selective inhibitor but is no longer in use as an antidepressant (Holmes et al. 2012). While other inhibitors included specific synthetic peptides (Niimi et al. 2012; Maurya et al. 2013), one interesting class of inhibitors included a class of macrocyclic lactones (milbemycins). Milbemycin derivatives (milbemycin oxims) are still widely used as antiparasitic agents in animals (Silva et al. 2013). Milbemycin oxims were used as drug efflux inhibitors not only in *C. albicans* but also in *C. glabrata*. Milbemycin oxims were also tested in animal models of infection with *C. albicans* and *C. glabrata*. Combination of flucoanzole with milbemycin oxims could reverse azole resistance with high efficacy, thus suggesting that these types of combination may have potential therapeutic utility (Silva et al. 2013).

The transcriptional complex that is required for ABC transporter regulation can itself be ratcheted by inhibitors. In a recent example, it was shown that the transcription factor CgPdr1 mediates upregulation of *CgCDR1* via the participation of the Mediator complex (Thakur et al. 2008). CgPdr1 interacts with a specific subunit of the Mediator (Gal11/Med15) and this interaction is necessary for *CgCDR1* upregulation (Thakur et al. 2008). In a recent study, a specific inhibitor of this interaction was identified (iKIX). This inhibitor blocked *CgCDR1* upregulation and therefore development of drug resistance. The inhibitor was found effective to reduce fungal burdens in a murine model of *C. glabrata* infection in combination with fluconazole (Nishikawa et al. 2016).

15.10 Resistance Mechanisms and Molecular Diagnostics

Accurate and early diagnostic of a fungal infection is known to be associated with favorable patient outcomes (Morrell et al. 2005). As far as antifungal resistance detection is concerned, the current methods implemented in diagnostic laboratories are mostly performed with the help of culture-dependent methods which are still time-consuming (Pfaller 2012). Having now acquired mutations in genes mediating drug resistance, DNA-based detection approaches of resistance mechanisms are feasible. DNA-based detection approaches use principally polymerase chain reactions (PCR) on nucleic acids (DNA) but can also use Nucleic Acid Sequence Based Amplification (NASBA)-based amplification–detection platforms that will rather target RNA (Arvanitis et al. 2014). In addition, real-time PCR using self-reporting fluorescent probes and molecular beacons allows the detection of single nucleotide polymorphisms associated with drug resistance (Perlin 2009). Molecular beacons are single-stranded oligonucleotide hybridization probes that form a stem-and-loop structure. This structure undergoes a structural change when bound to a target, which can be revealed by covalently linked fluorochromes. One additional advantage of molecular techniques using PCR-based approaches is that they can be multiplexed in order to detect multiple mutations in the same assay. Some examples of resistance mechanisms detection have been realized by the use of molecular beacon probes from reverse-transcribed mRNA on major drug resistance genes including ERG11, *MDR1*, *CDR1*, and *CDR2* (Park and Perlin 2005). At the same time, SNPs in ERG11 could be differentiated with molecular beacons. Clinical isolates displaying a range of azole susceptibilities were included in the analysis and fluconazole-resistant phenotype (MIC $\geq$ 64 μg/mL) was closely associated with overexpression of *CDR1*, *CDR2*, and *MDR1* along with four mutations in ERG11 (T229A, Y132F, S405F, G464S). Molecular beacons can be applied for the detection of echinocandin resistance in *C. albicans*, where *FKS1* mutations remain limited to regions of limited range (Balashov et al. 2006).

Several other methods such as of next-generation sequencing (NGS) have emerged over recent years due to the availability of benchtop platforms. This format makes these techniques suitable to clinical laboratories. In a recent study, 121 amplicons were designed to amplify known antifungal drug resistance genes and their mutations (Garnaud et al. 2015). Due to multiplexing capacities of the NGS technology, it was possible to include in single deep sequencing several individual *Candida* spp, some of which contain known resistance mechanisms. The expected mutations were recovered with high accuracy (Garnaud et al. 2015). Clinical samples with unknown resistance mechanisms were next assayed with the NGS approach and specific amplicons. The authors were able to recover as well known and still unknown mutations in the clinical samples (Garnaud et al. 2015). This pioneering study of molecular characterization in antifungal resistance from *Candida* spp. allowed early detection of mutated populations by combining in a single sequencing run multiple isolates and multiple resistance detections. It will be still challenging to introduce NGS approaches in the clinical laboratory, however, with the on-going progression of automation of these technologies, their routine application may become a reality in the next years or decades.

15.11 Conclusions and Perspectives

Chemotherapy will be faced with resistance issues, as it is observed in cancer, bacterial viral and antifungal therapies. Antifungal resistance, even if it not as serious as in other public health issues, has been an extraordinary exploratory tool to deepen our understanding of various cell functions in *C. albicans* and other fungal pathogens. Up to now, most of the resistance mechanisms have been solved to the molecular level. In the recent years however, reports on novel resistance profiles have appeared and, among them, the emergence of MDR is challenging not only for the researcher but also for the clinician, who sees alternatives therapeutic options becoming restricted. Currently, MDR is prominent in the species *C. glabrata* and the reasons behind MDR in this pathogen are being elucidated. *C. glabrata* is haploid and thus single genetic events are sufficient to result in drug resistance phenotypes. Very recent work suggests that some *C. glabrata* isolates exhibit intrinsic higher mutations rates than wild-type isolates, a feature that is known as hypermutator phenotype (Healey et al. 2016). This phenotype results from specific defects in the DNA repair machinery that is otherwise necessary to maintain the integrity of genomes during DNA replication. DNA repair defects generate genetic diversity, which may be important when the fungus encounter stressful environments, such as antifungal exposure. Now that the hypermutator phenotype is known from this yeast pathogen, several questions remain to be addressed, one of them being whether the hypermutator phenotype is the result of a recent clonal expansion in *C. glabrata*. Moreover, the existence of the hypermutator phenotype as a way to generate genetic diversity in fungal populations should be addressed in other types of fungal species.

Acknowledgements DS is supported by grants from the Swiss National Research Foundation No 31003A_146936.

References

Alastruey-Izquierdo A, Cuenca-Estrella M (2012) EUCAST and CLSI: how to assess in vitro susceptibility and clinical resistance. Curr Fungal Infect Rep 6:229–234. doi:10.1007/s12281-012-0100-3

Alexander BD, Johnson MD, Pfeiffer CD et al (2013) Increasing echinocandin resistance in *Candida glabrata*: clinical failure correlates with presence of *FKS* mutations and elevated minimum inhibitory concentrations. Clin Infect Dis 56:1724–1732. doi:10.1093/cid/cit136

Anderson TM, Clay MC, Cioffi AG et al (2014) Amphotericin forms an extramembranous and fungicidal sterol sponge. Nat Chem Biol 10:400–406. doi:10.1038/nchembio.1496

Arendrup MC, Cuenca-Estrella M, Lass-Flörl C, Hope WW (2013) Breakpoints for antifungal agents: an update from EUCAST focussing on echinocandins against *Candida* spp. and triazoles against *Aspergillus* spp. Drug Resist Updat 16:81–95. doi:10.1016/j.drup.2014.01.001

Arendrup MC, Cuenca-Estrella M, Lass-Flörl C, Hope WW (2014) EUCAST technical note on *Candida* and micafungin, anidulafungin and fluconazole. Mycoses 57:377–379. doi:10.1111/myc.12170

Arendrup MC, Perlin DS (2014) Echinocandin resistance: an emerging clinical problem? Curr Opin Infect Dis 27:484–492. doi:10.1097/QCO.0000000000000111

Arvanitis M, Anagnostou T, Fuchs BB et al (2014) Molecular and nonmolecular diagnostic methods for invasive fungal infections. Clin Microbiol Rev 27:490–526. doi:10.1128/CMR.00091-13

Asner SA, Giulieri S, Diezi M et al (2015) Acquired multidrug antifungal resistance in *Candida lusitaniae* during therapy. Antimicrob Agents Chemother 59: AAC.02204–15–8. doi:10.1128/AAC.02204-15

Balashov SV, Park S, Perlin DS (2006) Assessing resistance to the echinocandin antifungal drug caspofungin in *Candida albicans* by profiling mutations in *FKS1*. Antimicrob Agents Chemother 50:2058–2063. doi:10.1128/AAC.01653-05

Bonhomme J, d'Enfert C (2013) *Candida albicans* biofilms: building a heterogeneous, drug-tolerant environment. Curr Opin Microbiol 16:398–403. doi:10.1016/j.mib.2013.03.007

Brauner A, Fridman O, Gefen O, Balaban NQ (2016) Distinguishing between resistance, tolerance and persistence to antibiotic treatment. Nat Rev Micro 14:320–330. doi:10.1038/nrmicro.2016.34

Calabrese D, Bille J, Sanglard Dominique (2000) A novel multidrug efflux transporter gene of the major facilitator superfamily from *Candida albicans* (*FLU1*) conferring resistance to fluconazole. Microbiol (Reading, Engl) 146 (Pt 11): 2743–2754. doi:10.1099/00221287-146-11-2743

Chapeland-Leclerc F, Hennequin C, Papon N et al (2010) Acquisition of flucytosine, azole, and caspofungin resistance in *Candida glabrata* bloodstream isolates serially obtained from a hematopoietic stem cell transplant recipient. Antimicrob Agents Chemother 54:1360–1362. doi:10.1128/AAC.01138-09

Chau AS, Gurnani M, Hawkinson R et al (2005) Inactivation of sterol Delta 5,6-desaturase attenuates virulence in Candida albicans. Antimicrob Agents Chemother 49:3646–3651. doi:10.1128/AAC.49.9.3646-3651.2005

Cleveland AA, Farley MM, Harrison LH et al (2012) Changes in incidence and antifungal drug resistance in candidemia: results from population-based laboratory surveillance in Atlanta and Baltimore, 2008–2011. Clin Infect Dis 55:1352–1361

Costa C, Dias PJ, Sá-Correia I, Teixeira MC (2014) MFS multidrug transporters in pathogenic fungi: do they have real clinical impact? Front Physiol 5:197. doi:10.3389/fphys.2014.00197

Coste A, Turner V, Ischer F et al (2006) A mutation in Tac1p, a transcription factor regulating *CDR1* and *CDR2*, is coupled with loss of heterozygosity at chromosome 5 to mediate antifungal resistance in *Candida albicans*. Genetics 172:2139–2156. doi:10.1534/genetics.105.054767

Coste AT, Karababa M, Ischer F et al (2004) *TAC1*, transcriptional activator of *CDR* genes, is a new transcription factor involved in the regulation of *Candida albicans* ABC transporters *CDR1* and *CDR2*. Eukaryot Cell 3:1639–1652. doi:10.1128/EC.3.6.1639-1652.2004

Cowen LE (2013) The fungal Achilles' heel: targeting Hsp90 to cripple fungal pathogens. Curr Opin Microbiol 16:377–384. doi:10.1016/j.mib.2013.03.005

Cowen LE, Lindquist S (2005) Hsp90 potentiates the rapid evolution of new traits: drug resistance in diverse fungi. Science 309:2185–2189. doi:10.1126/science.1118370

Delarze E, Sanglard D (2015) Defining the frontiers between antifungal resistance, tolerance and the concept of persistence. Drug Resist Updat 23:12–19. doi:10.1016/j.drup.2015.10.001

Desai JV, Mitchell AP, Andes DR (2014) Fungal Biofilms, Drug Resistance, and Recurrent Infection. Cold Spring Harb Perspect Med 4:a019729–a019729. doi:10.1101/cshperspect.a019729

Desnos-Ollivier M, Bretagne S, Raoux D et al (2008) Mutations in the *fks1* gene in *Candida albicans*, *C. tropicalis*, and *C. krusei* correlate with elevated caspofungin MICs uncovered in AM3 medium using the method of the European committee on antibiotic susceptibility testing. Antimicrob Agents Chemother 52:3092–3098. doi:10.1128/AAC.00088-08

Desnos-Ollivier M, Moquet O, Chouaki T et al (2011) Development of echinocandin resistance in *Clavispora lusitaniae* during caspofungin treatment. J Clin Microbiol 49:2304–2306. doi:10.1128/JCM.00325-11

Dodgson AR, Dodgson KJ, Pujol C et al (2004) Clade-specific flucytosine resistance Is due to a single nucleotide change in the *FUR1* gene of *Candida albicans*. Antimicrob Agents Chemother 48:2223–2227. doi:10.1128/AAC.48.6.2223-2227.2004

Dunkel N, Blass J, Rogers PD, Morschhäuser J (2008a) Mutations in the multi-drug resistance regulator *MRR1*, followed by loss of heterozygosity, are the main cause of *MDR1* overexpression in fluconazole-resistant *Candida albicans* strains. Mol Microbiol 69:827–840. doi:10.1111/j.1365-2958.2008.06309.x

Dunkel N, Liu TT, Barker KS et al (2008b) A gain-of-function mutation in the transcription factor Upc2p causes upregulation of ergosterol biosynthesis genes and increased fluconazole resistance in a clinical *Candida albicans* isolate. Eukaryot Cell 7:1180–1190. doi:10.1128/EC.00103-08

Eddouzi J, Parker JE, Vale-Silva LA et al (2013) Molecular mechanisms of drug resistance in clinical *Candida* species isolated from Tunisian hospitals. Antimicrob Agents Chemother 57:3182–3193. doi:10.1128/AAC.00555-13

Edlind TD, Katiyar SK (2010) Mutational analysis of flucytosine resistance in *Candida glabrata*. Antimicrob Agents Chemother 54:4733–4738. doi:10.1128/AAC.00605-10

EUCAST-AFST (2008) EUCAST technical note on fluconazole. Clin Microbiol Infect 14:193–195. doi:10.1111/j.1469-0691.2007.01899.x

Ferrari S, Ischer F, Calabrese D et al (2009) Gain of function mutations in *CgPDR1* of *Candida glabrata* not only mediate antifungal resistance but also enhance virulence. PLoS Pathog 5:e1000268. doi:10.1371/journal.ppat.1000268

Fiori A, van Dijck P (2012) Potent synergistic effect of doxycycline with fluconazole against *Candida albicans* is mediated by interference with iron homeostasis. Antimicrob Agents Chemother 56:3785–3796. doi:10.1128/AAC.06017-11

Ford CB, Funt JM, Abbey D, et al (2015) The evolution of drug resistance in clinical isolates of *Candida albicans*. eLife 4:e00662. doi:10.7554/eLife.00662

Fothergill AW, Sutton DA, McCarthy DI, Wiederhold NP (2014) Impact of new antifungal breakpoints on antifungal resistance in *Candida* species. J Clin Microbiol 52:994–997. doi:10.1128/JCM.03044-13

Garnaud C, Botterel F, Sertour N, et al (2015) Next-generation sequencing offers new insights into the resistance of *Candida* spp. to echinocandins and azoles. J Antimicrob Chemother 1–10. doi:10.1093/jac/dkv139

Gaur M, Puri N, Manoharlal R et al (2008) MFS transportome of the human pathogenic yeast *Candida albicans*. BMC Genom 9:579. doi:10.1186/1471-2164-9-579

Gefen O, Balaban NQ (2009) The importance of being persistent: heterogeneity of bacterial populations under antibiotic stress. FEMS Microbiol Rev 33:704–717. doi:10.1111/j.1574-6976.2008.00156.x

Harrison BD, Hashemi J, Bibi M et al (2014) A tetraploid intermediate precedes aneuploid formation in yeasts exposed to fluconazole. PLoS Biol 12:e1001815. doi:10.1371/journal.pbio.1001815

Hayama K, Ishibashi H, Ishijima SA et al (2012) A d-octapeptide drug efflux pump inhibitor acts synergistically with azoles in a murine oral candidiasis infection model. FEMS Microbiol Lett 328:130–137. doi:10.1111/j.1574-6968.2011.02490.x

Healey KR, Zhao Y, Perez WB et al (2016) Prevalent mutator genotype identified in fungal pathogen *Candida glabrata* promotes multi-drug resistance. Nature Commun 7:1–10. doi:10.1038/ncomms11128

Holmes AR, Keniya MV, Ivnitski-Steele I et al (2012) The monoamine oxidase A inhibitor clorgyline is a broad-spectrum inhibitor of fungal ABC and MFS transporter efflux pump activities which reverses the azole resistance of *Candida albicans* and *Candida glabrata* clinical isolates. Antimicrob Agents Chemother 56:1508–1515. doi:10.1128/AAC.05706-11

Hope WW, Tabernero L, Denning DW, Anderson MJ (2004) Molecular mechanisms of primary resistance to flucytosine in *Candida albicans*. Antimicrob Agents Chemother 48:4377–4386. doi:10.1128/AAC.48.11.4377-4386.2004

Jensen RH, Astvad KMT, Silva LV et al (2015) Stepwise emergence of azole, echinocandin and amphotericin B multidrug resistance *in vivo* in *Candida albicans* orchestrated by multiple genetic alterations. J Antimicrob Chemother 70:1–5. doi:10.1093/jac/dkv140

Jensen-Pergakes KL, Kennedy MA, Lees ND et al (1998) Sequencing, disruption, and characterization of the *Candida albicans* sterol methyltransferase (*ERG6*) gene: drug susceptibility studies in *erg6* mutants. Antimicrob Agents Chemother 42:1160–1167

Jia N, Arthington-Skaggs B, Lee W et al (2002) *Candida albicans* sterol C-14 reductase, encoded by the *ERG24* gene, as a potential antifungal target site. Antimicrob Agents Chemother 46:947–957. doi:10.1128/AAC.46.4.947-957.2002

Kaur R, Castaño I, Cormack BP (2004) Functional genomic analysis of fluconazole susceptibility in the pathogenic yeast *Candida glabrata*: roles of calcium signaling and mitochondria. Antimicrob Agents Chemother 48:1600–1613. doi:10.1128/AAC.48.5.1600-1613.2004

Kelly S, Lamb D, Corran A, et al (1995) Mode of action and resistance to azole antifungals associated with the formation of 14alpha-methylergosta-8,24(28)-dien-3ß,6a-diol 207: 910–915. doi:10.1006/bbrc.1995.1272

Kohli A, Smriti Mukhopadhyay K et al (2002) In vitro low-level resistance to azoles in *Candida albicans* is associated with changes in membrane lipid fluidity and asymmetry. Antimicrob Agents Chemother 46:1046–1052. doi:10.1128/aac.46.4.1046-1052.2002

Lamb D, Kelly D, White T, Kelly S (2000) The R467 K amino acid substitution in *Candida albicans* sterol 14alpha-demethylase causes drug resistance through reduced affinity. Antimicrob Agents Chemother 44:63–67. doi:10.1128/aac.44.1.63-67.2000

Lamping E, Monk BC, Niimi K et al (2007) Characterization of three classes of membrane proteins involved in fungal azole resistance by functional hyperexpression in *Saccharomyces cerevisiae*. Eukaryot Cell 6:1150–1165. doi:10.1128/EC.00091-07

Lemke A, Kiderlen AF, Kayser O (2005) Amphotericin B. Appl Microbiol Biotechnol 68:151–162. doi:10.1007/s00253-005-1955-9

Lepak AJ, Andes DR (2014) Antifungal Pharmacokinetics and Pharmacodynamics. Cold Spring Harb Perspect Med 1–24. doi:10.1101/cshperspect.a019653

Lewis K (2006) Persister cells, dormancy and infectious disease. Nat Rev Micro 5:48–56. doi:10.1038/nrmicro1557

Li P, Seneviratne CJ, Alpi E et al (2015) Delicate metabolic control and coordinated stress response critically determine antifungal tolerance of *Candida albicans* biofilm Persisters. Antimicrob Agents Chemother 59:6101–6112. doi:10.1128/AAC.00543-15

Li R, Kumar R, Tati S, et al (2013) *C. albicans* Flu1 mediated efflux of salivary Histatin 5 reduces its cytosolic concentration and fungicidal activity. Antimicrob Agents Chemother. doi:10.1128/AAC.02295-12

Lortholary O, Renaudat C, Sitbon K et al (2014) Worrisome trends in incidence and mortality of candidemia in intensive care units (Paris area, 2002–2010). Intensive Care Med 40:1303–1312. doi:10.1007/s00134-014-3408-3

Luna-Tapia A, Kerns ME, Eberle KE, et al (2015). Trafficking through the late endosome significantly impacts *Candida albicans* tolerance of the azole antifungals. Antimicrob Agents Chemother AAC.04239–14. doi:10.1128/AAC.04239-14

Maebashi K, Kudoh M, Nishiyama Y et al (2002) A novel mechanism of fluconazole resistance associated with fluconazole sequestration in *Candida albicans* isolates from a myelofibrosis patient. Microbiol Immunol 46:317–326. doi:10.1111/j.1348-0421.2002.tb02702.x

Mansfield BE, Oltean HN, Oliver BG et al (2010) Azole drugs are imported by facilitated diffusion in *Candida albicans* and other pathogenic fungi. PLoS Pathog. doi:10.1371/journal.ppat.1001126

Marchetti O, Moreillon P, Entenza JM et al (2003) Fungicidal synergism of fluconazole and cyclosporine in *Candida albicans* Is not dependent on multidrug efflux transporters encoded by the *CDR1*, *CDR2*, *CaMDR1*, and *FLU1* Genes. Antimicrob Agents Chemother 47:1565–1570. doi:10.1128/AAC.47.5.1565-1570.2003

Marchetti O, Moreillon P, Glauser MP et al (2000) Potent synergism of the combination of fluconazole and cyclosporine in *Candida albicans*. Antimicrob Agents Chemother 44:2373–2381

Marr KA, Rustad TR, Rex JH, White TC (1999) The trailing end point phenotype in antifungal susceptibility testing is pH dependent. Antimicrob Agents Chemother 43:1383–1386

Martel CM, Parker JE, Bader O et al (2010a) A clinical isolate of *Candida albicans* with mutations in *ERG11* (encoding sterol 14alpha-demethylase) and *ERG5* (encoding C22 desaturase) is cross resistant to azoles and amphotericin B. Antimicrob Agents Chemother 54:3578–3583. doi:10.1128/AAC.00303-10

Martel CM, Parker JE, Bader O et al (2010b) Identification and characterization of four azole-resistant *erg3* mutants of *Candida albicans*. Antimicrob Agents Chemother 54:4527–4533. doi:10.1128/AAC.00348-10

Maubon D, Garnaud C, Calandra T et al (2014) Resistance of *Candida* spp. to antifungal drugs in the ICU: where are we now? Intensive Care Med 40:1–15. doi:10.1007/s00134-014-3404-7

Maurya IK, Thota CK, Verma SD et al (2013) Rationally designed transmembrane peptide mimics of the multidrug transporter protein Cdr1 act as antagonists to selectively block drug efflux and chemosensitize azole-resistant clinical isolates of *Candida albicans*. J Biol Chem 288:16775–16787. doi:10.1074/jbc.M113.467159

Mitchell KF, Zarnowski R, Sanchez H, et al (2015) Community participation in biofilm matrix assembly and function. Proc Natl Acad Sci U S A 201421437–17. doi:10.1073/pnas.1421437112

Miyazaki T, Inamine T, Yamauchi S et al (2010) Role of the Slt2 mitogen-activated protein kinase pathway in cell wall integrity and virulence in *Candida glabrata*. FEMS Yeast Res 10:343–352. doi:10.1111/j.1567-1364.2010.00611.x

Monk BC, Tomasiak TM, Keniya MV et al (2014) Architecture of a single membrane spanning cytochrome P450 suggests constraints that orient the catalytic domain relative to a bilayer. Proc Natl Acad Sci U S A 111:3865–3870. doi:10.1073/pnas.1324245111

Morio F, Loge C, Besse B et al (2010) Screening for amino acid substitutions in the *Candida albicans Erg11* protein of azole-susceptible and azole-resistant clinical isolates: new substitutions and a review of the literature. Diagn Microbiol Infect Dis 66:373–384. doi:10.1016/j.diagmicrobio.2009.11.006

Morio F, Pagniez F, Lacroix C et al (2012) Amino acid substitutions in the *Candida albicans* sterol $\Delta^{5,6}$-desaturase (Erg3p) confer azole resistance: characterization of two novel mutants with impaired virulence. J Antimicrob Chemother 67:2131–2138. doi:10.1093/jac/dks186

Morrell M, Fraser VJ, Kollef MH (2005) Delaying the empiric treatment of *Candida* bloodstream infection until positive blood culture results are obtained: a potential risk factor for hospital mortality. Antimicrob Agents Chemother 49:3640–3645. doi:10.1128/AAC.49.9.3640-3645.2005

Morschhäuser J, Barker KS, Liu TT et al (2007) The transcription factor Mrr1p controls expression of the *MDR1* efflux pump and mediates multidrug resistance in *Candida albicans*. PLoS Pathog 3:e164. doi:10.1371/journal.ppat.0030164

Mukherjee PK, Sheehan DJ, Hitchcock CA, Ghannoum MA (2005) Combination treatment of invasive fungal infections. Clin Microbiol Rev 18:163–194. doi:10.1128/CMR.18.1.163-194.2005

Mukhopadhyay K, Kohli A, Prasad R (2002) Drug susceptibilities of yeast cells are affected by membrane lipid composition. Antimicrob Agents Chemother. doi:10.1128/AAC.46.12.3695-3705.2002

Nett J, Lincoln L, Marchillo K et al (2007) Putative role of {beta}-1,3 glucans in *Candida albicans* biofilm resistance. Antimicrob Agents Chemother 51:510–520. doi:10.1128/AAC.01056-06

Niimi K, Harding DRK, Holmes AR et al (2012) Specific interactions between the *Candida albicans* ABC transporter Cdr1p ectodomain and a d-octapeptide derivative inhibitor. Mol Microbiol 85:747–767. doi:10.1111/j.1365-2958.2012.08140.x

Nishikawa JL, Boeszoermenyi A, Vale-Silva LA, et al (2016) Inhibiting fungal multidrug resistance by disrupting an activator–Mediator interaction. Nature 1–15. doi:10.1038/nature16963

O'Meara TR, Veri AO, Ketela T et al (2015) Global analysis of fungal morphology exposes mechanisms of host cell escape. Nature Commun 6:6741. doi:10.1038/ncomms7741

Park S, Kelly R, Kahn JN et al (2005) Specific substitutions in the echinocandin target Fks1p account for reduced susceptibility of rare laboratory and clinical *Candida* sp. isolates. Antimicrob Agents Chemother 49:3264–3273. doi:10.1128/AAC.49.8.3264-3273.2005

Park S, Perlin DS (2005) Establishing surrogate markers for fluconazole resistance in *Candida albicans*. Microb Drug Resist 11:232–238. doi:10.1089/mdr.2005.11.232

Paulsen IT, Sliwinski MK, Nelissen B et al (1998) Unified inventory of established and putative transporters encoded within the complete genome of *Saccharomyces cerevisiae*. FEBS Lett 430:116–125

Perlin DS (2011) Current perspectives on echinocandin class drugs. Future Microbiol 6:441–457. doi:10.2217/fmb.11.19

Perlin DS (2015) Echinocandin Resistance in *Candida*. Clin Infect Dis 61:S612–S617. doi:10.1093/cid/civ791

Perlin DS (2009) Antifungal drug resistance: do molecular methods provide a way forward? Curr Opin Infect Dis 22:568–573. doi:10.1097/QCO.0b013e3283321ce5

Perlin DS, Shor E, Zhao Y (2015) Update on antifungal drug resistance. Curr Clin Micro Rpt 2:84–95. doi:10.1007/s40588-015-0015-1

Pfaller MA (2012) Antifungal drug resistance: mechanisms, epidemiology, and consequences for treatment. Am J Med 125:S3–S13. doi:10.1016/j.amjmed.2011.11.001

Pfaller MA, Castanheira M, Diekema DJ et al (2010) Comparison of European Committee on Antimicrobial Susceptibility Testing (EUCAST) and Etest methods with the CLSI broth microdilution method for echinocandin susceptibility testing of *Candida* species. J Clin Microbiol 48:1592–1599. doi:10.1128/JCM.02445-09

Pfaller MA, Castanheira M, Messer SA et al (2014) Comparison of EUCAST and CLSI broth microdilution methods for the susceptibility testing of 10 Systemically active antifungal agents when tested against *Candida* spp. Diagn Microbiol Infect Dis 79:198–204. doi:10.1016/j.diagmicrobio.2014.03.004

Pfaller MA, Diekema DJ (2007) Epidemiology of invasive candidiasis: a persistent public health problem. Clin Microbiol Rev 20:133–163. doi:10.1128/CMR.00029-06

Pfaller MA, Diekema DJ (2012) Progress in antifungal susceptibility testing of *Candida* spp. using clinical and laboratory standards institute broth microdilution methods, 2010-2012. J Clin Microbiol 50:2846–2856. doi:10.1128/JCM.00937-12

Pfaller MA, Messer SA, Boyken L et al (2003) Variation in susceptibility of bloodstream isolates of *Candida glabrata* to fluconazole according to patient age and geographic location. J Clin Microbiol 41:2176–2179. doi:10.1128/JCM.41.5.2176-2179.2003

Pfaller MA, Messer SA, Hollis RJ et al (2009) Variation in susceptibility of bloodstream isolates of *Candida glabrata* to fluconazole according to patient age and geographic location in the United States in 2001 to 2007. J Clin Microbiol 47:3185–3190. doi:10.1128/JCM.00946-09

Prasad R, Goffeau A (2012) Yeast ATP-binding cassette transporters conferring multidrug resistance. Annu Rev Microbiol 66:39–63. doi:10.1146/annurev-micro-092611-150111

Ramage G, Saville SP, Thomas DP, Lopez-Ribot JL (2005) *Candida* biofilms: an update. Eukaryot Cell 4:633–638. doi:10.1128/EC.4.4.633-638.2005

Reinoso-Martín C, Schüller C, Schuetzer-Muehlbauer M, Kuchler K (2003) The yeast protein kinase c cell integrity pathway mediates tolerance to the antifungal drug caspofungin through activation of Slt2p mitogen-activated protein kinase signaling. Eukaryot Cell 2:1200–1210. doi:10.1128/EC.2.6.1200-1210.2003

Sanglard D, Coste A, Ferrari S (2009) Antifungal drug resistance mechanisms in fungal pathogens from the perspective of transcriptional gene regulation. FEMS Yeast Res 9:1029–1050. doi:10.1111/j.1567-1364.2009.00578.x

Sanglard D, Coste AT (2015) Activity of isavuconazole and other azoles against *Candida* clinical isolates and yeast model systems with known azole resistance mechanisms. Antimicrob Agents Chemother 60:229–238. doi:10.1128/AAC.02157-15

Sanglard D, Ischer F, Marchetti O et al (2003a) Calcineurin A of *Candida albicans*: involvement in antifungal tolerance, cell morphogenesis and virulence. Mol Microbiol 48:959–976. doi:10.1046/j.1365-2958.2003.03495.x

Sanglard D, Ischer F, Parkinson T et al (2003b) *Candida albicans* mutations in the ergosterol biosynthetic pathway and resistance to several antifungal agents. Antimicrob Agents Chemother 47:2404–2412. doi:10.1128/aac.47.8.2404-2412.2003

Sanglard D, Kuchler K, Ischer F et al (1995) Mechanisms of resistance to azole antifungal agents in *Candida albicans* isolates from AIDS patients involve specific multidrug transporters. Antimicrob Agents Chemother 39:2378–2386. doi:10.1128/aac.39.11.2378

Sanglard D, Odds FC (2002) Resistance of *Candida* species to antifungal agents: molecular mechanisms and clinical consequences. Lancet Infect Dis 2:73–85. doi:10.1016/s1473-3099(02)00181-0

Schwarzmüller T, Ma B, Hiller E et al (2014) Systematic phenotyping of a large-scale *Candida glabrata* deletion collection reveals novel antifungal tolerance genes. PLoS Pathog 10:e1004211–19. doi:10.1371/journal.ppat.1004211

Selmecki A, Forche A, Berman J (2006) Aneuploidy and isochromosome formation in drug-resistant *Candida albicans*. Science 313:367–370. doi:10.1126/science.1128242

Silva LV, Sanguinetti M, Vandeputte P et al (2013) Milbemycins: more than efflux inhibitors for fungal pathogens. Antimicrob Agents Chemother 57:873–886. doi:10.1128/AAC.02040-12

Singh SD, Robbins N, Zaas AK et al (2009) Hsp90 governs echinocandin resistance in the pathogenic yeast *Candida albicans* via calcineurin. PLoS Pathog 5:e1000532. doi:10.1371/journal.ppat.1000532.g009

Thakur JK, Arthanari H, Yang F et al (2008) A nuclear receptor-like pathway regulating multidrug resistance in fungi. Nature 452:604–609. doi:10.1038/nature06836

Tobudic S, Kratzer C, Lassnigg A, Presterl E (2012) Antifungal susceptibility of *Candida albicans* in biofilms. Mycoses 55:199–204. doi:10.1111/j.1439-0507.2011.02076.x

Vale-Silva LA, Coste AT, Ischer F et al (2012) Azole resistance by loss of function of the sterol $\Delta^{5,6}$-desaturase gene (*ERG3*) in *Candida albicans* does not necessarily decrease virulence. Antimicrob Agents Chemother 56:1960–1968. doi:10.1128/AAC.05720-11

Vandeputte P, Pineau L, Larcher G et al (2011) Molecular mechanisms of resistance to 5-fluorocytosine in laboratory mutants of *Candida glabrata*. Mycopathologia 171:11–21. doi:10.1007/s11046-010-9342-1

Vasicek EM, Berkow EL, Bruno VM et al (2014) Disruption of the transcriptional regulator Cas5 results in enhanced killing of *Candida albicans* by fluconazole. Antimicrob Agents Chemother 58:6807–6818. doi:10.1128/AAC.00064-14

Vediyappan G, Rossignol T, d'Enfert C (2010) Interaction of *Candida albicans* biofilms with antifungals: transcriptional response and binding of antifungals to beta-glucans. Antimicrob Agents Chemother 54:2096–2111. doi:10.1128/AAC.01638-09

Vermitsky J-P, Earhart KD, Smith WL et al (2006) Pdr1 regulates multidrug resistance in *Candida glabrata*: gene disruption and genome-wide expression studies. Mol Microbiol 61:704–722. doi:10.1111/j.1365-2958.2006.05235.x

Verstraeten N, Knapen WJ, Kint CI et al (2015) Obg and membrane depolarization are part of a microbial bet-hedging strategy that leads to antibiotic tolerance. Mol Cell 59:9–21. doi:10.1016/j.molcel.2015.05.011

Vincent BM, Lancaster AK, Scherz-Shouval R et al (2013) Fitness trade-offs restrict the evolution of resistance to amphotericin B. PLoS Biol 11:e1001692. doi:10.1371/journal.pbio.1001692

Walker LA, Gow NAR, Munro CA (2010) Fungal echinocandin resistance. Fungal Genet Biol 47:117–126. doi:10.1016/j.fgb.2009.09.003

Chapter 16
Unlocking the Therapeutic Potential of the Fungal Cell Wall: Clinical Implications and Drug Resistance

Chibuike Ibe, Louise A. Walker, Neil A.R. Gow and Carol A. Munro

Abstract The cell wall of *Candida albicans* is essential for cellular survival. The wall is a rich polysaccharide and mannoprotein complex. During wall synthesis, the cell wall components—glucan, chitin and mannoproteins—are assembled and stabilised in the cell wall space by the action of cell wall-associated enzymes and structural wall proteins. Cell surface proteins are also important in *C. albicans* virulence. The components of the cell wall are unique to fungi, thus making the wall an attractive drug target. Echinocandins are antifungal drugs that inhibit the synthesis of cell wall β-1,3-D-glucan and are fungicidal to *C. albicans*. Echinocandin treatment, at sub-inhibitory concentrations, can result in a compensatory increase in wall chitin content and triggers the expression of mannoproteins whose activities remodel the wall. *C. albicans* cells with thicker cell walls and elevated chitin levels are less susceptible to echinocandin therapy. The development of a molecule that non-competitively inhibits chitin synthesis and/or impairs cell wall integrity response regulatory pathways may impact the cell wall in such a way that reduces its compensatory ability. This type of molecule may have a synergistic effect in combined therapy with other cell wall inhibitors such as β-1,3-D-glucan-targeting antifungal drugs for managing candidiasis infections. The synthesis and localisation of mannoproteins at the cell surface also provide an attractive cell wall target for future antifungal agents.

16.1 Introduction

The high mortality rate of invasive candidiasis may largely be due to the increasing number of immunocompromised individuals and partly due to lack of approved prophylactic agents, poor diagnostics and lack of efficacious therapeutics (Brown et al. 2012a, b). One of the limitations of current antifungal drugs is the adverse side

---

C. Ibe · L.A. Walker · N.A.R. Gow · C.A. Munro (✉)
Aberdeen Fungal Group, Institute of Medical Sciences, University of Aberdeen,
Aberdeen AB252ZD, UK
e-mail: c.a.munro@abdn.ac.uk

R. Prasad (ed.), *Candida albicans: Cellular and Molecular Biology*,
DOI 10.1007/978-3-319-50409-4_16

effects in patients as a result of the common evolutionary ancestry between fungi and man. Consequently, the arsenal of antifungal drugs is limited to fungal targets that lack homologues in humans (Cowen and Steinbach 2008; Fairlamb et al. 2016).

The fungal cell wall and its components are unique to fungi; this makes them attractive targets for antifungal agents. The cell wall of fungi is composed of a coordinated assembly of polysaccharides and mannoproteins that function as an exoskeleton. The exoskeleton can be remodelled by a number of strictly regulated mechanisms that protect the protoplasm from harmful environmental agents and conditions (Munro and Richard 2012). The cell wall polysaccharides and glycoproteins are known as pathogen-associated molecular patterns (PAMPs) and can provoke immune responses during host invasion by *C. albicans*. The cell wall structure can also be altered enabling *C. albicans* to evade host immune responses to adapt to different niches at various stages of infection. Treatment with a cell wall inhibitor can disrupt wall structure leading to cell lysis and death. For example, the echinocandins, such as caspofungin, are antifungal drugs that target and inhibit the synthesis of the major cell wall structural polysaccharide β-1,3-D-glucan (Douglas et al. 1997), and are fungicidal to *C. albicans*. However, cell wall salvage responses that are activated when the wall is damaged trigger the expression and recruitment of proteins whose activities remodel the cell wall polysaccharides in the cell wall space, to maintain an intact wall during stress (Fonzi 1999; Gelis et al. 2012; Klis et al. 2009; Liu et al. 2005; Moreno-Ruiz et al. 2009; Munro 2013; Pardini et al. 2006; Plaine et al. 2008; Sorgo et al. 2011). One such cell wall remodelling activity leads to a thicker wall with a compensatory increase in wall chitin content (Walker et al. 2008). *C. albicans* strains with thicker cell walls and high chitin contents are less susceptible to caspofungin treatment (Imtiaz et al. 2012; Lee et al. 2012; Walker et al. 2008, 2013a). The mechanism(s) that underpins this response in *C. albicans* is poorly understood. For example, the synthesis and assembly of cell wall components, and cell wall remodelling in response to cell wall stressors is mediated by enzymes many of which are cell wall localised or plasma membrane associated. Many of these proteins, which may be potential targets for new antifungal agents, have yet to be studied in great detail.

16.2 *Candida albicans* Cell Wall

The fungal cell wall is a dynamic organelle. It is essential for cellular survival by providing the cell with mechanical strength to withstand fluctuations in osmotic conditions of the environment. At the same time the cell wall must retain sufficient plasticity to enable cell growth, cell division and morphogenesis. Thus the cell wall retains its basic architecture and executes its function by maintaining a delicate balance between wall rigidity and elasticity. Imaging of the cell wall ultrastructure by electron microscopy has identified two layers: an outer fibrillar layer and a translucent inner layer (Tokunaga et al. 1986). The highly crosslinked polysaccharide-rich inner layer is made of β-1,3-D- and β-1,6-D-glucans, and chitin (Fig. 16.1) (Kapteyn et al. 2000).

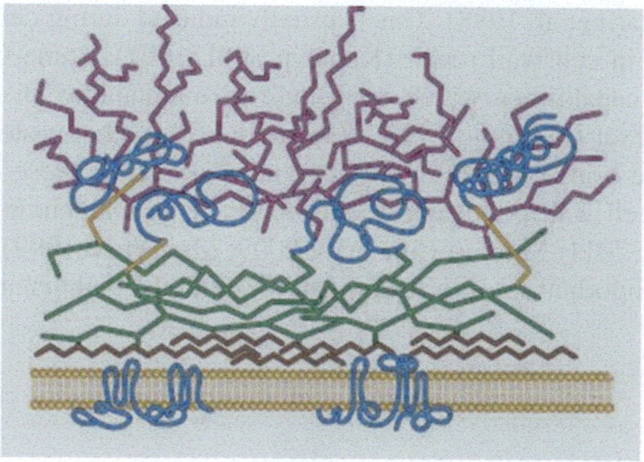

Fig. 16.1 *C. albicans* cell wall. The model shows the components and structural organisation of the *C. albicans* cell wall. The glycoprotein coat is made up of covalently attached glycosylphosphatidylinositol (GPI)-CWPs, proteins with internal repeats, PIR-CWPs that are attached via alkali-sensitive linkages; proteins coupled to the cell wall through disulphide bonds (reducing agent extractable, RAE-CWPs) and non-covalently linked (NCL)-CWPs (may be linked through hydrogen bonds). The GPI-CWPs are linked to β-1,6-D-glucans through GPI remnants. Pir-CWPs are attached to β-1,3-D-glucan through an transesterification reaction (Ecker et al. 2006). The cell wall may be strengthened by disulphide bridges between RAE-CWPs. In the inner polysaccharide-rich layer, the β-1,3-D-glucan network is linked to chitin and β-1,6-D-glucan. PIR-CWPs can also crosslink the β-1,3-D-glucan network thereby increasing the wall rigidity (De Groot et al. 2005). Adapted from Gow et al. (2011), Gow and Hube (2012)

The major structural polysaccharide β-1,3-D-glucan is moderately branched. It forms a stable association leading to a three-dimensional network that extends around the cell wall (Klis et al. 2001). Chitin molecules are covalently linked to the non-reducing ends of β-1,3-D-glucan (Fig. 16.1) through β-1,4-linkages (Kollar et al. 1995). Chitin cross-linking to β-1,3-D-glucan provides the wall with rigidity and strength. β-1,6-D-glucan molecules are soluble in water, highly branched and lack regular structure. They act as flexible linkers between the skeletal inner layer and the protein coat (Fig. 16.1). The protein coat determines the cell surface properties of the cell wall. It accounts for about 35–40% of the dry weight of the cell wall and is made up of mannoproteins that are heavily glycosylated. The *N*-linked mannans, attached to cell wall proteins, are phosphorylated on their β-mannan side chains (Shibata et al. 1995). The phosphoric acid residues give the glycoproteins the ability to bind other proteins, positively charged ions and dyes such as Alcian blue. A minor class of cell wall proteins (CWPs), proteins with internal repeats, PIR-CWPs (Toh-e et al. 1993), are tethered directly to β-1,3-D-glucan through alkali-sensitive linkages. These proteins can strengthen the cell wall matrix through the linkages in their internal repeats (Ecker et al. 2006). The major classes of CWPs are those modified with glycosylphosphatidylinositol (GPI) anchors. These proteins are covalently linked to β-1,6-D-glucan molecules through remnants of their GPI anchors (De Nobel and Lipke 1994), which are in turn covalently linked to β-1,3-D-glucans. β-1,6-D-glucan-linked GPI-CWPs may also be linked to chitin (Surarit et al. 1988). This is usually induced during cell wall stress and may play a role in cell wall repair (Kapteyn et al. 1997). Some CWPs possess dynamic bonding capabilities which contribute to maintaining an intact wall (Klis et al. 2010). For example, some GPI-CWPs are linked to other proteins in the wall through disulphide bridges. In *S. cerevisiae* and *C. albicans* these disulphide bridges can restrict cell wall permeability and may help in maintaining cell wall integrity (Laforet et al. 2011; Moreno-Ruiz et al. 2009; Mrsa et al. 1999). This indicates that CWPs play important roles in repair activities and the maintenance of cell wall architecture.

16.2.1 Cell Wall Chitin

Chitin is a linear homopolymer of more than 200 units of β-1,4-linked *N*-acetylglucosamine (GlcNac) residues. Chitin is synthesised by chitin synthase (Chs) enzymes encoded by *CHS* genes. All fungi studied to date possess at least one Chs enzyme. *C. albicans* has four chitin synthase genes, *CHS1*, *CHS2*, *CHS3* and *CHS8* (Munro et al. 2003; Mio et al. 1996; Chen-Wu et al. 1992). The Chs1 enzyme is essential for septum formation and cell wall integrity. Conditional *chs1* mutants do not form septa and eventually lyse (Munro et al. 2001). *CHS2* is expressed in the hyphal form of *C. albicans* and may also be involved in the formation of the primary septum in a dividing yeast cell (Gow et al. 1994). The most active chitin synthase enzyme as measured by in vitro assays is Chs2. *chs2*Δ null mutants have normal growth rate, chitin content and virulence (Gow et al. 1994). In vivo, chitin in both yeast and hyphae is synthesised mainly by Chs3, which is responsible for the synthesis of about 80-90% of total cellular chitin (Bulawa 1993; Sudoh et al. 1993). *chs3*Δ mutants are less virulent in a murine model of systemic candidiasis and have significantly reduced cell wall chitin content, with concurrent defects in cell wall integrity compared to the wild type (Bulawa 1993). *chs8*Δ mutants have 75% of wild-type chitin synthase activity, are hypersensitive to Calcofluor white, but have normal growth and chitin content (Munro et al. 2003). Chitin typically accounts for $\sim$1–2% of the yeast cell wall dry weight (Bulawa 1993); nevertheless, its synthesis is essential for cell viability (Munro et al. 2001). Chitin provides structural integrity to the cell wall and is essential for cell viability and is thus a potential target for antifungal agents. Polyoxins and nikkomycins are two known chitin synthase inhibitors that compete with UDP-GlcNAc for binding to the enzymes, due to their structural similarities. Despite their high in vitro activity, and evidence of efficacy in vivo (Becker et al. 1988), they were not developed further. This may be because their uptake into the fungal cytoplasm is limited due to competition with other substrates or they are susceptible to degradation by cellular proteases. There could also be variation in the affinity of the agents to the various chitin synthase proteins. Though the enzymes involved in chitin synthesis may have specialised functions, there is potential for functional redundancies between the proteins and under certain conditions they can be activated to generate alternative septa (Walker et al. 2013b). In addition, subtle differences, in structure between the proteins, may have complicated the development of efficient chitin synthase inhibitors over the years. For example, Sudoh et al. (2000) developed a Chs1-specific inhibitor, RO-09-3143, and showed that inhibition of Chs1 only arrests cell growth, but inhibition of Chs1 in *chs2Δ* deletion mutant is lethal. They concluded that simultaneous inhibition of Chs1 and Chs2 is lethal (Sudoh et al. 2000).

16.2.2 Cell Wall Glucans

16.2.2.1 β-1,3-D-Glucan

Cell wall glucan is the major polysaccharide found in *C. albicans* and *S. cerevisiae*, and constitutes about 50–65% of the cell wall dry weight. In these yeasts glucan is composed of glucose units linked with β-1,3- or β-1,6- glycosidic bonds. About 50–55% of wall glucan possesses β-1,3-linkages (Klis et al. 2002; Kollar et al. 1995). β-1,3-D-glucan is synthesised by β-1,3-D-glucan synthase an integral membrane protein, with a catalytic subunit, Fks1 (Douglas et al. 1994) and a regulatory subunit Rho1. The physical property of β-1,3-D-glucan is comparable to a flexible wire spring which explains the elastic nature of the cell wall (Klis et al. 2002, 2006). β-1,3-D-glucan has a pivotal role in cell wall architecture, providing the structural backbone for wall organisation and formation of cell wall matrix. It has been extensively pursued as an antifungal target as a means of preventing fungal cell growth through inhibiting cell wall formation. The echinocandins, a class of antifungal agents, were successfully developed that non-competitively inhibit β-1,3-D-glucan synthesis by inhibiting the Fks1 protein (Denning 2003). Resistance to the echinocandins occurs sporadically as fungal species can withstand the drug's activity by acquiring point mutations in two external "hotspot" domains of *FKS1* and/or through a compensatory increase in cell wall chitin content (Chapter 15; 22). β-1,3-D-glucan and chitin appear to have an overlapping function to provide mechanical support to the fungal cell wall, and maintain an intact wall (Douglas et al. 1997). Clinically efficacious antifungal agents, specifically targeting chitin synthase proteins to inhibit chitin synthesis, appear to be a missing piece in the clinical setting. Effective chitin synthase inhibitors would have the potential to be used synergistically with β-1,3-D-glucan inhibitors such as the echinocandins, to better manage invasive candidiasis, especially in breakthrough infections (Walker et al. 2008).

16.2.2.2 β-1,6-D-Glucan

β-1,6-D-glucan, like β-1,3-D-glucan, is a component of the cell wall of most fungi. In *C. albicans* and *S. cerevisiae*, β-1,6-D-glucan plays a central role in linking the GPI-CWPs to the cell wall matrix and thus it is vital in cell wall organisation and structure (Aimanianda et al. 2009; Kollar et al. 1995, 1997; Lesage and Bussey 2006; Lu et al. 1995; Ruiz-Herrera et al. 2006). The enzymatic pathway responsible for the synthesis of β-1,6-D-glucan is yet to be fully elucidated. Our understanding so far of β-1,6-D-glucan synthesis has come from studies on *S. cerevisiae* using killer toxin K1. Mutants deficient in β-1,6-D-glucan are resistant to the K1 toxin and this provides a correlation between the sensitivity to K1 toxin and β-1,6-D-glucan levels. Using this technique approximately ten genes have been identified as involved in the synthesis of β-1,6-D-glucan in *S. cerevisiae. KRE1, KRE5, KRE6, CNE1, CWH41/GLS1, ROT2,* and *SKN1* have been identified in the *C. albicans* genome as homologues of *S. cerevisiae* β-1,6-D-glucan synthesis genes (Shahinian and Bussey 2000). This suggests that β-1,6-D-glucan synthesis is similar in both organisms. The synthesis of β-1,6-D-glucan is thought to occur on the cell surface and β-1,6-D-glucan becomes crosslinked to the cell wall by transglycosidases and glycosylhydrolases (Aimanianda et al. 2009; Montijn et al. 1999). However, sequence and structural studies showed that some of the β-1,6-D-glucan synthesis proteins, such as Kre5 and Cwh41, are located in the endoplasmic reticulum (ER), while Kre6 and Skn1 are Golgi apparatus localised (Shahinian and Bussey 2000). Our limited knowledge of the synthesis of β-1,6-D-glucan and lack of any known β-1,6-D-glucan synthase protein have made the development of an antifungal drug against it difficult, despite its importance in cell wall organisation. However, research has shown that single or double knock-outs of genes associated with the synthesis of β-1,6-D-glucan has significant effect on the cell wall levels of β-1,6-D-glucan (Gilbert et al. 2010; Jiang et al. 1996).

16.2.3 Cell Wall Proteins

The outer layer of the cell wall of *C. albicans* is composed of mannoproteins that are heavily glycosylated. The availability of the *Candida* genome database (CGD) (http://www.candidagenome.org/), bioinformatics tools and the use of extremely sensitive mass spectrometry techniques have greatly advanced our knowledge of the nature and abundance of the cell wall proteome (De Groot et al. 2004; Gil-Bona et al. 2015; Richard and Plaine 2007). Proteomic analysis of *C. albicans* cell walls reveals covalently bound carbohydrate-active enzymes and adhesins (De Groot et al. 2004). CWPs are unique in their structure in that they generally contain an N-terminal signal peptide motif for localisation to the ER and secretory pathway. They are post-translationally modified with serine/threonine-linked *O*-manno-oligosaccharides and/or asparagine-linked *N*-glycans and may contain unique features such as a GPI-anchor attachment sequence and/or internal repeats (Klis et al. 2009; Ruiz-Herrera et al. 2006).

The proteins considered to be true covalently attached CWPs include the proteins with internal repeats (PIR)-, and GPI-CWPs (De Groot et al. 2005). The GPI-CWPs constitute about 88% of the total wall glycoprotein. In this paper we will focus on the biogenesis and localisation of the GPI-CWPs pointing to key enzymatic steps which may be suitable drug targets. CWPs can be split into groups based on their functional importance in the cell wall (Free 2013; Klis et al. 2010). Functionally important CWPs include (1) enzymes that participate in cell wall biogenesis. These include proteins with glycosylhydrolase, glycosyltransferase or transglucosidase activities such as Bgl2, Scw4, Crh11, Phr1 and Phr2. (2) Structural proteins with non-enzymatic activities such as agglutinins (e.g. Als1, Rbt1, Hwp1) (Dranginis et al. 2007) or β-1,3-D-glucan cross connectors (e.g. Pir1) are important for cell-to-cell interactions or wall integrity, respectively. Finally, cell surface proteins that are membrane-localised and function as mechano-sensors are primarily responsible for activating the cell wall integrity (CWI) pathways in response to cell wall damage. These proteins include Wsc1-3, Mid2 and Mtl1 (Heinisch et al. 2010; Levin 2011; Rodicio and Heinisch 2010).

16.2.4 Glycosylation of Cell Wall Proteins

Cell wall protein glycosylation is one of the most important post-translational modifications and has been extensively studied in *S. cerevisiae* (Jigami 2008). *C. albicans* mannosylation mutants with *O*- and *N*-mannosylation defects have been generated and the impact of the mutations on interactions with the host, immune response and cell wall integrity examined (Hall and Gow 2013). After synthesis proteins destined for the cell wall are extruded into the lumen of the ER. There the proteins are modified with 14 sugar oligosaccharides containing 2 GlcNAc, 9 mannose and 3 Glc residues. The synthesis of the oligosaccharide occurs in the ER. This involves the step-by-step addition of sugar residues to a dolichol lipid moiety by specific glycosyltransferases encoded by the *ALG* gene family (Aebi et al. 2010; Orlean 2012). The transferases use dolichol sugar intermediates and nucleotide sugars as substrates. The core oligosaccharides are transferred from the dolichol donors to the asparagine residues in the proteins. The target asparagine residues always occur in an asparagine-X-threonine/serine sequence, and with the exception of proline, X can be any amino acid. The transfer is mediated by the *N*-oligosaccharide transferase complex (OST) which catalyses the formation of glycosidic bonds between the NH_2 of the target asparagine residues and the first GlcNAc in the oligosaccharide chains (Kelleher and Gilmore 2006; Larkin and Imperiali 2011; Lennarz 2007; Schulz and Aebi 2009). Once the *N*-linked oligosaccharides have been attached to the proteins, they are further processed in the ER before being translocated to the Golgi. The processing and trimming in the ER and Golgi is responsible for the diversity of *N*-linked sugar components attached to the wall proteins (Helenius and Aebi 2004; Munro 2001). In *S. cerevisiae* the oligosaccharide $GlcNAc_2Man_9Glc_3$ is enzymatically processed in the ER to generate the triantennary core, $GlcNAc_2Man_8$, (Aebi et al. 2010; Quinn et al. 2009; Romero et al. 1997) which if properly folded, proteins bearing this glycan are exported to the Golgi. Deletion of any of the glycosidases (Mns1, Rot2 and Cwh41) responsible for trimming the oligosaccharide in *C. albicans* leads to reduced wall phosphomannan content, increased flocculation and attenuated virulence in a mouse model of systemic candidiasis (Mora-Montes et al. 2007). Misfolded proteins are trimmed and marked for destruction by the ER-associated protein degradation system

(Helenius and Aebi 2002). In the Golgi, mannan is further elaborated. Mannan elaboration begins with the addition of α-1,6-mannose to the α-1,3-mannose of the core glycan from GDP-mannose. The reaction is catalysed by the α-1,6-mannosyltransferase, Och1. In *C. albicans och1*Δ mutant strains have a temperature-sensitive growth defect. In addition, Δ*och1* mutants lack the α-1,6-linked polymannose backbone and show underglycosylation of *N*-acetylglucosaminidase leading to the absence of outer chain elongation. *C. albicans och1*Δ deletion mutants are hypersensitive to cell wall inhibitory agents and have a constitutively activated cell salvage pathway (Bates et al. 2006; Netea et al. 2006). The step-by-step addition of mannose by mannose polymerase I and II complex extends the α-1,6-mannose backbone. In *S. cerevisiae* mannose polymerase I contains Mnn9 and Van1 homologues and adds the first α-1,6-mannose to the Och1-derived α-1,6- mannose through the action of its Mnn9 subunit. Approximately, 10 α-1,6-linked mannose units are further added catalysed by Van1 (Hashimoto and Yoda 1997; Jungmann and Munro 1998; Jungmann et al. 1999; Rodionov et al. 2009; Stolz and Munro 2002). *Camnn9*Δ deletion mutants had growth defects, formed aggregates and showed characteristic phenotypes of cell wall defects with increased glucanase sensitivity. In addition, the mutant had a 50% reduction in mannan levels compared to the wild type (Southard et al. 1999). The Mnn10 and Mnn11 subunits of mannose polymerase II are responsible for the addition of around 60 α-1,6-linked mannose (Jungmann et al. 1999). Chain elongation is terminated with the addition of α-1,2-mannose when core *N*-glycan is formed.

Side branching α-1,2-linked mannose is added to the α-1,6-mannose backbone through the initiating α-1,2-mannose transferase activity of Mnn2, while Mnn5 adds the second α-1,2-mannose (Hall et al. 2013). *Camnn2*Δ deletion mutants showed low molecular weight, less complex short mannan fibrils (Hall and Gow 2013; Hall et al. 2013). *Camnn5*Δ deletion mutants had impaired hyphae formation (Bai et al. 2006). Both *Camnn2*Δ and *Camnn5*Δ had attenuated virulence in mice models of systemic candidiasis. In *S. cerevisiae* the addition of other α-1,2-mannose sugars is through the collective activity of Ktr3, Ktr2, Ktr1, Kre2/Mnt1 and Yur1 enzymes (Lussier et al. 1999), while Mnn1 adds α-1,3-mannose and blocks the elongation of α-1,2-mannose chains (Romero et al. 1999). The Mnn1-like protein family is made up of six members. Only the deletion of *C. albicans MNN14* resulted in attenuated virulence (Bates et al. 2013). The side branching mannose chains are further modified with mannose phosphate by the activities of Mnn6/Ktr6 and Mnn4. The mannose phosphate is a β-1,2-mannose moiety linked to the α-1,2-mannose side branching through a phosphodiester bond. This molecule adds negative charge to the cell surface of *C. albicans* and gives the ability to bind a cationic dye Alcian blue. It is also important for macrophage phagocytosis (McKenzie et al. 2010). The *C. albicans mnn4*Δ deletion mutant had impaired ability to bind Alcian blue and increased net hydrophobicity of the cell wall (Hobson et al. 2004; Singleton et al. 2005).

O-mannosylation begins in the lumen of the ER with the transfer of mannose from the dolichol phosphate mannose donors to wall proteins. The transferred mannose is attached through an α-linkage to the serine or threonine residues of the proteins. The serine and threonine at *N*-glycosylated sites are also potential targets for *O*-mannosylation (Orlean 2012). This initial process is catalysed by a family of protein *O*-mannosyltransferases (PMTs) (Hall and Gow 2013; Lussier et al. 1999; Prill et al. 2005). The α-mannose transferred to wall proteins can be further extended by transfer of mannose units from GDP-mannose by the Mnn1 and Ktr1 families in the Golgi (Lussier et al. 1999). In *C. albicans* the first and second α-1,2-linked mannoses are transferred by α-1,2-mannosyltransferases, Mnt1 and Mnt2, to the α-mannose forming a trisaccharide. Mnt1 and Mnt2 have been shown to have redundant functions. Significantly reduced attachment to epithelial cells was observed in both *C. albicans mnt1*Δ and *mnt2*Δ single mutants (Buurman et al. 1998; Munro et al. 2005). *C. albicans* cells with a *MNT1* and *MNT2* double gene knockout have truncated *O*-mannan, altered mannosyltransferase activity in vitro and attenuated virulence (Munro et al. 2005). This underlines the importance of *O*-mannosylated wall proteins in the pathogenesis of *C. albicans* infection. In *S. cerevisiae*, Mnn1 family members (Mnt3, Mnt2 and Mnn1) further extend the mannose chain by attaching up to two α-1,3-linked mannoses (Romero et al. 1999). The Mnn6 protein modifies the chain by tethering mannose phosphate to the second α-1,2-linked mannose in *S. cerevisiae* (Nakayama et al. 1998). The deletion of one *PMT* gene in *S. cerevisiae* is not lethal but in *C. albicans* deletion of *PMT2* has a lethal phenotype and mutants lacking single members of the 5-membered Pmt family have different phenotypes indicating specificity in the glycosylation of target proteins (Peltroche-Llacsahuanga et al. 2006; Prill et al. 2005; Rouabhia et al. 2005; Timpel et al. 2000). Pmt1 and Pmt4 are synthetically lethal (Prill et al. 2005). Pmt4 was found to mediate *O*-mannosylation during morphogenesis in specific environments. *C. albicans pmt6*Δ mutants have partial blockage of hyphal formation and defects in adherence to endothelial cells. *Capmt6*Δ and *Capmt1*Δ mutants are supersensitive to antifungal agents. *Ca*Pmt4 and *Ca*Pmt6 are required for full virulence in mice models of systemic candidiasis (Prill et al. 2005; Rouabhia et al. 2005; Timpel et al. 2000). *pmt*Δ mutants also have differential biofilm forming ability. A Pmt inhibitor, OGT2599, was found to inhibit early-stage biofilm formation in *C. albicans* (Peltroche-Llacsahuanga et al. 2006). In *S. cerevisiae* mutants with double gene deletion such as *pmt3*Δ *pmt2*Δ are osmotically fragile, but the knockout of some three gene combinations such as *pmt4*Δ *pmt3*Δ *pmt2*Δ has a lethal effect (Ecker et al. 2003; Lommel and Strahl 2009). These results demonstrate the importance of *O*-mannosylation in yeast cell wall assembly and cell survival. Arroyo et al. (2011) showed that inhibition of *O*-mannosylation (by deleting *PMT* genes) leads to activation of the CWI pathway and unfolded protein response. Inhibition of *N*-glycosylation (by inhibiting Alg7 with tunicamycin) leads to elevation of transcription of *PMT* genes (Travers et al. 2000). This finding suggests that the *O*- and *N*-mannosylation of cell wall proteins may have complimentary roles. The Pmt proteins have been proposed as suitable drug targets (Ernst and Prill 2001; Rouabhia et al. 2005). Pmt1 inhibitors, rhodanine-3-acetic acid derivatives, have been developed and have different morphological effects on *C. albicans*. These inhibitors can be used as a gateway for the search of potent antifungal agents (Orchard et al. 2004).

The gene products involved in each of the *O*- and *N*-mannosylation processes, in glycan synthesis, transfer and modification, have overlapping functions. Consequently, the disruption of any one gene may have no or only modest effect on the associated glycosylation step or pathway. Only simultaneous multiple gene disruption results in a severe phenotype and possibly cell death.

The majority of mannosyltransferases require metal ions (mainly Mn^{2+}) for function, making ion transport into the ER and Golgi an important factor for mannosylation. Pmr1 is required for the transport of divalent cations (Ca^{2+}/Mn^{2+}) into the Golgi. Deletion of *PMR1* in *C. albicans* truncates the branched *N*- and *O*-mannosylation (Bates et al. 2005). This could be due to the inhibition of multiple transferases due to insufficient Mn^{2+} in the Golgi. It is crucial that the initial steps of glycosylation of cell wall proteins occur in the ER before the proteins are exported to the Golgi for extension and branching. In the Golgi the *N*- and *O*-linked glycans can be extended into structures that contain about 200 and 5 mannose residues, respectively (Goto 2007; Orlean 2012; Yoko-o et al. 2003). The mannose utilised by the transferases in the Golgi is donated by GDP-mannose. GDP-mannose is synthesised in the cytoplasm and requires transport into the lumen of the Golgi by a unique sugar nucleotide transporter encoded by *VRG4*. Vrg4 catalyses the first rate-limiting step in the mannosylation of wall proteins in the Golgi. Partial loss of function of *CaVRG4* gene leads to mannosylation defects. Complete disruption of *CaVRG4* and *ScVRG4* results in cell death (Berninsone and Hirschberg 2000; Nishikawa et al. 2002). The *VRG4* gene is essential, fungal specific (i.e. does not have a mammalian homologue), and therefore presents a unique target for developing new antifungal drugs.

16.2.5 Glycosylphosphatidylinositol-Anchored Cell Wall Proteins

Protein modification has an important influence on protein structure and on the regulatory control of protein function. These modifications can also target a protein to its final destination. Most *O*- and *N*-glycosylated cell wall proteins also receive a GPI anchor during their passage through the secretory pathway (Orlean 2012). The GPI anchor serves to transiently attach the protein to the lumen of the ER and subsequently localise it to the plasma membrane and/or cell wall. Wall proteins destined to receive a preformed GPI anchor contain a unique signal sequence at their C-terminal (Eisenhaber and Eisenhaber 2007, 2010). The C-terminal signal sequence is recognised and cleaved by a GPI transamidase complex subunit GAA1/GPAA1 (Eisenhaber et al. 2014). This complex mediates the formation of an amide linkage between the newly formed C-terminal and the amino group of the ethanolamine of the GPI lipid moiety (Eisenhaber et al. 2014; Ferguson 1999; Kinoshita and Inoue 2000).

16.2.5.1 The Synthesis and Anchoring of Glycosylphosphatidylinositol Lipid Moiety to Cell Wall Proteins

The synthetic pathway for GPI lipids involves several enzyme-catalysed reactions. This multistep GPI synthetic pathway and the structure of GPI anchors has been well studied in mammals, protozoa and yeast (Tiede et al. 1999). The GPI anchors in these organisms share a common structure, $H_2N{-}CH_2{-}CH_2{-}PO_4\alpha6$Man α1,2-Man α1,6Man α1,4GlcNα1,6 *myo*inositol phospholipid. This structure is composed of a phosphatidylinositol, a linear trisaccharide with characteristic bonds between each sugar residue and phosphoethanolamine (PEtN). The side chains attached to the core structure and the lipid moiety of the anchor are responsible for the observed diversity in species. The synthesis of GPI anchors occurs in the ER and follows at least a 10-step process. It begins with the transfer of GlcNAc from UDP-GlcNAc to phosphatidylinositol and the formation of an α-1,6-linkage catalysed by UDP-GlcNAc transferase complex (Tiede et al. 2000; Watanabe et al. 1996). The phosphatidylinositol is attached to a ceramide- or diacylglycerol-based lipid. The reaction follows a deacetylation step to generate GlcN phosphatidylinositol (Watanabe et al. 1999). The *myo*-inositol ring is then acylated at position 2′ and the first mannose is added to position 4′ of GlcN by an α-1,4-linkage to produce mannose-GlcN-acyl phosphatidylinositol. These processes are catalysed by the acyltransferases, Gwt1 and Gpi14 (Maeda et al. 2001), respectively. The protein, Gwt1, has been studied as a potential antifungal drug target (Hata et al. 2011; Tsukahara et al. 2003; Wiederhold et al. 2015). The mannose is then modified with the addition of PEtN catalysed by the PEtN transferase Mcd4 (Wiedman et al. 2007). The phosphate of PEtN interacts with the position 6′ carbon of the mannose. The product, PEtN-Man-GlcN-acyl phosphatidylinositol, receives a second α-1,6-linked mannose residue and a third α-1,2-linked mannose residue (Canivenc-Gansel et al. 1998; Kang et al. 2005; Sato et al. 2007). The third mannose residue is modified with the addition of PEtN transferred from phosphatidylethanolamine to the position 6′ carbon to form (PEtN)Man-Man-(PEtN)Man-GlcN-acyl phosphatidylinositol (Taron et al. 2000). The PEtN at the third mannose residue is involved in the attachment to the recipient protein (Taron et al. 2000). The last step involves the addition of PEtN to the position 6′ carbon of the second mannose residue to form (PEtN)Man-(PEtN)Man-(PEtN)Man-GlcN-acyl phosphatidylinositol. The mature GPI anchor bearing three PEtNs is transferred to the protein with a C-terminus signal GPI anchoring sequence in a transamidation reaction. In this reaction the amino acid group of the PEtN on the third mannose acts as a nucleophile. The reaction is catalysed by a transamidase enzyme complex which cleaves a peptide bond near the C-terminus of the protein and attaches the GPI anchor by an amide linkage (Benghezal et al. 1996; Grimme et al. 2004; Hamburger et al. 1995). The entire process of GPI biosynthesis is catalysed by about 20 gene products which are organised into synthetic complexes in the ER membrane (Orlean 2012). Some of these proteins have well-defined enzymatic activity, while others may have a structural role or perform auxiliary functions. In *S. cerevisiae* and *C. albicans,* before the attachment of PEtN to the third mannose of the GPI lipid, a fourth mannose residue is added at position 2′ through an α-1,2-linkage. This reaction is catalysed by the Smp3 mannosyltransferase. Repression of *CaSMP3* expression leads to a defect in the addition of the fourth mannose and blockage of PEtN addition to the third mannose (Eisenhaber et al. 2003; Grimme et al. 2001; Grimme et al. 2004). GPI assembly, an activity critical for growth of *C. albicans,* is blocked leading to loss of cell viability (Eisenhaber et al. 2003; Grimme et al. 2001; Grimme et al. 2004; Kinoshita and Inoue 2000; Taron et al. 2000). In mammals the majority of GPI lipid moieties tethered to a protein contain three mannose residues (Grimme et al. 2001). The addition of a fourth mannose to the GPI anchor is not important in mammals but it is in fungi; therefore, Smp3 has been proposed as a potential antifungal target (Grimme et al. 2001). The transferase Mcd4 has also been proposed as a suitable drug target. An Mcd4 inhibitor, M720, was identified through genomics-based screening and found to be effective in a murine model of systemic candidiasis (Mann et al. 2015). Other molecules have been identified that significantly affect *C. albicans* growth by targeting the GPI anchor synthetic pathway (Hata et al. 2011; Tsukahara et al. 2003; Wiederhold et al. 2015), thus demonstrating that the process of GPI anchor synthesis provides a potential antifungal drug target.

16.2.5.2 Glycosylphosphatidylinositol Anchored Cell Wall Protein Localisation

After further processing of the GPI proteins in the ER and Golgi, they are transported via the secretory pathway and targeted to the plasma membrane. In *S. cerevisiae* and *C. albicans* a number of GPI proteins, such as Ecm331, remain localised to the plasma membrane (Mao et al. 2003; Mao et al. 2008). Other GPI proteins such as Pga59 and Pga62 are cleaved and targeted to the cell wall (Moreno-Ruiz et al. 2009). Cell wall localisation of GPI-anchored proteins depends on the cleavage of the lipid portion of the GPI moiety leaving a remnant with which the proteins are tethered to the wall. The sequence of amino acids immediately upstream is thought to be responsible for determining the final destination of GPI-anchored proteins (Frieman and Cormack 2003, Mao et al. 2008). The enzymes responsible for the cleavage and attachment of GPI-anchored proteins to the wall have been predicted to include the synthetically lethal GPI-anchored proteins, Dfg5 and Dcw1. Evidence now exists in *S. cerevisiae, C. albicans* and *Neurospora crassa* that the enzymes Dfg5 and Dcw1 are required for the incorporation of GPI-anchored proteins into the cell wall as well as the mannosyltransferase Och1 in *C. albicans* (Ao et al. 2015; Kitagaki et al. 2002; Maddi et al. 2012).

The availability of genome sequences has made it possible to predict the number of ORFs with GPI anchor signal sequences (De Groot et al. 2003; Eisenhaber et al. 2004; Richard and Plaine 2007). Many of the proteins predicted to be GPI-anchored have been shown to have enzymatic activities on substrates that are found within the wall itself and/or the environment. Their enzymatic activities are involved in the synthesis, assembly and remodelling of wall polymers as well as in virulence (Albrecht et al. 2006; De Boer et al. 2010; De Groot et al. 2005; Gelis et al. 2012; Klis et al. 2010; Munro 2010, 2013; Pardini et al. 2006; Schild et al. 2011; Sheppard et al. 2004; Staab et al. 2013). Some of these proteins may be suitable drug targets for developing new potent antifungal agents. So far less than 30% of the predicted GPI proteins in *C. albicans* have been functionally characterised. A large-scale analysis of GPI proteins (Plaine et al. 2008) and gene family targeted studies (Pardini et al. 2006; Calderon et al. 2010) are required to further understand the importance of these proteins in the cell wall.

16.2.5.3 Glycosylphosphatidylinositol-Anchored Proteins as Cell Wall Localised Virulence Factors

C. albicans is one of the most successful pathogens of humans adapting to numerous microenvironments (Biswas et al. 2007). The highly adaptive capability of *C. albicans* may be in part because of its ability to express different virulence factors at each stage of the infection, and in any infected niche, leading to the successful establishment of infection especially in an immunocompromised host (Calderone and Fonzi 2001; Kumamoto and Vinces 2005; Mayer et al. 2013). Thus, particular virulence factors such as cell wall adhesins may only be relevant at the initial stages of host colonisation and invasion. Attachment to host cells is required before any organism can cause infection. *C. albicans* attaches to host cells to colonise as a commensal and to invade as a pathogen (Gaur and Klotz 1997; Gow and Hube 2012). *C. albicans* cell surface proteins that function as adhesins mediate the process of attachment which promotes host–pathogen interactions (Staab et al. 2013). The adhesive properties of the cell wall proteins also mediate attachment to inanimate objects such as indwelling medical devices and promote cell-to-cell interactions amongst *C. albicans* cells, which are required for biofilms formation (Garcia et al. 2011; Verstrepen and Klis 2006). The various groups of *C. albicans* cell surface adhesins, invasins and proteinases may explain the versatility of the fungus in adapting to several niches in human hosts (Munro 2010). Some of these surface proteins are encoded by *SAP* and *ALS* gene families and *HWP1*.

The main group of adhesins in *C. albicans* is the agglutinin-like sequence (Als) surface glycoproteins encoded by the *ALS* genes. The *ALS* gene family encodes eight Als proteins (Calderone and Fonzi 2001, Gaur and Klotz 1997, Hoyer et al. 2008, Hoyer 2001). Proteins encoded by the *ALS* gene family have three basic characteristics—the N-terminus, the central domain and the C-terminus—which has varying degrees of similarity amongst the family members. The carboxylic end of these glycoproteins is attached to a GPI anchor which links them to the cell surface (Hoyer et al. 1995) whereas the amino end is involved with ligand binding (Hoyer et al. 2008; Sheppard et al. 2004). The peptide-binding ability at the amino end is conferred by the Ig-fold domain (Gaur and Klotz 2004). The Ig-fold mediates adhesion to fibronectin and cell-to-cell aggregation through Als to Als interaction (Nobile et al. 2008). Furthermore, just after the Ig-fold domain is the threonine-rich domain which contains an amyloid-forming sequence (Rauceo et al. 2006). The domain modulates the structure of the Ig-fold and enhances its adhesive properties. In all Als proteins, the threonine-rich domain is the most conserved. Furthermore, downstreams of the threonine-rich domain are tandem repeats which are responsible for non-specific binding to different surfaces (Frank et al. 2010, Lipke et al. 2014). Als proteins bind to different surfaces. Sheppard et al. (2004) heterologously expressed Als proteins in a non-adherent *S. cerevisiae* strain and observed that Als1, Als3 and Als5 proteins bind to all substrates tested, while Als6 and Als9 bind only to gelatin and laminin, respectively. Als7 did not bind to any of the tested surfaces. Since Als2 and Als4 could not be heterologously expressed, Zhao et al. (2005) used deletion mutants to demonstrate that while both are essential for attachment to endothelial cells only Als2 is essential for attachment to epithelial cells in a reconstituted human epithelium model. Als1 protein is required for *C. albicans* adherence to mice tongues (Kamai et al. 2002). Phan et al. (2007) used molecular models to show that Als3 binds to human epithelial cells through cadherins (transmembrane proteins involved in cell adhesion) and induces endocytosis in a manner similar to the mechanism by which host cells cadherins induce endocytosis. The Als3 protein has hyphal associated expression (Argimon et al. 2007). The *ALS* gene family members are known potent virulence factors and Als1 and Als3 have been shown to have the potential for use in the development of prophylactics. The recombinant N-terminus of *C. albicans* adhesin, rAls3p-N, protects mice against *C. albicans* infections by cell-mediated immune response. The vaccine has been shown to confer protection on immunocompetent mice from both systemic and vaginal candidiasis by significantly reducing fungal burden in the latter. Furthermore, in a corticosteroid-treated murine model of oropharyngeal candidiasis, the vaccine also reduced oral fungal burden, significantly. Human trials with the rAls3-N vaccine are ongoing (Ibrahim et al. 2006; Liu and Filler 2011; Spellberg et al. 2006).

Hyphal wall protein 1 (Hwp1) is an outer surface mannoprotein found on germ tubes and hyphal cells of *C. albicans* (Staab et al. 1999). The C-terminus contains a GPI anchor sequence which attaches the protein to the cell wall β-glucan resulting in surface exposure. The N-terminus containing the secretory signal sequence is rich in proline and glutamine amino acids, resembling a substrate for mammalian transglutaminase enzyme (Staab and Sundstrom 1998; Staab et al. 1999). Transglutaminase enzymes cross link proline-rich proteins to epithelial cells. It is the enzyme–substrate relationship between the Hwp1 N-terminus and the mammalian transglutaminase active site that enables the attachment of *C. albicans* to human buccal epithelial cells (Staab et al. 2013). *C. albicans* cells with homozygous gene deletion of *HWP1* show reduced ability to form hyphae and to bind to human buccal epithelial cells. This is consistent with the reduced virulence observed in a murine model of systemic candidiasis (Sundstrom et al. 2002; Tsuchimori et al. 2000). Hwp1 and Als proteins on cell surfaces can interact promoting cell-to-cell interaction that contributes to biofilm formation (Nobile et al. 2008). These studies show that Hwp1 protein is essential for *C. albicans* hyphal-specific attachment to epithelial cells and to other hyphal and yeast cells.

Overexpression of Hwp1 in *C. albicans* improved the adhesion to an in vivo catheter model and the protein has been suggested as a therapeutic target (Nobile et al. 2006b).

C. albicans has the ability to adhere to and form biofilms—a population of cells attached to one another and/or a surface surrounded by exopolysaccharide matrix—on indwelling medical devices which contribute to virulence and drug resistance (Sherry et al. 2014). Biofilm formation by *C. albicans* has been shown to be under the positive regulatory control of the transcription factor, Bcr1. Bcr1 regulates the expression of Als1, Als3 and Hwp1, which are involved in biofilm formation (Nobile and Mitchell 2005; Nobile et al. 2006a, b). These proteins plus Als2 are associated with various stages of biofilm formation in *C. albicans* (Green et al. 2004; Zhao et al. 2005), thus contributing to virulence.

To establish a successful infection with *C. albicans,* the organism must overcome host nutritional immunity to acquire nutrients including metals such as iron from the host (Almeida et al. 2009). Following the release of iron from lysed blood cells during candidiasis, *C. albicans* uses receptor proteins on the cell surface to take up iron in an endocytosis-mediated manner (Weissman et al. 2008). Als3, Rbt5, Rbt51/Pga10 and Pga7 are GPI-anchored cell wall proteins that have iron-binding receptors (Weissman and Kornitzer 2004). These proteins contain CFEM domains characterised by an eight-spaced cysteine residue sequence (Kulkarni et al. 2003). *C. albicans rbt5Δ* deletion mutants show reduced ability for iron uptake (Weissman and Kornitzer 2004). Rbt5 has hyphal-associated expression and this may explain why *C. albicans* hyphae bind more strongly to haemoglobin than yeasts cells. However, the *C. albicans rbt5Δ* mutants exhibited wild-type virulence (Braun et al. 2000). This could be due to redundancy in the functions of the iron-binding cell surface proteins.

Hydrolytic enzymes contribute to the pathogenesis of *C. albicans* infections. Of all the hydrolytic enzymes, the most extensively studied, are the secreted aspartyl proteinase (Sap) isoenzymes encoded by the *SAP* gene family (Naglik et al. 2003). This gene family encodes 10 Sap enzymes (Naglik et al. 2004; Odds 2008). Each mature Sap protein contains two aspartic acid residues conserved within the active site and a conserved cysteine residue which plays a structural role (Hube and Naglik 2001; Naglik et al. 2004; Naglik et al. 2003). Sap9 and Sap10 proteins are attached to the cell wall of *C. albicans* through C-terminal GPI anchored sequences, while Sap1–Sap8 proteins are secreted (Hube and Naglik 2001; Monod et al. 1994). These proteinases are differentially expressed under different conditions and this may be crucial for the overall success of *C. albicans* as both a commensal and a pathogen. For example, the yeast-associated Sap9-10 has a near neutral pH optimal activity, while the optimal pH for the hyphal-associated Sap4-6 is alkaline (Naglik et al. 2003; Schild et al. 2011). Sap proteins have been linked to the ability of *C. albicans* to adhere to and damage host tissue as well as the ability to evade the host immune response (Hube and Naglik 2001). Treatment of *C. albicans* with antifungal drugs induces expression levels of *SAP2* and *SAP9* genes (Copping et al. 2005). *C. albicans* can increase the expression level of certain *SAP* genes as a compensation for deleted *SAP* genes, suggesting possible redundancy in Sap protein function (Schaller et al. 1999). For example, *C. albicans* increased the expression level of *SAP5* as a compensatory response in the triple-gene deletion mutant *sap1-3*Δ (Naglik et al. 2008).

With the recognised importance of host–pathogen interactions in the development of candidiasis, our understanding of the main roles played by these GPI-anchored cell surface proteins is limited. Furthermore, the humoral and cell-mediated immune responses triggered by these proteins are still poorly understood. Thorough investigation of each of these GPI-anchored cell surface proteins may broaden the available option of protein candidates for the development of prophylactics, immunotherapies and therapeutics. It is possible to combine an immunogenic epitope from a selection of wall proteins as a multicomponent vaccine (Heilmann et al. 2012). Inhibiting or blocking these cell surface virulence factors may be useful in reducing the pathogenicity of *C. albicans* infection especially in infections with drug-resistant strains.

16.2.5.4 Glycosylphosphatidylinositol Cell Wall Proteins and Cell Wall Remodelling

Cell wall glycoproteins with enzymatic activities can function as glycosylhydrolases that cleave wall polymers. Glycosyltransferases or transglycosidases modify wall components by crosslinking wall polymers and fragments of polymers (Free 2013; Lombard et al. 2014; Munro 2013; Orlean 2012). The non-enzymatic or 'structural' glycoproteins may function as β-1,3-glucan cross connectors and as a scaffold for the attachment of other wall components (Yin et al. 2005). The concerted effort of these glycoproteins in constructing the wall during synthesis and assembly, and in remodelling the wall in response to stress, to continually maintain a balance between wall rigidity and plasticity makes the cell wall a dynamic organelle. Glycoproteins are synthesised and targeted to the cell wall in response to specific environmental conditions such as changes in pH, oxygen levels and incubation temperature (Ene et al. 2012a, b, 2015, Sosinska et al. 2008). GPI anchor proteins such as Pga31, Sap9 and Phr1 have been shown to be upregulated when *C. albicans* is grown in lactate but not in glucose (Ene et al. 2012b). Lactate-grown *C. albicans* is more adapted to osmotic stress and less sensitive to antifungals (Ene et al. 2012a, 2015). *C. albicans* grown on lactate is more strongly adherent to plastics and more virulent than glucose grown cells (Ene et al. 2012a). The carbon-source-induced alteration in osmotic tolerance was shown to be independent of the CWI pathway (Ene et al. 2012a), but rather by the alteration in the biophysical properties of the cell wall. However, this kind of alteration may also function as a compensatory mechanism to maintain an intact cell wall.

Generally, the CWI pathway or protein kinase C (Pkc) pathway is activated in response to damage to the cell wall (Levin 2011). In *S. cerevisiae* the signal is received by cell wall mechanosensors, which trigger activation of Pkc and the downstream MAP kinase cascade. Rlm1 is the transcription factor downstream of the pathway and regulates the expression of genes involved in cell wall construction and remodelling. Cell wall and plasma membrane-localised GPI-anchored proteins such as Sed1, Ccw14, Cwp1, Ecm331, Crh1, Dfg5 and Sap10 have been shown to be regulated by the protein kinase C pathway (Jung and Levin 1999; Krysan et al. 2005; Terashima et al. 2000). Ca^{2+}/calcineurin and the high osmolarity glycerol (HOG) pathways are also associated with cell wall salvage mechanisms in *C. albicans* (Munro et al. 2007). Caspofungin stress provokes compensatory chitin upregulation that reinforces the cell wall leading to drug tolerance. Yeast cells with elevated chitin content are less susceptible to wall perturbing agents such as caspofungin (Lee et al. 2012; Walker et al. 2013a) and glucanase (De Nobel et al. 2000). This has led to the proposal that combination therapies that block both glucan and chitin synthesis may be attractive to abrogate the acquisition of caspofungin tolerance and resistance. However, inhibiting the cell wall proteins which play central roles in cell wall remodelling leading to an altered cell wall architecture may also significantly impact upon cell wall integrity.

16.3 Cell Wall Inhibitory Agents and Fungal Resistance

Fungal cell wall components are absent from mammalian cells, which makes them and their synthesis promising targets for antifungal drugs. Damaging the cell wall leads to cell lysis and cell death. Although the cell wall promises to be a suitable target for antifungal drugs, only the echinocandins have been clinically approved. The echinocandins are the first successfully developed lipopeptides against fungal pathogens. They have lipid side chains coupled to a cyclic hexapeptide core. The lipid side chains are responsible for the antifungal activity (Odds et al. 2003). The echinocandins (caspofungin, micafungin and anidulafungin) exert their fungicidal effect by targeting and non-competitively inhibiting the β-1,3-D-glucan synthase catalytic subunit, Fks (Douglas et al. 1997). In vitro and in vivo experiments have shown that the echinocandins are fungicidal against *Candida* species. Echinocandins are active against invasive fungal infections and are used in some hospitals as the first line drug; however, there are now sporadic breakthrough infections in patients receiving echinocandin therapy (Arendrup et al. 2009; Brielmaier et al. 2008; Gamaletsou et al. 2014; Miller et al. 2006; Park et al. 2005; Pfeiffer et al. 2010). *C. albicans* strains that are resistant to echinocandins acquire single nucleotide polymorphisms (SNPs) in the *FKS1* gene (Park et al. 2005; Perlin 2007, 2011, 2015) (Chapter 15; 22). The SENTRY surveillance project has examined the prevalence of *FKS* mutations in echinocandin-resistant isolates and found that the majority of resistant isolates of the major pathogenic *Candida* species tested do not harbour these mutations suggesting that alternative mechanisms including cell wall remodelling may contribute to drug resistance (Castanheira et al. 2016). Cell wall remodelling has been shown to alter the susceptibility of *C. albicans* to echinocandin drugs (Lee et al. 2012; Walker et al. 2013a). In vitro treatment of *C. albicans* with sub-inhibitory concentrations of caspofungin induces increased chitin synthesis as a compensatory response, through activation of the

Pkc, calcineurin and HOG signalling pathways. *C. albicans* cells with elevated chitin content in the wall are less susceptible to caspofungin treatment (Walker et al. 2008). The observed elevation in cell wall chitin content appears to be an adaptive mechanism to maintain an intact cell wall and not a genetic alteration but remodelling of the cell wall may give the fungus a window of opportunity to acquire the point mutations that subsequently fix echinocandin resistance in the population. *C. albicans* with high cell wall chitin can tolerate echinocandin treatment in vivo in a murine model of systemic candidiasis (Lee et al. 2012). Some *C. albicans*-resistant clinical isolates have been found to have hot spot mutations plus a thicker wall with elevated wall chitin content (Ben-Ami et al. 2011; Imtiaz et al. 2012). The compensatory elevation of cell wall chitin content has also been associated with paradoxical growth (Stevens et al. 2004). This phenomenon refers to the ability of the fungus to grow at high drug concentrations. The results described above provide evidence indicating that as the number of *C. albicans* strains exposed to the echinocandins increases, the number of tolerant/resistant strains may rise (Perlin 2007; Pfaller 2012; Rueda et al. 2014). This may have a dramatic, negative effect, on the clinical outcome for candidiasis patients receiving echinocandin therapy in the near future. Similarly, these results also point to the urgent need for new effective antifungal agents. The compensatory increase in cell wall chitin content underpins the need for combination therapy for a broader spectrum of activity and/or more effective therapy.

Studies of the combinations of CWI pathway inhibition or chitin synthase inhibitors with echinocandin treatment have shown synergistic effects between the treatments (Walker et al. 2008). The combination of gene deletion leading to the inhibition of the calcineurin pathway or the use of calcineurin pathway inhibitors such as cyclosporine, with caspofungin treatment, has a synergistic effect against *C. albicans, A. fumigatus* and *C. neoformans* (Del Poeta et al. 2000; Kontoyiannis et al. 2003; Wiederhold et al. 2005). The combined treatment is also effective at inhibiting paradoxical growth of *C. albicans* in response to caspofungin treatment (Wiederhold et al. 2005). Similarly, the combination of a chitin synthase inhibitor such as nikkomycin Z with echinocandin treatment has synergistic effect against an echinocandin-resistant strain of *C. albicans* with an *FKS1* hot spot mutation (Walker et al. 2008).

Unfortunately, despite these intriguing results suggesting potential for synergistic effects, no chitin synthase inhibitors have been approved for clinical use. Unlike the azoles, absence of antagonism in combination with other antifungal agents provides evidence that combination therapy may be an important feature in echinocandin treatment which has not been fully exploited.

16.4 Old and New Components, Potential Targets for New Antifungal Molecules

Drug development for treatment of fungal pathogens is challenging as fungi share some biochemical pathways with their human hosts. The implication is that some fungal proteins involved in fundamental processes such as energy generation, stress response, protein localisation and gene regulation may have the same or similar functions in humans. Consequently, molecules that are able to inhibit fungal growth may be toxic to humans. This is why antifungal drug targets should ideally be specific to fungi. The cell wall is an obvious histological distinction between fungal and human cells. Because of this uniqueness, it is hoped that molecules of antifungal pharmacopoeia targeting fungal wall components, such as the echinocandins, will become staples as antibacterial wall-targeted drugs. Unfortunately, this has yet to be shown to be the case. The development of antifungal agents that target the cell wall has received great attention but thus far only limited success has been recorded. For example, the only class of antifungal agents that target the cell wall (the echinocandins) was discovered in the 1970s and took 30 years to be clinically approved (Roemer and Krysan 2014). Recently, other β-1,3-glucan inhibitors such as enfumafungin antibiotics have been identified. The lead molecule MK3118 (Fig. 16.2) is an orally active synthetic derivative of enfumafungin with in vitro antifungal activity (Roemer and Krysan 2014).

In vivo, MK3118 has been found to be active against *Candida* and *Aspergillus* species. Drug resistance to MK3118 was mapped to *FKS1,* although strains resistant to the echinocandins do not display cross resistance to MK3118, thus emphasising different mechanisms of action for both agents (Hector and Bierer 2011; Jimenez-Ortigosa et al. 2014; Pfaller et al. 2013). Another molecule, piperazinyl-pyridazinone, has also been identified and shown to inhibit β-1,3-glucan synthesis (Walker et al. 2011). Piperazinyl-pyridazinones possess both in vitro and in vivo antifungal activity. The piperazinyl-pyridazinones have in vitro activity against *Candida* and *Aspergillus* species and unlike the echinocandins, they are also active against *C. neoformans*, *Fusarium moniliforme* and *Trichophyton* species. In vivo, the piperazinyl-pyridazinones have only been effective against *Candida glabrata* in a murine model of systemic candidiasis (Walker et al. 2011). The mode of action of the piperazinyl-pyridazinone is different (i.e. it binds to a different region of the Fks1 protein) to that of the echinocandins. Echinocandin-resistant *C. albicans* was found to be susceptible to the piperazinyl-pyridazinones (Walker et al. 2011).

The unavailability of a specific protein whose catalytic activity has been directly linked to the synthesis of β-1,6-glucan has made it difficult to identify molecules that can directly inhibit the synthesis of this cell wall component. However, pyridobenzimidazole derivative (D75-4590) (Fig. 16.2) was found to target and inhibit Kre6 which is linked to the synthesis of β-1,6-glucan. Analysis of cell wall material from D75-4590-treated *C. albicans* cells showed a reduced level of β-1,6-glucan (Kitamura et al. 2009a, b). Since GPI-modified cell wall proteins are coupled to the

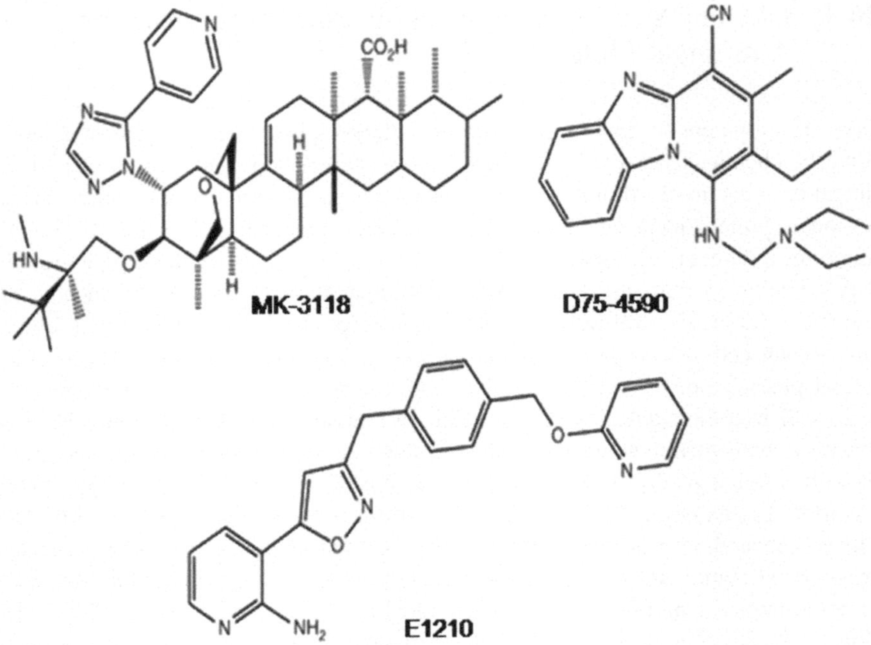

Fig. 16.2 Some candidate antifungal molecules in development. MK-3118 inhibits the synthesis of β-1,3-glucan and it is effective against *Candida* and *Aspergillus* species. D75-4590 is effective against *Candida* species. Its mode of action is through the inhibition of the synthesis of β-1,6-glucan. E1210 inhibits Gwt1 which is involved in the acylation of the inositol ring of GPI proteins. Adapted from Roemer and Krysan (2014)

cell wall by first attaching to β-1,6-glucan molecules, inhibition of β-1,6-glucans may lead to inhibition of the attachment of glycoproteins to the cell wall. Other derivatives of pyridobenzimidazole, such as D21-6076, were identified and found to have in vitro and in vivo activities against *C. albicans* (Kitamura et al. 2009a, b).

Anti-infective molecules have been identified which can interfere not with the cell wall components but with the processes that are essential in the synthesis of the cell wall, such as cell wall localisation of the GPI anchored wall glycoproteins. One is a pyridine-2-amine-based molecule, E1210 (Fig. 16.2). E1210 targets and inhibits the acylation of the inositol ring catalysed by Gwt1, an inositol acyltransferase. Depending on the strain background, deletion of *GWT1* is lethal, or leads to temperature sensitivity and slow growth. E1210 was effective at low concentrations against *Candida, Aspergillus, Fusarium* and *Scedosporium* species. It was also effective against caspofungin-resistant *C. albicans* (Hata et al. 2011, Tsukahara et al. 2003, Wiederhold et al. 2015). Another molecule, phenoxyacetanilide (gepinacin), was identified that could inhibit Gwt1 (McLellan et al. 2012). Gepinacin inhibited the growth of diverse yeasts and moulds. It was also found to inhibit the growth of *C. albicans* strains that are resistant to caspofungin, fluconazole or amphotericin B (McLellan et al. 2012). A Gwt1 inhibitor has been found to have synergistic effect with the Mcd4 inhibitor, M720 against *C. albicans* (Mann et al. 2015). Many of the GPI glycoproteins have roles in *C. albicans* virulence or have enzymatic activities in the cell wall, thus inhibiting their synthesis and localization to the cell wall could impact on virulence and cell wall biosynthesis significantly.

16.5 Perspectives

Fungal cell wall biosynthesis is a complex multistep process. The cell wall is a mutlipolymeric structure and its components are unique to and conserved in fungi. Consequently, known wall glycoproteins which act in a highly regulated manner to create and maintain a balance between wall rigidity and plasticity or that have roles in virulence have no human homologues. So far, even with our understanding of the cell wall biosynthetic processes, only the synthesis of β-1,3-glucan has been exploited as an antifungal target with the development of the echinocandins. Inhibition of β-1,3-glucan synthesis activates CWI pathways leading to cell wall remodelling. Cell wall remodelling, which results in an increase in chitin synthesis in addition to altered crosslinking of the cell wall polymers, is required by the fungus to adapt to external stress. The CWI pathway and chitin synthesis must be overcome in order to reduce the resistance to and increase the effectiveness of β-1,3-glucan-targeted therapies. CWI pathway and elevated chitin synthesis have the potential to be exploited as targets for antifungal agents. Several new β-1,3-glucan synthase inhibitors have been discovered. If they are clinically approved, they may share the same fate as the echinocandins, activating compensatory cell wall remodelling activities when used at sub-optimal concentrations. Thus a clinically approved agent that directly and non-competitively inhibits the synthesis of chitin may improve the clinical usefulness of β-1,3-glucan inhibitors in terms of spectrum of activity and effectiveness.

The glycosylation of cell wall proteins that is initiated and completed in the ER and the Golgi, respectively, is orchestrated by a group of transferases. These transferases, especially the essential enzyme, Vrg4, which catalyses the rate-limiting step in the glycosylation of wall proteins, are also suitable drug targets for developing new antifungal agents.

The synthesis and attachment of a GPI anchor to cell wall proteins represents important processes in cell wall biogenesis. All the enzymes involved in these steps are thought to be essential. The core steps in GPI anchor synthesis and attachment are conserved from fungi to human, but the unique steps, such as the addition of the fourth mannose, which is specifically present in fungi, may comprise a suitable drug target (Bowman and Free 2006). Antifungal agents targeting this fungal-specific step could have a significant impact on fungal growth and virulence.

Various hydrolases, transferases and transglycosidases function in the extracellular space to assemble and crosslink the wall components extruded into the wall space. The anticipated effective antifungal agents yet to be discovered and developed are those that target and inhibit these enzymes. Agents that inhibit the activities of these enzymes and block the modification of the cell wall polymers as well as their crosslinking to each other are believed to be excellent candidates for the development of new antifungal drugs (Gozalbo et al. 2004). Inhibitory molecules could block these targets without having to cross the plasma membrane and would inhibit important steps in cell wall construction in the extracellular space.

The cell wall remains one of the most attractive antifungal targets, yet it is a virtually an unexploited area of research compared to the bacterial cell wall in terms of clinically approved wall-targeted drugs. Even with recent successes in understanding cell wall biogenesis, more research is needed if more effective antifungal drug candidates are to earn clinical approval.

Acknowledgements We would like to acknowledge the support of the Wellcome Trust and MRC Centre for Medical Mycology (MR/N006364/1). The authors also thank the University of Aberdeen and Abia State University, Nigeria for Sponsorship.

References

Aebi M, Bernasconi R, Clerc S, Molinari M (2010) N-glycan structures: recognition and processing in the ER. Trends Biochem Sci 35:74–82

Aimanianda V, Clavaud C, Simenel C, Fontaine T, Delepierre M, Latge JP (2009) Cell wall beta-(1,6)-glucan of *Saccharomyces cerevisiae*: structural characterization and in situ synthesis. J Biol Chem 284:13401–13412

Albrecht A, Felk A, Pichova I, Naglik JR, Schaller M, de Groot P, Maccallum D, Odds FC, Schafer W, Klis F, Monod M, Hube B (2006) Glycosylphosphatidylinositol-anchored proteases of *Candida albicans* target proteins necessary for both cellular processes and host-pathogen interactions. J Biol Chem 281:688–694

Almeida RS, Wilson D, Hube B (2009) *Candida albicans* iron acquisition within the host. FEMS Yeast Res 9:1000–1012

Ao J, Chinnici JL, Maddi A, Free SJ (2015) The NLinked Outer Chain Mannans and the Dfg5p and Dcw1p Endo-alpha-1,6-Mannanases Are Needed for Incorporation of *Candida albicans* Glycoproteins into the Cell Wall. Eukaryot Cell 14:792–803

Arendrup MC, Garcia-Effron G, Buzina W, Mortensen KL, Reiter N, Lundin C, Jensen HE, Lass-Florl C, Perlin DS, Bruun B (2009) Breakthrough Aspergillus fumigatus and *Candida albicans* double infection during caspofungin treatment: laboratory characteristics and implication for susceptibility testing. Antimicrob Agents Chemother 53:1185–1193

Argimon S, Wishart JA, Leng R, Macaskill S, Mavor A, Alexandris T, Nicholls S, Knight AW, Enjalbert B, Walmsley R, Odds FC, Gow NA, Brown AJ (2007) Developmental regulation of an adhesin gene during cellular morphogenesis in the fungal pathogen *Candida albicans*. Eukaryot Cell 6:682–692

Arroyo J, Hutzler J, Bermejo C, Ragni E, Garcia-Cantalejo J, Botias P, Piberger H, Schott A, Sanz AB, Strahl S (2011) Functional and genomic analyses of blocked protein O-mannosylation in baker's yeast. Mol Microbiol 79:1529–1546

Bai C, Xu XL, Chan FY, Lee RT, Wang Y (2006) MNN5 encodes an iron-regulated alpha-1,2-mannosyltransferase important for protein glycosylation, cell wall integrity, morphogenesis, and virulence in *Candida albicans*. Eukaryot Cell 5:238–247

Bates S, MacCallum DM, Bertram G, Munro CA, Hughes HB, Buurman ET, Brown AJ, Odds FC, Gow NA (2005) *Candida albicans* Pmr1p, a secretory pathway P-type Ca^{2+}/Mn^{2+}-ATPase, is required for glycosylation and virulence. J Biol Chem 280(24):23408–23415

Bates S, Hughes HB, Munro CA, Thomas WP, MacCallum DM, Bertram G, Atrih A, Ferguson MA, Brown AJ, Odds FC, Gow NA (2006) Outer chain N-glycans are required for cell wall integrity and virulence of *Candida albicans*. J Biol Chem 281:90–98

Bates S, Hall RA, Cheetham J, Netea MG, MacCallum DM, Brown AJ, Odds FC, Gow NA (2013) Role of the *Candida albicans MNN1* gene family in cell wall structure and virulence. BMC Res Notes 6:294-0500-6-294

Becker JM, Marcus S, Tallock J, Miller D, Krainer E, Khare RK, Naider F (1988) Use of the chitin-synthesis inhibitor nikkomycin to treat disseminated candidiasis in mice. J Infect Dis 157 (1):212–214

Ben-Ami R, Garcia-Effron G, Lewis RE, Gamarra S, Leventakos K, Perlin DS, Kontoyiannis DP (2011) Fitness and virulence costs of *Candida albicans FKS1* hot spot mutations associated with echinocandin resistance. J Infect Dis 204:626–635

Benghezal M, Benachour A, Rusconi S, Aebi M, Conzelmann A (1996) Yeast Gpi8p is essential for GPI anchor attachment onto proteins. EMBO J 15:6575–6583

Berninsone PM, Hirschberg CB (2000) Nucleotide sugar transporters of the Golgi apparatus. Curr Opin Struct Biol 10:542–547

Biswas S, Van Dijck P, Datta A (2007) Environmental sensing and signal transduction pathways regulating morphopathogenic determinants of *Candida albicans*. Microbiol Mol Biol Rev 71:348–376

Bowman SM, Free SJ (2006) The structure and synthesis of the fungal cell wall. BioEssays 28:799–808

Braun BR, Head WS, Wang MX, Johnson AD (2000) Identification and characterization of *TUP1*-regulated genes in *Candida albicans*. Genetics 156:31–44

Brielmaier BD, Casabar E, Kurtzeborn CM, McKinnon PS, Ritchie DJ (2008) Early clinical experience with anidulafungin at a large tertiary care medical center. Pharmacotherapy 28:64–73

Brown GD, Denning DW, Levitz SM (2012a) Tackling human fungal infections. Science 336:647

Brown GD, Denning DW, Gow NA, Levitz SM, Netea MG, White TC (2012b) Hidden killers: human fungal infections. Sci Transl Med 4:165rv13

Bulawa CE (1993) Genetics and molecular biology of chitin synthesis in fungi. Annu Rev Microbiol 47:505–534

Buurman ET, Westwater C, Hube B, Brown AJ, Odds FC, Gow NA (1998) Molecular analysis of *Ca*Mnt1p, a mannosyl transferase important for adhesion and virulence of *Candida albicans*. Proc Natl Acad Sci USA 95:7670–7675

Calderon J, Zavrel M, Ragni E, Fonzi WA, Rupp S, Popolo L (2010) *PHR1*, a pH-regulated gene of *Candida albicans* encoding a glucan-remodelling enzyme, is required for adhesion and invasion. Microbiology 156:2484–2494

Calderone RA, Fonzi WA (2001) Virulence factors of *Candida albicans*. Trends Microbiol 9:327–335

Canivenc-Gansel E, Imhof I, Reggiori F, Burda P, Conzelmann A, Benachour A (1998) GPI anchor biosynthesis in yeast: phosphoethanolamine is attached to the alpha1,4-linked mannose of the complete precursor glycophospholipid. Glycobiology 8:761–770

Castanheira M, Messer SA, Rhomberg PR, Pfaller MA (2016) Antifungal susceptibility patterns of a global collection of fungal isolates: results of the SENTRY Antifungal Surveillance Program (2013). Diagn Microbiol Infect Dis 85:200–204

Chen-Wu JL, Zwicker J, Bowen AR, Robbins PW (1992) Expression of chitin synthase genes during yeast and hyphal growth phases of *Candida albicans*. Mol Microbiol 6:497–502

Copping VM, Barelle CJ, Hube B, Gow NA, Brown AJ, Odds FC (2005) Exposure of *Candida albicans* to antifungal agents affects expression of *SAP2* and *SAP9* secreted proteinase genes. J Antimicrob Chemother 55:645–654

Cowen LE, Steinbach WJ (2008) Stress, drugs, and evolution: the role of cellular signaling in fungal drug resistance. Eukaryot Cell 7:747–764

De Boer AD, de Groot PW, Weindl G, Schaller M, Riedel D, Diez-Orejas R, Klis FM, de Koster CG, Dekker HL, Gross U, Bader O, Weig M (2010) The *Candida albicans* cell wall protein Rhd3/Pga29 is abundant in the yeast form and contributes to virulence. Yeast 27:611–624

De Groot PW, Hellingwerf KJ, Klis FM (2003) Genome-wide identification of fungal GPI proteins. Yeast 20:781–796

De Groot PW, de Boer AD, Cunningham J, Dekker HL, de Jong L, Hellingwerf KJ, de Koster C, Klis FM (2004) Proteomic analysis of *Candida albicans* cell walls reveals covalently bound carbohydrate-active enzymes and adhesins. Eukaryot Cell 3:955–965

De Groot PW, Ram AF, Klis FM (2005) Features and functions of covalently linked proteins in fungal cell walls. Fungal Genet Biol 42:657–675

De Nobel H, Lipke PN (1994) Is there a role for GPIs in yeast cell-wall assembly? Trends Cell Biol 4:42–45

De Nobel H, Ruiz C, Martin H, Morris W, Brul S, Molina M, Klis FM (2000) Cell wall perturbation in yeast results in dual phosphorylation of the Slt2/Mpk1 MAP kinase and in an Slt2-mediated increase in *FKS2*-lacZ expression, glucanase resistance and thermotolerance. Microbiology 146(Pt 9):2121–2132

Del Poeta M, Cruz MC, Cardenas ME, Perfect JR, Heitman J (2000) Synergistic antifungal activities of bafilomycin A(1), fluconazole, and the pneumocandin MK-0991/caspofungin acetate (L-743,873) with calcineurin inhibitors FK506 and L-685,818 against *Cryptococcus neoformans*. Antimicrob Agents Chemother 44:739–746

Denning DW (2003) Echinocandin antifungal drugs. Lancet 362:1142–1151

Douglas CM, Foor F, Marrinan JA, Morin N, Nielsen JB, Dahl AM, Mazur P, Baginsky W, Li W, el-Sherbeini M (1994) The *Saccharomyces cerevisiae FKS1* (*ETG1*) gene encodes an integral membrane protein which is a subunit of 1,3-beta-D-glucan synthase. Proc Natl Acad Sci USA 91:12907–12911

Douglas CM, D'Ippolito JA, Shei GJ, Meinz M, Onishi J, Marrinan JA, Li W, Abruzzo GK, Flattery A, Bartizal K, Mitchell A, Kurtz MB (1997) Identification of the *FKS1* gene of *Candida albicans* as the essential target of 1,3-beta-D-glucan synthase inhibitors. Antimicrob Agents Chemother 41:2471–2479

Dranginis AM, Rauceo JM, Coronado JE, Lipke PN (2007) A biochemical guide to yeast adhesins: glycoproteins for social and antisocial occasions. Microbiol Mol Biol Rev 71:282–294

Ecker M, Mrsa V, Hagen I, Deutzmann R, Strahl S, Tanner W (2003) O-mannosylation precedes and potentially controls the N-glycosylation of a yeast cell wall glycoprotein. EMBO Rep 4:628–632

Ecker M, Deutzmann R, Lehle L, Mrsa V, Tanner W (2006) Pir proteins of *Saccharomyces cerevisiae* are attached to beta-1,3-glucan by a new protein-carbohydrate linkage. J Biol Chem 281:11523–11529

Eisenhaber B, Eisenhaber F (2007) Posttranslational modifications and subcellular localization signals: indicators of sequence regions without inherent 3D structure? Curr Protein Pept Sci 8:197–203

Eisenhaber B, Eisenhaber F (2010) Prediction of posttranslational modification of proteins from their amino acid sequence. Methods Mol Biol 609:365–384

Eisenhaber B, Maurer-Stroh S, Novatchkova M, Schneider G, Eisenhaber F (2003) Enzymes and auxiliary factors for GPI lipid anchor biosynthesis and post-translational transfer to proteins. BioEssays 25:367–385

Eisenhaber B, Eisenhaber F, Maurer-Stroh S, Neuberger G (2004) Prediction of sequence signals for lipid post-translational modifications: insights from case studies. Proteomics 4:1614–1625

Eisenhaber B, Eisenhaber S, Kwang TY, Gruber G, Eisenhaber F (2014) Transamidase subunit *GAA1/GPAA1* is a M28 family metallo-peptide-synthetase that catalyzes the peptide bond formation between the substrate protein's omega-site and the GPI lipid anchor's phosphoethanolamine. Cell Cycle 13:1912–1917

Ene IV, Heilmann CJ, Sorgo AG, Walker LA, de Koster CG, Munro CA, Klis FM, Brown AJ (2012a) Carbon source-induced reprogramming of the cell wall proteome and secretome modulates the adherence and drug resistance of the fungal pathogen *Candida albicans*. Proteomics 12:3164–3179

Ene IV, Adya AK, Wehmeier S, Brand AC, MacCallum DM, Gow NA, Brown AJ (2012b) Host carbon sources modulate cell wall architecture, drug resistance and virulence in a fungal pathogen. Cell Microbiol 14:1319–1335

Ene IV, Walker LA, Schiavone M, Lee KK, Martin-Yken H, Dague E, Gow NA, Munro CA, Brown AJ (2015) Cell wall remodeling enzymes modulate fungal cell wall elasticity and osmotic stress resistance. MBio 6:e00986–15

Ernst JF, Prill SK (2001) O-glycosylation. Med Mycol 39(Suppl 1):67–74

Fairlamb A, Gow NAR, Mathews KR, Waters AP (2016) Drug resistance in eukaryotic microorganisms. Nat Microbiol 1(7):16092

Ferguson MA (1999) The structure, biosynthesis and functions of glycosylphosphatidylinositol anchors, and the contributions of trypanosome research. J Cell Sci 112:2799–2809

Fonzi WA (1999). *PHR1* and *PHR2* of *Candida albicans* encode putative glycosidases required for proper cross-linking of beta-1,3- and beta-1,6-glucans. J Bacteriol 181:7070–7079

Frank AT, Ramsook CB, Otoo HN, Tan C, Soybelman G, Rauceo JM, Gaur NK, Klotz SA, Lipke PN (2010) Structure and function of glycosylated tandem repeats from *Candida albicans* Als adhesins. Eukaryot Cell 9:405–414

Free SJ (2013) Fungal cell wall organization and biosynthesis. Adv Genet 81:33–82

Frieman MB, Cormack BP (2003) The omega-site sequence of glycosylphosphatidylinositol-anchored proteins in *Saccharomyces cerevisiae* can determine distribution between the membrane and the cell wall. Mol Microbiol 50:883–896

Gamaletsou MN, Daikos GL, Walsh TJ, Perlin DS, Ortigosa CJ, Psaroulaki A, Pagoni M, Argyropoulou A, Nepka M, Perivolioti E, Kotsopoulou M, Perloretzou S, Marangos M, Kofteridis D, Grammatikou M, Goukos D, Petrikkos G, Sipsas NV (2014) Breakthrough candidaemia caused by phenotypically susceptible *Candida* spp. in patients with haematological malignancies does not correlate with established interpretive breakpoints. Int J Antimicrob Agents 44:248–255

Garcia MC, Lee JT, Ramsook CB, Alsteens D, Dufrene YF, Lipke PN (2011) A role for amyloid in cell aggregation and biofilm formation. PLoS ONE 6:e17632

Gaur NK, Klotz SA (1997) Expression, cloning, and characterization of a *Candida albicans* gene, *ALA1*, that confers adherence properties upon *Saccharomyces cerevisiae* for extracellular matrix proteins. Infect Immun 65:5289–5294

Gaur NK, Klotz SA (2004) Accessibility of the peptide backbone of protein ligands is a key specificity determinant in *Candida albicans* SRS adherence. Microbiology 150:277–284

Gelis S, de Groot PW, Castillo L, Moragues MD, Sentandreu R, Gomez MM, Valentin E (2012) Pga13 in *Candida albicans* is localized in the cell wall and influences cell surface properties, morphogenesis and virulence. Fungal Genet Biol 49:322–331

Gilbert NM, Donlin MJ, Gerik KJ, Specht CA, Djordjevic JT, Wilson CF, Sorrell TC, Lodge JK (2010) *KRE* genes are required for beta-1,6-glucan synthesis, maintenance of capsule architecture and cell wall protein anchoring in *Cryptococcus neoformans*. Mol Microbiol 76:517–534

Gil-Bona A, Parra-Giraldo CM, Hernaez ML, Reales-Calderon JA, Solis NV, Filler SG, Monteoliva L, Gil C (2015) *Candida albicans* cell shaving uncovers new proteins involved in cell wall integrity, yeast to hypha transition, stress response and host-pathogen interaction. J Proteomics 127(Pt B):340–351

Goto M (2007) Protein O-glycosylation in fungi: diverse structures and multiple functions. Biosci Biotechnol Biochem 71:1415–1427

Gow NA, Hube B (2012) Importance of the *Candida albicans* cell wall during commensalism and infection. Curr Opin Microbiol 15:406–412

Gow NA, Robbins PW, Lester JW, Brown AJ, Fonzi WA, Chapman T, Kinsman OS (1994) A hyphal-specific chitin synthase gene (*CHS2*) is not essential for growth, dimorphism, or virulence of *Candida albicans*. Proc Natl Acad Sci USA 91:6216–6220

Gow NA, van de Veerdonk FL, Brown AJ, Netea MG (2011) Candida albicans morphogenesis and host defence: discriminating invasion from colonization. Nat Rev Microbiol 10:112–122

Gozalbo D, Roig P, Villamon E, Gil ML (2004) *Candida* and candidiasis: the cell wall as a potential molecular target for antifungal therapy. Curr Drug Targets Infect Disord 4:117–135

Green CB, Cheng G, Chandra J, Mukherjee P, Ghannoum MA, Hoyer LL (2004) RT-PCR detection of *Candida albicans ALS* gene expression in the reconstituted human epithelium (RHE) model of oral candidiasis and in model biofilms. Microbiology 150:267–275

Grimme SJ, Westfall BA, Wiedman JM, Taron CH, Orlean P (2001) The essential Smp3 protein is required for addition of the side-branching fourth mannose during assembly of yeast glycosylphosphatidylinositols. J Biol Chem 276:27731–27739

Grimme SJ, Colussi PA, Taron CH, Orlean P (2004) Deficiencies in the essential Smp3 mannosyltransferase block glycosylphosphatidylinositol assembly and lead to defects in growth and cell wall biogenesis in *Candida albicans*. Microbiology 150:3115–3128

Hall RA, Gow NA (2013) Mannosylation in *Candida albicans*: role in cell wall function and immune recognition. Mol Microbiol 90:1147–1161

Hall RA, Bates S, Lenardon MD, Maccallum DM, Wagener J, Lowman DW, Kruppa MD, Williams DL, Odds FC, Brown AJ, Gow NA (2013) The Mnn2 mannosyltransferase family modulates mannoprotein fibril length, immune recognition and virulence of *Candida albicans*. PLoS Pathog 9:e1003276

Hamburger D, Egerton M, Riezman H (1995) Yeast Gaa1p is required for attachment of a completed GPI anchor onto proteins. J Cell Biol 129:629–639

Hashimoto H, Yoda K (1997) Novel membrane protein complexes for protein glycosylation in the yeast Golgi apparatus. Biochem Biophys Res Commun 241:682–686

Hata K, Horii T, Miyazaki M, Watanabe NA, Okubo M, Sonoda J, Nakamoto K, Tanaka K, Shirotori S, Murai N, Inoue S, Matsukura M, Abe S, Yoshimatsu K, Asada M (2011) Efficacy of oral E1210, a new broad-spectrum antifungal with a novel mechanism of action, in murine models of candidiasis, aspergillosis, and fusariosis. Antimicrob Agents Chemother 55:4543–4551

Hector RF, Bierer DE (2011) New beta-glucan inhibitors as antifungal drugs. Expert Opin Ther Pat 21:1597–1610

Heilmann CJ, Sorgo AG, Klis FM (2012) News from the fungal front: wall proteome dynamics and host-pathogen interplay. PLoS Pathog 8:e1003050

Heinisch JJ, Dupres V, Wilk S, Jendretzki A, Dufrene YF (2010) Single-molecule atomic force microscopy reveals clustering of the yeast plasma-membrane sensor Wsc1. PLoS ONE 5: e11104

Helenius A, Aebi M (2004) Roles of N-linked glycans in the endoplasmic reticulum. Annu Rev Biochem 73:1019–1049

Helenius J, Aebi M (2002) Transmembrane movement of dolichol linked carbohydrates during N-glycoprotein biosynthesis in the endoplasmic reticulum. Semin Cell Dev Biol 13:171–178

Hobson RP, Munro CA, Bates S, MacCallum DM, Cutler JE, Heinsbroek SE, Brown GD, Odds FC, Gow NA (2004) Loss of cell wall mannosylphosphate in *Candida albicans* does not influence macrophage recognition. J Biol Chem 279:39628–39635

Hoyer LL (2001) The *ALS* gene family of *Candida albicans*. Trends Microbiol 9:176–180

Hoyer LL, Scherer S, Shatzman AR, Livi GP (1995) *Candida albicans ALS1*: domains related to a *Saccharomyces cerevisiae* sexual agglutinin separated by a repeating motif. Mol Microbiol 15:39–54

Hoyer LL, Green CB, Oh SH, Zhao X (2008) Discovering the secrets of the *Candida albicans* agglutinin-like sequence (*ALS*) gene family–a sticky pursuit. Med Mycol 46:1–15

Hube B, Naglik J (2001) *Candida albicans* proteinases: resolving the mystery of a gene family. Microbiology 147:1997–2005

Ibrahim AS, Spellberg BJ, Avanesian V, Fu Y, Edwards JE Jr (2006) The anti-*Candida* vaccine based on the recombinant N-terminal domain of Als1p is broadly active against disseminated candidiasis. Infect Immun 74:3039–3041

Imtiaz T, Lee KK, Munro CA, Maccallum DM, Shankland GS, Johnson EM, Macgregor MS, Bal AM (2012) Echinocandin resistance due to simultaneous *FKS* mutation and increased cell wall chitin in a *Candida albicans* bloodstream isolate following brief exposure to caspofungin. J Med Microbiol 61:1330–1334

Jiang B, Sheraton J, Ram AF, Dijkgraaf GJ, Klis FM, Bussey H (1996) *CWH41* encodes a novel endoplasmic reticulum membrane N-glycoprotein involved in beta 1,6-glucan assembly. J Bacteriol 178:1162–1171

Jigami Y (2008) Yeast glycobiology and its application. Biosci Biotechnol Biochem 72:637–648

Jimenez-Ortigosa C, Paderu P, Motyl MR, Perlin DS (2014) Enfumafungin derivative MK-3118 shows increased *in vitro* potency against clinical echinocandin-resistant *Candida* species and *Aspergillus* species isolates. Antimicrob Agents Chemother 58:1248–1251

Jung US, Levin DE (1999) Genome-wide analysis of gene expression regulated by the yeast cell wall integrity signalling pathway. Mol Microbiol 34:1049–1057

Jungmann J, Munro S (1998) Multi-protein complexes in the cis Golgi of *Saccharomyces cerevisiae* with alpha-1,6-mannosyltransferase activity. EMBO J 17:423–434

Jungmann J, Rayner JC, Munro S (1999) The *Saccharomyces cerevisiae* protein Mnn10p/Bed1p is a subunit of a Golgi mannosyltransferase complex. J Biol Chem 274:6579–6585

Kamai Y, Kubota M, Kamai Y, Hosokawa T, Fukuoka T, Filler SG (2002) Contribution of *Candida albicans ALS1* to the pathogenesis of experimental oropharyngeal candidiasis. Infect Immun 70:5256–5258

Kang JY, Hong Y, Ashida H, Shishioh N, Murakami Y, Morita YS, Maeda Y, Kinoshita T (2005) PIG-V involved in transferring the second mannose in glycosylphosphatidylinositol. J Biol Chem 280:9489–9497

Kapteyn JC, Ram AF, Groos EM, Kollar R, Montijn RC, Van Den Ende H, Llobell A, Cabib E, Klis FM (1997) Altered extent of cross-linking of beta1,6-glucosylated mannoproteins to chitin in *Saccharomyces cerevisiae* mutants with reduced cell wall beta1,3-glucan content. J Bacteriol 179:6279–6284

Kapteyn JC, Hoyer LL, Hecht JE, Muller WH, Andel A, Verkleij AJ, Makarow M, Van Den Ende H, Klis FM (2000) The cell wall architecture of *Candida albicans* wild-type cells and cell wall-defective mutants. Mol Microbiol 35:601–611

Kelleher DJ, Gilmore R (2006) An evolving view of the eukaryotic oligosaccharyltransferase. Glycobiology 16:47R–62R

Kinoshita T, Inoue N (2000) Dissecting and manipulating the pathway for glycosylphos-phatidylinositol-anchor biosynthesis. Curr Opin Chem Biol 4:632–638

Kitagaki H, Wu H, Shimoi H, Ito K (2002) Two homologous genes, *DCW1* (YKL046c) and *DFG5*, are essential for cell growth and encode glycosylphosphatidylinositol (GPI)-anchored membrane proteins required for cell wall biogenesis in *Saccharomyces cerevisiae*. Mol Microbiol 46:1011–1022

Kitamura A, Higuchi S, Hata M, Kawakami K, Yoshida K, Namba K, Nakajima R (2009a) Effect of beta-1,6-glucan inhibitors on the invasion process of *Candida albicans*: potential mechanism of their *in vivo* efficacy. Antimicrob Agents Chemother 53:3963–3971

Kitamura A, Someya K, Hata M, Nakajima R, Takemura M (2009b) Discovery of a small-molecule inhibitor of {beta}-1,6-glucan synthesis. Antimicrob Agents Chemother 53:670–677

Klis FM, de Groot P, Hellingwerf K (2001) Molecular organization of the cell wall of *Candida albicans*. Med Mycol 39(Suppl 1):1–8

Klis FM, Mol P, Hellingwerf K, Brul S (2002) Dynamics of cell wall structure in *Saccharomyces cerevisiae*. FEMS Microbiol Rev 26:239–256

Klis FM, Boorsma A, De Groot PW (2006) Cell wall construction in *Saccharomyces cerevisiae*. Yeast 23:185–202

Klis FM, Sosinska GJ, de Groot PW, Brul S (2009) Covalently linked cell wall proteins of *Candida albicans* and their role in fitness and virulence. FEMS Yeast Res 9:1013–1028

Klis FM, Brul S, De Groot PW (2010) Covalently linked wall proteins in ascomycetous fungi. Yeast 27:489–493

Kollar R, Petrakova E, Ashwell G, Robbins PW, Cabib E (1995) Architecture of the yeast cell wall. The linkage between chitin and beta(1–>3)-glucan. J Biol Chem 270:1170–1178

Kollar R, Reinhold BB, Petrakova E, Yeh HJ, Ashwell G, Drgonova J, Kapteyn JC, Klis FM, Cabib E (1997) Architecture of the yeast cell wall. Beta(1–>6)-glucan interconnects mannoprotein, beta(1–>)3-glucan, and chitin. J Biol Chem 272:17762–17775

Kontoyiannis DP, Lewis RE, Osherov N, Albert ND, May GS (2003) Combination of caspofungin with inhibitors of the calcineurin pathway attenuates growth *in vitro* in *Aspergillus* species. J Antimicrob Chemother 51:313–316

Krysan DJ, Ting EL, Abeijon C, Kroos L, Fuller RS (2005) Yapsins are a family of aspartyl proteases required for cell wall integrity in *Saccharomyces cerevisiae*. Eukaryot Cell 4:1364–1374

Kulkarni RD, Kelkar HS, Dean RA (2003) An eight-cysteine-containing CFEM domain unique to a group of fungal membrane proteins. Trends Biochem Sci 28:118–121

Kumamoto CA, Vinces MD (2005) Contributions of hyphae and hypha-co-regulated genes to *Candida albicans* virulence. Cell Microbiol 7:1546–1554

Laforet L, Moreno I, Sanchez-Fresneda R, Martinez-Esparza M, Martinez JP, Arguelles JC, de Groot PW, Valentin-Gomez E (2011) Pga26 mediates filamentation and biofilm formation and is required for virulence in *Candida albicans*. FEMS Yeast Res 11:389–397

Larkin A, Imperiali B (2011) The expanding horizons of asparagine-linked glycosylation. Biochemistry 50:4411–4426

Lee KK, Maccallum DM, Jacobsen MD, Walker LA, Odds FC, Gow NA, Munro CA (2012) Elevated cell wall chitin in *Candida albicans* confers echinocandin resistance *in vivo*. Antimicrob Agents Chemother 56:208–217

Lennarz WJ (2007) Studies on oligosaccharyl transferase in yeast. Acta Biochim Pol 54:673–677

Lesage G, Bussey H (2006) Cell wall assembly in *Saccharomyces cerevisiae*. Microbiol Mol Biol Rev 70:317–343

Levin DE (2011) Regulation of cell wall biogenesis in *Saccharomyces cerevisiae*: the cell wall integrity signaling pathway. Genetics 189:1145–1175

Lipke PN, Ramsook C, Garcia-Sherman MC, Jackson DN, Chan CX, Bois M, Klotz SA (2014) Between Amyloids and Aggregation Lies a Connection with Strength and Adhesion. New J Sci 2014:815102

Liu TT, Lee RE, Barker KS, Lee RE, Wei L, Homayouni R, Rogers PD (2005) Genome-wide expression profiling of the response to azole, polyene, echinocandin, and pyrimidine antifungal agents in *Candida albicans*. Antimicrob Agents Chemother 49:2226–2236

Liu Y, Filler SG (2011) *Candida albicans* Als3, a multifunctional adhesin and invasin. Eukaryot Cell 10:168–173

Lommel M, Strahl S (2009) Protein O-mannosylation: conserved from bacteria to humans. Glycobiology 19:816–828

Lombard V, Golaconda Ramulu H, Drula E, Coutinho PM, Henrissat B (2014)The carbohydrate-active enzymes database (CAZy) in 2013.Nucleic Acids Res D490–5

Lu CF, Montijn RC, Brown JL, Klis F, Kurjan J, Bussey H, Lipke PN (1995) Glycosyl phosphatidylinositol-dependent cross-linking of alpha-agglutinin and beta 1,6-glucan in the *Saccharomyces cerevisiae* cell wall. J Cell Biol 128:333–340

Lussier M, Sdicu AM, Bussey H (1999) The *KTR* and *MNN1* mannosyltransferase families of *Saccharomyces cerevisiae*. Biochim Biophys Acta 1426:323–334

Maddi A, Fu C, Free SJ (2012) The *Neurospora crassa* dfg5 and dcw1 genes encode alpha-1,6-mannanases that function in the incorporation of glycoproteins into the cell wall. PLoS ONE 7:e38872

Maeda Y, Watanabe R, Harris CL, Hong Y, Ohishi K, Kinoshita K, Kinoshita T (2001) PIG-M transfers the first mannose to glycosylphosphatidylinositol on the lumenal side of the ER. EMBO J 20:250–261

Mann PA, McLellan CA, Koseoglu S, Si Q, Kuzmin E, Flattery A, Harris G, Sher X, Murgolo N, Wang H, Devito K, de Pedro N, Genilloud O, Kahn JN, Jiang B, Costanzo M, Boone C, Garlisi CG, Lindquist S, Roemer T (2015) Chemical genomics-based antifungal drug discovery: targeting glycosylphosphatidylinositol (GPI) precursor biosynthesis. ACS Infect Dis 1:59–72

Mao Y, Zhang Z, Wong B (2003) Use of green fluorescent protein fusions to analyse the N- and C-terminal signal peptides of GPI-anchored cell wall proteins in *Candida albicans*. Mol Microbiol 50:1617–1628

Mao Y, Zhang Z, Gast C, Wong B (2008) C-terminal signals regulate targeting of glycosylphosphatidylinositol-anchored proteins to the cell wall or plasma membrane in *Candida albicans*. Eukaryot Cell 7:1906–1915

Mayer FL, Wilson D, Hube B (2013) *Candida albicans* pathogenicity mechanisms. Virulence 4:119–128

McKenzie CG, Koser U, Lewis LE, Bain JM, Mora-Montes HM, Barker RN, Gow NA, Erwig LP (2010) Contribution of *Candida albicans* cell wall components to recognition by and escape from murine macrophages. Infect Immun 78:1650–1658

McLellan CA, Whitesell L, King OD, Lancaster AK, Mazitschek R, Lindquist S (2012) Inhibiting GPI anchor biosynthesis in fungi stresses the endoplasmic reticulum and enhances immunogenicity. ACS Chem Biol 7:1520–1528

Miller CD, Lomaestro BW, Park S, Perlin DS (2006) Progressive esophagitis caused by *Candida albicans* with reduced susceptibility to caspofungin. Pharmacotherapy 26:877–880

Mio T, Yabe T, Sudoh M, Satoh Y, Nakajima T, Arisawa M, Yamada-Okabe H (1996) Role of three chitin synthase genes in the growth of *Candida albicans*. J Bacteriol 178:2416–2419

Monod M, Togni G, Hube B, Sanglard D (1994) Multiplicity of genes encoding secreted aspartic proteinases in *Candida* species. Mol Microbiol 13:357–368

Montijn RC, Vink E, Muller WH, Verkleij AJ, Van Den Ende H, Henrissat B, Klis FM (1999) Localization of synthesis of beta1,6-glucan in *Saccharomyces cerevisiae*. J Bacteriol 181:7414–7420

Mora-Montes HM, Bates S, Netea MG, Diaz-Jimenez DF, Lopez-Romero E, Zinker S, Ponce-Noyola P, Kullberg BJ, Brown AJ, Odds FC, Flores-Carreon A, Gow NA (2007) Endoplasmic reticulum alpha-glycosidases of *Candida albicans* are required for N glycosylation, cell wall integrity, and normal host-fungus interaction. Eukaryot Cell 6:2184–2193

Moreno-Ruiz E, Ortu G, de Groot PW, Cottier F, Loussert C, Prevost MC, de Koster C, Klis FM, Goyard S, d'Enfert C (2009) The GPI-modified proteins Pga59 and Pga62 of *Candida albicans* are required for cell wall integrity. Microbiology 155:2004–2020

Mrsa V, Ecker M, Strahl-Bolsinger S, Nimtz M, Lehle L, Tanner W (1999) Deletion of new covalently linked cell wall glycoproteins alters the electrophoretic mobility of phosphorylated wall components of *Saccharomyces cerevisiae*. J Bacteriol 181:3076–3086

Munro C (2010) *Candida albican* cell wall mediated virulence. In: Ashbee RH, Bignell EM (eds) Pathogenic Yeast, The Yeast Handbook. Springer, Berlin, pp 69–95

Munro C, Richard ML (2012) The cell wall: glycoproteins, remodeling, and regulation. In: Calderone RA, Clancy CJ (eds) *Candida* and candidiasis, 2nd, Edition edn. ASM Press, Washington, DC, pp 197–223

Munro CA (2013) Chitin and glucan, the yin and yang of the fungal cell wall, implications for antifungal drug discovery and therapy. Adv Appl Microbiol 83:145–172

Munro CA, Winter K, Buchan A, Henry K, Becker JM, Brown AJ, Bulawa CE, Gow NA (2001) Chs1 of *Candida albicans* is an essential chitin synthase required for synthesis of the septum and for cell integrity. Mol Microbiol 39:1414–1426

Munro CA, Whitton RK, Hughes HB, Rella M, Selvaggini S, Gow NA (2003) *CHS8*-a fourth chitin synthase gene of *Candida albicans* contributes to *in vitro* chitin synthase activity, but is dispensable for growth. Fungal Genet Biol 40:146–158

Munro CA, Bates S, Buurman ET, Hughes HB, Maccallum DM, Bertram G, Atrih A, Ferguson MA, Bain JM, Brand A, Hamilton S, Westwater C, Thomson LM, Brown AJ, Odds FC, Gow NA (2005) Mnt1p and Mnt2p of *Candida albicans* are partially redundant alpha-1,2-mannosyltransferases that participate in O-linked mannosylation and are required for adhesion and virulence. J Biol Chem 280:1051–1060

Munro CA, Selvaggini S, de Bruijn I, Walker L, Lenardon MD, Gerssen B, Milne S, Brown AJ, Gow NA (2007) The PKC, HOG and Ca^{2+} signalling pathways co-ordinately regulate chitin synthesis in *Candida albicans*. Mol Microbiol 63:1399–1413

Munro S (2001) What can yeast tell us about N-linked glycosylation in the Golgi apparatus? FEBS Lett 498:223–227

Naglik J, Albrecht A, Bader O, Hube B (2004) *Candida albicans* proteinases and host/pathogen interactions. Cell Microbiol 6:915–926

Naglik JR, Rodgers CA, Shirlaw PJ, Dobbie JL, Fernandes-Naglik LL, Greenspan D, Agabian N, Challacombe SJ (2003) Differential expression of *Candida albicans* secreted aspartyl proteinase and phospholipase B genes in humans correlates with active oral and vaginal infections. J Infect Dis 188:469–479

Naglik JR, Moyes D, Makwana J, Kanzaria P, Tsichlaki E, Weindl G, Tappuni AR, Rodgers CA, Woodman AJ, Challacombe SJ, Schaller M, Hube B (2008) Quantitative expression of the *Candida albicans* secreted aspartyl proteinase gene family in human oral and vaginal candidiasis. Microbiology 154:3266–3280

Nakayama K, Feng Y, Tanaka A, Jigami Y (1998) The involvement of mnn4 and mnn6 mutations in mannosylphosphorylation of O-linked oligosaccharide in yeast *Saccharomyces cerevisiae*. Biochim Biophys Acta 1425:255–262

Netea MG, Gow NA, Munro CA, Bates S, Collins C, Ferwerda G, Hobson RP, Bertram G, Hughes HB, Jansen T, Jacobs L, Buurman ET, Gijzen K, Williams DL, Torensma R, McKinnon A, MacCallum DM, Odds FC, Van der Meer JW, Brown AJ, Kullberg BJ (2006) Immune sensing of *Candida albicans* requires cooperative recognition of mannans and glucans by lectin and Toll-like receptors. J Clin Invest 116:1642–1650

Nishikawa A, Poster JB, Jigami Y, Dean N (2002) Molecular and phenotypic analysis of *CaVRG4*, encoding an essential Golgi apparatus GDP-mannose transporter. J Bacteriol 184:29–42

Nobile CJ, Mitchell AP (2005) Regulation of cell-surface genes and biofilm formation by the *C. albicans* transcription factor Bcr1p. Curr Biol 15:1150–1155

Nobile CJ, Andes DR, Nett JE, Smith FJ, Yue F, Phan QT, Edwards JE, Filler SG, Mitchell AP (2006a) Critical role of Bcr1-dependent adhesins in *C. albicans* biofilm formation *in vitro* and *in vivo*. PLoS Pathog 2:e63

Nobile CJ, Nett JE, Andes DR, Mitchell AP (2006b) Function of *Candida albicans* adhesin Hwp1 in biofilm formation. Eukaryot Cell 5:1604–1610

Nobile CJ, Schneider HA, Nett JE, Sheppard DC, Filler SG, Andes DR, Mitchell AP (2008) Complementary adhesin function in *C. albicans* biofilm formation. Curr Biol 18:1017–1024

Odds FC (2008) Secreted proteinases and *Candida albicans* virulence. Microbiology 154:3245–3246

Odds FC, Brown AJ, Gow NA (2003) Antifungal agents: mechanisms of action. Trends Microbiol 11:272–279

Orchard MG, Neuss JC, Galley CM, Carr A, Porter DW, Smith P, Scopes DI, Haydon D, Vousden K, Stubberfield CR, Young K, Page M (2004) Rhodanine-3-acetic acid derivatives as inhibitors of fungal protein mannosyl transferase 1 (*PMT1*). Bioorg Med Chem Lett 14:3975–3978

Orlean P (2012) Architecture and biosynthesis of the *Saccharomyces cerevisiae* cell wall. Genetics 192:775–818

Pardini G, De Groot PW, Coste AT, Karababa M, Klis FM, de Koster CG, Sanglard D (2006) The *CRH* family coding for cell wall glycosylphosphatidylinositol proteins with a predicted transglycosidase domain affects cell wall organization and virulence of *Candida albicans*. J Biol Chem 281:40399–40411

Park S, Kelly R, Kahn JN, Robles J, Hsu MJ, Register E, Li W, Vyas V, Fan H, Abruzzo G, Flattery A, Gill C, Chrebet G, Parent SA, Kurtz M, Teppler H, Douglas CM, Perlin DS (2005) Specific substitutions in the echinocandin target Fks1p account for reduced susceptibility of rare laboratory and clinical *Candida* sp. isolates. Antimicrob Agents Chemother 49:3264–3273

Peltroche-Llacsahuanga H, Goyard S, d'Enfert C, Prill SK, Ernst JF (2006) Protein O-mannosyltransferase isoforms regulate biofilm formation in *Candida albicans*. Antimicrob Agents Chemother 50:3488–3491

Perlin DS (2007) Resistance to echinocandin-class antifungal drugs. Drug Resist Updat 10:121–130

Perlin DS (2011) Current perspectives on echinocandin class drugs. Future Microbiol 6:441–457

Perlin DS, Shor E, Zhao Y (2015) Update on antifungal drug resistance. Curr Clin Microbiol Rep 2:84–95

Pfaller MA (2012) Antifungal drug resistance: mechanisms, epidemiology, and consequences for treatment. Am J Med 125:S3–13

Pfaller MA, Messer SA, Motyl MR, Jones RN, Castanheira M (2013) Activity of MK-3118, a new oral glucan synthase inhibitor, tested against *Candida* spp. by two international methods (CLSI and EUCAST). J Antimicrob Chemother 68:858–863

Pfeiffer CD, Garcia-Effron G, Zaas AK, Perfect JR, Perlin DS, Alexander BD (2010) Breakthrough invasive candidiasis in patients on micafungin. J Clin Microbiol 48:2373–2380

Phan QT, Myers CL, Fu Y, Sheppard DC, Yeaman MR, Welch WH, Ibrahim AS, Edwards JE Jr, Filler SG (2007) Als3 is a *Candida albicans* invasin that binds to cadherins and induces endocytosis by host cells. PLoS Biol 5:e64

Plaine A, Walker L, Da Costa G, Mora-Montes HM, McKinnon A, Gow NA, Gaillardin C, Munro CA, Richard ML (2008) Functional analysis of *Candida albicans* GPI-anchored proteins: roles in cell wall integrity and caspofungin sensitivity. Fungal Genet Biol 45:1404–1414

Prill SK, Klinkert B, Timpel C, Gale CA, Schroppel K, Ernst JF (2005) *PMT* family of *Candida albicans*: five protein mannosyltransferase isoforms affect growth, morphogenesis and antifungal resistance. Mol Microbiol 55:546–560

Quinn RP, Mahoney SJ, Wilkinson BM, Thornton DJ, Stirling CJ (2009) A novel role for Gtb1p in glucose trimming of N-linked glycans. Glycobiology 19:1408–1416

Rauceo JM, De Armond R, Otoo H, Kahn PC, Klotz SA, Gaur NK, Lipke PN (2006) Threonine-rich repeats increase fibronectin binding in the *Candida albicans* adhesin Als5p. Eukaryot Cell 5:1664–1673

Richard ML, Plaine A (2007) Comprehensive analysis of glycosylphosphatidylinositol-anchored proteins in *Candida albicans*. Eukaryot Cell 6:119–133

Rodicio R, Heinisch JJ (2010) Together we are strong–cell wall integrity sensors in yeasts. Yeast 27:531–540

Rodionov D, Romero PA, Berghuis AM, Herscovics A (2009) Expression and purification of recombinant M-Pol I from *Saccharomyces cerevisiae* with alpha-1,6 mannosylpolymerase activity. Protein Expr Purif 66:1–6

Roemer T, Krysan DJ (2014) Antifungal drug development: challenges, unmet clinical needs, and new approaches. Cold Spring Harb Perspect Med 4:10.1101/cshperspect.a019703

Romero PA, Dijkgraaf GJ, Shahinian S, Herscovics A, Bussey H (1997) The yeast *CWH41* gene encodes glucosidase I. Glycobiology 7:997–1004

Romero PA, Lussier M, Veronneau S, Sdicu AM, Herscovics A, Bussey H (1999) Mnt2p and Mnt3p of *Saccharomyces cerevisiae* are members of the Mnn1p family of alpha-1,3-mannosyltransferases responsible for adding the terminal mannose residues of O-linked oligosaccharides. Glycobiology 9:1045–1051

Rouabhia M, Schaller M, Corbucci C, Vecchiarelli A, Prill SK, Giasson L, Ernst JF (2005) Virulence of the fungal pathogen *Candida albicans* requires the five isoforms of protein mannosyltransferases. Infect Immun 73:4571–4580

Rueda C, Cuenca-Estrella M, Zaragoza O (2014) Paradoxical growth of *Candida albicans* in the presence of caspofungin is associated with multiple cell wall rearrangements and decreased virulence. Antimicrob Agents Chemother 58:1071–1083

Ruiz-Herrera J, Elorza MV, Valentin E, Sentandreu R (2006) Molecular organization of the cell wall of *Candida albicans* and its relation to pathogenicity. FEMS Yeast Res 6:14–29

Sato K, Noda Y, Yoda K (2007) Pga1 is an essential component of Glycosylphosphatidylinositol-mannosyltransferase II of *Saccharomyces cerevisiae*. Mol Biol Cell 18:3472–3485

Schaller M, Korting HC, Schafer W, Bastert J, Chen W, Hube B (1999) Secreted aspartic proteinase (Sap) activity contributes to tissue damage in a model of human oral candidosis. Mol Microbiol 34:169–180

Schild L, Heyken A, de Groot PW, Hiller E, Mock M, de Koster C, Horn U, Rupp S, Hube B (2011) Proteolytic cleavage of covalently linked cell wall proteins by *Candida albicans* Sap9 and Sap10. Eukaryot Cell 10:98–109

Schulz BL, Aebi M (2009) Analysis of glycosylation site occupancy reveals a role for Ost3p and Ost6p in site-specific N-glycosylation efficiency. Mol Cell Proteomics 8:357–364

Shahinian S, Bussey H (2000) Beta-1,6-Glucan synthesis in *Saccharomyces cerevisiae* 35:477–489

Sheppard DC, Yeaman MR, Welch WH, Phan QT, Fu Y, Ibrahim AS, Filler SG, Zhang M, Waring AJ, Edwards JE Jr (2004) Functional and structural diversity in the Als protein family of *Candida albicans*. J Biol Chem 279:30480–30489

Sherry L, Rajendran R, Lappin DF, Borghi E, Perdoni F, Falleni M, Tosi D, Smith K, Williams C, Jones B, Nile CJ, Ramage G (2014) Biofilms formed by *Candida albicans* bloodstream isolates display phenotypic and transcriptional heterogeneity that are associated with resistance and pathogenicity. BMC Microbiol 14:182-2180-14-182

Shibata N, Ikuta K, Imai T, Satoh Y, Satoh R, Suzuki A, Kojima C, Kobayashi H, Hisamichi K, Suzuki S (1995) Existence of branched side chains in the cell wall mannan of pathogenic yeast, *Candida albicans*. Structure-antigenicity relationship between the cell wall mannans of *Candida albicans* and *Candida parapsilosis*. J Biol Chem 270:1113–1122

Singleton DR, Masuoka J, Hazen KC (2005) Surface hydrophobicity changes of two *Candida albicans* serotype B mnn4delta mutants. Eukaryot Cell 4:639–648

Sorgo AG, Heilmann CJ, Dekker HL, Bekker M, Brul S, de Koster CG, de Koning LJ, Klis FM (2011) Effects of fluconazole on the secretome, the wall proteome, and wall integrity of the clinical fungus *Candida albicans*. Eukaryot Cell 10:1071–1081

Sosinska GJ, de Groot PW, Teixeira de Mattos MJ, Dekker HL, de Koster CG, Hellingwerf KJ, Klis FM (2008) Hypoxic conditions and iron restriction affect the cell-wall proteome of *Candida albicans* grown under vagina-simulative conditions. Microbiology 154:510–520

Southard SB, Specht CA, Mishra C, Chen-Weiner J, Robbins PW (1999) Molecular analysis of the *Candida albicans* homolog of *Saccharomyces cerevisiae MNN9*, required for glycosylation of cell wall mannoproteins. J Bacteriol 181:7439–7448

Spellberg BJ, Ibrahim AS, Avanesian V, Fu Y, Myers C, Phan QT, Filler SG, Yeaman MR, Edwards JE Jr (2006) Efficacy of the anti-*Candida* rAls3p-N or rAls1p-N vaccines against disseminated and mucosal candidiasis. J Infect Dis 194:256–260

Staab JF, Sundstrom P (1998) Genetic organization and sequence analysis of the hypha-specific cell wall protein gene *HWP1* of *Candida albicans*. Yeast 14:681–686

Staab JF, Bradway SD, Fidel PL, Sundstrom P (1999) Adhesive and mammalian transglutaminase substrate properties of *Candida albicans* Hwp1. Science 283:1535–1538

Staab JF, Datta K, Rhee P (2013) Niche-specific requirement for hyphal wall protein 1 in virulence of *Candida albicans*. PLoS ONE 8:e80842

Stevens DA, Espiritu M, Parmar R (2004) Paradoxical effect of caspofungin: reduced activity against *Candida albicans* at high drug concentrations. Antimicrob Agents Chemother 48:3407–3411

Stolz J, Munro S (2002) The components of the *Saccharomyces cerevisiae* mannosyltransferase complex M-Pol I have distinct functions in mannan synthesis. J Biol Chem 277:44801–44808

Sudoh M, Nagahashi S, Doi M, Ohta A, Takagi M, Arisawa M (1993) Cloning of the chitin synthase 3 gene from *Candida albicans* and its expression during yeast-hyphal transition. Mol Gen Genet 241:351–358

Sudoh M, Yamazaki T, Masubuchi K, Taniguchi M, Shimma N, Arisawa M, Yamada-Okabe H (2000) Identification of a novel inhibitor specific to the fungal chitin synthase. Inhibition of chitin synthase 1 arrests the cell growth, but inhibition of chitin synthase 1 and 2 is lethal in the pathogenic fungus *Candida albicans*. J Biol Chem 275:32901–32905

Sundstrom P, Balish E, Allen CM (2002) Essential role of the *Candida albicans* transglutaminase substrate, hyphal wall protein 1, in lethal oroesophageal candidiasis in immunodeficient mice. J Infect Dis 185:521–530

Surarit R, Gopal PK, Shepherd MG (1988) Evidence for a glycosidic linkage between chitin and glucan in the cell wall of *Candida albicans*. J Gen Microbiol 134:1723–1730

Taron CH, Wiedman JM, Grimme SJ, Orlean P (2000) Glycosylphosphatidylinositol biosynthesis defects in Gpi11p- and Gpi13p-deficient yeast suggest a branched pathway and implicate gpi13p in phosphoethanolamine transfer to the third mannose. Mol Biol Cell 11:1611–1630

Terashima H, Yabuki N, Arisawa M, Hamada K, Kitada K (2000) Up-regulation of genes encoding glycosylphosphatidylinositol (GPI)-attached proteins in response to cell wall damage caused by disruption of *FKS1* in *Saccharomyces cerevisiae*. Mol Gen Genet 264:64–74

Tiede A, Bastisch I, Schubert J, Orlean P, Schmidt RE (1999) Biosynthesis of glycosylphosphatidylinositols in mammals and unicellular microbes. Biol Chem 380:503–523

Tiede A, Nischan C, Schubert J, Schmidt RE (2000) Characterisation of the enzymatic complex for the first step in glycosylphosphatidylinositol biosynthesis. Int J Biochem Cell Biol 32:339–350

Timpel C, Zink S, Strahl-Bolsinger S, Schroppel K, Ernst J (2000) Morphogenesis, adhesive properties, and antifungal resistance depend on the Pmt6 protein mannosyltransferase in the fungal pathogen *Candida albicans*. J Bacteriol 182:3063–3071

Toh-e A, Yasunaga S, Nisogi H, Tanaka K, Oguchi T, Matsui Y (1993) Three yeast genes, *PIR1*, *PIR2* and *PIR3*, containing internal tandem repeats, are related to each other, and *PIR1* and *PIR2* are required for tolerance to heat shock. Yeast 9:481–494

Tokunaga M, Kusamichi M, Koike H (1986) Ultrastructure of outermost layer of cell wall in *Candida albicans* observed by rapid-freezing technique. J Electron Microsc (Tokyo) 35:237–246

Travers KJ, Patil CK, Wodicka L, Lockhart DJ, Weissman JS, Walter P (2000) Functional and genomic analyses reveal an essential coordination between the unfolded protein response and ER-associated degradation. Cell 101:249–258

Tsuchimori N, Sharkey LL, Fonzi WA, French SW, Edwards JE Jr, Filler SG (2000) Reduced virulence of *HWP1*-deficient mutants of *Candida albicans* and their interactions with host cells. Infect Immun 68:1997–2002

Tsukahara K, Hata K, Nakamoto K, Sagane K, Watanabe NA, Kuromitsu J, Kai J, Tsuchiya M, Ohba F, Jigami Y, Yoshimatsu K, Nagasu T (2003) Medicinal genetics approach towards identifying the molecular target of a novel inhibitor of fungal cell wall assembly. Mol Microbiol 48:1029–1042

Verstrepen KJ, Klis FM (2006) Flocculation, adhesion and biofilm formation in yeasts. Mol Microbiol 60:5–15

Walker LA, Munro CA, de Bruijn I, Lenardon MD, McKinnon A, Gow NA (2008) Stimulation of chitin synthesis rescues *Candida albicans* from echinocandins. PLoS Pathog 4:e1000040

Walker LA, Gow NA, Munro CA (2013a) Elevated chitin content reduces the susceptibility of *Candida* species to caspofungin. Antimicrob Agents Chemother 57:146–154

Walker LA, Lenardon MD, Preechasuth K, Munro CA, Gow NA (2013b) Cell wall stress induces alternative fungal cytokinesis and septation strategies. J Cell Sci 126:2668–2677

Walker SS, Xu Y, Triantafyllou I, Waldman MF, Mendrick C, Brown N, Mann P, Chau A, Patel R, Bauman N, Norris C, Antonacci B, Gurnani M, Cacciapuoti A, McNicholas PM, Wainhaus S, Herr RJ, Kuang R, Aslanian RG, Ting PC, Black TA (2011) Discovery of a novel class of orally active antifungal beta-1,3-D-glucan synthase inhibitors. Antimicrob Agents Chemother 55:5099–5106

Watanabe R, Kinoshita T, Masaki R, Yamamoto A, Takeda J, Inoue N (1996) PIG-A and PIG-H, which participate in glycosylphosphatidylinositol anchor biosynthesis, form a protein complex in the endoplasmic reticulum. J Biol Chem 271:26868–26875

Watanabe R, Ohishi K, Maeda Y, Nakamura N, Kinoshita T (1999) Mammalian PIG-L and its yeast homologue Gpi12p are N-acetylglucosaminylphosphatidylinositol de-N-acetylases essential in glycosylphosphatidylinositol biosynthesis. Biochem J 339(Pt 1):185–192

Weissman Z, Kornitzer D (2004) A family of *Candida* cell surface haem-binding proteins involved in haemin and haemoglobin-iron utilization. Mol Microbiol 53:1209–1220

Weissman Z, Shemer R, Conibear E, Kornitzer D (2008) An endocytic mechanism for haemoglobin-iron acquisition in *Candida albicans*. Mol Microbiol 69:201–217

Wiederhold NP, Kontoyiannis DP, Prince RA, Lewis RE (2005) Attenuation of the activity of caspofungin at high concentrations against *Candida albicans*: possible role of cell wall integrity and calcineurin pathways. Antimicrob Agents Chemother 49:5146–5148

Wiederhold NP, Najvar LK, Fothergill AW, McCarthy DI, Bocanegra R, Olivo M, Kirkpatrick WR, Everson MP, Duncanson FP, Patterson TF (2015) The investigational agent E1210 is effective in treatment of experimental invasive candidiasis caused by resistant *Candida albicans*. Antimicrob Agents Chemother 59:690–692

Wiedman JM, Fabre AL, Taron BW, Taron CH, Orlean P (2007) *In vivo* characterization of the GPI assembly defect in yeast mcd4-174 mutants and bypass of the Mcd4p-dependent step in mcd4Delta cells. FEMS Yeast Res 7:78–83

Yin QY, de Groot PW, Dekker HL, de Jong L, Klis FM, de Koster CG (2005) Comprehensive proteomic analysis of *Saccharomyces cerevisiae* cell walls: identification of proteins covalently attached via glycosylphosphatidylinositol remnants or mild alkali-sensitive linkages. J Biol Chem 280:20894–20901

Yoko-o T, Wiggins CA, Stolz J, Peak-Chew SY, Munro S (2003) An N-acetylglucosaminyltransferase of the Golgi apparatus of the yeast *Saccharomyces cerevisiae* that can modify N-linked glycans. Glycobiology 13:581–589

Zhao X, Oh SH, Yeater KM, Hoyer LL (2005) Analysis of the *Candida albicans* Als2p and Als4p adhesins suggests the potential for compensatory function within the Als family. Microbiology 151:1619–1630

Chapter 17
Iron Acquisition in the Pathobiology of *Candida albicans*

Manjit Kumar Srivastav, Remya Nair and Krishnamurthy Natarajan

Abstract The host nutritional environment is a key determinant for survival of human pathogens including *C. albicans*. Iron is an essential micronutrient for the host as well as for *C. albicans*. However, iron sequestration in host proteins confers 'nutritional immunity'. In this chapter, we have reviewed our current understanding of the different iron uptake systems, the associated genes, and their transcriptional regulation in *C. albicans*. Besides, iron acquisition is essential for *C. albicans* virulence and iron homeostasis is also interlinked with lipid homeostasis and multidrug resistance. We have also provided an atlas of all annotated *C. albicans* iron acquisition genes with the attendant gene expression, and phenotype data including virulence.

Keywords Iron acquisition · *Candida albicans* · Virulence · Microarray profiling · Transcription factors · HAP complex · Sfu1 · Sef1 · Hap43/Cap2 · Lipid homeostasis · Multidrug resistance

17.1 Introduction

Iron, a micronutrient, is an essential cofactor for several metabolic steps including DNA synthesis and respiration. The bioavailability of iron in mammalian host, however, is highly restricted. Free iron is highly reactive and generates free radicals, including hydroxyl radicals and superoxide anions, which in turn have huge toxic effects on cells. Most of the iron in humans, close to about 1.8–2.3 g, is found in heme-bound form in hemoglobin, and about half that amount as storage iron and in

M.K. Srivastav · K. Natarajan (✉)
Laboratory of Eucaryotic Gene Regulation, School of Life Sciences, Jawaharlal Nehru University, New Delhi 110067, India
e-mail: nat0200@gmail.com

R. Nair
RGITBT, Bharati Vidyapeeth University, Pune, India

R. Nair
Amity University, Gurgaon, India

R. Prasad (ed.), *Candida albicans: Cellular and Molecular Biology*,
DOI 10.1007/978-3-319-50409-4_17

circulation. The storage iron is sequestered primarily in two proteins ferritin and transferrin. Ferritin has the highest amount of iron sequestered—about 4500 mol as ferric atoms per mole of ferritin. Although ferritin is largely intracellular, a small amount is also secreted. Transferrin on the other hand, is predominantly secreted by hepatocytes and other cell types. Each mole of transferrin coordinates two Fe^{3+} atoms with very high affinity ($K_d \sim 10^{-20}$ M). Thus *C. albicans* has to mobilize iron from these different host iron resources at different sites of infection and colonization. In this chapter, we will discuss our current understanding of iron acquisition mechanisms and its regulation in relation to the biology of *C. albicans* and indicate the gaps in our understanding of the iron acquisition mechanisms. We also provide a compendium of genome-scale studies that provided insights into the regulatory mechanisms governing expression of iron acquisition genes.

17.2 Iron Acquisition Pathways and the Machinery

C. albicans cells need to acquire iron from the host resources hemoglobin, ferritin, and transferrin, in addition to other sources of iron such as siderophores. This is accomplished through specific pathways that involve either cyclical reduction of Fe^{3+} to Fe^{2+} and oxidation back to Fe^{3+}, or direct uptake of Fe^{3+}-bound iron stores. Genetic studies of iron uptake genes in *C. albicans* have compared the growth properties or phenotypes and/or expression analysis in either iron deprivation or iron replete media conditions. For iron-deprived conditions, either defined limited iron medium such as Yeast Nitrogen Base w/o Ammonium Sulfate w/o Copper Sulfate w/o Ferric Chloride (MP Biomedicals LLC) or YPD medium containing one of the iron chelators, viz., bathophenanthroline disulfonic acid disodium salt hydrate (BPS) or 3-(2-Pyridyl)-5,6-diphenyl-1,2,4-triazine-p,p′-disulfonic acid monosodium salt hydrate (ferrozine) is added to the culture medium. Iron replete medium generally contained up to 100 μM of ferrous sulfate or ammonium iron(II) sulfate hexahydrate (ferrous ammonium sulfate) or ferric chloride in either YPD or the limited iron medium described above.

17.2.1 Iron Acquisition from Heme/Hemoglobin

As Candidaemia (blood-borne infection) is a serious life-threatening condition, we would discuss here the pathway and the molecular machinery involved in the acquisition of heme-bound iron from hemoglobin (Hb), as *C. albicans* can utilize Hb and hemin as sole sources of iron (Santos et al. 2003; Moors et al. 1992; Weissman et al. 2002). The heme/Hb-bound Fe^{2+} iron utilization was shown to be independent of the high-affinity iron uptake pathway components (Weissman et al. 2002; Santos et al. 2003), discussed in a later section. Although Hb is not normally predominant in free form, erythrocytes are the reservoir of Hb and heme. Red blood cells undergo rapid recycling through the reticuloendothelial pathway, but acute hemolysis also occurs under various pathological conditions, and Hb is released in circulation. The utilization of iron from hemoglobin also involves hemolysis by *C. albicans* (Watanabe et al. 1999; Manns et al. 1994; Favero et al. 2014). Hb seems to be assimilated preferentially by the hyphal form (Tanaka et al. 1997), since hyphal formation was induced by Hb and hemin (Pendrak and Roberts 2007; Casanova et al. 1997). Hb also induced the expression of a common receptor for *C. albicans* adhesion to extracellular matrix through interaction with laminin, fibronectin, and type IV collagen (Favero et al. 2014) thereby providing another avenue for *C. albicans* survival in the host.

Finally, Hb induced the expression of several *C. albicans* genes in vitro including *HMX1*, *HBR1* through *HBR4* (Pendrak et al. 2004a, b). The role of the *HBR* genes in Hb metabolism is not well understood. The *HMX1* gene codes for heme oxygenase, an activity conserved in mammalian cells as well, and *hmx1Δ* mutant was unable to grow in Hb-containing medium (Santos et al. 2003), as well as in iron-depleted medium (Pendrak et al. 2004a). Heme oxygenase activity leads to the metabolism of heme to release iron, and production of α-biliverdin and carbon monoxide (Pendrak et al. 2004a), two molecules that protect against oxidative killing and inflammation. Thus Hb utilization would provide adaptive advantage for *C. albicans* during bloodstream infection (Pendrak et al. 2004b).

A genetic surrogate screen for *C. albicans* genes that complemented a growth defect of a *Saccharomyces cerevisiae* mutant defective for Hb utilization led to identification of heme-binding cell wall genes *RBT5*, *RBT51/PGA10*, *WAP1/CSA1*, *CSA2*, and *PGA7* referred to as Hb receptor family (Weissman and Kornitzer 2004). Each of the five genes contains the CFEM domain (Weissman and Kornitzer 2004) incorporating eight cysteine residues, which could participate in reversible oxidation–reduction steps. *RBT5* and *RBT51* encode GPI-anchored heme and hemoglobin-binding proteins. Under iron starvation conditions, *RBT5* was highly induced and *rbt5*Δ/Δ mutation impaired utilization of heme and hemoglobin as iron sources. *RBT51*, but not *RBT5*, supported the growth of *S. cerevisiae* in medium containing hemoglobin as iron source (Weissman and Kornitzer 2004). Using Rhodamine-fused Hb, it was demonstrated that the fluorescent Hb bound to the cell surface and was rapidly internalized and eventually accumulated in the vacuole in a manner dependent on the Rbt5, Rbt51 CFEM proteins (Weissman et al. 2008). Furthermore, Hb trafficking required the endocytic pathway involving *VMA11* ATPase and components of the ESCRT machinery (Weissman et al. 2008). These studies established that Hb is a robust source of iron and is actively mobilized by elaborate cellular machinery.

17.2.2 Ferritin and Transferrin Iron Resources

As discussed in the previous section, transferrin and ferritin are rich sources of iron in the mammalian host. When *C. albicans* cells were starved for iron by pregrowth in medium containing an iron chelator bathophenanthroline disulfonate (BPS), supplementation of the medium with holotransferrin but not apotransferrin rescued the growth of iron-starved cells (Knight et al. 2005), indicating that Fe^{3+}-bound transferrin is utilized as an iron source. The utilization of this transferrin-bound iron is dependent on cell surface ferric reductase function, as would be discussed later Sect. 17.2.4. Studies involving co-culture of epithelial cells and *C. albicans* showed that the invasion of epithelial cells by *C. albicans* occurred only when the epithelial cells have high ferritin content (Almeida et al. 2008) and indeed *C. albicans* could grow in medium containing ferritin as the sole iron source. Moreover, ferritin bound to *C. albicans* only in the hyphal form but not yeast form (Almeida et al. 2008). Remarkably, *C. albicans* interaction with ferritin is dependent on the Als3 protein (Almeida et al. 2008). Als3 belongs to the agglutinin family of cell surface glycosylphosphatidylinositol-anchored glycoproteins (Liu and Filler 2011). Expression of *ALS3* was shown to be hyphal-specific and was observed in vivo during oral and systemic infections (Hoyer et al. 1998; Thewes et al. 2007; Zakikhany et al. 2007). Deletion of Als3 lead to strong reduction in the ability of *C. albicans* to bind ferritin and the strain grew poorly on agar plate with ferritin as sole source of iron. The affinity of Fe^{3+} iron to ferritin is drastically reduced under low pH conditions. Thus *C. albicans* can mobilize iron from ferritin through a specific cell surface receptor.

17.2.3 Iron Acquisition Through Siderophore

Siderophores are low molecular weight iron-chelating compounds of diverse chemical structures that are biosynthesized by certain bacteria and fungi under deficient iron conditions (Haas 2003; Andrews et al. 2003). Of the different siderophores, fungi produce the hydroxamate type that binds Fe^{3+} atoms at a very high affinity ($K_d \sim 10^{29}$M). Although early reports indicated that *C. albicans* could produce siderophores under iron-deficient conditions (Ismail et al. 1985; Holzberg and Artis 1983; Sweet and Douglas 1991), the *C. albicans* genome sequence data did not provide any leads into the existence of a siderophore biosynthetic pathway. Utilization of siderophores as a source of iron and the acquisition pathway was identified in *S. cerevisiae* (reviewed in Philpott and Protchenko 2008), involving two cell wall mannoprotein genes *FIT1-FIT3* and a family of four transporter genes *ARN1-ARN4* (Philpott and Protchenko 2008). *C. albicans* can also utilize iron from ferrichrome and ferrioxamine B siderophores (Hu et al. 2002). The *C. albicans* *SIT1/ARN1* transporter gene was identified by several laboratories (Lesuisse et al. 2002; Heymann et al. 2002; Hu et al. 2002; Ardon et al. 2001). The *SIT1/ARN1* gene is required for ferrichrome uptake, but not for ferrioxamine B uptake by *C. albicans* (Hu et al. 2002). Iron utilization from siderophores in *S. cerevisiae* takes place either through the reductase–permease system (Hu et al. 2002), releasing the Fe^{3+} chelate or by the internalization of the Fe^{3+}-siderophore complex (reviewed in Philpott and Protchenko 2008). The mechanistic details of Arn1/Sit1-mediated ferrichrome uptake in *S. cerevisiae* came from studies from the Philpott laboratory (Kim et al. 2002; Moore et al. 2003), wherein the receptor and transporter domains in Arn1/Sit1 coordinates siderophore intake by shuttling between endosome and plasma membrane.

Key differences in the siderophore acquisition pathway between *S. cerevisiae* and *C. albicans* include the absence of all *FIT1-FIT3* genes and the *ARN2-ARN3* genes in the *C. albicans* genome. Arn1/Sit1 was also shown to be plasma membrane-bound in *C. albicans*, and the elevated extracellular ferrichrome concentration, led to a cytoplasmic relocalization (Hu et al. 2002). The internalized siderophore is then trafficked into vacuole, but the mechanism of iron release from the siderophore is not understood. The utilization of iron from both the ferrichrome and the ferrixoxamine B siderophores requires the high-affinity iron permease genes *FTR1* and *FTR2* (Hu et al. 2002). Although *C. albicans* and *S. cerevisiae* does not produce siderophores, unlike *Aspergillus fumigatus*, heterologous siderophores produced by microflora at the site of infection could be used as an iron resource.

17.2.4 Iron Transport Machinery

Iron is predominant in natural sources in Fe^{3+} complexes as described in the preceding sections. The Fe^{3+} iron is also the most stable form compared to the highly reactive Fe^{2+} form, which upon auto-oxidation produces superoxide radicals. Thus the cellular strategies for iron uptake have to contend with the insoluble Fe^{3+} iron and the labile Fe^{2+} iron. As iron is essential for growth, cellular strategy for iron acquisition is critical under both iron-limiting (high-affinity uptake) as well as under iron-sufficient conditions (low-affinity uptake). The high-affinity iron transport system involves an essential iron redox mechanism whereby the external Fe^{3+} ion is first reduced to Fe^{2+} ion by cell surface-resident ferric reductase and the Fe^{2+} ion serves as a substrate for iron oxidase–permease enzyme complex (Philpott and Protchenko 2008; Kosman 2003). Our current understanding of the iron transport machinery in *C. albicans* has largely come from the elegant genetic studies in the yeast *S. cerevisiae*. The logic of iron uptake in *S. cerevisiae* (Kosman 2003; Radisky and Kaplan 1999; Philpott 2006; Philpott and Protchenko 2008; Van Ho et al. 2002) as well as the S. cerevisiae proteins involved in iron uptake has been comprehensively reviewed in the past (Kosman 2003; Philpott and Protchenko 2008). Therefore, we only provide a brief summary of the high-affinity uptake system in *S. cerevisiae*. Sequence comparison identified nine ferric reductase-like genes *FRE1-FRE7*, YGL160w, and YLR047c. Measurements of mRNA levels showed that only the *FRE1-FRE6* genes are regulated by iron levels (Georgatsou and Alexandraki 1999), although some of these six genes are also regulated by copper levels as would be discussed later. Interestingly, just the two Fre1p and Fre2p together account for >90% of the cell surface ferric reductase activity in *S. cerevisiae* (reviewed in Philpott and Protchenko 2008). The fungal ferric reductase enzymes are characterized by the ferric reductase domain (Pfam PF01794), cofactor FAD-binding domain (Pfam PF08022), electron donor NADPH-binding domain (Pfam PF08030), as well as multiple transmembrane domains. The ferric reductase activity itself resides in the transmembrane domain of ferric reductase protein; moreover, the fungal ferric reductases also contain additional transmembrane segments. The redox chemistry employs the electrons generated upon NADPH conversion to NADP and H^+, to reduce Fe^{3+} to Fe^{2+}. The reaction mechanism then involves the reoxidation of the soluble Fe^{2+} by the plasma membrane resident Fet3 multicopper oxidase, also called ferroxidase. The Fe^{3+} ion thus produced is transferred to the ferric permease Ftr1 by an elegant substrate-channeling mechanism identified previously (Kwok et al. 2006).

Next, we review the current literature on the *C. albicans* genes/putative genes of the three components of the high-affinity uptake system, viz., ferric reductase, oxidase (a copper-dependent enzyme), and iron permease (Almeida et al. 2008). Early studies to identify iron uptake pathway in *C. albicans* involved cloning genes by complementation and functional analyses in *S. cerevisiae* (Hammacott et al. 2000; Knight et al. 2002; Ziegler et al. 2011; Cheng et al. 2013). Although orthologs of the *S. cerevisiae* high-affinity uptake system genes are found in the *C. albicans* genome, only a limited number of functional studies have been carried out. Biochemical studies showed that *C. albicans* cells had substantially higher reductive iron uptake from transferrin as compared to *S. cerevisiae* (Knight et al. 2005). *C. albicans* genome sequence contains 17 putative ferric reductase-like genes (Binkley et al. 2014; Baek et al. 2008; Ibtissem et al. 2010). Therefore, we reanalyzed the translated protein sequence of all 17 putative ferric reductase-like genes for conserved protein domains using the SMART domain search (Letunic et al. 2015). We found that only 14 of the annotated ferric reductase-like ORFs (Table 17.1) contained all the three domains, viz., ferric reductase-transmembrane domain, the FAD domain, and the NAD domain, characteristic of the ferric reductase protein family.

Of these ferric reductase genes, functional analysis has been carried out only on three of them and would be described next. The *C. albicans CFL1* gene (Table 17.1) could complement the growth of *S. cerevisiae fre1Δ/Δ* mutant in iron-deficient medium (Hammacott et al. 2000). In *C. albicans*, the *CFL1* mRNA level was induced upon either iron or copper depletion. Interestingly, the *C. albicans cfl1Δ/Δ* mutant exhibited higher cell surface ferric reductase activity (Xu et al. 2014b). Moreover, the expression analysis showed that mRNA levels of other putative ferric reductase genes *FRP1* (orf19.5634), *CFL2* (orf19.1264) and *FRE10* (orf19.1415) were elevated in the *cfl1Δ/Δ* mutant independent of the external iron levels. A second ferric reductase gene *FRE10/CFL95* (orf19.1415) was initially cloned by complementation in *S. cerevisiae* (Knight et al. 2002). Deletion of the *FRE10* gene in *C. albicans* led to a large (~75%) decrease in cell

Table 17.1 Compendium of iron acquisition genes indicating their regulation of expression and virulence phenotype

C. albicans standard gene/systematic name	Alias	Growth and localization Upon iron depletion (Ref.)	Regulation (microarray)[a]				Virulence[b]
			−Fe versus +Fe			pH8 versus pH4	
			Chen et al. (2011)	Lan et al. (2004)	Hameed et al. (2008)	Bensen et al. (2004)	
Hemoglobin uptake							
CSA1/ C7_00090C_A	orf19.7114, *WAP1*	n.d.; CS (Lamarre et al. 2000)	Up	Up	Up	Up	n.d.
RBT5/ C4_00130W_A	orf19.5636	NC; CW & PM (Weissman and Kornitzer 2004; Weissman et al. 2008; Kuznets et al. 2014)	Up	Up	Up		NC (Braun et al. 2000)
HMX1/ C1_00350C_A	orf19.6773	NC; n.d. (Santos et al. 2003)	Up	Up	Up		Dn (Navarathna and Roberts 2010)
PGA10/ C4_00450C_A	orf19.5674, *RBT51*, *RBT8*, *CRW2*	NC; CW (Weissman and Kornitzer 2004; Weissman et al. 2008)	Up				NC (Perez et al. 2011)
CSA2/ C4_06920C_A	orf19.3117, *CRW1*	n.d.	Up				n.d.
PGA7/ C4_00120W_A	orf19.5635, *CRW3*, *RBT6*	n.d.; CW (Kuznets et al. 2014)	Up				Dn (Kuznets et al. 2014)
RBT4/ C1_07030C_A	orf19.6202, *PRY4*	n.d.		Dn			Dn (Braun et al. 2000; Rohm et al. 2013)

(continued)

Table 17.1 (continued)

C. albicans standard gene/systematic name	Alias	Growth and localization Upon iron depletion (Ref.)	Regulation (microarray)[a]				Virulence[b]
			−Fe versus +Fe			pH8 versus pH4	
			Chen et al. (2011)	Lan et al. (2004)	Hameed et al. (2008)	Bensen et al. (2004)	
Siderophore Uptake							
SIT1/ C2_08050C_A	orf19.2179, *ARN1*	Dn; P.M. (Santos et al. 2003; Hu et al. 2002)	Up	Up	Up	Up	NC (Heymann et al. 2002)
Ferritin uptake							
ALS3/ CR_07070C_A	orf19.1816, *ALS8*, *ALS2*, *ALS10*	Dn; CW (Almeida et al. 2008)					n.d.
Fe^{3+} to Fe^{2+} reductase							
CFL1/ C4_05770C_A	orf19.1263, *FRE1*	n.d.	Up	Up		Up	Dn (Xu et al. 2014b)
CFL2/ C4_05780C_A	orf19.1264, *FRE2*, *CFL97*	n.d.	Up	Up		Up	n.d.
CFL4/ C5_01360W_A	orf19.1932, *FRE5*, *FRE32*, *CFL94*	n.d.	Up	Up		Dn	n.d.
FRE10/ C4_04320W_A	orf19.1415, *CFL95*, *FRE1*, *RBT2*	n.d.	Up		Up	Dn	NC (Braun et al. 2000)
FRP1/ C4_00110C_A	orf19.5634, *FRE5*	Dn; n.d. (Liang et al. 2009b)	Up	Up	Up		n.d.

(continued)

Table 17.1 (continued)

C. albicans standard gene/systematic name	Alias	Growth and localization Upon iron depletion (Ref.)	Regulation (microarray)[a]				Virulence[b]
			−Fe versus +Fe			pH8 versus pH4	
			Chen et al. (2011)	Lan et al. (2004)	Hameed et al. (2008)	Bensen et al. (2004)	
FRP2/ C7_00100W_A	orf19.7112	n.d.	Up		Dn	Up	n.d.
CFL5/ C5_01380W_A	orf19.1930, *FRE31*, *FRE4*, *CFL90*	n.d.	Up	Up			n.d.
FRE9/ C2_05070W_A	orf19.3538, *CFL12*, *CFL8*	n.d.	Up			Up	n.d.
CFL11/ CR_06670W_A	orf19.701, *FRE8*, *CFL96*, *CFL9*	n.d.	Dn				n.d.
C7_00430W_A	orf19.7077, *FRE7*	n.d.			Dn		n.d.
FRE7/ CR_07290W_A	orf19.6139, *FRE3*, *CFL1*, *CFL93*	n.d.				Dn	n.d.
FRE3/ C4_05840W_A	orf19.1270, *CFL99*, *CFL3*	n.d.					n.d.
FRE30/ CR_07280W_A	orf19.6140	n.d.					n.d.
CR_07300W_A	orf19.6138, *CFL93*, *FRE6*, *FRE41*, *CFL6*	n.d.					n.d.

(continued)

Table 17.1 (continued)

C. albicans standard gene/systematic name	Alias	Growth and localization Upon iron depletion (Ref.)	Regulation (microarray)[a] −Fe versus +Fe			pH8 versus pH4	Virulence[b]
			Chen et al. (2011)	Lan et al. (2004)	Hameed et al. (2008)	Bensen et al. (2004)	
C1_11020W_A	orf19.2312, *FRE3*, *FRE11*, *FRE42*, *CFL1*	n.d.					n.d.
CR_06870C_A	orf19.1844, *FRE4*, *CFL91*, *FRE43*	n.d.					n.d.
C1_09780C_A	orf19.4843	n.d.					n.d.
Multicopper oxidase							
FET34/ C6_00440C_A	orf19.4215, *FET5*, *FET35*	NC; PM (Cheng et al. 2013; Ziegler et al. 2011)	Up	Up			Dn (Cheng et al. 2013)
FET3/ C6_00480C_A	orf19.4213, *FET33*, *FET31*	n.d.; ER[c] (Ziegler et al. 2011)	Up				n.d.
FET31/ C6_00460C_A	orf19.4211, *FET32*, *FET3*	n.d.; PM, ER[c] (Eck et al. 1999; Ziegler et al. 2011)					NC (Eck et al. 1999)
FET33/ C5_00460C_A	orf19.943, *FET397*	NC.; VM (Cheng et al. 2013; Ziegler et al. 2011)					NC (Cheng et al. 2013)
FET99/ C6_00470C_A	orf19.4212, *FET32*	n.d.; ER[c] (Ziegler et al. 2011)					n.d.

(continued)

Table 17.1 (continued)

C. albicans standard gene/systematic name	Alias	Growth and localization Upon iron depletion (Ref.)	Regulation (microarray)[a]				Virulence[b]
			−Fe versus +Fe			pH8 versus pH4	
			Chen et al. (2011)	Lan et al. (2004)	Hameed et al. (2008)	Bensen et al. (2004)	
Iron permease/transporter							
FTR1/ C1_14130W_A	orf19.7219	Dn; PM (Baek et al. 2008; Fang and Wang 2002; Wolf et al. 2010)	Up	Up	Up	Up	Dn (Ramanan and Wang 2000)
FTR2/ C1_14220C_A	orf19.3227	n.d.	Up	Dn		Up	NC (Ramanan and Wang 2000)
FTH1/ CR_01270C_A	orf19.4802	n.d.	Up	Up	Up	Up	n.d.
FTH2/ CR_01270C_A	orf19.3227, *FTR3*	n.d.					n.d.
Low affinity transporter							
SMF12/ C2_07160W_A	orf19.2270, *SMF1*, *SMF11*	n.d.				Up	n.d.
SMF3/ C2_00580C_A	orf19.2069, *SMF2*	NC; n.d. (Xu et al. 2014a)				Up	NC (Xu et al. 2014a)
C1_13840W_A	orf19.5022, *SMF2*, *SMF3*	n.d.					n.d.
C4_00990W_A	orf19.4690, *SMF11*, *SMF12*	n.d.					n.d.

(continued)

Table 17.1 (continued)

C. albicans standard gene/systematic name	Alias	Growth and localization Upon iron depletion (Ref.)	Regulation (microarray)[a]				Virulence[b]
			−Fe versus +Fe			pH8 versus pH4	
			Chen et al. (2011)	Lan et al. (2004)	Hameed et al. (2008)	Bensen et al. (2004)	
Copper/manganese transporter							
CCC2/ C5_03020W_A	orf19.4328	n.d.; Trans-Golgi (Weissman et al. 2002)	Up	Up		Dn	NC (Weissman et al. 2002)
CCC1/ C3_03710W_A	orf19.6948	NC; n.d. (Xu et al. 2014a)	Dn			Dn	NC (Xu et al. 2014a)
Mitochondrial iron transporter							
MRS4/ C2_08070C_A	orf19.2178, *MRS3*	Dn; MM (Xu et al. 2014a)					Dn (Xu et al. 2014a)
Transcription factors							
HAP2/ C1_07680W_A	orf19.1228	Dn; Nuc[d] (Hsu et al. 2013; Homann et al. 2009)	Up		Up		Dn (Hsu et al. 2013)
HAP3/ C4_01390W_A	orf19.4647, *HAP32*	Dn; Nuc[d] (Singh et al. 2011; Hsu et al. 2013; Homann et al. 2009)	Up	Up		Up	NC (Hsu et al. 2013)
HAP43/ C1_11210C_A	orf19.681, *CAP2*	Dn; Nuc (Singh et al. 2011; Hsu et al. 2013; Hsu et al. 2011; Chen et al. 2011; Homann et al. 2009)	Up	Up			Dn (Hsu et al. 2011; Singh et al. 2011)
SEF1/ CR_02190C_A	orf19.3753	Dn; Nuc (Chen and Noble 2012; Chen et al. 2011; Homann et al. 2009)	Up				Dn (Chen et al. 2011)

(continued)

Table 17.1 (continued)

C. albicans standard gene/systematic name	Alias	Growth and localization Upon iron depletion (Ref.)	Regulation (microarray)[a]				Virulence[b]
			−Fe versus +Fe			pH8 versus pH4	
			Chen et al. (2011)	Lan et al. (2004)	Hameed et al. (2008)	Bensen et al. (2004)	
SFU1/ C1_10020W_A	orf19.4869	NC; Nuc (Chen and Noble 2012; Chen et al. 2011; Lan et al. 2004; Singh et al. 2011; Homann et al. 2009)	Dn				NC (Chen et al. 2011)
HAP41/ C4_05110C_A	orf19.740	NC; Nuc[d] (Singh et al. 2011; Baek et al. 2008)	Dn				n.d.
HAP31/ CR_04290W_A	orf19.4647, *HAP3*	NC; Nuc[d] (Hsu et al. 2013; Singh et al. 2011)		Dn			NC (Hsu et al. 2013)
RIM101/ C1_14340C_A	orf19.7247, *HRM101*, *PRR2*	NC; Nuc[d] (Davis 2009; El Barkani et al. 2000; Ramon and Fonzi 2003; Nobile et al. 2008)				Up	Dn (Davis et al. 2000)
HAP42/ C2_01700C_A	orf19.1481	NC; Nuc[d] (Singh et al. 2011)					n.d.
HAP5/ C5_00940C_A	orf19.1973	NC; Nuc[d] (Singh et al. 2011; Hsu et al. 2013; Hsu et al. 2011; Baek et al. 2008; Homann et al. 2009)					Dn (Hsu et al. 2013)
AFT2/ C2_07170C_A	orf19.2272	NC; Nuc[d] (Liang et al. 2010; Xu et al. 2013)					Dn (Liang et al. 2010)

Genes in the various iron acquisition pathways were collated from CGD, and other published resources (Philpott and Protchenko 2008; Kosman 2003). The genes were grouped as per their functional activity and arranged based on their differential expression under iron deprivation conditions in microarray data
[a]Regulation: *Up* Upregulated; *Dn* Downregulated; Blank indicates No Data;
[b]Virulence: *NC* No change in growth; *Dn* Reduced Growth/Virulence; *n.d.* Not determined;
Localization: *PM* Plama membrane; *VM* Vacuolar membrane; *CW* Cell Wall; *CS* Cell surface; *MM* Mitochondrial membrane; *Nuc* Nuclear; [c]localization; [d]predicted in *S. cerevisiae*

surface ferric reductase activity and ~77% decrease in iron uptake (Knight et al. 2002, 2005; Braun et al. 2000). In addition to being a ferric reductase, Fre10 is also an iron-responsive cupric reductase (Jeeves et al. 2011). Fre10 has been shown to be maximally active at pH 4.4 as compared to that at pH 6.3–6.4 (Knight et al. 2005). Moreover, Fre10 is also required, although not essential, for iron uptake from transferrin (Knight et al. 2005). The expression of C7_00430W/orf19.7077 (originally named *FRE7*), a putative ferric reductase gene, was regulated by copper and ferric reductase activity was induced ~two-fold upon copper depletion (Woodacre et al. 2008). However, a homozygous mutant still had high levels of ferric reductase activity even under copper depleted conditions, indicating that this ferric reductase has a minor contribution to cell surface ferric reduction. The role of iron in the expression or in regulating the ferric reductase activity of C7_00430W/orf19.7077 is not understood. Despite the large number of ferric reductase genes found in the *C. albicans* genome, a comprehensive functional analysis of the rest of the genes has not been conducted. The role of iron in regulating the expression of the ferric reductase genes would be discussed later based on genome-wide microarray data.

The second component of the reductase pathway, the multicopper oxidase, also known as ferroxidase, converts ferrous ion to ferric ion (Kosman 2003). The *S. cerevisiae* genome encodes genes in the high-affinity uptake system *FET3* and *FET5* (in the vacoular membrane) and the low-affinity uptake *FET4* gene. The expression of each of *FET3*, *FET4* and *FET5* genes is induced under iron-deficient conditions. The *C. albicans* genome encodes five multicopper oxidase genes *FET3*, *FET31*, *FET33*, *FET34*, and *FET99* (Almeida et al. 2009). Expression of *FET34*, but not *FET33*, in an *S. cerevisiae fet3*Δ mutant led to substantial complementation of the iron-deficient growth defect (Cheng et al. 2013). Moreover, *C. albicans fet3*Δ/Δ mutation conferred partial growth defect under iron-deficient conditions, with *fet34*Δ/Δ mutation being more severely compromised for growth. Whereas Fet34::GFP was found on the plasma membrane, Fet33::GFP was found on the vacuolar membrane (Ziegler et al. 2011). Iron strongly repressed the transcription of *FET99* (Knight et al. 2002). Thus the multiple multicopper oxidase genes in *C. albicans* seem to perform redundantly in iron homeostasis.

In *S. cerevisiae*, biogenesis of Fet3, which is bound to four atoms of copper in its multicopper oxidase domain, occurs in trans-Golgi which requires P-type ATPase Ccc2 for copper transport. Two genes *CRP1* and *CCC2*, orthologs of the *S. cerevisiae CCC2* gene, have been identified and characterized in *C. albicans*. *CRP1* is specifically involved in copper detoxification, while Ccc2 is required for biogenesis of Fet3. Ccc2 was localized into trans-Golgi while Crp1 localizes to plasma membrane. Due to the copper requirement for the oxidase activity, the intracellular copper transporter Ccc2 is essential for this reductive pathway (Weissman et al. 2002; Almeida et al. 2008). Thus copper and iron homeostasis are intimately linked.

The third component of the reductive pathway is iron permease. The permease forms a complex with multicopper oxidase and transfers Fe^{3+} into the cell (Almeida et al. 2009). *C. albicans* genome encodes four iron permease genes *FTR1*, *FTR2*, *FTH1*, and *FTH2* (Almeida et al. 2009). All the four protein sequences contain an iron-binding motif Arg-Glu-Gly-Leu-Glu, which is essential for the permease activity (Fang and Wang 2002). Under iron deprivation conditions, high-affinity iron permease *FTR1* was induced and essential for growth (Ramanan and Wang 2000). The *ftr1Δ* null mutant exhibited severe growth defect in iron deprivation conditions (Almeida et al. 2008). Therefore, both under in vivo and in vitro conditions, the *FTR1* gene is an essential component of the reductive pathway. Expression of *FTR1* and *FTR2*, with ~87% sequence identity, in the *S. cerevisiae ftr1Δ* mutant rescued its low-iron-deficient growth defect (Ramanan and Wang 2000). Unlike the *FTR1* gene, the expression of the *FTR2* gene is induced under high iron condition (Ramanan and Wang 2000). Furthermore, *ftr1*Δ/Δ but not *ftr2*Δ/Δ mutants exhibited growth defect in low-iron media (Ramanan and Wang 2000). The *FTR1* gene is also required for high-affinity iron uptake from transferrin, along with *FRE10* and *FET3* genes.

The vacuole is a special organelle that retains high amount of iron stores. Indeed a dedicated iron transport system exists at the vacuolar membrane including the ferric reductase Fre6, multicopper oxidase Fet5, and the iron permease Fth1 in *S. cerevisiae* (Philpott and Protchenko 2008). Moreover, the transporter Ccc1 is required for iron import into the vacuole (Puig et al. 2005), and *CCC1* expression is controlled by the transcription factor Yap5 (Li et al. 2008). An analogous machinery for vacuolar iron homeostasis has not yet been identified in *C. albicans*.

17.2.5 Direct Ferrous Iron Uptake (Low Affinity Uptake System)

In yeast, the ferrous iron has also been shown to be directly transported into cells through divalent metal ion transporters. Unlike the ferric reductases, these transporters are not specific for ferrous iron, and can also transport Cu^{2+}, Mn^{2+}, and often Ni^{2+}. The *S. cerevisiae* genes *SMF1*, *SMF2*, and *SMF3* encode homologues of the mammalian Nramp protein (Portnoy et al. 2000). Whereas Smf1 and Smf2 are involved in Mn^{2+} transport, Smf1 can also transport Fe^{2+} ion (Portnoy et al. 2000; Cohen et al. 2000). Smf3 was localized to vacuolar membrane indicating that Smf3 is required for mobilization of iron reserves from the vacuoles (Portnoy et al. 2000; Philpott and Protchenko 2008). Moreover, *SMF3* expression was downregulated by iron, and *SMF3* deletion caused iron starvation. *C. albicans* genome encodes four Nramp-like genes with homology to the yeast *SMF* genes (Table 17.1). Biochemical fractionation showed that *CCC1* and *SMF3* are both vacuolar; whereas deletion of *CCC1* gene led to reduction in vacuolar iron content and deletion of *SMF3* gene led to higher vacuolar iron content indicating that Smf3 functions antagonistic to the Ccc1 transporter (Xu et al. 2014a).

17.3 Transcriptional Regulation of Iron Homeostasis in *C. albicans*

As discussed in the preceding sections, mRNAs encoding the various components of the iron uptake systems are upregulated upon iron deprivation in *C. albicans*, similar to their counterparts in *S. cerevisiae*. In *S. cerevisiae*, the transcription factor Aft1 has a predominant role in transcriptional upregulation of iron acquisition genes (Fig. 17.1). Moreover, an Aft1 paralog Aft2 also has partially nonoverlapping role in regulation of iron acquisition gene expression (reviewed in Kaplan and Kaplan 2009). Besides, the mRNA levels of genes involved in iron utilization genes or that increase iron utilization have been shown to be under the negative regulation of Aft1/Aft2-induced genes *CTH1* and *CTH2*, encoding mRNA-binding proteins leading to instability of target mRNAs (Fig. 17.1) (reviewed in Vergara and Thiele 2008). The Aft1/Aft2- and Cth1/Cth2-regulated genes, collectively known as iron regulon, account for bulk of the genes in the different iron acquisition pathways and certain genes encoding iron requiring proteins such as iron–sulfur (Fe–S) cluster proteins.

17.3.1 Iron-Responsive Transcription Factors

The major point of difference in the control of iron homeostasis between *S. cerevisiae* and *C. albicans* (reviewed in Noble 2013; Blankenship and Mitchell

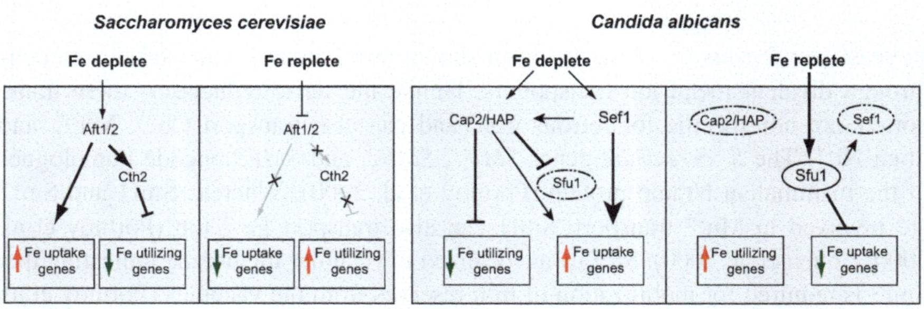

Fig. 17.1 Comparision of Iron Homeostasis Regulation in *C. albicans* and *S. cerevisise*. In *S. cerevisiae*, the Aft1 and Aft2 transcription factors upregulate expression of iron uptake genes, and that of the negative regulator Cth2 upon iron deprivation. The Cth2 protein binds to the mRNA of genes encoding proteins that require iron and target them for degradation. Under iron replete conditions, Aft1-Aft2 is inoperative, and therefore the negative control by Cth2 is absent. In *C. albicans* upon iron deprivation, two transcription factors the Cap2/Hap43-HAP complex and Sef1 are induced, which then upregulate iron uptake gene expression. Cap2/Hap43 also repressed the expression of the negative regulator Sfu1. Cap2/Hap43-HAP complex also directly repressed the expression of genes encoding iron utilization proteins. Under iron replete conditions, the Cap2/Hap43-HAP complex and Sef1 expression is lost. Consequently, Sfu1 is expressed, which then repressed the expression of iron uptake genes. Sfu1 also interacts with Sef1 under iron replete conditions, and sequesters any residual Sef1 and targets Sef1 for degradation

2011) involves a complete rewiring of the transcription factors (Fig. 17.1). Whereas an ortholog of *S. cerevisiae AFT1* gene is absent in *C. albicans*, the *AFT2* ortholog was neither required for *C. albicans* growth in iron-limited medium nor the cell surface ferric reductase activity and cellular iron content was significantly impaired by *aft2Δ* mutation (Liang et al. 2010; Xu et al. 2013). Moreover, orf19.5334, the sequence ortholog of *CTH2* in *C. albicans* was not required for growth under iron deprivation conditions and for regulation of *ACO1* mRNA level (Singh et al. 2011). These studies indicated that the paradigm for iron homeostasis regulation identified in *S. cerevisiae* was not operative in *C. albicans*. Indeed studies from several fungal genomes have revealed a complete transcriptional rewiring of the regulation of expression of iron acquisition genes and genes involved in iron homeostasis.

Functional genomics approaches in *C. albicans* led to the discovery of Sfu1 (Lan et al. 2004), HAP transcription regulators (Singh et al. 2011; Chen et al. 2011; Hsu et al. 2011; Baek et al. 2008) and the Cys_6Zn_2 transcriptional activator Sef1 as key transcription factors controlling iron homeostasis (Lan et al. 2004; Chen et al. 2011; Homann et al. 2009). The *C. albicans* Cap2/Hap43, orthologous to the fungal *hapX* factors, is a bipartite transcription factor bearing bZIP domain and the Hap4-like domain, and interacts biochemically with the Hap5 to a form a regulatory complex (Singh et al. 2011). The *C. albicans* Hap5 protein contains two evolutionarily conserved domains, the histone H2A-like histone-fold domain and a Hap4-interacting domain, and the Hap3 protein contains histone H2B-like histone-fold domain (Singh et al. 2011).

Mutational studies showed that *HAP5*, *HAP3*, *HAP2*, and *HAP43/CAP2* are required for *C. albicans* growth under iron deprivation conditions. Genetic studies showed that both the Hap4-like (HAP4L) and the bZIP bipartite domains are required for Cap2/Hap43 function (Singh et al. 2011). Although the complex formation by the HAP proteins has not been demonstrated, the regulation of expression and their requirement for iron-limited growth indicates the interaction between Hap2, Hap3, and Hap5. The role of *HAP43/CAP2* in iron homeostasis gene regulation is not completely understood. *HAP43* is required for induction of several iron acquisition genes and the *HAP3* regulator (Singh et al. 2011; Chen et al. 2011). Hap43 was also shown to be recruited to *FRP2* promoter (Chen et al. 2011), a ferric reductase gene induced upon iron deprivation. Paradoxically, *HAP43* was required for repression of iron utilizing gene promoters and that of the expression of the *SFU1* gene (Singh et al. 2011; Chen et al. 2011) by recruitment to the *SFU1* promoter (Chen et al. 2011) under conditions of iron deprivation. Moreover, as *C. albicans* encodes multiple *HAP3* and *HAP4* genes, the *HAP* genes participate in the control of multiple pathways in addition to regulation of iron homeostasis (Hsu et al. 2013).

Under iron replete conditions, expression of iron uptake genes is repressed by the negative regulator Sfu1, a GATA family zinc finger transcription factor (Chen et al. 2011; Lan et al. 2004), presumably to offset iron overload in cells (Fig. 17.1). In *S. pombe*, the Sfu1 ortholog Fep1 interacts with Tup11, a general transcriptional repressor protein, in high iron conditions to repress iron uptake pathway genes in

S. pombe (Pelletier et al. 2007). Moreover, Sfu1 interacts with Sef1 in high iron conditions leading to Sef1 degradation (Chen and Noble 2012).

Under iron deprivation conditions, however, the expression of iron acquisition genes was induced by the Sef1 regulator, through a two-way control (Fig. 17.1). First through activation of the Cap2/Hap43 expression, which in turn is a repressor of Sfu1 (Chen et al. 2011; Singh et al. 2011), and by direct activation of iron acquisition gene expression (Chen et al. 2011). The Cap2/Hap43 regulator also performs a dual role during iron deprivation conditions (Fig. 17.1)—promoter binding and repression of *SFU1* expression, and by activating expression of several iron uptake genes (Singh et al. 2011; Chen et al. 2011), through promoter binding as shown for orf19.7112/*FRP2* (Chen et al. 2011). Moreover, the activation of iron acquisition genes also requires the promoter binding of the Hap5 subunit of the Hap2-Hap3-Hap5 complex (Singh et al. 2011). Consistent with the differential roles for these transcriptional regulators, their expression is also controlled in a manner dependent on the iron availability. The expression of *CAP2/HAP43* and *HAP3* (Singh et al. 2011) and that of *SEF1* (Chen et al. 2011) is highly induced upon iron deprivation and is repressed under iron replete condition (Fig. 17.1). Moreover, the Sef1 factor also can cause flavin secretion, a likely metabolic response to iron limitation (Schillig and Morschhäuser 2013). A central issue in the regulatory paradigm is to understand the mechanism of sensing of iron status by the *C. albicans* cells to activate the different transcription factor responses. In this context, Sfu1 interacts with Sef1 under iron replete conditions to sequester from nucleus and target for degradation (Fig. 17.1) in a manner dependent on the Ssn3 kinase (Chen and Noble 2012). Moreover, Fe^{3+} atom, along with Zn^{2+} is required for activation of the DNA binding activity of the Sfu1 ortholog in *Histoplasma capsulatum* (Chao et al. 2008), indicating a role for direct metal sensing as a key regulatory step.

17.3.2 *Iron Homeostasis Transcriptome*

As discussed in the preceding sections, bulk of the genes annotated to be related to iron acquisition pathways in *C. albicans* have not been studied experimentally in *C. albicans*. Therefore to probe the involvement of these genes, here we have compiled the transcriptome data from published microarray studies that compared mRNA levels between iron replete versus iron-depleted conditions (Chen et al. 2011; Lan et al. 2004; Hameed et al. 2008). Upon iron deprivation, ~515 genes (Chen et al. 2011) and ~470 genes (Lan et al. 2004) were differentially expressed, and at least 94 genes (10.5%) were common to both datasets, even though the media and the iron deprivation conditions employed in the two studies were not the same. These iron deprivation responsive differentially expressed genes belonged to various biological processes including pathogenesis, transport, response to stress, cellular homeostasis, cellular respiration, etc.

All genes in the hemoglobin uptake pathway, with the exception of *RBT4*, were induced upon iron deprivation in at least one of the three datasets (Table 17.1). The expression of *SIT1*, the siderophore transporter was induced in all three datasets, but mRNA level of *ALS3* involved in ferritin uptake was not significantly altered upon iron deprivation in these datasets. As discussed in an earlier section, although *C. albicans* genome contains 17 ferric reductase-like genes. However, only a limited number of genes have been studied for their role in iron uptake and cellular localization (Table 17.1). The microarray data showed that the expression of *CFL1*, *CFL2*, *CFL4*, *FRE10*, *FRP1*, *FRP2*, *CFL5*, and *FRE9* are upregulated upon iron deprivation (Table 17.1). While *CFL11* was downregulated, rest of the genes were either not differentially expressed or no data was available (Table 17.1). Of the five multicopper oxidase genes, the microarray data contained *FET34* and *FET3* that were upregulated upon iron deprivation. Among the high-affinity iron permease genes, *FTR1*, *FTR2*, and *FTH1* were upregulated (Table 17.1). Whereas the expression of *CCC2*, a key gene in the high-affinity uptake pathway, was upregulated, the expression of *CCC1* was downregulated (Table 17.1). None of the predicted low-affinity ferrous transporter genes were differentially expressed upon iron deprivation (Table 17.1), although *SMF12* and *SMF3* were upregulated in alkaline pH8 compared to pH4 (Table 17.1).

Among the transcription factors, the expression of *HAP2*, *HAP3*, *HAP43*, and *SEF1* were upregulated, and the expression *SFU1*, *HAP41*, and *HAP31* were downregulated upon iron deprivation. The expression of *HAP3* and *SFU1* were upregulated in alkaline pH (Table 17.1). *C. albicans* genome also encodes *SEF2*, a gene paralogous to *SEF1*, whose deletion caused copper-sensitive growth defect, but not BPS-sensitive growth defect (Homann et al. 2009). Interestingly, *SEF2* expression is upregulated by Sef1 and downregulated by Hap43 (Chen et al. 2011). The *CaSEF1* ortholog in *S. cerevisiae* has no role in growth under iron deprivation conditions (Chen et al. 2011) (Fig. 17.1).

Remarkably, the iron transcriptome data showed differential expression of several genes not directly involved in iron acquisition. Microarray profiling of the *hap43*Δ/Δ with reference to the wild-type under iron deprivation conditions identified ~181 genes that were commonly up- or downregulated in the two datasets (Singh et al. 2011; Chen et al. 2011). These 181 genes should form the core of the *HAP43*-dependent genes as the two microarray studies been conducted under different culture conditions to elicit iron deprivation. Although a large number of genes were dependent on either *HAP43* or *SEF1* for their up- or downregulation during iron deprivation (Chen et al. 2011), a substantial fraction of genes were dependent on both *HAP43* and *SEF1* for their upregulation (18%) or downregulation (7%) under iron deprivation conditions. These genome-wide microarray data showed that *SEF1* and *HAP43* were required for regulation of distinct as well as a common set of genes during iron deprivation response. The complete impact of these transcriptomic data is yet to be fully realized since the genome-wide Sef1 and Hap43 ChIP-Chip data yielded only a limited number of genes bound by these transcription factors (Chen et al. 2011).

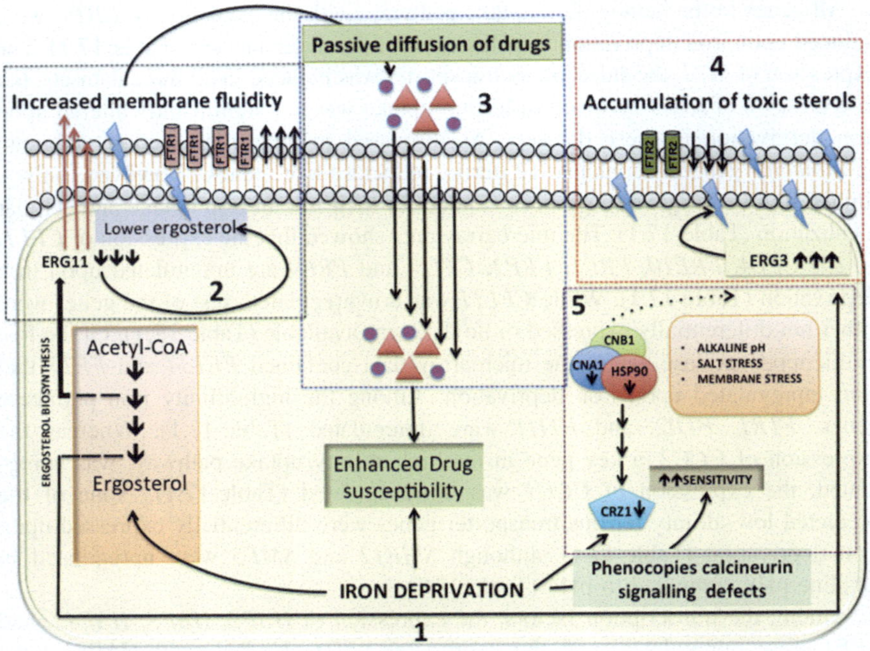

Fig. 17.2 Schematic diagram highlighting the events related to iron homeostasis and MDR in *C. albicans*. Iron deprivation renders *C. albicans* cells hypersusceptible to drugs. Ergosterol biosynthesis pathway genes *ERG11* and *ERG3* are reciprocally regulated upon iron deprivation (*1*). Decrease in *ERG11* results in lowered ergosterol levels and elevated membrane fluidity (*2*) Increased passive diffusion of drugs (*3*). Upregulation of *ERG3* results in accumulation of toxic sterols (*4*). Iron deprivation phenocopies calcineurin signaling mutants resulting in enhanced sensitivity toward alkaline pH, salt, and membrane stress (*5*). Also, the calcinuerin pathway genes, *CRZ1*, *HSP90*, and *CMP1* are downregulated upon iron deprivation

17.4 Signaling Pathways and Iron Homeostasis

17.4.1 pH Response Pathway

As indicated before, the environmental pH has a profound effect on iron solubility. Even an increase of pH unit 1.0 causes rapid equilibrium shift from the soluble ferrous to the insoluble ferric form. The *RIM101* pathway is required for pH sensing and is also required for an appropriate transcriptional response to changes in the environmental pH (Bensen et al. 2004). In alkaline pH, iron acquisition genes were found to be upregulated indicating that alkaline pH mimics iron starvation in a manner dependent on *RIM101*. Moreover, *C. albicans rim101Δ/Δ* mutant was impaired for growth under iron deprivation conditions (Bensen et al. 2004). Rim101 is a zinc finger transcriptional repressor. As indicated in Table 17.1, Rim101 is required for upregulation of the reductive iron uptake pathway genes *FRP1*, *FRE1*, *FRE2*, *FRE7*, and *FRE9* under alkaline pH conditions (Baek et al.

2008; Bensen et al. 2004; Liang et al. 2009a). Indeed Rim101 has been shown to directly bind to the *FRP1* promoter to mediate pH-dependent transcriptional control (Baek et al. 2008). Bulk of the iron-responsive genes was also regulated by Rim101 (Table 17.1). It is noteworthy that the expression of ferritin receptor *ALS3* is regulated by Rim101 in iron-poor environment (Nobile et al. 2008).

17.4.2 HOG1 Pathway

Under iron replete conditions, the expression of four iron uptake genes *FRE10*, *RBT5*, *FTR1*, and *FET34* were derepressed in cells lacking *HOG1* encoding the mitogen-activated protein kinase (MAPK) (Enjalbert et al. 2006). Moreover, in iron-limited medium the protein levels of two of the five multicopper ferroxidases Fet3 and Fet34 were elevated as evident from the *C. albicans* cell wall proteome profiling (Kaba et al. 2013). Moreover, Hog1 activation through its phosphorylation is elevated under high iron conditions compared to iron-deprived condition (Kaba et al. 2013). Consistent with a role for Hog1 in iron homeostasis, *hog1Δ* and the MAP kinase kinase *pbs2Δ* mutants had elevated levels of the multicopper ferroxidase proteins, and cell surface ferric reductase activity in high iron conditions (Kaba et al. 2013). It is unclear as to how the Hog1 signaling integrates with the transcriptional program to repress iron acquisition under high iron conditions.

17.4.3 Calcium and Calcineurin Pathway

The role of the Ca^{2+}-calcineurin pathway has been investigated in *C. albicans*. Calcineurin is an evolutionarily conserved protein phosphatase, and is a heterdimeric complex composed of the catalytic subunit calcineurin A (Cmp1/Cna1) and the regulatory Ca^{2+} binding subunit calcineurin B (Cnb1). Calcineurin is activated by Ca^{2+}-Calmodulin, and the activated calcineurin is required for a variety of cellular stress responses and for survival and pathogenesis of *C. albicans* (Bader et al. 2003, 2006; Liu et al. 2015; Steinbach et al. 2007; Yu et al. 2015). The connection of calcineurin pathway to iron homeostasis was uncovered when iron deprivation led to downregulation of *CMP1/CNA1* expression as well as that of the *HSP90* chaperone, and the downstream transcription factor *CRZ1* (Hameed et al. 2011). Moreover, the phenotypic defects of *cmp1*Δ and *cnb1*Δ calcineurin mutants, viz., susceptibility to alkaline pH, membrane, and salt stress were specifically enhanced in iron-deprived medium (Hameed et al. 2011). However, the molecular mechanism of how calcineurin pathway controls iron homeostasis is yet to be understood.

17.5 Link Between Iron Homeostasis and Redox Status and Hypoxia

C. albicans encounters oxygen-poor environment in host tissues. This hypoxic conditions lead to activation of genes involved in ergosterol (*ERG* genes) biosynthesis, fatty acid, iron metabolism, cell wall structure, glycolysis, and fermentation, while respiration genes are repressed in *C. albicans*. Under hypoxic conditions, 44 iron-responsive genes showed differential expression, of which 15 were upregulated and 29 were downregulated (Synnott et al. 2010), including *FRP1*, *FTR2*, *FRP2*, *FRE9*, *FRE10*, *FTR1*, *FET34*, *RBT5*, and *SIT1* that were upregulated and *FRE7* and *FTR2* were downregulated. Ribonucleotide reductase genes *RNR22* and *RNR21* as well as heme biosynthetic pathway genes *HEM13* and *HEM14* were also upregulated under hypoxic conditions. While *SEF1* expression was upregulated, *SFU1* expression was downregulated under hypoxic conditions. Another transcription factor Upc2 has been shown to be required for upregulation of *FRE7*, *FTR2*, and *FET34* expression indicating an overlap between the transcriptional response to hypoxia and iron limitation (Sosinska et al. 2008; Synnott et al. 2010).

17.6 Iron Acquisition and Filamentation

Iron deprivation conditions elicited *C. albicans* hyphal induction at 37 °C, in a manner that could be reversed by the addition of excess iron (Hameed et al. 2008). Moreover, the *ftr1*Δ and *ccc2*Δ mutants displayed a constitutive filamentous phenotype, which, remarkably, can be reversed by supplementation of iron (Hameed et al. 2008). mRNA expression analysis showed that *EFG1* mRNA level was upregulated under iron-deficient conditions, and *ftr1*Δ and *ccc2*Δ mutations also led to comparable increase in *EFG1* mRNA level, consistent with the constitutive filamentation in the two mutants. Indeed *efg1*Δ mutation abrogated filamentation under both iron-replete as well as iron-deplete conditions (Hameed et al. 2008). Moreover, *TPK1* encoding protein kinase A that acts upstream to *EFG1* was required for the filamentation response in iron-deficient medium (Hameed et al. 2008). There also appears to be a differential role for ferric reductase genes in *C. albicans*, as the expression of *FRE10* and *FRE7* transcript levels is lowered in hyphal stage as compared to yeast form, leading to decrease in ferric reductase activity but not cupric reductase activity (Jeeves et al. 2011).

The cellular morphology also differentially impacts the ability of *C. albicans* to bind hemoglobin as hyphal form seems to adhere more efficiently than the yeast form to hemoglobin-coated matrix (Tanaka et al. 1997). Moreover, the expression of *RBT5* and *WPA1*, two genes required for *C. albicans* growth in hemoglobin–iron-containing medium, was induced in filamentous condition as compared to that in the yeast form (Braun et al. 2000). Haemin increased the filamentation in the absence of *hmx1*. Iron deprivation, haemin, and temperature shift from 30–37 °C also induced the expression of *CaHMX1* (Santos et al. 2003).

17.7 Iron Acquisition in Mammalian Host and Virulence

C. albicans encounters iron-poor environments in the host. Therefore to combat the hostile environment, effective modulation of iron acquisition and homeostasis is of paramount importance for the pathogen survival. As discussed in the preceding sections, *C. albicans* genome encodes a large number of ferric reductase-like genes, ferrooxidase genes and iron permease genes, and therefore it is essential to identify the specific components and relative contributions required for virulence attribute in the host. Accordingly, in this section, we would review the literature regarding the in vivo expression of the iron acquisition genes, and summarize the published work on the requirement of the iron acquisition genes during *C. albicans* infection in mouse disseminated Candidiasis model.

Laser ablation inductively coupled mass spectrometry was used to examine iron deposition in normal versus *C. albicans* infected mouse kidney samples (Potrykus et al. 2013). Iron levels were found to be redistributed from the cortex to the medulla in the infected kidneys, and this iron redistribution appears to be due to an enhanced accumulation of ferritin-iron and the globin polypeptides in the renal medulla (Potrykus et al. 2013). Concomitantly, the level of heme oxygenase HO-1, an iron metabolizing enzyme, in spleen was decreased indicating a splenic malfunction. Using laser capture microdissected kidney samples of mice infected with *C. albicans*, mRNA levels for key iron uptake genes were analyzed quantitative real-time PCR. The expression of several high-affinity iron uptake gene expression was elevated in the *C. albicans* cells recovered from the site of infection. During early infection (at 12 h postinfection), expression of *ALS3* and *CSA1* was enhanced, while during late infection stage *ALS3*, *CSA1*, *CSA2*, *HMX1*, and *PGA10* expression was elevated. However, *FTR1* expression was enhanced both at early and late stage of infection. Thus the mRNA-level measurements suggested that the expression of a specific subset of iron acquisition genes could be modulated in vivo in response to iron limitation consequent to the iron redistribution from the cortex to the medulla. To examine the time course of expression of iron acquisition genes in kidney from infected mice, Xu et al. (2015) employed a highly sensitive, multiplexed mRNA measurement platform called nCounter from Nanostring Technologies. This study identified *CSA1*, *CSA2*, *SIT1*, *FTR1*, and *FRP2* and *RBT1* genes that were induced from 12 h to 48 h postinfection (Xu et al. 2015). Of the 231 known or predicted transcription factor genes analyzed, *HAP3*, *HAP43*, and *SEF1* were expressed as early as 12 h postinfection (Xu et al. 2015) indicating that induction of these key low-iron-induced transcription factors led to the upregulation of the downstream iron acquisition gene expression.

From among the 42 genes annotated to participate in iron acquisition that have been compiled here, virulence data was available for only about one-third of the genes from mouse systemic infection model (Table 17.1). Among these genes, null mutations in *HMX1*, *PGA7*, *RBT4*, *CFL1*, *FET34*, and *FTR1* impaired virulence to various extents. However, *CCC2*, *FTR2*, *FET31*, *FET33*, *FRE10*, *SIT1*, *PGA10*, and *RBT5* deletions did not alter the virulence of *C. albicans* in the mouse systemic infection model. The *ccc2*Δ/Δ mutant showed moderate reduction in pathogenesis in a mouse model of systemic infection, but this reduction was found only at low cell inoculum (Almeida et al. 2008; Weissman et al. 2002). Of the 11 transcription factor genes *HAP2*, *HAP43/CAP2*, *SEF1*, *RIM101*, *HAP5*, and *AFT2* null mutants had defective virulence, whereas *HAP3*, *SFU1*, and *HAP31* mutants showed wild-type virulence phenotype (Table 17.1).

The systemic infection model employed in the studies summarized above provides an outcome of the complex virulence process involving clearance from the blood stream to establishment of systemic infection in various tissues, with kidney being the predominant organ colonized by *C. albicans* in this infection model. Thus a virulence defect does not generally provide an insight into the actual step of the pathogenesis. Accordingly, *C. albicans* pathogenesis has also been studied using alternative model systems including reconstituted human epithelium (Heymann et al. 2002), and an infectivity model in mice (Noble et al. 2010). Although the *sit1*Δ/Δ mutant showed no virulence defect in systemic infection model (Table 17.1), the mutant showed defective infection in reconstituted human epithelium (Heymann et al. 2002; Hu et al. 2002). While the *ftr1*Δ/Δ null mutant was unable to damage oral epithelial cells and was avirulent in mouse model of systemic infection (Almeida et al. 2008), the *ftr2*Δ/Δ mutant showed wild-type phenotype (Ramanan and Wang 2000). Of the several mutants analyzed in the iron acquisition pathways, only *sef1*Δ/Δ, *rbt4*Δ/Δ (both defective in systemic infection), and *cfl4*Δ/Δ mutants were defective in the Noble et al. mouse infectivity study (Noble et al. 2010). Using a Keratitis model of *C. albicans* infection in rabbit cornea of the eye, *rbt1*Δ/Δ and *rbt4*Δ/Δ mutants do not form hyphae, although formed robust hyphae in vitro, and completely lost the ability of corneal invasion (Braun et al. 2000). Both the *rbt1*Δ/Δ and *rbt4*Δ/Δ mutants were also attenuated in their virulence in systemic mouse infection model (Braun et al. 2000). Together these studies have yielded a limited number of iron acquisition genes that are expressed during experimental infection in mouse kidney and those that are required for virulence.

17.8 Iron Homeostasis and Multidrug Resistance

C. albicans continually evolves mechanisms to curb the effects of antifungal drugs. The gradual evolution of frequently encountered multidrug resistance (MDR) represents one such strategy (Shapiro et al. 2011). MDR is a multifactorial phenomenon manifested via two broad categories in *C. albicans*. The first category encompasses mechanisms to escape the impact of antifungal drugs on the cell. This comprises alterations in the drug target that prevents drug binding or overexpression of drug efflux pumps. The second category includes mechanisms that allow the cell to cope with drug-induced stress. For instance, the induction of chaperones and signal transduction cascades dedicated to sense and counteract various stresses and also metabolic alterations that reduce the toxicity of the drug (Morschhauser 2010).

The underpinnings of the impact of iron homeostasis on MDR phenomena in *C. albicans* comes from several studies (Prasad et al. 2006; Vasicek et al. 2014). For instance, iron deprivation by chemical chelators resulted in highly susceptible *C. albicans* cells toward a variety of drugs. This elevated drug susceptibility in iron-poor media is not restricted to *C. albicans*, since other *Candida* species such as *C. kefyr*, *C. glabrata*, *C. tropicalis*, *C. krusei*, *C. dubliniensis*, and *C. parapsilosis* also manifest enhanced drug susceptibility. Notably, the extent of iron-dependent susceptibility is variable among different species of *Candida* (Prasad et al. 2006). Additionally, the anti-*Candida* activity of lactoferrin, one of the nonspecific host defense factors present in saliva is contributed by its ability to bind and sequester environmental iron (Kuipers et al. 1999).

The link between high-affinity iron transporters and iron deprivation-induced drug susceptibility was found as *ftr1Δ*, *ftr2Δ*, *ftr1Δ ftr2Δ*, and *ccc2Δ* mutants showed increased susceptibility to drugs even in the absence of external chemical iron chelators. Interestingly, the drug susceptible phenotype elicited by these mutants could be reversed upon supplementation of the media with extracellular iron, pointing to the direct roles of iron in drug sensitivity (Prasad et al. 2006).

The overexpression of drug efflux pumps is one of the predominant mechanisms governing MDR in *C. albicans*. In this context, drug-resistant clinical isolates overexpress efflux pump proteins, but do not seem to play a perceptible role in iron-induced drug susceptibility. When assayed under iron deprivation conditions, null mutants of efflux pump-encoding genes such as *cdr1Δ*, *cdr2Δ*, *camdr1Δ*, *flu1Δ*, and *cdr1Δ cdr2Δ* do not display any further increase in drug susceptibility in iron-deprived conditions. Also, iron deprivation does not impact the expression of genes encoding these major drug efflux pumps (Prasad et al. 2006).

17.9 Cellular Iron Impacts Lipid Homeostasis

Past studies have established that any imbalance in lipid composition is associated with changes in the physical state of the membrane leading to enhanced diffusion of drugs and susceptibility (Mukhopadhyay et al. 2002; Kohli et al. 2002). Consequently, any interruptions within membrane microdomain major constituents, ergosterol or sphingolipids, result in selective loss of trafficking of ABC multidrug transporter leading to a nonfunctional protein. While investigating iron deprivation-induced imbalances in lipids using high throughput MS-based lipid profiling, a significant decrease in the total sterol content was observed. Not only did low-iron levels lead to a significant decrease in ergosterol levels but also in some of the intermediates of ergosterol biosynthetic pathway (Hameed et al. 2011). Low level of ergosterol was also associated with iron acquisition mutants (*ftr1Δ*, *ccc2Δ*) (Prasad et al. 2006).

Genome-wide transcriptome analysis confirmed that iron levels do impact ergosterol biosynthetic pathway. For instance, two important genes of ergosterol biosynthetic pathway, *ERG11* and *ERG3* displayed reciprocal regulation upon iron deprivation. The downregulation of *ERG11* correlated well with the lower levels of ergosterol in iron-deprived cells. The upregulation of *ERG3* was reflected in increased accumulation of toxic sterols, which acts synergistically with the drugs in iron-limited conditions thus, leading to enhance drug susceptibilities (Prasad et al. 2006).

In-depth analysis of the lipidome of iron-deprived *C. albicans* by the Prasad laboratory revealed that iron deprivation not only affects the ergosterol content but also results in imbalances in sphingolipids (SLs) and phosphoglycerides. Although the total SL content of iron-deprived cells remained constant, certain SL species, for instance, inositolphosphorylceramide (IPC) levels were significantly lowered. In comparison, only minor fluctuations in phosphoglyceride composition were noted upon iron deprivation. The various classes of phosphoglycerides viz., phosphatidyl choline (PC), phosphatidyl ethanolamine (PE), phosphatidyl inositol (PI) remained broadly unchanged, except the levels of phosphatidyl glycerol (PG), which was significantly reduced (Hameed et al. 2011).

17.10 Closing Remarks

In this chapter, we have summarized the molecular machinery of iron acquisition and its requirement for the pathobiology of *C. albicans*. Genetic and biochemical studies have shown that *C. albicans* can acquire iron from siderophore, ferritin, and hemoglobin. The *C. albicans* genome contains genes that encode high-affinity iron uptake system comprising ferric reductase, multicopper oxidase and iron permease, and siderophore, ferritin, and hemoglobin uptake genes. The in vivo functional relevance of many of these genes has been established largely through studies in mouse infection model. Despite the large effort from numerous laboratories, a comprehensive evaluation of the contributions of each iron acquisition gene is yet to be realized. Moreover, with the recent adoption of newer technologies such as CRISPR-Cas9 for gene deletion (Vyas et al. 2015; Min et al. 2016), nCounter for highly sensitive gene expression measurement (Xu et al. 2015), and Laser capture microdissection and tissue imaging (Potrykus et al. 2013) in *C. albicans* research, the full scope of the iron acquisition genes to iron homeostasis and *C. albicans* virulence would be realized.

References

Almeida RS, Brunke S, Albrecht A, Thewes S, Laue M, Edwards JE, Filler SG, Hube B (2008) The hyphal-associated adhesin and invasin Als3 of *Candida albicans* mediates iron acquisition from host ferritin. PLoS Pathog 4(11):e1000217

Almeida RS, Wilson D, Hube B (2009) Candida albicans iron acquisition within the host. FEMS Yeast Res 9(7):1000–1012

Andrews SC, Robinson AK, Rodriguez-Quinones F (2003) Bacterial iron homeostasis. FEMS Microbiol Rev 27(2–3):215–237

Ardon O, Bussey H, Philpott C, Ward DM, Davis-Kaplan S, Verroneau S, Jiang B, Kaplan J (2001) Identification of a Candida albicans ferrichrome transporter and its characterization by expression in Saccharomyces cerevisiae. J Biol Chem 276(46):43049–43055

Bader T, Bodendorfer B, Schroppel K, Morschhauser J (2003) Calcineurin is essential for virulence in Candida albicans. Infect Immun 71(9):5344–5354

Bader T, Schroppel K, Bentink S, Agabian N, Kohler G, Morschhauser J (2006) Role of calcineurin in stress resistance, morphogenesis, and virulence of a Candida albicans wild-type strain. Infect Immun 74(7):4366–4369

Baek YU, Li M, Davis DA (2008) Candida albicans ferric reductases are differentially regulated in response to distinct forms of iron limitation by the Rim101 and CBF transcription factors. Eukaryot Cell 7(7):1168–1179

Bensen ES, Martin SJ, Li M, Berman J, Davis DA (2004) Transcriptional profiling in Candida albicans reveals new adaptive responses to extracellular pH and functions for Rim101p. Mol Microbiol 54(5):1335–1351. doi:10.1111/j.1365-2958.2004.04350.x

Binkley J, Arnaud M, Inglis D, Skrzypek M, Shah P, Wymore F, Binkley G, Miyasato S, Simison M, Sherlock G (2014) The candida genome database: the new homology information page highlights protein similarity and phylogeny. Nucleic Acids Res 42 (Database issue): D711–716

Blankenship JR, Mitchell AP (2011) Candida albicans adds more weight to iron regulation. Cell Host Microbe 10(2):93–94

Braun BR, Head WS, Wang MX, Johnson AD (2000) Identification and characterization of TUP1-regulated genes in Candida albicans. Genetics 156(1):31–44

Casanova M, Cervera AM, Gozalbo D, Martínez JP (1997) Hemin induces germ tube formation in Candida albicans. Infect Immun 65(10):4360–4364

Chao LY, Marletta MA, Rine J (2008) Sre1, an iron-modulated GATA DNA-binding protein of iron-uptake genes in the fungal pathogen histoplasma capsulatum. Biochemistry 47(27):7274–7283. doi:10.1021/bi800066s

Chen C, Noble SM (2012) Post-transcriptional regulation of the Sef1 transcription factor controls the virulence of Candida albicans in its mammalian host. PLoS Pathog 8(11):e1002956. doi:10.1371/journal.ppat.1002956

Chen C, Pande K, French SD, Tuch BB, Noble SM (2011) An iron homeostasis regulatory circuit with reciprocal roles in Candida albicans commensalism and pathogenesis. Cell Host Microbe 10(2):118–135

Cheng X, Xu N, Yu Q, Ding X, Qian K, Zhao Q, Wang Y, Zhang B, Xing L, Li M (2013) Novel insight into the expression and function of the multicopper oxidases in Candida albicans. Microbiology 159(6):1044–1055. doi:10.1099/mic.0.065268-0

Cohen A, Nelson H, Nelson N (2000) The family of SMF metal ion transporters in yeast cells. J Biol Chem 275(43):33388–33394

Davis D, Edwards JE Jr, Mitchell AP, Ibrahim AS (2000) Candida albicans RIM101 pH response pathway is required for host-pathogen interactions. Infect Immun 68(10):5953–5959

Davis DA (2009) How human pathogenic fungi sense and adapt to pH: the link to virulence. Curr Opin Microbiol 12(4):365–370

Eck R, Hundt S, Hartl A, Roemer E, Kunkel W (1999) A multicopper oxidase gene from Candida albicans: cloning, characterization and disruption. Microbiology 145(Pt 9):2415–2422

El Barkani A, Kurzai O, Fonzi WA, Ramon A, Porta A, Frosch M, Muhlschlegel FA (2000) Dominant active alleles of RIM101 (PRR2) bypass the pH restriction on filamentation of Candida albicans. Mol Cell Biol 20(13):4635–4647

Enjalbert B, Smith DA, Cornell MJ, Alam I, Nicholls S, Brown AJ, Quinn J (2006) Role of the Hog1 stress-activated protein kinase in the global transcriptional response to stress in the fungal pathogen Candida albicans. Mol Biol Cell 17(2):1018–1032. doi:10.1091/mbc.E05-06-0501

Fang HM, Wang Y (2002) Characterization of iron-binding motifs in Candida albicans high-affinity iron permease CaFtr1p by site-directed mutagenesis. Biochem J 368(Pt 2):641–647

Favero D, Furlaneto-Maia L, Franc EJG, Peggau H, Goes MCF (2014) Hemolytic factor production by clinical isolates of Candida albicans. Curr Microbiol 68:161–166

Georgatsou E, Alexandraki D (1999) Regulated expression of the Saccharomyces cerevisiae Fre1p/Fre2p Fe/Cu reductase related genes. Yeast 15(7):573–584

Haas H (2003) Molecular genetics of fungal siderophore biosynthesis and uptake: the role of siderophores in iron uptake and storage. Appl Microbiol Biotechnol 62(4):316–330

Hameed S, Dhamgaye S, Singh A, Goswami SK, Prasad R (2011) Calcineurin signaling and membrane lipid homeostasis regulates iron mediated multidrug resistance mechanisms in *Candida albicans*. PLoS ONE 6(4):e18684. doi:10.1371/journal.pone.0018684

Hameed S, Prasad T, Banerjee D, Chandra A, Mukhopadhyay CK, Goswami SK, Lattif AA, Chandra J, Mukherjee PK, Ghannoum MA, Prasad R (2008) Iron deprivation induces EFG1-mediated hyphal development in Candida albicans without affecting biofilm formation. FEMS Yeast Res 8(5):744–755

Hammacott JE, Williams PH, Cashmore AM (2000) Candida albicans CFL1 encodes a functional ferric reductase activity that can rescue a Saccharomyces cerevisiae fre1 mutant. Microbiology 146(Pt 4):869–876

Heymann P, Gerads M, Schaller M, Dromer F, Winkelmann G, Ernst JF (2002) The siderophore iron transporter of Candida albicans (Sit1p/Arn1p) mediates uptake of ferrichrome-type siderophores and is required for epithelial invasion. Infect Immun 70(9):5246–5255

Holzberg M, Artis WM (1983) Hydroxamate siderophore production by opportunistic and systemic fungal pathogens. Infect Immun 40(3):1134–1139

Homann OR, Dea J, Noble SM, Johnson AD (2009) A phenotypic profile of the Candida albicans regulatory network. PLoS Genet 5(12):e1000783

Hoyer LL, Payne TL, Bell M, Myers AM, Scherer S (1998) Candida albicans ALS3 and insights into the nature of the ALS gene family. Curr Genet 33:451–459

Hsu PC, Chao CC, Yang CY, Ye YL, Liu FC, Chuang YJ, Lan CY (2013) Diverse Hap43-independent functions of the Candida albicans CCAAT-binding complex. Eukaryot Cell 12:804–815

Hsu PC, Yang CY, Lan CY (2011) Candida albicans Hap43 is a repressor induced under low-iron conditions and is essential for iron-responsive transcriptional regulation and virulence. Eukaryot Cell 10(2):207–225. doi:10.1128/EC.00158-10

Hu CJ, Bai C, Zheng XD, Wang YM, Wang Y (2002) Characterization and functional analysis of the siderophore-iron transporter CaArn1p in Candida albicans. J Biol Chem 277(34):30598–30605

Ibtissem G, Frédérique B, Pierre G, Sandrine G, SILAR P (2010) The Nox/Ferric reductase/Ferric reductase-like families of Eumycetes. Fungal Biology 114:766–777

Ismail A, Bedell GW, Lupan DM (1985) Siderophore production by the pathogenic yeast, Candida albicans. Biochem Biophys Res Commun 130(2):885–891

Jeeves RE, Mason RP, Woodacre A, Cashmore AM (2011) Ferric reductase genes involved in high-affinity iron uptake are differentially regulated in yeast and hyphae of Candida albicans. Yeast 28:629–644

Kaba HE, Nimtz M, Muller PP, Bilitewski U (2013) Involvement of the mitogen activated protein kinase Hog1p in the response of Candida albicans to iron availability. BMC Microbiol 13:16. doi:10.1186/1471-2180-13-16

Kaplan CD, Kaplan J (2009) Iron acquisition and transcriptional regulation. Chem Rev 109 (10):4536–4552. doi:10.1021/cr9001676

Kim Y, Yun CW, Philpott CC (2002) Ferrichrome induces endosome to plasma membrane cycling of the ferrichrome transporter, Arn1p, in saccharomyces cerevisiae. EMBO J 21(14):3632–3642

Knight SA, Lesuisse E, Stearman R, Klausner RD, Dancis A (2002) Reductive iron uptake by Candida albicans: role of copper, iron and the TUP1 regulator. Microbiology 148(Pt 1):29–40

Knight SA, Vilaire G, Lesuisse E, Dancis A (2005) Iron acquisition from transferrin by Candida albicans depends on the reductive pathway. Infect Immun 73(9):5482–5492

Kohli A, Smriti NFN, Mukhopadhyay K, Rattan A, Prasad R (2002) In vitro low-level resistance to azoles in candida albicans is associated with changes in membrane lipid fluidity and asymmetry. Antimicrob Agents Chemother 46(4):1046–1052

Kosman DJ (2003) Molecular mechanisms of iron uptake in fungi. Mol Microbiol 47(5):1185–1197

Kuipers ME, de Vries HG, Eikelboom MC, Meijer DK, Swart PJ (1999) Synergistic fungistatic effects of lactoferrin in combination with antifungal drugs against clinical Candida isolates. Antimicrob Agents Chemother 43(11):2635–2641

Kuznets G, Vigonsky E, Weissman Z, Lalli D, Gildor T, Kauffman SJ, Turano P, Becker J, Lewinson O, Kornitzer D (2014) A relay network of extracellular heme-binding proteins drives C. albicans iron acquisition from hemoglobin. PLoS Pathog 10(10):e1004407. doi:10.1371/journal.ppat.1004407

Kwok EY, Severance S, Kosman DJ (2006) Evidence for iron channeling in the Fet3p-Ftr1p high-affinity iron uptake complex in the yeast plasma membrane. Biochemistry 45(20):6317–6327. doi:10.1021/bi052173c

Lamarre C, Deslauriers N, Bourbonnais Y (2000) Expression cloning of the Candida albicans CSA1 gene encoding a mycelial surface antigen by sorting of Saccharomyces cerevisiae transformants with monoclonal antibody-coated magnetic beads. Mol Microbiol 35(2):444–453

Lan CY, Rodarte G, Murillo LA, Jones T, Davis RW, Dungan J, Newport G, Agabian N (2004) Regulatory networks affected by iron availability in Candida albicans. Mol Microbiol 53 (5):1451–1469

Lesuisse E, Knight SA, Camadro JM, Dancis A (2002) Siderophore uptake by Candida albicans: effect of serum treatment and comparison with Saccharomyces cerevisiae. Yeast 19(4):329–340

Letunic I, Doerks T, Bork P (2015) SMART: recent updates, new developments and status in 2015. Nucleic Acids Res 43:D257–D260

Li L, Bagley D, Ward DM, Kaplan J (2008) Yap5 is an iron-responsive transcriptional activator that regulates vacuolar iron storage in yeast. Mol Cell Biol 28(4):1326–1337

Liang Y, Gui L, Wei DS, Zheng W, Xing LJ, Li MC (2009a) Candida albicans ferric reductase FRP1 is regulated by direct interaction with Rim101p transcription factor. FEMS Yeast Res 9 (2):270–277

Liang Y, Zheng W, Wei D, Xing L, Li M (2009b) Function of ferric reductase FRP1 gene in Candida albicans. Wei Sheng Wu Xue Bao 49(3):337–342

Liang Y, Wei D, Wang H, Xu N, Zhang B, Xing L, Li M (2010) Role of Candida albicans Aft2p transcription factor in ferric reductase activity, morphogenesis and virulence. Microbiology 156(Pt 10):2912–2919. doi:10.1099/mic.0.037978-0

Liu S, Hou Y, Liu W, Lu C, Wang W, Sun S (2015) Components of the calcium-calcineurin signaling pathway in fungal cells and their potential as antifungal targets. Eukaryot Cell 14 (4):324–334. doi:10.1128/EC.00271-14

Liu Y, Filler SG (2011) *Candida albicans* Als3, a multifunctional adhesin and invasin. Eukaryot Cell 10:168–173

Manns JM, Mosse DM, Buckley HR (1994) Production of a hemolytic factor by Candida albicans. Infect Immun 62:5154–5156

Min K, Ichikawa Y, Woolford CA, Mitchell AP (2016) *Candida albicans* gene deletion with a transient CRISPR-Cas9 system. mSphere 1(3):e00130–00116. doi:10.1128/mSphere.00130-16
Moore RE, Kim Y, Philpott CC (2003) The mechanism of ferrichrome transport through Arn1p and its metabolism in Saccharomyces cerevisiae. Proc Natl Acad Sci USA 100(10):5664–5669
Moors MA, Stull TL, Blank KJ, Buckley HR, Mosser DM (1992) A role for complement receptor-like molecules in iron acquisition by Candida albicans. J Exp Med 175(6):1643–1651
Morschhauser J (2010) Regulation of multidrug resistance in pathogenic fungi. Fungal Genet Biol 47(2):94–106. doi:10.1016/j.fgb.2009.08.002
Mukhopadhyay K, Kohli A, Prasad R (2002) Drug susceptibilities of yeast cells are affected by membrane lipid composition. Antimicrob Agents Chemother 46(12):3695–3705
Navarathna DH, Roberts DD (2010) Candida albicans heme oxygenase and its product CO contribute to pathogenesis of candidemia and alter systemic chemokine and cytokine expression. Free Radic Biol Med 49(10):1561–1573. doi:10.1016/j.freeradbiomed.2010.08.020
Nobile CJ, Solis N, Myers CL, Fay AJ, Deneault JS, Nantel A, Mitchell AP, Filler SG (2008) Candida albicans transcription factor Rim101 mediates pathogenic interactions through cell wall functions. Cell Microbiol 10(11):2180–2196
Noble SM (2013) Candida albicans specializations for iron homeostasis: from commensalism to virulence. Curr Opin Microbiol. doi:10.1016/j.mib.2013.09.006
Noble SM, French S, Kohn LA, Chen V, Johnson AD (2010) Systematic screens of a Candida albicans homozygous deletion library decouple morphogenetic switching and pathogenicity. Nat Genet 42(7):590–598
Pelletier B, Mercier A, Durand M, Peter C, Jbel M, Beaudoin J, Labbe S (2007) Expression of Candida albicans Sfu1 in fission yeast complements the loss of the iron-regulatory transcription factor Fep1 and requires Tup co-repressors. Yeast 24(10):883–900
Pendrak ML, Chao MP, Yan SS, Roberts DD (2004a) Heme oxygenase in Candida albicans is regulated by hemoglobin and is necessary for metabolism of exogenous heme and hemoglobin to alpha-biliverdin. J Biol Chem 279(5):3426–3433
Pendrak ML, Yan SS, Roberts DD (2004b) Sensing the host environment: recognition of hemoglobin by the pathogenic yeast Candida albicans. Arch Biochem Biophys 426(2):148–156
Pendrak ML, Roberts DD (2007) Hemoglobin is an effective inducer of hyphal differentiation in Candida albicans. Med Mycol 45(1):61–71
Perez A, Ramage G, Blanes R, Murgui A, Casanova M, Martinez JP (2011) Some biological features of Candida albicans mutants for genes coding fungal proteins containing the CFEM domain. FEMS Yeast Res 11(3):273–284. doi:10.1111/j.1567-1364.2010.00714.x
Philpott CC (2006) Iron uptake in fungi: a system for every source. Biochim Biophys Acta 1763 (7):636–645
Philpott CC, Protchenko O (2008) Response to iron deprivation in Saccharomyces cerevisiae. Eukaryot Cell 7(1):20–27
Portnoy ME, Liu XF, Culotta VC (2000) Saccharomyces cerevisiae expresses three functionally distinct homologues of the nramp family of metal transporters. Mol Cell Biol 20(21):7893–7902
Potrykus J, Stead D, MacCallum DM, Urgast DS, Raab A, van Rooijen N, Jr F, Brown AJP (2013) Fungal iron availability during deep seated candidiasis is defined by a complex interplay involving systemic and local events. PLoS Pathog 9(10):e1003676. doi:10.1371/journal.ppat.1003676
Prasad T, Chandra A, Mukhopadhyay CK, Prasad R (2006) Unexpected link between iron and drug resistance of Candida spp.: iron depletion enhances membrane fluidity and drug diffusion, leading to drug-susceptible cells. Antimicrob Agents Chemother 50(11):3597–3606
Puig S, Askeland E, Thiele DJ (2005) Coordinated remodeling of cellular metabolism during iron deficiency through targeted mRNA degradation. Cell 120(1):99–110
Radisky D, Kaplan J (1999) Regulation of transition metal transport across the yeast plasma membrane. J Biol Chem 274(8):4481–4484

Ramanan N, Wang Y (2000) A high-affinity iron permease essential for Candida albicans virulence. Science 288(5468):1062–1064

Ramon AM, Fonzi WA (2003) Diverged binding specificity of Rim101p, the Candida albicans ortholog of PacC. Eukaryot Cell 2(4):718–728

Rohm M, Lindemann E, Hiller E, Ermert D, Lemuth K, Trkulja D, Sogukpinar O, Brunner H, Rupp S, Urban CF, Sohn K (2013) A family of secreted pathogenesis-related proteins in Candida albicans. Mol Microbiol 87(1):132–151. doi:10.1111/mmi.12087

Santos R, Buisson N, Knight S, Dancis A, Camadro JM, Lesuisse E (2003) Haemin uptake and use as an iron source by *Candida albicans*: role of CaHMX1-encoded haem oxygenase. Microbiology 149(Pt 3):579–588

Schillig R, Morschhäuser J (2013) Analysis of a fungus-specific transcription factor family, the Candida albicans zinc cluster proteins, by artificial activation. Mol Microbiol 89(5):1003–1017. doi:10.1111/mmi.12327

Shapiro RS, Robbins N, Cowen LE (2011) Regulatory circuitry governing fungal development, drug resistance, and disease. Microbiol Mol Biol Rev 75(2):213–267. doi:10.1128/mmbr.00045-10

Singh RP, Prasad HK, Sinha I, Agarwal N, Natarajan K (2011) Cap2-HAP complex is a critical transcriptional regulator that has dual but contrasting roles in regulation of iron homeostasis in Candida albicans. J Biol Chem 286(28):25154–25170

Sosinska GJ, de Groot PW, Teixeira de Mattos MJ, Dekker HL, de Koster CG, Hellingwerf KJ, Klis FM (2008) Hypoxic conditions and iron restriction affect the cell-wall proteome of Candida albicans grown under vagina-simulative conditions. Microbiology 154(Pt 2):510–520

Steinbach WJ, Reedy JL, Cramer RA Jr, Perfect JR, Heitman J (2007) Harnessing calcineurin as a novel anti-infective agent against invasive fungal infections. Nat Rev Microbiol 5(6):418–430. doi:10.1038/nrmicro1680

Sweet SP, Douglas LJ (1991) Effect of iron concentration on siderophore synthesis and pigment production by Candida albicans. FEMS Microbiol Lett 64(1):87–91

Synnott JM, Guida A, Mulhern-Haughey S, Higgins DG, Butler G (2010) Regulation of the hypoxic response in Candida albicans. Eukaryot Cell 9:1734–1746

Tanaka WT, Nakao N, Mikami T, Matsumoto T (1997) Hemoglobin is utilized by Candida albicans in the hyphal form but not yeast form. Biochem Biophys Res Commun 232(2):350–353

Thewes S, Kretschmar M, Park H, Schaller M, Filler SG, Hube B (2007) In vivo and ex vivo comparative transcriptional profiling of invasive and non-invasive Candida albicans isolates identifies genes associated with tissue invasion. Mol Microbiol 63(6):1606–1628

Van Ho A, Ward DM, Kaplan J (2002) Transition metal transport in yeast. Annu Rev Microbiol 56:237–261. doi:10.1146/annurev.micro.56.012302.160847

Vasicek EM, Berkow EL, Flowers SA, Barker KS, Rogers PD (2014) UPC2 is universally essential for azole antifungal resistance in Candida albicans. Eukaryot Cell 13(7):933–946. doi:10.1128/ec.00221-13

Vergara SV, Thiele DJ (2008) Post-transcriptional regulation of gene expression in response to iron deficiency: co-ordinated metabolic reprogramming by yeast mRNA-binding proteins. Biochem Soc Trans 36(5):1088–1090. doi:10.1042/BST0361088

Vyas VK, Barrasa MI, Fink GR (2015) A *Candida albicans* CRISPR system permits genetic engineering of essential genes and gene families. Sci Adv 1(3):e1500248. doi:10.1126/sciadv.1500248

Watanabe T, Takano M, Murakami M, Tanaka H, Matsuhisa A, Nakao N, Mikami T, Suzuki M, Matsumoto T (1999) Characterization of a haemolytic factor from *Candida albicans*. Microbiology 145(Pt 3):689–694. doi:10.1099/13500872-145-3-689

Weissman Z, Kornitzer D (2004) A family of Candida cell surface haem-binding proteins involved in haemin and haemoglobin-iron utilization. Mol Microbiol 53(4):1209–1220

Weissman Z, Shemer R, Conibear E, Kornitzer D (2008) An endocytic mechanism for haemoglobin-iron acquisition in Candida albicans. Mol Microbiol 69(1):201–217

Weissman Z, Shemer R, Kornitzer D (2002) Deletion of the copper transporter CaCCC2 reveals two distinct pathways for iron acquisition in Candida albicans. Mol Microbiol 44(6):1551–1560

Wolf JM, Johnson DJ, Chmielewski D, Davis DA (2010) The Candida albicans ESCRT pathway makes Rim101-dependent and -independent contributions to pathogenesis. Eukaryot Cell 9 (8):1203–1215. doi:10.1128/EC.00056-10

Woodacre A, Mason RP, Jeeves RE, Cashmore AM (2008) Copper-dependent transcriptional regulation by Candida albicans Mac1p. Microbiology 154(Pt 5):1502–1512

Xu N, Cheng X, Yu Q, Qian K, Ding X, Liu R, Zhang B, Xing L, Li M (2013) Aft2, a novel transcription regulator, is required for iron metabolism, oxidative stress, surface adhesion and hyphal development in Candida albicans. PLoS ONE 8(4):e62367. doi:10.1371/journal.pone.0062367

Xu N, Dong Y, Cheng X, Yu Q, Qian K, Mao J, Jia C, Ding X, Zhang B, Chen Y, Zhang B, Xing L, Li M (2014a) Cellular iron homeostasis mediated by the Mrs4–Ccc1–Smf3 pathway is essential for mitochondrial function, morphogenesis and virulence in Candida albicans. Biochim Biophys Acta 1843:629–639

Xu N, Qian K, Dong Y, Chen Y, Yu Q, Zhang B, Xing L, Li M (2014b) Novel role of the Candida albicans ferric reductase gene CFL1 in iron acquisition, oxidative stress tolerance, morphogenesis and virulence. Res Microbiol 165(3):252–261. doi:10.1016/j.resmic.2014.03.001

Xu W, Solis NV, Ehrlich RL, Woolford CA, Filler SG, Mitchell AP (2015) Activation and alliance of regulatory pathways in C. albicans during mammalian infection. PLoS Biol 13(2):e1002076. doi:10.1371/journal.pbio.1002076

Yu S-J, Chang Y-L, Chen Y-L (2015) Calcineurin signaling: lessons from Candida species. FEMS Yeast Res 15(4). doi:10.1093/femsyr/fov016

Zakikhany K, Naglik JR, Schmidt-Westhausen A, Holland G, Schaller M, Hube B (2007) In vivo transcript profiling of *Candida albicans* identifies a gene essential for interepithelial dissemination. Cell Microbiol 9:2938–2954

Ziegler L, Terzulli A, Gaur R, McCarthy R, Kosman DJ (2011) Functional characterization of the ferroxidase, permease high-affinity iron transport complex from Candida albicans. Mol Microbiol 81(2):473–485. doi:10.1111/j.1365-2958.2011.07704.x

Chapter 18
Structure–Function Analyses of Multidrug Transporters

Erwin Lamping, Golnoush Madani, Hee Ji Lee, Masakazu Niimi and Richard D. Cannon

Abstract Proteins catalysing the transport of molecules across biological membranes are vital for organisms in all kingdoms of life. These proteins are needed for the uptake of nutrients and the efflux of signalling molecules and toxic compounds. In the human opportunistic pathogen *Candida albicans,* efflux proteins can translocate antifungal drugs, such as fluconazole, and confer drug resistance. There are two main families of membrane proteins involved in drug transport, the major facilitator superfamily (MFS) and the ATP-binding cassette (ABC) proteins. Both types of protein possess multiple membrane spanning α-helices in transmembrane domains (TMDs), and ABC proteins, in addition, contain cytosolic nucleotide-binding domains (NBDs) involved in ATP hydrolysis. ABC proteins, and to a lesser extent MFS proteins, have broad substrate specificities that are determined by the structure and arrangement of the transmembrane α-helices. The multidrug transporter most often associated with drug resistance of *C. albicans* clinical isolates is ABC protein Cdr1. This is a pleiotropic drug resistance (PDR) ABC protein with domain arrangement NBD_1-TMD_1-NBD_2-TMD_2 and unique large extracellular loops between transmembrane α-helices that may be important for pump function. There is no crystal structure for Cdr1, but X-ray structures of related proteins indicate that unique interactions between conserved elements of the TMDs and the NBDs are important for pump function.

E. Lamping · G. Madani · H.J. Lee · M. Niimi · R.D. Cannon (✉)
Sir John Walsh Research Institute, Faculty of Dentistry, University of Otago, PO Box 56, Dunedin 9054, New Zealand
e-mail: richard.cannon@otago.ac.nz

E. Lamping
e-mail: erwin.lamping@otago.ac.nz

M. Niimi
e-mail: masa.niimi@otago.ac.nz

M. Niimi
Mycology Unit, Department of Microbiology, Faculty of Medicine, Chulalongkorn University, Bangkok, Thailand

R. Prasad (ed.), *Candida albicans: Cellular and Molecular Biology*,
DOI 10.1007/978-3-319-50409-4_18

Keywords ATP-binding cassette (ABC) · Major facilitator superfamily (MFS) · Chemosensitization · Suppressor mutation · Drug resistance

18.1 Introduction

Fungi possess a variety of integral membrane proteins, the functions of which are to transport substances across biological membranes (Perlin et al. 2014). They are present in many cellular membranes, including vacuolar, mitochondrial, peroxisomal and endoplasmic reticulum membranes, with many important transporters located in the plasma membrane. The plasma membrane is a vital permeability barrier for cellular homeostasis. This barrier stops the free flow of ions, small molecules and macromolecules across the membrane and maintains chemical and electrochemical gradients between the inside and the outside of the fungal cell. In order to transport molecules across this barrier, the fungal cell utilises integral membrane proteins. If the molecule moves down a concentration gradient facilitated by membrane protein channels or pores, then the transport protein simply accelerates passive diffusion, and no energy is required for this process. In many instances, however, molecules are transported against a concentration gradient, and energy is required.

Energy-dependent transport is carried out predominantly by proteins belonging to one of two large protein families, either the ATP-binding cassette (ABC) transporter family, or the major facilitator superfamily (MFS). ABC transporters are primary active transporters that use the energy released by ATP hydrolysis to power transport, whereas MFS proteins are secondary active transporters that utilise electrochemical gradients (e.g. Na^+ or H^+ gradients established by primary active transporters) across the membrane as their energy source. Fungi generate many compounds that are toxic to the cell unless effluxed such as secondary metabolites or endpoints of metabolic pathways including ethanol and organic acids. Fungi also encounter xenobiotics from their environment, which are kept outside the cell by multiple efflux proteins in plasma membranes. Multidrug efflux proteins transport a variety of substrates, and the range of substrates for different pumps can overlap. It is interesting to note that fungi that inhabit a variety of environmental niches, such as *Cryptococcus neoformans,* contain a large repertoire of 33 ABC (Kovalchuk and Driessen 2010) and 150–170 MFS proteins (Janbon et al. 2014). Another example of species that inhabit a variety of niches are *Aspergillus* spp. with 50–70 ABC proteins (Kovalchuk and Driessen 2010) and 275–356 MFS transporters (Coleman and Mylonakis 2009). In contrast, the fungus *Candida albicans*, which is almost entirely associated with mammalian hosts, possesses a more limited range of 26 ABC (Kovalchuk and Driessen 2010) and 85–95 MFS transporters (Coleman and Mylonakis 2009; Gaur et al. 2008), possibly because of its exposure to a smaller range of xenobiotics.

A complication of treating fungal infections with antifungal drugs is that fungal genomes encode transporters that are able to export these drugs out of the cells thereby rendering the fungi resistant. It has been found that the strong selective pressure exerted by antifungal drugs can select for mutations in transcriptional regulators that result in over-expression of ABC or MFS transporters. For example, the high-level azole resistance of *C. albicans* has been shown to be due to the over-expression of MFS protein Mdr1 or, most frequently, the over-expression of the ABC proteins Cdr1 and Cdr2 (Cannon et al. 2009). Over-expression of Mdr1 can be caused by gain-of-function mutations in transcription factor Mrr1 (Dunkel et al. 2008), whereas the over-expression of Cdr1 and Cdr2 is frequently caused by gain-of-function mutations in transcription factor Tac1 (Coste et al. 2007; Coste et al. 2004). This chapter describes the multidrug transporters of *C. albicans*, focusing on the clinically more significant ABC proteins, and current knowledge of structure–function relationships for these proteins.

18.2 Multidrug Efflux Transporters of *C. albicans*

18.2.1 MFS Transporters

MFS transporters are integral membrane proteins with, most commonly, either 12 or 14 transmembrane spans (TMSs); exceptions do exist with one MFS transporter predicted to have 6 and another small MFS family predicted to have 24 TMSs (Law et al. 2008; Reddy et al. 2012). *C. albicans* possesses an estimated 95 MFS transporters (Gaur et al. 2008) that can be divided into 17 separate phylogenetic lineages. The four major MFS families are the SP (sugar porter), the DHA1 (drug: H^+ antiporter 1), the DHA2 and the ACS (anion:cation symporter) family with 22, 22, 9 and 16 representatives, respectively (Gaur et al. 2008). SP-, DHA1- and ACS-type transporters are predicted to have 12 TMSs and DHA2 members have 14 TMSs. The *C. albicans* genome encodes between 26 (Costa et al. 2014) and 31 (Gaur et al. 2008) DHA1 and DHA2 MFS transporters. Few of these ORFs have been characterised and only one, Mdr1, is thought to be a clinically relevant MFS multidrug efflux pump. *C. albicans* Mdr1, a DHA1 protein (Fig. 18.1 and Table 18.1) also called Ben^r, was originally identified by its ability to confer benomyl and methotrexate resistance on *Saccharomyces cerevisiae* (Fling et al. 1991). It prefers generally less hydrophobic and less branched substrates than *C. albicans* Cdr1 (Puri et al. 2010), and it transports compounds such as fluconazole (FLC) and ketoconazole (KTZ), but not miconazole (MCZ) or itraconazole (ITC) (Lamping et al. 2007). High levels of *MDR1* expression have been detected in certain azole-resistant clinical isolates of *C. albicans* (Sanglard et al. 1995; White et al. 2002; Wirsching et al. 2000). In general, MFS multidrug efflux transporters efflux a narrower range of substrates than ABC multidrug efflux proteins (Puri et al. 2010; Sa-Correia et al. 2009). Mdr1 expression is up-regulated in clinical isolates by gain-of-function mutations in transcription factor Mrr1 (Dunkel et al. 2008; Morschhauser et al. 2007). For further details on the transcriptional control of

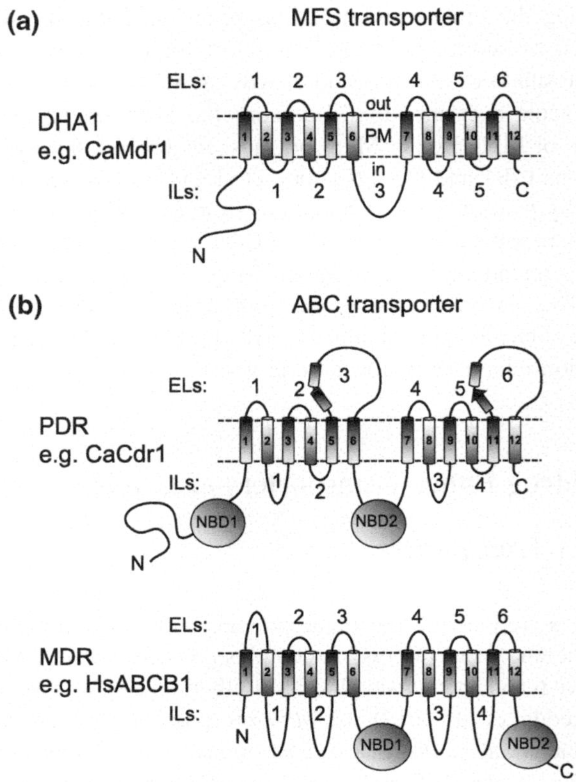

Fig. 18.1 Basic topologies of MFS, and PDR and MDR ABC transporters. *Cylinders* indicate α-helices, the *arrow* indicates a *β*-sheet, *ovals* indicate cytosolic NBDs, and the plasma membrane (PM) separating the inside (*in*) from the outside (*out*) of the cell is represented by two *dashed lines*. **a** General topology of DHA1 MFS transporters such as *C. albicans* Mdr1 (Ca Mdr1). DHA1 MFS transporters typically comprise ~500 amino acids including 12 TMSs that are connected by short intracellular loops (ILs) or extracellular loops (ELs). The core MFS transporter fold consists of four TMS repeats (i.e. TMS1-3, TMS4-6, TMS7-9, TMS10-12) that originated from two separate tandem-duplications of a three TMS ancestor; i.e. TMS1/4/7/10, TMS2/5/8/11 and TMS3/6/9/12 share the same TMS ancestor (also see Fig. 18.2a). **b** The prototypic fungal PDR family multidrug efflux transporters *S. cerevisiae* Pdr5 and *C. albicans* Cdr1 (Ca Cdr1; top panel), are characterised by a topology with the NBDs NBD1 and NBD2 preceding TMD1 (TMS1-6) and TMD2 (TMS7-12), respectively. All other ABC transporter families, including the MDR family such as the human multidrug efflux pump HsABCB1 (Hs ABCB1; bottom panel), exhibit the typical ABC transporter topology with the TMDs preceding the respective NBDs

C. albicans MFS transporters see the next chapter. Another DHA1 MFS transporter with a potential role in drug resistance is Flu1 (Table 18.1). Although there is no evidence of Flu1 over-expression in azole-resistant clinical isolates, disruption of *FLU1* in *C. albicans* caused susceptibility to mycophenolic acid but it had no effect on FLC sensitivity (Calabrese et al. 2000). Flu1 is thought to efflux the human fungicidal peptide histatin 5 (Hst 5), although Hst 5 did not induce over-expression of *FLU1* mRNA (Li et al. 2013).

Two other genes with homology to DHA1 MFS transporters, *NAG3* (also known as *TMP1*) and *NAG4* (also known as *TMP2*), may also be involved in drug transport (Table 18.1). When these genes in *C. albicans* are disrupted the cells become more susceptible to cycloheximide (CHX) (Yamada-Okabe and Yamada-Okabe 2002) but their over-expression has not been associated with clinical drug resistance.

18.2.2 ABC Transporters

As their name suggests, ABC proteins contain nucleotide-binding domains (NBDs). These domains contain highly conserved amino acid motifs and are characterised by the Walker A and the Walker B motifs (Walker et al. 1982) and the hallmark ABC signature motif or C-loop. The 26 *C. albicans* ABC proteins (Braun et al. 2005) can be classified into the following subgroups: PDR (pleiotropic drug resistance), MDR (multidrug resistance), MRP (multidrug resistance-associated protein), ALDp (adrenoleukodystrophy protein), EF3 (elongation factor EF-3) and RL1 (RNAse L inhibitor) (Gaur et al. 2005). The EF3 and RL1 ABC proteins lack TMSs and so, despite the ability to hydrolyse ATP, are not involved in transport.

The two ABC proteins most often associated with drug resistance in clinical *C. albicans* isolates are Cdr1 and Cdr2 (Maebashi et al. 2001; Sanglard et al. 1995; White 1997) (Table 18.1). These are 'full-size' ABC transporters in that they contain two homologous halves each with a NBD and a TMD (Cannon et al. 2009). The two TMDs each contain six TMSs and together they are thought to form the substrate transport channel. Cdr1 and Cdr2 are PDR-type ABC proteins with an inverted order of the domains $[NBD\text{-}TMD]_2$—only found in fungi and plants (Crouzet et al. 2006)—as opposed to the typical $[TMD\text{-}NBD]_2$ arrangement found in all other ABC proteins (Fig. 18.1) (Lamping et al. 2010).

Cdr1 and Cdr2 have a broad substrate specificity and, when heterologously expressed in *S. cerevisiae,* they confer resistance to FLC, KTZ, MCZ, ITC, rhodamine 6G (R6G), CHX and cerulenin (CER) among many other compounds (Puri et al. 2010; Tanabe et al. 2011). Both Cdr1 and Cdr2 are often co-expressed at elevated levels in azole-resistant clinical *C. albicans* isolates because their promoters contain drug-responsive elements $CGGN_4CGG$ (de Micheli et al. 2002; Liu et al. 2007); the sequence recognised by transcription factor Tac1 (Coste et al. 2004). These azole-resistant isolates possess gain-of-function mutations in Tac1 (Coste et al. 2007; Coste et al. 2004). The use of Cdr1- and Cdr2-specific antibodies, Cdr1-specific inhibitors, and gene deletions, has shown that Cdr1 contributes more to FLC efflux than Cdr2 in FLC-resistant clinical isolates (Holmes et al. 2008; Tsao et al. 2009).

Table 18.1 *C. albicans* MFS and ABC proteins that may be involved in multidrug resistance

Transporter type	ORF[a]	Gene	Number of amino acids	MW (kDa)	Potential substrates[b]	Evidence for role in drug transport	
						in vitro	in vivo
MFS[c]	orf19.5604	*MDR1*	564	62.9	FLC, KTZ, BEN, CHX	(Fling et al. 1991) (Lamping et al. 2007)	(Wirsching et al. 2000) (Sanglard et al. 1995) (White et al. 2002) (Morschhauser et al. 2007)
	orf19.6577	*FLU1*	610	67.6	FLC, CHX, MPH, Hst5	(Calabrese et al. 2000)	
	orf19.2158	*NAG3*	561	62.6	CHX	(Yamada-Okabe and Yamada-Okabe 2002)	
	orf19.2160	*NAG4*	581	64.6	CHX	(Yamada-Okabe and Yamada-Okabe 2002)	
ABC	orf19.6000	*CDR1*	1501	170	FLC, KTZ, MCZ, ITC, R6G, CHX, CER	(Lamping et al. 2007)	(Maebashi et al. 2001) (Sanglard et al. 1995) (Tsao et al. 2009) (White 1997)
	orf19.5958	*CDR2*	1499	169	FLC, KTZ, MCZ, ITC, R6G, CHX, CER	(Lamping et al. 2007)	(Maebashi et al. 2001) (Tsao et al. 2009) (White et al. 2002)
	orf19.1783	*YOR1*	1488	168		(Znaidi et al. 2008)[d]	

[a]ORF identifier from *Candida* Genome Database: http://www.candidagenome.org/
[b]FLC, fluconazole; KTZ, ketoconazole; BEN, benomyl; CHX, cycloheximide; MPH, mycophenolic acid; Hst5, histatin 5; MCZ, miconazole; ITC, itraconazole; R6G, rhodamine 6G; CER, cerulenin
[c]all MFS transporters are DHA1 type consisting of 12 predicted TMSs
[d]expression of *YOR1* (and *CDR1* and *MDR1*) induced by Upc2, substrates are unknown

18.3 Topology of *C. albicans* Multidrug Efflux Transporters

18.3.1 MFS Transporter Mdr1

C. albicans Mdr1 has 12 predicted TMSs connected by hydrophilic loops (Fig. 18.1a) (Pasrija et al. 2007). Unlike other MFS transporters such as human glucose transporters GLUT1, -3 or -5 or *Escherichia coli* lactose transporter LacY, multidrug efflux transporter EmrD, and tetracycline transporter TetA, most fungal MFS transporters, including *S. cerevisiae* Flr1 and *C. albicans* Mdr1, Flu1, Nag3 and Nag4, have a ~100–160 amino acid N-terminal extension (Fig. 18.1a; data not shown), the function of which remains to be determined. Early topology studies of DHA1 transporters using reporter fusions revealed that the N- and C-termini of these proteins are cytoplasmic (Law et al. 2008). There is a degree of homology between the N- and C-terminal halves of MFS proteins indicating that they may have arisen by duplication of an ancestor protein with 6 TMSs (Paulsen and Skurray 1993). These two halves are separated by the extended cytoplasmic intracellular loop 3 (IL3; Fig. 18.1a), which permits a relatively large, rigid, movement between the two domains during the proposed 'rocker-switch'-type substrate translocation through the centre of the transporter (Law et al. 2008; Yan 2013; Yan 2015). Further structural and functional analyses revealed that an even earlier duplication of a 3-TMS ancestor possibly gave rise to this 6 TMS precursor protein. Thus, the 12 TMS core of most MFS transporters consists of four inverted 3-TMS repeats (Fig. 18.1a) (Quistgaard et al. 2016; Yan 2015).

18.3.2 ABC Transporters Cdr1 and Cdr2

All ABC transporters consist of four core sub-domains: two TMDs that provide a channel for the passage of compounds across the lipid bilayer, and two cytoplasmic NBDs that provide the necessary energy for active transport (Fig. 18.1b). Prokaryotes commonly express these individual sub-domains as separate polypeptide chains, whereas eukaryotes usually express half-size transporters (TMD-NBD) that form functional homo- or hetero-dimers, or full-size transporters with all four sub-domains $[TMD\text{-}NBD]_2$ expressed as a single polypeptide (Fig. 18.1b) (Hollenstein et al. 2007a, b; Jones et al. 2009). As mentioned above, PDR transporters like Cdr1 and Cdr2 are the only ABC sub-group with an inverted domain arrangement where the two NBDs precede their respective TMDs $[NBD\text{-}TMD]_2$ (Fig. 18.1b). They also have a ~150 amino acid N-terminal extension rich in hydrophilic (S, T, N, Q) and charged (D, E, R, K) residues (Lamping et al. 2010), similar to the fungal MFS transporters mentioned above. This N-terminal domain may be important for protein–protein interactions that regulate PDR transporter function and/or expression. Our analysis of 244 PDR sequences also predicted four small (EL1, -2, -4, -5) and two large ELs (EL3, -6) with PDR family-defining helical and beta-sheet motifs near the exits of TMS5 and TMS11, and two small (IL2, -4) and two large ILs (IL1, -3) connecting 12 predicted TMSs (Fig. 18.1b) (Lamping et al. 2010). Plant and human PDR transporters exhibit the same topology with the same conserved EL3 and EL6 helical and beta-sheet motifs near the exits of TMS5 and TMS11 (Fig. 18.1b; unpublished data). MDR ABC proteins, like human ABCB1 (also known as P-glycoprotein or Pgp), also have 12 TMSs but, unlike PDR transporters such as Cdr1 and Cdr2, they have equally large IL1-IL4 that extend deep into the cytosol, and they do not have large ELs (Fig. 18.1b). As discussed below, this has significant structural and functional consequences for these different types of ABC exporter.

18.4 Structural Analyses of Multidrug Transporters

The structure of proteins, and protein complexes, provides pivotal clues about how proteins fulfil their function. There are several techniques that can generate structural information at atomic or near-atomic resolution—two of the most frequently used techniques being X-ray crystallography and cryo-electron microscopy (CryoEM) (Delmar et al. 2015; Ford et al. 2009; Vinothkumar 2015). There are hundreds of thousands of structures of soluble proteins deposited in the protein data bank (pdb; http://www.rcsb.org/pdb/home/home.do), but high-resolution structures for integral membrane proteins are much harder to obtain and account for less than 1% of all deposited structures. The amphiphilic nature of membrane proteins makes it very difficult to obtain enough mono-disperse protein and to produce the well-ordered three-dimensional (3-D) crystals that are required for X-ray crystallography. However, recent advances in CryoEM technology (i.e. improved detector technology and software analysis tools (Cheng et al. 2015; Delmar et al. 2015; Henderson 2015; Nannenga et al. 2013; Rigaud et al. 2000; Vinothkumar 2015)), which do not require well-ordered 3-D crystals for the structural analysis of proteins, has helped to increase the pace for solving membrane protein structures, including structures for MFS and ABC transporters.

18.4.1 MFS Transporters

There are currently 64 solved structures for 23 MFS transporters representing 10 different MFS transporter families and that display all four major pump conformations (inward open—inward occluded—outward occluded—outward open) of the recently proposed ‘clamp and switch model’ for MFS transporter function (Quistgaard et al. 2016).

It would appear that all MFS transporters, irrespective of their substrate specificities and transport coupling mechanisms, share a common structural fold known as the MFS fold (Yan 2013). The canonical MFS fold comprises 12 TMSs which are organised into two discreetly folded N- and C-terminal domains displaying a two-fold pseudosymmetry (Yan 2013). Each domain further divides into a pair of inverted 3 + 3 TMS repeats, the basic functional unit of MFS proteins (Yan 2013). Interestingly, the corresponding TMSs (i.e. TMS1, -4, -7, -10; TMS2, -5, -8, -11; and TMS3, -6, -9, -12; blue, magenta and green, respectively; Fig. 18.2a) of two 3-TMS repeats are always positioned next to each other in opposite transmembrane orientations (Fig. 18.2a). The substrate-translocation cavity at the centre of the transporter is enclosed by the blue helices (TMS1, -4, -7 and -10; Fig. 18.2a). Major interactions between TMS1 and TMS7 near the extracellular space and between TMS4 and TMS10 near the cytosol are thought to control access of solutes to the substrate-binding cavity (Nomura et al. 2015; Quistgaard et al. 2016; Yan 2015). Some residues of the magenta helices (TMS2, -5, -8 and -11) that typically mediate the interface between the N- and C-terminal domains (Fig. 18.2a), also face the substrate-binding cavity and may participate in substrate binding/transport and coupling of substrate transport to the symport or antiport of H^+ or other chemiosmotic solutes. The green helices (TMS3, -6, -9 and -12; Fig. 18.2a) are positioned on the outside of the protein, mostly stabilising MFS transporter structure. Most MFS transporters also have a conserved hallmark A-motif Gx_3(D/E)(R/K)xG [x](R/K)(R/K) (x indicates any residue) between TMS2 and -3 and/or TMS8 and -9 (Quistgaard et al. 2016). These motifs are important for gating interactions in the four different MFS pump conformations (Quistgaard et al. 2016; Yan 2013; Yan 2015). Some drug: H^+ MFS antiporter families, including *C. albicans* Mdr1, also have a conserved C-motif Gx_3GPx_2G near the end of TMS5 (Paulsen et al. 1996; Paulsen and Skurray 1993), which has been proposed to be involved in linking H^+ translocation to export as it is only conserved among export proteins (Paulsen and Skurray 1993).

18.4.2 ABC Transporters

ABC transporters are divided into three types of importers (type I, type II and energy-coupling factor (ECF) importers; Fig. 18.3), only found in prokaryotes, and two types of exporters (type I and type II exporters; Fig. 18.2b, c), found in all kingdoms of life. While the NBDs have generally well-conserved structures, their TMD folds vary significantly, as demonstrated in the centre and bottom panels in Figs. 18.2b, c and 18.3—possibly a reflection of their large substrate diversity (Jones et al. 2009; Kerr et al. 2010; Locher 2004; Rees et al. 2009). Many ABC transporters possess additional domains responsible for regulating transporter activity, necessary for protein–protein interactions or important for protein targeting to the correct membrane. Type I and type II ABC importers, for example, require an additional high-affinity substrate-binding protein component (yellow domains in Fig. 18.3a, b) that delivers a specific ligand to the transporter, while others have

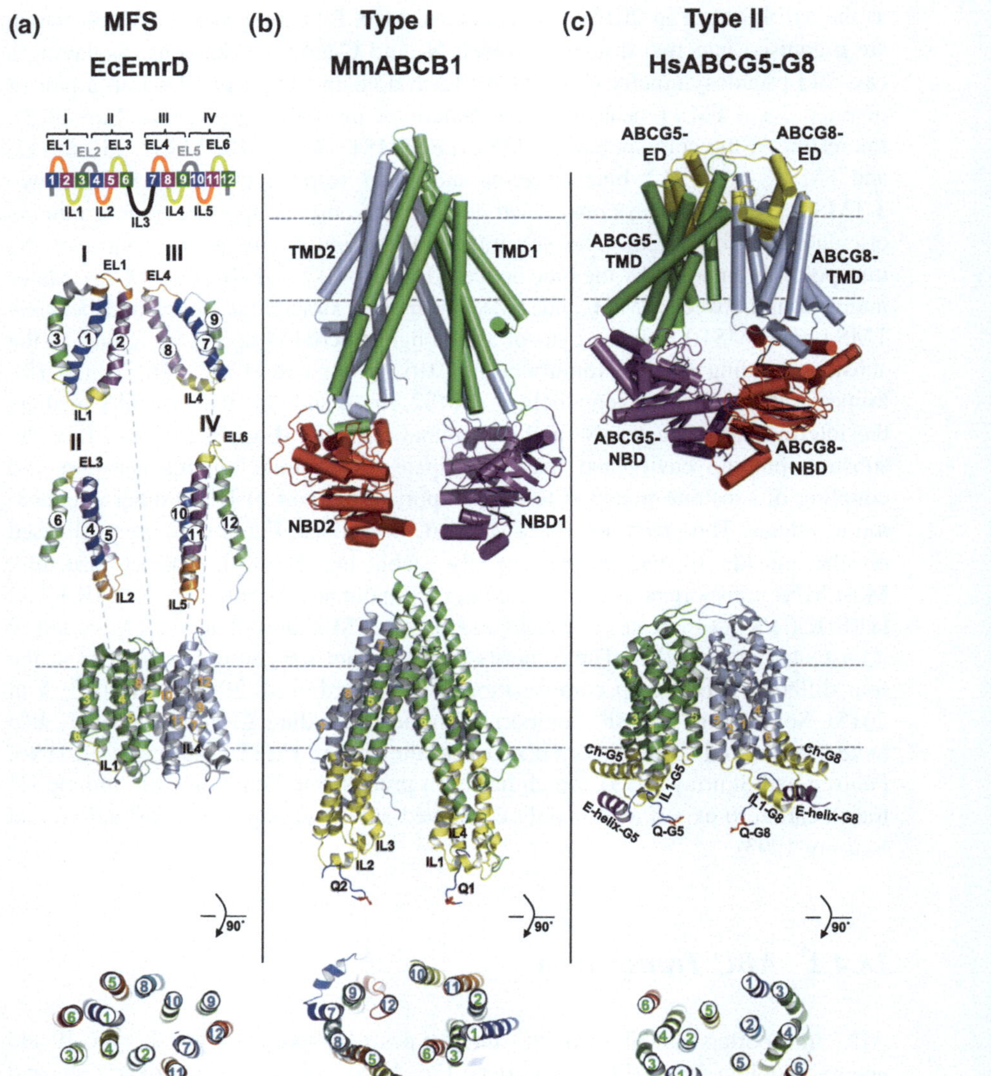
(a)
MFS
EcEmrD
(b)
Type I
MmABCB1
(c)
Type II
HsABCG5-G8
EL1
EL2
EL3
EL4
EL5
EL6
IL1
IL2
IL3
IL4
IL5
TMD2
TMD1
NBD2
NBD1
Q2
Q1
ABCG5-ED
ABCG8-ED
ABCG5-TMD
ABCG8-TMD
ABCG5-NBD
ABCG8-NBD
Ch-G5
Ch-G8
IL1-G5
IL1-G8
E-helix-G5
E-helix-G8
Q-G5
Q-G8
90°

◀**Fig. 18.2** Structures of a typical MFS transporter and type I and type II ABC exporters. All structures were prepared with PyMOL Molecular Graphics System. The lipid bilayer is indicated with dashed grey lines. **a** *E. coli* multidrug efflux transporter EcEmrD (pdb: 2GFP; (Yin et al. 2006)), like *C. albicans* Mdr1 an H^+/drug antiporter. Because all MFS transporters share the same core fold, it is likely that *C. albicans* Mdr1 folds like EcEmrD. The *top* panel in **a** shows how a typical MFS transporter contains four core subunits. Subunit I comprises TMS1/2/3, subunit II TMS4/5/6, subunit III TMS7/8/9 and subunit IV comprises TMS10/11/12. MFS transporters possibly experienced two separate gene duplications of a common 3 TMS ancestor. Thus, TMSs with the same colour (*blue*, *magenta* or *yellow*) are equivalent and ELs and ILs with the same colour (*orange*, *yellow* or *grey*) are also equivalent. The centre panel in **a** shows the entire EcEmrD molecule with the N-terminal half (TMS1-6) in *green*, the C-terminal half (TMS7-12) in *pale blue*, the IL3 linker between TMD1 and TMD2 in *black* and the conserved IL1 and IL4 highlighted in *yellow*. **b** and **c** Nucleotide-free conformations of type I and type II ABC exporters. **b** Type I *Mus musculus* multidrug efflux pump MmABCB1 (pdb: 3G5U; (Aller et al. 2009)); and **c** Type II *Homo sapiens* cholesterol ABC exporter HsABCG5-G8 (pdb: 5DO7; (Lee et al. 2016)). MmABCB1 is a single polypeptide, whereas HsABCG5-G8 is a heterodimer of two separate polypeptide chains. Their N- and C-terminal TMDs are *green* and *pale blue* and their N- and C-terminal NBDs *magenta* and *red*, respectively. The *centre* panels in **b** and **c** highlight the major contact points between the TMDs (ILs and coupling helices (Ch) in *yellow*) and the Q-loops (*dark blue*; conserved Qs are shown as *red* sticks) of the NBDs for type I (**b**) and type II (**c**) ABC exporters. The four ILs of type I exporters provide 'asymmetric' contact with the two NBD Q-loops (centre panel), whereas IL1 and a conserved positively charged cytosolic Ch preceding TMS1 provide 'symmetric' contact with their respective NBDs; i.e. the Ch and IL1 of ABCG5 and ABCG8 are in contact with the Q-loop and a conserved E-helix (*magenta*; *centre panel*) of their own NBDs (*centre right*). Visible TMS helices are numbered in *yellow* (N-terminal TMD) and *orange* (C-terminal TMD). The *bottom panels* show cross-sections through the centres of the TMDs for MFS (**a**), type I (**b**) and type II (**c**) ABC transporters viewed from the extracellular side after tilting the TMDs depicted in the *centre panels* forward by 90°. The N- and C-terminal TMDs are rainbow colour-coded and numbered in *green* (N-terminal TMD) or *blue* (C-terminal TMD), respectively. Each transporter has a unique TMS topology with unique arrangements of TMSs surrounding a centrally located substrate-binding pocket

N-terminal TMD extensions that cause transporter dimerization and/or help target the transporter to its correct destination.

Due to the lack of any type II exporter structure, *S. cerevisiae* Pdr5 was initially modelled on the structures of type I exporters (Rutledge et al. 2011). However, the first type II ABC exporter structure for the open conformation of the human cholesterol transporter HsABCG5-G8 (Fig. 18.2c) has recently been solved (Lee et al. 2016). Perhaps not surprisingly, given the unique features characterising PDR transporters (Lamping et al. 2010), HsABCG5-G8 exhibited a unique TMD fold (bottom of Fig. 18.2c), quite different from that of type I exporter MmABCB1 (bottom of Fig. 18.2b). Lee and colleagues noticed a unique TMD-NBD cis-interface with novel connecting-helices (Ch; yellow) and E-helices (magenta) just before TMS1 and after the Q-loop of ABCG5 and ABCG8, respectively (centre panel; Fig. 18.2c). Conserved salt-bridges stabilised cis contacts between the TMDs (Ch and IL1) and NBDs (Q-loop and E-helix) of HsABCG5-G8 (centre panel in Fig. 18.2c) (Lee et al. 2016). We previously reported PDR transporter-specific positively charged helices just before TMS1 and TMS7 (Lamping et al. 2010) that align with the two Chs of ABCG5 and ABCG8 (data not shown), and we can

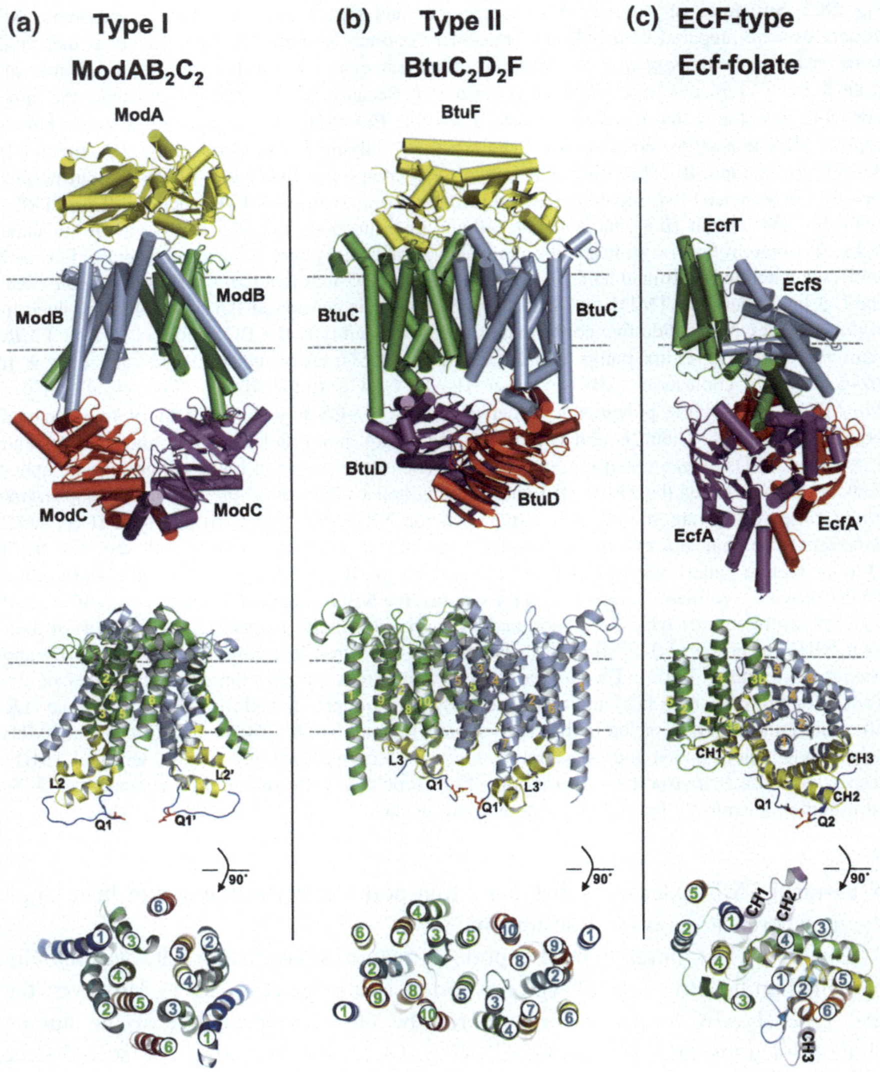

confirm that all fungal PDR transporters also have the conserved E-helices adjacent to their respective Q-loops (Fig. 18.4b) suggesting that Pdr5 and Cdr1 have the same structural/functional constraints as HsABCG5-G8. Also of significant interest was how the unique extracellular domains of HsABCG5-G8 fold. The previously identified PDR transporter EL motifs (Lamping et al. 2010) are visualised in ABCG5 (PDRA = ^{557}I-F^{565}; PDRB = ^{569}K-F^{580}) and ABCG8 (EL6 motif = ^{584}W-S^{591}; EL6 helix = ^{595}F-S^{609}) as four separate helices (yellow; top

◄**Fig. 18.3** Structures of type I, type II and ECF-type ABC importers. Nucleotide-free conformations of three ABC importers with the same annotations used in Fig. 18.2b, c. **a** Type I *Archaeoglobus fulgidus* molybdate ABC importer $ModAB_2C_2$ (pdb: 2ONK; (Hollenstein et al. 2007a, b)). **b** Type II *E. coli* vitamin B12 ABC importer $BtuC_2D_2F$ (pdb: 2QI9; (Hvorup et al. 2007)). **c** *Lactobacillus brevis* energy-coupling factor (EFC) ABC importer Ecf-folate transporter (pdb: 4HUQ; (Xu et al. 2013)). $ModAB_2C_2$ and $BtuC_2D_2F$ each consist of five separate domains that are encoded by three separate genes, the periplasmic molybdate and vitamin B12 binding proteins ModA and BtuF (*yellow*), the N- (*green*) and C-terminal (*pale blue*) TMDs ModB and BtuC and the N- (*magenta*) and C-terminal (*red*) NBDs ModC and BtuD, respectively. In contrast, the *L. brevis* Ecf-folate transporter consists of four domains, each encoded by a separate gene: EcfA (*magenta*) and EcfA' (*red*) are the N- and C-terminal NBDs, and EcfT (*green*) and the folate binding protein EcfS (*pale blue*) are the N- and C-terminal TMDs. The centre panels highlight the major contact points of the TMDs with the Q-loops of the NBDs. ModB-IL2 (L2 and L2'; *left*) and BtuC-IL3 (L3 and L3'; *centre*) provide the major contact points with the Q-loops of ModC and BtuD, respectively, ECF importers like the *L. brevis* Ecf-folate transporter, however, interact with the Q-loops of EcfA and EcfA' via three intracellular coupling helices (CH1, CH2 and CH3) that are part of the large EcfT-IL2 (*right*). The *bottom* panels highlight the unique TMD folds of these three different types of ABC importer. Note how EcfS is toppled over towards EcfT—this conformation reflects the structure of the Ecf-folate transporter after substrate release

panel of Fig. 18.2c). Both helices of each monomer bend towards each other and form a critical part of the TMD by dipping into the central binding cavity of the transporter (top panel; Fig. 18.2c).

Most of the current knowledge about the transport and energy coupling of ABC exporters relates to studies with type I ABC exporters. *S. aureus* Sav1866, homolog of one of the best-studied multidrug efflux pumps, human Pgp, was one of the first type I ABC exporters for which the structure was solved (Dawson and Locher 2006; Dawson and Locher 2007). The structures obtained were for the closed conformation with the substrate-binding cavity accessible to the extracellular space. Most other type I ABC exporter structures solved since then have been for open conformations with the substrate-binding cavity open to the cytosol (Fig. 18.2b). These structures revealed important insights into how type I ABC exporters couple ATP-binding and hydrolysis with substrate transport. Typically, the first IL of each TMD (IL1 and IL3) is in close contact with its proximal NBD Q-loop (Q-loop1 and Q-loop2, respectively; Fig. 18.2b centre), whereas the second loop (IL2 and IL4) reaches across the opposite NBD by contacting Q-loop2 and Q-loop1, respectively (centre panel of Fig. 18.2b). Computer simulations of the transport cycle of type I exporters suggested large, rigid, conformational changes of the NBDs upon ATP-binding that move the two NBDs close together and the cylindrical, barrel-like, structure formed by the two TMDs, especially TMS4, -6, -10 and -11 that surround a central substrate-binding cavity (bottom panel of Fig. 18.2b), becomes exposed to the outer leaflet and the extracellular space (Aller et al. 2009; Dawson and Locher 2006; Jin et al. 2012; Li et al. 2014; Prajapati and Sangamwar 2014; Ravna et al. 2009; Rosenberg et al. 2003; Velamakanni et al. 2008; Ward et al. 2007; Ward et al. 2013). This twisting motion of the TMDs induced by movement of the hinge-like contacts between IL1-4 with the Q-loops of NBD1 and NBD2 (Fig. 18.2b) causes helix rotation (*Helix Rotation Model*) (Gutmann et al.

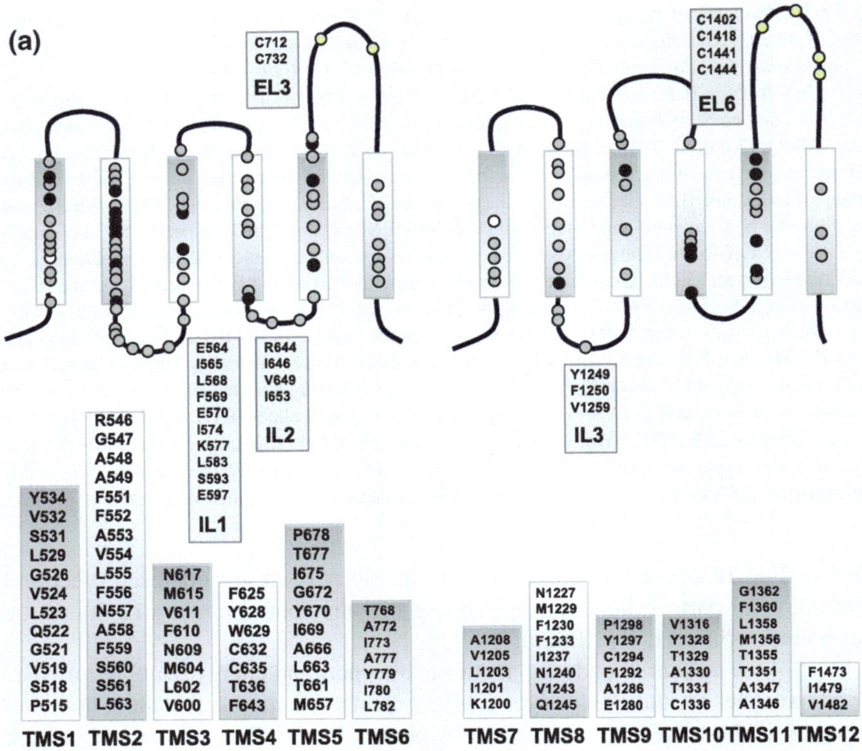

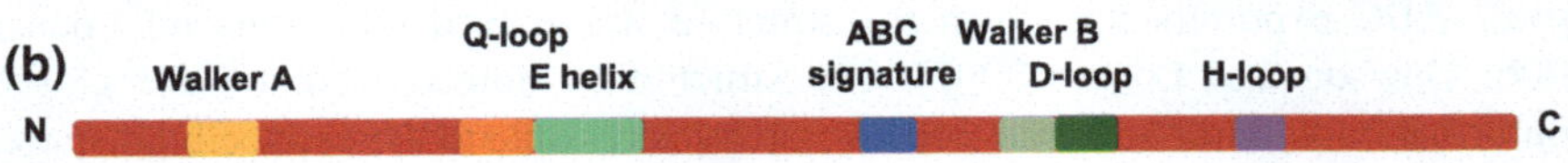

NBD1

	Walker A1*	Q-loop1*	E-helix1	ABC1	Walker B1*	D-loop1	H-loop1*
CaCdr1	^{187}GRPGAGC	^{236}SAETDVH	^{248}VGDTLE	^{303}VSGGE	^{323}IQCWDN	^{329}ATRGLD	^{361}YQCSQ
ScPdr5	^{193}GRPGSGC	^{242}NAEADVH	^{254}VFETLV	^{309}VSGGE	^{329}FQCWDN	^{335}ATRGLD	^{367}YQCSQ

NBD2

	Walker A2	Q-loop2	E-helix2	ABC2*	Walker B2	D-loop2	H-loop2
CaCdr1	^{895}GASGAGK	^{940}VQQQDVH	^{952}VREALQ	^{1001}LNVEQ	^{1022}LLFLDE	^{1028}PTSGLD	^{1059}HQPSA
ScPdr5	^{905}GASGAGK	^{949}CQQQDLH	^{961}VRESLR	^{1010}LNVEQ	^{1031}LVFLDE	^{1037}PTSGLD	^{1068}HQPSA

◂**Fig. 18.4** Summary of Cdr1 mutagenesis studies. **a** Cdr1 TMD residues that have been shown to be important for Cdr1 efflux pump function (Kolaczkowski et al. 2013; Rawal et al. 2013; Shah et al. 2015) are identified. TMSs are shown as light grey boxes and residues that selectively affected transport of some, but not all, Cdr1 test substrates are shown as *grey circles*, and residues with severe functional defects (i.e. no drug efflux and reduced ATPase activity) are shown as *black circles*. A list of these residues is shown underneath the corresponding TMS or next to the corresponding IL. Residues G521 (*centre* of TMS1) and A1208 (*centres* of TMS7) that, when mutated, improved efflux of small compounds such as resazurin, an Snq2 substrate (Kolaczkowski et al. 2013), are shown as *white circles* and the six conserved EL3 and EL6 cysteines (Lamping et al. 2010) are indicated as *yellow circles*. **b** A graphical representation of the conserved Cdr1/Pdr5 NBD motifs highlighted in different colours with a list of the Walker A, Walker B, Q-loop, E-helix, D-loop, H-loop and ABC signature motif residues for Cdr1 and Pdr5 shown underneath. Red residues are critical for ATP-binding and/or hydrolysis as discussed in the text. * = non-canonical NBD motifs typical for PDR transporters

2010) of TMSs surrounding the substrate-binding cavity and thus changes the affinity of substrates so that they can be released to the cell exterior after transporter closure. Like efflux pump substrates, inhibitors also bind to the substrate-binding cavity (Aller et al. 2009) but, unlike Pgp substrates that can induce the ATPase activity of the efflux pump up to 100-fold by helping bring the two NBDs closer together and thus increase the affinity for ATP and induce their own transport—also referred to as the *Induced Fit Model*—inhibitors seem to keep the NBDs apart and freeze the transporter in an open conformation (Aller et al. 2009; Gutmann et al. 2010; Ma and Biggin 2013). ATP hydrolysis is believed to then reset the transporter in the inward facing conformation so that another transport cycle can begin. The *Switch Model* proposed that after ATP hydrolysis the two NBDs separate by >20 Å (Higgins and Linton 2004), while others suggested that the large gap separating the NBDs in the open conformation of most type I ABC exporter structures is not a biologically viable state. Some studies have shown that the NBDs indeed remain in contact at all times during an alternating ATP-binding and hydrolysis cycle (George and Jones 2012; Jones and George 2009; Sauna et al. 2007). In the *Constant Contact Model,* the ATP-bound active site is closed while the second ATP-binding site is empty and slightly further apart (George and Jones 2012). After transport, ATP hydrolysis and the release of Pi and ADP, the empty low-affinity site switches to a high-affinity site and another ATP hydrolysis cycle begins (Fribourg et al. 2014; George and Jones 2012; Ravna et al. 2009).

However, despite all these valuable contributions, details of the ATP kinetics of type I ABC exporters remain unresolved. Electron microscopy single particle analysis of *E. coli* MsbA and Pgp revealed remarkable differences between the two pumps, not only in their conformational changes in the presence of nucleotides and substrates, but also in the separation between their two NBDs. Therefore, perhaps more than one model is required to describe type I ABC exporter function (Moeller et al. 2015).

18.5 Structure–Function Activity Relationships Deduced from *C. albicans* Multidrug Transporter Mutagenesis

18.5.1 MFS Transporter Mdr1

Structural and functional analyses of *C. albicans* Mdr1 have indicated that TMS5 and IL3 residues are critical for drug transport. Alanine substitutions of 21 TMS5 residues showed clustering of functionally critical residues, including all conserved C-motif residues (G251, G255, P256, G259), to one face of TMS5 (Pasrija et al. 2007). The functional importance of IL3 residues was also established by site-directed mutagenesis (SDM) (Mandal et al. 2012). This region contains a number of charged residues, half of which were critical for Mdr1 function (Mandal et al. 2012). Yan proposed a model whereby IL3 helices provide a latch that strengthens the intracellular gate in the outward open conformation of GLUT1 and related SPs (Yan 2015). *C. albicans* Mdr1 also has remnants of the 'hallmark' A-motif (174-SPMSENAIFGRT-185; Paulsen et al. 1996) with three of the six conserved residues (i.e. E178, G183 and R184) found between TMS2 and -3, and, like most other MFS transporters, it has a conserved D235 at the end of TMS4 and a conserved 296-P-E-(T/S) motif at the cytosolic end of TMS6 (Nomura et al. 2015; Paulsen et al. 1996). Based on existing structures (Nomura et al. 2015; Quistgaard et al. 2016; Yan 2013; Yan 2015), Mdr1 R184 is likely to form important salt-bridges with D235 and E297 at different stages of the transport cycle. The critical contribution of E297 to Mdr1 drug transport was confirmed by Mandal and colleagues (Mandal et al. 2012).

An important question is why does Mdr1 have a wider substrate specificity than most other MFS pump homologues? MFS proteins often have conserved gating residues lining the central binding cavity, which are responsible for their substrate specificity. For example, a single-point mutation Q166E in TMS5 of the substrate-binding cavity of GLUT5 was enough to switch the substrate-binding preference from fructose to glucose (Nomura et al. 2015). Similarly, the *E. coli* multidrug efflux pump EmrD has two long helical regions composed of TMS4, -5, -10, -11 and slightly extended IL2 and IL4 that, together with a run of three positively charged residues at the end of TMS4, are proposed to act as a substrate-specificity filter (Yin et al. 2006). Another example of the importance of TMS1, -4, -7, -10 residues (blue helices on Fig. 18.2a) and, in particular, residues of the proton-dependent oligopeptide transporter (POT) family-defining motif $Ex_2ERFxYY$ of TMS1, in substrate selectivity, proton binding and oligopeptide symport was revealed by the structural analysis of the *Streptococcus thermophilus* POT oligopeptide symporter PepTSt (Solcan et al. 2012). These structural and mechanistic principles appear to hold true for most, if not all, MFS transporters, and, in the absence of an Mdr1 structure, they provide a model for how *C. albicans* multidrug efflux pump Mdr1 works.

18.5.2 ABC Transporter Cdr1

18.5.2.1 TMDs and ELs

Early random mutagenesis studies with *S. cerevisiae* Pdr5 identified several residues that were critical for drug transport. It was not clear, however, how these residues, which were distributed throughout the molecule, caused reduced efflux of some or all drugs and/or affected the ATPase activity of Pdr5 (Egner et al. 2000; Egner et al. 1998; Tutulan-Cunita et al. 2005). Both MDR and PDR ABC exporters appear to have multiple overlapping substrate-binding sites encompassing a large substrate-binding cavity within their TMDs, with Pdr5 and Cdr1 possessing at least three separate substrate-binding sites (Golin et al. 2003; Golin et al. 2007; Puri et al. 2009; Shukla et al. 2003). There are, however, differences in the preferred substrates of ABC proteins. Pgp for instance prefers hydrophobic substrates with hydrogen-bond acceptor groups (Seelig and Landwojtowicz 2000), whereas size (optimum ~ 200–225 Å^3), but not hydrophobicity, seemed the major determinant for a good Pdr5 substrate (Golin et al. 2003; Golin et al. 2007). Cdr1, on the other hand, seemed to prefer hydrophobic substrates and molecular-branching (Puri et al. 2010).

In the absence of a high-resolution structure, bioinformatics and biochemistry are the best ways to identify residues involved in drug binding and/or transport. Heterologous expression of efflux pumps in *S. cerevisiae* provides many advantages for studying structure–function relationships. *S. cerevisiae* can be easily genetically manipulated and a vast mutant collection is available. Cdr1/Pdr5 variants can be rapidly generated and expressed at high levels at the correct cellular location (Lamping and Cannon 2010; Niimi et al. 2012; Tanabe et al. 2011). In addition, one can quickly select for naturally occurring suppressor mutants of transport defective variants by isolating resistant colonies that appear after prolonged incubation at 30 °C on plates containing high (Niimi et al. 2012) concentrations of toxic efflux pump substrates (Golin and Ambudkar 2015). SDM of all 252 Cdr1 TMS residues indicated that almost half of them (~42%) affected substrate transport and/or ATPase activity (Fig. 18.4a) (Rawal et al. 2013), making interpretation of the effects of mutations of individual residues very challenging. Interestingly, most residues affecting substrate transport (grey circles; Fig. 18.4a) and/or ATPase activity (black circles; Fig. 18.4a) were located in TMS1, -2, -5, -8 and -11, whereas TMS4, -6, -10 and -12 contributed least to substrate transport and/or ATPase activity (Fig. 18.4a) (Rawal et al. 2013). Given this result, it was perhaps a little surprising that only ~ 18% of all IL mutants affected Cdr1 drug transport and none of them affected the Cdr1 ATPase activity (Fig. 18.4a) (Shah et al. 2015). The observed phenotypes of TMS5 mutant residues affecting Cdr1 pump function did not relate to which face of the helix they belonged to, the degree of conservation of the residues or the nature of their side chains (Puri et al. 2009; Rawal et al. 2013). In contrast, TMS11 mutations of both Cdr1 and Pdr5 that affected drug efflux were mainly clustered on the hydrophilic face of the helix, suggesting a direct contribution of TMS11 to a centrally located drug-binding cavity (Egner et al. 2000; Kueppers et al. 2013; Rawal et al. 2013; Saini et al. 2005; Shukla et al. 2004).

The role of EL residues in protein assembly, cell surface localisation and substrate specificity has been highlighted in a random mutagenesis study of Pdr5 (Egner et al. 1998). A single substitution of one of the six conserved EL3/EL6 cysteines (yellow residues in Fig. 18.4a) of fungal PDR transporters (Lamping et al. 2010), C1427Y, caused mislocalization of the mutant Pdr5 to the endoplasmic reticulum. The equivalent residue in Cdr1, C1418, was found to be essential for pump function (Shukla et al. 2003). These studies highlighted the importance of the conserved extracellular cysteines that possibly form three extracellular disulfide bonds that stabilise the structure of the highly conserved extracellular domains of PDR transporters (Lamping et al. 2010). Molecular mapping and biochemical characterization of more than 50 Pdr5/Cdr1 suppressor mutants that had lost their ability to respond to efflux pump inhibitors RC21v3, a Cdr1 specific D-octapeptide inhibitor (Niimi et al. 2012), macrolides such as FK506 (Egner et al. 1998; Kralli and Yamamoto 1996; Lamping et al. 2007; Tanabe et al. 2011) and milbemycins (Lamping et al. 2007; Lee et al. 2001; Niimi et al. 2012; Silva et al. 2013), depsipeptides enniatin (Hiraga et al. 2005; Holmes et al. 2008; Lamping et al. 2007; Lee et al. 2001; Tanabe et al. 2011) and beauvericin (Lee et al. 2001; Tanabe et al. 2011) further highlighted the importance of the conserved EL3 and EL6 in Cdr1/Pdr5 efflux pump function ((Lamping et al. 2014) and unpublished results). How these extracellular domains contribute to type II exporter function is an important question that needs to be answered in order to understand how PDR multidrug efflux pumps work.

18.5.2.2 NBDs and ILs

The highly conserved NBDs are defined by the Walker A and the Walker B motifs (Walker et al. 1982) and the hallmark ABC signature motif which are critical for ATP-binding and hydrolysis (Fig. 18.4b; (Prasad and Goffeau 2012)). Typically, ABC transporters have two canonical ATP-binding sites, also referred to as composite nucleotide-binding domains (CNBD), comprising Walker A and Walker B motifs of one NBD and the ABC signature motif of the other. CNBD1 and CNBD2 of most ABC proteins are believed to alternatingly bind and hydrolyse ATP (see Sect. 18.4.2). However, PDR transporters have a high, non-inducible, basal ATPase activity (Ernst et al. 2008) and they are characterised by one canonical CNBD2 (i.e. typical Walker A2, Walker B2 and ABC1 signature motifs; Fig. 18.4b) and one non-canonical CNBD1 with its Walker A1 K residue replaced by a C (C193 in Cdr1), its Walker B1 catalytic E replaced by an N (N328 in Cdr1), the conserved H-loop H of NBD1 replaced by a Y (Y361 in Cdr1) and the C-loop G of ABC2 replaced by a E (E1004 in Cdr1; red residues in Fig. 18.4b).

SDM of C199 of Walker A1 and K911 of Walker A2 (Fig. 18.4b) indicated that the Walker A2 K911, but not the Walker A1 C199, was critical for Pdr5 ATPase activity and drug transport (Ernst et al. 2008). SDM of Cdr1 Walker B1 (D327,

N328), Walker B2 (D1026, E1027) as well as all five ABC1 (VSGGE) and ABC2 (LNVEQ; Fig. 18.4b) residues found that all four Walker B residues (Rai et al. 2006) and ABC residues S304, G306 and E307 of ABC1 and the equivalent ABC2 residues N1002 and E1004, but not Q1005 (Kumar et al. 2010), were critical for both the ATPase activity as well as drug efflux, indicating the importance of both CNBDs in ATP-binding and/or hydrolysis (Kumar et al. 2010). In contrast, alanine substitution of the Pdr5 D-loop residues D340A and D1042A (red residues; Fig. 18.4b) 'uncoupled' the ATPase activity from drug transport by affecting drug efflux but not the ATPase activity of Pdr5, indicating that D-loop residues are important for the crosstalk between the NBDs and the TMDs of PDR transporters (Furman et al. 2013). Surprisingly, mutation of the conserved H-loop H1068A had no effect on Pdr5 ATPase activity, instead it selectively abolished R6G transport without affecting the transport of any other substrates (Ernst et al. 2008). This led to the *Kinetic Substrate Selection* model for Pdr5 transport, whereby the authors proposed that the H1068A mutation changed the transport kinetics in such a way that only the transport of 'slow' substrates such as R6G was impaired because the average time it took 'slow' substrates to enter or diffuse into the transporter far exceeded the time the mutant transporter was open for substrates to enter (Ernst et al. 2010).

A study of the Pdr5 S558Y (Cdr1 A548; Fig. 18.4a) mutation at the top of TMS2 provided important clues about how TMS2 may be critical for the crosstalk between the NBDs and the TMDs (Sauna et al. 2008). The phenotype of yeast with this mutation was drug hypersensitivity despite wild-type ATPase activity or, in other words, this mutant had a pump defect because its ATPase activity could not be translated into the conformational changes needed for drug efflux. This defect could be recovered by second-site suppressor mutations in Q-loop1 (N242 K and E244G; Fig. 18.4b) and the centre of IL1 (S597I, S597T (Cdr1 S587)) (Ananthaswamy et al. 2010). Similarly, second-site Pdr5 N242 K suppressor mutants with improved CHX transport were found in the IL2-TMS5 (M649I, V656L, A666G) region and just C-terminal of ABC2 (K1016I) (Ananthaswamy et al. 2010). The same V656L second-site suppressor mutation was also found in a search for improved CHX transport (i.e. increased CHX resistance) of yeast over-expressing wild-type Pdr5 (Downes et al. 2013). Alanine substitution of Pdr5 V656, however, caused the Pdr5 ATPase activity to become uncoupled from drug efflux and led to drug hypersensitivity of the strain over-expressing the mutant protein. Taken together, these data support a model whereby Pdr5 IL2 may possibly be part of a cis-transmission interface between TMD1 and NBD1, unlike MmABCB1 where IL2 is a critical part of a trans-transmission interface by being in direct contact with Q-loop2 of NBD2 (Fig. 18.2b). The critical importance of the TMS2-IL1 region for the crosstalk between the NBDs and TMDs of PDR transporters was further supported by the Cdr1 SDM studies described in Sect. 18.5.2.1 with 7 of 24 functionally critical Cdr1 TMD residues (black residues; Fig. 18.4a) located in TMS2 and almost all remaining TMS2 residues affecting efflux, and more than 50% of all IL residues affecting Cdr1 drug efflux were located in IL1 (Fig. 18.4a) (Rawal et al. 2013; Shah et al. 2015).

Taken together, these studies suggest a critical role for the deviant CNBD1 in intra-domain signalling of ATP-binding at the canonical CNBD2 (Ananthaswamy et al. 2010; Furman et al. 2013) which causes the NBDs to move closer together and, in turn, leads to large conformational changes at the TMDs followed by drug release in the closed conformation of the transporter. In addition, TMS2-IL1, and possibly IL2, seem to play a critical role in the cis-transmission crosstalk between the NBDs and the TMDs. Despite all these findings, however, it is still not clear whether CNBP1 can also bind and/or hydrolyze ATP, or whether it has evolved an alternative 'regulatory' function and is only able to bind—possibly at all times—but not hydrolyze ATP (Furman et al. 2013; Gupta et al. 2011; Prasad and Goffeau 2012).

18.6 Model of Fungal PDR Transporters

It is evident that there has been a plethora of studies dissecting Cdr1/Pdr5 structure and function. The vast amount of biochemical data accumulated over the past two decades combined with PDR transporter phylogeny and the first type II ABC exporter structure, HsABCG5-G8, provide evidence for the following tentative model of how PDR transporters function. Cdr1/Pdr5 TMS1, -2, -5, -7, 8, -11 (yellow to red; Fig. 18.5c) form a large central substrate-binding cavity with TMS3, -4, -6, -9, -10, -12 (green to blue; Fig. 18.5c) stabilising the protein during the transport cycle. Salt-bridges help stabilise critical contacts between the positively charged Ch near TMS1 and IL1, Q-loop1 and E-helix1 of NBD1. The homologous parts of the C-terminal half provide similar cis interactions between NBD2 and TMD2. CNBD1 is possibly bound to ATP at all times while CNBD2 provides the energy required for the large conformational changes associated with drug efflux by binding and hydrolysing ATP. Openings between TMS1 and TMS11 (gate 1; Fig. 18.5c) and TMS5 and TMS7 (gate 2; Fig. 18.5c) may provide gates for substrates to enter, in contrast to type I exporters where substrates are thought to enter between TMS4 and TMS6 and TMS10 and TMS12 (Fig. 18.5a). The extracellular domain also plays a critical part in this process, as evidenced by phylogeny and the HsABCG5-G8 structure. How these extracellular domains contribute to drug efflux, however, remains to be explored.

18.7 Concluding Remarks

C. albicans possesses many membrane transport proteins which perform vital cellular functions. ABC and, to a lesser extent, MFS transporters have broad substrate specificities that can include antifungal drugs and xenobiotics. Increased expression of pumps, particularly Cdr1, is selected for by growth in the presence of drugs and confers drug resistance. Because of their promiscuity, the normal

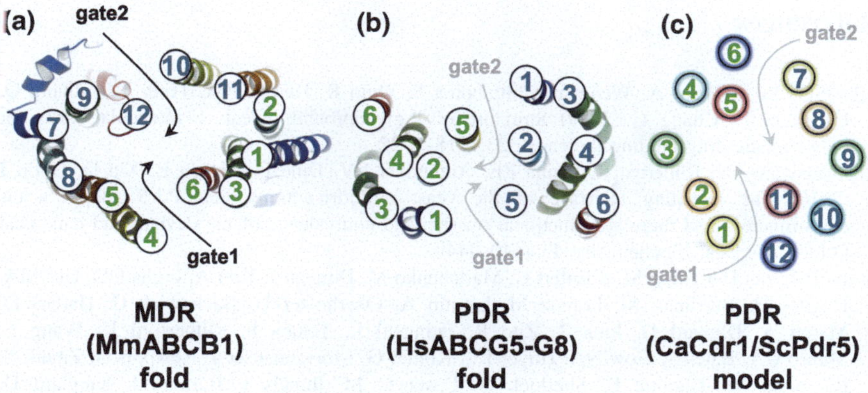

Fig. 18.5 Fungal PDR transporters are predicted to fold like HsABCG5-G8. The TMD folds for MmABCB1 (**a**) and HsABCG5-G8 (**b**) are the same as shown in Fig. 18.2b, c. This figure shows how MDR transporters (**a**) have TMS4, -5, -6 and TMS10, -11, -12 lining the substrate-binding pocket, whereas the PDR transporter homolog HsABCG5-G8 (**b**) has TMS1, -2, and -5 of ABCG5 (*left half*) and ABCG8 (*right half*) lining the substrate-binding pocket. Annotations are the same as in Fig. 18.2**c**. The predicted TMS topology for Cdr1 and Pdr5 (Lamping et al. 2014) (**c**). This topology model was created by careful consideration of all published and unpublished data relating to the structure and function of Cdr1 and Pdr5. The colours of TMSs: *red*, *orange*, *yellow*, *green*, *turquois* and *blue* indicate homologous N- (*lighter* colours) and C-terminal (*darker* colours) TMS pairs from the most to the least 'important' (i.e. the most important TMSs had the most number of residues affecting Cdr1/Pdr5 ATPase activity and/or drug efflux; also see Fig. 18.4a). Proposed (**a**) and possible gates (**b**; **c**) where pump substrates may enter the transporter from the inner leaflet of the lipid bilayer are indicated with *black* (**a**) and *grey* (**b**; **c**) *arrows*. Note how the predicted PDR TMS topology closely resembles that of ABCG5-G8 (**b**) but has little semblance to type I exporters (**a**)

biological function of multidrug efflux pumps is an unanswered question. In vitro and in vivo studies have demonstrated that inhibition of Cdr1 can sensitise azole-resistant *C. albicans* clinical isolates to FLC and ITC. Further advances in the determination of the 3D structures of multidrug transporters will help design more potent specific and broad spectrum pump inhibitors and may also elucidate the reaction cycle for an important class of membrane transporters. Molecular modelling and site-directed mutagenesis is beginning to reveal that the arrangement of external loops, transmembrane helices and interactions between these helices and the NBDs are important for pump function and are potential drug targets.

Acknowledgements The authors acknowledge financial support from the Marsden Fund of the Royal Society of New Zealand (UOO1305) and a research grant from Chulalongkorn University, Thailand.

References

Aller SG, Yu J, Ward A, Weng Y, Chittaboina S, Zhuo R, Harrell PM, Trinh YT, Zhang Q, Urbatsch IL, Chang G (2009) Structure of P-glycoprotein reveals a molecular basis for poly-specific drug binding. Science 323:1718–1722

Ananthaswamy N, Rutledge R, Sauna ZE, Ambudkar SV, Dine E, Nelson E, Xia D, Golin J (2010) The signaling interface of the yeast multidrug transporter Pdr5 adopts a cis conformation, and there are functional overlap and equivalence of the deviant and canonical Q-loop residues. Biochemistry 49:4440–4449

Braun BR, van Het Hoog M, d'Enfert C, Martchenko M, Dungan J, Kuo A, Inglis DO, Uhl MA, Hogues H, Berriman M, Lorenz M, Levitin A, Oberholzer U, Bachewich C, Harcus D, Marcil A, Dignard D, Iouk T, Zito R, Frangeul L, Tekaia F, Rutherford K, Wang E, Munro CA, Bates S, Gow NA, Hoyer LL, Kohler G, Morschhauser J, Newport G, Znaidi S, Raymond M, Turcotte B, Sherlock G, Costanzo M, Ihmels J, Berman J, Sanglard D, Agabian N, Mitchell AP, Johnson AD, Whiteway M, Nantel A (2005) A human-curated annotation of the *Candida albicans* genome. PLoS Genet 1:36–57

Calabrese D, Bille J, Sanglard D (2000) A novel multidrug efflux transporter gene of the major facilitator superfamily from *Candida albicans* (*FLU1*) conferring resistance to fluconazole. Microbiology 146(Pt 11):2743–2754

Cannon RD, Lamping E, Holmes AR, Niimi K, Baret PV, Keniya MV, Tanabe K, Niimi M, Goffeau A, Monk BC (2009) Efflux-mediated antifungal drug resistance. Clin Microbiol Rev 22:291–321

Cheng Y, Grigorieff N, Penczek PA, Walz T (2015) A primer to single-particle cryo-electron microscopy. Cell 161:438–449

Coleman JJ, Mylonakis E (2009) Efflux in fungi: la piece de resistance. PLoS Pathog 5:e1000486

Costa C, Dias PJ, Sa-Correia I, Teixeira MC (2014) MFS multidrug transporters in pathogenic fungi: do they have real clinical impact? Front Physiol 5:197

Coste A, Selmecki A, Forche A, Diogo D, Bougnoux ME, d'Enfert C, Berman J, Sanglard D (2007) Genotypic evolution of azole resistance mechanisms in sequential *Candida albicans* isolates. Eukaryot Cell 6:1889–1904

Coste AT, Karababa M, Ischer F, Bille J, Sanglard D (2004) *TAC1*, transcriptional activator of *CDR* genes, is a new transcription factor involved in the regulation of *Candida albicans* ABC transporters *CDR1* and *CDR2*. Eukaryot Cell 3:1639–1652

Crouzet J, Trombik T, Fraysse AS, Boutry M (2006) Organization and function of the plant pleiotropic drug resistance ABC transporter family. FEBS Lett 580:1123–1130

Dawson RJ, Locher KP (2006) Structure of a bacterial multidrug ABC transporter. Nature 443:180–185

Dawson RJ, Locher KP (2007) Structure of the multidrug ABC transporter Sav 1866 from Staphylococcus aureus in complex with AMP-PNP. FEBS Lett 581:935–938

de Micheli M, Bille J, Schueller C, Sanglard D (2002) A common drug-responsive element mediates the upregulation of the *Candida albicans* ABC transporters *CDR1* and *CDR2*, two genes involved in antifungal drug resistance. Mol Microbiol 43:1197–1214

Delmar JA, Bolla JR, Su CC, Yu EW (2015) Crystallization of membrane proteins by vapor diffusion. Methods Enzymol 557:363–392

Downes MT, Mehla J, Ananthaswamy N, Wakschlag A, Lamonde M, Dine E, Ambudkar SV, Golin J (2013) The transmission interface of the *Saccharomyces cerevisiae* multidrug transporter Pdr5: Val-656 located in intracellular loop 2 plays a major role in drug resistance. Antimicrob Agents Chemother 57:1025–1034

Dunkel N, Blass J, Rogers PD, Morschhauser J (2008) Mutations in the multi-drug resistance regulator *MRR1*, followed by loss of heterozygosity, are the main cause of *MDR1* overexpression in fluconazole-resistant *Candida albicans* strains. Mol Microbiol 69:827–840

Egner R, Bauer BE, Kuchler K (2000) The transmembrane domain 10 of the yeast Pdr5p ABC antifungal efflux pump determines both substrate specificity and inhibitor susceptibility. Mol Microbiol 35:1255–1263

Egner R, Rosenthal FE, Kralli A, Sanglard D, Kuchler K (1998) Genetic separation of FK506 susceptibility and drug transport in the yeast Pdr5 ATP-binding cassette multidrug resistance transporter. Mol Biol Cell 9:523–543

Ernst R, Kueppers P, Klein CM, Schwarzmueller T, Kuchler K, Schmitt L (2008) A mutation of the H-loop selectively affects rhodamine transport by the yeast multidrug ABC transporter Pdr5. Proc Natl Acad Sci U S A 105:5069–5074

Ernst R, Kueppers P, Stindt J, Kuchler K, Schmitt L (2010) Multidrug efflux pumps: substrate selection in ATP-binding cassette multidrug efflux pumps–first come, first served? FEBS J 277:540–549

Fling ME, Kopf J, Tamarkin A, Gorman JA, Smith HA, Koltin Y (1991) Analysis of a *Candida albicans* gene that encodes a novel mechanism for resistance to benomyl and methotrexate. Mol Gen Genet 227:318–329

Ford RC, Kamis AB, Kerr ID, Callaghan R (2009) The ABC transporters: structural insights into drug transport. Transporters as Drug Carriers: Struct Funct Substrates 44:3

Fribourg PF, Chami M, Sorzano CO, Gubellini F, Marabini R, Marco S, Jault JM, Levy D (2014) 3D cryo-electron reconstruction of BmrA, a bacterial multidrug ABC transporter in an inward-facing conformation and in a lipidic environment. J Mol Biol 426:2059–2069

Furman C, Mehla J, Ananthaswamy N, Arya N, Kulesh B, Kovach I, Ambudkar SV, Golin J (2013) The deviant ATP-binding site of the multidrug efflux pump Pdr5 plays an active role in the transport cycle. J Biol Chem 288:30420–30431

Gaur M, Choudhury D, Prasad R (2005) Complete inventory of ABC proteins in human pathogenic yeast, *Candida albicans*. J Mol Microbiol Biotechnol 9:3–15

Gaur M, Puri N, Manoharlal R, Rai V, Mukhopadhayay G, Choudhury D, Prasad R (2008) MFS transportome of the human pathogenic yeast *Candida albicans*. BMC Genom 9:579

George AM, Jones PM (2012) Perspectives on the structure-function of ABC transporters: the switch and constant contact models. Prog Biophys Mol Biol 109:95–107

Golin J, Ambudkar SV (2015) The multidrug transporter Pdr5 on the 25th anniversary of its discovery: an important model for the study of asymmetric ABC transporters. Biochem J 467:353–363

Golin J, Ambudkar SV, Gottesman MM, Habib AD, Sczepanski J, Ziccardi W, May L (2003) Studies with novel Pdr5p substrates demonstrate a strong size dependence for xenobiotic efflux. J Biol Chem 278:5963–5969

Golin J, Ambudkar SV, May L (2007) The yeast Pdr5p multidrug transporter: how does it recognize so many substrates? Biochem Biophys Res Commun 356:1–5

Gupta RP, Kueppers P, Schmitt L, Ernst R (2011) The multidrug transporter Pdr5: a molecular diode? Biol Chem 392:53–60

Gutmann DA, Ward A, Urbatsch IL, Chang G, van Veen HW (2010) Understanding polyspecificity of multidrug ABC transporters: closing in on the gaps in ABCB1. Trends Biochem Sci 35:36–42

Henderson R (2015) Overview and future of single particle electron cryomicroscopy. Arch Biochem Biophys 581:19–24

Higgins CF, Linton KJ (2004) The ATP switch model for ABC transporters. Nat Struct Mol Biol 11:918–926

Hiraga K, Yamamoto S, Fukuda H, Hamanaka N, Oda K (2005) Enniatin has a new function as an inhibitor of Pdr5p, one of the ABC transporters in *Saccharomyces cerevisiae*. Biochem Biophys Res Commun 328:1119–1125

Hollenstein K, Dawson RJ, Locher KP (2007) Structure and mechanism of ABC transporter proteins. Curr Opin Struct Biol 17:412–418

Hollenstein K, Frei DC, Locher KP (2007) Structure of an ABC transporter in complex with its binding protein. Nature 446:213–216

Holmes AR, Lin YH, Niimi K, Lamping E, Keniya M, Niimi M, Tanabe K, Monk BC, Cannon RD (2008) ABC transporter Cdr1p contributes more than Cdr2p does to fluconazole efflux in fluconazole-resistant *Candida albicans* clinical isolates. Antimicrob Agents Chemother 52:3851–3862

Hvorup RN, Goetz BA, Niederer M, Hollenstein K, Perozo E, Locher KP (2007) Asymmetry in the structure of the ABC transporter-binding protein complex BtuCD-BtuF. Science 317:1387–1390

Janbon G, Ormerod KL, Paulet D, Byrnes EJ 3rd, Yadav V, Chatterjee G, Mullapudi N, Hon CC, Billmyre RB, Brunel F, Bahn YS, Chen W, Chen Y, Chow EW, Coppee JY, Floyd-Averette A, Gaillardin C, Gerik KJ, Goldberg J, Gonzalez-Hilarion S, Gujja S, Hamlin JL, Hsueh YP, Ianiri G, Jones S, Kodira CD, Kozubowski L, Lam W, Marra M, Mesner LD, Mieczkowski PA, Moyrand F, Nielsen K, Proux C, Rossignol T, Schein JE, Sun S, Wollschlaeger C, Wood IA, Zeng Q, Neuveglise C, Newlon CS, Perfect JR, Lodge JK, Idnurm A, Stajich JE, Kronstad JW, Sanyal K, Heitman J, Fraser JA, Cuomo CA, Dietrich FS (2014) Analysis of the genome and transcriptome of *Cryptococcus neoformans* var. *grubii* reveals complex RNA expression and microevolution leading to virulence attenuation. PLoS Genet 10:e1004261

Jin MS, Oldham ML, Zhang Q, Chen J (2012) Crystal structure of the multidrug transporter P-glycoprotein from *Caenorhabditis elegans*. Nature 490:566–569

Jones PM, George AM (2009) Opening of the ADP-bound active site in the ABC transporter ATPase dimer: evidence for a constant contact, alternating sites model for the catalytic cycle. Proteins 75:387–396

Jones PM, O'Mara ML, George AM (2009) ABC transporters: a riddle wrapped in a mystery inside an enigma. Trends Biochem Sci 34:520–531

Kerr ID, Jones PM, George AM (2010) Multidrug efflux pumps: the structures of prokaryotic ATP-binding cassette transporter efflux pumps and implications for our understanding of eukaryotic P-glycoproteins and homologues. FEBS J 277:550–563

Kolaczkowski M, Sroda-Pomianek K, Kolaczkowska A, Michalak K (2013) A conserved interdomain communication pathway of pseudosymmetrically distributed residues affects substrate specificity of the fungal multidrug transporter Cdr1p. Biochim Biophys Acta 1828:479–490

Kovalchuk A, Driessen AJ (2010) Phylogenetic analysis of fungal ABC transporters. BMC Genomics 11:177

Kralli A, Yamamoto KR (1996) An FK506-sensitive transporter selectively decreases intracellular levels and potency of steroid hormones. J Biol Chem 271:17152–17156

Kueppers P, Gupta RP, Stindt J, Smits SH, Schmitt L (2013) Functional impact of a single mutation within the transmembrane domain of the multidrug ABC transporter Pdr5. Biochemistry 52:2184–2195

Kumar A, Shukla S, Mandal A, Shukla S, Ambudkar SV, Prasad R (2010) Divergent signature motifs of nucleotide binding domains of ABC multidrug transporter, CaCdr1p of pathogenic *Candida albicans*, are functionally asymmetric and noninterchangeable. Biochim Biophys Acta 1798:1757–1766

Lamping E, Baret PV, Holmes AR, Monk BC, Goffeau A, Cannon RD (2010) Fungal PDR transporters: Phylogeny, topology, motifs and function. Fungal Genet Biol 47:127–142

Lamping E, Cannon RD (2010) Use of a yeast-based membrane protein expression technology to overexpress drug resistance efflux pumps. Methods Mol Biol 666:219–250

Lamping E, Monk BC, Niimi K, Holmes AR, Tsao S, Tanabe K, Niimi M, Uehara Y, Cannon RD (2007) Characterization of three classes of membrane proteins involved in fungal azole resistance by functional hyperexpression in *Saccharomyces cerevisiae*. Eukaryot Cell 6:1150–1165

Lamping E, Niimi M, Niimi K, Tanabe K, Cannon RD (2014) How do the transmembrane domains of PDR transporters like Cdr1p fold. Paper presented at the conference celebrating 20 years of *CDR1* research, New Delhi, India, 4–8 January 2014

Law CJ, Maloney PC, Wang DN (2008) Ins and outs of major facilitator superfamily antiporters. Annu Rev Microbiol 62:289–305

Lee JY, Kinch LN, Borek DM, Wang J, Wang J, Urbatsch IL, Xie XS, Grishin NV, Cohen JC, Otwinowski Z, Hobbs HH, Rosenbaum DM (2016) Crystal structure of the human sterol transporter ABCG5/ABCG8. Nature 533:561–564

Lee MD, Galazzo JL, Staley AL, Lee JC, Warren MS, Fuernkranz H, Chamberland S, Lomovskaya O, Miller GH (2001) Microbial fermentation-derived inhibitors of efflux-pump-mediated drug resistance. Farmaco 56:81–85

Li J, Jaimes KF, Aller SG (2014) Refined structures of mouse P-glycoprotein. Protein Sci 23:34–46

Li R, Kumar R, Tati S, Puri S, Edgerton M (2013) *Candida albicans* flu1-mediated efflux of salivary histatin 5 reduces its cytosolic concentration and fungicidal activity. Antimicrob Agents Chemother 57:1832–1839

Liu TT, Znaidi S, Barker KS, Xu L, Homayouni R, Saidane S, Morschhauser J, Nantel A, Raymond M, Rogers PD (2007) Genome-wide expression and location analyses of the *Candida albicans* Tac1p regulon. Eukaryot Cell 6:2122–2138

Locher KP (2004) Structure and mechanism of ABC transporters. Curr Opin Struct Biol 14:426–431

Ma J, Biggin PC (2013) Substrate versus inhibitor dynamics of P-glycoprotein. Proteins 81:1653–1668

Maebashi K, Niimi M, Kudoh M, Fischer FJ, Makimura K, Niimi K, Piper RJ, Uchida K, Arisawa M, Cannon RD, Yamaguchi H (2001) Mechanisms of fluconazole resistance in *Candida albicans* isolates from Japanese AIDS patients. J Antimicrob Chemother 47:527–536

Mandal A, Kumar A, Singh A, Lynn AM, Kapoor K, Prasad R (2012) A key structural domain of the *Candida albicans* Mdr1 protein. Biochem J 445:313–322

Moeller A, Lee SC, Tao H, Speir JA, Chang G, Urbatsch IL, Potter CS, Carragher B, Zhang Q (2015) Distinct conformational spectrum of homologous multidrug ABC transporters. Structure 23:450–460

Morschhauser J, Barker KS, Liu TT, Bla BWJ, Homayouni R, Rogers PD (2007) The transcription factor Mrr1p controls expression of the *MDR1* efflux pump and mediates multidrug resistance in *Candida albicans*. PLoS Pathog 3:e164

Nannenga BL, Iadanza MG, Vollmar BS, Gonen T (2013) Overview of electron crystallography of membrane proteins: crystallization and screening strategies using negative stain electron microscopy. In: Coligan John E et al Current protocols in protein science Chapter 17: Unit17 15

Niimi K, Harding DR, Holmes AR, Lamping E, Niimi M, Tyndall JD, Cannon RD, Monk BC (2012) Specific interactions between the *Candida albicans* ABC transporter Cdr1p ectodomain and a D-octapeptide derivative inhibitor. Mol Microbiol 85:747–767

Nomura N, Verdon G, Kang HJ, Shimamura T, Nomura Y, Sonoda Y, Hussien SA, Qureshi AA, Coincon M, Sato Y (2015) Structure and mechanism of the mammalian fructose transporter GLUT5. Nature 526:397–401

Pasrija R, Banerjee D, Prasad R (2007) Structure and function analysis of CaMdr1p, a major facilitator superfamily antifungal efflux transporter protein of *Candida albicans*: identification of amino acid residues critical for drug/H+ transport. Eukaryot Cell 6:443–453

Paulsen IT, Brown MH, Littlejohn TG, Mitchell BA, Skurray RA (1996) Multidrug resistance proteins QacA and QacB from *Staphylococcus aureus*: membrane topology and identification of residues involved in substrate specificity. Proc Natl Acad Sci U S A 93:3630–3635

Paulsen IT, Skurray RA (1993) Topology, structure and evolution of two families of proteins involved in antibiotic and antiseptic resistance in eukaryotes and prokaryotes–an analysis. Gene 124:1–11

Perlin MH, Andrews J, Toh SS (2014) Essential letters in the fungal alphabet: ABC and MFS transporters and their roles in survival and pathogenicity. Adv Genet 85:201–253

Prajapati R, Sangamwar AT (2014) Translocation mechanism of P-glycoprotein and conformational changes occurring at drug-binding site: Insights from multi-targeted molecular dynamics. Biochim Biophys Acta 1838:2882–2898

Prasad R, Goffeau A (2012) Yeast ATP-binding cassette transporters conferring multidrug resistance. Annu Rev Microbiol 66:39–63

Puri N, Gaur M, Sharma M, Shukla S, Ambudkar SV, Prasad R (2009) The amino acid residues of transmembrane helix 5 of multidrug resistance protein CaCdr1p of *Candida albicans* are involved in substrate specificity and drug transport. Biochim Biophys Acta 1788:1752–1761

Puri N, Prakash O, Manoharlal R, Sharma M, Ghosh I, Prasad R (2010) Analysis of physico-chemical properties of substrates of ABC and MFS multidrug transporters of pathogenic *Candida albicans*. Eur J Med Chem 45:4813–4826

Quistgaard EM, Low C, Guettou F, Nordlund P (2016) Understanding transport by the major facilitator superfamily (MFS): structures pave the way. Nat Rev Mol Cell Biol 17:123–132

Rai V, Gaur M, Shukla S, Shukla S, Ambudkar SV, Komath SS, Prasad R (2006) Conserved Asp327 of walker B motif in the N-terminal nucleotide binding domain (NBD-1) of Cdr1p of *Candida albicans* has acquired a new role in ATP hydrolysis. Biochemistry 45:14726–14739

Ravna AW, Sylte I, Sager G (2009) Binding site of ABC transporter homology models confirmed by ABCB1 crystal structure. Theor Biol Med Model 6:20

Rawal MK, Khan MF, Kapoor K, Goyal N, Sen S, Saxena AK, Lynn AM, Tyndall JD, Monk BC, Cannon RD, Komath SS, Prasad R (2013) Insight into pleiotropic drug resistance ATP-binding cassette pump drug transport through mutagenesis of Cdr1p transmembrane domains. J Biol Chem 288:24480–24493

Reddy VS, Shlykov MA, Castillo R, Sun EI, Saier MH Jr (2012) The major facilitator superfamily (MFS) revisited. FEBS J 279:2022–2035

Rees DC, Johnson E, Lewinson O (2009) ABC transporters: the power to change. Nat Rev Mol Cell Biol 10:218–227

Rigaud J-L, Chami M, Lambert O, Levy D, Ranck J-L (2000) Use of detergents in two-dimensional crystallization of membrane proteins. Biochim Biophys Acta (BBA)-Biomembranes 1508:112–128

Rosenberg MF, Kamis AB, Callaghan R, Higgins CF, Ford RC (2003) Three-dimensional structures of the mammalian multidrug resistance P-glycoprotein demonstrate major conformational changes in the transmembrane domains upon nucleotide binding. J Biol Chem 278:8294–8299

Rutledge RM, Esser L, Ma J, Xia D (2011) Toward understanding the mechanism of action of the yeast multidrug resistance transporter Pdr5p: a molecular modeling study. J Struct Biol 173:333–344

Sa-Correia I, dos Santos SC, Teixeira MC, Cabrito TR, Mira NP (2009) Drug: H+ antiporters in chemical stress response in yeast. Trends Microbiol 17:22–31

Saini P, Prasad T, Gaur NA, Shukla S, Jha S, Komath SS, Khan LA, Haq QM, Prasad R (2005) Alanine scanning of transmembrane helix 11 of Cdr1p ABC antifungal efflux pump of *Candida albicans*: identification of amino acid residues critical for drug efflux. J Antimicrob Chemother 56:77–86

Sanglard D, Kuchler K, Ischer F, Pagani JL, Monod M, Bille J (1995) Mechanisms of resistance to azole antifungal agents in *Candida albicans* isolates from AIDS patients involve specific multidrug transporters. Antimicrob Agents Chemother 39:2378–2386

Sauna ZE, Bohn SS, Rutledge R, Dougherty MP, Cronin S, May L, Xia D, Ambudkar SV, Golin J (2008) Mutations define cross-talk between the N-terminal nucleotide-binding domain and transmembrane helix-2 of the yeast multidrug transporter Pdr5: possible conservation of a signaling interface for coupling ATP hydrolysis to drug transport. J Biol Chem 283:35010–35022

Sauna ZE, Kim IW, Nandigama K, Kopp S, Chiba P, Ambudkar SV (2007) Catalytic cycle of ATP hydrolysis by P-glycoprotein: evidence for formation of the E. S reaction intermediate with ATP-gamma-S, a nonhydrolyzable analogue of ATP. Biochemistry 46:13787–13799

Seelig A, Landwojtowicz E (2000) Structure-activity relationship of P-glycoprotein substrates and modifiers. Eur J Pharm Sci 12:31–40

Shah AH, Rawal MK, Dhamgaye S, Komath SS, Saxena AK, Prasad R (2015) Mutational analysis of intracellular loops identify cross talk with nucleotide binding domains of yeast ABC transporter Cdr1p. Sci Rep 5:11211

Shukla S, Ambudkar SV, Prasad R (2004) Substitution of threonine-1351 in the multidrug transporter Cdr1p of *Candida albicans* results in hypersusceptibility to antifungal agents and threonine-1351 is essential for synergic effects of calcineurin inhibitor FK520. J Antimicrob Chemother 54:38–45

Shukla S, Saini P, Smriti Jha S, Ambudkar SV, Prasad R (2003) Functional characterization of *Candida albicans* ABC transporter Cdr1p. Eukaryot Cell 2:1361–1375

Silva LV, Sanguinetti M, Vandeputte P, Torelli R, Rochat B, Sanglard D (2013) Milbemycins: more than efflux inhibitors for fungal pathogens. Antimicrob Agents Chemother 57:873–886

Solcan N, Kwok J, Fowler PW, Cameron AD, Drew D, Iwata S, Newstead S (2012) Alternating access mechanism in the POT family of oligopeptide transporters. EMBO J 31:3411–3421

Tanabe K, Lamping E, Nagi M, Okawada A, Holmes AR, Miyazaki Y, Cannon RD, Monk BC, Niimi M (2011) Chimeras of *Candida albicans* Cdr1p and Cdr2p reveal features of pleiotropic drug resistance transporter structure and function. Mol Microbiol 82:416–433

Tsao S, Rahkhoodaee F, Raymond M (2009) Relative contributions of the *Candida albicans* ABC transporters Cdr1p and Cdr2p to clinical azole resistance. Antimicrob Agents Chemother 53:1344–1352

Tutulan-Cunita AC, Mikoshi M, Mizunuma M, Hirata D, Miyakawa T (2005) Mutational analysis of the yeast multidrug resistance ABC transporter Pdr5p with altered drug specificity. Genes to cells: devoted to molecular & cellular mechanisms 10:409–420

Velamakanni S, Yao Y, Gutmann DA, van Veen HW (2008) Multidrug transport by the ABC transporter Sav 1866 from *Staphylococcus aureus*. Biochemistry 47:9300–9308

Vinothkumar KR (2015) Membrane protein structures without crystals, by single particle electron cryomicroscopy. Curr Opin Struct Biol 33:103–114

Walker JE, Saraste M, Runswick MJ, Gay NJ (1982) Distantly related sequences in the alpha- and beta-subunits of ATP synthase, myosin, kinases and other ATP-requiring enzymes and a common nucleotide binding fold. EMBO J 1:945–951

Ward A, Reyes CL, Yu J, Roth CB, Chang G (2007) Flexibility in the ABC transporter MsbA: alternating access with a twist. Proc Natl Acad Sci U S A 104:19005–19010

Ward AB, Szewczyk P, Grimard V, Lee CW, Martinez L, Doshi R, Caya A, Villaluz M, Pardon E, Cregger C, Swartz DJ, Falson PG, Urbatsch IL, Govaerts C, Steyaert J, Chang G (2013) Structures of P-glycoprotein reveal its conformational flexibility and an epitope on the nucleotide-binding domain. Proc Natl Acad Sci U S A 110:13386–13391

White TC (1997) Increased mRNA levels of *ERG16*, *CDR*, and *MDR1* correlate with increases in azole resistance in *Candida albicans* isolates from a patient infected with human immunodeficiency virus. Antimicrob Agents Chemother 41:1482–1487

White TC, Holleman S, Dy F, Mirels LF, Stevens DA (2002) Resistance mechanisms in clinical isolates of *Candida albicans*. Antimicrob Agents Chemother 46:1704–1713

Wirsching S, Michel S, Morschhauser J (2000) Targeted gene disruption in *Candida albicans* wild-type strains: the role of the MDR1 gene in fluconazole resistance of clinical *Candida albicans* isolates. Mol Microbiol 36:856–865

Xu K, Zhang M, Zhao Q, Yu F, Guo H, Wang C, He F, Ding J, Zhang P (2013) Crystal structure of a folate energy-coupling factor transporter from *Lactobacillus brevis*. Nature 497:268–271

Yamada-Okabe T, Yamada-Okabe H (2002) Characterization of the *CaNAG3*, *CaNAG4*, and *CaNAG6* genes of the pathogenic fungus *Candida albicans*: possible involvement of these genes in the susceptibilities of cytotoxic agents. FEMS Microbiol Lett 212:15–21

Yan N (2013) Structural advances for the major facilitator superfamily (MFS) transporters. Trends Biochem Sci 38:151–159

Yan N (2015) Structural biology of the major facilitator superfamily transporters. Annu Rev Biophys 44:257–283

Yin Y, He X, Szewczyk P, Nguyen T, Chang G (2006) Structure of the multidrug transporter EmrD from *Escherichia coli*. Science 312:741–744

Znaidi S, Weber S, Al-Abdin OZ, Bomme P, Saidane S, Drouin S, Lemieux S, De Deken X, Robert F, Raymond M (2008) Genomewide location analysis of *Candida albicans* Upc2p, a regulator of sterol metabolism and azole drug resistance. Eukaryot Cell 7:836–847

Chapter 19
MDR1 and Its Regulation

Joachim Morschhäuser

Abstract *Candida albicans* infections are commonly treated with fluconazole, an antifungal drug that inhibits ergosterol biosynthesis. *C. albicans* can acquire resistance to fluconazole by various mechanisms, one of which is the constitutive overexpression of the *MDR1* gene, encoding a multidrug efflux pump of the major facilitator superfamily. While Mdr1 is known for its role in drug resistance, the physiological function of this transporter has not been elucidated. This chapter summarizes what is currently known about the transcriptional control of *MDR1* expression.

19.1 The Discovery of *MDR1* as a Multidrug Efflux Pump

In fungi, two types of efflux pumps that are localized in the cytoplasmic membrane and transport toxic substances out of the cell are known: The ABC transporters, which use ATP as the energy source to drive transport, and the major facilitators, which are energized by the proton gradient across the cell membrane (see Chap. 18). *MDR1* encodes a member of the latter class of transporters. It was identified in a screen for genes that, upon overexpression in the heterologous host *Saccharomyces cerevisiae*, conferred resistance to the antimitotic drug benomyl, to which *C. albicans* is highly resistant, and therefore originally termed *BEN*^{r} (Fling et al. 1991). Subsequent studies found that the gene also mediates resistance to various other, structurally unrelated toxic compounds (Ben-Yaacov et al. 1994). Disruption of the gene in a *C. albicans* strain in which it was strongly expressed resulted in increased susceptibility of the mutants to several of these inhibitors, but did not decrease benomyl resistance, and it was renamed *MDR1*, for multidrug resistance (Goldway et al. 1995). *MDR1* from *C. albicans* was the first multidrug efflux pump identified in a fungal pathogen.

J. Morschhäuser (✉)
Institut für Molekulare Infektionsbiologie, Universität Würzburg, Josef-Schneider-Str. 2, 97080 Würzburg, Germany
e-mail: joachim.morschhaeuser@uni-wuerzburg.de

R. Prasad (ed.), *Candida albicans: Cellular and Molecular Biology*,
DOI 10.1007/978-3-319-50409-4_19

19.2 The Role of *MDR1* in Fluconazole Resistance

The first indication that *MDR1* also mediates resistance to clinically relevant antifungal drugs was obtained by Sanglard and coworkers. They found that a fluconazole-resistant *C. albicans* isolate that was recovered from an AIDS patient with oropharyngeal candidiasis after prolonged therapy with this drug strongly overexpressed *MDR1*, whereas *MDR1* expression was undetectable in previous, more susceptible isolates from the same patient and in isolates from other patients (Sanglard et al. 1995). The same study also showed that *MDR1* overexpression in *S. cerevisiae* conferred increased fluconazole resistance. Similar investigations performed in different laboratories also observed *MDR1* upregulation in fluconazole-resistant clinical *C. albicans* isolates, indicating that overexpression of this efflux pump is an important mechanism of fluconazole resistance, besides other resistance mechanisms such as overexpression of the ABC transporters *CDR1* and *CDR2* or mutations in the drug target (White 1997; Franz et al. 1998, 1999; Lopez-Ribot et al. 1998; Perea et al. 2001; Martinez et al. 2002; Saidane et al. 2006). *MDR1* overexpression correlated with reduced intracellular fluconazole accumulation, which was reversed upon treatment of the cells with the respiratory inhibitor sodium azide, indicating that Mdr1 is an energy-dependent drug efflux pump (Sanglard et al. 1995; Albertson et al. 1996). Mdr1 has a narrower substrate spectrum than the ABC transporters Cdr1 and Cdr2, and among clinically useful antifungal agents, *MDR1* overexpression causes resistance only to fluconazole and voriconazole, but not other azoles or different antimycotic drugs (Sanglard et al. 1995; Niimi et al. 2006; Wakiec et al. 2007).

Initial attempts to genetically prove that *MDR1* mediates fluconazole resistance in *C. albicans* failed, because deletion of *MDR1* in *C. albicans* laboratory strains did not result in hypersusceptibility of the mutants to the drug (Sanglard et al. 1996; Morschhäuser et al. 1999). This is explained by the fact that the parental strains used in these studies exhibited negligible *MDR1* expression levels. The development of a recyclable, dominant selection marker enabled the generation of *mdr1*Δ mutants from *MDR1*-overexpressing, fluconazole-resistant clinical isolates of both *C. albicans* and *C. dubliniensis* and confirmed that *MDR1* upregulation indeed contributed to the increased drug resistance of these strains (Wirsching et al. 2000b, 2001).

19.3 Regulation of *MDR1* Expression

In fluconazole-susceptible *C. albicans* isolates, *MDR1* is expressed at very low levels, such that *MDR1* mRNA is not or barely detectable by Northern hybridization, and no upregulation of *MDR1* expression in the presence of fluconazole was observed in such studies (Albertson et al. 1996; Franz et al. 1998; Gupta et al. 1998). Similar results were also obtained in DNA microarray analyses or when *GFP* or *RLUC* were used as reporters of *MDR1* expression (Wirsching et al. 2000a;

Harry et al. 2005; Sellam et al. 2009b; Schneider and Morschhäuser 2015). A more sensitive detection of *MDR1* transcript levels by RT-qPCR showed that the very low basal *MDR1* expression levels in drug-susceptible strains are indeed elevated in the presence of fluconazole (Sellam et al. 2009a; Vasicek et al. 2014). However, the fluconazole-induced *MDR1* upregulation is insufficient to confer increased drug resistance, because deletion of *MDR1* in fluconazole-susceptible *C. albicans* strains does not result in hypersusceptibility to the drug (Sanglard et al. 1996; Morschhäuser et al. 1999; Schubert et al. 2011a). Some other toxic compounds induce *MDR1* expression much more strongly (Gupta et al. 1998; Harry et al. 2005). Especially, benomyl and H_2O_2 have been frequently used to study the involvement of *cis*-regulatory sequences and *trans*-acting factors in the transcriptional control of *MDR1* expression (Karababa et al. 2004; Harry et al. 2005; Hiller et al. 2006a, b; Rognon et al. 2006; Morschhäuser et al. 2007; Schubert et al. 2008, 2011a, b; Mogavero et al. 2011; Sasse et al. 2011, 2012; Ramírez-Zavala et al. 2014; Schneider and Morschhäuser 2015).

19.4 Transcription Factors Controlling *MDR1* Expression

Several transcription factors are involved in the regulation of *MDR1* expression. The basic region leucine zipper (bZip) transcription factor Cap1 was identified by its ability to confer fluconazole resistance when *CAP1* was overexpressed in *S. cerevisiae* (Alarco et al. 1997). This phenotype was mediated by the upregulation of *FLR1*, the closest *MDR1* homolog in *S. cerevisiae*. Production of a C-terminally truncated, hyperactive form of Cap1 in *C. albicans* caused *MDR1* upregulation and increased resistance to fluconazole and other Mdr1 substrates (Alarco and Raymond 1999; Schubert et al. 2011a). Like its counterpart Yap1 in *S. cerevisiae*, Cap1 is involved in the response to oxidative stress in *C. albicans* (Alarco and Raymond 1999; Zhang et al. 2000). In the presence of oxidants, the formation of disulfide bonds within the C-terminal domain of Yap1/Cap1 causes conformational changes that prevent its interaction with the nuclear exportin Crm1, resulting in nuclear retention and transcriptional activation (Moye-Rowley 2003). Deletion of *CAP1* abolished the inducibility of *MDR1* expression by H_2O_2 and reduced its induction by benomyl, which also causes oxidative stress (Rognon et al. 2006; Znaidi et al. 2009; Schubert et al. 2011a). Cap1 binds to the *MDR1* promoter in vivo, where it recruits the Ada2 subunit of the SAGA coactivator complex to stimulate transcription in response to oxidative stress (Sellam et al. 2009a; Znaidi et al. 2009; Ramírez-Zavala et al. 2014). Cap1 is therefore required to promote *MDR1* upregulation by certain environmental stimuli. However, Cap1 as well as Ada2 are dispensable for the constitutive *MDR1* overexpression in fluconazole-resistant *C. albicans* strains (Alarco and Raymond 1999; Rognon et al. 2006; Schubert et al. 2011a; Ramírez-Zavala et al. 2014).

The zinc cluster transcription factor Mrr1 is a central regulator of *MDR1* expression. *MRR1* was identified as a gene that was coordinately upregulated with

MDR1 in fluconazole-resistant *C. albicans* isolates. Deletion of *MRR1* in these isolates abolished *MDR1* overexpression and drug resistance, and *MRR1* was also essential for the upregulation of *MDR1* by the inducers benomyl and H_2O_2 (Morschhäuser et al. 2007). All *MDR1*-overexpressing, fluconazole-resistant *C. albicans* isolates that have been examined to date contain gain-of-function mutations in *MRR1* that result in constitutive activity of the transcription factor, and the same is true for *MDR1*-overexpressing strains of the related species *C. dubliniensis* and *C. parapsilosis* (Dunkel et al. 2008a; Schubert et al. 2008; Eddouzi et al. 2013; Lohberger et al. 2014; Branco et al. 2015). Mrr1 cooperates with Mcm1, a MADS box transcription factor that also binds to the *MDR1* promoter, to induce *MDR1* expression (Riggle and Kumamoto 2006; Lavoie et al. 2008; Mogavero et al. 2011). Mcm1 is essential for the constitutive *MDR1* overexpression in strains containing a gain-of-function mutation in Mrr1 and is also important for benomyl-induced *MDR1* upregulation, which depends on Mrr1. In contrast, Mcm1 is dispensable for the induction of *MDR1* expression by H_2O_2, which is mediated by Cap1 in cooperation with Mrr1 (Mogavero et al. 2011).

Another zinc cluster transcription factor, Upc2, also binds to the *MDR1* promoter and acts as either a positive or negative regulator of *MDR1* expression, depending on the conditions (Znaidi et al. 2008). Upc2 upregulates the expression of ergosterol biosynthesis (*ERG*) genes upon ergosterol depletion, for example in the presence of fluconazole, and gain-of function mutations in Upc2 cause constitutive *ERG* gene overexpression and increased fluconazole resistance in many clinical *C. albicans* isolates (Silver et al. 2004; MacPherson et al. 2005; Dunkel et al. 2008b; Heilmann et al. 2010; Hoot et al. 2011; Flowers et al. 2012). Hyperactive forms of Upc2 also cause a slight upregulation of *MDR1* expression, which, however, is not sufficient to contribute to fluconazole resistance (Dunkel et al. 2008b; Znaidi et al. 2008; Schubert et al. 2011a; Flowers et al. 2012). Interestingly, the moderate upregulation of *MDR1* in the presence of fluconazole mentioned above did not depend on Upc2 (Vasicek et al. 2014). However, when Upc2 was artificially targeted to the Mrr1 binding site in the *MDR1* promoter by replacing its DNA-binding domain with that of Mrr1, this resulted in high fluconazole-inducible *MDR1* expression levels and strongly increased fluconazole resistance (Schneider and Morschhäuser 2015). This observation indicates that if *C. albicans* acquired the ability to efficiently induce the expression of *MDR1* and other Mrr1 target genes by fluconazole, the fungus would become intrinsically resistant to the drug.

19.5 Other Potential Regulators of *MDR1* Expression

Rep1 is a protein of unknown function that contains a putative DNA-binding domain and has been reported to be a negative regulator of *MDR1* expression. Deletion of *REP1* resulted in increased *MDR1* expression, especially in the presence of the inducing compound 4–nitroquinoline *N*-oxide (4–NQO), and slightly increased fluconazole resistance (Chen et al. 2009). Cph1, which regulates the response of *C. albicans* to mating pheromone and is also involved in the induction of filamentous growth under some conditions, has been proposed to be a negative regulator of *MDR1* expression (Lo et al. 2015). Deletion of *CPH1* resulted in increased 4–NQO-induced *MDR1* transcript levels, but did not derepress *MDR1* expression in the absence of an inducer. Cph1 is not known to bind to the *MDR1* promoter, and it remains possible that *cph1*Δ mutants are simply hypersensitive to 4–NQO, resulting in a more efficient *MDR1* induction by the compound. *MDR1* promoter deletion analyses did not identify *cis*-regulatory sequences whose absence would result in constitutive activity (Harry et al. 2005; Hiller et al. 2006b; Rognon et al. 2006). Therefore, although some negative regulatory factors may limit the inducibility of *MDR1* expression, relief from repression is not sufficient to result in constitutive *MDR1* upregulation, and efficient *MDR1* induction depends on positive regulators such as Mrr1 and Cap1.

19.6 Conclusion

MDR1 overexpression is known as one of several mechanisms of fluconazole resistance in *C. albicans*. Up to now, *MDR1* overexpression in fluconazole-resistant clinical isolates has been invariably linked to gain-of-function mutations in the transcription factor Mrr1. Of note, Mdr1 contributes only partially to the increase in fluconazole resistance caused by a hyperactive Mrr1, because Mrr1 can mediate resistance even in the absence of the efflux pump, albeit at a reduced level (Schubert et al. 2011a). Although the target genes of Mrr1 have been identified (Morschhäuser et al. 2007; Schubert et al. 2011a), deletion studies so far failed to uncover which of the other genes that are upregulated by hyperactive forms of the transcription factor contribute to drug resistance (Kusch et al. 2004; Schillig and Morschhäuser 2013). The physiological function of Mdr1 also is still unknown, and apart from Mrr1, other transcription factors such as Cap1 and Upc2 may have an important role in the appropriate regulation of *MDR1* expression in response to intracellular and environmental signals.

Acknowledgements I thank Christina Popp for critical reading of the manuscript. Work in my laboratory is supported by the Deutsche Forschungsgemeinschaft (DFG grants MO 846/6, MO 846/7, and Transregio 124).

References

Alarco AM, Raymond M (1999) The bZip transcription factor Cap1p is involved in multidrug resistance and oxidative stress response in *Candida albicans*. J Bacteriol 181:700–708

Alarco AM, Balan I, Talibi D, Mainville N, Raymond M (1997) AP1-mediated multidrug resistance in *Saccharomyces cerevisiae* requires *FLR1* encoding a transporter of the major facilitator superfamily. J Biol Chem 272:19304–19313

Albertson GD, Niimi M, Cannon RD, Jenkinson HF (1996) Multiple efflux mechanisms are involved in *Candida albicans* fluconazole resistance. Antimicrob Agents Chemother 40:2835–2841

Ben-Yaacov R, Knoller S, Caldwell GA, Becker JM, Koltin Y (1994) *Candida albicans* gene encoding resistance to benomyl and methotrexate is a multidrug resistance gene. Antimicrob Agents Chemother 38:648–652

Branco J, Silva AP, Silva RM, Silva-Dias A, Pina-Vaz C, Butler G, Rodrigues AG, Miranda IM (2015) Fluconazole and voriconazole resistance in *Candida parapsilosis* is conferred by gain-of-function mutations in *MRR1* transcription factor gene. Antimicrob Agents Chemother 59:6629–6633

Chen CG, Yang YL, Tseng KY, Shih HI, Liou CH, Lin CC, Lo HJ (2009) Rep1p negatively regulating *MDR1* efflux pump involved in drug resistance in *Candida albicans*. Fungal Genet Biol 46:714–720

Dunkel N, Blaß J, Rogers PD, Morschhäuser J (2008a) Mutations in the multi-drug resistance regulator *MRR1*, followed by loss of heterozygosity, are the main cause of *MDR1* overexpression in fluconazole-resistant *Candida albicans* strains. Mol Microbiol 69:827–840

Dunkel N, Liu TT, Barker KS, Homayouni R, Morschhäuser J, Rogers PD (2008b) A gain-of-function mutation in the transcription factor Upc2p causes upregulation of ergosterol biosynthesis genes and increased fluconazole resistance in a clinical *Candida albicans* isolate. Eukaryot Cell 7:1180–1190

Eddouzi J, Parker JE, Vale-Silva LA, Coste A, Ischer F, Kelly S, Manai M, Sanglard D (2013) Molecular mechanisms of drug resistance in clinical *Candida* species isolated from Tunisian hospitals. Antimicrob Agents Chemother 57:3182–3193

Fling ME, Kopf J, Tamarkin A, Gorman JA, Smith HA, Koltin Y (1991) Analysis of a *Candida albicans* gene that encodes a novel mechanism for resistance to benomyl and methotrexate. Mol Gen Genet 227:318–329

Flowers SA, Barker KS, Berkow EL, Toner G, Chadwick SG, Gygax SE, Morschhäuser J, Rogers PD (2012) Gain-of-function mutations in *UPC2* are a frequent cause of *ERG11* upregulation in azole-resistant clinical isolates of *Candida albicans*. Eukaryot Cell 11:1289–1299

Franz R, Kelly SL, Lamb DC, Kelly DE, Ruhnke M, Morschhäuser J (1998) Multiple molecular mechanisms contribute to a stepwise development of fluconazole resistance in clinical *Candida albicans* strains. Antimicrob Agents Chemother 42:3065–3072

Franz R, Ruhnke M, Morschhäuser J (1999) Molecular aspects of fluconazole resistance development in *Candida albicans*. Mycoses 42:453–458

Goldway M, Teff D, Schmidt R, Oppenheim AB, Koltin Y (1995) Multidrug resistance in *Candida albicans*: disruption of the *BEN*^{r} gene. Antimicrob Agents Chemother 39:422–426

Gupta V, Kohli A, Krishnamurthy S, Puri N, Aalamgeer SA, Panwar S, Prasad R (1998) Identification of polymorphic mutant alleles of *CaMDR1*, a major facilitator of *Candida albicans* which confers multidrug resistance, and its in vitro transcriptional activation. Curr Genet 34:192–199

Harry JB, Oliver BG, Song JL, Silver PM, Little JT, Choiniere J, White TC (2005) Drug-induced regulation of the *MDR1* promoter in *Candida albicans*. Antimicrob Agents Chemother 49:2785–2792

Heilmann CJ, Schneider S, Barker KS, Rogers PD, Morschhäuser J (2010) An A643T mutation in the transcription factor Upc2p causes constitutive *ERG11* upregulation and increased fluconazole resistance in *Candida albicans*. Antimicrob Agents Chemother 54:353–359

Hiller D, Sanglard D, Morschhäuser J (2006a) Overexpression of the *MDR1* gene is sufficient to confer increased resistance to toxic compounds in *Candida albicans*. Antimicrob Agents Chemother 50:1365–1371

Hiller D, Stahl S, Morschhäuser J (2006b) Multiple *cis*-acting sequences mediate upregulation of the *MDR1* efflux pump in a fluconazole-resistant clinical *Candida albicans* isolate. Antimicrob Agents Chemother 50:2300–2308

Hoot SJ, Smith AR, Brown RP, White TC (2011) An A643 V amino acid substitution in Upc2p contributes to azole resistance in well-characterized clinical isolates of *Candida albicans*. Antimicrob Agents Chemother 55:940–942

Karababa M, Coste AT, Rognon B, Bille J, Sanglard D (2004) Comparison of gene expression profiles of *Candida albicans* azole-resistant clinical isolates and laboratory strains exposed to drugs inducing multidrug transporters. Antimicrob Agents Chemother 48:3064–3079

Kusch H, Biswas K, Schwanfelder S, Engelmann S, Rogers PD, Hecker M, Morschhäuser J (2004) A proteomic approach to understanding the development of multidrug-resistant *Candida albicans* strains. Mol Genet Genomics: MGG 271:554–565

Lavoie H, Sellam A, Askew C, Nantel A, Whiteway M (2008) A toolbox for epitope-tagging and genome-wide location analysis in Candida albicans. BMC Genom 9:578

Lo HJ, Tseng KY, Kao YY, Tsao MY, Lo HL, Yang YL (2015) Cph1p negatively regulates *MDR1* involved in drug resistance in *Candida albicans*. Int J Antimicrob Agents 45:617–621

Lohberger A, Coste AT, Sanglard D (2014) Distinct roles of *Candida albicans* drug resistance transcription factors *TAC1*, *MRR1*, and *UPC2* in virulence. Eukaryot Cell 13:127–142

Lopez-Ribot JL, McAtee RK, Lee LN, Kirkpatrick WR, White TC, Sanglard D, Patterson TF (1998) Distinct patterns of gene expression associated with development of fluconazole resistance in serial *Candida albicans* isolates from human immunodeficiency virus-infected patients with oropharyngeal candidiasis. Antimicrob Agents Chemother 42:2932–2937

MacPherson S, Akache B, Weber S, De Deken X, Raymond M, Turcotte B (2005) *Candida albicans* zinc cluster protein Upc2p confers resistance to antifungal drugs and is an activator of ergosterol biosynthetic genes. Antimicrob Agents Chemother 49:1745–1752

Martinez M, Lopez-Ribot JL, Kirkpatrick WR, Bachmann SP, Perea S, Ruesga MT, Patterson TF (2002) Heterogeneous mechanisms of azole resistance in *Candida albicans* clinical isolates from an HIV-infected patient on continuous fluconazole therapy for oropharyngeal candidosis. J Antimicrobial Chemother 49:515–524

Mogavero S, Tavanti A, Senesi S, Rogers PD, Morschhäuser J (2011) Differential requirement of the transcription factor Mcm1 for activation of the *Candida albicans* multidrug efflux pump *MDR1* by its regulators Mrr1 and Cap1. Antimicrob Agents Chemother 55:2061–2066

Morschhäuser J, Michel S, Staib P (1999) Sequential gene disruption in *Candida albicans* by FLP-mediated site-specific recombination. Mol Microbiol 32:547–556

Morschhäuser J, Barker KS, Liu TT, Blaß-Warmuth J, Homayouni R, Rogers PD (2007) The transcription factor Mrr1p controls expression of the *MDR1* efflux pump and mediates multidrug resistance in *Candida albicans*. PLoS Pathog 3:e164

Moye-Rowley WS (2003) Regulation of the transcriptional response to oxidative stress in fungi: similarities and differences. Eukaryot Cell 2:381–389

Niimi K, Maki K, Ikeda F, Holmes AR, Lamping E, Niimi M, Monk BC, Cannon RD (2006) Overexpression of *Candida albicans CDR1*, *CDR2*, or *MDR1* does not produce significant changes in echinocandin susceptibility. Antimicrob Agents Chemother 50:1148–1155

Perea S, Lopez-Ribot JL, Kirkpatrick WR, McAtee RK, Santillan RA, Martinez M, Calabrese D, Sanglard D, Patterson TF (2001) Prevalence of molecular mechanisms of resistance to azole antifungal agents in *Candida albicans* strains displaying high-level fluconazole resistance isolated from human immunodeficiency virus-infected patients. Antimicrob Agents Chemother 45:2676–2684

Ramírez-Zavala B, Mogavero S, Schöller E, Sasse C, Rogers PD, Morschhäuser J (2014) SAGA/ADA complex subunit Ada2 is required for Cap1- but not Mrr1-mediated upregulation of the *Candida albicans* multidrug efflux pump *MDR1*. Antimicrob Agents Chemother 58:5102–5110

Riggle PJ, Kumamoto CA (2006) Transcriptional regulation of *MDR1*, encoding a drug efflux determinant, in fluconazole-resistant *Candida albicans* strains through an Mcm1p binding site. Eukaryot Cell 5:1957–1968

Rognon B, Kozovska Z, Coste AT, Pardini G, Sanglard D (2006) Identification of promoter elements responsible for the regulation of *MDR1* from *Candida albicans*, a major facilitator transporter involved in azole resistance. Microbiology 152:3701–3722

Saidane S, Weber S, De Deken X, St-Germain G, Raymond M (2006) *PDR16*-mediated azole resistance in *Candida albicans*. Mol Microbiol 60:1546–1562

Sanglard D, Kuchler K, Ischer F, Pagani JL, Monod M, Bille J (1995) Mechanisms of resistance to azole antifungal agents in *Candida albicans* isolates from AIDS patients involve specific multidrug transporters. Antimicrob Agents Chemother 39:2378–2386

Sanglard D, Ischer F, Monod M, Bille J (1996) Susceptibilities of *Candida albicans* multidrug transporter mutants to various antifungal agents and other metabolic inhibitors. Antimicrob Agents Chemother 40:2300–2305

Sasse C, Schillig R, Dierolf F, Weyler M, Schneider S, Mogavero S, Rogers PD, Morschhäuser J (2011) The transcription factor Ndt80 does not contribute to Mrr1-, Tac1-, and Upc2-mediated fluconazole resistance in *Candida albicans*. PLoS ONE 6:e25623

Sasse C, Schillig R, Reimund A, Merk J, Morschhäuser J (2012) Inducible and constitutive activation of two polymorphic promoter alleles of the *Candida albicans* multidrug efflux pump *MDR1*. Antimicrob Agents Chemother 56:4490–4494

Schillig R, Morschhäuser J (2013) Analysis of a fungus-specific transcription factor family, the *Candida albicans* zinc cluster proteins, by artificial activation. Mol Microbiol 89:1003–1017

Schneider S, Morschhäuser J (2015) Induction of *Candida albicans* drug resistance genes by hybrid zinc cluster transcription factors. Antimicrob Agents Chemother 59:558–569

Schubert S, Rogers PD, Morschhäuser J (2008) Gain-of-function mutations in the transcription factor *MRR1* are responsible for overexpression of the *MDR1* efflux pump in fluconazole-resistant *Candida dubliniensis* strains. Antimicrob Agents Ch 52:4274–4280

Schubert S, Barker KS, Znaidi S, Schneider S, Dierolf F, Dunkel N, Aid M, Boucher G, Rogers PD, Raymond M, Morschhäuser J (2011a) Regulation of efflux pump expression and drug resistance by the transcription factors Mrr1, Upc2, and Cap1 in *Candida albicans*. Antimicrob Agents Chemother 55:2212–2223

Schubert S, Popp C, Rogers PD, Morschhäuser J (2011b) Functional dissection of a *Candida albicans* zinc cluster transcription factor, the multidrug resistance regulator Mrr1. Eukaryot Cell 10:1110–1121

Sellam A, Askew C, Epp E, Lavoie H, Whiteway M, Nantel A (2009a) Genome-wide mapping of the coactivator Ada2p yields insight into the functional roles of SAGA/ADA complex in *Candida albicans*. Mol Biol Cell 20:2389–2400

Sellam A, Tebbji F, Nantel A (2009b) Role of Ndt80p in sterol metabolism regulation and azole resistance in *Candida albicans*. Eukaryot Cell 8:1174–1183

Silver PM, Oliver BG, White TC (2004) Role of *Candida albicans* transcription factor Upc2p in drug resistance and sterol metabolism. Eukaryot Cell 3:1391–1397

Vasicek EM, Berkow EL, Flowers SA, Barker KS, Rogers PD (2014) UPC2 is universally essential for azole antifungal resistance in *Candida albicans*. Eukaryot Cell 13:933–946

Wakiec R, Prasad R, Morschhäuser J, Barchiesi F, Borowski E, Milewski S (2007) Voriconazole and multidrug resistance in *Candida albicans*. Mycoses 50:109–115

White TC (1997) Increased mRNA levels of *ERG16*, *CDR*, and *MDR1* correlate with increases in azole resistance in *Candida albicans* isolates from a patient infected with human immunodeficiency virus. Antimicrob Agents Chemother 41:1482–1487

Wirsching S, Michel S, Köhler G, Morschhäuser J (2000a) Activation of the multiple drug resistance gene *MDR1* in fluconazole-resistant, clinical *Candida albicans* strains is caused by mutations in a trans-regulatory factor. J Bacteriol 182:400–404

Wirsching S, Michel S, Morschhäuser J (2000b) Targeted gene disruption in *Candida albicans* wild-type strains: the role of the *MDR1* gene in fluconazole resistance of clinical *Candida albicans* isolates. Mol Microbiol 36:856–865

Wirsching S, Moran GP, Sullivan DJ, Coleman DC, Morschhäuser J (2001) *MDR1*-mediated drug resistance in *Candida dubliniensis*. Antimicrob Agents Chemother 45:3416–3421

Zhang X, De Micheli M, Coleman ST, Sanglard D, Moye-Rowley WS (2000) Analysis of the oxidative stress regulation of the *Candida albicans* transcription factor, Cap1p. Mol Microbiol 36:618–629

Znaidi S, Weber S, Al-Abdin OZ, Bomme P, Saidane S, Drouin S, Lemieux S, De Deken X, Robert F, Raymond M (2008) Genomewide location analysis of *Candida albicans* Upc2p, a regulator of sterol metabolism and azole drug resistance. Eukaryot Cell 7:836–847

Znaidi S, Barker KS, Weber S, Alarco AM, Liu TT, Boucher G, Rogers PD, Raymond M (2009) Identification of the *Candida albicans* Cap1p regulon. Eukaryot Cell 8:806–820

Chapter 20
Insights into *Candida* Lipids

Rajendra Prasad, Sudhanshu Shukla and Ashutosh Singh

Abstract Although lipid metabolic pathways are fairly well established in yeast, our knowledge of lipid compositional profile, particularly in pathogenic species, is rather limited. Fungal lipids are important on two accounts; first, they possess lipids, particularly sphingolipids, which are unique to *Candida* species and are absent in mammalian host hence are novel drug targets. Second, the functionality of some of the multidrug resistance (MDR) export proteins is dependent upon optimal lipid environment implying their role in clinical drug resistance. The comprehensive high-throughput lipidomics combined with genetic approaches applied to human pathogenic diploid *C. albicans* has started providing insight into mysteries surrounded around this important class of biomolecules. Recent studies already revealed functional interactions between lipids, virulence, and MDR determinants in *Candida*. This chapter reviews some of the recent advances in the field and highlights the role of lipids involved in cross talks between different cellular circuits that impact the acquisition of MDR in *Candida*.

20.1 Introduction

The complexity and dynamic nature of the "lipidome" of eukaryotic cells are increasingly apparent. Taking into account the diversity within many lipid classes created by variability in such factors as acyl chain substitutions, degree of desaturation, hydroxylation, and phosphorylation, the number of molecular species in

R. Prasad (✉)
Amity Institute of Integrative Sciences and Health, Amity University, Haryana, India
e-mail: rp47jnu@gmail.com; rprasad@ggn.amity.edu

R. Prasad · S. Shukla
Amity Institute of Biotechnology, Amity University, Haryana, India
e-mail: sshukla99@gmail.com

A. Singh
Department of Biochemistry, Lucknow University, Lucknow, India
e-mail: ashutosh.singh29@gmail.com

R. Prasad (ed.), *Candida albicans: Cellular and Molecular Biology*,
DOI 10.1007/978-3-319-50409-4_20

eukaryotic cells exceeds over thousand. Such estimates of the diversity of the lipidome do not even account for the various water-soluble turnover products of lipid metabolism, many of which are vital signaling molecules themselves (Wenk 2005). Superimposed with this diversity in lipids, there are host of specialized lipids, which serve to modify proteins and to target and anchor them to specific membranes (van Meer et al. 2008).

Lipids primarily drive their importance as being integral part of membrane structure where lipid bilayer provides matrix of hydrophobic environment for the embedded proteins to perform dedicated functions. Interestingly, while it would have been feasible to form a lipid membrane that could act as a physical barrier from a single lipid component, the cell invests significant resources in generating a zoo of lipids for the optimal performance of membrane proteins (Shevchenko and Simons 2010; Wenk 2010). The predominant eukaryotic membrane lipids including those of yeasts are glycerophospholipids, sphingolipids (SLs), and sterols. The head group of glycerophospholipids (GPLs) can vary, as can the bonds linking the hydrocarbon chains to glycerol, as can the fatty acids (FAs), which differ in length and degree of saturation. Also, the SLs have the combinatorial propensity to create diversity by different ceramide backbones and, above all, more than 500 different carbohydrate structures, which make up the head groups of the glycosphingolipids (Futerman and Hannun 2004). While mammalian cell membranes predominantly contain cholesterol and ergosterol for yeast membranes, respectively, hundreds of different lipid species of GPLs and SLs classes provide diversity to their composition. The diversity in steroids is generated by the addition of different functional groups to the steroid nucleus. Major lipid structures of fungi are depicted in Fig. 20.1.

20.2 *Candida* Lipids

Lipids play a wide variety of roles in pathogenic fungi and are also considered as potential drug targets. Compared to other yeasts where lipid metabolic pathways are fairly well established (Gasper et al. 2007), the lipid compositional profile, particularly in *Candida* species, are not well understood (Mahmoudabadi et al. 2001; van Meer et al. 2008). Nonetheless, conventional methods used thus far have resulted in identification of major classes of lipids of *Candida*. Expectedly, *Candida* has been shown to possess a mixture of typical eukaryotic cell lipids (Mahmoudabadi et al. 2001). The recent introduction of high-throughput analytical tools has accelerated our ability to analyze and quantify not only most of the lipids but could also quantitate the molecular species which arise due to variation in fatty acid composition, their position of esterification, and degree of unsaturation (Ejsing et al. 2009; Guan and Wenk 2006). Thus, in *C. albicans*, nine major GPL classes namely PC, PE, phosphatidyl serine (PS), phosphatidyl glycerol (PG), phosphatidyl

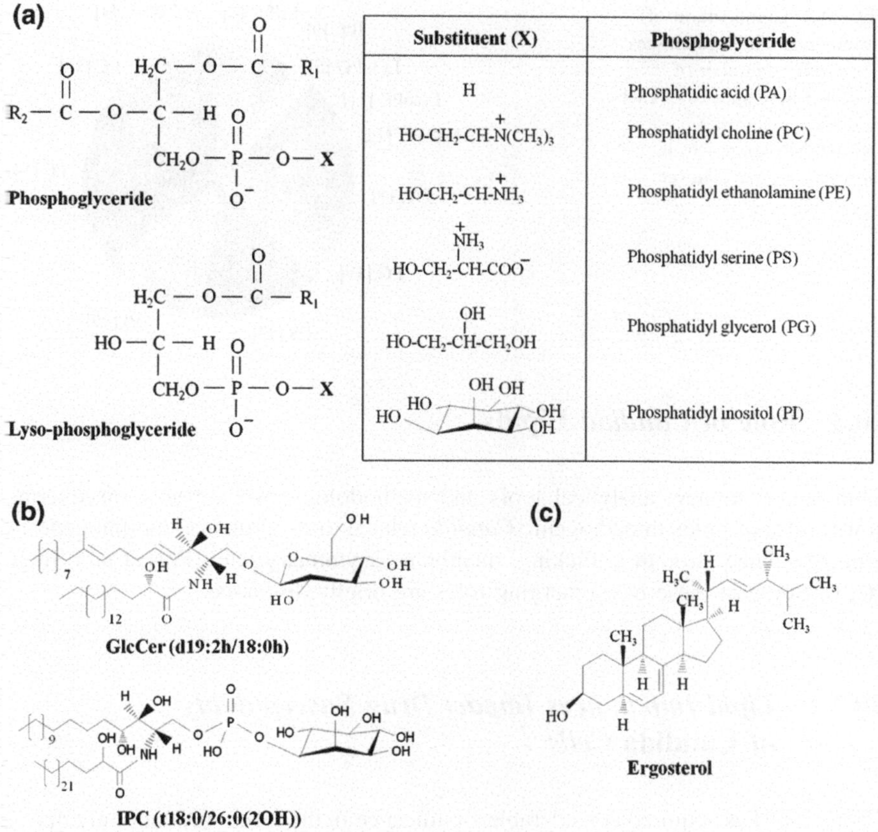

Fig. 20.1 Structure of major lipids in fungi. **a** Glycerophospholipid structure. A glycerol backbone is ester linked to FA at the sn-1 (R1) and sn-2 (R2) with a polar head group at the sn-3 position. Subclasses are based on the polar head group (X). **b** Structure of 2 key sphingolipids in fungi namely d19:2/18:0h glucosylceramide and t18:0/26:0 inositolphosphorylceramide. These lipid species are known to play a key role in survival of fungi at alkaline and acidic pH. **c** Structure of ergosterol

inositol (PI), phosphatidic acid (PA), lysoPG, lysoPE, lysoPC (depicted in Fig. 20.2); four major groups of SL, ceramides, inositolphosphorylceramide (IPC), mannosylinositolphosphorylceramide (MIPC), and $M(IP)_2C$; neutral lipids like sterols, diacylglycerol (DAG) and triacylglycerol (TAG), have been identified (Singh et al. 2010).

In contrast to GPLs where no major differences among *Candida* species exist, SL class is different showing variation in composition as well as in molecular species. For instance, the abundance of many SL species is distinct in *C. albicans* compared to non-albicans strain like *C. glabrata* or nonpathogenic yeast *S. cerevisеae* (Ejsing et al. 2009; Singh et al. 2010). However, the physiological consequences of these differences are not evident (Singh et al. 2010).

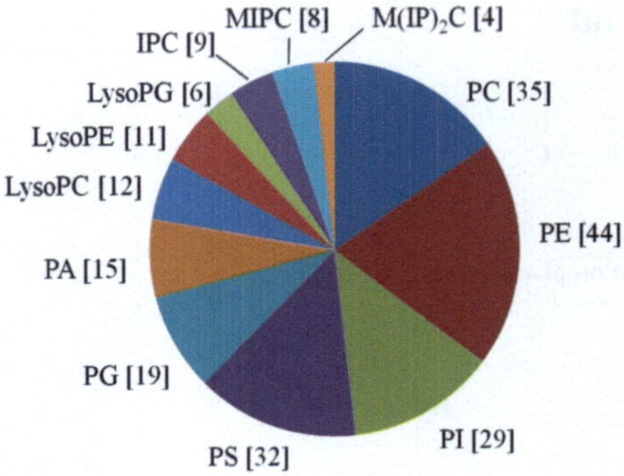

Fig. 20.2 Composition of various phospholipid classes in *Candida*. Number of identified lipid species of each lipid class, using the ESI-MS/MS approach, is shown in *square bracket*

20.3 Role of *Candida* Lipids

With access to new analytical tools and methodology, we are now discovering newer roles of lipids in pathogenic *Candida* related to its virulence, morphogenesis, drug resistance, protein trafficking, membrane asymmetry, etc. (Prasad and Singh 2013). Some of these new emerging roles are briefly discussed.

20.3.1 Lipid Imbalances Impact Drug Susceptibility of Candida *Cells*

C. albicans has acquired considerable significance in the recent past not only due to the severity of their infections but also for their ability to develop tolerance against antifungal drugs particularly those target lipids or its metabolism (Cannon et al. 1998; Maesaki et al. 1999). The most commonly encountered clinical drug resistance is against frequently used azoles, which target ergosterol metabolism. Not surprisingly, the most predominate mechanisms of azole resistance include point mutations or overexpression of the drug target. The decrease in the import of drugs and an increased efflux of drugs are also important strategies mastered commonly in clinical multidrug resistance (MDR) by the resistant organisms. The development of drug tolerance is gradual which can be the result of a single or combination of several mechanisms, however, the efflux protein mediated extrusion of noxious compounds from the cell is one of the most frequently used strategies for resistance to cytotoxic drugs in *Candida* cells. For this, *Candida* cells are geared up with two kinds of major drug exporters belonging to ABC or MFS super-families (discussed in Chap. 18).

These membrane multidrug exporters are localized within plasma membrane of *Candida* cells embedded within the hydrophobic core of lipids. It has been observed that imbalances in membrane lipids impact drug susceptibilities. This was more evident by the observation that any disruption in ergosterol or SL biosynthetic pathways either by mutations or by the treatment of specific inhibitors, results in hypersusceptibility toward drugs, implying a close link between these lipids and MDR in *C. albicans* (Mukhopadhyay et al. 2004; Prasad et al. 2005; Pasrija et al. 2005a, b, 2008). What emerges from these observations point to a preferential localization of some of the major drug transporters within ergosterol and sphingolipid rich micro-domains in plasma membrane (discussed below).

20.3.2 ABC Multidrug Transporter and Microdomians of Membrane

In *Candida* cells, membrane rafts are highly enriched in SLs and ergosterol and are well characterized by their resistance to detergent solubility as detergent resistant fractions (DRM) providing hydrophobic surface to stabilize the interaction with the large hydrophobic proteins (Pasrija et al. 2005b, 2008; Simons and Sampaio 2011). The ABC and MFS multidrug efflux proteins show different propensities toward lipids. For instance, ABC transporter, Cdr1 protein is preferentially localized within the "lipid rafts" while no such preference to lipids or the domains was evident for Mdr1 protein belonging to MFS superfamily. Reinforcing these observations, it is elegantly demonstrated that the disruption of SL or ergosterol biosynthetic pathway led to massive protein trafficking defect resulting in mislocalization of Cdr1 protein, thereby affecting its functionality. In contrast, the localization and functionality of sister MFS transporter, Mdr1 protein remained properly localized within the plasma membrane and fully functional (Pasrija et al. 2008).

20.3.3 Lipidomics of Azole Susceptible (AS) and Resistant (AR) Clinical Isolates

The adaptation of *C. albicans* to tolerate antifungals is accompanied with many specific and global changes in lipid homeostasis. This was revealed from a recent study, where a detailed lipidomics of several genetically matched (isogenic) as well as select sequential AS and AR clinical isolates of *C. albicans,* were performed (Singh et al. 2011, 2012, 2013). These studies not only provided a comprehensive evaluation of lipids as the determinants of drug resistance but also showed that though each AR isolate possessed a characteristic lipid composition, there were consistent differences in several lipid classes. Development of azole tolerance also impelled the remodeling of molecular species of lipids.

Notably, the lipidomic response of matched pair isolates was associated with simultaneous overproduction of efflux pump membrane proteins pointing to a possible common regulatory mechanism between the two phenomena. Such a common link has already been observed in case of *S. cerevisiae* and *C. glabrata,* where genes encoding efflux pumps, such as ScPdr5 or CgCdr1 and CgCdr2, play an important role in regulating lipid levels (Shahi and Moye-Rowley 2009). There are not enough instances to confirm coregulation of lipids and MDR in *Candida*, although Psd1 has been found to be upregulated in some AR clinical isolates (Singh et al. 2012).

20.3.4 Cross talk Between Lipids, Mitochondria, Cell Wall (CW) Integrity and Virulence

A close link between lipid homeostasis, mitochondrial membrane dysfunction, and CW integrity is demonstrated by a recent, lipidomics study of clinical AR isolates of *C. albicans* (Singh et al. 2012). The study revealed gradual corrective changes in lipidome coinciding with the development of fluconazole tolerance. In that study, a decrease in mitochondrial membrane action potential could be correlated to the PG levels and CW integrity in several genetically matched clinical AR isolates of *C. albicans* (Singh et al. 2012). In *S. cereviseae*, the Pdr pathway has been shown as a link between the mitochondrial membrane structure and lipids, however, no such mechanism has been identified in *Candida* (Shingu-Vazquez and Traven 2011).

20.3.5 MDR Proteins as Lipid Translocators

Membranes of *Candida* display asymmetric distribution of phospholipids wherein the aminophosphoglycerides such as PE and PS are discreetly located on the inner leaflet of the lipid bilayer as compared to other lipids including PC which show no preferential distribution and are localized on both the outer and inner leaflets of bilayer (Smriti et al. 2002; Shukla et al. 2006). The asymmetrical distribution of membrane lipids is very specific, and its loss results in various physiological consequences in *Candida* (Smriti et al. 2002; Shukla et al. 2006). The lipid asymmetry is maintained by membrane bound phospholipid translocators and some of them include ABC drug extrusion pump proteins (Smriti et al. 2002; Shukla et al. 2006; Prasad et al. 2016). The nulls of multidrug ABC transporters, Cdr1 and Cdr2 proteins show defective energy dependent phospholipid translocation between the two leaflets of lipid bilayers (Smriti et al. 2002). The phospholipid translocation could also be elegantly demonstrated by functionally reconstituted purified Cdr1 protein (Shukla et al. 2006). What emerges from these studies suggest that some of the ABC transporters involved in clinical drug resistance also act as general phospholipid translocators where they flop GPLs from inner leaflet to outer leaflet by sharing common drug binding sites of the protein (Smriti et al. 2002; Shukla et al. 2006).

Recently, functional characterization of a MRP subfamily member Mlt1 revealed that it is required to transport PC into vacuoles of *C. albicans,* in an energy dependent manner and its absence affects lipid homeostasis. Interestingly, PC transporter Mlt1 which impacts lipid homeostasis also affects plethora of phenotypes such as delayed endocytosis, defects in hyphae and biofilm, susceptibility to drugs, protease activity/secretion, together culminating in attenuated virulence of *Candida* cells (Khandelwal et al. 2016).

20.3.6 Lipids in Hyphae and Biofilms

The morphological switch from yeast to hyphae form or to assemble biofilms is related to *Candida* cell's ability to invade, evade host immune response, and thus influence its virulence (Chandra et al. 2007). Notably, the expression profiling of *C. albicans* biofilms revealed differential regulation of lipid biosynthesis genes (Yeater et al. 2007; Lattif et al. 2008). A lipidomics study of the planktonic and biofilm forms of *C. albicans* cells, points to a remodeling of GPL and SL composition between the two forms. Any interruption in SL (raft lipid constituent) biosynthesis abolished biofilm formation in *C. albicans* (Lattif et al. 2011). Thus, there are strong evidences to point that lipids are crucial for the development of hyphae and development of biofilms in *C. albicans*.

20.3.7 Lipids as Virulence Factors

Recently, a screening of homozygous deletion mutant library showed that glucosylceramide biosynthesis is required for virulence in *C. albicans* (Noble et al. 2010). Mutants of *HSX11* (glucosyltransferase), Orf19.260 (or Sld1, sphingolipid desaturase) and Orf19.4831 (SL methyltransferases/cyclopropane synthases), which are involved in biosynthesis of glucosylceramide (Fig. 20.3a) and *HET1* (SL transfer protein), showed attenuated virulence and proliferation in *C. albicans*. The glucosylceramide pathway is *Candida* specific and is not found in other fungi like *S. cerevisiae* and *Schizosaccharomyces pombe*. Interestingly, the virulence effecter glucosylceramide acts independent of morphogenetic switching in *C. albicans* (Noble et al. 2010). In a separate study, *CHO1* (PS synthase) and *PSD1/PSD2* (PS decarboxylases) gene mutants, which are defective in PS and PE biosynthesis, respectively, also show attenuated virulence of *C. albicans* in a mouse model (Chen et al. 2010).

In this context, ergosterol is also shown to be critical for proper functionality of the vacuolar ATPases for the maintenance of vacuolar pH and intracellular ion homeostasis in *Candida* cells. The amiodarone susceptible disruptant of *ERG* and *VMA* genes, failed to acidify vacuoles and could not grow on non-fermentable carbon source, became susceptible to calcofluor white, demonstrated attenuated virulence in murine models of Candidiasis (Zhang et al. 2010).

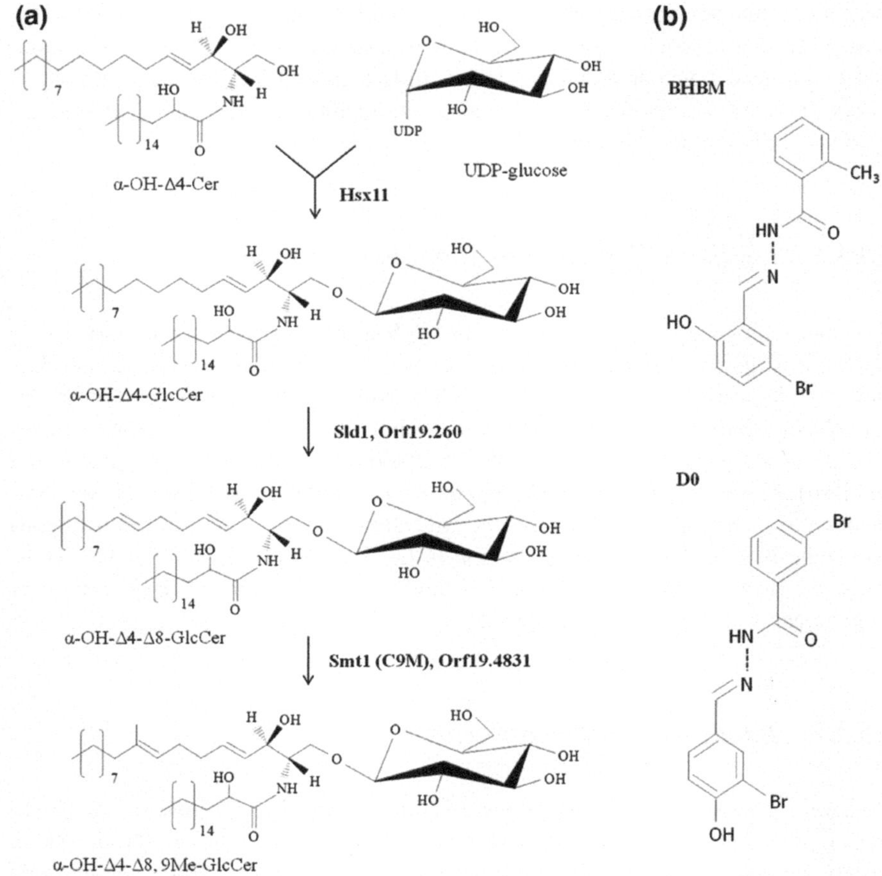

Fig. 20.3 **a** Partial scheme of glucosylceramide biosynthetic pathway in *C. albicans*. The glucosylceramide biosynthesis in *C. albicans* is facilitated by the enzymes Hsx11 (glucosyltransferase), orf19.260 (or Sld1, sphingolipid desaturase) and orf19.4831 (or Smt1, SL C9-methyltransferase/cyclopropane synthase). **b** Structures of BHBM and D0, the inhibitors of glucosylceramide biosynthesis

20.4 Current Approaches for Lipid Detection in *Candida*

The radiolabelled and fluorescent precursors used earlier to metabolically label lipids in tissues and cells were only the available methods to measure and quantify lipids. However, now with the advent of high resolution and analytical power, in particular, mass spectrometry and liquid chromatography have begun providing interesting insights that hold great future for lipidomic research. Multiple mass spectrometry platforms have been developed for simultaneous structural and quantitative analyses of lipids. These include the electrospray ionization (ESI), fast atom bombardment (FAB), matrix-assisted laser desorption/ionization (MALDI) coupled with mass detectors like tandem mass spectrometry (MS/MS), multiple-stage mass spectrometry (MS^n), time-of-flight (TOF), hybrid quadrupole time-of-flight mass spectrometer (QqTOF MS), and many more (Köfeler et al. 2012). Among these, ESI-based MS/MS or commonly known as triple quadrupole mass spectrometers have been the most extensively used instruments in routine analysis of fungal lipids. For example, analyses of polar phospholipids and its molecular species were performed based on the unique head group mass of each class of *Candida* lipids (Singh et al. 2010). However, due to low mass resolution in quadrupole MS, it is difficult to gather precise structural information of the analyzed molecular lipid species in complex lipid samples, specially the unknown lipids. However, it is a cost-effective technique that is widely used to analyze and quantify known lipid species for which a pure standard is available. It is possible to gather some information regarding the fatty acyls attached to the lipid species and their position (*sn*-1 or *sn*-2) using product ion scanning in ESI-MS/MS. Improved data can be obtained when ESI-MS/MS are coupled with high performance liquid chromatography (HPLC) platforms. For more correct identification of lipid species, hybrid MS are now available that have high mass resolution and therefore can provide the accurate mass of the lipid species (Ejsing et al. 2009). Although the quality of data gathered by a hybrid MS is much better but a higher cost of running, time required per sample, and extensive data analysis are major challenges with this technique. Only recently, MS methods for analyzing lipids from fungi are being developed. In the context of *Candida*, it is already enhancing our awareness and enabling in recognizing fresh roles of lipids in its cellular functions and pathogenesis. Lipidomics, therefore promise an additional layer of information to omics data, enabling identification of new roles of lipids.

20.5 Sphingolipids as Antifungal Drug Targets

Over the years, SLs and their biosynthetic pathways have shown great promise for its exploitation as novel drug target. Fungal SLs play key physiological roles in their survival and pathogenesis, possess unique structures synthesized by fungal specific enzymes as compared to human host. Notably, many pathogenic fungi share common structure of SLs species, making them targets for broad spectrum antifungals (Del Poeta et al. 2014).

One of the prime examples of sphingolipids being targeted to kill fungi is the glucosylceramide (GlcCer) biosynthetic pathway. In fungi, GlcCer are implicated in extracellular growth (alkali tolerance) and virulence. The fungal GlcCer is quite unique and is composed of Δ4-Δ8,9-methyl sphingosine backbone trans linked to 2-hydroxy-octadecanoic acid at C3 position (Figs. 20.1c and 20.3a). On the other hand, the mammalian GlcCer is composed of Δ4-sphingosine backbone attached to hexadecanoic acid. This structural difference has been exploited to identify a new class of antifungals designated as "hydrazycins" targeting fungal GlcCer. Hydrazycin compounds, namely [N′-(3-bromo-4-hydroxybenzylidene)-2-methylbenzohydrazide (BHBM) and 3-bromo-N′-(3-bromo-4-hydroxybenzylidene) benzohydrazide (D0)] (Fig. 20.3b), exhibit potent antifungal activity in both in vitro and mouse model studies. Both BHBM and D0 have shown very promising activity against a range of human pathogenic fungi including *C. albicans* (Mor et al. 2015). The mechanistic basis of BHBM activity has been explained by Mor et al. in *C. neoformans*. Treatment with BHBM causes depletion in GlcCer content of the cell. Furthermore, BHBM has four identifiable gene targets namely *APL5*, *COS111*, *MKK1*, and *STE2*, which have direct role in cell cycle progression and vesicular transport. Therefore, the antifungal potency in fungi is believed to be caused by severe defects in vesicular trafficking resulting accumulation of intracellular vesicles and altered cellular morphology (Mor et al. 2015). This is still an emerging area which warrants understanding of SL biochemical pathways and development of high-throughput compound screens for improved drugs and therapy.

Acknowledgements The work from authors (RP) laboratory discussed has been supported in part by grants from the Department of Biotechnology (BT/PR11158/BRB/10/640/2008, BT/PR13641/Med/29/175/2010, BT/PR14879/BRB10/885/2010, BT/01/CEIB/10/III/12).

References

Cannon RD, Fischer FJ, Niimi K, Niimi M, Arisawa M (1998) Drug pumping mechanisms in *Candida albicans*. Nihon Ishinkin Gakkai Zasshi 39:73–78

Chandra J, McCormick TS, Imamura Y, Mukherjee PK, Ghannoum MA (2007) Interaction of *Candida albicans* with adherent human peripheral blood mononuclear cells increases *C. albicans* biofilm formation and results in differential expression of pro- and anti-inflammatory cytokines. Infect Immun 75:2612–2620

Chen YL, Montedonico AE, Kauffman S, Dunlap JR, Menn FM, Reynolds TB (2010) Phosphatidylserine synthase and phosphatidylserine decarboxylase are essential for cell wall integrity and virulence in *Candida albicans*. Mol Microbiol 75(5):1112–1132. doi:10.1111/j.1365-2958.2009.07018.x

Del Poeta M, Nimrichter L, Rodrigues ML, Luberto C (2014) Synthesis and biological properties of fungal glucosylceramide. PLoS Pathog 10(1):e1003832. doi:10.1371/journal.ppat.1003832

Ejsing CS, Sampaio JL, Surendranath V, Duchoslav E, Ekroos K, Klemm RW, Simons K, Shevchenko A (2009) Global analysis of the yeast lipidome by quantitative shotgun mass spectrometry. Proc Natl Acad Sci U S A. 106(7):2136–2141. doi:10.1073/pnas.0811700106

Futerman AH, Hannun YA (2004) The complex life of simple sphingolipids. EMBO Rep 5 (8):777–782

Gaspar ML, Aregullin MA, Jesch SA, Nunez LR, Villa-García M, Henry SA (2007) The emergence of yeast lipidomics. Biochim Biophys Acta 1771:241–254

Guan XL, Wenk MR (2006) Mass spectrometry-based profiling of phospholipids and sphingolipids in extracts from Saccharomyces cerevisiae. Yeast 23:465–477

Khandelwal NK, Kaemmer P, Förster TM, Singh A, Coste AT, Andes DR, Hube B, Sanglard D, Chauhan N, Kaur R, d'Enfert C, Mondal AK, Prasad R (2016) Pleiotropic effects of the vacuolar ABC transporter MLT1 of *Candida albicans* on cell function and virulence. Biochem J 473(11):1537–1552. doi:10.1042/BCJ20160024

Köfeler HC, Fauland A, Rechberger GN, Trötzmüller M (2012) Mass spectrometry based lipidomics: an overview of technological platforms. Metabolites. 2(1):19–38. doi:10.3390/metabo2010019

Lattif AA, Chandra J, Chang J, Liu S, Zhou G, Chance MR, Ghannoum MA, Mukherjee PK (2008) Proteomic and pathway analyses reveal phase-dependent over-expression of proteins associated with carbohydrate metabolic pathways in *Candida albicans* biofilms. Open Proteomics 1:5–26

Lattif AA, Mukherjee PK, Chandra J, Roth MR, Welti R, Rouabhia M, Ghannoum MA (2011) Lipidomics of *Candida albicans* biofilms reveals phase-dependent production of phospholipid molecular classes and role for lipid rafts in biofilm formation. Microbiology 157(11):3232–3242. doi:10.1099/mic.0.051086-0

Maesaki S, Marichal P, Vanden Bossche H, Sanglard D, Kohno S (1999) Rhodamine 6G efflux for the detection of CDR1-overexpressing azole-resistant *Candida albicans* strains. J Antimicrob Chemother 44:27–31

Mahmoudabadi AZ, Boote V, Drucker DB (2001) Characterization of polar lipids of oral isolates of *Candida*, *Pichia* and *Saccharomyces* by fast atom bombardment mass spectrometry (FAB MS). J Appl Microbiol 90:668–675

Mor V, Rella A, Farnoud AM, Singh A, Munshi M, Bryan A, Naseem S, Konopka JB, Ojima I, Bullesbach E, Ashbaugh A, Linke MJ, Cushion M, Collins M, Ananthula HK, Sallans L, Desai PB, Wiederhold NP, Fothergill AW, Kirkpatrick WR, Patterson T, Wong LH, Sinha S, Giaever G, Nislow C, Flaherty P, Pan X, Cesar GV, de Melo Tavares P, Frases S, Miranda K, Rodrigues ML, Luberto C, Nimrichter L, Del Poeta M (2015) Identification of a new class of antifungals targeting the synthesis of fungal sphingolipids. MBio. 6(3):e00647. doi:10.1128/mBio.00647-15

Mukhopadhyay K, Prasad T, Saini P, Pucadyil TJ, Chattopadhyay A, Prasad R (2004) Membrane sphingolipid-ergosterol interactions are important determinants of multidrug resistance in *Candida albicans*. Antimicrob Agents Chemother 48(5):1778–1787

Noble SM, French S, Kohn LA, Chen V, Johnson AD (2010) Systematic screens of a *Candida albicans* homozygous deletion library decouple morphogenetic switching and pathogenicity. Nat Genet 42(7):590–598. doi:10.1038/ng.605

Pasrija R, Krishnamurthy S, Prasad T, Ernst JF, Prasad R (2005a) Squalene epoxidase encoded by ERG1 affects morphogenesis and drug susceptibilities of *Candida albicans*. J Antimicrob Chemother 55:905–913

Pasrija R, Prasad T, Prasad R (2005b) Membrane raft lipid constituents affect drug susceptibilities of *Candida albicans*. Biochem Soc Trans 33:1219–1223

Pasrija R, Panwar SL, Prasad R (2008) Multidrug transporters CaCdr1p and CaMdr1p of *Candida albicans* display different lipid specificities: both ergosterol and sphingolipids are essential for targeting of CaCdr1p to membrane rafts. Antimicrob Agents Chemother 52:694–704

Prasad R, Singh A (2013) Lipids of *Candida albicans* and their role in multidrug resistance. Curr Genet 59(4):243–250. doi:10.1007/s00294-013-0402-1

Prasad T, Saini P, Gaur NA, Vishwakarma RA, Khan LA, Haq QM, Prasad R (2005) Functional analysis of CaIPT1, a sphingolipid biosynthetic gene involved in multidrug resistance and morphogenesis of *Candida albicans*. Antimicrob Agents Chemother 49(8):3442–3452

Prasad R, Rawal MK, Shah AH (2016) *Candida* efflux ATPases and antiporters in clinical drug resistance. Adv Exp Med Biol 892:351–376. doi:10.1007/978-3-319-25304-6_15

Shahi P, Moye-Rowley WS (2009) Coordinate control of lipid composition and drug transport activities is required for normal multidrug resistance in fungi. Biochim Biophys Acta 1794:852–859

Shevchenko A, Simons K (2010) Lipidomics: coming to grips with lipid diversity. Nat Rev Mol Cell Biol 11(8):593–598. doi:10.1038/nrm2934

Shingu-Vazquez M, Traven A (2011) Mitochondria and fungal pathogenesis: drug tolerance, virulence, and potential for antifungal therapy. Eukaryot Cell 10:1376–1383

Shukla S, Rai V, Banerjee D, Prasad R (2006) Characterization of Cdr1p, a major multidrug efflux protein of *Candida albicans*: purified protein is amenable to intrinsic fluorescence analysis. Biochemistry 45(7):2425–2435

Simons K, Sampaio JL (2011) Membrane organization and lipid rafts. Cold Spring Harb Perspect Biol 3(10):a004697. doi:10.1101/cshperspect.a004697

Singh A, Prasad R (2011) Comparative lipidomics of azole sensitive and resistant clinical isolates of *Candida albicans* reveals unexpected diversity in molecular lipid imprints. PLoS ONE 6(4): e19266. doi:10.1371/journal.pone.0019266

Singh A, Prasad T, Kapoor K, Mandal A, Roth M, Welti R, Prasad R (2010) Phospholipidome of *Candida*: each species of *Candida* has distinctive phospholipid molecular species. OMICS 14 (6):665–677. doi:10.1089/omi.2010.0041

Singh A, Yadav V, Prasad R (2012) Comparative lipidomics in clinical isolates of *Candida albicans* reveal crosstalk between mitochondria, cell wall integrity and azole resistance. PLoS ONE 7(6):e39812. doi:10.1371/journal.pone.0039812

Singh A, Mahto KK, Prasad R (2013) Lipidomics and *in vitro* azole resistance in *Candida albicans*. OMICS 17(2):84–93. doi:10.1089/omi.2012.0075

Smriti Krishnamurthy S, Dixit BL, Gupta CM, Milewski S, Prasad R (2002) ABC transporters Cdr1p, Cdr2p and Cdr3p of a human pathogen *Candida albicans* are general phospholipid translocators. Yeast 19(4):303–318

van Meer G, Voelker DR, Feigenson GW (2008) Membrane lipids: where they are and how they behave. Nat Rev Mol Cell Biol 9:112–124. doi:10.1038/nrm2330

Wenk MR (2005) The emerging field of lipidomics. Nat Rev Drug Discov. 4:594–610

Wenk MR (2010) Lipidomics: new tools and applications. Cell 143(6):888–895. doi:10.1016/j.cell.2010.11.033

Yeater KM, Chandra J, Cheng G, Mukherjee PK, Zhao X, Rodriguez-Zas SL, Kwast KE, Ghannoum MA, Hoyer LL (2007) Temporal analysis of *Candida albicans* gene expression during biofilm development. Microbiology 153:2373–2385

Zhang YQ, Gamarra S, Garcia-Effron G, Park S, Perlin DS, Rao R (2010) Requirement for ergosterol in V-ATPase function underlies antifungal activity of azole drugs. PLoS Pathog 6 (6):e1000939. doi:10.1371/journal.ppat.1000939

Chapter 21
Molecular Targets for Anticandidal Chemotherapy

Sławomir Milewski

Abstract A relatively small number of anticandidal chemotherapeutics used in clinical practice is at least in part consequence of a limited number of their molecular targets: ergosterol in the membrane, lanosterol demethylase, β(1→3) glucan synthase, and DNA/RNA biosynthesis. Much more potential novel targets have been revealed by the comparative genomic studies identifying essential genes unique for *Candida albicans* or resulted from recognition of biomolecules targeted by newly discovered antifungals of natural or synthetic origin. The most promising of them are proteins/enzymes involved in biosynthesis of the cell wall components: chitin synthase, *O*-mannosyl tranferase, α-1,2 mannosyltransferases, and inositol acylase, in biosynthesis of membrane lipids (Eq. inositolphosphoceramide synthase), in methonine biosynthesis, in translation and posttranslation processing (fungal-specific elongation factors, *N*-mirystoyl transferase) and in signaling pathways. Target candidates have been also identified among proteins participating in expression of *C. albicans* virulence factors: filamentation, biofilm formation, and production of secretory hydrolytic enzymes.

Keywords Antifungal chemotherapy · Targets · Selective toxicity

21.1 Introduction

Disseminated (invasive) fungal infections remain as one of the major problems in modern chemotherapy. Although invasive mycoses are much less frequent than the superficial ones, they are associated with unacceptably high mortality rates. The estimated number of cases is more than 2 million/year worldwide, with over 1.5 million deaths (Brown et al. 2012). In fact, more people die of the invasive fungal diseases than of tuberculosis or malaria. Many species of fungi are responsible for

S. Milewski (✉)
Department of Pharmaceutical Technology and Biochemistry, Gdańsk University of Technology, 11/12 Narutowicza St., 80-1233 Gdańsk, Poland
e-mail: slamilew@pg.gda.pl

R. Prasad (ed.), *Candida albicans: Cellular and Molecular Biology*,
DOI 10.1007/978-3-319-50409-4_21

invasive infections. More than 90% of all reported fungal-related deaths result from species that belong to one of the four genera: *Cryptococcus*, *Candida*, *Aspergillus*, and *Pneumocystis* (Brown et al. 2012). *Candida* species are the most common fungal etiological agents of life-threatening invasive infections in immunocompromised hosts, transplant recipients and patients hospitalized in intensive care units. These species are also the fourth most common cause of nosocomial (hospital-acquired) bloodstream infections (Wisplinghoff et al. 2004). Among a dozen of human pathogenic *Candida* species, *C. albicans* is an etiological agent most often causing disseminated candidiasis (Pfaller and Diekema 2007).

The high mortality rate of invasive candidiases and other disseminated mycoses is due to the several factors, including shortcomings of diagnosing, a limited number of effective antifungal chemotherapeutics, and increasing fungal resistance to available medicines. The arsenal of drugs used in the clinical treatment of disseminated candidiases comprises 10 compounds/preparations, with only four molecular targets (Table 21.1). None of these antifungals fulfills all criteria of the "ideal" anticandidal drug. Such compound should exhibit a broad spectrum of activity, no association with intrinsic or acquired resistance, fungicidal mode of action, good pharmacokinetic and pharmacodynamic profile, no mammalian toxicity or side effects, and low cost (Wong et al. 2014).

The most important feature of a good chemotherapeutic agent is its proper selective toxicity, i.e., the highest possible antimicrobial activity and the lowest possible mammalian toxicity. Analyzing properties of drugs used in different areas of antimicrobial chemotherapy, one may enumerate five major cases of molecular bases advantageous for good selective toxicity: (1) target, essential for the pathogen, is absent from the host; (2) substantial difference in structure and/or properties

Table 21.1 Antifungal agents/preparations used in clinical treatment of invasive candidiasis

Antifungal	Molecular target	Major drawbacks
Amphotericin B	Ergosterol	High mammalian toxicity
Liposomal Amphotericin B	Ergosterol	Cost
5-fluorocytosine	Thymidylate synthase	Narrow spectrum, fungistatic, resistance
Fluconazole	Cytochrome P_{450} dependent lanosterol demethylase	Fungistatic, resistance
Itraconazole	Cytochrome P_{450} dependent lanosterol demethylase	Resistance, low bioavailability
Voriconazole	Cytochrome P_{450} dependent lanosterol demethylase	Side effects, resistance
Posaconazole	Cytochrome P_{450} dependent lanosterol demethylase	Cost, resistance
Caspofungin	$\beta(1\rightarrow3)$glucan synthase	Narrow spectrum
Micafungin	$\beta(1\rightarrow3)$glucan synthase	Narrow spectrum
Anidulafungin	$\beta(1\rightarrow3)$glucan synthase	Narrow spectrum

of the target present in the pathogen and in the host, giving rise to the differential drug: target affinity; (3) selective uptake of a drug by the pathogen but not host cells; (4) specific activation of the prodrug in the pathogen but not in host cells; (5) qualitatively different consequences of target inhibition in pathogen and host cells. Obviously option (1) is the most promising, provided a proper agent acting as a selective target inhibitor is found but the other four can be also exploited. Unfortunately, compared to other microbial pathogens, fungi are evolutionarily close to humans, what considerably limits the scope of molecular targets constituting a proper basis for selective toxicity. Despite this, extensive efforts aimed at further exploitation of known targets and at finding novel targets are being continued. In this chapter, a survey of targets for known anticandidal drugs and other targets, considered promising for evaluation of novel drug candidates, is presented.

21.2 Target Search

Antifungal targets have been previously found mainly from the studies on mechanisms of action of newly discovered antifungal agents of natural or synthetic origin. Nowadays, in the postgenomic era, the major tools used to identify novel drug targets are comparative genomics, genome-wide transcriptome analysis (expression profiling at the RNA level using DNA microarrays), or proteome analysis (expression profiling at the protein level using two-dimensional gel electrophoresis). Comparative genomics approach is one of them that finds differences between two organisms at the molecular level. *C. albicans*, whose genome was sequenced in 2004, offers its use for genome-wide screening of potential targets. The "-omic" approaches allow to identify genes/proteins that are unique for the fungal pathogen. In one of such studies, 57 potential drug targets, based on 55 genes experimentally confirmed as essential for *C. albicans* or *Aspergillus fumigatus* and other 2 genes (*kre2* and *erg6*) relevant for fungal survival within the host were selected. Further selection according to the six criteria: (1) be essential or relevant for fungi survival; (2) be present in all analyzed pathogens, therefore allowing a broad spectrum of drug action; (3) be absent in the human genome, therefore avoiding unwanted side effects; (4) be preferentially an enzyme and have the potential for assay ability; (5) not be auxotrophic, thereby avoiding host provision of the necessary substrate for the blocked pathway; and (6) have a cellular localization potentially accessible to the drug activity, afforded 4 genes: *trr1* (coding for cytoplasmic thioredoxin reductase), *rim8* (coding for the protein involved in the proteolytic activation of a transcriptional factor in response to alkaline pH), *kre2* (encoding α-1,2-mannosyltransferase), and *erg6* (coding for Δ^{24}-sterol C-methyltransferase) (Abadio et al. 2011). In another study, 1,618 proteins out of 14,633 total proteins present in *C. albicans* proteome were identified as unique (42 functionally known, 1,566 hypothetical, and 10 unknown), so that finally 52 were selected as potential antifungal drug targets unique to *C. albicans* (Tripathi et al. 2014).

Good antifungal targets are typically considered to be either gene products essential for viability or so-called virulence factors. Large-scale essential gene identification in *C. albicans* is possible through a number of alternative approaches, including antisense-based gene inactivation (De Backer et al. 2001), homozygote null mutants (Bruno et al. 2006), transposon-based heterozygote screens for hypomorphs (Uhl et al. 2003), conditional promoter replacement (CPR, Nakayama et al. 2000), and gene replacement and conditional expression (GRACE, Roemer et al. 2003). For a virulence factor to be established, the corresponding knockout strain should be shown to be avirulent when assayed in an appropriate animal models, usually murine but also invertebratal (Pukkila-Worley et al. 2009; Chamilos et al. 2009); reintroduction of the wild-type allele should restore virulence.

Genomics-based strategies for discovery of novel antifungal drug targets have allowed to identify several novel potential targets but have thus far produced no agents in clinical development.

21.3 Established and Proposed Antifungal Targets

Targets for the established antifungal chemotherapeutics and for putative antifungal agents are proteins, mostly enzymes, components of the cell membrane or of the cell wall.

21.3.1 Components of the Cell Membrane and Their Biosynthesis

More than a half of antifungal chemotherapeutics used in clinics target a sterol component of the fungal cell membrane or an enzyme catalyzing one of the steps in its biosynthesis. Moreover, some other enzymes of this pathway are the targets for antifungals used in therapy of topical mycoses and a few enzymes catalyzing particular steps in biosynthesis of other lipids present in the fungal cell membrane have ben proposed as novel targets.

21.3.1.1 Ergosterol in the Membrane

Mammalian cell membranes contain cholesterol, which in their fungal counterparts is substituted by ergosterol. The latter is a molecular target for polyene macrolide antifungal antibiotics, of which Amphotericin B, Nystatin, Pimaricin are approved drugs but only the first of them is used for the treatment of invasive fungal infections. Amphotericin B (AmB) is known as a “golden standard” of antifungal chemotherapy. Indeed AmB demonstrates most of the features expected for the "ideal" antifungal, including a fungicidal action, broad antifungal spectrum, and lack of fungal resistance. The only (but important) drawback is substantial mammalian toxicity, especially nephrotoxicity, which is a consequence of mechanism of biological action.

Currently there are two major hypotheses on mechanism of antifungal action of AmB. According to the "barrel-stave-pore" mechanism (Fig. 21.1a, b), AmB molecules form complexes with ergosterol in the membrane and a few such complexes (4–12) self-assemble to barrel-stave-like transmembrane pores, giving rise to leakage of low molecular weight cell components, including ions (Gray et al. 1990). The actual structure of AmB-induced channels is not known but has been modeled by molecular dynamics (Bagiński et al. 2002). In alternative mechanism, called a "sterol sponge" model (Fig. 21.1c, d), AmB extracts ergosterol from the hydrophobic interior of fungal membrane and deposits it on the cell's surface as single complexes or a "pile" of complexes (Palacios et al. 2012; Anderson et al. 2014). In both mechanisms, binding of AmB to ergosterol is the first step but the antibiotic also binds to cholesterol in mammalian cell membranes. This is only a slightly higher affinity to ergosterol than to cholesterol that constitutes a molecular basis of selective toxicity of AmB. In consequence, a minimal concentration at which AmB is toxic to mammalian cells is only 5–10 times higher than the minimal fungicidal concentration. The fungistatic effect of AmB action is linked to antibiotic-induced ion (especially potassium) leakage, while the fungicidal action, observed at higher concentrations, is due to the apoptosis-like effect and peroxidation of membrane lipids (Chen et al. 1978; Phillips et al. 2003).

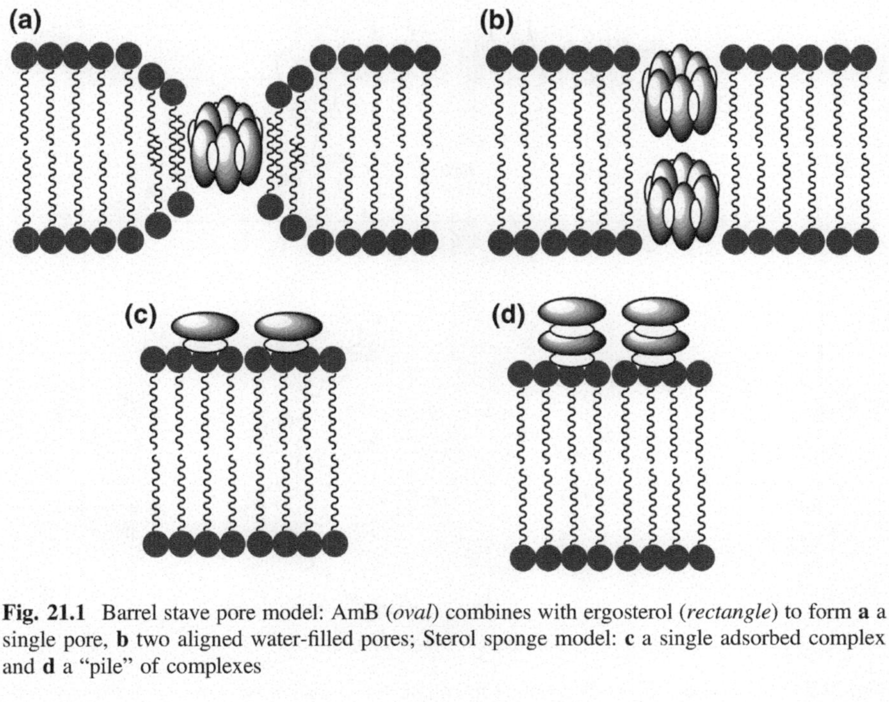

Fig. 21.1 Barrel stave pore model: AmB (*oval*) combines with ergosterol (*rectangle*) to form **a** a single pore, **b** two aligned water-filled pores; Sterol sponge model: **c** a single adsorbed complex and **d** a "pile" of complexes

AmB, as other polyene macrolides, is poorly soluble in aqueous solutions. Monomeric form of AmB exists in water at concentrations below 10^{-7} M. At higher concentrations, AmB undergoes complex processes of self-association, formation of dimers and soluble oligomers. Finally, at concentrations higher than 10^{-5} M insoluble aggregates are observed. It was shown that water-soluble aggregates of AmB are toxic to erythrocytes and fungal cells, while the monomers are toxic only to fungal cells (Legrand et al. 1992).

Substantial mammalian toxicity of AmB is diminished in AmBisome™, a liposomal preparation of this antibiotic (Moen et al. 2009); however its cost is high. It is believed that selective toxicity of AmB can be improved by a proper chemical modification of the antibiotic molecule. Most of them concerned the carboxyl functionality or the amino sugar substituent. A number of derivatives of AmB were obtained but only some of them, especially those modified at the mycosamine residue, exhibited improved selective toxicity. Substantial improvement was observed for *N*-D-ornithyl- AMB (Wright et al. 1982), *N*-methyl-*N*-D-fructopyranosylamphotericin B methyl ester (MF-AME, Grzybowska et al. 1997), or *N*-piperidinepropionyl amphotericin B methyl ester (PAME, Hac-Wydro et al. 2005). *N,N*-di-(3-aminopropyl) AmB exhibited much higher than AmB antifungal activity and slightly lower mammalian toxicity (Volmer et al. 2010). Truly spectacular effects, i.e., elimination of mammalian in vitro toxicity with retention or at most minimal decrease of antifungal activity, have been recently achieved for the AmB derivatives shown in Fig. 21.2: 2'-deoxyAmB (Wilcock et al. 2013), AmB urea derivatives (Davis et al.

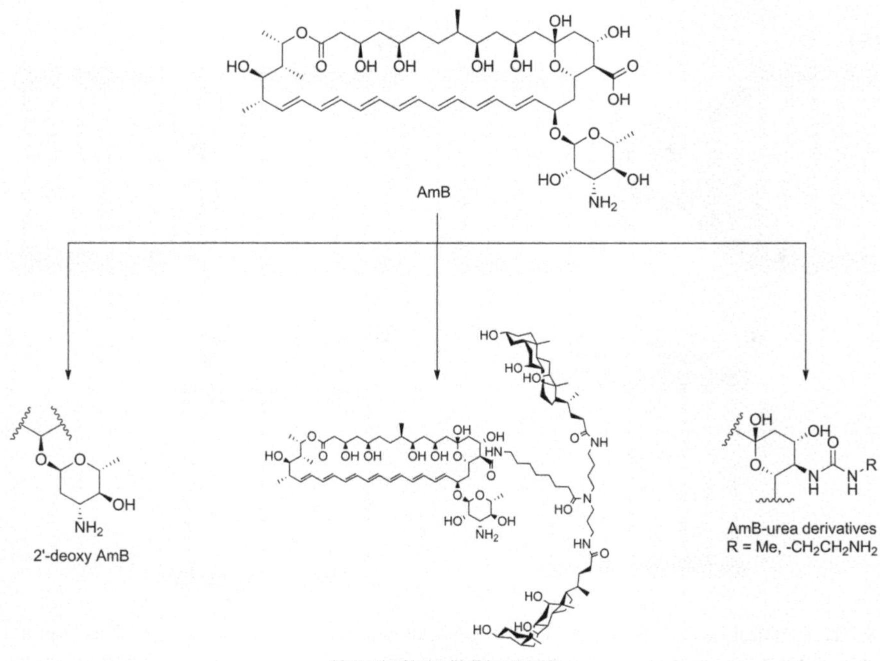

Fig. 21.2 Modifications of AmB resulting in complete elimination of mammalian toxicity with retention of antifungal activity

2015), and conjugates of AmB with "molecular umbrellas" (Janout et al. 2015). In the case of the 2'-deoxy- and urea derivatives, elimination of mammalian toxicity was a consequence of subtle structural modifications resulting in substantial decrease of affinity to cholesterol, whereas conjugation to molecular umbrella prevented formation of water-soluble aggregates of AmB. Substantially improved selective toxicity of AmB urea derivatives was confirmed also in vivo, in the murine model of disseminated candidiasis (Davis et al. 2015).

21.3.1.2 Ergosterol Biosynthesis

Presence of ergosterol in fungal cell membrane is essential for survival. Ergosterol deficient yeast mutants are not viable unless the growth medium is supplemented with proper sterol (Parks and Casey 1995). Inhibition of ergosterol biosynthesis may therefore give rise to hampered cell growth and, under some circumstances, to the fungicidal effect. With ergosterol depleted and replaced with unusual sterols, the normal permeability and fluidity of the fungal membrane is altered, with secondary consequences for membrane-bound enzymes, such as those involved in cell wall synthesis.

The biochemical pathway of ergosterol biosynthesis shown in Fig. 21.3 consists of several steps. Most of the initial steps are identical with those of the cholesterol biosynthetic pathway in humans. Some of the steps are the targets for antifungals, namely these catalyzed by squalene epoxidase, lanosterol 14α-demethylase Δ^{14} reductase, and $\Delta^7 \rightarrow \Delta^8$ isomerase.

Imidazole and triazole derivatives ("azole antifungals") are the largest class of antifungal agents in clinical use. The principal molecular target of azole antifungals is lanosterol 14α-demethylase ($P450_{14DM}$, CYP51), encoded by the *ERG11* gene, which catalyzes the oxidative removal of the 14α-methyl group of lanosterol and/or eburicol in fungi by a typical P450 monooxygenase activity but in some fungal species, they can also inhibit the subsequent Δ^{22}-desaturase step (Kelly et al. 1997). $P450_{14DM}$ occurs in different kingdoms, such as fungi, higher plants, and animals, with the same metabolic role, i.e., removal of the 14-methyl group of sterol precursors such as lanosterol, obtusifoliol, ihydrolanosterol, and 24(28)-methylene-24,25-dihydrolanosterol, and this is the only known P450 distributed widely in eukaryotes with essentially the same metabolic role. This protein contains an iron protoporphyrin moiety located at the active site, and the antifungal azoles bind to the iron atom via a nitrogen atom in the imidazole or triazole ring. The remainder of the azole molecule binds to the apoprotein in a manner dependent on the individual azole's structure. The exact conformation of the active site differs between fungal species and amongst the many mammalian P450 monooxygenases (Vanden Bossche et al. 1995). The precise nature of the interaction between each azole molecule and each kind of P450 therefore determines the extent of the azole's inhibitory effect in different fungal species. So far, the only crystal structure of a Cyp51p molecule and its complex with azole antifungals to have been published is for the one from *Mycobacterium tuberculosis* (Podust et al. 2001), although the 3D models of *C. albicans* Cyp51p complexed with azole antifungals were built (Ji et al.

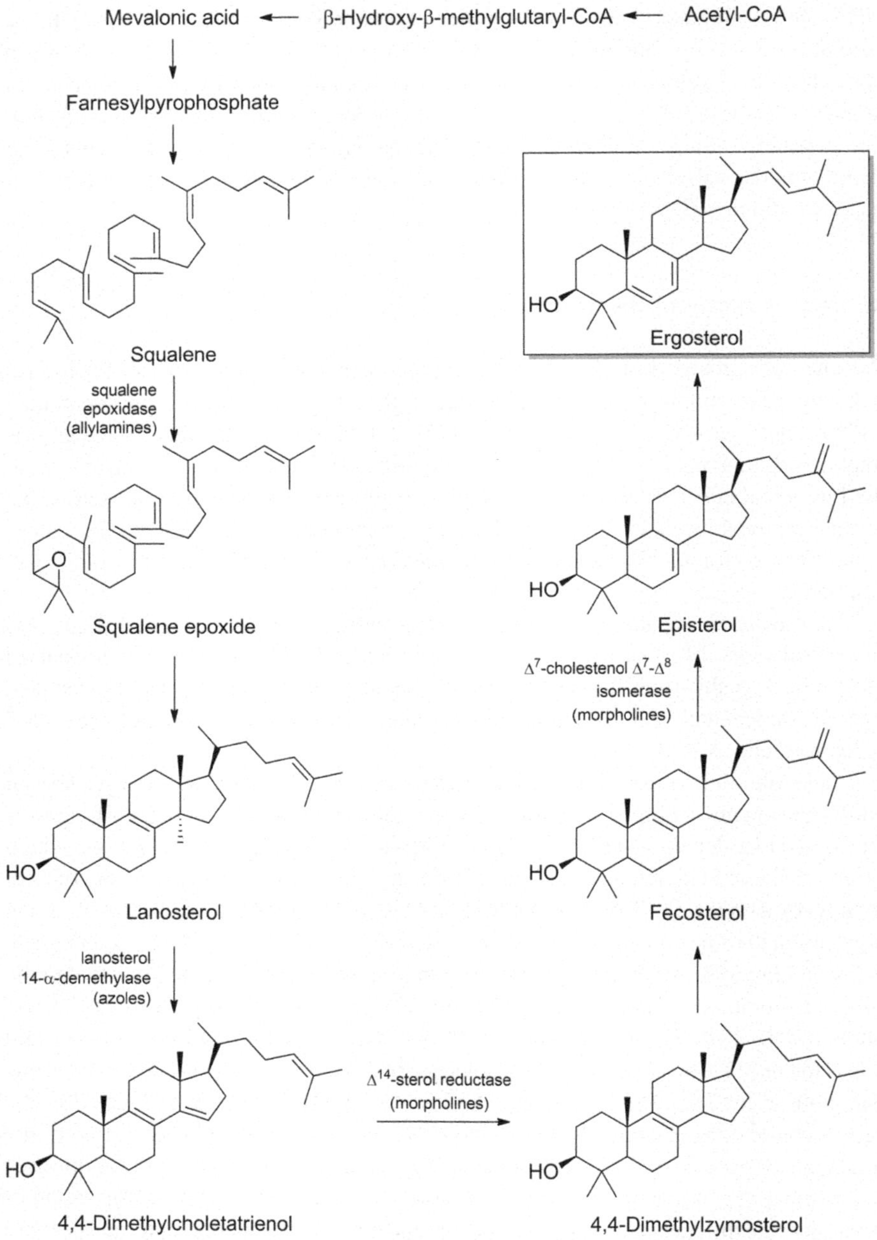

Fig. 21.3 Pathway of ergosterol biosynthesis in fungi with indicated targets for antifungals

2000; Xiao et al. 2004). Amino acid sequences of fungal and mammalian enzymes are 38% identical.

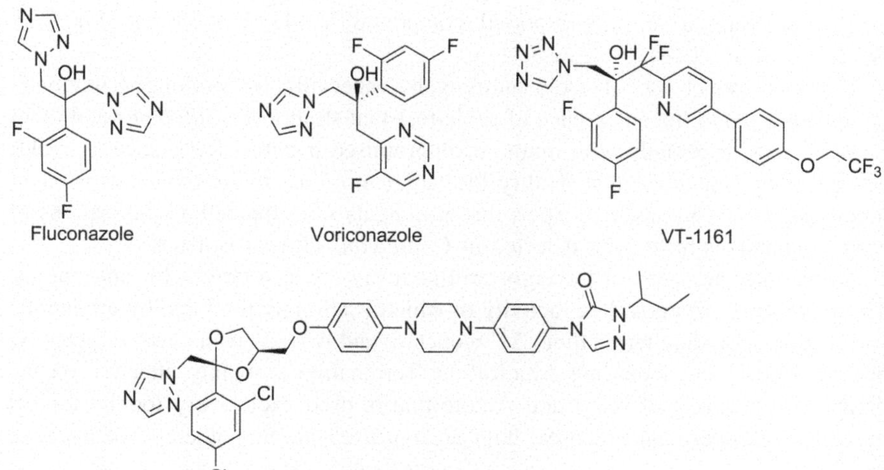

Fig. 21.4 Inhibitors of lanosterol 14α-demethylase ($P450_{14DM}$) used in antifungal chemotherapy (fluconazole, itraconazole, and voriconazole) and drug candidate VT-1161

Nowadays, triazole antifungals (Fig. 21.4) constitute the most important group of antifungal chemotherapeutics. The first azole antifungal used in chemotherapy of disseminated candidiasis was an imidazole derivative ketoconazole (Heeres et al. 1979) but this drug had several drawbacks, such as poor oral absorption, low activity in immunocompromised patients, and severe side effects. Other azole antifungals introduced later belong to the triazole family. Fluconazole became available for use by clinicians in 1990 and provided many advantages over the use of imidazoles, especially good absorption through the gastrointestinal tract, good pharmacokinetic properties as well as broad spectrum of activity (Brammer et al. 1990). Intraconazole is preferentially used for the treatment of aspergilloses and voriconazole and posaconazole (new generation triazoles) are the newest clinical drugs of the series. Albaconazole, efinaconazole, ravuconazole, and isavuconazole have been designed to overcome the deficiencies of their parent drugs (Chiou et al. 2000; Potoski and Brown 2002; Peyton et al. 2015).

The selective toxicity of azole antifungals results from the difference in consequences of $P450_{14DM}$ inhibition in fungi and mammals. In fungi, the resulting accumulation and metabolism of 14α-methylated sterol species leads to the synthesis of toxic compounds, which are unable to successfully replace ergosterol. Substitution of cholesterol with 14α-methylated sterols in mammalian membranes has little if any effect on membrane functionality, and moreover humans are able to compensate cholesterol deficiency by its supply from the diet. Side effects of chemotherapy with azole antifungals are predominantly linked to inhibition of off-target human cytochromes P450 (CYPs), such as CYP3A4, CYP2C9, CYP2C19, and CYP46 (Nivoix et al. 2008). In this respect, an interesting new option could be novel inhibitors of fungal $P450_{14DM}$, not interacting with human

CYPs, including a recently reported compound VT-1161 shown in Fig. 21.4 (Warrilow et al. 2014).

The widespread use of azole antifungals, especially for prolonged treatment periods, has led to the emergence of azole-resistant strains of *C. albicans* and other *Candida* spp., especially with immunocompromised patients. Resistance to azoles can occur by mutations that modify the target molecule or by overexpression of membrane efflux pumps that export the antifungals from the cell. Combinations of both mechanisms have been detected in *C. albicans* clinical isolates.

Some other enzymes of the ergosterol pathway are also targets for antifungals. These are squalene epoxidase, activity of which is strongly inhibited by allylamine antifungals, including terbinafine, Δ^{14} reductase and $\Delta^7 \rightarrow \Delta^8$ isomerase, targeted by phenylmorpholines, including Amorolfine. Terbinafine is mainly effective on the dermatophyte group of fungi and Amorolfine is used exclusively for the topical treatment of superficial mycoses; both are not used in chemotherapy of invasive candidiases.

21.3.1.3 Sfingolipid Biosynthesis

Sfingolipids are essential components of fungal and mammalian cell membranes. Although structures of these lipids of fungal and mammalian origin are similar, the biosynthetic pathways are different. The difference concerns the last steps, i.e., conversion of sphingosine to sphingolipids. The first step of the fungal version of sfingolipid biosynthetic pathway is catalyzed by inositolphosphoceramide (IPC) synthase. The enzyme, located in Golgi, transfers the phoshoinositol group from phosphatidyl inositol (PI) to the 1-hydroxyl group of phytoceramide. This enzyme was proposed as a target for antifungals after discovery of strong inhibitors of this enzyme: aureobasidin A, khafrefungin, and galbonolide A (Sugimoto et al. 2004; Aeed et al. 2009). Unfortunately, these cyclic peptides are very good substrates for fungal ABC-type drug transporters (Ogawa et al. 1998), so that these agents are prone to multidrug resistance. Moreover, the *AUR1* gene of *S. cerevisiae* encodes a protein that is necessary for IPC synthase activity and confers resistance to the antifungal drug aureobasidin A when mutated (Nagiec et al. 1997). Aur1p may thus represent a new target for screening for IPC synthase inhibitor compounds. Some antifungal target potential was suggested also for serine palmitoyl transferase, catalyzing the committed step in sphingolipid biosynthesis (Yamaji-Hasegawa et al. 2005) but selective toxicity of inhibitors of this enzyme known so far is questionable.

21.3.1.4 Fatty Acid Δ^9 Desaturase

The *OLE1* gene encoding fatty acid Δ^9 desaturase in *C. albicans* is essential for viability and virulence in a murine model of systemic candidiasis even when the animals were supplemented with a high fat diet (Xu et al. 2009). The fungal fatty acid desaturases contain a cytochrome b_5 domain fused to the carboxyl termini, which is absent in the mammalian homologs (Martin et al. 2007). Therefore, compounds targeting the fungi-specific part of the enzyme may become promising antifungal drug candidates.

21.3.2 *Cell Wall and Biosynthesis of Its Components*

One of the very few obvious qualitative differences between fungal and mammalian cells is presence of the cell wall in the former but in the latter. The fungal cell wall is first of all an osmoprotective barrier and inhibition of its biosynthesis or functionality may lead to the fungicidal effect, similar to the bactericidal effect of β-lactams or vancomycin. Three major polysaccharide components of the fungal cell wall are glucan, chitin, and mannan (as mannoproteins). The relative proportions of these constituents in various fungi differ markedly but in *Candida* species β(1→3)glucan is the most abundant (>50%) and chitin is a minor component (1–5%). A few molecular targets for anticandial chemotherapy have been identified in pathways of β(1→3)glucan, chitin, and mannoptotein biosynthesis and assembly.

21.3.2.1 Biosynthesis of Precursors of Cell Wall Polysaccharides

Two out of three components of the fungal cell wall, namely chitin and mannoproteins contain N-acetyl-D-glucosamine (GlcNAc), incorporated into their structures from the common precursor, UDP-GlcNAc. The cytosolic four-step pathway of UDP-GlcNAc biosynthesis (shown as a part of Fig. 21.5) is a branch of glycolysis (Milewski et al. 2006) and is initiated by formation of D-glucosamine-6-phosphate from D-fructose-6-phosphate and L-glutamine, catalyzed by glucosamine-6-phosphate synthase. The enzyme is very abundant and is present in almost all living organisms, except for some protozoans (Milewski, 2002) The *GFA1* gene encoding GlcN-6-P synthase is essential for fungi (Whelan and Ballou 1975). Inhibition of GlcN-6-P synthase activity in *Candida* spp. results in a fungicidal effect due to the hampered biosynthesis of chitin and mannoprotein, as was shown for the antifungal antibiotic tetaine (Milewski et al. 1986). On the other hand, a short-term inhibition of GlcN-6-P synthase in mammals is not lethal due to the slow turnover of mammalian amino sugar-containing macromolecules and very rapid turnover of mRNA encoding the enzyme (Bates et al. 1966). This difference in consequences of GlcN-6-P synthase inhibition constitutes a basis for exploitation of this enzyme as a target for antifungal chemotherapy. Several GlcN-6-P synthase inhibitors have been proposed, with L-glutamine analogs, N^3-acyl-L-2,3-diaminopropanoic acid derivatives as especially promising. Oligopeptides incorporating N^3-(4-methoxyfumaroyl)-L-2,3-diaminopropanoic acid (FMDP) demonstrated high in vitro and in vivo anticandidal activity and very low, if any, mammalian toxicity (Milewski et al. 1988; Andruszkiewicz et al. 1990), thus confirming target validity but their further development was limited by frequent fungal-specific resistance. Activity of GlcN-6-P synthase is also inhibited by antibiotic kanosamine (3-amino-3-deoxy-D-glucose) and a transition state analog 2-amino-2-deoxy-D-glucitol-6-phosphate (ADGP), however both exhibit poor antifungal activity (Janiak and Milewski 2001; Janiak et al. 2002) Search for other inhibitors of GlcN-6-P synthase as anticandidal drug candidates is being continued (Melcer et al. 2007; Zhao et al. 2013; Bahare et al. 2015).

Another step of cell wall polysaccharide precursor biosynthesis proposed as a potential target is isomerization of D-fructose-6-phosphate to D-mannose-6-phosphate (Fig. 21.5). This reaction initiates a cytosolic pathway leading to GDP-mannose, a sugar nucleotide precursor of mannan present in fungal cell wall mannoproteins. A respective homologous enzyme exists in humans but human glycoproteins are not as mannose-rich as the fungal ones, so that consequences of phosphomannose isomerase inhibition in fungal cells could be deeper. The Fru-6-P to Man-6-P isomerization is catalyzed by Zn-dependent phosphomannose isomerase. The *PMI1* gene encoding *C. albicans* phosphomannose isomerase is essential (Smith et al. 1995). Strong inhibitors of this enzyme are known but they do not exhibit anticandidal activity due to the inability of crossing the cell membrane (Desvergnes et al. 2012; Foret et al. 2009).

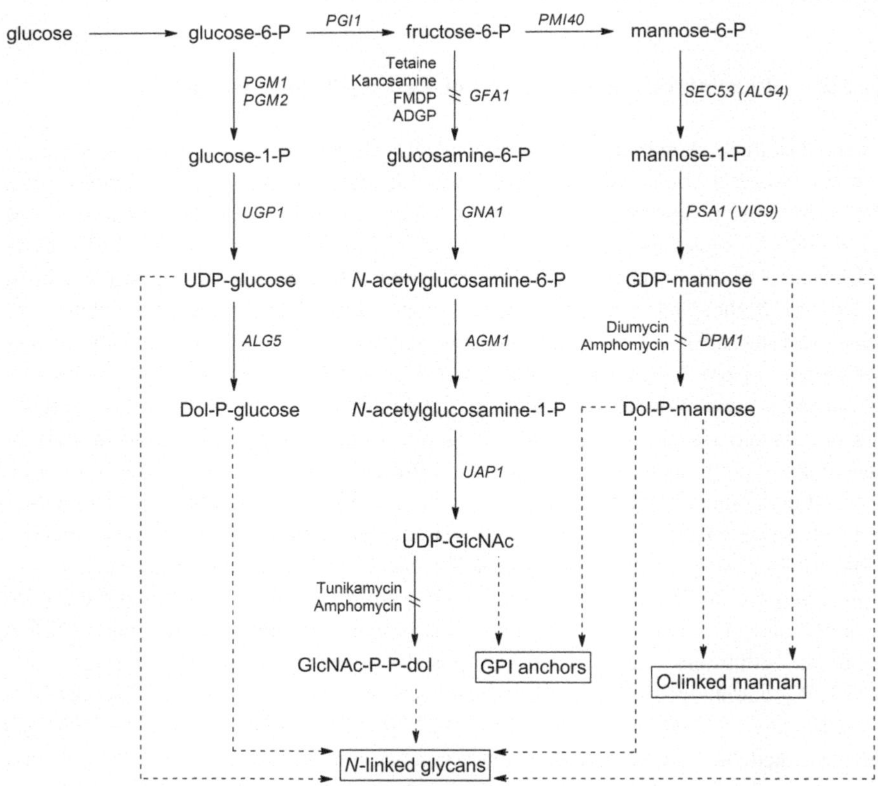

Fig. 21.5 Biosynthesis of precursors for protein-linked oligosaccharides

Antibiotics tunikamycin, amphomycin, and diumycin inhibit activity of enzymes catalyzing formation of dolichol phosphate bound sugars in pathways of biosynthesis of precursors for protein-linked oligosaccharides (Fig. 21.5) but none of them displayed acceptable selective toxicity (Elbein 1984).

21.3.2.2 Biosynthesis of O-Linked Mannans

Mannoproteins are the nonrigid components of the fungal cell wall. Some of them located in the inner part, intermeshed with the glucan network, play a structural role as a filling cementing the wall. Those exposed on the surface are important for adhesion of candidal cells to human epithelium which is a starting point for effective colonization and for intercellular recognition and interaction. In this functional aspect, they are similar to human glycoproteins but are structurally different for their exceptionally high mannose content, especially located in the O-linked oligosaccharide fragments (Fig. 21.6).

Protein O-mannosylation is essential for fungi including *C. albicans*, where reduced O-mannosylation results in attenuated virulence and diminished adherence to mammalian epithelial tissues (Ernst and Prill 2001). O-mannosylated proteins occur also in mammals; they are relatively abundant in brain. Defects in O-mannosylation of brain proteins may lead to neuromuscular diseases, for example, congenital muscular dystrophies (Willer et al. 2003). However, biosynthetic pathways of O-mannosylation in fungi and in mammals are somewhat different. The fungal version is shown in Fig. 21.7. The first mannose derived from Dolichol-P-Man is added in endoplasmic reticulum in two steps. The second of them, i.e., O-mannosylation is catalyzed by a set of Dol-P-Man: protein O-mannosyl transferases Pmt1-6p, with the most important Pmt1p and Pmt2p and

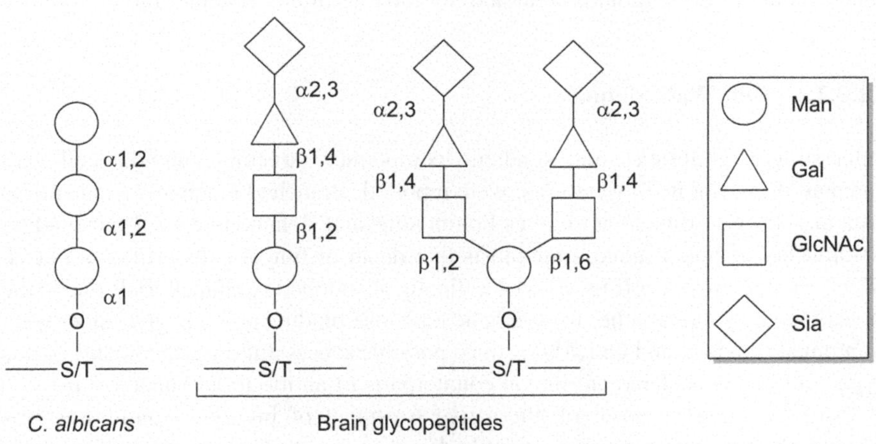

Fig. 21.6 Comparison of structures of candidal and mammalian O-mannosyl glycans. Man—mannose; Gal—galactose; GlcNAc—N-acetylglucosamine; Sia—sialic acid

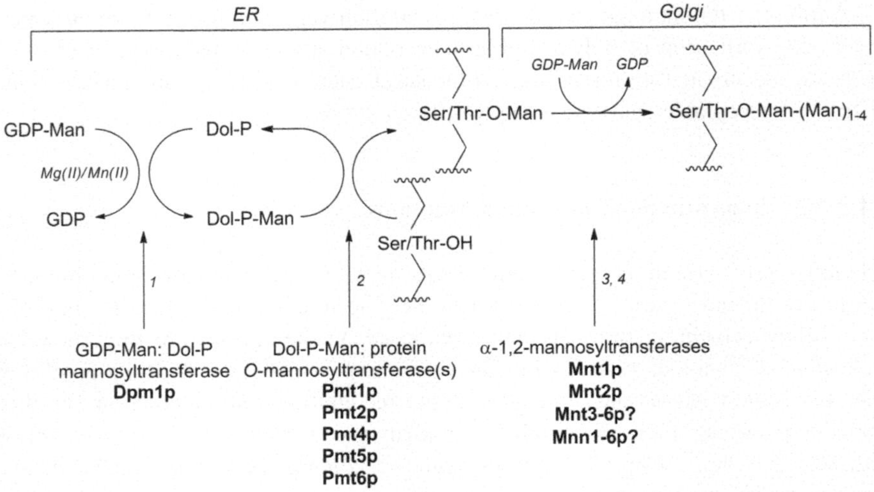

Fig. 21.7 O-mannosylation pathway in *C. albicans*. Dol-P—dolichol phosphate; Man—mannose; ER—endoplasmic reticulum

these enzymes were indicated as potential targets for antifungals. Derivatives of rhodanine-3-acetic acid, strong inhibitors of Pmt1p from *C. albicans*, exhibited also high growth inhibitory activity in vitro (Orchard et al. 2004). High target potential could be also attributed to α-1,2 mannosyltransferases Mnt1p and Mnt2p, responsible for addition of further Man residues in Golgi (Munro et al. 2005), since such enzymes are absent in mammals. Moreover, the *kre2* gene encoding α-1,2-mannosyltransferase in *S. cerevisiae* was one of the four genes selected in the comparative genomics studies aimed at identification of the most promising antifungal drug targets (Abadio et al. 2011). However, no specific inhibitors of these enzymes demonstrating antifungal activity have been described so far.

21.3.2.3 Cell Wall Mannan

Interestingly, antibiotics and synthetic compounds targeting candidal cell wall mannan itself, not its biosynthesis, were reported. Branched mannan was identified as a target for antifungal antibiotics Pramicidins and Benanomycins. Their binding to branched mannan induces apoptosis-like death of fungal cells (Hiramoto et al. 2003). Benanomycin A (Fig. 21.8) was finally abandoned in clinical trials due to the hepatic toxicity but searches for synthetic mannose-binding quinone glycosides were continued (Igarashi and Oki 2004), since possible successful drug candidates of this type could be considered antifungal counterparts of antibacterial vancomycin.

An interesting proposal of alternative treatment of invasive candidioses was immunotherapy with recombinant antibodies directed at *Candida* cell wall proteins. One of them are heat shock proteins Hsp70 and Hsp90. Mycograb™, a recombinant

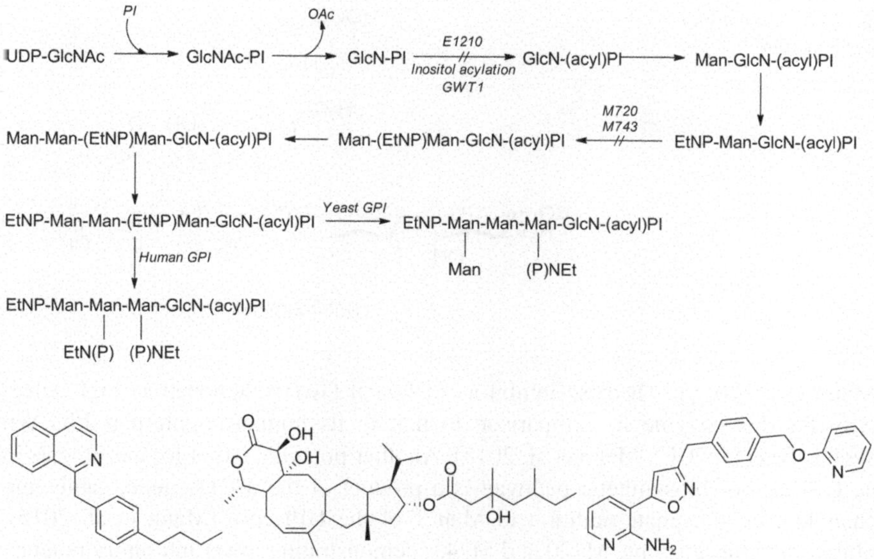

BIQ M743 E1210

Fig. 21.8 Biosynthesis of GPI anchor and some inhibitors of this pathway. PI—phosphatidylinositol; EtNP—ethanolamine phosphate; Man—mannose; GlcN—glucosamine

antibody derived from anti-Hsp90 antibody entered clinical trials and demonstrated synergy in combination with Fluconazole and Amphotericin B against a number of *Candida* spp, although was not finally approved for clinical use (Bugli et al. 2013).

21.3.2.4 Biosynthesis of GPI Anchor

Some of the cell wall mannoproteins are linked to the cell membrane by glycosylphosphatidylinositol (GPI), a complex glycolipid structure that acts as a membrane anchor. In *C. albicans*, complete GPI anchors are required for full morphogenesis, virulence, and resistance to macrophages (Richard et al. 2002). The GPI anchors are present in different eukaryotic cells and while the GPI core glycan is conserved in all organisms, many differences in additional modifications to GPI structures and biosynthetic pathways are known (de Macedo et al. 2003). Some of the enzymes catalyzing particular steps of the GPI anchor biosynthetic pathway have been proposed as molecular targets for antifungal chemotherapy (Fig. 21.9). One of them is an inositol acylase encoded in *C. albicans* by the *GWT1* gene (Umemura et al. 2003; Tsukahara et al. 2003). Its validity was first confirmed by 1-(4-butylbenzyl) isoquinoline BIQ (Tsukuhara et al. 2003). Further screening for Gwt1p inhibitors resulted in discovery of compound E1210, demonstrating high antifungal in vitro and in vivo activity (Hata et al. 2011), which has recently entered antifungal clinical development. Several other inhibitors have been also reported

Fig. 21.9 Structure of Benanomycin A

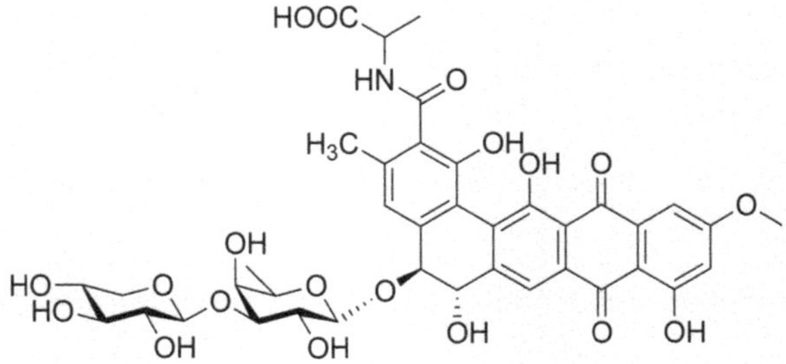

(Mann et al. 2015). All these inhibitors of fungal Gwt1p demonstrate high selectivity for this enzyme in comparison to that to its human counterpart PIG-Wp (Watanabe et al. 2012, Mann et al. 2015). Another potential target for antifungals in the GPI anchor biosynthetic pathway is a product of the *MCD4* gene, catalyzing ethanolamine phosphate addition to Man 1 of the GPI core (Mann et al. 2015). Inhibitors of this enzyme, M720 and M743, demonstrating lower inhibitory potency against its human counterpart PIG-Np, have been developed (Mann et al. 2015). The last step in the pathway of GPI formation in yeast (addition of the fourth mannose residue) is different than that in mammals (addition of another ethanolamine phosphate), so that the yeast enzyme is a potential target but its inhibitors demonstrating antifungal activity are not known.

21.3.2.5 Chitin Biosynthesis

Glucan and chitin are the skeletal components of the candidal cell wall. Although the former is the major constituent, chitin, a homopolymer of N-acetyl-D-glucosamine, is also essential for cell survival and effective inhibition of chitin biosynthesis results in fungicidal effect. Since chitin is not present in mammals, inhibition of its biosynthesis has been considered a safe and selective option for development of anticandidal agents. Chitin formation can be hampered upon inhibition of a chitin precursor (UDP-GlcNAc) biosynthesis in cytosol or of the GlcNAc polymerizing, cell membrane-located enzyme, chitin synthase. This enzyme accepts the GlcNAc units delivered in UDP-GlcNAc from the cytosolic side of the cell membrane and adds them to the growing chitin chain, extruded on the outer face of the membrane. Indeed, chitin synthase has been for many years considered one of the most promising targets for antifungals. Unfortunately, in human pathogenic fungi there is no single enzyme catalyzing chitin formation. *C. albicans* has four chitin synthases, Chs1, Chs2, Chs3, and Chs8. Although bulk chitin (80%) is synthesized by Chs3, this is Chs1, which is an essential enzyme responsible for formation of primary septum and Chs2 is the chitin synthase exhibiting the highest activity measured in vitro, although this is rather a repair enzyme (Munro 2013). Chs8 synthesizes long chitin fibrils arranged tangentially in the septum but contributes only a small proportion of in vitro Chs activity (Lenardon et al. 2007). Such a multiplicity of potential antifungal target is disadvantageous since the successful drug candidate should be an effective inhibitor of all isoforms.

A number of chitin synthase inhibitors are known, including substrate or bisubstrate analogs, transition state analogs, and noncompetitive inhibitors, but many of them do not exhibit antifungal activity (Behr 2011; Chaudhary et al. 2013). Among those demonstrating antifungal efficacy are oligopeptide antibiotics, Polyoxins and Nikkomycins, structural analogs of UDP-GlcNAc and thus competitive inhibitors of chitin synthase. Nikkomycin Z (Fig. 21.10) has been considered the most promising drug candidate, due to its fungicidal mode of action, a broad spectrum of antifungal activity, and good in vivo efficacy in murine candidiasis, cryptococcosis, histoplasmosis, and blastomycosis (Hector et al. 1990; Chapman et al. 1992) but stalled in Phase I clinical studies. Several attempts of chemical modification of Nikkomycins aimed at improvement of their antifungal properties, especially uptake by fungal cells, were not particularly successful (Ruiz-Herrera and San-Blas 2003). Interestingly, Nikkomycins *X* and *Z* demonstrated strong synergistic effect in combination with glucan synthase inhibitors and with inhibitors of ergosterol biosynthesis (Hector and Braun 1986; Milewski et al. 1991). Antifungal synergy with azole-type antifungals was also found for another chitin synthase inhibitor of natural origin, synerazol (Fig. 21.10), but this compound did not exhibit anticandidal in vivo activity (Ando et al. 1991). Synthetic RO-09-3143 (Fig. 21.10), resulting from the screening for potential chitin synthase inhibitors, appeared a very

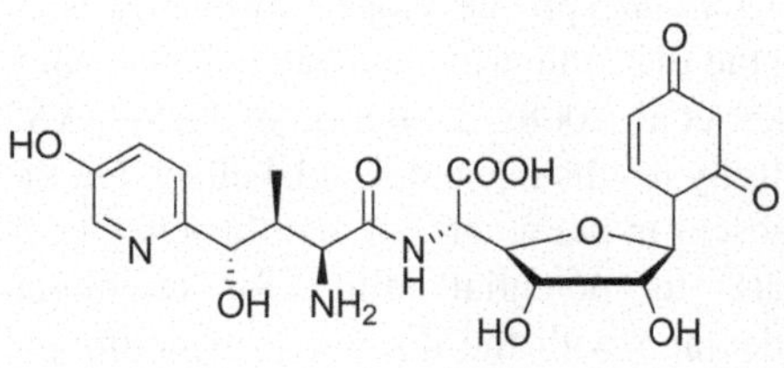

Nikkomycin Z

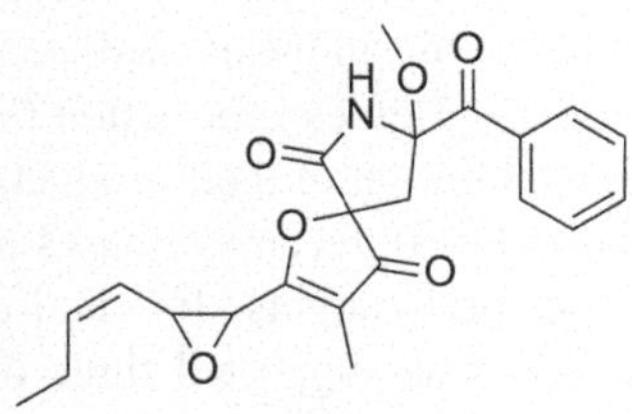

Synerazol

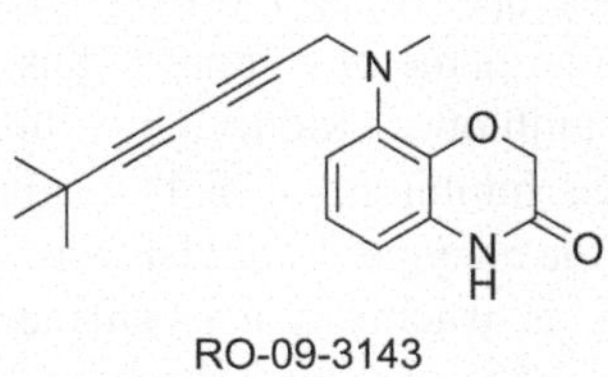

RO-09-3143

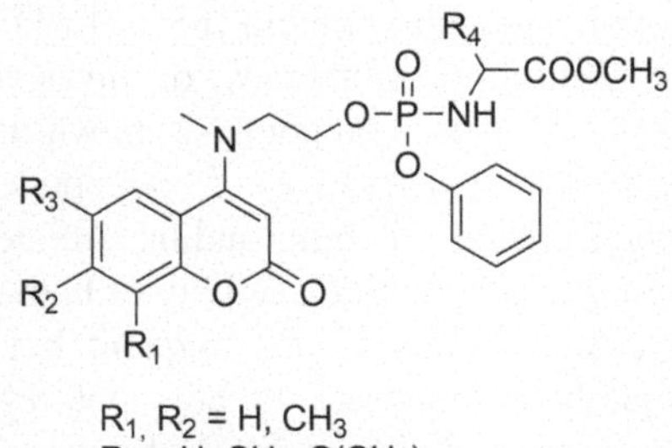

$R_1, R_2 = H, CH_3$
$R_3 = H, CH_3, C(CH_3)_3$
$R_4 = H, CH_3$, Benzyl, $CH(CH_3)_2$

Phosphoroamidate derivatives of coumarin

Fig. 21.10 Selected inhibitors of chitin synthase

strong noncompetitive inhibitor of Chs1 but not Chs2 and therefore was fungicidal only in a *chs2* null mutant background, while it was fungistatic for wild-type *C. albicans* cells (Sudoh et al., 2000). This result raises doubts whether searching for antifungal drug candidates among noncompetitive inhibitors of chitin synthase is a promising direction, since the enzyme inhibitory activity, specific for only one type of Chs isozymes, is not advantageous. The recently proposed phoshoramidate derivatives of coumarin (Fig. 21.10) could be an interesting novel option but nothing is known about their in vivo antifungal efficacy (Ji et al. 2016).

21.3.2.6 Glucan Biosynthesis

The major skeletal component of candidal cell wall is β(1→3)glucan. Its physiological role in fungal cell can be therefore related to that of peptidoglycan in bacteria. Rigidity provided to the cell by a dense β(1→3)glucan layer is additionally strengthened by covalent attachment to chitin.

The process of glucan synthesis and assembly is only partly known—much better in *S. cerevisiae* and rather poorly in *Candida*. The key polymerization of glucose units derived from the UDP-glucose is catalyzed by the membrane-located enzyme β(1→3)glucan synthase. This hetero oligomeric protein is composed of a transmembrane catalytic subunit Fks and a regulatory subunit, the GTP-binding protein Rho1, associated with Fks at the inner membrane side (Douglas 2001). In most fungi at least two genes encoding Fks are present; in *S. cerevisiae* there are three versions of Fks, with two of them, Fks1 and Fks2 being synthetically lethal (Lesage et al. 2004). Although several *Candida* spp. developed mechanisms of enhanced chitin biosynthesis mediated via the PKC integrity pathway compensating defects in β(1→3)glucan formation (Walker et al. 2008), disruption of *FKS1*/*FKS2* or effective inhibition of their products activity results in candidacidal effect. On the other hand β(1→3)glucan synthases are absent in mammals, so that this enzyme is one of the best targets for antifungals. Its potential utility for antifungal chemotherapy was suggested three decades ago with the discovery of antifungal antibiotics targeting Fks, papulacandins, and aculeacins (Hector 1993).

Nowadays, three inhibitors of β(1→3)glucan synthase: caspofungin, micafungin, and anidulafungin (Fig. 21.11) are parenterally administered drugs, used in the clinical chemotherapy of invasive candidiases and aspergilloses (Denning 2003). All these compounds, known under a common name of echinocandins, are semisynthetic derivatives of lipopeptide antifungal antibiotics produced by different fungal species. Echinocandins are noncompetitive inhibitors of β(1→3)glucan synthase, able to block activity of both isoforms of the enzyme. Candidal resistance to echinocandins is not frequent but if happens, is usually a consequence of single-base mutations in hot spot regions of the *FKS* sequence (Perlin 2011). Another group of promising inhibitors of β(1→3)glucan synthase are derivatives of enfumafungin (Fig. 21.11), a triterpene glycoside natural product. These compounds, demonstrating antifungal activity similar to that of echinocandins, but orally bioavailable, are currently under development (Apgar et al. 2015).

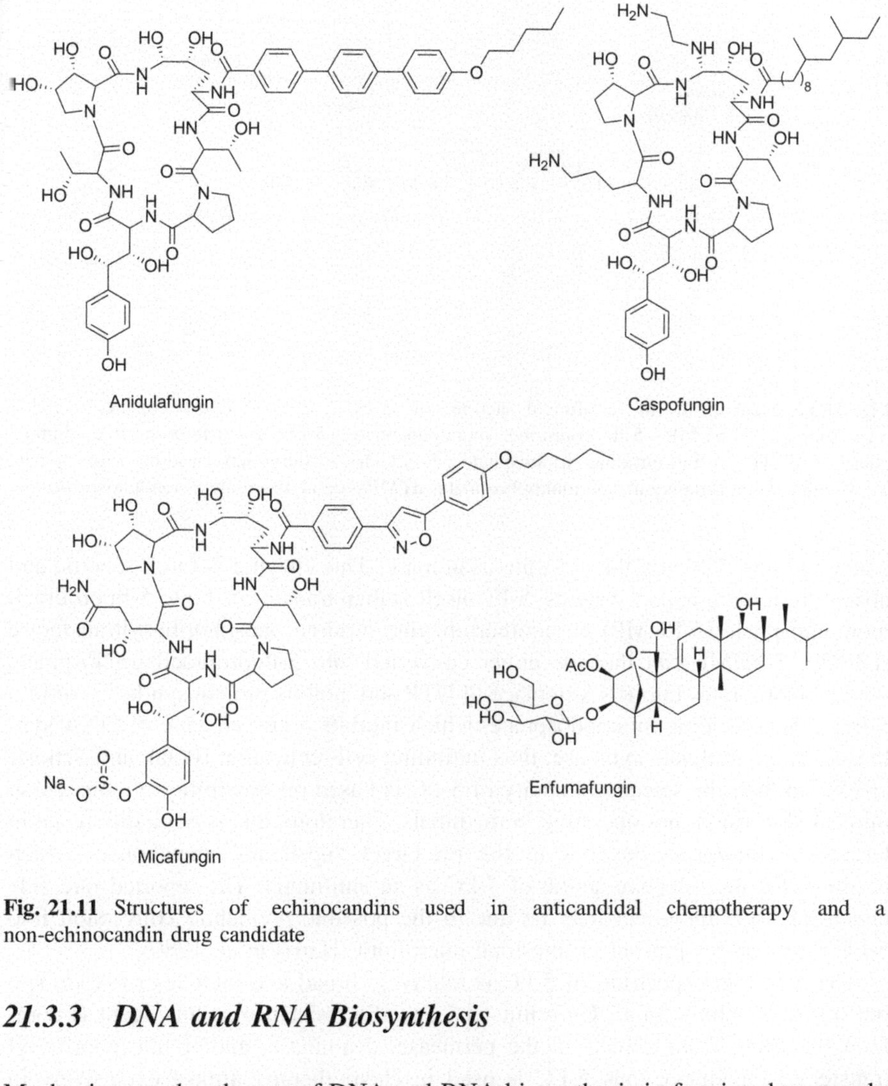

Fig. 21.11 Structures of echinocandins used in anticandidal chemotherapy and a non-echinocandin drug candidate

21.3.3 DNA and RNA Biosynthesis

Mechanisms and pathways of DNA and RNA biosynthesis in fungi and mammals are thought to be very similar, if not identical, so that one cannot expect possible targets for antifungal chemotherapy there. However, one of the antifungal chemotherapeutics is a prodrug, specifically converted in fungal cells to an active inhibitor of nucleic acid biosynthesis. This is 5-fluorocytosine (5-FC), synthesized originally as a potential antitumor agent, never used in antitumor chemotherapy but successfully applied for the treatment of systemic candidiasis and of cryptococcal meningitis (Vermes et al. 2000).

Mechanism of antifungal action of 5-FC is shown in Fig. 21.12. This fluoropyrimidine rapidly enters the fungal cells through specific transmembrane carriers, such as cytosine permease or pyrimidine transporters and is subsequently

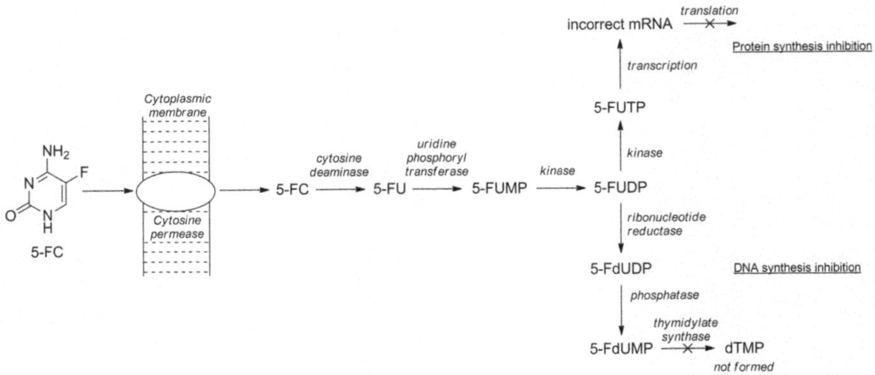

Fig. 21.12 Mechanism of antifungal action of 5-FC. 5-FC—5-fluorocytosine; 5-FU—5-fluorouracil; 5-FUMP—5-fluorouridine monophosphate; 5-FUDP—5-fluorouridine diphosphate; 5-FUTP—5-fluorouridine triphosphate; 5-FdUDP—5-fluorodeoxyuridine diphosphate; 5-FdUMP—5-fluorodeoxyuridine monophosphate; dTMP—deoxythymidine monophosphate

converted into 5-FU by the cytosine deaminase. This enzyme is fungi-specific and absent from mammalian systems. 5-FU itself is then transformed into 5-fluorouracil monophosphate (5-FUMP) by another enzyme, uridine phosphoribosyltransferase (UPRT). 5-FUMP can then be either converted into 5-fluorouracil triphosphate, which incorporates into RNA in place of UTP and inhibits protein synthesis, or into 5-fluorodeoxyuridine monophosphate, which inhibits a key enzyme of DNA synthesis, the thymidylate synthase, thus inhibiting cell replication (Polak and Scholer 1975). So that, the selective toxicity of 5-FC is based on possibility of conversion into 5-FU in fungi, not operating in mammals. Therefore, this is not a difference in target structure/properties but in the pre-target "upstream" metabolism which accounts for the selective action of 5-FC as an antifungal. The reported rare side effects of 5-FC treatment may be due to the possible premature conversion into 5-FU by microbes present in intestinal microflora (Harris et al. 1986).

The antifungal spectrum of 5-FC is relatively broad and includes *Candida* spp but not filamentous fungi. Unfortunately, specific resistance to this agent is common (resulting from defects in the permease, deaminase, and/or phosphoribosyl transferase enzymes), thus 5-FC is used in chemotherapy almost exclusively in combination with Amphotericin B (Banerjee et al. 2014).

21.3.4 Amino Acid Biosynthesis

Nine out of twenty proteinogenic amino acids, namely Phe, Val, Thr, Trp, Ile, Met, Leu, Lys, and His, are regarded as essential for humans and must be acquired from the diet. The human pathogenic fungi have developed the own pathways of biosynthesis of all amino acids essential for humans. Several steps of these pathways are catalyzed by enzymes that are absent from mammalian cells and unique for fungi, so that are potential targets for antifungal chemotherapy. A number of inhibitors of these enzymes exhibiting antifungal activity have been reported.

Obviously, usefulness of enzymes of amino acid biosynthetic pathways as potential targets for antifungal chemotherapy might be questionable. Requirement of fungal cells for a particular amino acid, biosynthesis of which is inhibited by any enzyme inhibitor, may be satisfied by the exogenous supply provided by amino acids and peptides present in human serum, what obviously cancels the possible chemotherapeutic effect of the inhibitor. From this point of view, the fungi-specific pathways of methionine and tryptophan biosynthesis seem to be the most promising targets, since the human serum levels of these two essential amino acids are particularly low (<20 μM). Moreover, *Candida* mutants auxotrophic for L-methionine were avirulent in animal models (Aoki et al. 1995; Suliman et al. 2007).

21.3.4.1 Aspartate-Derived Pathways of Threonine, Methionine, and Isoleucine Biosynthesis

L-Threonine, L-isoleucine, and L-methionine belong to the so-called aspartate family and are synthesized through the pathways absent in mammals. The common initial part of these pathways is a three-step conversion of L-aspartate to L-homoserine, catalyzed by aspartate kinase Hom3p, aspartate semialdehyde dehydrogenase Hom2p, and homoserine dehydrogenase Hom6p. Two further steps catalyzed by homoserine kinase Thr1p and threonine synthase Thr4p afford finally L-threonine. L-methionine is formed from L-homoserine through L-homocysteine, which is S-methylated in the final step by methionine synthase Met6p. L-isoleucine derives from L-threonine (four further steps). It was shown that deletion of *THR1*, *THR4,* or *MET6* attenuated virulence of *C. albicans* (Suliman et al. 2007). Deletion of *HOM6* causes translational arrest in *C. albicans* cells grown under amino acid starvation conditions and reduces cell adhesion to polystyrene (Tsai et al. 2016). Moreover deletion of *THR1* or *MET6* or inhibition of activity of Thr1p or Met6p results in intracellular accumulation of toxic intermediates, L-homoserine or L-homocysteine (Kingsbury and McCusker 2010a, b, c). Accumulation of L-homocysteine additionally disturbs ergosterol biosynthesis (Parks and Casey 1995).

Effective antifungal agents were found among inhibitors of homoserine dehydrogenase, Hom6p. A natural compound, (S)-2-amino-4-oxo-5-hydroxypentanoic acid (Fig. 21.13), known as an antibiotic RI-331, is an enzyme-assisted suicide inhibitor of homoserine dehydrogenase (Yamaki et al. 1990; Jacques et al. 2003). This compound is active against medically important yeasts of the *Candida* genus. Furthermore, it was shown effective in the treatment of systemic murine candidiasis, being well tolerated in mice (Yamaguchi et al. 1988). High antifungal in vitro activity was found for the phenolic compounds resulting from the high-throughput screen of a library of small molecules, however there were doubts, whether Hom6p was in fact their primary target (Ejim et al. 2004).

Activity of threonine synthase, Thr4p, was inhibited by L-(Z)-2-amino-5-phosphono-3-pentenoic acid, known as Rhizocticin A (Fig. 21.13). This compound is transported to the fungal cells by the oligopeptide transport system and cleaved inside by intracellular peptidases to release an active inhibitor of Thr4p, which leads to the growth inhibitory effect (Kugler et al. 1990). Inhibitors of homoserine kinase Thr1p demonstrating antifungal activity were also reported (De Pascale et al. 2011a, b).

The methionine branch of the aspartate pathway seems especially promising due to the well-documented attenuated virulence and growth defects of *Candida* strains auxotrophic for L-methionine (Aoki et al. 1995; Suliman et al. 2007). At least two enzymes catalyzing particular steps of this pathway were proposed as targets, namely homoserine transacetylase Met2p (Nazi et al. 2007) and methionine synthase Met6p (Suliman et al. 2007). Activity of the former is inhibited by a natural substance, Ebelactone A and its synthetic derivatives but their antifungal activity was moderate (De Pascale et al. 2011a). Surprisingly enough, inhibitors of methionine synthase Met6p demonstrating antifungal activity have not been reported yet. Although Met6p is present also in mammalian systems, this is a cobalamin-dependent enzyme, while the fungal Met6p uses 5-methyl-THF as a methyl donor (Banerjee and Matthews 1990; Prasannan et al. 2009), so that it seems that differences between the fungal and mammalian versions of methionine synthase may be exploited in search for highly selective antifungal agents. Another compound strongly affecting methionine biosynthesis was an antibiotic azoxybacillin (Fig. 21.13) demonstrating good antifungal in vitro activity in minimal media (Aoki et al. 1994). This activity was due to the interference with expression of genes encoding ATP sulfurylase, homoserine transacetylase, and sulfite reductase (Aoki et al. 1996), possibly by binding to an unidentified transcriptional factor controlling expression of genes of sulfur assimilation and methionine biosynthesis. Azoxybacillin itself exhibited very low antifungal activity in an animal infection

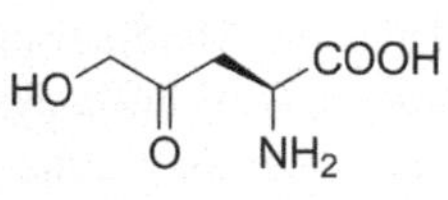

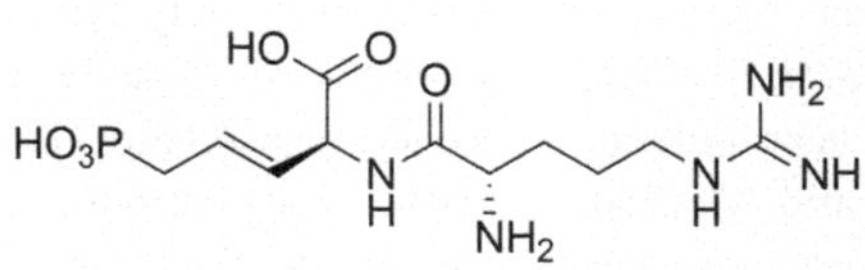

RI-331 Rhizocticin

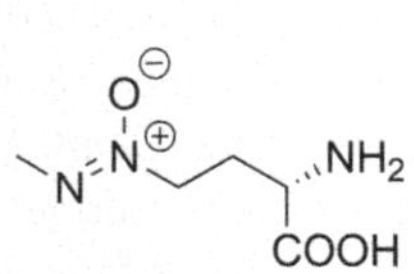

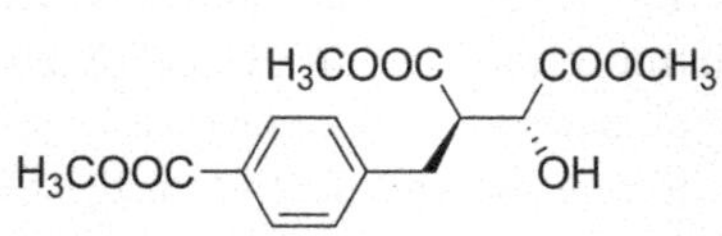

Azoxybacilin (*2R,3S*)-3-(*p*-carboxybenzyl)malate trimethyl ester

Fig. 21.13 Selected inhibitors of amino acid biosynthesis

model that could be only slightly improved for benzyl and t-butyl esters of this compound (Aoki et al. 1995, 1996).

21.3.4.2 Lysine Biosynthesis

Bacteria and fungi have evolved two different pathways of lysine biosynthesis: L, L-diaminopimelate pathway in bacteria and α-aminoadipate pathway (α-AA) in fungi. Enzymes catalyzing three initial steps of the α-AA pathway, namely homocitrate synthase (HCS), homoaconitase (HA), and homoisocitrate dehydrogenase (HID) are specific for higher fungi and as such, seemed to be proper candidates for antifungal targets. Some inhibitors of these enzymes have been reported: thiahomocitrate is a very strong inhibitor of HID but lacking antifungal activity (Yamamoto and Eguchi 2008) while (2R, 3S)-3-(*p*-carboxybenzyl) malate trimethyl ester (Fig. 21.13) targets the same enzyme and exhibits some antifungal activity in minimal media (Gabriel et al. 2013). Nevertheless, *C. albicans* mutants with disrupted genes encoding HS or HID demonstrated virulence similar to that of the wild-type cells and no growth defect in lysine-containing media (Kur et al. 2010; Gabriel et al. 2014), thus suggesting that these enzymes are not promising targets for antifungal chemotherapy. The main reason for little if any effect of lysine auxotrophy on *Candida* survival under in vivo conditions seems to be a relatively high level of L-lysine in human serum, well above the threshold limit ~ 0.1 mM determined in vitro in minimal media (Gabriel et al. 2014).

21.3.4.3 Biosynthesis of Branched-Chain Amino Acids

Usefulness of the branched-chain amino acids pathway (Leu and Val) as a source of potential targets for antifungal chemotherapy is controversial. Several inhibitors of some of the enzymes of this pathway are known and especially acetolactate synthase Ile2p was shown to be a target of several structurally different classes of inhibitors widely used as herbicides, particularly the sulfonylureas (including sulfometuron methyl), imidazolinones, and sulfonanilides (Grandoni et al. 1998). Deletion of the *ILV1* gene coding for threonine deaminase and of the *ILV2* gene encoding acetohydroxyacid (acetolactate) synthase resulted in attenuated virulence of *C. albicans* in a murine model of infection (Kingsbury and McCusker 2010b) but on the other hand, the *ILV5* gene encoding enzyme catalyzing the second step in the common part of the branched-chain amino acid biosynthesis was shown to be nonessential for *C. albicans* virulence in a murine infection model (Becker et al. 2010). Significant antifungal in vitro activity against *C. albicans* exhibited some triazolo-pyrimidine-sulfonamides targeting Ilv2p, however this effect could be bypassed through supplementation with exogenous branched-chain amino acids or by the addition of serum to the medium (Richie et al. 2013).

21.3.4.4 Biosynthesis of Aromatic Amino Acids

Although the enzymes of the aromatic amino acid pathway (with shikimate as the major common intermediate) from bacteria and plants have been extensively studied, their fungal counterparts are rather poorly characterized. Plant DAHP synthase is a target for glyphosate (*N*-phosphomethylglycine), a well-known herbicide but inhibitors of the fungal enzyme with potential application in antifungal chemotherapy are not known. Disruption of the *ARO3* and *ARO4* genes encoding catalytically redundant 3-deoxy-D-arabinoheptulosonate-7-phosphate (DAHP) synthases catalyzing the first committed step of aromatic amino acids biosynthesis in *C. albicans* resulted in auxotrophy for Phe, Tyr, and Trp, and the growth impairment could be only in part rescued by supplementation of the growth medium with 5 mM aromatic amino acids (Sousa et al. 2002). Search for the inhibitors of the tryptophan branch of the aromatic pathway as potential antifungal drugs seems especially promising since the L-Trp level in serum is low.

21.3.4.5 Histidine Biosynthesis

Recently, it was shown that blocking histidine synthesis in *A. fumigatus* leads to reduced growth and attenuated virulence (Alcazar-Fuoli 2016) thus confirming a potential usefulness of enzymes catalyzing particular steps in histidine biosynthetic pathway as possible targets for antifungal chemotherapy. Previously, it was suggested that imidazole glycerol phosphate synthase His7p could be an attractive target (Rivalta et al. 2012), however effective inhibitors exhibiting antifungal activity are still to be discovered.

21.3.5 Protein Biosynthesis

A few proteins involved in protein biosynthesis or posttranslational processing were proposed as targets for anticandidal chemotherapy, namely two of the aminoacyl-tRNA synthases, one of the elongations factors and an enzyme catalyzing one of the posttranslational protein modifications. None of them resulted from any directed screening or rational choice but all were identified as targets of newly discovered antifungal antibiotics. This is not surprising, since any fundamental differences between fungal and mammalian translation that could constitute a molecular basis for selective toxicity are not known.

21.3.5.1 Aminoacyl-tRNA Synthetases

Aminoacyl-tRNA synthetases (aaRSa) catalyze attachment of amino acids to their cognate tRNAs to produce the aminoacyl tRNAs (also aa-tRNA or charged tRNA) that are the substrates for translation. The aaRSs enzymes are not only responsible for producing the raw materials for protein biosynthesis, but also for ensuring the fidelity of translation. These enzymes are found in all living organisms and most organisms contain at least 20 different aaRSs. All aaRSs have a pivotal role in translation of mRNA, and thus are of vital importance, however it might be expected that a great homology exists between respective aaRS from different organisms, so that selectivity of a potential inhibitor may be hard to achieve. Nevertheless, surprisingly great number of aaRSs have been proposed as targets for antimicrobial or antiparasitic chemotherapy (Vondenhoff and Van Aerschot 2011; Pham et al. 2014). In addition, mupirocin, inhibitor of bacterial isoleucyl-tRNA synthetase (IleRS), is used as a drug (Bactroban™) for topical treatment of *Staphylococcus aureus* (Nakama et al. 2001). The same aaRS in its fungal version is a target for cispentacin, a cyclic β-amino acid produced by *Bacillus cereus* and *Streptomyces setonii,* which exhibited in vivo activity against *C. albicans* infection in mice (Oki et al. 1989). Through active transport by the proline-specific permease this compound accumulates in yeast cells up to 200-fold of the extracellular concentration (Capobianco et al. 1993). Icofungipen, previously known as BAY 10-8888 (Fig. 21.14), was discovered through a program directed toward a more potent derivative of cispentacin (Ziegelbauer et al. 1998). Good clinical efficacy and safety were observed in phase I and II clinical trials, although low mycologic eradication rates were observed in HIV-positive patients. To this end, higher dosage may be desirable (Ochsner et al. 2007). Another fungal aaRS indicated as a potential target was leucyl-tRNA synthetase (LeuRS) inhibited by 5-fluoro-1,3-dihydro-1-hydroxy-2,1-benzoxaborole AN2690, that entered clinical trials (Rock et al. 2007; Seiradake et al. 2009).

21.3.5.2 Elongation Factors

Although protein synthesis is a universal process in living cells, it has always been considered as one of the more attractive targets for the development of

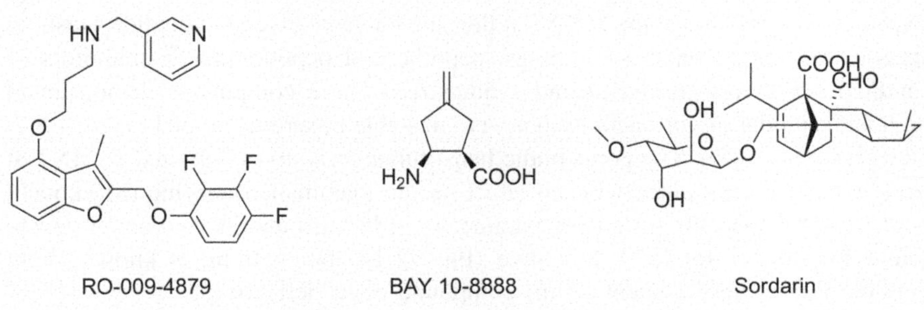

Fig. 21.14 Inhibitors of protein biosynthesis or posttranslational processing

antimicrobial agents and some well-known antibacterials, including erythromycin, tetracycline, and chloramphenicol are inhibitors of translation, exploiting structural differences between prokaryotic and eukaryotic translational machineries. It is known that the fungal translation has also exploitable differences relative to its mammalian counterpart, including the two soluble protein factors: elongation factor 3 (eEF3), which is absent from mammalian cells (Triana-Alonso et al. 1995), and eEF2, which is functionally distinct from its mammalian counterpart (Domínguez and Martín 1998). A screening program for inhibitors of *C. albicans* protein synthesis in vitro established in the 90s of the previous century resulted in discovery of a new antibiotic class called sordarins (Kinsman et al. 1998; Dominguez et al. 1999).

Sordarins (structure of one of them is shown in Fig. 21.14) bind to the eEF2, stabilizing the EF2/ribosome complex via a likely additional interaction with ribosomal subunit protein (Søe et al. 2006), and preventing the final step of translocation in protein synthesis. Despite eEF2 being highly conserved among eukaryotes, sordarin and its derivatives interact exclusively with a very specific region of eEF2, which is fungi-unique, particularly to *Candida* spp. (Chakraborty et al. 2013). Several synthetic sordarin derivatives were obtained, exhibiting improved in vivo antifungal activity in animal models and low cytotoxicity (Aviles et al. 2000; Herreros et al. 2001; Kamai et al. 2005). High specificity of sordarins for the fungal target and the relative ease with which new variants of these agents can be synthesized hold promise for positive future developments with this series.

Surprisingly enough, compounds targeting eEF3, with concomitant antifungal activity, have not been discovered so far, so that eEF3 is still a potential target.

21.3.5.3 N-Myristoyltransferase

Myristoyl-CoA:protein N-mirystoyltransferases (NMTs) are the family of enzymes catalyzing transfer of myristate from myristoyl-CoA (MYA) to the N-terminal glycine of proteins, following the removal of the initiator methionine residue of growing polypeptide chain. NMTs are essential for growth and survival of human pathogenic fungi (Weinberg et al. 1995). Fungal and human NMTs share high sequence conservation at the MYA binding site but have divergent peptide substrate specificities. Based on these findings, peptidic and peptidomimetic inhibitors of candidal NMT were designed and synthesized. These compounds demonstrated high anticandidal in vitro activity but were unstable in serum (Sikorski et al. 1997; Devadas et al. 1997). The nonpeptidic benzofuran derivatives inhibited CaMNT at nanomolar concentrations, with no effect on the mammalian enzyme (Masubuchi et al. 2003). Especially promising parameters of in vitro and in vivo activity were found for the RO-09-4879 derivative (Fig. 21.14) but nothing is known about possible further development of this compound or its derivatives.

21.3.6 Signaling Pathways

A number of various signaling pathways, responsible mainly for response to environmental stimuli, stress and adaptation to changing conditions, operate in candidal cells. Although most of the genes encoding components of the signaling systems are not essential, products of these genes are very often important for fungal virulence. Some of these pathways or at least some of their components are fungi-specific, so that they are considered potential antifungal targets.

21.3.6.1 pH Signaling

An ability to respond to a wide pH range in broadly fluctuating environmental conditions allows human pathogenic fungi to cope with rapid changes in their extracellular settings (Davis 2009). The primary pH response is mediated by the PacC/Rim signaling pathway (Xu et al. 2004). In *C. albicans* this pathway is required for many functions associated with pathogenesis and virulence (Davis 2003). The pathway is fungi-specific, does not exist in mammalian cells, and components of this system have been proposed as possible targets for antifungal chemotherapy (Cornet and Gaillardin 2014).

21.3.6.2 PKC1-Dependent Cell Wall Integrity Pathway (CWIP)

Integrity of the cell wall during fungal growth is essential for survival and adaptation to environmental stresses. Calcium-dependent protein kinase (PKC1)-dependent cell wall integrity pathway is the main signaling system for sensing cell wall perturbations under many cell stress conditions. The signaling activity of this pathway is initiated at the cell surface through sensor–transducer proteins and is transmitted to downstream effectors through Rom2p-Rho1p complexes culminating with the activation of the MAPK/Slt2p. The antifungal natural product cercosporamide is a highly selective and potent Pkc1p inhibitor (Sussman et al. 2004) that in combination with caspofungin has a synergistic effect against *C. albicans.*

21.3.6.3 Calcium Signaling Pathways

Calcium homeostasis is associated with numerous physiological processes in *C. albicans*, such as stress responses, virulence, hyphal development, and adhesion (Liu et al. 2015). One of the regulators of calcium homeostasis, calcineurin (CN), has been identified as a virulence factor, mycelial growth regulator, and drug resistance mediator in *Candida* spp. (Chen et al. 2011; Brand et al. 2007). A number of compounds, including calcium channel blockers, calmodulin-binding agents, calcineurin effectors, and inhibitors of the calcium-triggered secretory systems, exhibit antifungal activity when used alone or in combination with antifungal drugs by interfering with components in the calcium signaling pathway (Liu et al. 2015). Interference in calcineurin-mediated signaling was observed for the well-known household medicine and spice curcumin, demonstrating antifungal activity (Kumar et al. 2014).

21.3.6.4 Two-Component Signaling Systems

The two-component signal transduction pathways are based on the transfer of phosphoryl groups among their components (phosphorelays) and are one of the primary means by which microorganisms sense and respond to environmental cues. The fungal versions of this pathway involves three proteins: histidine kinase, histidine phosphotransferase, and response regulator protein. In *C. albicans* at least four such signaling systems have been identified (Shor and Chauhan 2015). Two-component signal transduction pathways present attractive targets for antifungal drug discovery because they exist in fungi but not in mammalian cells. Moreover, they were shown to be important for virulence (Yamada-Okabe et al. 1999).

The mitogen-activated protein kinase (MAPK) pathways are one of the most important eukaryotic signal networks allowing adaptation to environmental changes. Four such pathways have been identified in *C. albicans*: the Mkc1 pathway, the Cek1 pathway, the Cek2 pathway, and the high-osmolarity glycerol (HOG) pathway, and shown to be involved in invasive hyphal growth, morphogenesis, biogenesis of the cell wall and the stress response (Román et al. 2009). The HOG pathway is composed of a two-component-system (TCS)-like phosphorelay system and the Hog1-type mitogen-activated protein kinase cascade. The TCS is not present in mammals (Chauhan and Calderone 2008) and possibility of selective inhibition of Hog1 in *Saccharomyces cerevisiae* by 4- and 5-substituted 1,2,3-triazoles was demonstrated (Dinér et al. 2011), thus confirming the target potential of this pathway.

21.3.6.5 TOR Signaling Pathway

The targets of rapamycin (TOR) proteins are members of a ubiquitous family of signaling proteins that have been discovered by mutations that control cell growth by conferring resistance to the growth inhibitory effect of the potent antifungal agent rapamycin. The TOR pathway which plays an important role in *C. albicans* morphogenesis and virulence (Zacchi et al. 2010) is composed of two functionally distinct protein complexes TORC1 and TORC2 and the former is specifically inhibited by rapamycin (Loewith et al. 2002). At least two components of TOR, Sit4p and Tco89p, do not have mammalian counterparts (Betz and Hall 2013), what makes them perfect antifungal target candidates.

21.3.7 Virulence Factors and Virulence-Regulating Transcription Factors

A number of virulence factors have been identified in human pathogenic fungi. Some of them are ubiquitous but most are species-specific. Principal virulence factors of *C. albicans* are yeast-to-mycelia morphological transformation, ability to biofilm formation and production of some secretory hydrolytic enzymes, especially aspartic proteases and phospholipases. Targeting the virulence factors or virulence-regulating transcription factors have been proposed as novel antifungal strategies (Pierce and Lopez-Ribot 2013; Bahn 2015), alternative to those aimed at cell viability. Although any novel antifungals targeting virulence factors would possibly exhibit narrow spectrum of activity, their expected advantage should be much weaker selective pressure for the development of fungal resistance (Clatworthy et al. 2007). On the other hand, it is doubtful whether virulence-inhibiting small molecule inhibitors would be effective in immunocompromised patients, who constitute the major group of the population prone to Candida infections. It seems therefore that virulence-inhibiting drugs could rather be used for prophylactic or an adjuvant therapy or as a pretreatment for medical devices that are at high risk of Candida colonization.

Searching for novel targets related to biofilm formation in *C. albicans* and developing the corresponding specific drugs are especially important in light of the growing number of life-threatening disseminated candidiases resulting from the catheter-based infections. It is known that *C. albicans* biofilm formation is controlled by six transcriptional regulators, of which the Bcr1p and Efg1p were proposed as a possible antifungal targets (Nobile et al. 2006; Lassak et al. 2011). On the other hand, several small molecule compounds inhibiting biofilm formation by *C. albicans* cells were found in the high-throughput screenings (LaFleur et al. 2011; Siles et al. 2013) but their actual targets were not identified.

Filamentation of a dimorphic fungus *C. albicans* not only represents a virulence trait per se, but is also coordinately regulated with other virulence factors, which are associated with cellular morphology. It is clear that yeast-to-mycelia transition is a complex process, as signaling pathways may converge on separate or identical transcription factors and transcription factors may converge themselves on common target genes to trigger expression of hypha-specific genes (Sudbery 2011). Definitely, there is no single gene controlling morphological transition but *UME6* and *NRG1* were identified as especially important (Saville et al. 2006), so that they could be considered potential anticandial targets.

A number of small molecule compounds were reported as inhibitors of *C. albicans* filamentation but probably most of them act nonspecifically (Shareck and Belhumeur 2011). On the other hand, 21 specific inhibitors from the high-throughput screening were found but their effectiveness was confirmed only in vitro (Toenjes et al. 2005; Midkiff et al. 2011). The in vivo validation is in this case important, since, for example, exogenous administration of farnesol, which in vitro prevents the *C. albicans* yeast-to-mycelium conversion, surprisingly increased virulence in the murine model of disseminated candidiasis (Navarathna et al. 2007).

Definitely inhibitors of *C. albicans* secretory aspartic proteases of phospholipases are unlikely to be fungicidal agents but could support antifungal chemotherapy (Sousa dos Santos 2010). A number of inhibitors of C. albicans aspartic proteases (Sap) are known, including pepstatin A, but their antifungal potential as single agents is low (Santos and Braga-Silva 2013). The phospholipase inhibitor ebelactone B and the proteinase inhibitor pepstatin A were reported to reduce the damage to human tissues caused by Candida species (Trofa et al. 2009; Gácser et al. 2007).

21.3.8 Other Targets

A few proteins, mainly enzymes, not belonging to any group specified above, have been also proposed as novel possible anticandidal targets.

21.3.8.1 Histone Deacetylase

Histone deacetylases (HDACs) are a family of enzymes which deacetylate N^{ω}-acetyl-lysine residues on core histones and other cellular proteins, playing an important role in gene regulation and also in the control of cell proliferation and motility. Compound MGCD290, an inhibitor of Hos2p histone deacetylase exhibited synergistic growth inhibitory effect in combination with azole antifungals against opportunistic fungal pathogens, including *C. albicans* (Pfaller et al. 2009).

21.3.8.2 Topoisomerase 1

Topoisomerase 1 (Top1p) is essential for cell survival and is a virulence factor for some fungi and its deletion in *C. albicans* induces slow cellular growth and aberrant cell morphology (Fostel et al. 1992). The fungal *TOP1* gene has a considerable amount of coding sequence not present in human homologs (Stewart et al. 1996), thus suggesting differences in their protein structure and presenting the possibility of exploring these topoisomerases as drug targets.

21.3.8.3 Vacuolar-Type H^+-ATPase (V-ATPase)

Pathogenic fungi require optimum V-ATPase function for secretion of virulence factors, induction of stress response pathways, hyphal morphology, and homeostasis of pH and other cations in order to successfully survive within and colonize the host. Thus, compounds that inhibit the V-ATPase directly, or indirectly by modulating the membrane milieu, can profoundly affect fungal viability and virulence (Zhang and

Rao 2012), what justifies a systematic screen for fungal-specific V-ATPase inhibitors with potential application in antifungal chemotherapy.

21.3.8.4 Glyoxalate Cycle—Isocitrate Lyase

Presence of the glyoxylate cycle (absent in mammalian host) enables *C. albicans* to survive in nutrient-limited host niches, so that the key enzymes of this cycle, isocitrate lyase Iclp and malate synthase Mlsp could be potential antifungal targets. Antifungal activity of some inhibitors of *C. albicans* Iclp was demonstrated (Cheah et al. 2014; Bae et al. 2015) what paves the way for new inhibitor design and seems a good starting point for search for the more effective compounds.

21.4 Conclusions

The extensive antifungal target search efforts have not resulted yet in extention of a very limited set of defined targets for clinical antifungal chemotherapeutics. Genomic-based searches revealed several potential targets but none of them has been validated in terms of finding clinically useful inhibitors of these targets. On the other hand, some well-understood, fungal-specific targets, such as chitin synthase, have so far eluded chemical exploitation. It seems obvious that generation of new antifungal drugs against *C. albicans* should be a joint effort by pharmacologists, pathologists, chemists, and medical scientists. Exploitation of pathogenicity factors as antifungal targets may lead to discovery of novel agents but rather for antifungal prophylaxis or for support of direct chemotherapy. Hopefully, generation of new antifungal drugs will be achieved in the near future with continued progress in recognition of novel targets for antifungal chemotherapy.

References

Abadio AKR, Kioshima ES, Teixeira MM, Martins NF, Maigret B, Felipe MSS (2011) Comparative genomics allowed the identification of drug targets against human fungal pathogens. BMC Genom 12:75

Aeed PA, Young CL, Nagiec MM, Elhammer AP (2009) Inhibition of inositol phosphorylceramide synthase by the cyclic peptide aureobasidin A. Antimicrob Agents Chemother 53:496–504

Alcazar-Fuoli L (2016) Amino acid biosynthetic pathways as antifungal targets for fungal infections. Virulence. doi:10.1080/21505594.2016.1169360

Anderson TM, Clay MC, Cioffi AG, Diaz KA, Hisao GS, Tuttle MD, Nieuwkoop AJ, Comellas G, Maryum N, Wang S, Uno BE, Wildeman EL, Gonen T, Rienstra CM, Burke MD (2014) Amphotericin forms an extramembranous and fungicidal sterol sponge. Nat Chem Biol 10:400–406

Ando O, Satake H, Nakajima M, Sato A, Nakamura T, Kinoshita T, Furuya K, Haneishi T (1991) Synerazol, a new antifungal antibiotic. J Antibiot 44:382–389

Andruszkiewicz R, Milewski S, Zieniawa T, Borowski E (1990) Anticandidal properties of N^3-(4-methoxyfumaroyl)-L-2,3-diaminopropanoic acid oligopeptides. J Med Chem 33:132–135

Aoki Y, Kamiyama T, Fujii T, Yamamoto M, Ohwada J, Arisawa M (1995) Design of an antifungal methionine inhibitor not antagonized by methionine. Biol Pharm Bull 18:1267–1271

Aoki Y, Kongoh M, Nakamura M Fujii T, Yamazaki T, Shimada H, Arisawa M (1994) A new methionine antagonist that has antifungal activity: mode of action. Jpn J Antibiot 40:127–132

Aoki Y, Yamamoto M, Hosseini-Mazinani SM, Koshikawa N, Sugimoto K, Arisawa M (1996) Antifungal azoxybacilin exhibits activity by inhibiting gene expression of sulfite reductase. Antimicrob Agents Chemother 40:127–132

Apgar JM, Wilkening RR, Greenlee ML, Balkovec JM, Flattery AM, Abruzzo GK, Galgoci AM, Giacobbe RA, Gill CJ, Hsu MJ, Liberator P, Misura AS, Motyl M, Nielsen Kahn J, Powles M, Racine F, Dragovic J, Habulihaz B, Fan W, Kirwan R, Lee S, Liu H, Mamai A, Nelson K, Peel M (2015) Novel orally active inhibitors of β-1,3-glucan synthesis derived from enfumafungin. Bioorg Med Chem Lett 25:5813–5818

Aviles P, Falcoz C, San Roman R, Gargallo-Viola D (2000) Pharmacokinetics-pharmacodynamics of a sordarin derivative (GM 237354) in a murine model of lethal candidiasis. Antimicrob Agents Chemother 44:2333–2340

Bae M, Kim H, Moon K, Nam SJ, Shin J, Oh KB, Oh DC (2015) Mohangamides A and B, new dilactone-tethered pseudo-dimeric peptides inhibiting *Candida albicans* isocitrate lyase. Org Lett 17:712–715

Bagiński M, Resat H, Borowski E (2002) Comparative molecular dynamics simulations of amphotericin B-cholesterol/ergosterol membrane channels. Biochim Biophys Acta 1567:63–78

Bahare RS, Ganguly S, Choowongkomon K, Seetaha S (2015) Synthesis, HIV-1 RT inhibitory, antibacterial, antifungal and binding mode studies of some novel N-substituted 5-benzylidine-2,4-thiazolidinediones. DARU J Pharm Sci 23:6

Bahn Y-S (2015) Exploiting fungal virulence-regulating transcription factors as novel antifungal drug targets. PLoS Pathog 11:e1004936

Banerjee D, Burkard L, Panepinto JC (2014) Inhibition of nucleotide biosynthesis potentiates the antifungal activity of amphotericin B. PLoS ONE 9:e87246

Banerjee RV, Matthews RG (1990) Cobalamin-dependent methionine synthase. FASEB J 4:1450–1459

Bates CJ, Adams WR, Handschumacher RE (1966) Control of the formation of uridine diphospho-N-acetyl-hexosamine and glycoprotein synthesis in rat liver. J Biol Chem 241:1705–1712

Becker J, Kauffman S, Hauser M, Huang L, Lin M, Sillaots S, Jiang B, Xu D, Roemer T (2010) Pathway analysis of *Candida albicans* survival and virulence determinants in a murine infection model. Proc Natl Acad Sci USA 107:22044–22049

Behr JB (2011) Chitin synthase, a fungal glycosyltransferase that is a valuable antifungal target. Chimia 65:1–5

Betz C, Hall MN (2013) Where is mTOR and what is it doing there? J Cell Biol 203:563–574

Brammer KW, Farrow PR, Faulkner JK (1990) Pharmacokinetics and tissue penetration of fluconazole in humans. Rev Infect Dis 12(suppl. 3):S318–S326

Brand A, Shanks S, Duncan VM, Yang M, Mackenzie K, Gow NA (2007) Hyphal orientation of Candida albicans is regulated by a calcium dependent mechanism. Curr Biol 17:347–352

Brown GD, Denning DW, Gow NAR, Levitz SM, Netea MG, White TC (2012) Hidden killers: Human fungal infections. Sci Transl Med 4:165rv13

Bruno VM, Kalachikov S, Subaran R, Nobile CJ, Kyratsous C et al (2006) Control of the *C. albicans* cell wall damage response by transcriptional regulator Cas5. PLoS Pathog 2:e21. doi:10.1371/journal.ppat.0020021

Bugli F, Cacaci M, Martini C, Torelli R, Posteraro B, Sanguinetti M, Paroni Sterbini F (2013) Human monoclonal antibody-based therapy in the treatment of invasive candidiasis. Clin Develop Immunol 2013:403121

Capobianco JO, Zakula D, Coen ML, Goldman RC (1993) Anti-Candida activity of cispentacin: the active transport by amino acid permeases and possible mechanisms of action. Biochim Biophys Res Commun 190:1037–1044

Chakraborty B, Mukherjee R, Sengupta J (2013) Structural insights into the mechanism of translational inhibition by the fungicide sordarin. J Comput Aided Mol Des 27:173–184

Chamilos G, Nobile CJ, Bruno VM, Lewis RE, Mitchell AP, Kontoyiannis DP (2009) *Candida albicans* Cas5, a regulator of cell wall integrity, is required for virulence in murine and toll mutant fly models. J Infect Dis 200:152–157

Chapman T, Kinsman O, Houston J (1992) Chitin biosynthesis in *Candida albicans* grown in vitro and in vivo and its inhibition by nikkomycin Z. Antimicrob Agents Chemother 36:1909–1914

Chauhan N, Calderone R (2008) Two-component signal transduction proteins as potential drug targets in medically important fungi. Infect Immun 76:4795–4803

Chaudhary PM, Tupe SG, Deshpande MV (2013) Chitin synthase inhibitors as antifungal agents. Mini Rev Med Chem 13:222–236

Cheah H-L, Lim V, Sandai D (2014) Inhibitors of the glyoxylate cycle enzyme ICL1 in *Candida albicans* for potential use as antifungal agents. PLoS ONE 9:e95951

Chen WC, Chou DL, Feingold DS (1978) Dissociation between ion permeability and lethal action of polyene antibiotics on *Candida albicans*. Antimicrob Agents Chemother 13:914–917

Chen YL, Brand A, Morrison EL, Silao FG, Bigol UG, Malbas FF Jr, Nett JE, Andes DR, Solis NV, Filler SG, Averette A, Heitman J (2011) Calcineurin controls drug tolerance, hyphal growth, and virulence in *Candida dubliniensis*. Eukaryot Cell 10:803–819

Chiou C, Groll A, Walsh T (2000) New drugs and novel targets for treatment of invasive fungal infections in patients with cancer. Oncologist 5:120–135

Clatworthy AE, Pierson E, Hung DT (2007) Targeting virulence: a new paradigm for antimicrobial therapy. Nat Chem Biol 3:541–548

Cornet M, Gaillardin C (2014) pH Signaling in human fungal pathogens: a new target for antifungal strategies. Eukaryot Cell 13:342–352

Davis D (2003) Adaptation to environmental pH in *Candida albicans* and its relation to pathogenesis. Curr Genet 44:1–7

Davis DA (2009) How human pathogenic fungi sense and adapt to pH: the link to virulence. Curr Opin Microbiol 12:365–370

Davis SA, Vincent BM, Endo MM, Whitesell L, Marchillo K, Andes DR, Lindquist S, Burke MD (2015) Nontoxic antimicrobials that evade drug resistance. Nat Chem Biol 11:481–487

De Backer MD, Nelissen B, Logghe M, Viaene J, Loonen I, Vandoninck S, de Hoogt R, Dewaele S, Simons FA, Verhasselt P, Vanhoof G, Contreras R, Luyten WH (2001) An antisense-based functional genomics approach for identification of genes critical for growth of *Candida albicans*. Nat Biotechnol 19:235–241

De Macedo CS, Shams-Eldin H, Smith TK, Schwarz RT, Azzouz N (2003) Inhibitors of glycosyl-phosphatidylinositol anchor biosynthesis. Biochimie 85:465–472

De Pascale G, Griffiths EJ, Shakya T, Nazi I, Wright GD (2011a) Identification and characterization of new inhibitors of fungal homoserine kinase. ChemBioChem 12:1179–1182

De Pascale G, Nazi I, Harrison PH, Wright GD (2011b) β-Lactone natural products and derivatives inactivate homoserine transacetylase, a target for antimicrobial agents. J Antibiot (Tokyo) 64:483–487

Denning DW (2003) Echinocandin antifungal drugs. Lancet 362:1142–1151

Desvergnes S, Courtiol-Legourd S, Daher R, Dabrowski M, Salmon L, Therisod M (2012) Synthesis and evaluation of malonate-based inhibitors of phosphosugar-metabolizing enzymes: Class II fructose-1,6-bis-phosphate aldolases, type I phosphomannose isomerase, and phosphoglucose isomerase. Bioorg Med Chem 20:1511–1520

Devadas B, Freeman SK, Zupec ME, Lu HF, Nagarajan SR, Kishore NS, Lodge JK, Kuneman DW, McWherter CA, Vinjamoori DV, Getman DP, Gordon JI, Sikorski JA (1997) Design and synthesis of novel imidazole-substituted dipeptide amides as potent and selective inhibitors of *Candida albicans* myristoylCoA:protein N-myristoyltransferase and identification of related tripeptide inhibitors with mechanism-based antifungal activity. J Med Chem 40:2609–2625

Dinér P, Veide Vilg J, Kjellen J, Migdal I, Andersson T, Gebbia M, Giaever G, Nislow C, Hohmann S, Wysocki R, Tamas MJ, Grotli M (2011) Design, synthesis, and characterization of a highly effective Hog1 inhibitor: a powerful tool for analyzing MAP kinase signaling in yeast. PLoS ONE 6:e20012

Dominguez JM, Gomez-Lorenzo MG, Martin JJ (1999) Sordarin inhibits fungal protein synthesis by blocking translocation differently to fusidic acid. J Biol Chem 274:22423–22427

Domínguez JM, Martín JJ (1998) Identification of elongation factor 2 as the essential protein targeted by sordarins in *Candida albicans*. Antimicrob Agents Chemother 42:2279–2283

Douglas CM (2001) Fungal beta(1,3)-d-glucan synthesis. Med Mycol 39(Suppl. 1):55–66

Ejim L, Mirza IA, Capone C, Nazi I, Jenkins S, Chee GL, Berghuis AM, Wright GD (2004) New phenolic inhibitors of yeast homoserine dehydrogenase. Bioorg Med Chem 12:3825–3830

Elbein AD (1984) Inhibitors of the biosynthesis and processing of N-linked oligosaccharides. CRC Crit Rev Biochem 16:21–49

Ernst JF, Prill SK-H (2001) O-glycosylation. Med Mycol 39(suppl. 1):67–74

Foret J, de Courcy B, Gresh N, Piquemal J-P, Salmon L (2009) Synthesis and evaluation of non-hydrolyzable D-mannose 6-phosphate surrogates reveal 6-deoxy-6-dicarboxymethyl-D-mannose as a new strong inhibitor of phosphomannose isomerases. Bioorg Med Chem 17:7100–7107

Fostel JM, Montgomery DA, Shen LL (1992) Characterization of DNA topoisomerase I from *Candida albicans* as a target for drug discovery. Antimicrob Agents Chemother 36:2131–2138

Gabriel I, Kur K, Laforce-Nesbitt SS, Pulickal AS, Bliss JM, Milewski S (2014) Phenotypic consequences of *LYS4* gene disruption in *Candida albicans*. Yeast 31:299–308

Gabriel I, Vetter ND, Palmer DRJ, Milewska MJ, Wojciechowski M, Milewski S (2013) Homoisocitrate dehydrogenase from *Candida albicans*: properties, inhibition, and targeting by an antifungal pro-drug. FEMS Yeast Res 13:143–155

Gácser A, Schafer W, Nosanchuk JS, Salomon S, Nosanchuk JD (2007) Virulence of *Candida parapsilosis*, *Candida orthopsilosis*, and *Candida metapsilosis* in reconstituted human tissue models. Fungal Genet Biol 44:1336–1341

Grandoni JA, Marta PT, Schloss JV (1998) Inhibitors of branched chain amino acid biosynthesis as potential antituberculosis agents. J Antimicrob Chemother 42:475–482

Gray KC, Brajtburg J, Powderly WG, Kobayashi GS, Medoff G (1990) Amphotericin B: current understanding of mechanisms of action. Antimicrob Agents Chemother 34:183–188

Grzybowska J, Sowiński P, Gumieniak J, Zieniawa T, Borowski E (1997) *N*-methyl-*N*-D-fructopyranosylamphotericin B methyl ester, new amphotericin B derivative of low toxicity. J Antibiot 50:709–711

Hac-Wydro K, Dynarowicz-Łątka P, Grzybowska J, Borowski E (2005) N-(1-piperidinepropionyl) amphotericin B methyl ester (PAME)—a new derivative of the antifungal antibiotic amphotericin B: searching for the mechanism of its reduced toxicity. J Colloid Interface Sci 287:476–484

Harris BE, Manning BW, Federle TW, Diasio RB (1986) Conversion of 5-fluorocytosine to 5-fluorouracil by human intestinal microflora. Antimicrob Agents Chemother 29:44–48

Hata K, Horii T, Miyazaki M, Watanabe NA, Okubo M, Sonoda J, Nakamoto K, Tanaka K, Shirotori S, Murai N, Inoue S, Matsukura M, Abe S, Yoshimatsu K, Asada M (2011) Efficacy of oral E1210, a new broad-spectrum antifungal with a novel mechanism of action, in murine models of candidiasis, aspergillosis, and fusariosis. Antimicrob Agents Chemother 55:4543–4551

Hector RF (1993) Compounds active against cell walls of medically important fungi. Clin Microbiol Rev 6:1–21

Hector RF, Braun PC (1986) Synergistic action of nikkomycins X and Z with papulacandin B on whole cells and regenerating protoplasts of *Candida albicans*. Antimicrob Agents Chemother 29:389–394

Hector RF, Zimmer BL, Pappagianis D (1990) Evaluation of Nikkomycins X and Z in murine models of coccidioidomycosis, histoplasmosis, and blastomycosis. Antimicrob Agents Chemother 34:587–593

Heeres J, Backx LJJ, Mostmans JH, Van Cutsem J (1979) Antimycotic imidazoles. Part 4. Synthesis and antifungal activity of ketoconazole, a new potent orally active broad spectrum antifungal agent. J Med Chem 22:1003–1005

Herreros E, Almela MJ, Lozano S, Gomez de las Heras F, Gargallo-Viola D (2001) Antifungal activities and cytotoxicity studies of six new azasordarins. Antimicrob Agents Chemother 45:3132–3139

Hiramoto F, Nomura N, Furumai T, Oki T, Igarashi Y (2003) Apoptosis-like cell death of Saccharomyces cerevisiae induced by a mannose-binding antifungal antibiotic, pradimicin. J Antibiot (Tokyo) 56:768–772

Igarashi Y, Oki T (2004) Mannose-binding quinone glycoside, MBQ: potential utility and action mechanism. Adv Appl Microbiol 54:147–166

Jacques S, Mirza A, Ejim L, Koteva K, Hughes DW, Green K, Kinach R, Honek JF, Lai HK, Berghuis AM, Wright GD (2003) Enzyme-assisted suicide: molecular basis for the antifungal activity of 5-hydroxy-4-oxonorvaline by potent inhibition of homoserine dehydrogenase. Chem Biol 10:989–995

Janiak A, Cybulska B, Szlinder-Richert J, Borowski E, Milewski S (2002) Facilitated diffusion of glucosamine-6-phosphate synthase inhibitors enhances their antifungal activity. Acta Biochim Polon 49:77–86

Janiak A, Milewski S (2001) Mechanism of antifungal action of kanosamine. Med Mycol 39:401–408

Janout V, Schell WA, Thevenin D, Yu Y, Perfect JR, Regen SL (2015) Taming amphotericin B. Bioconjugate Chem 26:2021–2024

Ji H, Zhang W, Zhou Y, Zhang M, Zhu J, Song Y, Lü J (2000) A three-dimensional model of lanosterol 14alpha-demethylase of *Candida albicans* and its interaction with azole antifungals. J Med Chem 43:2493–2505

Ji Q, Ge Z, Ge Z, Chen K, Wu H, Liu X, Huang Y, Yuan L, Yang X, Liao F (2016) Synthesis and biological evaluation of novel phosphoramidate derivatives of coumarin as chitin synthase inhibitors and antifungal agents. Eur J Med Chem 108:166–176

Kamai Y, Kakuta M, Shibayama T, Fukuoka T, Kuwahara S (2005) Antifungal activities of R-135853, a sordarin derivative, in experimental candidiasis in mice. Antimicrob Agents Chemother 49:52–56

Kelly SL, Lamb DC, Baldwin BC, Corran AJ, Kelly DE (1997) Characterization of Saccharomyces cerevisiae cyp61, sterol & Δ22-desaturase, and inhibition by azole antifungal agents. J Biol Chem 272:9986–9988

Kingsbury JM, McCusker JH (2010a) Fungal homoserine kinase (thr1Δ) mutants are attenuated in virulence and die rapidly upon threonine starvation and serum incubation. Eukaryot Cell 9:729–737

Kingsbury JM, McCusker JH (2010b) Homoserine toxicity in *Saccharomyces cerevisiae* and *Candida albicans* homoserine kinase (thr1Δ) mutants. Eukaryot Cell 9:717–728

Kingsbury JM, McCusker JH (2010c) Cytocidal amino acid starvation of *Saccharomyces cerevisiae* and *Candida albicans* acetolactate synthase (ilv2Δ) mutants is influenced by the carbon source and rapamycin. Microbiology 156:929–939

Kinsman OS, Chalk PA, Jackson HC, Middleton RF, Shuttleworth A, Rudd BAM, Jones CA, Noble HM, Wildman HG, Dawson MJ, Lynn S, Hayes MV (1998) Isolation and characterisation of an antifungal antibiotic (GR 135402) with protein synthesis inhibition. J Antibiot 51:41–49

Kugler M, Loeffler W, Rapp C, Kern A, Jung G (1990) Rhizocticin A, an antifungal phosphono-oligopeptide of *Bacillus subtilis* ATCC 6633: biological properties. Arch Microbiol 153:276–281

Kumar A, Dhamgaye S, Maurya IK, Singh A, Sharma M, Prasad R (2014) Curcumin targets cell wall integrity via calcineurin-mediated signaling in *Candida albicans*. Antimicrob Agents Chemother 58:167–175

Kur K, Gabriel I, Morschhäuser J, Barchiesi F, Spreghini E, Milewski S (2010) Disruption of homocitrate synthase genes in *Candida albicans* affects growth but not virulence. Mycopathologia 170:397–402

LaFleur MD, Lucumi E, Napper AD, Diamond SL, Lewis K (2011) Novel high-throughput screen against *Candida albicans* identifies antifungal potentiators and agents effective against biofilms. J Antimicrob Chemother 66(4):820–826

Lassak T, Schneider E, Bussmann M, Kurtz D, Manak JR, Srikantha T, Soll DR, Ernst JF (2011) Target specificity of the *Candida albicans* Efg1 regulator. Mol Microbiol 82:602–618

Legrand P, Romero EA, Cohen BE, Bolard B (1992) Effects of aggregation and solvent on the toxicity of amphotericin B to human erythrocytes. Antimicrob Agents Chemother 36:2518–2522

Lenardon MD, Whitton RK, Munro CA, Marshall D, Gow NA (2007) Individual chitin synthase enzymes synthesize microfibrils of differing structure at specific locations in the *Candida albicans* cell wall. Mol Microbiol 66:1164–1173

Lesage G, Sdicu AM, Menard P, Shapiro J, Hussein S, Bussey H (2004) Analysis of beta-1,3-glucan assembly in *Saccharomyces cerevisiae* using a synthetic interaction network and altered sensitivity to caspofungin. Genetics 167:35–49

Liu S, Hou Y, Liu W, Lu C, Wang W, Sund S (2015) Components of the calcium-calcineurin signaling pathway in fungal cells and their potential as antifungal targets. Eukaryot Cell 14:324–334

Loewith R, Jacinto E, Wullschleger S, Lorberg A, Crespo JL, Bonenfant D, Oppliger W, Jenoe P, Hall MN (2002) Two TOR complexes, only one of which is rapamycin sensitive, have distinct roles in cell growth control. Mol Cell 10:457–468

Mann PA, McLellan CA, Koseoglu S, Si Q, Kuzmin E, Flattery A, Harris G, Sher X, Murgolo N, Wang H, Devito K, de Pedro N, Genilloud O, Kahn JN, Jiang B, Costanzo M, Boone C, Garlisi CG, Lindquist S, Roemer T (2015) Chemical genomics-based antifungal drug discovery: Targeting glycosylphosphatidylinositol (GPI) precursor biosynthesis. ACS Infect Dis 1:59–72

Martin CE, Oh CS, Jiang Y (2007) Regulation of long chain unsaturated fatty acid synthesis in yeast. Biochim Biophys Acta 1771:271–285

Masubuchi M, Ebiike H, Kawasaki K, Sogabe S, Morikami K, Shiratori Y, Tsujii S, Fujii T, Sakata K, Hayase M, Shindoh H, Aoki Y, Ohtsuka T, Shimma N (2003) Synthesis and biological activities of benzofuran antifungal agents targeting fungal N-myristoyltransferase. Bioorg Med Chem 11:4463–4478

Melcer A, Łącka I, Gabriel I, Wojciechowski M, Liberek B, Wiśniewski A, Milewski S (2007) Rational design of *N*-alkyl derivatives of 2-amino-2-deoxy-D-glucitol-6P as antifungal agents. Bioorg Med Chem Lett 17:6602–6606

Midkiff J, Borochoff-Porte N, White D, Johnson DI (2011) Small molecule inhibitors of the *Candida albicans* budded-to-hyphal transition act through multiple signaling pathways. PLoS ONE 6:e25395

Milewski S (2002) Glucosamine-6-phosphate synthase-the multi-facets enzyme. Biochim Biophys Acta 1597:173–192

Milewski S, Chmara H, Andruszkiewicz R, Borowski E, Zaremba M, Borowski J (1988) Antifungal peptides with novel inhibitors of glucosamine-6-P synthase. Drugs Exptl Clin Res 14:466–477

Milewski S, Chmara H, Borowski E (1986) Antibiotic tetaine—a selective inhibitor of chitin and mannoprotein biosynthesis in *Candida albicans*. Arch Microbiol 145:234–240

Milewski S, Gabriel I, Olchowy J (2006) Enzymes of UDP-GlcNAc biosynthesis in yeast. Yeast 23:1–14

Milewski S, Mignini F, Borowski E (1991) Synergistic action of nikkomycin X/Z with azole antifungals on *Candida albicans*. J Gen Microbiol 13:2155–2161

Moen MD, Lyseng-Williamson KA, Scott LJ (2009) Liposomal amphotericin B: a review of its use as empirical therapy in febrile neutropenia and in the treatment of invasive fungal infections. Drugs 69:361–392

Munro CA (2013) Chitin and glucan, the Yin and Yang of the fungal cell wall, implications for antifungal drug discovery and therapy. Adv Appl Microbiol 83:145–172

Munro CA, Bates S, Buurman ET, Hughes HB, Maccallum DM, Bertram G, Atrih A, Ferguson MA, Bain JM, Brand A, Hamilton S, Westwater C, Thomson LM, Brown AJ, Odds FC, Gow NA (2005) Mnt1p and Mnt2p of Candida albicans are partially redundant alpha-1,2-mannosyltransferases that participate in O-linked mannosylation and are required for adhesion and virulence. J Biol Chem 280:1051–1060

Nagiec MM, Nagiec EE, Baltisberger JA, Wells GB, Wells GB, Lester RL, Dickson RC (1997) Sphingolipid synthesis as a target for antifungal drugs. Complementation of the inositol phosphorylceramide synthase defect in a mutant strain of *Saccharomyces cerevisiae* by the *AuR1* gene. J Biol Chem 272:9809–9817

Nakama T, Nureki O, Yokoyama S (2001) Structural basis for the recognition of isoleucyl-adenylate and an antibiotic, mupirocin, by isoleucyl-tRNA synthetase. J Biol Chem 276:47387–47393

Nakayama H, Mio T, Nagahashi S, Kokado M, Arisawa M, Aoki Y (2000) Tetracycline-regulatable system to tightly control gene expression in the pathogenic fungus *Candida albicans*. Infect Immun 68:6712–6719

Navarathna DH, Hornby JM, Krishnan N, Parkhurst A, Duhamel GE, Nickerson KW (2007) Effect of farnesol on a mouse model of systemic candidiasis, determined by use of a *DPP3* knockout mutant of *Candida albicans*. Infect Immun 75:1609–1618

Nazi I, Scott A, Sham A, Rossi L, Williamson PR, Kronstad JW, Wright GD (2007) Role of homoserine transacetylase as a new target for antifungal agents. Antimicrob Agents Chemother 51:1731–1736

Nivoix Y, Leveque D, Herbrecht R, Koffel J-C, Beretz L, Ubeaud-Sequier G (2008) The enzymatic basis of drug-drug interactions with systemic triazole antifungals. Clin Pharmacokinet 47:779–792

Nobile CJ, Andes DR, Nett JE, Smith FJ, Yue F, Phan QT, Edwards JE, Filler SG, Mitchell AP (2006) Critical role of Bcr1-dependent adhesins in *C. albicans* biofilm formation in vitro and in vivo. PLoS Pathog 2:e63

Ochsner UA, Sun X, Jarvis T, Critchley I, Janjic N (2007) Aminoacyl-tRNA synthetases: essential and still promising targets for new anti-infective agents. Expert Opin Investing Drugs 16:573–593

Ogawa A, Hashida-Okado T, Endo M, Yoshioka H, Tsuruo T, Takesako K, Kato I (1998) Role of ABC transporters in aureobasidin A resistance. Antimicrob Agents Chemother 42:755–761

Oki T, Hirano M, Tomatsu K, Numata K, Kamei H (1989) Cispentacin, a new antifungal antibiotic. II. In vitro and in vivo antifungal activities. J Antibiot (Tokyo) 42:1756–1762

Orchard MG, Neuss JC, Galley CM, Carr A, Porter DW, Smith P, Scopes DI, Haydon D, Vousden K, Stubberfield CR, Young K, Page M (2004) Rhodanine-3-acetic acid derivatives as inhibitors of fungal protein mannosyl transferase 1 (PMT1). Bioorg Med Chem Lett 14:3975–3978

Palacios DS, Dailey I, Endo MM, Uno BE, Wilcock BC, Burke MD (2012) Amphotericin primarily kills yeast by simply binding ergosterol. Proc Natl Acad Sci USA 109:2234–2239

Parks L, Casey W (1995) Physiological implications of sterol biosynthesis in yeast. Annu Rev Microbiol 49:95–116

Perlin DS (2011) Current perspectives on echinocandin class drugs. Future Microbiol 6:441–457

Peyton LR, Gallagher S, Hashemzadeh M (2015) Triazole antifungals: a review. Drugs Today (Barc). 51:705–718

Pfaller MA, Diekema DJ (2007) Epidemiology of invasive candidiasis: A persistent public health problem. Clin Microbiol Rev 20:133–163

Pfaller MA, Messer SA, Georgopapadakou N, Martell LA, Besterman JM, Diekema DJ (2009) Activity of MGCD290, a Hos2 histone deacetylase inhibitor, in combination with azole antifungals against opportunistic fungal pathogens. J Clin Microbiol 47:3797–3804

Pham JS, Dawson KL, Jackson KE, Lim EE, Pasaje CF, Turner KE, Ralph SA (2014) Aminoacyl-tRNA synthetases as drug targets in eukaryotic parasites. Int J Parasitol Drugs Drug Resist 4:1–13

Phillips AJ, Sudbery I, Ramsdale M (2003) Apoptosis induced by environmental stresses and amphotericin B in *Candida albicans*. Proc Nat Acad Sci USA 100:14327–14332

Pierce CG, Lopez-Ribot JL (2013) Candidiasis drug discovery and development: new approaches targeting virulence for discovering and identifying new drugs. Expert Opin Drug Discov 8:1117–1126

Podust LM, Poulos TL, Waterman MR (2001) Crystal structure of cytochrome P450 14 alpha-sterol demethylase (CYP51) from *Mycobacterium tuberculosis* in complex with azole inhibitors. Proc Natl Acad Sci USA 98:3068–3073

Polak A, Scholer HJ (1975) Mode of action of 5-fluorocytosine and mechanisms of resistance. Chemotherapy 21:113–130

Potoski BA, Brown J (2002) The safety of voriconazole. Clin Infect Dis 35:1273–1275

Prasannan P, Suliman HS, Robertus JD (2009) Kinetic analysis of site-directed mutants of methionine synthase from *Candida albicans*. Biochem Biophys Res Commun 382:730–734

Pukkila-Worley R, Peleg AY, Tampakakis E, Mylonakis E (2009) *Candida albicans* hyphal formation and virulence assessed using a *Caenorhabditis elegans* infection mode. Eukaryot Cell 8:1750–1758

Richard M, Ibata-Ombetta S, Dromer F, Bordon-Pallier F, Jouault T, Gaillardin C (2002) Complete glycosylphosphatidylinositol anchors are required in *Candida albicans* for full morphogenesis, virulence and resistance to macrophages. Mol Microbiol 44:841–853

Richie DL, Thompson KV, Studer C, Prindle VC, Aust T, Riedl R, Estoppey D, Tao J, Sexton JA, Zabawa T, Drumm J, Cotesta S, Eichenberger J, Schuierer S, Hartmann N, Movva NR, Tallarico JA, Ryder NS, Hoepfner D (2013) Identification and evaluation of novel acetolactate synthase inhibitors as antifungal agents. Antimicrob Agents Chemother 57:2272–2280

Rivalta I, Sultan MM, Lee N, Manley GA, Loria JP, Batista VS (2012) Allosteric pathways in imidazole glycerol phosphate synthase. PNAS 109:1428–1436

Rock FL, Mao W, Yaremchuk A, Tukalo M, Crepin T, Zhou H, Zhang YK, Hernandez V, Akama T, Baker SJ, Plattner JJ, Shapiro L, Martinis SA, Benkovic SJ, Cusack S, Alley M (2007) An antifungal agent inhibits an aminoacyl-tRNA synthetase by trapping tRNA in the editing site. Science 316:1759–1761

Roemer T, Jiang B, Davison J, Ketela T, Veillette K, Breton A, Tandia F, Linteau A, Sillaots S, Marta C, Martel N, Veronneau S, Lemieux S, Kauffman S, Becker J, Storms R, Boone C, Bussey H (2003) Large-scale essential gene identification in *Candida albicans* and applications to antifungal drug discovery 50:167–181

Román E, Alonso-Monge R, Gong Q, Li D, Calderone R, Pla J (2009) The Cek1 MAPK is a short-lived protein regulated by quorum sensing in the fungal pathogen *Candida albicans*. FEMS Yeast Res 9:942–955

Ruiz-Herrera J, San-Blas G (2003) Chitin synthesis as target for antifungal drugs. Curr Drug Targets Infect Disord 3:77–91

Santos AL, Braga-Silva LA (2013) Aspartic protease inhibitors: effective drugs against the human fungal pathogen *Candida albicans*. 13:155–162

Saville SP, Lazzell AL, Bryant AP, Fretzen A, Monreal A, Solberg EO, Monteagudo C, Lopez-Ribot JL, Milne GT (2006) Inhibition of filamentation can be used to treat disseminated candidiasis. Antimicrob Agents Chemother 50:3312–3316

Seiradake E, Mao W, Hernandez V, Baker SJ, Plattner JJ, Alley MR, Cusack S (2009) Crystal structures of the human and fungal cytosolic leucyl-tRNA synthetase editing domains: a structural basis for the rational design of antifungal benzoxaboroles. J Mol Biol 390:196–207

Shareck J, Belhumeur P (2011) Modulation of morphogenesis in *Candida albicans* by various small molecules. Eukaryot Cell 10:1004–1012

Shor E, Chauhan N (2015) A case for two-component signaling systems as antifungal drug targets. PLoS Pathog 11:e1004632

Sikorski JA, Devadas B, Zupec ME, Freeman SK, Brown DL, Lu HF, Nagarajan S, Mehta PP, Wade AC, Kishore NS, Bryant ML, Getman DP, McWherter CA, Gordon JI (1997) Selective peptidic and peptidomimetic inhibitors of *Candida albicans* myristoylCoA: protein N-myristoyltransferase: a new approach to antifungal therapy. Biopolymers 43:43–71

Siles SA, Srinivasan A, Pierce CG, Lopez-Ribot JL, Ramasubramanian AK (2013) High-throughput screening of a collection of known pharmacologically active small compounds for identification of *Candida albicans* biofilm inhibitors. Antimicrob Agents Chemother 57:3681–3687

Smith DJ, Proudfoot AE, Detiani M, Wells TN, Payton MA (1995) Cloning and heterologous expression of the *Candida albicans* gene *PMI 1* encoding phosphomannose isomerase. Yeast 11:301–310

Søe R, Mosley RT, Justice M, Nielsen-Kahn J, Shastry M, Merrill AR, Andersen GR (2006) Sordarin derivatives induce a novel conformation of the yeast ribosome translocation factor eEF2. J Biol Chem 282:657–666

Sousa S, McLaughlin M, Pereira S, Van Horn S, Knowlton R, Brown JR, Nicholas RO, Livi GP (2002) The *ARO4* gene of *Candida albicans* encodes a tyrosine-sensitive DAHP synthase: evolution, functional conservation and phenotype of Aro3p-Aro4p-deficient mutants. Microbiology 148:1291–1303

Sousa dos Santos AL (2010) HIV aspartyl protease inhibitors as promising compounds against *Candida albicans*. World J Biol Chem 1:21–30

Stewart L, Ireton GC, Champoux JJ (1996) The domain arrangement of human topoisomerase I. J Biol Chem 271:7602–7608

Sudbery PE (2011) Growth of *Candida albicans* hyphae. Nat Rev Microbiol 9:737–748

Sudoh M, Yamazaki T, Masubuchi K, Taniguchi M, Shimma N, Arisawa M, Yamada-Okabe H (2000) Identification of a novel inhibitor specific to the fungal chitin synthase; inhibition of chitin synthase 1 arrests the cell growth, but that of chitin synthase 1 and 2 is lethal in the pathogenic fungus *Candida albicans*. J Biol Chem 275:32901–32905

Sugimoto Y, Sakoh H, Yamada K (2004) IPC synthase as a useful target for antifungal drugs. Curr Drug Targets Infect Disord 4:311–322

Suliman HS, Appling DR, Robertus JD (2007) The gene for cobalamin-independent methionine synthase is essential in *Candida albicans*: a potential antifungal target. Arch Biochem Biophys 467:218–226

Sussman A, Huss K, Chio LC, Heidler S, Shaw M, Ma D, Zhu G, Campbell RM, Park TS, Kulanthaivel P, Scott JE, Carpenter JW, Strege MA, Belvo MD, Swartling JR, Fischl A, Yeh WK, Shih C, Ye XS (2004) Discovery of cercosporamide, a known antifungal natural product, as a selective Pkc1 kinase inhibitor through high-throughput screening. Eukaryot Cell 3:932–943

Toenjes KA, Munsee SM, Ibrahim AS, Jeffrey R, Edwards JE Jr, Johnson DI (2005) Small-molecule inhibitors of the budded-to-hyphal form transition in the pathogenic yeast *Candida albicans*. Antimicrob Agents Chemother 49:963–972

Triana-Alonso FJ, Chakraburtty K, Nierhaus KH (1995) The elongation factor 3 unique in higher fungi and essential for protein biosynthesis is an E site factor. J Biol Chem 270:20473–20478

Tripathi H, Lugman S, Meena A, Khan F (2014) Genomic identification of potential targets unique to *Candida albicans* for the discovery of antifungal agent. Curr Drug Targets 5:136–149

Trofa D, Agovino M, Stehr F, Schafer W, Rykunov D, Fiser A, Hamari Z, Nosanchuk JD, Gacser A (2009) Acetylsalicylic acid (aspirin) reduces damage to reconstituted human tissues infected with Candida species by inhibiting extracellular fungal lipases. Microbes Infect 11:1131–1139

Tsai PW, Chien CY, Yeh YC, Tung L, Chen HF, Chang TH, Lan CY (2016) *Candida albicans* Hom6 is a homoserine dehydrogenase involved in protein synthesis and cell adhesion. J Microbiol Immunol Infect. doi:10.1016/j.jmii.2016.03.001

Tsukahara K, Hata K, Nakamoto K, Sagane K, Watanabe NA, Kuromitsu J, Kai J, Tsuchiya M, Ohba F, Jigami Y, Yoshimatsu K, Nagasu T (2003) Medicinal genetics approach towards identifying the molecular target of a novel inhibitor of fungal cell wall assembly. Mol Microbiol 48:1029–1042

Uhl MA, Biery M, Craig N, Johnson AD (2003) Haploinsufficiency-based large-scale forward genetic analysis of filamentous growth in the diploid human fungal pathogen *C. albicans*. EMBO J 22:2668–2678

Umemura M, Okamoto M, Nakayama K, Sagane K, Tsukahara K, Hata K, Jigami Y (2003) *GWT1* gene is required for inositol acylation of glycosylphosphatidylinositol anchors in yeast. J Biol Chem 278:23639–23647

Vanden Bossche H, Koymans L, Moereels H (1995) P450 inhibitors of use in medical treatment: focus on mechanisms of action. Pharmacol Ther 67:79–100

Vermes A, Guchelaar HJ, Dankert J (2000) Flucytosine: a review of its pharmacology, clinical indications, pharmacokinetics, toxicity and drug interactions. J Antimicrob Chemother 46:171–179

Volmer AA, Szpilman AM, Carreira EM (2010) Synthesis and biological evaluation of amphotericin B derivatives. Nat Prod Rep 27:1329–1349

Vondenhoff GH, Van Aerschot A (2011) Aminoacyl-tRNA synthetase inhibitors as potential antibiotics. Eur J Med Chem 46:5227–5236

Walker LA, Munro CA, de Bruijn I, Lenardon MD, McKinnon A, Gow NA (2008) Stimulation of chitin synthesis rescues *Candida albicans* from echinocandins. PLoS Pathog 4:e1000040

Warrilow AGS, Hull CM, Parker JE, Garvey EP, Hoekstra WJ, Moore, WR, Schotzinger RJ, Kelly DE, Kelly SL (2014) The clinical candidate VT-1161 is a highly potent inhibitor of *Candida albicans* CYP51 but fails to bind the human enzyme. 58:7121–7127

Watanabe NA, Miyazaki M, Horii T, Sagane K, Tsukahara K, Hata K (2012) E1210, a new broad-spectrum antifungal, suppresses *Candida albicans* hyphal growth through inhibition of glycosylphosphatidylinositol biosynthesis. Antimicrob Agents Chemother 56:960–971

Weinberg RA, McWherter CA, Freeman SK, Wood DC, Gordon JI, Lee SC (1995) Genetic studies reveal that myristoylCoA:protein N-myristoyltransferase is an essential enzyme in *Candida albicans*. Mol Microbiol 16:241–250

Whelan WL, Ballou CE (1975) Sporulation in D-glucosamine auxotrophs of *Saccharomyces cerevisiae*: meiosis with defective ascospore wall formation. J Bacteriol 124:1545–1557

Wilcock BC, Endo MM, Uno BE, Burke MD (2013) C_2'-OH of amphotericin B plays an important role in binding the primary sterol of human cells but not yeast cells. J Am Chem Soc 135:8488–8491

Willer T, Valeroy MC, Tanner W, Crucesy J, Strahl S (2003) O-mannosyl glycans: from yeast to novel associations with human disease. Curr Opinion Struct Biol 13:621–630

Wisplinghoff H, Bischoff T, Tallent SM, Seifert H, Wenzel RP, Edmond MB (2004) Nosocomial bloodstream infections in US hospitals: Analysis of 24,179 cases from a prospective nationwide surveillance study. Clin Infect Dis 39:309–317

Wong SSW, Samaranayake LP, Seneviratne CJ (2014) In pursuit of the ideal antifungal agent for *Candida* infections: high-throughput screening of small molecules. Drug Discov Today 19:1721–1730

Wright JJK, Albarella JA, Krepski LR, Loebenberg D (1982) *N*-aminoacyl derivatives of polyene macrolide antibiotics and their esters. J Antibiot 35:911–914

Xiao L, Madison V, Chau AS, Loebenberg D, Palermo RE, McNicholas PM (2004) Three-dimensional models of wild-type and mutated forms of cytochrome P450 14alpha-sterol demethylases from *Aspergillus fumigatus* and *Candida albicans* provide insights into posaconazole binding. Antimicrob Agents Chemother 48:568–574

Xu D, Sillaots S, Davison J, Hu W, Jiang B, Kauffman S, Martel N, Ocampo P, Oh C, Trosok S, Veillette K, Wang H, Yang M, Zhang L, Becker J, Martin CE, Roemer T (2009) Chemical genetic profiling and characterization of small-molecule compounds that affect the biosynthesis of unsaturated fatty acids in *Candida albicans*. J Biol Chem 284:19754–19764

Xu W, Smith FJJ, Subaran R, Mitchell AP (2004) Multivesicular body-ESCRT components function in pH response regulation in *Saccharomyces cerevisiae* and *Candida albicans*. Mol Biol Cell 15:5528–5537

Yamada-Okabe T, Mio T, Ono N, Kashima Y, Matsui M, Arisawa M, Yamada-Okabe H (1999) Roles of three histidine kinase genes in hyphal development and virulence of the pathogenic fungus *Candida albicans*. J Bacteriol 181:7243–7247

Yamaguchi H, Uchida K, Hiratani T, Nagate T, Watanabe N, Omura S (1988) RI-331, a new antifungal antibiotic. Ann N Y Acad Sci 544:188–190

Yamaji-Hasegawa A, Takahashi A, Tetsuka Y, Senoh Y, Kobayashi T (2005) Fungal metabolite sulfamisterin suppresses sphingolipid synthesis through inhibition of serine palmitoyltransferase. Biochemistry 44:268–277

Yamaki H, Yamaguchi M, Imamura H, Suzuki H, Nishimura T, Saito H, Yamaguchi H (1990) The mechanism of antifungal action of (S)-2-amino-4-oxo-5-hydroxypentanoic acid, RI-331: the inhibition of homoserine dehydrogenase in *Saccharomyces cerevisiae*. Biochem Biophys Res Commun 168:837–843

Yamamoto T, Eguchi T (2008) Thiahomoisocitrate: a highly potent inhibitor of homoisocitrate dehydrogenase involved in the α-aminoadipate pathway. Bioorganic Med Chem 16:3372–3376

Zacchi LF, Gomez-Raja J, Davis DA (2010) Mds3 regulates morphogenesis in *Candida albicans* through the TOR pathway. Mol Cell Biol 30:3695–3710

Zhang Y, Rao R (2012) The V-ATPase as a target for antifungal drugs. Curr Protein Pept Sci 13:134–140

Zhao H, Zhou M, Duan L, Wang W, Zhang J, Wang D, Liang X (2013) Efficient synthesis and anti-fungal activity of oleanolic acid oxime esters. Molecules 18:3615–3629

Ziegelbauer K, Babczinski P, Schonfeld W (1998) Molecular mode of action of the antifungal beta-amino acid BAY 10-8888. Antimicrob Agents Chemother 42:2197–2205

Chapter 22
Antifungals

David S. Perlin

Abstract Each year, fungi are responsible for millions of superficial and systemic infections that all require effective antifungal therapy. After more than six decades of development, existing antifungal agents represent relatively few targets and chemical classes, which include agents that alter membrane function (azole and polyene), cell wall structure (echinocandin) and DNA and protein synthesis (pyrimidine analog). Overall, these agents display a range of chemical spectrum, pharmacokinetics and pharmacodynamics, toxicity and bioavailability. A new generation of agents is needed to overcome some chemical limitations and meet the challenges posed by expanding prophylaxis and empiric therapy along with selection of drug resistant strains. The challenging history of antifungal development serves as a foundation for new discovery.

22.1 Introduction

Invasive fungal infections have increased over the past few decades resulting in nearly 1.4 deaths annually (Brown et al. 2012). Some fungal diseases are severe like cryptococcal meningitis and invasive aspergillosis, while others are recurrent like *Candida* vaginitis or oral candidiasis. The most serious fungal infections are often a consequence of serious underlying health conditions like asthma, AIDS, diabetes, cancer, organ transplantation, and corticosteroid therapies (Brown et al. 2012). These infections require timely and appropriate antifungal therapy. Regrettably, invasive fungal diseases are often diagnosed late in the disease process, which delays effective therapy and increases morbidity and mortality. It is recognized that

D.S. Perlin (✉)
Public Health Research Institute, New Jersey Medical School – Rutgers Biomedical and Health Sciences, Newark, NJ, USA
e-mail: perlinds@njms.rutgers.edu

R. Prasad (ed.), *Candida albicans: Cellular and Molecular Biology*,
DOI 10.1007/978-3-319-50409-4_22

more than 90% of fungal-associated deaths result from *Cryptococcus*, *Candida*, *Aspergillus*, and *Pneumocystis* species (Brown et al. 2012). This clinical recognition of the importance of fungal infections has led to a dramatic rise in the use of antifungal agents for treatment and prevention. Unfortunately, treatment options for invasive fungal infections are highly limited, as existing antifungal drugs represent only a few chemical classes including the polyene, azole, echinocandin, and antimetabolite agents. These agents prevent cell growth or kill growing cells at therapeutic dosages by interfering with fundamental cellular processes including cell membrane function (azoles and polyenes), cell wall integrity (echinocandins), and DNA and protein synthesis (antimetabolite) (Fig. 22.1). In addition to their antimicrobial properties, antifungal agents at low levels can also reduce fungal virulence and modulate the host immune response (Mihu et al. 2014). Understanding the specific chemical classes, their mode of action and basic pharmacology provides a window into the evolution of modern antifungal drug therapy, as well as needs and opportunities for new development. This review will focus on the development and basic drug properties of the major antifungal drug classes. Properties such as acquired drug resistance and biofilm formation that impact drug action are addressed in other chapters.

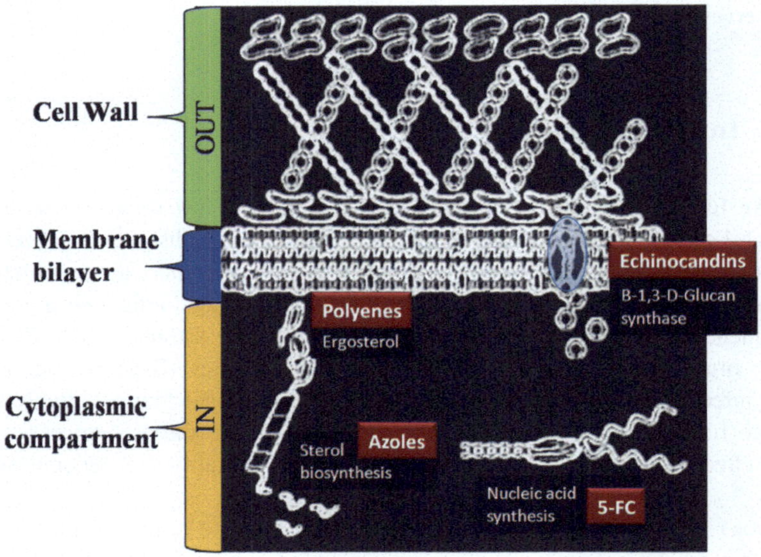

Fig. 22.1 Schematic showing prominent cellular targets and corresponding antifungal drugs

22.2 Polyenes

The polyene antifungal agents are a subgroup of macrolides that are poly-unsaturated and contain one or more sequences of alternating double and single carbon–carbon bonds. The most prominent members are nystatin, natamycin, and amphotericin B (Fig. 22.2). These agents bind the membrane sterol ergosterol to form concentration-dependent channels that disrupt cellular function often resulting in cell lysis. Due to their mechanism of action, they are fungicidal and are active against a wide range of yeasts and molds. Amphotericin B (AmB), first isolated from *Streptomyces nodosus* in 1955 (Dutcher 1968), was the cornerstone for the treatment of systemic fungal infections for many decades (Dodds et al. 2000). The formation of aqueous pores by AmB provides a basis for its fungicidal behavior. However, there is increasing evidence indicating that AmB also forms nonaqueous (cation-selective) channels at concentrations below the threshold at which aqueous pores are formed (Cohen 2010). The nonaqueous channels formed by AmB have a hydrophilic water-filled core that allows the permeation of monovalent cations such as K^+ or Na^+, but are not able to span the bilayer (Cohen 1992, 2010) This membrane perturbing behavior has a broad effect on cell killing, resistance and known immune-inducing responses (Sau et al. 2003). The non-aqueous membrane perturbation appears to affect small lipid-anchored Ras GTPases, which are implicated in AmB-induced yeast apoptosis and lethal oxidative damage via the cAMP–PKA and MAPK pathway. Deletion of RAS genes or certain downstream function can reduce AmB-mediated ROS production and ROS-scavenging agents blocks AmB-lethal activity and decrease Ras signaling (Grahl et al. 2015; Phillips et al. 2003).

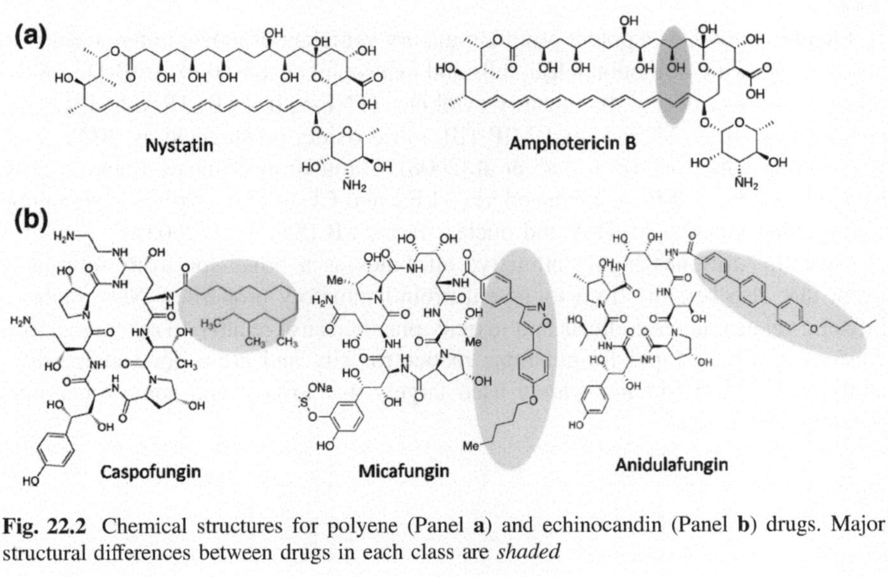

Fig. 22.2 Chemical structures for polyene (Panel **a**) and echinocandin (Panel **b**) drugs. Major structural differences between drugs in each class are *shaded*

AMB has been a workhorse antifungal agent due to its broad-spectrum activity and role in treating endemic and opportunistic mycoses. It is poorly water soluble and is formulated in a detergent micelle as a deoxycholate-soluble salt of AmB (AmB deoxycholate (AmB-D)), which has been used for intravenous infusion since 1953 (Dutcher 1968). It has a wide distribution within the body, including the brain, which makes it useful to help treat cryptococcal meningitis. Acquired resistance is rarely observed and generally results from selection of less susceptible yeasts or rare molds (Ellis 2002). The parenteral formulation is associated with well-known adverse effects including fever, chilling, vomiting, headache, and nausea during administration and nephrotoxicity (Hamill 2013). Infusion-related adverse effects, along with prominent nephrotoxicity related to binding of AmB to cholesterol, have relegated it to second or third-line status behind the azoles and echinocandin drugs. Lipid preparations of amphotericin B, amphotericin B lipid complex (ABLC) and liposomal amphotericin B (L-AmB), have significantly reduced nephrotoxicity relative to AmB-deoxycholate formulations (Hamill 2013; Linden 2003). Yet, there remains residual nephrotoxicity that is generally similar between ABLC and L-AmB in patients (Safdar et al. 2010).

AMB has long been assumed to be a concentration-dependent fungicidal agent, as a wide range of in vitro and in vivo studies have demonstrated concentration-dependent killing versus *Candida* species (Andes et al. 2001; Groll et al. 2000). Dosing regimens that maximize peak drug concentrations (C_{max}/MIC) appear to be more effective than dosing strategies that maximize overall exposure (AUC/MIC) or threshold exposures (time above MIC). This is true for candidiasis (Andes et al. 2001; Andes 2003; Takemoto et al. 2006) and invasive pulmonary aspergillosis (Wiederhold et al. 2006). The lipid formulations L-AmB and ABLC appear somewhat less active relative to conventional AmB-D (Andes et al. 2006), although they show similar pharmacodynamic properties (Andes et al. 2006; Takemoto et al. 2006).

Finally, AmB-D is a potent proinflammatory stimulant of innate immune cells. It has been shown that mononuclear cells and neutrophils exposed to AmB-D rapidly release a number of proinflammatory cytokines (TNF-α, IL-6, IL-1RA, and IL-1β), chemokines (IL-8, MCP-1, and MIP-1β), nitric oxide, prostaglandins, ROIs, and intercellular adhesions (Ben-Ami et al. 2008). Stimulation of innate immune cells by AmB is believed to be mediated via TLR2 and CD14 and involves a signaling cascade that includes MyD88 and nuclear factor κB (Sau et al. 2003).

Overall, AmB has highly attractive attributes as a broad-spectrum antifungal agent that also benefits from its potent proinflammatory properties. Nevertheless, infusion-related adverse effects and toxicity plague its use relative to other available antifungal agents. Lipid formulations reduce toxicity and are a good alternative, although they are still not widely used largely due to cost and some remaining potential for toxicity.

22.3 Azoles

The azole antifungal agents are among the largest and most widely used antifungal agents with millions of patients each year using these agents to treat superficial or systemic mycoses. The azole compounds block the biosynthesis of the membrane sterol ergosterol by binding to and inhibiting the cytochrome P450 enzyme Lanosterol 14 α-demethylase (Cyp51A or Erg11). This leads to ergosterol depletion and accumulation of toxic sterol compounds (Odds et al. 2003). Azole action is fungistatic with most yeast and fungicidal with molds like *A. fumigatus.* Azole class drugs remain a critical therapeutic tool and often preferred option to treat a wide range of invasive mycoses (Pappas et al. 2016). Azoles comprise a broad chemical class of molecules with a wide range of structural signatures. Recently, the first full-length structure of a fungal cytochrome P450 enzyme (CYP51) was determined by X-ray crystallography of Erg11p from the yeast *Saccharomyces cerevisiae* (Monk et al. 2014). The enzyme has a common P450 fold with α-helices and β-sheets comprising the catalytic domain and the heme cofactor forming the active site (Fig. 22.3). The substrate channel contacts the surface of the lipid bilayer. In the active site, the sixth position of the octahedral coordinate geometry of the iron may be occupied by a nitrogen in either the imidazole ring or the triazole ring of the

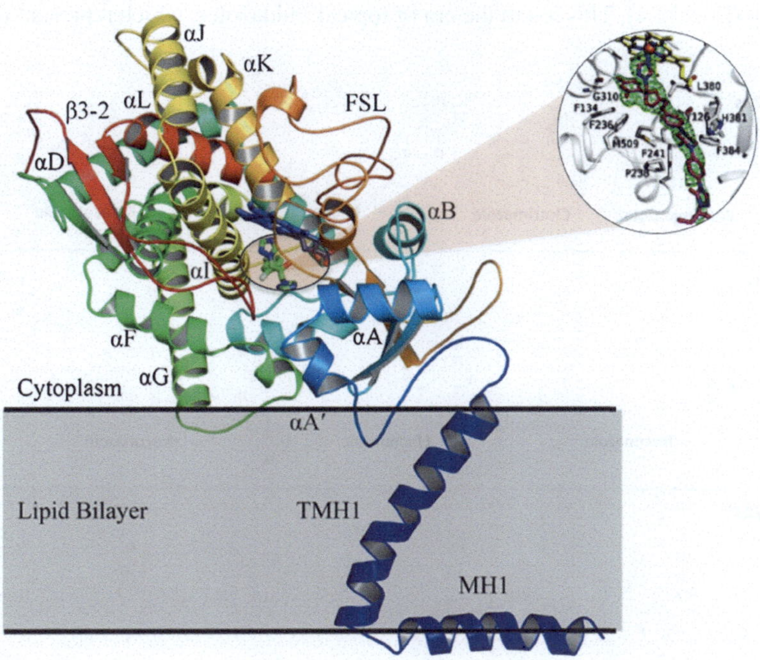

Fig. 22.3 The structure of Erg11 showing bound triazole drugs. The full structure of Erg11p from *Saccharomyces cerevisiae* is shown with bound substrate in active site with its heme moiety and enlargement of the same region (*inset*) with bound itraconazole per Sagatova et al. (2015)

azole drugs (Sagatova et al. 2015). It has been proposed that the specific interaction of azoles with its molecular target occurs via a stepwise mechanism in which displacement of a water molecule from the sixth binding site of the heme creates a pentacoordinated complex with the azole nitrogen and the heme iron (Balding et al. 2008). In high-resolution structures, the triazole ring of fluconazole is coordinated to the iron within the heme cofactor (Sagatova et al. 2015) (Fig. 22.3). The triazole head groups in the smaller azole drugs fluconazole and voriconazole are presumed to coordinate with heme iron and displace O_2, and their unmodified difluorinated head groups occupy the space engaged by first sterol ring of lanosterol and the sterol analog estriol. The long tail of itraconazole fills the entry channel and is presumed to exclude all but one water molecule (Fig. 22.3). Homology overlay modeling of Cyp51A and Erg11 using related high-resolution structural models have been used to describe the impact of specific amino acid substitutions on the interaction of triazole drugs with the target enzyme (Xiao et al. 2004). The most prominent resistance-conferring substitutions alter the apparent interaction of drug with the heme cofactor (Monk et al. 2014; Sagatova et al. 2015).

The history of azole development is instructive to understand the evolution of this modern antifungal class. The first reported antifungal activity of an azole compound was in the 1940s for the heterocyclic aromatic organic compound benzimidazole (Woolley 1944). In the 1960s, Merck introduced substituted benzimidazole compounds as antihelminthic drugs but also possessing antifungal activity (Fig. 22.4). This led to the era of topical imidazoles, which represent one of

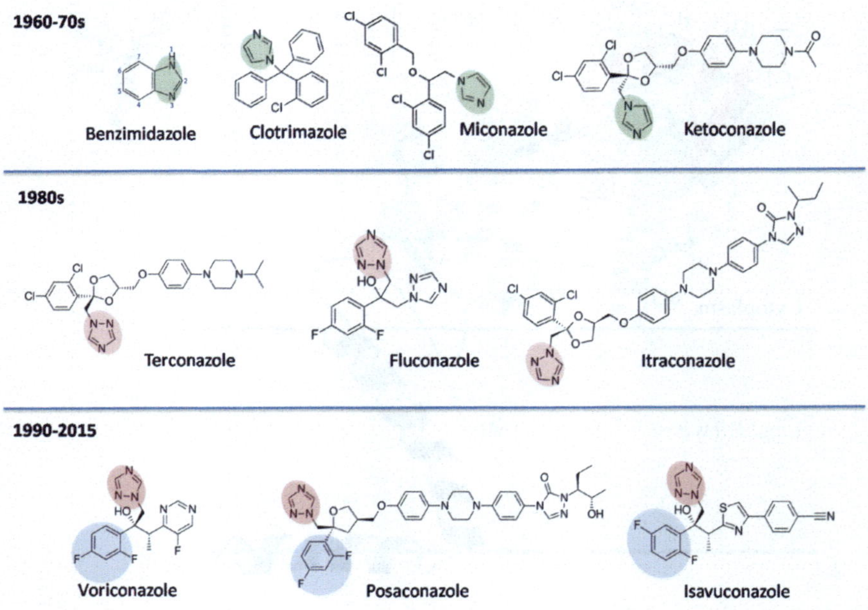

Fig. 22.4 Classes of azole and trizole drugs by decade of development. The azole and triazole cores (*red shading*) are shown, as well as the different difluorophenyl substituents (*blue shading*) for the newer highly active triazole drugs

the two major classes of azole derivatives. Clotrimazole was among the first imidazole anti-fungals developed. It was highly active on a range of fungi including dermatophytes, pathogenic yeasts, and filamentous dimorphic fungi. It was and remains particularly useful for dermatophytic infections and superficial mucosal yeast infections such as vaginitis. Miconazole synthesized by Janssen Pharmaceutica in 1969 was one of the first substituted imidazoles developed for systemic antifungal therapy with broad antifungal activity on yeasts, dermatophytes, and some filamentous molds like *Aspergillus*. Ketoconazole was developed in 1977 as an orally bioavailable imidazole with a broad spectrum for superficial and systemic mycoses. However, it suffered from toxicity and it promoted colonization with *C. glabrata* and *A. fumigatus* (Cauwenbergh 1986). Overall, the imidazoles tended to be limited in systemic therapeutic efficacy and have been relegated largely to superficial yeast infections.

The second major class and most widely used antifungal agents are the triazoles, which display a broader antifungal spectrum with reduced toxicity when compared with the imidazole drugs. Terconazole developed by Janssen Pharmaceutica in 1983 was the first triazole drug marketed for human use. It was followed by the modern era of triazole antifungal agents with the introduction of **fluconazole** by Pfizer in 1990 (Fig. 22.4). Orally active, fluconazole's spectrum of activity includes most *Candida* species, *Cryptococcus neoformans*, some dimorphic fungi, and dermatophytes. The discovery of fluconazole and modern triazoles was a major landmark since it was the first antifungal drug to be developed that could be used both orally for superficial infections or intravenously for more serious systemic infections. Fluconazole is orally active with a high degree of systemic bioavailability that can be used to treat many systemic or mucosal *Candida* infections due to *C. albicans* or other susceptible species (Sheehan et al. 1999). It is largely inactive against *Aspergillus* species and is not recommended for infections due to *C. krusei,* which are inherently resistant or *C. glabrata,* which rapidly develops resistance.

Fluconazole crosses the blood–brain barrier and distributes well into cerebrospinal fluid with high levels relative to serum, and it has a long serum half-life ($\sim$24 h), which facilitates once-daily dosing (Brammer et al. 1990). These features make it valuable to treat cryptococcal meningits when used in combination with 5-fluorocytosine or AmB. Finally, it is minimally metabolized with >80% eliminated unchanged in the urine. **Itraconazole**, an orally active triazole invented by Janssen Pharmaceutica in 1984, is a more potent triazole with activity against a wide range of yeasts and molds (Fromtling 1988). It has good oral absorption, high tissue distribution with tissue levels exceeding that of plasma with a long elimination half-life ($\sim$24 h in humans) and degradation to inactive metabolites. It has minimal penetration through the blood–brain barrier, which precludes its use for cryptococcal infections.

Despite widespread use, first-generation triazole agents proved to have a number of clinically important limitations related to limited spectrum, proclivity for resistance development, adverse drug–drug interactions, suboptimal pharmacokinetics, and toxicity (Maertens 2004). The second-generation triazoles, **voriconazole,**

posaconazole, and **isavuconazole** were developed to address many of these issues (Kale and Johnson 2005). All possess a difluorophenyl substituent ring in addition to a triazole moiety (Fig. 22.4). They have greater potency and possess increased activity against fluconazole-resistant yeasts and *Aspergillus* spp. However, resistance to azoles tends to be class-specific and many *Candida* and *Aspergillus* strains with inherent or acquired resistance to first-generation triazole agents (e.g., fluconazole) are often cross-resistant to the second-generation agents (Sanglard and Odds 2002).

Voriconazole, a closely related chemical derivative of fluconazole, was approved in 2002 by the FDA. It has expanded spectrum of activity against a variety of yeasts and filamentous fungi (Pearson et al. 2003), and it can be administered either orally or parenterally. It exhibits fungicidal activity against *Aspergillus* species and is recommended as first-line therapy for patients with invasive aspergillosis (Kale and Johnson 2005). It is approved for treatment of fluconazole-resistant invasive *Candida* infections, candidemia in non-neutropenic patients, disseminated candidiasis in skin, abdomen, kidney, bladder wall, and wounds; esophageal candidiasis and serious infections caused by *Scedosporium apiospermum* and *Fusarium* species in patients intolerant of or refractory to other therapy (Diekema et al. 2003; Pfaller et al. 2003). Like other triazoles, the 24 h area under the concentration-time curve (AUC)/MIC ratio is the critical pharmacokinetic parameter associated with treatment efficacy (Andes et al. 2003). Voriconazole is rapidly absorbed following oral administration with 90% bioavailability in adults. Protein binding is only 48–60%, which facilitates free drug distributions and high penetration into tissue and central nervous system. Voriconazole undergoes extensive conversion with nearly 80% of its metabolites eliminated renally. It undergoes extensive hepatic-metabolism, primarily via CYP2C19 (Pearson et al. 2003) and exhibits nonlinear pharmacokinetics in adults; drug–drug interactions occur through inhibition of CYP2C9, CYP2C19, and CYP3A4 isoenzymes (Pearson et al. 2003).

Posaconazole, developed initially by Schering-Plough, is a structural analog of itraconazole that was approved for marketing in the United States in 2006 (Schiller and Fung 2007). It has broad-spectrum activity against many yeasts and filamentous fungi, including *Candida* species, *Cryptococcus neoformans*, *Aspergillus* species, and certain rare and cryptic molds. It is approved for the prophylaxis of invasive fungal infections in patients with neutropenia or in haematopoietic stem cell transplant recipients undergoing high-dose immunosuppressive therapy for graft-versus-host disease, and for the treatment of fungal infections (Schiller and Fung 2007). Posaconazole is available as an oral suspension and is highly distributed to various body sites, including bone, central nervous system, and eye. It has low solubility in aqueous and acidic media and its absorption is significantly dependent upon food intake with maximum plasma concentration reached in 5–8 h following a single dose oral administration. Posaconazole is highly protein bound (>98%), mostly to albumin. It has a large mean volume of distribution after oral administration, which suggests extravascular distribution and penetration into intracellular spaces. Posaconazole primarily circulates in plasma and then is widely distributed to the tissues and slowly eliminated. Posaconazole is not metabolized to a significant extent through the cytochrome P450 enzyme system (CYP) and also has no effect on the key CYP isoenzymes. Because it is excreted mostly as unchanged drug in the feces (77%), posaconazole can be administered to patients with poor renal function (Schiller and Fung 2007). Posaconazole has well-established efficacy and tolerability as an oral suspension and can be administered three or four times daily (Moore et al. 2015). However, pharmacokinetic studies have found that the oral suspension had problematic bioavailability due to gastric acid (Moore et al. 2015). A gastro-resistant tablet and intravenous solution were developed to address the pharmacokinetic limitations, and both were found to achieve higher mean plasma concentrations than the original oral suspension (McKeage 2015).

Isavuconazole is a new extended-spectrum water-soluble triazole, codeveloped by Astellas and Basilea as an intravenous and oral broad-spectrum antifungal with activity against *Aspergillus* spp. and Mucorales such as *Rhizopus oryzae* and *Mucormycetes* species (Seyedmousavi et al. 2015; Kovanda et al. 2016), and *Cryptococcus* species (Falci and Pasqualotto 2013). It was approved in March 2015 by the U.S. FDA for the treatment of invasive aspergillosis and invasive mucormycosis (McCormack 2015). An isavuconazonium prodrug is rapidly cleaved into the active component isavuconazole after oral or IV administration (Rybak et al. 2015). Isavuconazole is extensively distributed; it is highly protein bound (>99%) predominantly to albumin with a mean plasma half-life of 130 h. The absolute bioavailability of isavuconazole following oral administration is 98% with no significant prodrug observed in plasma. Isavuconazole appears to be associated with less toxicity than voriconazole (Rybak et al. 2015; Miceli and Kauffman 2015).

The triazole drugs are a foundation component of modern antifungal therapy. Their broad use for empiric therapy and prophylaxis has been somewhat limited by the selection of inherently less susceptible yeast and mold species (Pfaller and Diekema 2004; Malani and Kauffman 2007), and by the acquisition of well-defined resistance mechanisms including target site modification and upregulation of drug efflux transporters in otherwise susceptible species (Sanglard and Odds 2002; Cowen et al. 2015). (Reviewed in other chapters.) Drug–drug interactions are a problem and the triazoles have been hindered by significant inter- and intrapatient pharmacokinetic variability (Allen et al. 2015). Large differences in the pharmacokinetics of critically ill patients for triazoles, including voriconazole and posaconazole, have prompted a need for therapeutic drug monitoring (TDM) (Jager et al. 2016). For example, it has been shown that patients with therapeutic drug monitoring during voriconazole treatment showed a more favorable response compared to the non-TDM group (Park et al. 2012). Nevertheless, the triazoles remain the most widely prescribed antifungal agents over the past 3 decades.

22.4 Echinocandins

The echinocandin drugs anidulafungin, caspofungin, and micafungin are the first cell wall active antifungal drugs. They inhibit glucan synthase, which is responsible for the biosynthesis of β-1,3-D-glucan, a principal structural component of fungal cell walls (Onishi et al. 2000). The three known chemical classes of natural product inhibitors of 1,3-β-D-glucan synthesis (Douglas 2001) include the lipopeptides (e.g., echinocandins and arborcandins), the glycolipid papulacandins, and the terpenoids (e.g., enfumafungins). All glucan synthase inhibitors are noncompetitive with their biosynthetic substrate UDP-glucose. The echinocandins are the only approved class for clinical application. They are semi-synthetic cyclic hexapeptides with an amide-linked fatty acyl side chain (Hammond 1993) (Fig. 22.2). A prominent feature of the echinocandins that contributed to their development was the potent in vivo animal model activity against infections due to *C. albicans* (Bartizal et al. 1992) and *Pneumocystis jiroveci* (Schmatz et al. 1991). This led to medicinal chemistry efforts at Merck, Eli Lilly, and Fujisawa (Astellas) and the development of the echinocandin drugs caspofungin, anidulafungin (Vicuron and Pfizer), and micafungin, respectively (Wiederhold and Lewis 2003). The three drugs share overall chemical structure but vary in specific members of the hexapeptide ring and the amide-linked side chains (Fig. 22.2). The U.S. Food and Drug Administration has approved echinocandin drugs for the treatment of esophageal and invasive candidiasis, including candidemia, empirical therapy in febrile neutropenic patients and prophylaxis in patients undergoing hematopoietic stem cell transplantation (HSCT) (Turner et al. 2006; Perlin 2011). The first in-class drug caspofungin was also approved for salvage therapy for patients with invasive aspergillosis refractory to conventional therapy (Walsh et al. 2008). Echinocandin drugs show in vitro fungicidal activity against susceptible *Candida* spp. (Ernst et al. 1999; Barchiesi et al. 2005), although they are fungistatic against molds in which they alter cell morphology, cell wall composition, and organization (Bowman et al. 2002, 2006). The echinocandins are largely inactive against *Zygomycetes, Cryptococcus* species, or *Fusarium* species. As echinocandin drugs have a distinct mechanism of action specific against glucan synthase, they are highly effective against yeasts with reduced susceptibility to azoles (Bachmann et al. 2002; Pfaller et al. 2003; Niimi et al. 2006) or to strains of *C. albicans* with acquired azole resistance due to target site mutations in Erg11 or overexpression of major drug efflux transporters. Furthermore, the echinocandins are also active against some *Candida* biofilms (Kuhn et al. 2002; Bachmann et al. 2003; Ferreira et al. 2009; Simitsopoulou et al. 2013); although, this is not a uniform property. The echinocandins have an excellent therapeutic index with a low potential for renal or hepatic toxicity or serious drug–drug interactions (Chen et al. 2011; Kofla and Ruhnke 2011). Despite a common mechanism of action, the three echinocandins vary in metabolism, half-lives, drug interaction profiles, and pharmacodynamic targets (Lepak and Andes 2011; Nett and Andes 2016; Lepak and Andes 2015; Andes et al. 2010). Ultimately, this mandates somewhat different strategies for drug dosing in patients.

Echinocandin drugs are only administered through an intravenous route owing to their high molecular weights and poor solubility, including weak absorption by the gastrointestinal tract. Echinocandins also show concentration-dependent (C_{max}) killing against *Candida* species and a persistent antifungal effect (Nett and Andes 2016; Hope et al. 2007). Fungicidal efficacy correlates with the ratio of AUC:MIC (Lepak and Andes 2011, 2015; Nett and Andes 2016).

All echinocandin drugs exhibit a high degree of binding to plasma proteins (>99%) and distribute minimally to the brain and eye. As the echinocandin drugs are not primarily metabolized by cytochrome P450 enzyme systems, they display fewer drug–drug interactions relative to other types of antifungal drugs. Caspofungin is characterized by nonlinear pharmacokinetics in which tissue distribution accounts for an initial phase of declining plasma levels, followed by gradual release from extravascular tissues (Lepak and Andes 2015; Gumbo et al. 2007). Caspofungin degrades spontaneously and is metabolized hepatically via hydrolysis and N-acetylation to inactive metabolites; less than 2% of active drug is excreted in urine. It has a long net terminal half-life of 27–50 h. Micafungin shows linear elimination pharmacokinetics, producing an adult terminal half-life of ~15 h (Boucher et al. 2004; Lepak and Andes 2015; Wiederhold and Lewis 2007). It is also metabolized hepatically by hydroxylation, arylsulfatase and catechol O-methyltransferase. In contrast to the other echinocandin drugs, anidulafungin is not metabolized but rather undergoes slow spontaneous degradation with a terminal half-life of 40–50 h; less than 1% of unchanged drug is excreted in the urine (Boucher et al. 2004)

A new echinocandin, CD101, is being developed by Cidara with similar mechanism of action to other echinocandin drugs (James et al. 2015). However, owing to some novel medicinal chemistry, it is slowly metabolized and has a very long plasma half-life with little toxicity, which would enable it to be dosed at a much higher level than other echinocandin class drugs. It is currently in early phase clinical trials.

SCY078 is a non-echinocandin enfumafungin-derived glucan synthase inhibitor being developed by Scynexis that is orally bioavailable (Apgar et al. 2015). It displays the same spectrum of antifungal activity as echinocandin class drugs, with somewhat lower pharmacodynamic targets. It is active against some, but not all, echinocandin resistant strains of *Candida* species with acquired *FKS* mutations (Jimenez-Ortigosa et al. 2014; Pfaller et al. 2013) suggesting a close interaction site with echinocandins on glucan synthase. It is presently in mid-phase clinical trials for both oral and intravenous formulations.

Drug resistance has emerged in recent years, and has been reported in most *Candida* species. However, resistance is most prominent among *C. glabrata*. The mechanism of resistance resulting in therapeutic failure is well characterized and involves modification of the catalytic subunits of glucan synthase encoded by the *FKS* genes (Perlin 2015; Arendrup and Perlin 2014).

22.5 Flucytosine (5-FC)

Flucytosine (5-FC) is an orally administered class of nucleoside (pyrimidine) analog used to treat systemic infections due to *Cr. neoformans* and less frequently due to invasive *Candida* species (Bennet 1977). It is a synthetic fluorinated analog of cytosine that is deaminated to 5-fluorouracil (5-FU) and then converted to 5-fluorodeoxyuridylic acid monophosphate, a noncompetitive inhibitor of thymidylate synthase. This metabolic product interferes with DNA synthesis; and it can also incorporate into RNA blocking protein synthesis. Flucytosine has a limited clinical spectrum and it is frequently administered in combination with either AmB or fluconazole because they tend to act synergistically, and the multiple targets limit development of drug resistance, which is common with 5-FC monotherapy (Yao et al. 2014). Combination therapy with 5-FC and AmB is recommended for initial management of severe cryptococcal pneumonia and meningoencephalitis (Day et al. 2013; Loyse et al. 2013; Yao et al. 2014). Resistance to 5-FC results from impaired cellular uptake due to mutation of the cytosine permease, as well as from metabolic defects involving mutations in cytosine deaminase or uracil phosphoribosyl transferase (Vanden Bossche et al. 1998; Hope et al. 2004).

5-FC is highly bioavailable following oral administration showing nearly 90% adsorption and peak serum levels achieved within 1–2 h. As it has low serum protein binding, it tends to be widely distributed among tissues and compartments with drug levels in major organs of spleen, heart, liver, kidney, and lung equivalent to that found in the serum, and it has penetration into bone, vertebrae, peritoneal, and synovial fluids (Vermes et al. 2000). Importantly, there is excellent distribution of 5-FC in cerebrospinal fluid, which are 60–80% of serum levels. Unlike echinocandin drugs, more than 70–90% of 5-FC is unchanged and is eliminated in the urine (Vanden Bossche et al. 1998).

22.6 Immunologic Benefits of Antifungal Therapy

The interplay of the immune system with antifungal agents, both directly and indirectly, plays a critical role in controlling fungal infections. As already described, antifungal drugs can upregulate innate recognition receptors that differently modulate antifungal effector functions (Yamaguchi et al. 1993; Ben-Ami et al. 2008). Azole-mediated depletion of membrane ergosterol renders cells more susceptible to both oxidative and nonoxidative damage by phagocytes. Fluconazole promotes a protective Th1 response (Cenci et al. 1998) and voriconazole induces expression of Toll-like receptor 2 nuclear factor kB activity and TNF-α in human monocytes (Mihu et al. 2014; Simitsopoulou et al. 2008). Lipid formulation of AmB enhance antihyphal activity of human phagocytes, PMNs or MNCs (Wolf et al. 1991); although, it is not associated with reactive oxygen production suggesting a critical role for other nonoxidative innate immune mechanisms (Wolf et al. 1991; Dotis et al. 2008). The echinocandins display a broad capacity to enable *C. albicans* and *A. fumigatus* to stimulate TLR upregulation through cell wall remodeling and glucan unmasking (Wheeler and Fink 2006; Lamaris et al. 2008). Exposure of *C. albicans* to low levels of caspofungin results in unmasking of β-glucan, which enables the fungus to illicit proinflammatory cytokine release from macrophages (Wheeler and Fink 2006; Beyda et al. 2015). Furthermore, caspofungin or micafungin mediated β-glucan exposure in hyphae from a variety of pathogenic molds is associated with increased expression of dectin-1 in neutrophils and TNF-α release by macrophages (Hohl 2009; Hohl et al. 2008). Overall, β-glucan unmasking by echinocandins and enhancement of PMN activity support an immunopharmacologic mode of action for this class of drugs (Beyda et al. 2015).

22.7 Perspective

All current antifungal agents have some deficiencies in spectra, pharmacokinetics, and/or drug–drug interactions, which limit their effectives in treating invasive fungal infections. It has been more than six decades since the introduction of amphotericin B to treat fungal infections. Since that time, only three new classes of drugs (azoles, echinocandins, and antimetabolite) have been approved. The global increase in the number of immunosuppressed patients due to underlying disease and/or modern medical intervention continues to drive the rise in fungal-associated disease, which has increased pressure for widespread prophylaxis and empiric drug use. In concert, the ever-expanding use of antifungal agents has led to emergence of resistance, which limits or in some cases eliminates effective therapy and is a major cause for concern. Therefore, there remains an urgent need to find safe, effective, rapidly fungicidal, broad-spectrum antifungal agents (Pitman et al. 2011). In this context, there has been renewed interest by biopharmaceutical companies to develop antifungal agents and at least ten new drug candidates against old or new targets are in development (Perfect 2015).

References

Allen D, Wilson D, Drew R, Perfect J (2015) Azole antifungals: 35 years of invasive fungal infection management. Expert Rev Anti Infect Ther 13(6):787–798

Andes D (2003) In vivo pharmacodynamics of antifungal drugs in treatment of candidiasis. Antimicrob Agents Chemother 47(4):1179–1186

Andes D, Stamsted T, Conklin R (2001) Pharmacodynamics of amphotericin B in a neutropenic-mouse disseminated-candidiasis model. Antimicrob Agents Chemother 45 (3):922–926

Andes D, Marchillo K, Stamstad T, Conklin R (2003) In vivo pharmacokinetics and pharmacodynamics of a new triazole, voriconazole, in a murine candidiasis model. Antimicrob Agents Chemother 47(10):3165–3169

Andes D, Safdar N, Marchillo K, Conklin R (2006) Pharmacokinetic-pharmacodynamic comparison of amphotericin B (AMB) and two lipid-associated AMB preparations, liposomal AMB and AMB lipid complex, in murine candidiasis models. Antimicrob Agents Chemother 50(2):674–684

Andes D, Diekema DJ, Pfaller MA, Bohrmuller J, Marchillo K, Lepak A (2010) In vivo comparison of the pharmacodynamic targets for echinocandin drugs against Candida species. Antimicrob Agents Chemother 54(6):2497–2506

Apgar JM, Wilkening RR, Greenlee ML, Balkovec JM, Flattery AM, Abruzzo GK, Galgoci AM, Giacobbe RA, Gill CJ, Hsu MJ, Liberator P, Misura AS, Motyl M, Nielsen Kahn J, Powles M, Racine F, Dragovic J, Habulihaz B, Fan W, Kirwan R, Lee S, Liu H, Mamai A, Nelson K, Peel M (2015) Novel orally active inhibitors of beta-1,3-glucan synthesis derived from enfumafungin. Bioorg Med Chem Lett 25(24):5813–5818

Arendrup MC, Perlin DS (2014) Echinocandin resistance: an emerging clinical problem? Curr Opin Infect Dis 27(6):484–492

Bachmann SP, Patterson TF, Lopez-Ribot JL (2002) In vitro activity of caspofungin (MK-0991) against Candida albicans clinical isolates displaying different mechanisms of azole resistance. J Clin Microbiol 40(6):2228–2230

Bachmann SP, Ramage G, VandeWalle K, Patterson TF, Wickes BL, Lopez-Ribot JL (2003) Antifungal combinations against Candida albicans biofilms in vitro. Antimicrob Agents Chemother 47(11):3657–3659

Balding PR, Porro CS, McLean KJ, Sutcliffe MJ, Marechal JD, Munro AW, de Visser SP (2008) How do azoles inhibit cytochrome P450 enzymes? A density functional study. J Phys Chem A 112(50):12911–12918

Barchiesi F, Spreghini E, Tomassetti S, Arzeni D, Giannini D, Scalise G (2005) Comparison of the fungicidal activities of caspofungin and amphotericin B against Candida glabrata. Antimicrob Agents Chemother 49(12):4989–4992

Bartizal K, Abruzzo G, Trainor C, Krupa D, Nollstadt K, Schmatz D, Schwartz R, Hammond M, Balkovec J, Vanmiddlesworth F (1992) In vitro antifungal activities and in vivo efficacies of 1,3-b-D- glucan synthesis inhibitors L-671,329, L-646,991, tetrahydroechinocandin B, and L-687,781, a papulacandin. Antimicrob Agents Chemother 36(8):1648–1657

Ben-Ami R, Lewis RE, Kontoyiannis DP (2008) Immunocompromised hosts: immunopharmacology of modern antifungals. Clin Infect Dis 47(2):226–235

Bennet JE (1977) Flucytosine. Ann Intern Med 86(3):319–321

Beyda ND, Liao G, Endres BT, Lewis RE, Garey KW (2015) Innate inflammatory response and immunopharmacologic activity of micafungin, caspofungin, and voriconazole against wild-type and FKS mutant Candida glabrata isolates. Antimicrob Agents Chemother 59 (9):5405–5412

Boucher HW, Groll AH, Chiou CC, Walsh TJ (2004) Newer systemic antifungal agents: pharmacokinetics, safety and efficacy. Drugs 64(18):1997–2020

Bowman JC, Hicks PS, Kurtz MB, Rosen H, Schmatz DM, Liberator PA, Douglas CM (2002) The antifungal echinocandin caspofungin acetate kills growing cells of Aspergillus fumigatus in vitro. Antimicrob Agents Chemother 46(9):3001–3012

Bowman JC, Abruzzo GK, Flattery AM, Gill CJ, Hickey EJ, Hsu MJ, Kahn JN, Liberator PA, Misura AS, Pelak BA, Wang TC, Douglas CM (2006) Efficacy of caspofungin against Aspergillus flavus, Aspergillus terreus, and Aspergillus nidulans. Antimicrob Agents Chemother 50(12):4202–4205

Brammer KW, Farrow PR, Faulkner JK (1990) Pharmacokinetics and tissue penetration of fluconazole in humans. Rev Infect Dis 12(Suppl 3):S318–326

Brown GD, Denning DW, Gow NA, Levitz SM, Netea MG, White TC (2012) Hidden killers: human fungal infections. Sci Transl Med 4(165):165rv113

Cauwenbergh G (1986) Prophylaxis of mycotic infections in immunocompromised patients: a review of 27 reports and publications. Drugs Exp Clin Res 12(5):419–427

Cenci E, Mencacci A, Del Sero G, Fe d'Ostiani C, Montagnoli C, Bacci A, Bistoni F, Romani L (1998) Induction of protective Th1 responses to Candida albicans by antifungal therapy alone or in combination with an interleukin-4 antagonist. J Chemother 10(2):160–163
Chen SC, Slavin MA, Sorrell TC (2011) Echinocandin antifungal drugs in fungal infections: a comparison. Drugs 71(1):11–41
Cohen BE (1992) A sequential mechanism for the formation of aqueous channels by amphotericin B in liposomes. The effect of sterols and phospholipid composition. Biochim Biophys Acta 1108(1):49–58
Cohen BE (2010) Amphotericin B membrane action: role for two types of ion channels in eliciting cell survival and lethal effects. J Membr Biol 238(1–3):1–20
Cowen LE, Sanglard D, Howard SJ, Rogers PD, Perlin DS (2015) Mechanisms of antifungal drug resistance. Cold Spring Harb Perspect Med 5(7):a019752
Day JN, Chau TT, Lalloo DG (2013) Combination antifungal therapy for cryptococcal meningitis. N Engl J Med 368(26):2522–2523
Diekema DJ, Messer SA, Hollis RJ, Jones RN, Pfaller MA (2003) Activities of caspofungin, itraconazole, posaconazole, ravuconazole, voriconazole, and amphotericin B against 448 recent clinical isolates of filamentous fungi. J Clin Microbiol 41(8):3623–3626
Dodds ES, Drew RH, Perfect JR (2000) Antifungal pharmacodynamics: review of the literature and clinical applications. Pharmacotherapy 20(11):1335–1355
Dotis J, Simitsopoulou M, Dalakiouridou M, Konstantinou T, Panteliadis C, Walsh TJ, Roilides E (2008) Amphotericin B formulations variably enhance antifungal activity of human neutrophils and monocytes against Fusarium solani: comparison with Aspergillus fumigatus. J Antimicrob Chemother 61(4):810–817
Douglas CM (2001) Fungal beta(1,3)-D-glucan synthesis. Med Mycol 39(Suppl 1):55–66
Dutcher JD (1968) The discovery and development of amphotericin B. Dis Chest 54(Suppl 1):296–298
Ellis D (2002) Amphotericin B: spectrum and resistance. J Antimicrob Chemother 49(Suppl 1):7–10
Ernst EJ, Klepser ME, Ernst ME, Messer SA, Pfaller MA (1999) In vitro pharmacodynamic properties of MK-0991 determined by time-kill methods. Diagn Microbiol Infect Dis 33 (2):75–80
Falci DR, Pasqualotto AC (2013) Profile of isavuconazole and its potential in the treatment of severe invasive fungal infections. Infect Drug Resist 6:163–174
Ferreira JA, Carr JH, Starling CE, de Resende MA, Donlan RM (2009) Biofilm formation and effect of caspofungin on biofilm structure of Candida species bloodstream isolates. Antimicrob Agents Chemother 53(10):4377–4384
Fromtling RA (1988) Overview of medically important antifungal azole derivatives. Clin Microbiol Rev 1(2):187–217
Grahl N, Demers EG, Lindsay AK, Harty CE, Willger SD, Piispanen AE, Hogan DA (2015) Mitochondrial activity and Cyr1 are key regulators of Ras1 activation of C. albicans virulence pathways. PLoS Pathog 11(8):e1005133
Groll AH, Giri N, Petraitis V, Petraitiene R, Candelario M, Bacher JS, Piscitelli SC, Walsh TJ (2000) Comparative efficacy and distribution of lipid formulations of amphotericin B in experimental Candida albicans infection of the central nervous system. J Infect Dis 182 (1):274–282
Gumbo T, Drusano GL, Liu W, Kulawy RW, Fregeau C, Hsu V, Louie A (2007) Once-weekly micafungin therapy is as effective as daily therapy for disseminated candidiasis in mice with persistent neutropenia. Antimicrob Agents Chemother 51(3):968–974
Hamill RJ (2013) Amphotericin B formulations: a comparative review of efficacy and toxicity. Drugs 73(9):919–934
Hammond M (1993) Chemical and structure activity studies on the echinocandin lipopeptides. Cutaneous Antifungal Agents. J. Rippon and R. Fromtling. Marcel Dekker, Inc., New York, N. Y., pp 395–420

Hohl TM (2009) Stage-specific innate immune recognition of Aspergillus fumigatus and modulation by echinocandin drugs. Med Mycol 47(Suppl 1):S192–198
Hohl TM, Feldmesser M, Perlin DS, Pamer EG (2008) Caspofungin modulates inflammatory responses to Aspergillus fumigatus through stage-specific effects on fungal beta-glucan exposure. J Infect Dis 198(2):176–185
Hope WW, Shoham S, Walsh TJ (2007) The pharmacology and clinical use of caspofungin. Expert Opin Drug Metab Toxicol 3(2):263–274
Hope WW, Tabernero L, Denning DW, Anderson MJ (2004) Molecular mechanisms of primary resistance to flucytosine in Candida albicans. Antimicrob Agents Chemother 48(11):4377–4386
Jager NG, van Hest RM, Lipman J, Taccone FS, Roberts JA (2016) Therapeutic drug monitoring of anti-infective agents in critically ill patients. Expert Rev Clin Pharmacol 1–19
James K, Laudeman C, Malkar N, Krishnan R, Polowy K (2015) Structure-activity relationship of a series of echinocandins and the discovery of CD101, a highly stable and soluble, once-weekly novel echinocandin. In: ICAAC2015, San Diego
Jimenez-Ortigosa C, Paderu P, Motyl MR, Perlin DS (2014) Enfumafungin derivative MK-3118 shows increased in vitro potency against clinical echinocandin-resistant Candida Species and Aspergillus species isolates. Antimicrob Agents Chemother 58(2):1248–1251
Kale P, Johnson LB (2005) Second-generation azole antifungal agents. Drugs Today (Barc) 41 (2):91–105
Kofla G, Ruhnke M (2011) Pharmacology and metabolism of anidulafungin, caspofungin and micafungin in the treatment of invasive candidosis: review of the literature. Eur J Med Res 16 (4):159–166
Kovanda LL, Maher R, Hope WW (2016) Isavuconazonium sulfate: a new agent for the treatment of invasive aspergillosis and invasive mucormycosis. Expert Rev Clin Pharmacol 1–11
Kuhn DM, George T, Chandra J, Mukherjee PK, Ghannoum MA (2002) Antifungal susceptibility of Candida biofilms: unique efficacy of amphotericin B lipid formulations and echinocandins. Antimicrob Agents Chemother 46(6):1773–1780
Lamaris GA, Lewis RE, Chamilos G, May GS, Safdar A, Walsh TJ, Raad II, Kontoyiannis DP (2008) Caspofungin-mediated beta-glucan unmasking and enhancement of human polymorphonuclear neutrophil activity against Aspergillus and non-Aspergillus hyphae. J Infect Dis 198(2):186–192
Lepak AJ, Andes DR (2011) Antifungal PK/PD considerations in fungal pulmonary infections. Semin Respir Crit Care Med 32(6):783–794
Lepak AJ, Andes DR (2015) Antifungal pharmacokinetics and pharmacodynamics. Cold Spring Harb Perspect Med 5(5):a019653
Linden PK (2003) Amphotericin B lipid complex for the treatment of invasive fungal infections. Expert Opin Pharmacother 4(11):2099–2110
Loyse A, Bicanic T, Jarvis JN (2013) Combination antifungal therapy for cryptococcal meningitis. N Engl J Med 368(26):2522
Maertens JA (2004) History of the development of azole derivatives. Clin Microbiol Infect 10 (Suppl 1):1–10
Malani AN, Kauffman CA (2007) Changing epidemiology of rare mould infections: implications for therapy. Drugs 67(13):1803–1812
McCormack PL (2015) Isavuconazonium: first global approval. Drugs 75(7):817–822
McKeage K (2015) Posaconazole: a review of the gastro-resistant tablet and intravenous solution in invasive fungal infections. Drugs 75(4):397–406
Miceli MH, Kauffman CA (2015) Isavuconazole: A new broad-spectrum triazole antifungal agent. Clin Infect Dis 61(10):1558–1565
Mihu MR, Pattabhi R, Nosanchuk JD (2014) The impact of antifungals on toll-like receptors. Front Microbiol 5:99
Monk BC, Tomasiak TM, Keniya MV, Huschmann FU, Tyndall JD, O'Connell JD 3rd, Cannon RD, McDonald JG, Rodriguez A, Finer-Moore JS, Stroud RM (2014) Architecture of a

single membrane spanning cytochrome P450 suggests constraints that orient the catalytic domain relative to a bilayer. Proc Natl Acad Sci U S A 111(10):3865–3870

Moore JN, Healy JR, Kraft WK (2015) Pharmacologic and clinical evaluation of posaconazole. Expert Rev Clin Pharmacol 8(3):321–334

Nett JE, Andes DR (2016) Antifungal agents: spectrum of activity, pharmacology, and clinical indications. Infect Dis Clin North Am 30(1):51–83

Niimi K, Maki K, Ikeda F, Holmes AR, Lamping E, Niimi M, Monk BC, Cannon RD (2006) Overexpression of Candida albicans CDR1, CDR2, or MDR1 does not produce significant changes in echinocandin susceptibility. Antimicrob Agents Chemother 50(4):1148–1155

Odds FC, Brown AJ, Gow NA (2003) Antifungal agents: mechanisms of action. Trends Microbiol 11(6):272–279

Onishi J, Meinz M, Thompson J, Curotto J, Dreikorn S, Rosenbach M, Douglas C, Abruzzo G, Flattery A, Kong L, Cabello A, Vicente F, Pelaez F, Diez MT, Martin I, Bills G, Giacobbe R, Dombrowski A, Schwartz R, Morris S, Harris G, Tsipouras A, Wilson K, Kurtz MB (2000) Discovery of novel antifungal (1,3)-beta-D-glucan synthase inhibitors. Antimicrob Agents Chemother 44(2):368–377

Pappas PG, Kauffman CA, Andes DR, Clancy CJ, Marr KA, Ostrosky-Zeichner L, Reboli AC, Schuster MG, Vazquez JA, Walsh TJ, Zaoutis TE, Sobel JD (2016) Clinical practice guideline for the management of candidiasis: 2016 update by the infectious diseases Society of America. Clin Infect Dis 62(4):e1–50

Park WB, Kim NH, Kim KH, Lee SH, Nam WS, Yoon SH, Song KH, Choe PG, Kim NJ, Jang IJ, Oh MD, Yu KS (2012) The effect of therapeutic drug monitoring on safety and efficacy of voriconazole in invasive fungal infections: a randomized controlled trial. Clin Infect Dis 55 (8):1080–1087

Pearson MM, Rogers PD, Cleary JD, Chapman SW (2003) Voriconazole: a new triazole antifungal agent. Ann Pharmacother 37(3):420–432

Perfect J (2015) New Antifungal Pipeline. In: ICAAC 2015, San Diego

Perlin DS (2011) Current perspectives on echinocandin class drugs. Future Microbiol 6(4):441–457

Perlin DS (2015) Mechanisms of echinocandin antifungal drug resistance. Ann N Y Acad Sci 1354:1–11

Pfaller MA, Diekema DJ (2004) Rare and emerging opportunistic fungal pathogens: concern for resistance beyond Candida albicans and Aspergillus fumigatus. J Clin Microbiol 42(10):4419–4431

Pfaller MA, Diekema DJ, Messer SA, Boyken L, Hollis RJ, Jones RN, G. International Fungal Surveillance Participant (2003) In vitro activities of voriconazole, posaconazole, and four licensed systemic antifungal agents against Candida species infrequently isolated from blood. J Clin Microbiol 41(1):78–83

Pfaller MA, Messer SA, Boyken L, Rice C, Tendolkar S, Hollis RJ, Diekema DJ (2003) Caspofungin activity against clinical isolates of fluconazole-resistant Candida. J Clin Microbiol 41(12):5729–5731

Pfaller MA, Messer SA, Motyl MR, Jones RN, Castanheira M (2013) Activity of MK-3118, a new oral glucan synthase inhibitor, tested against Candida spp. by two international methods (CLSI and EUCAST). J Antimicrob Chemother 68(4):858–863

Phillips AJ, Sudbery I, Ramsdale M (2003) Apoptosis induced by environmental stresses and amphotericin B in Candida albicans. Proc Natl Acad Sci U S A 100(24):14327–14332

Pitman SK, Drew RH, Perfect JR (2011) Addressing current medical needs in invasive fungal infection prevention and treatment with new antifungal agents, strategies and formulations. Expert Opin Emerg Drugs

Rybak JM, Marx KR, Nishimoto AT, Rogers PD (2015) Isavuconazole: pharmacology, pharmacodynamics, and current clinical experience with a new triazole antifungal agent. Pharmacotherapy 35(11):1037–1051

Safdar A, Ma J, Saliba F, Dupont B, Wingard JR, Hachem RY, Mattiuzzi GN, Chandrasekar PH, Kontoyiannis DP, Rolston KV, Walsh TJ, Champlin RE, Raad II (2010) Drug-induced nephrotoxicity caused by amphotericin B lipid complex and liposomal amphotericin B: a review and meta-analysis. Medicine (Baltimore) 89(4):236–244

Sagatova AA, Keniya MV, Wilson RK, Monk BC, Tyndall JD (2015) Structural insights into binding of the antifungal drug fluconazole to Saccharomyces cerevisiae Lanosterol 14alpha-Demethylase. Antimicrob Agents Chemother 59(8):4982–4989

Sanglard D, Odds FC (2002) Resistance of Candida species to antifungal agents: molecular mechanisms and clinical consequences. Lancet Infect Dis 2(2):73–85

Sau K, Mambula SS, Latz E, Henneke P, Golenbock DT, Levitz SM (2003) The antifungal drug amphotericin B promotes inflammatory cytokine release by a Toll-like receptor- and CD14-dependent mechanism. J Biol Chem 278(39):37561–37568

Schiller DS, Fung HB (2007) Posaconazole: an extended-spectrum triazole antifungal agent. Clin Ther 29(9):1862–1886

Schmatz DM, Powles M, McFadden DC, Pittarelli LA, Liberator PA, Anderson JW (1991) Treatment and prevention of *Pneumocystis carinii* pneumonia and further elucidation of the *P. carinii* life cycle with 1,3-b-glucan synthesis inhibitor L-671,329. J Protozool 38(6):151S–153S

Seyedmousavi S, Verweij PE, Mouton JW (2015) Isavuconazole, a broad-spectrum triazole for the treatment of systemic fungal diseases. Expert Rev Anti Infect Ther 13(1):9–27

Sheehan DJ, Hitchcock CA, Sibley CM (1999) Current and emerging azole antifungal agents. Clin Microbiol Rev 12(1):40–79

Simitsopoulou M, Roilides E, Paliogianni F, Likartsis C, Ioannidis J, Kanellou K, Walsh TJ (2008) Immunomodulatory effects of voriconazole on monocytes challenged with Aspergillus fumigatus: differential role of Toll-like receptors. Antimicrob Agents Chemother 52 (9):3301–3306

Simitsopoulou M, Peshkova P, Tasina E, Katragkou A, Kyrpitzi D, Velegraki A, Walsh TJ, Roilides E (2013) Species-specific and drug-specific differences in susceptibility of Candida biofilms to echinocandins: characterization of less common bloodstream isolates. Antimicrob Agents Chemother 57(6):2562–2570

Takemoto K, Yamamoto Y, Ueda Y (2006) Evaluation of antifungal pharmacodynamic characteristics of Am Bisome against Candida albicans. Microbiol Immunol 50(8):579–586

Turner MS, Drew RH, Perfect JR (2006) Emerging echinocandins for treatment of invasive fungal infections. Expert Opin Emerg Drugs 11(2):231–250

Vanden Bossche H, Dromer F, Improvisi I, Lozano-Chiu M, Rex JH, Sanglard D (1998) Antifungal drug resistance in pathogenic fungi. Med Mycol 36(Suppl 1):119–128

Vermes A, Guchelaar HJ, Dankert J (2000) Flucytosine: a review of its pharmacology, clinical indications, pharmacokinetics, toxicity and drug interactions. J Antimicrob Chemother 46 (2):171–179

Walsh TJ, Anaissie EJ, Denning DW, Herbrecht R, Kontoyiannis DP, Marr KA, Morrison VA, Segal BH, Steinbach WJ, Stevens DA, van Burik JA, Wingard JR, Patterson TF, Infectious Diseases Society America (2008) Treatment of aspergillosis: clinical practice guidelines of the Infectious Diseases Society of America. Clin Infect Dis 46(3):327–360

Wheeler RT, Fink GR (2006) A drug-sensitive genetic network masks fungi from the immune system. PLoS Pathog 2(4):e35

Wiederhold NP, Lewis JS 2nd (2007) The echinocandin micafungin: a review of the pharmacology, spectrum of activity, clinical efficacy and safety. Expert Opin Pharmacother 8(8):1155–1166

Wiederhold NP, Lewis RE (2003) The echinocandin antifungals: an overview of the pharmacology, spectrum and clinical efficacy. Expert Opin Investig Drugs 12(8):1313–1333

Wiederhold NP, Tam VH, Chi J, Prince RA, Kontoyiannis DP, Lewis RE (2006) Pharmacodynamic activity of amphotericin B deoxycholate is associated with peak plasma concentrations in a neutropenic murine model of invasive pulmonary aspergillosis. Antimicrob Agents Chemother 50(2):469–473

Wolf JE, Stein SH, Little KD, Abegg AL, Little JR (1991) Amphotericin B selectively stimulates macrophages from high responder mouse strains. Immunopharmacol Immunotoxicol 13 (3):221–235

Woolley DW (1944) Some biological effects produced by benzimidazole and their reversal by purines. J Biol Chem 152:225–232

Xiao L, Madison V, Chau AS, Loebenberg D, Palermo RE, McNicholas PM (2004) Three-dimensional models of wild-type and mutated forms of cytochrome P450 14alpha-sterol demethylases from Aspergillus fumigatus and Candida albicans provide insights into posaconazole binding. Antimicrob Agents Chemother 48(2):568–574

Yamaguchi H, Abe S, Tokuda Y (1993) Immunomodulating activity of antifungal drugs. Ann N Y Acad Sci 685:447–457

Yao ZW, Lu X, Shen C, Lin DF (2014) Comparison of flucytosine and fluconazole combined with amphotericin B for the treatment of HIV-associated cryptococcal meningitis: a systematic review and meta-analysis. Eur J Clin Microbiol Infect Dis 33(8):1339–1344

Chapter 23
Using Yeast to Discover Inhibitors of Multidrug Efflux in *Candida albicans*

Brian C. Monk and Mikhail V. Keniya

Abstract Fungal infections range from mild readily treated conditions such athlete's foot or thrush to life-threatening illnesses that affect immunocompromised or immunosuppressed patients. The antimycotic azole drugs, which target the enzyme lanosterol 14α-demethylase required for ergosterol biosynthesis, have provided therapy for a wide range of fungal infections over several decades. However, significant resistance is conferred by the removal of azole drugs from fungal cells via multidrug efflux pumps belong to the ABCG class of ATP-binding cassette transporters and a DHA1 class major facilitator superfamily efflux pump. This chapter will describe how the genetically tractable and biochemically well-characterised yeast *Saccharomyces cerevisiae* has provided a platform for the discovery and testing of novel inhibitors of these two classes of drug efflux pump in fungal pathogens of humans.

23.1 Introduction

The remarkable genetic diversity of fungal pathogens, the plasticity of their genomes in during xenobiotic threat, the multifaceted relationships but biological similarities of these eukaryotes with the human host, as well as economic considerations make the expansion of the existing armamentarium of antifungal drugs a daunting challenge. Since the sequencing of the model yeast *Saccharomyces cerevisiae* (Goffeau et al. 1996) the genomic era has produced an array of information, insights and tools that assist antifungal discovery and development. Despite these advances and the entry of a few new drug candidates into clinical trials (Denning and Bromley 2015), application in the clinic of new classes of broad-spectrum antifungals that are not susceptible to drug resistance remains a distant goal. Instead, better preventative strategies, improved diagnostics and increased potency of existing antifungal medications are needed to minimise the

B.C. Monk (✉) · M.V. Keniya
Department of Oral Sciences, University of Otago, Dunedin, New Zealand
e-mail: brian.monk@otago.ac.nz

R. Prasad (ed.), *Candida albicans: Cellular and Molecular Biology*,
DOI 10.1007/978-3-319-50409-4_23

incidence of life-threatening infection in severely debilitated patients and, where possible, avoid the emergence of drug resistance, particularly multidrug resistance. This chapter will reflect on the role that the genetically tractable and biochemically well-characterised yeast *S. cerevisiae* has played in identifying prototype drugs that can be used to block efflux-mediated multidrug resistance in fungal pathogens of humans.

The fungal kingdom is estimated to comprise up to 5.1 million species (Hawkesworth 2001; O'Brien et al. 2005; Blackwell 2011), of which fewer than 1% have been described and under 0.01% cause superficial or invasive infections requiring medical treatment (Pfaller and Diekema 2007, 2010). Physical barriers such as the skin plus active protection at epithelial and mucosal surfaces limit normal colonisation to an even smaller group of fungal pathogens. The commensals *Candida albicans* and *Candida glabrata* usually live harmlessly on human skin, in the mouth and the gastrointestinal, respiratory and reproductive tracts. Inhibition of host salivary flow (due to dry mouth induced by Sjogren's disease, drug treatments or head and neck irradiation) or abrasion of the oral epithelial lining by ill-fitting dentures, especially in the elderly, can lead to oral thrush or denture stomatitis caused by *C. albicans*. Antibiotic treatments that eliminate bacterial infections may destroy much of the endogenous microflora while immune deficiency can degrade protective responses on mucosal surfaces. Oral or vaginal thrush can then arise due to the overgrowth of commensal fungal species. Considered by many to be an opportunist, *C. glabrata* is less pathogenic than *C. albicans*. However, *C. glabrata* gives rise to systemic infections that are more likely to be fatal than those caused by *C. albicans* and a much higher proportion of its clinical isolates show intrinsic resistance to the azole drugs (Pfaller 2012). Most other pathogenic fungi that cause opportunistic infections in humans live in soil or in association with plants, e.g. *Aspergillus* and *Cryptococcus* species, respectively. These organisms are usually cleared by immune-competent individuals and are not normally transmitted from person to person. *Candida parapsilosis* is an exception because it colonises the skin and its spread is linked to poor hand hygiene among some health professionals, a problem that is particularly common in aged care environments (Pfaller and Diekema 2010). Contaminated medical devices are another common source of exogenous infection (Asmundsdottir et al. 2008; Pfaller and Diekema 2010). Superficial infections caused by *C. albicans* and the dermatophytes are usually readily treated with relatively inexpensive antifungal agents such as terbinafine, nystatin, the imidazoles and over-the-counter preparations of fluconazole (FLC). In contrast, life-threatening invasive fungal infections (IFIs) associated with serious underlying disease such as AIDS and modern medical techniques that degrade natural defences are an important concern that is aggravated when multidrug resistance (MDR) is involved (Pfaller and Diekema 2007, 2010).

MDR in yeast and fungi generally involves drug efflux pumps that can confer dramatically (~50–1000-fold) reduced susceptibility to structurally diverse xenobiotics (Balzi and Goffeau 1995; Prasad et al. 1995; Prasad and Goffeau 2012; Kolaczkowski et al. 1996; Coste et al. 2004, 2006, 2009; Cannon et al. 2009; Lamping et al. 2010). Of particular importance in the clinic is the reduced susceptibility of significant fungal pathogens to the widely used and well-tolerated azole antifungals that target lanosterol 14α-demethylase, a key enzyme in the ergosterol biosynthesis pathway (Sanglard and Billie 2002; Monk and Goffeau 2008; Sanglard and Billie 2002; Sanglard et al. 1998; Monk et al. 2014). The triazole drugs FLC and itraconazole (ITC) and, the more recently introduced voriconazole (VCZ) and posaconazole (PCZ), have now served for decades as frontline antifungals. Despite the increasingly widespread use of the echinocandin drugs that have been marketed since about 2000, azole drugs are still used selectively in the clinic to treat infections caused by commensal overgrowth on mucous membranes and to provide prophylaxis against or to treat potentially lethal IFIs, especially those involving *Candida* and *Aspergillus* species. The acquisition of MDR is recognised as a multistep process that exploits the fungistatic nature of the azole drugs (White 1997; Cowen et al. 2002; Ford et al. 2015). For example, a series of clinical isolates of *C. albicans* showing increasing resistance to FLC was a common outcome of repeated or long-term treatment of recurrent oropharyngeal candidiasis in AIDS patients. Homeostatic responses via stress response pathways sustain cell viability of some cells sufficiently long in the presence of these xenobiotics to enable genetic change at the chromosomal and nucleotide levels that can increase the number of target molecules, reduce the affinity of the target for drugs or incur mutations in transcription factors that drive the overexpression of drug efflux pumps (Cannon et al. 2009). In the case of the azole drugs, their intracellular concentration and their ability to inactivate lanosterol 14α-demethylase are reduced. Some of the key molecules that contribute to this outcome are shown in the Fig. 23.1.

23.2 Azole Tolerance

Exposure to drugs that modify the cell surface or the cell wall in *S. cerevisiae* or *C. albicans* can rapidly induce tolerance to growth-inhibitory drug concentrations via the activity of stress response pathways that, in some important cases, invoke a calcium spike (Sanglard et al. 2003). The stress response pathways link cell surface receptors with key protective modulators such as the regulatory phosphatase calcineurin and a network of related factors, including the heat-shock protein Hsp90. This response was detected experimentally because the immunosuppressants cyclosporin A and FK506 act synergistically with FLC (Sanglard et al. 2003) by binding with cyclophilin A (Cyp1p) or Rbp1p, respectively. These complexes inhibit calcineurin activity by binding at its subunit A and B interface, thereby blocking the binding of Ca^{2+}-calmodulin and restricting access of substrate to the active site (Cruz et al. 2002; Sanglard et al. 2003). In contrast, cyclosporin A and FK506 are also both substrates and inhibitors of the ABC transporter pumps that increase efflux of FLC in response to genetic modification of transcription factors (Saini et al. 2005). The transcription factor Crz1p is one of several factors active downstream of calcineurin that contribute to azole tolerance (Cowen et al. 2006;

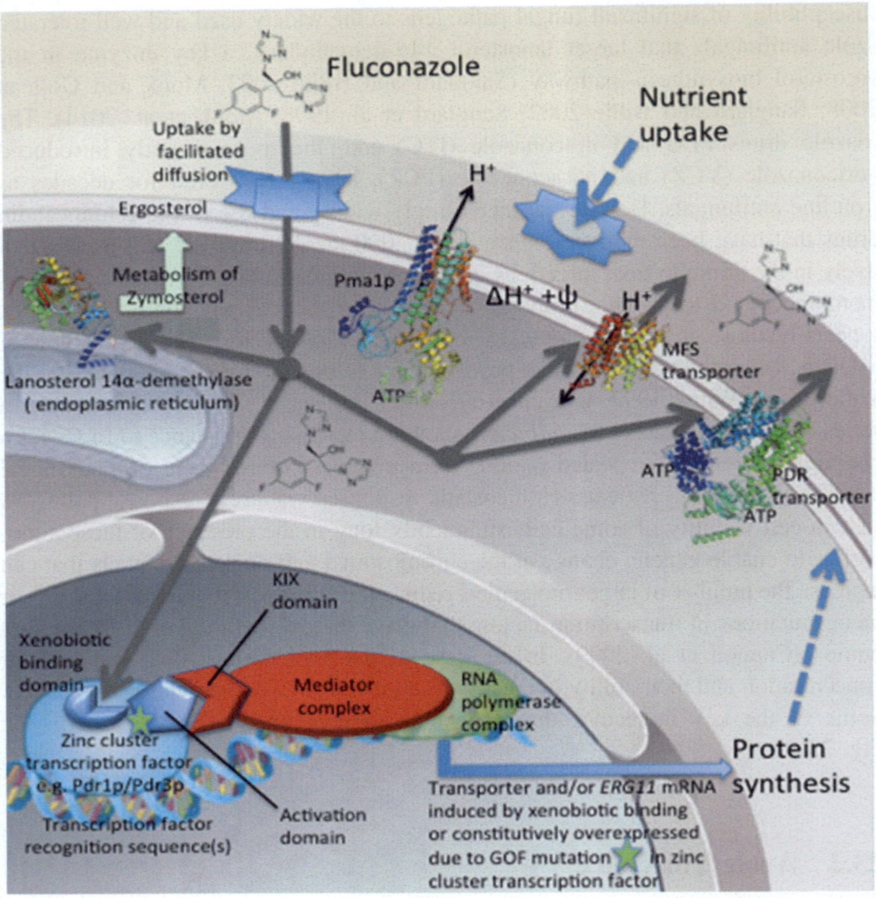

Fig. 23.1 Multidrug resistance in *S. cerevisiae* and *C. albicans*. Fluconazole is shown as a representative of the azole drugs. It enters the cell by facilitated diffusion and inhibits ergosterol biosynthesis by binding in the active site of its target lanosterol 14α-demethylase (Erg11p). It is effluxed from the cell by either an MFS transporter (Mdr1p in *C. albicans*) or the PDR ABC transporters (Pdr5p in *S. cerevisiae* or Cdrp1p and Cdr2p in *C. albicans*), with both types of efflux pump powered indirectly by the plasma membrane proton pump (Pma1p). Azoles and a range of other compounds bind to the xenobiotic binding site of zinc cluster transcriptional regulators resident at specific recognition sequences on the genome. The interaction of the transcriptional regulator with the multisubunit mediator and RNA polymerase complexes, depending on the organism, can induce the synthesis of the ABC and MFS transporters and/or Erg11p. Along with chromosomal changes such as aneuploidy and loss of heterozygosity, this confers tolerance to azole drugs. Intermediate- or high-level resistance is induced when gain-of-function (GOF) mutations occur in transcriptional regulators that constitutively upregulate mRNA synthesis of their target genes. All protein structures shown are homology models except for *S. cerevisiae* Erg11p

Cowen and Lindquist 2005; Karababa et al. 2006; Onyewu et al. 2004). This calcineurin substrate is required for calcium-induced transcription of the β(1,3)-glucan synthase subunit Gsc1p (also known as Fks1p), which may play a role in the repair of cell wall damage, and for the vacuolar Ca^{2+}-ATPase pump Pmc1p, which may dampen the calcium spike-induced activity of the system (Sanglard et al. 2003). Although fungal calcineurin is considered too similar to its human homologue to be an antifungal target, the effects of drugs that modify indirectly the activity of this key mediator of the fungal stress response suggest that targets in the pathways regulating calcineurin activity might help augment the activity of existing antifungals (Heitman 2005). For example, Hsp90, which stabilises calcineurin and is inhibited by geldamycin, is required for rapid selection and maintenance of azole-resistant isolates that arise due to alterations in the ergosterol biosynthetic pathway, but not those involving the more slowly induced drug efflux pumps (Cowen et al. 2006; Cowen and Lindquist 2005). Hsp90 is reported to potentiate adaptation to various unrelated cytotoxic compounds via the induction of aneuploidy (Cowen and Lindquist 2005; Chen et al. 2012). This usually involves the transient duplication of chromosomes or chromosomal regions that carry genes responsible for azole resistance and is a rapid and reversible strategy for adaption to stress, e.g. during treatment with antifungals like FLC (Berman 2016). This type of plasticity is common in fungi including pathogenic fungi. For example, FLC or VCZ are used as maintenance treatments of *Cryptococcus* infection after primary treatment with amphotericin B and 5-flucytosine. Some strains of *Cryptococcus neoformans* and *Cryptococcus gattii* are innately heteroresistant to azole drugs. In response to drug stress subpopulations of these organisms duplicate and transiently retain whole chromosomes or chromosomal fragments that carry genes responsible for azole resistance (Sionov et al. 2013).

Subpopulations of fungal cells with tolerance sufficient to survive azole treatment can undergo more stable genetic modifications that generate constitutive forms of azole resistance (Akins 2005; Sanglard and Billie 2002). The progressive stepwise acquisition of increasingly resistant genetically determined azole resistance phenotypes in *C. albicans* involves multiple mechanisms that give low-, intermediate- and high-level resistance (White 1997; White et al. 1998). While azole resistance of commensal *C. albicans* strains can emerge during long-term exposure to azole drugs in immunosuppressed patients, there is no evidence of this occurring with over-the counter purchases of FLC used to treat superficial infections in immune-competent individuals (Cross et al. 2000; Mathema et al. 2001). Antifungal resistance emerges during antifungal prophylaxis of immunocompromised patients and from the selection of innately resistant organisms generated during previous exposure to azole drugs.

Several fungal species and strains show significant resistance to the azole drugs (Pfaller and Diekema 2007). The ARTEMIS Worldwide survey of the FLC susceptibility of clinical isolates between 1994 and 2004 indicated that significant resistance determined in a standardised agar diffusion test was shown by $\leq 3\%$ of *C. albicans* strains and most of the less frequently isolated *Candida* species. The most important exceptions showing FLC resistance at higher frequency were

C. parasilosis (~5%), *C. glabrata* (~18%) and *C. krusei* (~70%). Of these species, *C. glabrata* and *C. krusei* gave the largest increases in the reported frequency of FLC-resistant isolates during the study. The bulk of *A. fumigatus* strains have intrinsic high-level resistance to FLC but fortunately most strains of *C. krusei* and *A. fumigatus* can be treated with ITC or VCZ. Clinical isolates of *C. glabrata* showing high-level or dose-dependent resistance to FLC are usually cross resistant to the extended spectrum azoles such as VCZ, PCZ and ITC. A discussion of genetically determined azole resistance in *C. albicans* and some other fungal pathogens is presented below as a prelude to description of the discovery of chemosensitizers that circumvent efflux-based resistance to azole drugs.

23.3 Genetically Determined Azole Resistance

23.3.1 Low-Level Azole Resistance

Low-level resistance of *C. albicans* to individual azole antifungals or subgroups of these drugs is due to the development of point mutations that reduce drug binding by the drug target lanosterol 14α-demethylase (i.e. CaErg11p) while broad-spectrum low-level azole resistance is likely to result from its overexpression caused by aneuploidy (Sagatova et al. 2016; Akins 2005; Sanglard and Bille 2002; White et al. 1998). The mutations detected in Erg11p may be single or multiple and these most often occur in three hot spots located within amino acids 105–165, 266–287 and 405–488 of the *C. albicans* primary sequence (Becher and Wirsel 2012; Morio et al. 2010; Parker et al. 2014). Of the 140 different CaErg11p amino acid substitutions found in clinical isolates of *C. albicans*, a much smaller subset has been shown to confer azole resistance (Becher and Wirsel 2012). A useful test of whether an *ERG11* mutation confers azole resistance is to show that the mutant enzyme confers resistance on an azole-sensitive *S. cerevisiae* test strain (Sanglard et al. 1998; Sagatova et al. 2016). Resistance can also be accompanied by promoter changes that compensate for reduced enzyme activity caused by single or multiple mutations. In *Aspergillus fumigatus* a single or a pair of mutations in the Cyp51A together with tandem duplications in the promoter give rise to resistance to azole drugs (e.g. tandem repeat 34 (TR_{34})/L98H and TR_{46}/Y121F/T289A). This type of resistance is of increasing concern to clinicians and has been attributed to environmental exposure to agricultural azoles in Britain, the Netherlands, France and more recently in China, India and the Iran (Azevedo et al. 2015; Chowdhary et al. 2013; Snelders et al. 2008, 2009; Liu et al. 2015; Mohammadi et al. 2016).

23.3.2 Intermediate- and High-Level Azole Resistance

In fungal systems, intermediate- and high-level azole resistance involves the development of constitutive efflux pump overexpression mediated by gain-of-function (GOF) mutations in transcription factors (See Fig. 23.1). In higher organisms GOF mutations in transcription factors also can increase the expression of drug efflux pumps, e.g. during MDR in cancer cells (Eckford and Sharom 2009; Moitra et al. 2011; Chen et al. 2012). While low-level resistance in fungi can be treated effectively with more potent azole drugs, it is not surprising that infections associated with intermediate- and high-level azole-resistant fungi significantly reduce survival rates and result in extended hospitalisation compared with azole susceptible clinical isolates (Pfaller 2012).

A common cause of intermediate- or high-level azole resistance in fungi is the overexpression of plasma membrane transport proteins that pump azoles out of cells, thus reducing intracellular azole concentrations below the levels at which Erg11p is inhibited (Perea et al. 2001; White et al. 1998). There are two main classes of efflux pumps responsible for fungal multidrug resistance, each with a different pumping mechanism and energy source (Cannon et al. 2009). ATP-binding cassette (ABC) proteins are primary transporters that use energy from the hydrolysis of ATP to complete their reaction cycle. Major facilitator superfamily (MFS) pumps are secondary transporters that translocate substrates by utilising the electrochemical gradient ($\Delta H^{+} + \psi$) across the cell membrane generated by the electrogenic plasma membrane proton pumping ATPase (Pma1p). Both classes of drug efflux pump are integral membrane proteins with distinctive functional domains: ABC transporters contain nucleotide-binding domains (NBDs) while both ABC and MFS pumps contain transmembrane domains (TMDs) that confer substrate specificity. In *C. albicans*, the expression of ABC transporters CaCdr1p and CaCdr2p and the MFS transporter CaMdr1p (Ben^{r}p) have been implicated in the resistance of clinical isolates to azoles (Maebashi et al. 2001; Perea et al. 2001; Sanglard et al. 1995; White 1997). The expression of the ABC pumps in *C. albicans* has been more frequently associated with resistance than the expression of Mdr1p, and the expression of Cdr1p appears to be the predominant contributor to clinically significant azole resistance (Holmes et al. 2008; Tsao et al. 2009).

Although VCZ appears susceptible to the same mechanisms of resistance as FLC (see below) its potency may mask azole resistance in some fungi. Intermediate-level azole resistance to FLC plus a limited set of other azole drugs (VCZ and KTC) seems to be due to overexpression of CaMdr1p, while high-level resistance to the entire set of current azole drugs occurs with the overexpression of CaCdr1p and CaCdr2p. CaCdr2p appears to be the product of recent gene duplication (Holmes et al. 2006). Its contribution to efflux-mediated azole resistance, despite the induction of expression by $\sim$50-fold, appears to be less important than CaCdr1p, which also shows a significant level of expression in the absence of azole drugs (Cannon et al. 2009; Holmes et al. 2008).

The ABCG transporters in the Pleiotropic Drug Resistance (PDR) family provide the most clinically significant mechanisms of resistance in both *C. albicans* and *C. glabrata*. In *C. albicans*, the overexpression can be correlated with mutations of the *TAC1* transcriptional regulator that are rendered homozygous (Coste et al. 2006). In *C. glabrata,* high-level azole resistance is caused by expression of the ABC proteins Cdr1p and Cdr2p (also called Pdh1p) (Izumikawa et al. 2003; Miyazaki et al. 1998; Sanglard et al. 2001) determined by GOF mutations in Cg*PDR1*, the single orthologue of the *S. cerevisiae PDR1*/*PDR3* transcriptional regulator pair (Vermitsky et al. 2006). The Cdr proteins pump a variety of substrates, including the azoles (Lamping et al. 2007) but not the echinocandins (Niimi et al. 2006). In *C. neoformans,* the ABC proteins CneAfr1p and CneMdr1p are the only efflux pumps that have been linked to antifungal drug resistance (Sanguinetti et al. 2006; Thornewell et al. 1997). In *A. fumigatus*, gene disruption studies have implicated *CDR1B* (abcB) and possibly *CDR1B* (abcA) in resistance to multiple azoles in different genetic backgrounds and both genes are induced in the presence of azole drugs (Paul and Moye-Rowley 2014). There is no direct evidence for a significant contribution by MFS pumps to MDR in clinical isolates of *C. glabrata*, *C. neoformans* or *A. fumigatus*, even though the expression of such pumps is induced by exposure to azole drugs (Nascimento et al. 2003).

The stepwise acquisition of azole resistance detected in *C. albicans* clinical isolates from individual patients has been used to understand the genetic plasticity of fungal pathogens and provided pairs of genetically related isolates used in screens for molecules that target the overexpressed MFS and ABC transporters conferring intermediate and high-level azole resistance. For example, the G5 strain of *C. albicans* is a daughter clinical isolate showing azole resistance primarily due to the overexpression of CaCdr1p that is not found in its parental G4 strain previously isolated from an AIDS patient. Similarly, the F5 strain is a daughter clinical isolate showing FLC resistance due to the overexpression of Mdr1p not found in its parental isolate F2. These similar pairs of strains been used to validate the mode of action of putative inhibitors of fungal drug efflux pumps that have been identified in screens using heterologous expression of targeted drug efflux pumps (Keniya et al. 2015; Niimi et al. 2012; Nim et al. 2014; Rawal et al. 2014).

Despite the comparative superiority of its pharmacokinetic properties, the fungistatic activity of FLC means that the clearance of pathogenic fungi relies ultimately on the host immune response. The occurrence of FLC-resistant clinical isolates of *C. albicans* became a significant problem in the early 90s among immune-compromised AIDS patients receiving FLC prophylaxis. This problem is now rare where high activity anti-retroviral therapy (HAART) is used to treat HIV patients. In the United States only 1–3% of *C. albicans* clinical isolates now show high-level resistance to FLC (Pfaller 2012). This emphasises the importance of the immune system in clearing fungal infections during treatment with fungistatic drugs. While HAART his very successful in partially restoring the immune response of HIV patients, the commencement of this therapy can be lethal for some patients infected with *Cryptococcus neoformans* due to a hyperactive immune response, known as immune reconstitution inflammatory syndrome, directed against the pathogen (Singh and Perfect 2007).

The increased frequency of the selection or emergence of clinical isolates that are intrinsically highly resistant to the azole drugs has made the treatment of fungal infections more complex. The expression of genes encoding drug efflux pumps often correlates with the azole resistance found in clinical isolates of C. *albicans* (Rogers and Barker 2003; White 1997; Maebashi et al. 2001) and with an increased frequency of *C. glabrata* (Bennett et al. 2004; Sanglard et al. 1999) and *C. neoformans* (Posteraro et al. 2003; Sanguinetti et al. 2006) clinical isolates. In many settings *C. glabrata* infections now occur at comparable or greater frequencies than *C. albicans* infections, with ~18% of *C. glabrata* clinical isolates being highly resistant to azole drugs due to the overexpression of the drug efflux pumps CgCdr1p and CgCdr2p, and with CaCdr1p dominating the response. Although *C. krusei* infections are rare compared to *C. albicans* and *C. glabrata,* <80% of its clinical isolates are innately resistant to azole drugs. This resistance is thought to be due to the presence of multiple copies of CkErg11p as well as the expression of the drug efflux pump CkAbc1p (Katiyar and Edlind 2001; Lamping et al. 2009). Some filamentous fungi possess separate *CYP51* genes (*CYP51A* and *CYP51B*) found in two distinct gene clusters (Ferreira et al. 2005; Mellado et al. 2001). Gene knock-out experiments have shown that *A. fumigatus* Cyp51Ap but not Cyp51Bp is responsible for the innate resistance to FLC and KTC. As the echinocandins are not fungicidal against *A. fumigatus*, VCZ and PCZ are used to treat such infections because they strongly inhibit Cyp51A. The prevalence of VCZ-resistant *A. fumigatus* isolates is low and resistance is usually due to point mutations in Cyp51Ap (Chen et al. 2005; da Silva Ferreira et al. 2004; Nascimento et al. 2003). Finally, the rapid acquisition of resistance to both azole and echinocandin drugs on exposure of *C. glabrata* to either of these distinct drug classes is a sobering reminder of the plasticity of a fungal genome on exposure to xenobiotics (Healey et al. 2016). Mutations in the mismatch repair gene *MSH2* give a mutator phenotype that allows this haploid organism to develop a higher propensity to breakthough antifungal treatment and develop MDR in a mouse colonisation model.

23.4 General Features of MDR in *C. albicans* and Related Systems

The mechanisms responsible for MDR in *C. albicans* have a long evolutionary history. For example, pumps belonging to the ATP-binding cassette (ABC) and major facilitator superfamily (MFS) are found widely in both bacteria and eukaryotes and many are import compounds required for biosynthesis or export xenobiotics including antibacterials (Piddock 2006). The ABC transporters that confer azole resistance in *C. albicans* (Cdr1p and Cdr2p) belong to the pleiotropic drug resistance (PDR) family (Balzi and Goffeau 1995; Prasad et al. 1995; Prasad and Goffeau 2012). They are full-length homodimeric members of the ABCG family of transporters, which also includes homodimeric and heterodimic half-length ABC transporters (Cannon et al. 2009; Lamping et al. 2010). CaMdr1p, the single MFS transporter implicated in clinically significant antifungal resistance in *C. albicans*, belongs to the Drug:H^+ Antiporter (DHA1) class of efflux pumps (reviewed in Redhu et al. 2016). MFS transporters have yet to be implicated in the clinically relevant azole resistance of other fungal pathogens of humans.

23.4.1 Fungal ABC Transporter Structure and Function

ABC transporters are found in all living organisms. Most ABC proteins are membrane proteins and many are thought to transport a variety of substances across membranes. The functional unit of ABC transporters consists of a cytoplasmic nucleotide-binding domain (NBD) and a transmembrane domain (TMD) (Dean 2002; Bouige et al. 2002; Dassa and Bouige 2001; Higgins 2001; Rees et al. 2009; Taglicht and Michaelis 1998). Half-size transporters have one such unit while full-size transporters have two. Sequencing of entire fungal genomes has revealed the repertoire of ABC proteins in a range of species (Verrier et al. 2008; Bouige et al. 2002; Dassa and Bouige 2001). The *S. cerevisiae* genome (Goffeau et al. 1996) was found to encode 29–30 ABC proteins (Decottignies and Goffeau 1997; Taglicht and Michaelis 1998). Each ABC protein contains at least one NBD but a subset lacks a predicted TMD. This subset comprises proteins involved in DNA repair or other ATP-requiring functions. *C. albicans* is predicted to have a similar number (27) of ABC proteins as *S. cerevisiae* (Gaur et al. 2005) and *C. glabrata* approximately two-thirds that number (18) (Gaur et al. 2005; Dujon et al. 2004). Much larger numbers of ABC proteins are found in *A fumigatus* (49) and *C. neoformans* (54) (Nierman et al. 2005; Loftus et al. 2005). The ABC proteins of *S. cerevisiae* have been classified into six classes: PDR (pleiotropic drug resistance), MDR (multidrug resistance), MRP (multidrug resistance-associated protein), RLI (RNase L inhibitor), ALDP (adrenoleukodystrophy protein) and YEF3 (yeast elongation factor EF-3) (Taglicht and Michaelis 1998). Of these, the PDR, MDR and MRP are most often associated with antifungal resistance. Xenobiotic efflux mediated by ABC transporters in fungi is carried out by full-sized ABC transporters and the major pumps involved in the azole resistance of *C. albicans, C. glabrata, C. krusei, A. fumigatus* and *C. neoformans* are all in the PDR class of the ABCG family. The domain arrangement of most full-sized eukaryotic ABC transporters is $[TMD\text{-}NBD]_2$ (Bauer et al. 1999; Bouige et al. 2002; Dean 2002; Gaur et al. 2005; Tusnady et al. 2006; Verrier et al. 2008). Fungal PDR transporters have a reverse topology $[NBD\text{-}TMD]_2$ and plant PDR transporters are the only other class of full-size ABC transporters with the same reverse topology (Verrier et al. 2008; Crouzet et al. 2006; Jasinski et al. 2003; van den Brule and Smart 2002).

It is widely accepted that substrate transport by PDR ABC transporters involves interactions between the membrane bound and cytosolic domains. The pairs of

NBDs carry out cooperative ATP binding and hydrolysis, while the pairs of TMDs containing 6 putative α-helical transmembrane segments (TMS1-12) determine specificity for efflux substrates.

The interdomain interactions are normally tightly coupled with substrate binding, ATP binding and ATP hydrolysis. Bacterial ABC transporters usually fully couple substrate transport with ATP hydrolysis despite being multiunit entities made up of separate NBDs and TMDs. The 'ATP switch model' (Higgins and Linton 2004) still reflects current thinking on the efflux mechanism of ABC transporters. The first step is the binding of substrate to a high-affinity binding site in the TMD that is usually considered to be open to the cytosol or the inner leaflet (IL) of the membrane. Binding of efflux substrate triggers a conformational change in the NBDs via formation of new contacts between each TMD and its cognate NBD. These allow NBD dimerization in the presence of ATP. The resultant conformational change opens the substrate-binding pocket to the extracellular space and the substrate is released. After substrate release, hydrolysis of the bound ATP returns the transporter to its internally open conformation and the transport cycle can be repeated. Thus, ATP binding, and not its hydrolysis, is considered the power stroke of substrate transport. While the molecular details of this model have yet to be elucidated (Jones and George 2004), current knowledge suggests that specific inhibitors would affect particular stages of the reaction cycle.

Both human ABCB1 and ABCC1, the prototype MDR- and MRP-type transporters, respectively, are associated with multidrug resistance of cancer cells (Dean 2002). Their innate basal ATPase activity is stimulated several-fold in the presence of substrate. Fungal PDR transporters are not stimulated significantly by the presence of xenobiotic substrates and a fully uncoupled transport mechanism for the archetypal fungal PDR transporter Pdr5p has been suggested (Ernst et al. 2008). In contrast, Sauna and colleagues (Sauna et al. 2008) have proposed coupling between TMS2 and NBD1 via contact between the Q-loop of nucleotide-binding domain 1 and intracellular loop 1 in *S. cerevisiae* Pdr5p (ScPDR5p).

Fungal PDR transporters, like many other ABC pumps, appear truly pleiotropic, with several hundred substrates reported for ScPdr5p, the orthologue of *C. albicans* Cdr1p (Kolaczkowski et al. 1996, 1998; Prasad et al. 1995). Many substrates appear to be chemically unrelated, but most are hydrophobic yet sufficiently charged to facilitate oriented binding within a large hydrophobic-binding site. This site is thought to be capable of attracting multiple substrates from the IL of the plasma membrane or from the cytoplasm. Golin and colleagues have demonstrated that ScPdr5p binds the substrates rhodamine 6G (R6G), cycloheximide and triazole drugs via sites that, at best, partially overlap, and that substrate size may be a key characteristic (Golin et al. 2003, 2007). The presence of separate non-interacting binding sites and/or multiple efflux pump-mediated pathways across the lipid bilayer is consistent with the finding that efflux of ScPdr5p substrate R6G is not inhibited in the presence of the azole substrate FLC. The ATPase activity of ScPdr5p is not activated (or inhibited) by FLC, unlike mammalian ABC transporters that are activated by their substrates. For example, the human sterol transporter ABCG5/ABCG8 is a member of the ABCG family and the ATPase activity of the reconstituted recombinant enzyme is stimulated by bile acids (Lee et al. 2016). Although fungal PDR transporters such as ScPdr5p and CaCdr1p are associated with multidrug resistance, their biological functions are probably related to membrane biogenesis and lipid homeostasis (Shahi and Moye-Rowley 2009). This could mean that the 'uncoupled' ATPase activity claimed for Pdr5p (Ernst et al. 2008) and found in other PDR transporters may reflect coupling to the transport of endogenous pump substrate(s) such as phospholipids, sphingolipids or sterols.

Crystal structures of the *S. aureus* multidrug transporter Sav1866 (Dawson and Locher 2006), the mouse homolog of human ABCB1 (Aller et al. 2009) and ABCG5/ABCG8 (Lee et al. 2016) are consistent with existing models for the transport mechanism. These transporters have a central cavity sufficiently large to bind many different substrates, and possibly more than one substrate simultaneously. An intriguing question is how can multidrug transporters sequester, with sufficient affinity, a diversity of chemically unrelated hydrophobic and charged compounds from the cytoplasm or lipid bilayer? Alanine scanning mutagenesis of the transmembrane domain of *C. albicans* Cdr1p overexpressed in *S. cerevisiae* (Rawal et al. 2013) suggested that the differential effect on drug binding and transport by substitutions in individual TMD residues is consistent with a large polyspecific drug-binding pocket containing various vestibules that differentially bind FLC and ITC, as predicted in earlier studies that indicated the presence of at least three drug-binding sites in Pdr5p (Golin et al. 2007).

23.4.2 Fungal MFS Transporter Structure and Function

MFS efflux pumps are widely distributed in nature, from bacteria to humans, and organisms generally have multiple gene copies. The MFS efflux pumps in the DHA1 class transport substrates via an antiport mechanism that use membrane potential, including the proton gradient generated by the plasma membrane proton pump. The numbers of DHA1 transporters vary considerably among *S. cerevisiae* and pathogenic fungi—12 transporters in *S. cerevisiae*, 9 in *C. neoformans*, 10 in *C. glabrata,* 18 in *C. albicans*, 28 in *C. parasilosis* and 54 in *A. fumigatus,* but most are uncharacterised (Costa et al. 2014). Of these AQR1, AZR1, FLR1, QDR1, QDR2, YHK8 in *S. cerevisiae*, *MDR1* in *C. albicans*, and *AQR1*, *TPO3* and *QDR2* in *C. glabrata* have been reported to confer resistance or be upregulated by one or more imidazole or azole drugs. It has been suggested that the clinical impact of MFS pumps on MDR is far from fully investigated (Costa et al. 2014).

C. albicans Mdr1p (CaMdr1p) is a 564 amino acid integral membrane protein proposed to contain two sets of six transmembrane helices. Each set of helices is interconnected by internal and external loops and the two halves of the molecule are linked together by a cytoplasmic loop. The N-terminal set of helices is thought to enable proton translocation and the C-terminal set determine substrate recognition (Prasad and Rawal 2014). The A-motif between helices 2 and 3, which is regarded as characteristic of 12 transmembrane helix MFS pumps, is poorly represented in

CaMdr1p while the B-motif located in transmembrane helix 4 is strongly conserved and thought to be involved in energy coupling (Paulsen and Skurray 1993). The efflux pump has an antiporter motif G(X6)G(X3)GP(X2)GP(X2)G in transmembrane segment 5 that is very similar to the motif G(X8)G(X3)GP(X2)GG required for H^+ antiporter activity in bacterial MFS drug/H^+ antiporters. CaMdr1p has therefore thought to efflux substrates in exchange for one or more protons. The first X-ray crystal structures of eukaryotic MFS pumps have been published. These include a fungal phosphate transporter PiPT (Pedersen et al. 2013), the plant nitrate transporter NRT1.1 (Sun et al. 2014; Parker and Newstead 2014) and the human glucose transporter GLUT1 (Deng et al. 2014). However, none of these pumps export xenobiotics and they all have very low similarity to CaMdr1p. These crystal structures have been used to suggest an induced transition fit model for transport (Quistgaard et al. 2016). This involves an occluded state, with the binding of both protons and the effluxed substrate driving antiport.

Insight into a possible drug-binding cavity in MFS drug transporters comes from the crystal structures of the *S. aureus* multidrug-binding transcription regulator QacR with six different structurally diverse drugs bound (reviewed by (Schumacher and Brennan 2003)). Similar to ABCB1 (Aller et al. 2009), this structure reveals a sizeable drug-binding pocket that includes multiple drug-binding mini-pockets. The drug-binding pocket is exceptionally rich in aromatic residues and contains negatively charged amino acids that can neutralise positively charged substrates.

23.4.3 Regulation of Drug Efflux-Mediated MDR in C. albicans

Gene disruption of Zn_2Cys_6 zinc cluster encoding genes in the region of the *C. albicans* mating type locus has shown that deletion of the transcriptional regulator *TAC1* conferred increased azole susceptibility and depressed expression of *CDR1* (Rustad et al. 2002; Coste et al. 2004). In addition to a requirement for *TAC1* in normal expression of Cdr1p, increased *TAC1* copy number due to aneuploidy and changes in the Tac1p primary sequence associated with the loss of heterozygosity have been found to enhance *TAC1* function (Coste et al. 2006, 2007). Like *S. cerevisiae,* GOF mutations in the zinc cluster transcriptional regulator *CgPDR1* upregulate the *C. glabrata* drug efflux pumps CgCdr1p and CgCdr2p.

Azole-resistant clinical isolates of *C. albicans* have been described that require the overproduction of *MDR1* to confer azole resistance (White 1997; Wirsching et al. 2000a). CaMdr1p, an orthologue of Flu1p in *S. cerevisiae*, transports a range of structurally unrelated compounds such as benomyl, cerulenin, brefeldin A, FLC, VCZ but not all azole drugs (White et al. 1998; Morschhauser et al. 2007; Hiller et al. 2006; Rognon et al. 2006; Niimi et al. 2006). *MDR1* expression is regulated by several transcription factors that bind to elements in the *MDR1* promoter (Vandeputte et al. 2012; Schubert et al. 2011a, b). It is positively regulated by transcription factors including Mcm1p, Mrr1p and the Upc2p (Riggle and Kumamoto 2006; Dunkel et al. 2008a, b) and negatively regulated by Cap1p (Alarco and Raymond 1999) and Cph1p (Lo et al. 2015). In *C. albicans,* GOF mutations in *MRR1* result in Mdr1p overexpression and drug resistance (Dunkel et al. 2008a). This property has been demonstrated by modifying drug susceptible clinical isolates, while deletion of *MDR1* reverses the drug resistance phenotype (Wirsching et al. 2000b; Hiller et al. 2006). Such experiments imply that Mrr1p is the key regulator of Mdr1p-mediated drug resistance. Genetic analysis shows the transcriptional activity of this zinc cluster protein can be activated by single amino acid GOF substitutions similar to those found in other zinc cluster proteins, while chromatin immunoprecipitation experiments are consistent with autoregulation of *MRR1* expression. GOF mutations in *UPC2* are known to upregulate *MDR1* expression (Dunkel et al. 2008b) and lead to pronounced azole resistance via upregulation of CaErg11p. It has been suggested that the regulation of both *TAC1* and *MRR1* in *C. albicans* resembles the positive autoregulation found for *S. cerevisiae PDR3* (Paul and Moye-Rowley 2014).

23.5 Pump Systems and Fungal MDR

The systems that contribute to multidrug resistance in fungi might seem to have evolved as protective responses to xenobiotics in the environment. However, the export of xenobiotics from bacteria occurs predominantly via the MFS pumps and not the ABC transporters (Fluman and Bibi 2009). The main role of bacterial ABC transporters appears to be the uptake of nutrients, although some bacteria use ABC transporters to export the compounds such as the toxic peptides they produce to reduce competition by other microorganisms (Claverys et al. 2007). Some experimental evidence suggests that the *S. cerevisiae* Pdr5p and the *C. albicans* Cdr1p and Cdr2p may function as floppases and *C. albicans* Cdr3p as a flippase that directionally transport phospholipids in order to maintain bilayer asymmetry (Smriti et al. 2002; Shukla et al. 2007). A caveat is that the phospholipid analogues used in these studies contained fluorescent (nitrobenoxadiazole) substituents that may be transport substrates.

The individual drug efflux pumps responsible for MDR are not essential in fungal systems, i.e. their deletion in *C. albicans,* or the deletion of their orthologues in *S. cerevisiae* does not result in cell death in non-selective medium. While such results might be due to the presence of compensating orthologues, the overexpression of individual efflux pumps confers resistance (i.e. they become essential) in the presence of their xenobiotic substrates. If the function of ABC transporters such as CaCdr1p is to maintain membrane lipid asymmetry, this property might protect against the xenobiotic uptake or provide charged gradients that allow slightly positively charged xenobiotics to move across the membrane outside the body of a transporter. Both of these notions seem unlikely as the import of azole drugs is not mediated by passive diffusion but by facilitated diffusion in *C. albicans* and *A.*

fumigatus (Esquivel et al. 2015; Mansfield et al. 2010) while individual amino acid substitutions within CaCdr1p differentially affect susceptibility to particular azoles, consistent with a large multi-vestibule drug-binding site (Rawal et al. 2013). Furthermore, electron density consistent with cholesterol has been found within an internal-binding site in the membrane sector of the crystal structure of the ABCG5/ABCG8 transporter heterodimer (Lee et al. 2016).

Primary (ATP-requiring) and the secondary (electrochemical potential-dependent) transporters consume significant amounts of energy to efflux drugs. In order to avoid an inordinate fitness cost, efflux should to be responsive to environmental change (e.g. activated when a xenobiotic or stressor is present) and tightly regulated (i.e. turned off when not needed). Consistent with this idea, the activity of efflux pumps is regulated via both homeostatic mechanisms including stress response pathways that involve enzyme phosphorylation (Wada et al. 2002), as well as by transcription factor-mediated activation of gene expression that is induced by xenobiotics (Thakur et al. 2008). In addition, cells can adapt using genetically determined upregulation of efflux pumps after extended periods of stress though genetic modification, e.g. via gross chromosomal rearrangements, aneuploidy, loss of heterozygosity or GOF modification of transcription factors (Ford et al. 2015; Morschhauser 2016). Finally, a different form of MDR appears to be associated with the formation of biofilms, an important habit adopted by *C. albicans* during infection and contamination of biomedical devices (Ramage et al. 2006). Mutants of *C. albicans* carrying single and double deletion mutations in *CaCDR1*, *CaCDR2* and *CaMDR1* or *CaCDR1*/*CaCDR2* and *CaMDR1*/*CDR1* are hypersusceptible to FLC when planktonic, but show a resistant phenotype during biofilm growth (Ramage et al. 2002). Thus, upregulation of these drug efflux pumps may not be important in the resistance of biofilms to azole drugs.

Our knowledge of antifungal resistance mechanisms and the need to improve the efficacy of the current repertoire of cheap antifungal drugs, especially the widely used, well-tolerated and broad-spectrum azole drugs, strongly suggests the value of investment in the discovery of compounds that can overcome multidrug resistance by inhibiting the development or function of drug efflux pathways. By analogy to chemosensitizers that block the resistance of tumour cells to anticancer therapeutics (Saraswathy and Gong 2013), where resistance often involves upregulation of genes controlling efflux pumps, chemosensitizers that block efflux pump function can be expected to make pathogenic fungi such as *C. albicans* more sensitive to antifungals that are efflux pump substrates, i.e. by increasing the intracellular concentrations of these antifungals. This approach has the potential to increase antifungal potency in target cells and thereby reduce the amount of antifungal drug required in the clinic, diminish side effects caused by the metabolism of antifungals, make some fungistatic drugs fungicidal, delay the emergence of drug resistance and improve the treatment outcomes of patients infected with multidrug-resistant organisms. The remainder of this chapter will describe why and how the model yeast *S. cerevisiae* can host the discovery of direct and indirect inhibitors of multidrug efflux in *C. albicans*. We also suggest Erg11p inhibitors that are not efflux pumps substrates could overcome this problem.

23.5.1 S. cerevisiae *Is Widely Used for Genetic, Biochemical and Physiological Studies*

Yeasts have made many important contributions to our understanding of biology, from individual enzyme reactions to systems biology. They continue to provide informative experimental models for human health and disease and have generated numerous techniques applicable to more complex systems. The baker's yeast *S. cerevisiae* is an invaluable model organism for drug discovery, even though it has only a core of the properties that characterise either higher eukaryotes or more complex fungal systems that extend to the moulds and mushrooms. Completion of the *S. cerevisiae* genome sequence of ~6200 open reading frames in 1996 (Goffeau et al. 1996) created the expectation that a new age of antifungal discovery based on novel targets would follow once the genomes of the most prominent fungal pathogens and the human genome were sequenced and annotated. Knowledge of the biochemistry, physiology and genetics of this generally regarded as safe budding yeast and sophisticated curation of the *S. cerevisiae* genome sequence underpin the comparative genomics of eukaryotes. In addition, a series of functional genomics and proteomics techniques have made the yeast genome and yeast-based experimental systems powerful, accessible and widely used.

As exemplified by the *S. cerevisiae* genome, realisation of the full value of genomic information requires ongoing sequence upgrading and annotation. In March 2016 the Candida Genome Database (http://www.candidagenome.org) Assembly 22, version s06-m0-r01 indicated 1583 verified ORFs, 152 ORFs that were considered dubious and 4483 uncharacterized ORFs. However, at least one-third of the *S. cerevisiae* ORFs and almost two-third of the *C. albicans* ORFs are still classified as unknown.

The *S. cerevisiae* genome sequence has facilitated the development and application of functional genomic techniques on a genome-wide scale. These techniques include genome-wide microarrays, two-hybrids clone sets, protein chips and epitope- and florescent protein-tagged yeast strains (Parsons et al. 2004; Tong and Boone 2006; Giaever et al. 2004; Giaever and Nislow 2014; Lee et al. 2014). Some of these techniques have been applied successfully to other fungi, including *C. albicans* (Roemer et al. 2003). Many of the studies with *S. cerevisiae,* together with relevant genetic and biochemical literature, can be accessed via databases such as the *Saccharomyces* genome database (SGD, http://www.yeastgenome.org). Genome-wide studies of genetic interactions, gene expression, protein–protein interactions, and protein abundance and localization in *S. cerevisiae* are a starting point for systems biology modelling of fungi. This integrative approach is helping ascribe the roles to many genes of unknown function.

23.6 Commonality and Differences Among Eukaryotes and Fungal Pathogens

Antifungal specificity in medicine depends on the divergence of pathogenic fungi from humans. The fungal and animal kingdoms separated over 800 million years ago and only ~45% of the 6200 predicted *S. cerevisiae* proteins show amino acid sequence homology with at least one human protein, i.e. homology at a BLAST E-value of $<10^{-10}$ (Parsons et al. 2006). The most important fungal pathogens of humans, including most of the fungal species for which complete or near complete genome sequences are available, are found in the *Saccharomyces*, *Candida*, *Fusarium*, *Aspergillus*, *Coccidioides–Histoplasma*, *Cryptococcus* and *Pneumocystis* genetic clusters of the Ascomycota and Basidiomycota (Galagan et al. 2007). Within the Ascomycetes, the phylogenetic distance between members of the hemiascomycetes, which includes *S. cerevisiae*, *C. glabrata* and *C. albicans*, is equivalent to that between the chordates, e.g. men and fish. There are even greater phylogenetic distances between hemiascomycetes and the Eurotiomycetes pathogens *Histoplasma capsulatum, Coccidioides immitis* and *Aspergillus* species, or between the Ascomycota and the Basidiomycota pathogen *Cryptococcus neoformans* (Fitzpatrick et al. 2006). Synteny between haploid genomes within the *Aspergillus* cluster (*A. fumigatus*, *A. niger*, *A. terreus* and the model organism *A. nidulans*) and the average of only 68% amino acid identity between *A. nidulans*, *A. fumigatus* and *A. oryzae* protein pairs are consistent with rapid evolutionary divergence since its emergence 200 million years ago (Galagan et al. 2005; Wortman et al. 2006). The evolution of the leucine-tRNA CUG codon to specify serine separates most *Candida* species from other members of the hemiascomycetous group (Fitzpatrick et al. 2006). All the major pathogenic *Candida* species appear to be diploid or aneuploid apart from *C. glabrata*, which is haploid and has a physiology, genome and codon usage akin to the model yeast *S. cerevisiae.*

C. albicans and *S. cerevisiae* are thought to have diverged about 140-330 million years ago (Seoighe et al. 2000; Shields and Wolfe 1997). After a whole genome duplication (WGD) event about 100 million years ago in an ancestor most closely related to the *Kluyveromyces* lineage (Wolfe and Shields 1997), the evolution of *S. cerevisiae* and *C. glabrata* involved substantial gene loss. These processes introduced many syntenic breakpoints and inversions in the *S. cerevisiae* and *C. glabrata* genomes, but syntenic blocks representing represent 88% of the two genomes are maintained (Fischer et al. 2006). As in *Aspergillus* species, and expected with the acquisition of new gene functions for secondary metabolism or adaption to ecological niches, synteny is not conserved in subtelomeric regions. Comparison between representative hemiascomycetes of the proportions of putative essential genes among syntenic orthologues and among all orthologues at increasing phylogenetic distances found that essential genes tend to remain adjacent during evolution; a trend that was maintained even at large evolutionary distances where massive chromosomal change has occurred (Fischer et al. 2006). Furthermore, different rates of macro- and micro-genomic rearrangement were associated with different genomic lineages, and the least stable genomes amongst the hemiascomycetous yeast were those of the fungal pathogens tested, i.e. *C. albicans* and *C. glabrata* (Fischer et al. 2006). Chromosomal duplication and aneuploidy of selected regions of the *C. albicans* genome during the acquisition of drug resistance by some strains provides additional genetic plasticity (Selmecki et al. 2006). The evolution of fungal genomes is an ongoing and dynamic process that responds to environmental conditions, but there is also considerable selective pressure to maintain a core set of essential genes.

The likelihood of discovering broad-spectrum drugs that target proteins encoded by essential fungal genes depends on the balance between the heterogeneous divergence of the major fungal pathogens from the human host and the maintenance of structure–function relationships in a core set of essential proteins. While only 10–25% of fungal proteins shared by ten fungal genomes (including four from the fungal pathogens *C. glabrata*, *C. albicans*, *A. fumigatus* and *C. neoformans*) have human homologs that are >40% identical, genes specifying proteins with core essential functions are much less divergent (Liu et al. 2006). About 20% of the genes in the fungal genomes tested thus far (*S. cerevisiae* and *C. albicans*) are essential but these are overlapping rather than identical sets of genes. Among the 1250 essential *S. cerevisiae* proteins, a core set of only $\sim$250 were also encoded in all ten fungal genomes at >40% identity. As more fungal genomes are sequenced this core set is likely to further diminish in number. Of the core set of essential proteins, 80% showed >40% homology with human proteins. This may reflect the finding that essential proteins have greater numbers of protein–protein interactions than other proteins and that their expression has the strongest response to growth rate (Castrillo et al. 2007).

The antifungal drugs used therapeutically are from a limited number of structural classes. They are directed against few molecular targets and impact on either the plasma membrane, the cell wall or RNA- and DNA-based macromolecules. Sanglard and Billie have reviewed the mechanisms of these drugs for *Candida* species (Sanglard and Billie 2002). The use of existing targets and drugs in the clinic, and their associated problems, provide standards against which new developments should be judged. The existing antifungals, including over-the-counter medications such as the FLC, provide highly effective treatments for the immune sufficient patients without causing significant drug resistance (Mathema et al. 2001). Drug resistance, especially MDR mediated by drug efflux pumps, is an important and growing threat for the immune-compromised and patients whose barriers to disseminated infection have been reduced or disabled.

23.7 *S. cerevisiae* as a Surrogate in Drug Discovery

S. cerevisiae is an invaluable experimental surrogate for more genetically complex and less understood fungal pathogens, including *C. albicans*. It has helped dissect processes such as fungal dimorphism (Gimeno et al. 1992; Palecek et al. 2002), the nature of sex in *C. albicans* (Hull et al. 2000) the formation of biofilms on prosthetic devices (Reynolds and Fink 2001) and the molecular basis of azole resistance in fungal pathogens (Sanglard et al. 1995, 1998). It is a genetically malleable model for investigating drug targets, including studies of the mode of action of antifungal drugs (Chen et al. 2007) and identifying target-associated metabolic pathways (Imoto et al. 2003). Comparative genomic, functional genomic and proteomic tools enable more efficient identification of gene function, the visualisation of the transcriptional responses to environmental factors and xenobiotics, and the determination of the cellular distribution, interaction networks and roles of proteins that impact on the lifestyles of fungal pathogens and their relationship with the human host (Parsons et al. 2006). Such studies help understand the on- and off-target effects of existing and novel antifungals, the molecular basis of antifungal resistance, and identify and prioritise new targets for antifungal discovery and development. Although the value *S. cerevisiae* in identifying antifungals is exemplified by the discovery and characterization of the echinocandin class of drug (Onishi et al. 2000), novel classes of antifungal agent that affect previously untapped targets identified primarily from genomic information have yet to appear in the clinic. If the experience with antibiotic discovery is replicated, it may be some time before this expectation is realised (Payne et al. 2007). In the meantime, it seems pragmatic to improve existing drugs by understanding their mode of action, determining how particular mutations affect drug binding and finding ways to circumvent their efflux from the cell.

23.8 Use of Yeast Deletion Mutants to Identify Drug Targets

Collections of systematic yeast deletion mutants with unique, readily identifiable molecular bar codes are making a significant contribution to antifungal discovery (Giaever et al. 2002). One collection comprises a set of heterozygous diploid strains each containing a deletion in one allele for each of the ~6200 yeast genes. Other sets comprise homozygous deletion strains, and *MATa* strains and *MATα* strains deleted for each of ~5000 non-essential genes. Building on earlier genetic studies, ~1200 genes in *S. cerevisiae* that are essential for growth and/or viability have been identified. This foreshadowed the ability to identify and prioritise new targets in pathogenic fungi using the rationale that the deletion of an essential gene should be functionally equivalent to the inactivation of a protein target by a fungicidal drug (see below for some important caveats) (Giaever et al. 2004; Giaever and Nislow 2014). Furthermore, if an essential protein is inactivated by an antifungal with high affinity, the MIC for the drug and the concentration of its protein target should be related. This also applies to the efflux pumps that become essential in the presence of a xenobiotic substrate. Since drug tolerance developed on exposure to azoles may alter the physiology of cells, it may be preferable to apply a counterscreen involving a related host strain that tolerates xenobiotic exposure due to the expression of an alternative class of drug efflux pump instead of a null host.

As the heterozygous deletion strains are diploid for all alleles except the test locus, and because only about 3% of the *S. cerevisiae* genes show haploinsufficiency under standard growth conditions (Deutschbauer et al. 2005), hypersensitization of strains by chemicals can be used in fitness tests to link a compound directly to its target or other mode of action-related genes (Baetz et al. 2004; Giaever et al. 2004; Lum et al. 2004). This form of chemical-genetic mapping allows target identification, drug screening and mode of action studies. The approach is sensitive and economical because the response of pools of deletion strains to individual drugs at semi-inhibitory (sub-MIC) concentrations can be quantitated using PCR amplification of strain-specific genome bar codes and microarray analysis. Furthermore, chemically induced haploinsufficiency for non-essential genes can identify synthetic-lethal interactions or proteins in the same cellular pathway as antifungal targets (Fleming et al. 2002; Giaever et al. 2004; Hanway et al. 2002). The hypersensitivity of yeast deletion mutants to compounds to which MDR confers resistance has been used to identify ergosterol biosynthetic enzymes, drug efflux pumps and sequestration via the vacuolar ATPase as contributors to this important phenotype (Parsons et al. 2004). However, this approach is not applicable where families of proteins or even unrelated proteins have analogous functions.

The overexpression of a specific drug efflux pump from the *PDR5* locus in a yeast genetic background in which other efflux pumps have been deleted provides a cellular template for the discovery of inhibitors of drug efflux pumps in a strain that is hypersensitive to xenobiotics (Nakamura et al. 2001). In addition, the xenobiotic-dependent induction of reporters at the *PDR5* locus provides tools to discover inhibitors of transcription factor-related drug resistance. Thus *S. cerevisiae* allows mapping and exploitation for drug discovery of compound–target, compound–pathway, synthetic lethal target–effector interactions and multidrug resistance phenomena. Finally, the proteins involved in such phenomena do not exist alone—most physically interact with other proteins and this should only occur in appropriate cellular compartments.

23.9 Is the Protein Interactome Important for Drug Efflux?

Most studies of the yeast protein interactome have used either the yeast two-hybrid (YTH) system to identify soluble intranuclear bait–prey couplings or MS to identify components of affinity-tagged soluble or detergent solubilised protein co-complexes (AP-MS) (Babu et al. 2012; Paumi et al. 2009). All three approaches detect functional and regulatory interactions that can be annotated through ‘guilt by association’. However, the identification of networks with different topological and biological properties is method dependent (Jensen and Bork 2008). These problems can be minimised using a derivative of the membrane yeast two-hybrid (MYTH) system that reconstitutes a split ubiquitin to report on interactions between membrane proteins presented as baits and their preys in situ (Snider et al. 2013). Reconstitution of a quasi-native ubiquitin activates genes allowing growth on selective medium that can be verified using a lacZ reporter, while the expected bait localization is confirmed using a YFP attached to the bait (iMYTH) (Paumi et al. 2009; Snider et al. 2013). The phenotypic analysis used to assess the impact of selected yeast ABC transporter bait constructs on cell function required a bait–prey complex for ubiquitin reconstitution. Bimolecular fluorescent complementation using a split YFP confirmed bait–prey interactions and identified the subcellular localization of the bait–prey complex.

Although ABC transporters have extramembranous nucleotide-binding domains involved in ATP hydrolysis and transmembrane domains that determine the substrate specificity of transport, their function may be modulated by interactions with other proteins. Of 30 yeast proteins identified as containing the nucleotide-binding domains characteristic of ABC proteins in *S. cerevisiae*, only 19 could be used as iMYTH baits because they contain transmembrane domains consistent with ABC transporter function plus N- or C-terminal sequences predicted to be accessible from the cytosol. Using 18 of these baits, 537 binary interactions involving 366 soluble and membrane proteins, including many new interactions, were identified (Snider et al. 2013). Bimolecular fluorescent complementation and/or AP-MS experiments validated 81% of the interactions detected by iMYTH.

Snider et al. provided evidence of subfamily-specific physical interactions dominated by either soluble or membrane proteins and interactions affecting function independent of mRNA expression levels. Gene deletions revealed the importance of ABC–ABC interactions among the ten-member ABCG subfamily involved in multidrug efflux and sterol uptake. Only the anaerobiosis-induced sterol transporter Aus1p, the stress-responsive transporter Pdr15p and the half-sized transporter Adp1p did not directly interact with least one of the other seven members of the ABCG family, although these proteins link indirectly with Pdr5p, Pdr10p and the YOL075c protein via other membrane proteins. It was suggested that physiologically relevant protein–protein interactions affect overall ABC transporter function, including the maintenance of transport promiscuity in *S. cerevisiae*. In contrast, little is known about the nature or impact of protein–protein interactions that occur among the ABCG proteins in fungal pathogens. We assume that if the *Candida* ABCG homologues (i.e. CgCdr1p, CgCdr2p, CaCdr1p and CaCdr2p) heterologously overexpressed in *S. cerevisiae* behave in ways similar to their overexpressed native yeast counterpart (i.e. Pdr5p), compounds that affect the function of these drug efflux pumps should behave similarly in the fungal pathogen. We are encouraged, for example, by results from chemogenomic analysis and known drug synergies found with both *S. cerevisiae* and *C. albicans* that involve azole drugs and chemosensitizers that affect drug efflux (see below).

23.10 Chemosensitization

Stagnation in the development of novel antifungal agents has highlighted the need to improve the efficacy of existing antimycotics. Chemosensitization of fungal systems, like efforts designed to block the drug efflux mechanisms in cancer cells, uses compounds that should have minimal effects on growth, but which in combination with an antimycotic increase antifungal efficacy (Campbell et al. 2012). This approach is increasingly likely to accepted by the pharmaceutical industry because of the advantage, if not necessity, of drug combination formulations to assure efficacy and economic product lifetime in the market, e.g. antibacterials such as augmentin and current multidrug treatments of AIDS and malaria. Antifungal chemosensitizers act by increasing the potency of antimycotics, thereby reducing antifungal dosages, potentially to levels that minimise both production costs and side effects. In some cases a fungistatic drug such as FLC can be converted into a fungicidal drug. This approach should therefore limit, but not necessarily exclude, the development of drug resistance. Chemosensitizers could also be used salvage patients by counteracting pre-existing target- and efflux pump-based azole resistance. However, FDA approval is a very significant economic hurdle for chemosensitizer–antifungal combinations because both the antifungal and chemosensitizer must be shown to act specifically and their individual and collective mechanisms of action understood. Chemogenomic profiling should be helpful in this regard.

There are several mechanisms by which a chemosensitizer can be envisaged to improve the efficacy of azole drugs in fungal cells:

1. Enhancement of azole uptake via modulation of membrane composition, cell permeability or facilitated diffusion.
2. Restriction of the ATP needed to power efflux by ABC drug transporters.
3. Reduction of electrochemical potential needed to power efflux by MFS drug transporters.
4. Direct inhibition of ABC drug efflux pumps.
5. Direct inhibition of MFS drug efflux pumps.
6. Inhibition of the expression of ABC and/or MFS drug efflux pumps.
7. Down-regulation of Erg11p overexpression.

Chemogenomic profiling measures the cellular response to individual or pairs of inhibitory compounds by determining the fitness of genome-wide libraries of suitable *S. cerevisiae* deletion strains. Chemogenomic analysis of *S. cerevisiae* supports some of the ideas described above and this yeast provides the physiological and biochemical tools needed to evaluate each of these mechanisms and to screen for chemosensitizers likely to be effective in pathogenic fungi. For example, chemogenomic profiles obtained for FLC found gene deletions such as the *PDR5* drug efflux pump and members of the SAGA and mediator complexes that transcribe this pump (Jansen et al. 2009) sensitised the system, consistent with the observation that the calcineurin and Pdr5p inhibitor FK506 (Tacrolimus) enhanced synergistically the activity of FLC in *S. cerevisiae* and rendered FLC cidal on *C. albicans*. A subsequent chemogenomic study identified small molecule compounds that perturbed the plasma membrane, including those affecting sphingolipid metabolisms, across multiple fungal pathogens (Spitzer et al. 2011). Sphingolipids are known to be important for plasma membrane function, including the activity of the plasma membrane proton pump (Pma1p) which determines the electrochemical potential of this membrane, the uptake of nutrients and indirectly the activity of ABC and MFS transporters.

23.11 The Yeast AD Expression System

S. cerevisiae provides a compatible host system that enables study of drug efflux responsible for MDR and better opportunity to create robust screens that identify inhibitors of this activity than in the more complex genetic systems of fungi such as *C. albicans*. The laboratory of Andre Goffeau developed an *S. cerevisiae* host strain (AD1-8u$^-$) that gave constitutive hyperexpression *S. cerevisiae* drug efflux pumps Pdr5p, Snq2p and Yor1p from the genomic *PDR5* locus (Decottignies et al. 1994, 1995, 1998). The host strain is deleted of 7 major ATP-binding cassette (ABC) transporters and the autoregulatory *PDR3* transcriptional regulator. It contains the mutant GOF Pdr1-3p transcriptional regulator that acts globally to deliver a series of functional membrane proteins to the plasma membrane (DeRisi et al. 2000). A strong drug resistance and efflux phenotype was achieved by homologous hyperexpression of Yor1p or Pdr5p in this yeast strain with depleted pump activity (Decottignies et al. 1994, 1998). The host is ultrasensitive to many xenobiotics due to the deletion of 7 ABC transporters (6 domiciled in the plasma membrane and one in the vacuole). We recognised that this system should enable the stable heterologous expression of drug efflux pumps and other proteins. We found this to be the case and have shown that several proteins were targeted to membranes or compartments (Lamping et al. 2007; Nakamura et al. 2001). The expression system was refined initially by preparing transformation cassettes in a series of modified plasmids (Lamping et al. 2007; Nakamura et al. 2001). The PDR pump to be studied was directionally cloned into plasmid pABC3 (GenBank accession number DQ903883.1), or in derivatives allowing C-terminal fusions (green fluorescent protein (GFP) or monomeric red fluorescent protein (mRFP) or tags (His, Cys, FLAG/His or His/Cys), downstream of the *S. cerevisiae PDR5* promoter. The cloned PDR gene was then excised from the plasmid as a transformation cassette containing a *URA3* or *HIS1* marker, flanked by *PDR5* upstream and downstream sequences. The system was improved using recombinant PCR or overlapping DNA fragments to provide transformation cassettes. Each approach requires the use of a high fidelity polymerase such as KOD (Lamping et al. 2007) and DNA sequencing to confirm the identity of the inserted gene. Upon transformation of the AD host to Ura$^+$ or His$^+$, the flanking sequences direct the integration of the cassette at the *PDR5* locus (Lamping et al. 2007). The reliability of obtaining the correct transformant has been enhanced by making the host auxotrophic requirements for histidine and uracil more robust by deleting the open reading frames for these genes, i.e. deletion of chromosomal *URA3 and HIS1* in strain AD2Δ prevents spontaneous reversion of the auxotropic requirement or integration of the transformation cassette at these loci (Sagatova et al. 2015).

The insertion of a single copy of the gene of interest into the genomic *PDR5* locus usually ensures a stable phenotype that may not be possible with plasmid-based systems. The inserted gene is constitutively hyperexpressed due to the GOF *Pdr1-3p* transcriptional regulator which acts on the *PDR5* promoter and also induces the expression of a network of genes required for membrane protein synthesis and correct protein trafficking within the yeast cell (Balzi and Goffeau 1995; DeRisi et al. 2000). The deletion of seven endogenous ABC pumps in the AD strain makes it hypersensitive to antifungals and many other xenobiotics.

The AD host strain has been used to hyperexpress membrane proteins implicated in azole resistance in pathogenic fungi (Lamping et al. 2007). It has enabled comparison of the contributions to azole resistance by different protein classes and by individual alleles encoding drug targets and efflux pumps (Holmes et al. 2006), the identification drug efflux pump substrates (Lamping et al. 2007) and analysis of the functional effects of the phosphorylation of the dominant efflux pumps from *C. glabrata* (Wada et al. 2002, 2005). It helped demonstrate that drug efflux pumps do not mediate clinically significant resistance to the echinocandin drugs (Niimi et al. 2006). The expression system has also been used to assess the function and determine the X-ray crystal structure of the azole drug target Erg11p from *S. cerevisiae* (Monk et al. 2014; Sagatova et al. 2015), *C. glabrata* and *C. albicans* (B. C. Monk and M.V. Keniya, unpublished observations). More recently the system has visualised the structural and functional effects of mutations in ScErg11p that confer azole resistance (Sagatova et al. 2016), thereby mimicking phenomena found in strains of *C. albicans*, *C. glabrata*, *A. fumigatus* and some agricultural pathogens.

The AD system has been used to express functional PDR pumps from *C. albicans*, *Candida glabrata*, *Candida krusei* and *Cryptococcus neoformans*, the *C. albicans* MFS pump Mdr1p and human ABC pump ABCB1 (P-glycoprotein) (Lamping et al. 2007, 2009). Mass spectrometry of tryptic fragments obtained from overexpressed protein bands detected in SDS-PAGE of suitable membrane fractions obtained by differential centrifugation of yeast homogenates provides a final check of heterologous protein expression (Niimi et al. 2002). This approach can, for example, detect the difference in the sequences of overexpressed CaCdr1p and CaCdr2p. Heterologous expression of efflux pumps ranged from 29% of plasma membrane protein for CaCdr1p to 3.2% for ABCB1 (Lamping et al. 2007). This highlights a limitation of the system. In general, the further the genetic distance of the source of the pump from *S. cerevisiae*, the lower the level of expression obtained.

The AD system enables robust whole-cell and in vitro assays with high signal to noise ratios that can be used to screen for chemosensitizers of the drug efflux pumps that mediate azole resistance (Lamping et al. 2007; Niimi et al. 2004). When CaCdr1p was hyperexpressed in AD, it increased the resistance of AD to azole antifungals by 400–1000-fold (Lamping et al. 2007). Chemosensitizers with broad-spectrum activity against a range of azole-resistant *Candida* species were expected increase the potency of azole drugs in both resistant and susceptible strains. This approach has enabled whole-cell screens to identify PDR pump inhibitors from multiple sources, including peptides in combinatorial library that act from the outside of the cell and thus not susceptible to efflux-mediated resistance (Niimi et al. 2004, 2012). It has also allowed identification of compounds from combinatorial libraries that inhibit drug efflux mediated by a fungal MFS transporter (Keniya et al. 2015). Such efflux pump inhibitors may help minimise the selection of azole-resistant commensals or intrinsically azole-resistant opportunistic fungi in at risk immunosuppressed patients.

23.12 How Might ABC and MFS Efflux Pump Inhibitors Act?

Inhibitors of PDR pumps could work in several different ways. They might be pump pseudosubstrates, competitively inhibiting either ATP hydrolysis or substrate transport. They could be non-competitive inhibitors that 'lock' the pump part way through its reaction cycle by binding a substrate exit site, the NBD(s) or other sequences that modulate inter-subunit interactions or the conformation changes required for the reaction cycle. Inhibitors that bind to the NBDs (e.g. nucleotide analogues) may lack specificity, as NBDs are highly conserved between ABC transporters. However, several motifs found in the NBDs of fungal ABC transporters show significant divergence from the consensus motifs of ABC transporters (Prasad and Goffeau 2012).

Inhibitors of MFS transport are most likely to be competitive inhibitors that mimic efflux pump substrates and these could act from either side of the plasma membrane. Non-competitive inhibitors might also block the reaction cycle by binding outside the active site in ways that limit conformational change.

Compounds that inhibit generation of the plasma membrane electrochemical gradient required for nutrient uptake would be expected to inhibit indirectly drug efflux mediated by either MFS or ABC transporters. This would be achieved by inhibiting the plasma membrane proton pump or using ionophores to collapse the electrochemical gradient. For example, the surface active D-decapeptide BM2 (D-NH_2-RRRFWWFRRR-$CONH_2$) is a potent inhibitor of *S. cerevisiae* and *C. albicans* Pma1p that indirectly blocks azole resistance mediated by CaCdr1p or CaMdr1p (Monk et al. 2005). In contrast, ionophores are likely to be too non-specific in the human host.

23.13 Discovery of Inhibitors of Antifungal Efflux

Screening systems have been used to identify a number of fungal efflux pump inhibitors. These include isonitrile (Yamamoto et al. 2005), FK506 and its homologue FK520 (Lamping et al. 2007; Nim et al. 2014), enniatins (Hiraga et al. 2005), beauvericin (Tanabe et al. 2011), milbemycins (Lee et al. 2001; Silva et al. 2013), sulphated sterols (Digirolamo et al. 2009), capisterones (Li et al. 2006), derivatives of cerulenin (Diwischek et al. 2009), chalcones (Lacka et al. 2015), phenothiazine (Kolaczkowski et al. 2003, 2009), ibuprofen (Ricardo et al. 2009), farnesol (Sharma and Prasad 2011), three synthetic peptides (Niimi et al. 2004, 2012; Maurya et al. 2013), jatrophanes (Rawal et al. 2014), unnarmicin A and C (Tanabe et al. 2007, 2011), compounds with a squarile core (Keniya et al. 2015) and the out-of-patent drug clorgyline (Holmes et al. 2012). Hiraga et al. (Hiraga et al. 2005) were the first publish the application of a yeast system to identify inhibitors of drug efflux by fungal ABCG defflux pumps. The host strain (Δsyr1/erg3 Δpdr5 Δsnq2) was 16- and 4-fold more sensitive to the Pdr5p substrates cycloheximide and cerulenin, respectively, than the parental wild-type strain. Overexpression of Pdr5p from a plasmid in this host strain conferred strong resistance compared with a strain carrying the empty vector. These strains provided a screen that detected chemosensitization to the Pdr5p substrates cycloheximide or cerulenin by FK506 or cyclosporin A and showed that Pdr5p-dependent R6G efflux was blocked by FK506. The same screen was used to identify enniatin, a compound from a *Fusarium* that inhibited Pdr5p and appeared to be less toxic than FK506 (Hiraga et al. 2005). At least ten compounds including KN20 (Niimi et al. 2004), RC21v3 (Niimi et al. 2012), clorgyline (Holmes et al. 2012), beauvericin (205) and unnarmicin A and C (Tanabe et al. 2007, 2011), FK520 (Nim et al. 2014), jatrophanes (Rawal et al. 2014) and milbemycins (Lee et al. 2001) have been shown to be Cdr1p inhibitors using the *S. cerevisiae* AD expression system (Lamping et al. 2007; Nakamura et al. 2001). The ability to express separately a variety ABC and MFS efflux pumps, including the human ABC transporter Pgp (Lamping et al. 2007) in a hypersensitive yeast host, has provided screens and counterscreens to drive inhibitor specificity plus secondary screens to evaluate inhibitor spectrum. Counterscreens using an alternative target under comparable test conditions can be used to eliminate false positives in the presence of a xenobiotic (Keniya et al. 2015; Niimi et al. 2004, 2012).

Several assays that measure pump function can be used to identify pump inhibitors. As fungal PDR transporters are not essential for growth, the deletion of one or more PDR genes, or the transcriptional regulators *PDR1* and *PDR3*, is not lethal. Thus, efflux pump inhibition alone does not give a growth defect phenotype. While *S. cerevisiae* AD strains overexpressing a functional fungal PDR or CaMdr1p pump can grow in medium containing antifungals that are pump substrates, the null host strain cannot grow under these conditions. If a compound by itself does not affect growth but inhibits the appropriate efflux pump, the cells will grow poorly, fail to grow or even die in the presence of an antifungal substrate of the efflux pump. This is the basis of the chemosensitization assay used in both low- and intermediate-throughput screens. The efflux of fluorescent substrates, such as R6G and rhodamine-123 (R123) catalysed by PDR transporters, and Nile Red catalysed by both PDR and MFS transporters, can also be measured (Holmes et al. 2006; Ivnitski-Steele et al. 2009; Keniya et al. 2015). This has enabled the use of the glucose- and time-dependent efflux of R6G or Nile Red in medium-throughput screens for pump inhibitors employing a microtiter plate spectrofluorimeter. The cell-associated fluorescence of energised cells has been adapted to high-throughput screens (HTS). The cells are preloaded with R6G or Nile Red under de-energised conditions. When energised with glucose, cells with active PDR pumps have low fluorescence, while those with the pump inhibited are highly fluorescent. These properties can be detected by flow cytometric analysis of cells serially aspirated from wells of microtitre plates, where each well contains a different test compound. Having identified a putative PDR pump inhibitor in a primary screen of a compound library, it is important to confirm that the inhibitor targets the pump directly and specifically. For example, a fluorescence-based HTS will identify antifungal inhibitors that indirectly affect efflux pumps, as well as pump specific inhibitors. Secondary assays can also be used to distinguish competitive pump inhibitors that share efflux pump substrate-binding sites from inhibitors that block the reaction cycle, i.e. that inhibit the ATPase activity of ABC transporters.

The application of secondary screens that test specificity or identify off-target effects is important because several inhibitors of fungal pumps have been reported as inhibiting Pgp, including curcuminoids (Limtrakul et al. 2004), phenothiazine derivatives (Maki and Dey 2006), propafenone (Schmid et al. 1999), chalcones (Liu et al. 2008), beauvericin (Dornetshuber et al. 2009), enniatin (Dornetshuber et al. 2009) and FK506 (Suzuki et al. 2010). The AD system has provided a range of secondary assays that use whole *S. cerevisiae* cells or membrane fractions for in vitro assays to validate and further analyse hits.

1. Whole-cell susceptibility assays use either liquid minimum inhibitory concentration (MIC) assays or agar-based diffusion assays to confirm that potential pump inhibitors alone do not affect growth.
2. Chemosensitization assays use solid or liquid media to show compounds that inhibit the efflux of surrogate fluorescent substrates also reverse resistance to clinically important substrates such as FLC. Agarose diffusion assays use a sub-MIC concentration of efflux pump substrate or no substrate in a medium overlaid with an agarose layer containing a strain expressing the target or other efflux pumps. Filter paper discs containing the putative chemosensitizer, its carrier medium or a control chemosensitizer are laid on top of the agarose. The resultant zones of inhibition are used to determine whether the test compound is a chemosensitizer, provide a visual estimate of its relative potency and help determine its specificity (Niimi et al. 2004, 2012). Checkerboard chemosensitization assays—essentially liquid MIC assays in which the concentrations of both a pump inhibitor and an antifungal substrate are varied—are used to calculate the fractional inhibitory concentration index (FICI), which quantifies the

degree of synergism between the two compounds (Niimi et al. 2004; Holmes et al. 2008). Defining the chemosensitization of a range of antifungal substrates by pump inhibitors helps elucidate the nature of the pump inhibition. Inclusion of strains expressing alternative ABC and MFS pumps or mammalian ABC pumps as counterscreens indicates whether target specificity is acceptable. If it is not, use of these inhibitors might give side effects.

3. In order to determine that a hit directly inhibits efflux function, its effect on energy-dependent efflux of a fluorescent substrate, such as R6G or Nile Red, into the culture supernatant is measured using a microtitre filter plate method (Holmes et al. 2008). This is particularly important where the primary screen employs flow cytometry-based identification of putative inhibitors, as changes in whole-cell fluorescence may not be energy dependent and may reflect cell death or intracellular sequestration rather than inhibition of efflux.
4. In vitro secondary assays are a critical complement to whole-cell assays. Putative inhibitors are tested for their effects on the ATPase activities of membrane preparations from *S. cerevisiae* strains expressing the efflux pump. With a few exceptions (e.g. CaCdr2p), the ATPase activity of ABC transporters is vanadate, oligomycin and FK506 sensitive and can be distinguished from the oligomycin resistant ATPase activity of the dominant plasma membrane protein proton pumping ATPase. It is important to demonstrate that efflux pump 'hits' inhibit the efflux of multiple substrates plus the ATPase activity of the transporter. For example, FK506 inhibits FLC and R6G transport by ScPdr5p and the ATPase activity of the enzyme. The in vitro ATPase assay can also be used to indicate whether inhibitors act competitively or non-competitively in relation to nucleotide (ATP) binding (Niimi et al. 2004).

23.13.1 Low–Medium-Throughput Screens for Inhibitors of Drug Efflux

We developed a low–medium-throughput approach to antifungal discovery designed to overcome antifungal resistance in key pathogenic fungi by inhibiting drug efflux pumps either directly or indirectly via accessible target sites at the fungal cell surface. This involved a biologically stable, surface-targeting resource of chemical and conformational diversity that was provided by a D-octapeptide library containing an N-terminal combinatorial D-pentapeptide component (Monk and Harding 2005). The surface-targeting feature of the D-octapeptides used the observation that a TRITC-labelled model D-peptide containing a C-terminal amidated triarginine motif was excluded from yeast cells and appeared to associate with cell wall phosphomannan. This was mimicked in a manually synthesised a 1.85×10^6 member D-octapeptide combinatorial library of the form D–NH_2–A–B–X_1–X_2–X_3–R–R–R–$CONH_2$, where the amino acids A and B are known for each pool, X_1, X_2 and X_3 may be any of 18 amino acids except glycine and cysteine, and R is arginine. The combinatorial library comprised an 18 × 18 array of 324 peptide pools, with each pool theoretically containing 5832 peptides (Monk et al. 2005; Niimi et al. 2004). A medium-throughput screen of these pools, depending on the number and complexity of the primary and secondary screens applied, took up to a month. The peptide pool that best met the screening criteria was selected for deconvolution to identify the active principal. This involved synthesising and screening 18 derivatives at each position to sequentially identify X_1, X_2 and X_3. The manual peptide resynthesis required at each position took about a month and the subsequent primary and secondary screens required about a week per position. Once the primary sequence of the active principal was identified by deconvolution, the peptide was resynthesized, purified by HPLC and its activity confirmed. This process generally took another 2 months and included more comprehensive mode of action and toxicity studies. Initial studies successfully identified BM2 as an inhibitor of the plasma membrane proton pump Pma1p (Monk et al. 2005).

A chemosensitizer of *S. cerevisiae* Pdr5p was identified by screening the D-octapeptide combinatorial library in a 96-well microtitre plate format using with the GOF *pdr1-3* mutation to drive Pdr5p overexpression in an *S. cerevisiae* strain deleted of five other ABC transporters and the transcriptional regulator *PDR3* (Niimi et al. 2004). Peptide pools at a set D-octapeptide concentration (~50 μg/ml) that did not affect growth yield in the absence of FLC, but in the presence of FLC at 0.25 × MIC (50 μg/ml) blocked yeast growth, were identified. The chemosensitizing pools were titrated to identify the most potent pools and these were assayed for inhibition of the overexpressed ScPdr5p using the vanadate- and oligomycin-sensitive ATPase activity of plasma membrane preparations at pH 7.0. Deconvolution of the selected pool, including the use of HPLC-purified candidate polypeptides, identified the D-octapeptide derivative KN20 (D-NH_2-NWWKVRRR-$CONH_2$ + Mtr), a non-competitive inhibitor (at 4 μM) of in vitro ATPase activity of Pdr5p and a potential chemosensitizer (at 40 μM) of FLC efflux by Pdr5p. Mtr (4-methoxy-2,3,6-trimethylbenzensulfonyl) is a chemical blocking agent used in the synthesis of the peptide library. A single Mtr substituent on the peptide backbone was required for chemosensitization and inhibition of ATPase activity. Although KN20 had attributes expected of a chemosensitizer of its target efflux pump, and showed chemosensitization of CaCdr1p and CaCdr2p, the peptide concentrations required for chemosensitization also permeabilized yeast cells to R6G (Niimi et al. 2004). For example, TRITC-labelled KN20 preferentially bound the plasma membrane of yeast cells overexpressing Pdr5p, but chemosensitization appeared indirect and at least partially mediated through non-lethal permeabilization of the plasma membrane.

The non-specific chemosensitization obtained with KN20 caused a re-evaluation of our approach to chemosensitizer discovery and the introduction of a counter-screen that took into account more subtle effects in cell growth caused by the presence of azole drug. A primary chemosensitization screen of the combinatorial peptide library was undertaken using *S. cerevisiae* AD overexpressing CaCdr1p (AD/CaCDR1) in an agarose diffusion growth assay in the presence and absence of

FLC at 0.25 × MIC (75 μg/ml). *S. cerevisiae* AD cells overexpressing the MFS transporter CaMdr1p (AD/CaMDR1) in the presence or absence of FLC (20 μg/ml) provided the counterscreen needed to avoid non-selective chemosensitization to FLC. An in vitro secondary screen of plasma membrane oligomycin-sensitive ATPase activity of CaCdr1p in a 96-well microtitre plate format was used to confirm specificity for the target enzyme. These screens identified the CaCdr1p-specific chemosensitizer RC21 D-NH_2-FFKWQRRR-$CONH_2$ (Niimi et al. 2012). An Mtr derivative of the D-octapeptide (at 1.5 μM) chemosensitized strain AD/CaCDR1 and inhibited CaCdr1p ATPase activity with a 50% inhibitory concentration of ~1.5 μM. The inhibitory peptide, denoted RC21v1, was found to be a stereospecific inhibitor of CaCdr1p, i.e. D-RC21v1 but not L-RC21v1-chemosensitized CaCdr1p. D-RC21v1 failed to chemosensitize AD strains overexpressing other PDR drug efflux pumps or the human ABCB1 drug efflux pump to FLC. The stereospecificity of RC21v1 indicated that its target was likely to be a protein and its specificity for CaCdr1p but not CaCdr2p helped demonstrate that the CaCdr1p drug efflux pump is the dominant contributor to drug efflux in FLC-resistant clinical isolates of *C. albicans* overexpressing both Cdr1p and Cdr2p (Holmes et al. 2008). RC21v1 was also found to be a highly specific inhibitor of R6G efflux from *S. cerevisiae* and *C. albicans* cells overexpressing CaCdr1p. The Mtr modification required for activity of the peptide derivative was found to be in the side chain of in the most amino terminal arginine within the C-terminal triarginine motif and did not involve a modification of the tryptophan side chain or any of the other arginines. This knowledge enabled the synthesis of RC21v3 (D-NH_2-FFKWQR(Mtr)RR-$CONH_2$) which rendered an azole-resistant *C. albicans* isolate that overexpresses CaCdr1p susceptible to azole drugs in a mouse model of oral candidiasis (Hayama et al. 2012).

An advantage of heterologous expression of drug targets in *S. cerevisiae* is the tractability of its genetics. For example, exposure of *S. cerevisiae* AD cells overexpressing an efflux pump to an efflux pump antifungal substrate and an efflux pump inhibitor selects for suppressor mutations with resistance to the pump inhibitor. Often these mutations are within the efflux pump gene, and the location of these mutations can give insight into the mechanism of pump inhibition. Agarose diffusion assays were used to find chemosensitizer suppressor mutants that gave stable resistance to RC21v3 in the presence of FLC (Niimi et al. 2012). For all 12 yeast suppressor mutants analysed, SDS-PAGE analysis and measurements of CaCdr1p-specific ATPase activity showed that each strain overexpressed near control amounts of functional CaCdr1p in the plasma membrane. DNA sequence analysis demonstrated that each suppressor mutant involved a single nucleotide substitution within the CaCdr1p open reading frame. Of the six mutations identified, five introduced a single positive charge into CaCdr1p that were mapped to surface-exposed extracellular sites on the enzyme in an homology model of CaCdr1p based on a model of *S. cerevisiae* Pdr5p (Rutledge et al. 2011). The other mutation introduced a large aromatic group buried near the extracellular end of transmembrane segment 5 that could modify the closed conformation of the enzyme. While the recent structure of ABCG5/ABCG8 (Lee et al. 2016) suggests some significant limitations in the Pdr5p and Cdr1p homology models (Niimi et al. 2012; Rutledge et al. 2011), CaCdr1p was unambiguously identified as the molecular target of RC21. Some of the chemosensitization suppressor mutants had another valuable attribute. Specific groups of the suppressor mutants were found to be resistant to known drug-like chemosensitizers of CaCdr1p including the immunosuppressant FK506, the depsipeptide enniatin and the macrolides milbemycin β9 and α11 (Niimi et al. 2012). The suppressor mutations in CaCdr1p provided a provisional map of amino acids that affected the binding, either directly or indirectly, of these chemosensitizers. We therefore consider that peptide derivatives such as RC21v3 and the chemosensitizer suppressor mutants it induces provide a model and tools for the ongoing discovery of fungal-specific chemosensitizers.

The yeast screening system has been used to identify chemosensitizers in crude extracts from marine-derived fungi and bacteria. AD strains overexpressing the drug efflux pumps ScPdr5p, CaCdr1p, CgCdr1p, CgCdr2p and CaMdr1p were used to identify the cyclodepsipeptides unnarmicin A and C (Tanabe et al. 2007). These compounds chemosensitized to FLC all the yeast strains expressing ABC transporters but not those expressing CaMdr1p. They were also found to inhibit R6G efflux by yeast cells overexpressing CaCdr1p at low micromolar concentrations, the oligomycin-sensitive ATPase activity of CaCdr1p in membrane preparations at submicromolar concentration and sensititize azole-resistant clinical isolates of *C. albicans* to FLC.

Nim et al. used AD1-8u$^-$ strains expressing a series of 252 alanine substitutions in the membrane sector of overexpressed CaCdr1p to study the direct chemosensitizing action of FK520, an analogue of the FK506 (Nim et al. 2014). Tests with FLC, ketoconazole and miconazole suggested these substrates have localised interactions with specific transmembrane helices that were either dependent or independent of the azole drug tested. Consistent with a previous study by Rawal et al., the drug-dependent interactions suggested that each drug binds to significantly different parts of a large drug-binding pocket, while the drug-independent interactions for azole drug implied that a group of amino acids makes contact with a similar part of each drug, i.e. the azole head group (Rawal et al. 2013). This systematic mutagenesis approach is a powerful way to assess drug and chemosensitizer interactions that will have greater value once a high-resolution X-ray structure of CaCdr1p is obtained.

Screens using CaCdr1p or CaMdr1p overexpressed in the AD1-8u$^-$ host were used to identify three jatrophanes from *Euphorbia squamosa* that significantly inhibited the efflux of Nile Red (Rawal et al. 2014). All three compounds (deacetylserrulatin B, euphosquamosin A and euphosquamosin C) were toxic for the host cells at ~2.3 μM with deacetylserrulatin B and euphosquamosin C appearing to be effluxed by both CaCdr1p and CaMdr1p. All three compounds showed synergy with FLC, with deacetylserrulatin B and euphosquamosin C showing strong synergy (FICIs $\leq$ 0.06) and euphosquamosin A showing modest synergy (FICI = 0.46). The strong synergy with FLC of low concentrations (0.3 μM) of deacetylserrulatin B toward CaCdr1p suggested a higher efficiency than most previously reported inhibitors, e.g. disulfiram (Holmes et al. 2008; Shukla et al. 2004), the D-octapeptide KN20 (Niimi et al. 2004), enniatins (Hiraga et al. 2005), unnarmicins (Tanabe et al. 2007), FK506 (Lamping et al. 2007), curcumin (Sharma et al. 2009), milbemycins (Silva et al. 2013) and chlorgyline (Holmes et al. 2012). However, its growth-inhibitory activity at concentrations only 10-fold higher disqualifies deacetylserrulatin B as a chemosensitizer. Deacetylserrulatin B also inhibited CaMdr1p in the presence of FLC at 10-fold lower concentrations than chlorgyline (Holmes et al. 2012; Rawal et al. 2014). The effects of the three compounds on clinical isolates of *C. albicans* expressing efflux pumps have yet to be described.

Maurya et al. prepared a series of fluorescently tagged transmembrane peptide mimetics (TMPMs) corresponding the 12 transmembrane helices of CaCdr1p (Maurya et al. 2013). The yeast system was used to show that a select group of the TMPMs bound to the surface of cells overexpressing CaCdr1p and chemosensitized these cells to FLC. Alanine scanning mutants of CaCdr1p in the AD host system indicated that TMPM8 specifically associates with one face of TMS8. TMPM8 was found to chemosensitize the *C. albicans* Gu5 strain, which overexpresses CaCdr1p, to the CaCdr1p substrates FLC, anisomycin and cycloheximide. The non-toxic nature of the TMPMs and their ability to re-sensitise the Gu5 strain to FLC in a mouse model of disseminated candidiasis indicate the potential of this approach.

Although significant commercial effort has been directed towards the discovery of inhibitors of drug efflux mediated by bacterial MFS pumps such as the Mex-Opr multidrug efflux pumps responsible for fluoroquinolone resistance in clinical isolates of *Pseudomonas aeruginosa* (Lomovskaya and Bostian 2006), limited effort has been directed towards the discovery of inhibitors of the fungal MFS drug efflux pumps.

A set of structurally related low molecular mass compounds synthesised using combinatorial chemistry of a cyclobutene-dione (squarile) core was screened using a *S. cerevisiae* AD strain overexpressing the MFS pump CaMdr1p, with a strain overexpressing the ABC transporter CaCdr1p serving as a counterscreen (Keniya et al. 2015). Although five compounds showed CaMdr1p-specific chemosensitization to FLC at low μM concentrations, only one specifically chemosensitized yeast overexpressing CaMdr1p to FLC in a pH-dependent manner, acted synergistically with FLC in liquid MIC assays, and inhibited Nile Red efflux by CaMdr1p but not CaCdr1p, and was not toxic to human cells. It also chemosensitized to FLC, the *C. albicans* FR2 strain, which overexpresses CaMdr1p but not CaCdr1p. The discovery of a first in class small molecule inhibitor of CaMdr1p demonstrates the power of the yeast-based screening system.

Nim et al. (2016) used Nile Red efflux and checkerboard chemosensitisation assays with FLC in the AD system to identify two macrocylic diterpenes, euphomelliferene and euphomelliferene B, as synergistic chemosensitisers of CaMdr1p-dependent FLC efflux in *S. cerevisiae* (FICI values of 0.07). However, these compounds gave only additive (FICI values of 0.6–0.7) rather than synergistic chemosensitization of the Mdr1p overexpressing F5 strain of *C. albicans* to FLC.

This result was explained by the possible presence of ABC transporters in strain F5 that negated synergy with CaMdr1p.

23.13.2 *High-Throughput Screens for Inhibitor of Drug Efflux*

High-throughput screens allow rapid testing of large numbers of compounds, usually in microtitre plates with robotic control. Flow cytometry-based assays (Sklar et al. 2007) of fluorescently tagged whole cells in suspension can be adapted readily for automated HTS. Quantification of ABC transporter activity is particularly amenable to fluorescence-based assays using substrates including R6G, R123 (Holmes et al. 2006), fluorescein diacetate (Kolaczkowski et al. 2009) and tetramethylrosamine (A.R. Holmes and R.D. Cannon, unpublished observations). This allows application of HTS to the discovery of efflux pump inhibitors that chemosensitize resistant cells to existing antifungals. Single substrates have been used to search for fungal PDR pump inhibitors and a multiplex HTS developed that simultaneously assays for inhibitors to multiple pumps (see below). The choice of fluorescent substrate in HTS is important. Ideally the fluorescent substrate will show the same transport properties as a clinically important substrate, e.g. FLC. For multiplex screens, the fluorescent compound must be a substrate of each pump tested. Our initial HTS for inhibitors of CaCdr1p used R6G as the fluorescent substrate. Enniatin B is an inhibitor of CaCdr1p that reverses FLC resistance of both *C. albicans* clinical isolates and AD/CaCDR1, and it inhibits R6G efflux by strain AD/CaCDR1 (Holmes et al. 2008). While R6G is not an ideal surrogate for FLC because it does not block FLC efflux, enniatin B can be used as a positive (inhibitor) control in dye-based HTS. Inhibitors of CaCdr1p, CaCdr2p and CaMdr1p have been sought using a triplex HTS. As R6G is not a substrate of CaMdr1p, Nile Red was used as the fluorescent multiplex substrate because it is pumped by all three pumps and its efflux by strain AD/CaCDR1 is inhibited by enniatin B (Ivnitski-Steele et al. 2009). Nile Red has the added advantage in whole-cell, flow cytometry-based HTS that only cell-associated Nile Red exhibits fluorescence. It was also found that the accumulation of both R6G and Nile Red in the yeast cells could be measured in the HyperCyt® flow cytometry system and that *S. cerevisiae* AD strains expressing individual *C. albicans* efflux pumps were easily distinguished from the negative-control AD/pABC3 strain in single-strain assays. In trial HTS, the Prestwick Chemical Library (PCL; Illkirch, France; a collection of 1200 off-patent small molecule drugs) was screened for inhibitors of CaCdr1p or CaCdr2p using flow cytometric analysis of R6G accumulation in strains AD/CaCDR1 and AD/CaCDR2 (Holmes et al. 2012). Nine compounds were identified in the primary screen as inhibitors of CaCdr1p or CaCdr2p. Seven of these were active against one or the other transporter. Disulfiram, a hit that was specific to CaCdr2p, had been reported previously to chemosensitize a

Cdr2p-expressing strain to FLC (Holmes et al. 2008). The finding that the hit econazole was active against the Cdr2p-expressing strain demonstrated that compounds identified in screens for efflux inhibition may also be pump substrates that compete with the fluorescent substrate. However, the clinical use of a pump substrate as an efflux pump inhibitor may not be ideal because many pump substrates induce pump expression. Induction of CaCdr1p and CaCdr2p expression would be an undesirable attribute of a pump inhibitor selected to reverse efflux-induced resistance. Hence, secondary assays were used to measure the sensitivity of strains to hits alone plus their chemosensitization of these strains to antifungal efflux pump substrates.

Two hits inhibited both classes of efflux pump. One of these hits, ebselen, is a known antifungal with activity against the *C. albicans* plasma membrane ATPase (Billack et al. 2009) and therefore was considered to act indirectly on the efflux pumps like the Pma1p inhibitor BM2 (Monk et al. 2005). The monoamine oxidase A inhibitor (MAOI) clorgyline was found to inhibit by over 90% the activity of both CaCdr1p and CaCdr2p (Holmes et al. 2012). Secondary in vitro assays confirmed inhibition of pump-mediated efflux by clorgyline. It not only reversed the FLC resistance of *S. cerevisiae* strains expressing other fungal ABC transporters individually (*C. glabrata* Cdr1p or *C. krusei* Abc1p) and the *C. albicans* MFS transporter Mdr1p but also chemosensitized these recombinant strains to other azoles (ITC and miconazole). Importantly, clorgyline showed synergy with FLC against FLC-resistant *C. albicans* clinical isolates and a *C. glabrata* strain, and inhibited R6G efflux from a FLC-resistant *C. albicans* clinical isolate. Clorgyline is a therefore broad-spectrum inhibitor of two classes of fungal efflux pumps that act synergistically with azoles against azole-resistant *C. albicans* and *C. glabrata* strains.

A trial multiplex assay using a portion of the Prestwick compound library has identified three hits: the antifungal tolnaftate, the fungal efflux pump substrate antimycin A and the nematode ABC transporter substrate ivermectin. When individual pump-expressing strains were screened separately, the same three hits identified in the multiplex were also found. This validated the multiplex screen for fungal PDR pump inhibitors. The flow cytometry assay was developed further as a multiplex assay in which three "sentinel" strains (expressing *C. albicans* Cdr1p, Cdr2p or Mdr1p), each tagged with a different fluorophore, were included in each well (Strouse et al. 2010). This screen was applied to the Molecular Libraries Small Molecule Repository of 329,018 compounds. A total of 357 compounds were cherry-picked for further analysis using strains expressing individual pumps and for dose-response determinations. Although broad-spectrum hits were confirmed, none of the compounds inhibited efflux pumps at submicromolar concentrations. Interestingly, at least three of the hits were also identified in a related HTS that used growth chemosensitization of a *C. albicans* strain to FLC its basis.

Both Nile Red and R6G have been found to be compatible with whole-cell spectrophotometer-based assays that do not require a filtration step. Pump activity in the presence of 20 mM glucose for cells preloaded R6G has enabled the screening of a library containing 12,401 compounds and the 15 hits that inhibited efflux await further test (A.R. Holmes and R.D. Cannon, unpublished observations).

23.14 Discovery of Inhibitors of the Transcriptional Regulator Pdr1p and the Mediator Complex

The Pdr1p transcriptional regulator is a zinc cluster protein that binds constitutively as a homodimer, or as a heterodimer with Pdr3p, to the one or more copies of Pdr1p/Pdr3p response element (PDRE) sequences found in the promoters of the PDR-related genes it regulates (Fardeau et al. 2007; Mamnun et al. 2002; Katzmann et al. 1996). Analogous but non-identical proteins are found in *C. glabrata* and other fungal pathogens including *C. albicans*. For example, CgPdr1p appears to be a structural and functional composite of ScPdr1p and ScPdr3p with some additional functions unique to *C. glabrata* (Vermitsky et al. 2006; Vermitsky and Edlind 2004). In *C. albicans TAC1* is a transcriptional regulator of *CDR1* and *CDR2* expression, MRR1 regulates *MDR1* expression, and *UPC2* regulates *ERG11* expression. Several GOF mutations present in hyperactive alleles of these zinc cluster transcription factors result in the overexpression of target genes, but of these only the GOF mutations in *UPC2* deleteriously affect virulence (Lohberger et al. 2014).

Numerous compounds, including some substrates of drug efflux pumps, activate Pdr1p. These include the azole drugs, sterols and rifampicin. For example, radioactive FLC binds to the central regions of Pdr1p and Pdr3p and FLC also induces Pdr1p/Pdr3p-dependent gene expression (Thakur et al. 2008). This knowledge was used to discover inhibitors of Pdr1p by screening 2540 compounds from the National Cancer Institute (USA) Representative Library, i.e. 821 compounds from the Mechanistic Diversity set, 1600 from the Diversity Set III and 119 from the Natural Products Set II. Compounds that were non-toxic for the yeast AD strain and reduced significantly the functional expression of a Ura3p-GFP reporter expressed from the *PDR5* locus under the control of wild-type *PDR1* in the presence of FLC at 2x MIC were considered as putative antagonists of Pdr1p (M.V. Keniya and B.C. Monk, unpublished observations). A GOF *pdr1-3* mutant and a *ΔPDR1* variant of the same test strain served as negative and positive controls, respectively. FLB-12, an antagonist of the pregnane-X-receptor (Venkatesh et al. 2011), provided a positive control.

Hits selected in the primary screen were re-evaluated using rifampicin, an inducer of Pdr1p-mediated transcription that does not affect yeast growth (M.V. Keniya and B.C. Monk, unpublished observations). Antagonists reduced the fluorescence caused by the rifampicin-dependent expression of the Ura3p-GFP reporter. Hits selected from the primary and secondary screens were tested against strains with an attenuated version of the apoptosis-inducing mouse BCL-1 (BAX) ORF containing a frame shift. An apoptotic response required the *PDR1*-dependent induction of BCL-1 mRNA expression from the *PDR5* locus in amounts that yield a functional reporter due to a low frequency of frame-shift reversal. The strains were tested using peptone-containing medium, which naturally contains sterols that induce Pdr1p, or synthetic defined medium containing the Pdr1p activators FLC or rifampicin (Venkatesh et al. 2011). Putative antagonists of Pdr1p gave cell growth in the presence of a Pdr1p inducer.

Four shortlisted hits were tested using a *PDR1* strain expressing either an ABC (CaCdr1p) or MFS (CaMdr1p) efflux pump from the *PDR5* locus. Only one of these compounds was found to chemosensitize the yeast cells to FLC, indicating that it inhibited transcription and was not a substrate of either drug efflux pump. However, when this compound was tested at 25–50 μM against *C. glabrata* strain CBS138, a clinical isolate that overexpresses the CgCdr1p drug efflux pump in the presence of FLC, it caused increased resistance to FLC after 24 h but had no effect on its MIC_{80} for FLC after 48 h (~400 μM).

These results suggest that while antagonists of transcriptional regulators that affect drug efflux pumps can be identified, most are likely to be drug efflux pump substrates and/or compounds with limited spectrum of action and thus unlikely to be suitable for clinical application. More recently an in vitro fluorescence polarisation assay that exploited the physical interaction between the CgPdr1p activation domain and the CgGal11A KIX domain of the Mediator complex was used to screen 140,000 compounds (Nishikawa et al. 2016). This approach together with downstream cell-based assays using *S. cerevisiae* as a surrogate and *C. glabrata* strains containing wild-type or constitutively active CgPdr1p, identified compound iKIX1 which inhibited the CgPdr1p-CgGal11 KIX domain interaction with an apparent inhibition constant of 18 μM. Importantly, iKIX1 re-sensitised azole-resistant *C. glabrata* in vitro and in animal models for disseminated and urinary tract *C. glabrata* infection. The effect of iKIX1 on fungal pathogens other than *C. glabrata* is not known.

23.15 Structural Analysis of Antifungal Targets

Antifungal discovery, like many areas of antimicrobial discovery, has been hindered by a paucity of structural information on existing and potential antifungal targets. X-ray crystallography of protein crystals can be used to determine the binding of known ligands, understand the impact of mutations and applied in screens to identify the binding of drug-like fragments or novel compounds (for a recent review see Tamay-Cach et al. 2016). The identification of the pharmacophore for a target (the features that define the binding of a ligand to a biological molecule) can increase the efficiency of virtual screening, facilitate the design of drugs and improve drug development. An important priority for antifungal research should therefore be to increase structural knowledge of established antifungal targets such as lanosterol 14α-demethylase, glucan synthase, squalene monooxygenase, the plasma membrane proton pump and the drug efflux pumps CaCdr1p, CgCdr1p and CaMdr1p. The recent structural resolution of ScErg11p, in complex with a wide range of azole drugs (Monk et al. 2014; Sagatova et al. 2015, 2016) as well as CaErg11p and CgErg11p in complex with ITC (M.V. Keniya and B.C. Monk, unpublished observations) has enabled virtual screening campaigns and the opportunity to identify new classes of antifungal agents. Both *C. albicans* squalene monooxygenase (Favre and Ryder 1997) and the plasma membrane proton pump (Keniya et al. 2013) have been functionally overexpressed in *S. cerevisiae*, making possible the purification and structural analysis of these membrane proteins. The availability of inhibitors of these enzymes increases the likelihood that high-resolution structures can be obtained. In contrast, while glucan synthase (Fks1p) can be purified from yeast (Inoue et al. 1995), assayed in vitro (Shedletzky et al. 1997) and different classes of inhibitor are available (Hector and Bierer 2011), structural studies on this enzyme are hindered because of uncertainties about its subunit composition (Edlind and Katiyar 2004) and the natural biosynthesis of small quantities of the enzyme localised to sites of membrane expansion in the plasma membrane (Sanchez-Leon and Riquelme 2015). The recent X-ray structure of the ABCG5/ABCG8 heterodimer obtained using lipid cuboid phase technology is an important achievement (Lee et al. 2016). However, the crystal structure is at low resolution (3.9 Å) and gives only a single conformation of the enzyme. For CaCdr1p and CaMdr1p, the purification of active forms of these transporters and their X-ray crystallographic resolution with substrates and known inhibitors should enable more directed approaches to the discovery of fungal-specific chemosensitizers. The use of cryoelectron microscopy, which has advanced significantly since 2-D crystals of ScPdr5p were obtained (Ferreira-Pereira et al. 2003), may provide a way forward if X-ray crystallography remains problematic, as evidenced by the recently obtained structure of a heterodimeric *Thermus thermophilus* ABC transporter (Kim et al. 2015). While many of the compounds identified as inhibitors of drug efflux pumps may not find use in the clinic, they may have value by driving their targets into conformations that assist crystallisation and structural determination at resolutions that enable structure-directed drug discovery. As a caveat, it should be noted that while crystal structures are available for the putative and soluble antifungal drug targets translation elongation factor 2 and N-myristoyl transferase, and these targets have yielded the sorderins and a range of N-myristoyl transferase inhibitors, respectively, none of these compounds have broad-spectrum antifungal activity.

The purification of multi-milligramme quantities of these proteins, preferably in monodisperse form, is required for crystallisation and structural resolution by X-ray crystallography, or by nuclear magnetic resonance in the case of proteins or protein domains <30 kDa. A good understanding of the physiological role, regulation and post-translational processing of the target proteins is an essential prerequisite for most structural studies. In some instances the native protein can be crystallised, but it is often necessary to over-express a genetically engineered, affinity-tagged protein so micro-heterogeneity can be minimised. The problems associated with protein crystallisation are more complex for membrane proteins, which make up approximately 70% of the drug targets used in medicine and all the antifungal targets discussed in this chapter. The ability to functionally express these targets and drug efflux pumps in *S. cerevisiae* provides several approaches needed to implement structure–function-led drug discovery that minimises opportunity for the drug resistance. For example, it is now possible to develop an antifungal discovery strategy that seeks inhibitors of fungal lanosterol 14α-demethylase that are not substrates of the drug efflux pumps involved in azole resistance. Based on our finding that inhibitors of MDR render the fungistatic azole drugs fungicidal (Niimi et al. 2012), we predict that this new class of demethylase inhibitors will be fungicides.

References

Akins RA (2005) An update on antifungal targets and mechanisms of resistance in *Candida albicans*. Med Mycol 43(4):285–318

Alarco AM, Raymond M (1999) The bZip transcription factor Cap1p is involved in multidrug resistance and oxidative stress response in *Candida albicans*. J Bacteriol 181(3):700–708

Aller SG, Yu J, Ward A, Weng Y, Chittaboina S, Zhuo R, Harrell PM, Trinh YT, Zhang Q, Urbatsch IL, Chang G (2009) Structure of P-glycoprotein reveals a molecular basis for poly-specific drug binding. Science 323(5922):1718–1722

Asmundsdottir LR, Erlendsdottir H, Haraldsson G, Guo H, Xu J, Gottfredsson M (2008) Molecular epidemiology of candidemia: evidence of clusters of smoldering nosocomial infections. Clin Infect Dis 47(2):e17–24. doi:10.1086/589298

Azevedo MM, Faria-Ramos I, Cruz LC, Pina-Vaz C, Rodrigues AG (2015) Genesis of azole antifungal resistance from agriculture to clinical settings. J Agric Food Chem 63(34):7463–7468. doi:10.1021/acs.jafc.5b02728

Babu M, Vlasblom J, Pu S, Guo X, Graham C, Bean BD, Burston HE, Vizeacoumar FJ, Snider J, Phanse S, Fong V, Tam YY, Davey M, Hnatshak O, Bajaj N, Chandran S, Punna T, Christopolous C, Wong V, Yu A, Zhong G, Li J, Stagljar I, Conibear E, Wodak SJ, Emili A, Greenblatt JF (2012) Interaction landscape of membrane-protein complexes in *Saccharomyces cerevisiae*. Nature 489(7417):585–589. doi:10.1038/nature11354

Baetz K, McHardy L, Gable K, Tarling T, Reberioux D, Bryan J, Andersen RJ, Dunn T, Hieter P, Roberge M (2004) Yeast genome-wide drug-induced haploinsufficiency screen to determine drug mode of action. Proc Natl Acad Sci USA 101(13):4525–4530

Balzi E, Goffeau A (1995) Yeast multidrug resistance: the PDR network. J Bioenerg Biomembr 27 (1):71–76

Bauer BE, Wolfger H, Kuchler K (1999) Inventory and function of yeast ABC proteins: about sex, stress, pleiotropic drug and heavy metal resistance. Biochim Biophys Acta 1461(2):217–236

Becher R, Wirsel SG (2012) Fungal cytochrome P450 sterol 14alpha-demethylase (CYP51) and azole resistance in plant and human pathogens. Appl Microbiol Biotechnol 95(4):825–840. doi:10.1007/s00253-012-4195-9

Bennett JE, Izumikawa K, Marr KA (2004) Mechanism of increased fluconazole resistance in *Candida glabrata* during prophylaxis. Antimicrob Agents Chemother 48(5):1773–1777

Berman J (2016) Ploidy plasticity: a rapid and reversible strategy for adaptation to stress. FEMS Yeast Res 16(3). doi:10.1093/femsyr/fow020

Billack B, Santoro M, Lau-Cam C (2009) Growth inhibitory action of ebselen on fluconazole-resistant *Candida albicans*: role of the plasma membrane H+-ATPase. Microb Drug Resist 15(2):77–83. doi:10.1089/mdr.2009.0872

Blackwell M (2011) The fungi: 1, 2, 3 … 5.1 million species? Am J Bot 98(3):426–438. doi:10.3732/ajb.1000298

Bouige P, Laurent D, Piloyan L, Dassa E (2002) Phylogenetic and functional classification of ATP-binding cassette (ABC) systems. Curr Protein Pept Sci 3(5):541–559

Campbell BC, Chan KL, Kim JH (2012) Chemosensitization as a means to augment commercial antifungal agents. Front Microbiol 3:79. doi:10.3389/fmicb.2012.00079

Cannon RD, Lamping E, Holmes AR, Niimi K, Baret PV, Keniya MV, Tanabe K, Niimi M, Goffeau A, Monk BC (2009) Efflux-mediated antifungal drug resistance. Clin Microbiol Rev 22(2):291–321. doi:10.1128/CMR.00051-08

Castrillo JI, Zeef LA, Hoyle DC, Zhang N, Hayes A, Gardner DC, Cornell MJ, Petty J, Hakes L, Wardleworth L, Rash B, Brown M, Dunn WB, Broadhurst D, O'Donoghue K, Hester SS, Dunkley TP, Hart SR, Swainston N, Li P, Gaskell SJ, Paton NW, Lilley KS, Kell DB, Oliver SG (2007) Growth control of the eukaryote cell: a systems biology study in yeast. J Biol 6(2):4

Chen J, Li H, Li R, Bu D, Wan Z (2005) Mutations in the cyp51A gene and susceptibility to itraconazole in *Aspergillus fumigatus* serially isolated from a patient with lung aspergilloma. J Antimicrob Chemother 55(1):31–37

Chen SH, Sheng CQ, Xu XH, Jiang YY, Zhang WN, He C (2007) Identification of Y118 amino acid residue in *Candida albicans* sterol 14alpha-demethylase associated with the enzyme activity and selective antifungal activity of azole analogues. Biol Pharm Bull 30(7):1246–1253

Chen G, Bradford WD, Seidel CW, Li R (2012) Hsp90 stress potentiates rapid cellular adaptation through induction of aneuploidy. Nature 482(7384):246–250. doi:10.1038/nature10795

Chowdhary A, Kathuria S, Xu J, Meis JF (2013) Emergence of azole-resistant *Aspergillus fumigatus* strains due to agricultural azole use creates an increasing threat to human health. PLoS Pathog 9(10):e1003633. doi:10.1371/journal.ppat.1003633

Claverys JP, Martin B, Havarstein LS (2007) Competence-induced fratricide in streptococci. Mol Microbiol 64(6):1423–1433. doi:10.1111/j.1365-2958.2007.05757.x

Costa C, Dias PJ, Sa-Correia I, Teixeira MC (2014) MFS multidrug transporters in pathogenic fungi: do they have real clinical impact? Front Physiol 5:197. doi:10.3389/fphys.2014.00197

Coste AT, Karababa M, Ischer F, Bille J, Sanglard D (2004) TAC1, transcriptional activator of CDR genes, is a new transcription factor involved in the regulation of *Candida albicans* ABC transporters CDR1 and CDR2. Eukaryot Cell 3(6):1639–1652. doi:10.1128/EC.3.6.1639-1652.2004

Coste A, Turner V, Ischer F, Morschhauser J, Forche A, Selmecki A, Berman J, Bille J, Sanglard D (2006) A mutation in Tac1p, a transcription factor regulating CDR1 and CDR2, is coupled with loss of heterozygosity at chromosome 5 to mediate antifungal resistance in *Candida albicans*. Genetics 172(4):2139–2156

Coste A, Selmecki A, Forche A, Diogo D, Bougnoux ME, d'Enfert C, Berman J, Sanglard D (2007) Genotypic evolution of azole resistance mechanisms in sequential *Candida albicans* isolates. Eukaryot Cell 6(10):1889–1904. doi:10.1128/EC.00151-07

Coste AT, Crittin J, Bauser C, Rohde B, Sanglard D (2009) Functional analysis of cis- and trans-acting elements of the *Candida albicans* CDR2 promoter with a novel promoter reporter system. Eukaryot Cell 8(8):1250–1267. doi:10.1128/EC.00069-09

Cowen LE, Lindquist S (2005) Hsp90 potentiates the rapid evolution of new traits: drug resistance in diverse fungi. Science 309(5744):2185–2189

Cowen LE, Anderson JB, Kohn LM (2002) Evolution of drug resistance in *Candida albicans*. Annu Rev Microbiol 56:139–165. doi:10.1146/annurev.micro.56.012302.160907

Cowen LE, Carpenter AE, Matangkasombut O, Fink GR, Lindquist S (2006) Genetic architecture of Hsp90-dependent drug resistance. Eukaryot Cell 5(12):2184–2188

Cross EW, Park S, Perlin DS (2000) Cross-Resistance of clinical isolates of *Candida albicans* and *Candida glabrata* to over-the-counter azoles used in the treatment of vaginitis. Microb Drug Resist 6(2):155–161. doi:10.1089/107662900419474

Crouzet J, Trombik T, Fraysse AS, Boutry M (2006) Organization and function of the plant pleiotropic drug resistance ABC transporter family. FEBS Lett 580(4):1123–1130

Cruz MC, Goldstein AL, Blankenship JR, Del Poeta M, Davis D, Cardenas ME, Perfect JR, McCusker JH, Heitman J (2002) Calcineurin is essential for survival during membrane stress in *Candida albicans*. EMBO J 21(4):546–559

da Silva Ferreira ME, Capellaro JL, dos Reis ME, Malavazi I, Perlin D, Park S, Anderson JB, Colombo AL, Arthington-Skaggs BA, Goldman MH, Goldman GH (2004) In vitro evolution of itraconazole resistance in *Aspergillus fumigatus* involves multiple mechanisms of resistance. Antimicrob Agents Chemother 48(11):4405–4413

Dassa E, Bouige P (2001) The ABC of ABCS: a phylogenetic and functional classification of ABC systems in living organisms. Res Microbiol 152(3–4):211–229

Dawson RJ, Locher KP (2006) Structure of a bacterial multidrug ABC transporter. Nature 443 (7108):180–185

Dean M (2002) The human ATP-binding cassette (ABC) transporter superfamily. NCBI; available online at the NCBI Bookshelf: http://www.ncbi.nlm.nih.gov/bookshelf/br.fcgi?book=mono_001&part=A137

Decottignies A, Goffeau A (1997) Complete inventory of the yeast ABC proteins. Nat Genet 15 (2):137–145. doi:10.1038/ng0297-137

Decottignies A, Kolaczkowski M, Balzi E, Goffeau A (1994) Solubilization and characterization of the overexpressed PDR5 multidrug resistance nucleotide triphosphatase of yeast. J Biol Chem 269(17):12797–12803

Decottignies A, Lambert L, Catty P, Degand H, Epping EA, Moye-Rowley WS, Balzi E, Goffeau A (1995) Identification and characterization of SNQ2, a new multidrug ATP binding cassette transporter of the yeast plasma membrane. J Biol Chem 270(30):18150–18157

Decottignies A, Grant AM, Nichols JW, de Wet H, McIntosh DB, Goffeau A (1998) ATPase and multidrug transport activities of the overexpressed yeast ABC protein Yor1p. J Biol Chem 273 (20):12612–12622

Deng D, Xu C, Sun P, Wu J, Yan C, Hu M, Yan N (2014) Crystal structure of the human glucose transporter GLUT1. Nature 510(7503):121–125. doi:10.1038/nature13306

Denning DW, Bromley MJ (2015) Infectious disease. How to bolster the antifungal pipeline. Science 347(6229):1414–1416. doi:10.1126/science.aaa6097

DeRisi J, van den Hazel B, Marc P, Balzi E, Brown P, Jacq C, Goffeau A (2000) Genome microarray analysis of transcriptional activation in multidrug resistance yeast mutants. FEBS Lett 470(2):156–160

Deutschbauer AM, Jaramillo DF, Proctor M, Kumm J, Hillenmeyer ME, Davis RW, Nislow C, Giaever G (2005) Mechanisms of haploinsufficiency revealed by genome-wide profiling in yeast. Genetics 169(4):1915–1925

Digirolamo JA, Li XC, Jacob MR, Clark AM, Ferreira D (2009) Reversal of fluconazole resistance by sulfated sterols from the marine sponge *Topsentia* sp. J Nat Prod 72(8):1524–1528. doi:10.1021/np900177m

Diwischek F, Morschhauser J, Holzgrabe U (2009) Cerulenin analogues as inhibitors of efflux pumps in drug-resistant *Candida albicans*. Arch Pharm (Weinheim) 342(3):150–164. doi:10.1002/ardp.200800160

Dornetshuber R, Heffeter P, Sulyok M, Schumacher R, Chiba P, Kopp S, Koellensperger G, Micksche M, Lemmens-Gruber R, Berger W (2009) Interactions between ABC-transport proteins and the secondary *Fusarium* metabolites enniatin and beauvericin. Mol Nutr Food Res 53(7):904–920. doi:10.1002/mnfr.200800384

Dujon B, Sherman D, Fischer G, Durrens P, Casaregola S, Lafontaine I, De Montigny J, Marck C, Neuveglise C, Talla E, Goffard N, Frangeul L, Aigle M, Anthouard V, Babour A, Barbe V, Barnay S, Blanchin S, Beckerich JM, Beyne E, Bleykasten C, Boisrame A, Boyer J, Cattolico L, Confanioleri F, De Daruvar A, Despons L, Fabre E, Fairhead C, Ferry-Dumazet H,

Groppi A, Hantraye F, Hennequin C, Jauniaux N, Joyet P, Kachouri R, Kerrest A, Koszul R, Lemaire M, Lesur I, Ma L, Muller H, Nicaud JM, Nikolski M, Oztas S, Ozier-Kalogeropoulos O, Pellenz S, Potier S, Richard GF, Straub ML, Suleau A, Swennen D, Tekaia F, Wesolowski-Louvel M, Westhof E, Wirth B, Zeniou-Meyer M, Zivanovic I, Bolotin-Fukuhara M, Thierry A, Bouchier C, Caudron B, Scarpelli C, Gaillardin C, Weissenbach J, Wincker P, Souciet JL (2004) Genome evolution in yeasts. Nature 430(6995):35–44

Dunkel N, Blass J, Rogers PD, Morschhauser J (2008a) Mutations in the multi-drug resistance regulator MRR1, followed by loss of heterozygosity, are the main cause of MDR1 overexpression in fluconazole-resistant *Candida albicans* strains. Mol Microbiol 69(4):827–840. doi:10.1111/j.1365-2958.2008.06309.x

Dunkel N, Liu TT, Barker KS, Homayouni R, Morschhauser J, Rogers PD (2008b) A gain-of-function mutation in the transcription factor Upc2p causes upregulation of ergosterol biosynthesis genes and increased fluconazole resistance in a clinical *Candida albicans* isolate. Eukaryot Cell 7(7):1180–1190. doi:10.1128/EC.00103-08

Eckford PD, Sharom FJ (2009) ABC efflux pump-based resistance to chemotherapy drugs. Chem Rev 109(7):2989–3011. doi:10.1021/cr9000226

Edlind TD, Katiyar SK (2004) The echinocandin "target" identified by cross-linking is a homolog of Pil1 and Lsp1, sphingolipid-dependent regulators of cell wall integrity signaling. Antimicrob Agents Chemother 48(11):4491

Ernst R, Kueppers P, Klein CM, Schwarzmueller T, Kuchler K, Schmitt L (2008) A mutation of the H-loop selectively affects rhodamine transport by the yeast multidrug ABC transporter Pdr5. Proc Natl Acad Sci USA 105(13):5069–5074

Esquivel BD, Smith AR, Zavrel M, White TC (2015) Azole drug import into the pathogenic fungus *Aspergillus fumigatus*. Antimicrob Agents Chemother 59(6):3390–3398. doi:10.1128/AAC.05003-14

Fardeau V, Lelandais G, Oldfield A, Salin H, Lemoine S, Garcia M, Tanty V, Le Crom S, Jacq C, Devaux F (2007) The central role of PDR1 in the foundation of yeast drug resistance. J Biol Chem 282(7):5063–5074. doi:10.1074/jbc.M610197200

Favre B, Ryder NS (1997) Cloning and expression of squalene epoxidase from the pathogenic yeast *Aspergillus fumigatus*. Gene 189(1):119–126

Ferreira ME, Colombo AL, Paulsen I, Ren Q, Wortman J, Huang J, Goldman MH, Goldman GH (2005) The ergosterol biosynthesis pathway, transporter genes, and azole resistance in *Aspergillus fumigatus*. Med Mycol 43(Suppl 1):S313–319

Ferreira-Pereira A, Marco S, Decottignies A, Nader J, Goffeau A, Rigaud JL (2003) Three-dimensional reconstruction of the *Saccharomyces cerevisiae* multidrug resistance protein Pdr5p. J Biol Chem 278(14):11995–11999. doi:10.1074/jbc.M212198200

Fischer G, Rocha EP, Brunet F, Vergassola M, Dujon B (2006) Highly variable rates of genome rearrangements between hemiascomycetous yeast lineages. PLoS Genet 2(3):e32

Fitzpatrick DA, Logue ME, Stajich JE, Butler G (2006) A fungal phylogeny based on 42 complete genomes derived from supertree and combined gene analysis. BMC Evol Biol 6:99

Fleming JA, Lightcap ES, Sadis S, Thoroddsen V, Bulawa CE, Blackman RK (2002) Complementary whole-genome technologies reveal the cellular response to proteasome inhibition by PS-341. Proc Natl Acad Sci USA 99(3):1461–1466

Fluman N, Bibi E (2009) Bacterial multidrug transport through the lens of the major facilitator superfamily. Biochim Biophys Acta 1794(5):738–747. doi:10.1016/j.bbapap.2008.11.020

Ford CB, Funt JM, Abbey D, Issi L, Guiducci C, Martinez DA, Delorey T, Li BY, White TC, Cuomo C, Rao RP, Berman J, Thompson DA, Regev A (2015) The evolution of drug resistance in clinical isolates of Candida albicans Elife 4:e00662. doi:10.7554/eLife.00662

Galagan JE, Calvo SE, Cuomo C, Ma LJ, Wortman JR, Batzoglou S, Lee SI, Basturkmen M, Spevak CC, Clutterbuck J, Kapitonov V, Jurka J, Scazzocchio C, Farman M, Butler J, Purcell S, Harris S, Braus GH, Draht O, Busch S, D'Enfert C, Bouchier C, Goldman GH, Bell-Pedersen D, Griffiths-Jones S, Doonan JH, Yu J, Vienken K, Pain A, Freitag M, Selker EU, Archer DB, Penalva MA, Oakley BR, Momany M, Tanaka T, Kumagai T, Asai K, Machida M, Nierman WC, Denning DW, Caddick M, Hynes M, Paoletti M, Fischer R,

Miller B, Dyer P, Sachs MS, Osmani SA, Birren BW (2005) Sequencing of Aspergillus nidulans and comparative analysis with A. fumigates and A. oryzae. Nature 438(7071):1105–1115

Galagan JE, Henn MR, Ma L-J, Cuomo CA, Biren B (2007) Genomics of the fungal kingdom: insights into eukaryotic biology. Genome Res 15:1620–1631

Gaur M, Choudhury D, Prasad R (2005) Complete inventory of ABC proteins in human pathogenic yeast, Candida albicans. J Mol Microbiol Biotechnol 9(1):3–15

Giaever G, Nislow C (2014) The yeast deletion collection: a decade of functional genomics. Genetics 197(2):451–465. doi:10.1534/genetics.114.161620

Giaever G, Chu AM, Ni L, Connelly C, Riles L, Veronneau S, Dow S, Lucau-Danila A, Anderson K, Andre B, Arkin AP, Astromoff A, El-Bakkoury M, Bangham R, Benito R, Brachat S, Campanaro S, Curtiss M, Davis K, Deutschbauer A, Entian KD, Flaherty P, Foury F, Garfinkel DJ, Gerstein M, Gotte D, Guldener U, Hegemann JH, Hempel S, Herman Z, Jaramillo DF, Kelly DE, Kelly SL, Kotter P, LaBonte D, Lamb DC, Lan N, Liang H, Liao H, Liu L, Luo C, Lussier M, Mao R, Menard P, Ooi SL, Revuelta JL, Roberts CJ, Rose M, Ross-Macdonald P, Scherens B, Schimmack G, Shafer B, Shoemaker DD, Sookhai-Mahadeo S, Storms RK, Strathern JN, Valle G, Voet M, Volckaert G, Wang CY, Ward TR, Wilhelmy J, Winzeler EA, Yang Y, Yen G, Youngman E, Yu K, Bussey H, Boeke JD, Snyder M, Philippsen P, Davis RW, Johnston M (2002) Functional profiling of the *Saccharomyces cerevisiae* genome. Nature 418(6896):387–391

Giaever G, Flaherty P, Kumm J, Proctor M, Nislow C, Jaramillo DF, Chu AM, Jordan MI, Arkin AP, Davis RW (2004) Chemogenomic profiling: identifying the functional interactions of small molecules in yeast. Proc Natl Acad Sci U S A 101(3):793–798

Gimeno CJ, Ljungdahl PO, Styles CA, Fink GR (1992) Unipolar cell divisions in the yeast *S. cerevisiae* lead to filamentous growth: regulation by starvation and RAS. Cell 68(6):1077–1090

Goffeau A, Barrell BG, Bussey H, Davis RW, Dujon B, Feldmann H, Galibert F, Hoheisel JD, Jacq C, Johnston M, Louis EJ, Mewes HW, Murakami Y, Philippsen P, Tettelin H, Oliver SG (1996) Life with 6000 genes. Science 274(5287):546, 563–547

Golin J, Ambudkar SV, Gottesman MM, Habib AD, Sczepanski J, Ziccardi W, May L (2003) Studies with novel Pdr5p substrates demonstrate a strong size dependence for xenobiotic efflux. J Biol Chem 278(8):5963–5969. doi:10.1074/jbc.M210908200

Golin J, Ambudkar SV, May L (2007) The yeast Pdr5p multidrug transporter: how does it recognize so many substrates? Biochem Biophys Res Commun 356(1):1–5. doi:10.1016/j.bbrc.2007.02.011

Hanway D, Chin JK, Xia G, Oshiro G, Winzeler EA, Romesberg FE (2002) Previously uncharacterized genes in the UV- and MMS-induced DNA damage response in yeast. Proc Natl Acad Sci U S A 99(16):10605–10610

Hawkesworth D (2001) The magnitude of fungal diversity: the 1.5 million species estimate revisited. Mycol Res 109:1422–1432

Hayama K, Ishibashi H, Ishijima SA, Niimi K, Tansho S, Ono Y, Monk BC, Holmes AR, Harding DR, Cannon RD, Abe S (2012) A D-octapeptide drug efflux pump inhibitor acts synergistically with azoles in a murine oral candidiasis infection model. FEMS Microbiol Lett 328(2):130–137. doi:10.1111/j.1574-6968.2011.02490.x

Healey KR, Zhao Y, Perez WB, Lockhart SR, Sobel JD, Farmakiotis D, Kontoyiannis DP, Sanglard D, Taj-Aldeen SJ, Alexander BD, Jimenez-Ortigosa C, Shor E, Perlin DS (2016) Prevalent mutator genotype identified in fungal pathogen *Candida glabrata* promotes multi-drug resistance. Nat Commun 7:11128. doi:10.1038/ncomms11128

Hector RF, Bierer DE (2011) New beta-glucan inhibitors as antifungal drugs. Expert Opin Ther Pat 21(10):1597–1610. doi:10.1517/13543776.2011.603899

Heitman J (2005) Cell biology. A fungal Achilles' heel. Science 309(5744):2175–2176

Higgins CF (2001) ABC transporters: physiology, structure and mechanism—an overview. Res Microbiol 152(3–4):205–210

Higgins CF, Linton KJ (2004) The ATP switch model for ABC transporters. Nat Struct Mol Biol 11(10):918–926

Hiller D, Sanglard D, Morschhauser J (2006) Overexpression of the MDR1 gene is sufficient to confer increased resistance to toxic compounds in *Candida albicans*. Antimicrob Agents Chemother 50(4):1365–1371. doi:10.1128/AAC.50.4.1365-1371.2006

Hiraga K, Yamamoto S, Fukuda H, Hamanaka N, Oda K (2005) Enniatin has a new function as an inhibitor of Pdr5p, one of the ABC transporters in *Saccharomyces cerevisiae*. Biochem Biophys Res Commun 328(4):1119–1125. doi:10.1016/j.bbrc.2005.01.075

Holmes AR, Tsao S, Ong SW, Lamping E, Niimi K, Monk BC, Niimi M, Kaneko A, Holland BR, Schmid J, Cannon RD (2006) Heterozygosity and functional allelic variation in the *Candida albicans* efflux pump genes CDR1 and CDR2. Mol Microbiol 62(1):170–186

Holmes AR, Lin YH, Niimi K, Lamping E, Keniya M, Niimi M, Tanabe K, Monk BC, Cannon RD (2008) ABC transporter Cdr1p contributes more than Cdr2p does to fluconazole efflux in fluconazole-resistant *Candida albicans* clinical isolates. Antimicrob Agents Chemother 52(11):3851–3862. doi:10.1128/AAC.00463-08

Holmes AR, Keniya MV, Ivnitski-Steele I, Monk BC, Lamping E, Sklar LA, Cannon RD (2012) The monoamine oxidase A inhibitor clorgyline is a broad-spectrum inhibitor of fungal ABC and MFS transporter efflux pump activities which reverses the azole resistance of *Candida albicans* and *Candida glabrata* clinical isolates. Antimicrob Agents Chemother 56(3):1508–1515. doi:10.1128/AAC.05706-11

Hull CM, Raisner RM, Johnson AD (2000) Evidence for mating of the "asexual" yeast *Candida albicans* in a mammalian host. Science 289(5477):307–310

Imoto S, Savoie CJ, Aburatani S, Kim S, Tashiro K, Kuhara S, Miyano S (2003) Use of gene networks for identifying and validating drug targets. J Bioinform Comput Biol 1(3):459–474

Inoue SB, Takewaki N, Takasuka T, Mio T, Adachi M, Fujii Y, Miyamoto C, Arisawa M, Furuichi Y, Watanabe T (1995) Characterization and gene cloning of 1,3-beta-D-glucan synthase from *Saccharomyces cerevisiae*. Eur J Biochem 231(3):845–854

Ivnitski-Steele I, Holmes AR, Lamping E, Monk BC, Cannon RD, Sklar LA (2009) Identification of Nile red as a fluorescent substrate of the *Saccharomyces cerevisiae* ATP-binding cassette transporters Cdr1p and Cdr2p and the major facilitator superfamily transporter Mdr1p. Anal Biochem 394(1):87–91. doi:10.1016/j.ab.2009.07.001

Izumikawa K, Kakeya H, Tsai HF, Grimberg B, Bennett JE (2003) Function of *Candida glabrata* ABC transporter gene, PDH1. Yeast 20(3):249–261

Jansen G, Lee AY, Epp E, Fredette A, Surprenant J, Harcus D, Scott M, Tan E, Nishimura T, Whiteway M, Hallett M, Thomas DY (2009) Chemogenomic profiling predicts antifungal synergies. Mol Syst Biol 5:338. doi:10.1038/msb.2009.95

Jasinski M, Ducos E, Martinoia E, Boutry M (2003) The ATP-binding cassette transporters: structure, function, and gene family comparison between rice and Arabidopsis. Plant Physiol 131(3):1169–1177

Jensen LJ, Bork P (2008) Biochemistry. Not comparable, but complementary. Science 322 (5898):56–57. doi:10.1126/science.1164801

Jones PM, George AM (2004) The ABC transporter structure and mechanism: perspectives on recent research. Cell Mol Life Sci 61(6):682–699

Karababa M, Valentino E, Pardini G, Coste AT, Bille J, Sanglard D (2006) CRZ1, a target of the calcineurin pathway in *Candida albicans*. Mol Microbiol 59(5):1429–1451

Katiyar SK, Edlind TD (2001) Identification and expression of multidrug resistance-related ABC transporter genes in *Candida krusei*. Med Mycol 39(1):109–116

Katzmann DJ, Hallstrom TC, Mahe Y, Moye-Rowley WS (1996) Multiple Pdr1p/Pdr3p binding sites are essential for normal expression of the ATP binding cassette transporter protein-encoding gene PDR5. J Biol Chem 271(38):23049–23054

Keniya MV, Cannon RD, Nguyen A, Tyndall JD, Monk BC (2013) Heterologous expression of *Candida albicans* Pma1p in *Saccharomyces cerevisiae*. FEMS Yeast Res 13(3):302–311. doi:10.1111/1567-1364.12035

Keniya MV, Fleischer E, Klinger A, Cannon RD, Monk BC (2015) Inhibitors of the *Candida albicans* major facilitator superfamily transporter Mdr1p responsible for fluconazole resistance. PLoS ONE 10(5):e0126350. doi:10.1371/journal.pone.0126350

Kim J, Wu S, Tomasiak TM, Mergel C, Winter MB, Stiller SB, Robles-Colmanares Y, Stroud RM, Tampe R, Craik CS, Cheng Y (2015) Subnanometre-resolution electron cryomicroscopy structure of a heterodimeric ABC exporter. Nature 517(7534):396–400. doi:10.1038/nature13872

Kolaczkowski M, van der Rest M, Cybularz-Kolaczkowska A, Soumillion JP, Konings WN, Goffeau A (1996) Anticancer drugs, ionophoric peptides, and steroids as substrates of the yeast multidrug transporter Pdr5p. J Biol Chem 271(49):31543–31548

Kolaczkowski M, Kolaczowska A, Luczynski J, Witek S, Goffeau A (1998) In vivo characterization of the drug resistance profile of the major ABC transporters and other components of the yeast pleiotropic drug resistance network. Microb Drug Resist 4(3):143–158. doi:10.1089/mdr.1998.4.143

Kolaczkowski M, Michalak K, Motohashi N (2003) Phenothiazines as potent modulators of yeast multidrug resistance. Int J Antimicrob Agents 22(3):279–283

Kolaczkowski M, Kolaczkowska A, Motohashi N, Michalak K (2009) New high-throughput screening assay to reveal similarities and differences in inhibitory sensitivities of multidrug ATP-binding cassette transporters. Antimicrob Agents Chemother 53(4):1516–1527. doi:10.1128/AAC.00956-08

Lacka I, Konieczny MT, Bulakowska A, Kodedova M, Gaskova D, Maurya IK, Prasad R, Milewski S (2015) Chemosensitization of multidrug resistant *Candida albicans* by the oxathiolone fused chalcone derivatives. Front Microbiol 6:783. doi:10.3389/fmicb.2015.00783

Lamping E, Monk BC, Niimi K, Holmes AR, Tsao S, Tanabe K, Niimi M, Uehara Y, Cannon RD (2007) Characterization of three classes of membrane proteins involved in fungal azole resistance by functional hyperexpression in *Saccharomyces cerevisiae*. Eukaryot Cell 6 (7):1150–1165

Lamping E, Ranchod A, Nakamura K, Tyndall JD, Niimi K, Holmes AR, Niimi M, Cannon RD (2009) Abc1p is a multidrug efflux transporter that tips the balance in favor of innate azole resistance in *Candida krusei*. Antimicrob Agents Chemother 53(2):354–369. doi:10.1128/AAC.01095-08

Lamping E, Baret PV, Holmes AR, Monk BC, Goffeau A, Cannon RD (2010) Fungal PDR transporters: phylogeny, topology, motifs and function. Fungal Genet Biol 47(2):127–142. doi:10.1016/j.fgb.2009.10.007

Lee MD, Galazzo JL, Staley AL, Lee JC, Warren MS, Fuernkranz H, Chamberland S, Lomovskaya O, Miller GH (2001) Microbial fermentation-derived inhibitors of efflux-pump-mediated drug resistance. Farmaco 56(1–2):81–85

Lee AY, St Onge RP, Proctor MJ, Wallace IM, Nile AH, Spagnuolo PA, Jitkova Y, Gronda M, Wu Y, Kim MK, Cheung-Ong K, Torres NP, Spear ED, Han MK, Schlecht U, Suresh S, Duby G, Heisler LE, Surendra A, Fung E, Urbanus ML, Gebbia M, Lissina E, Miranda M, Chiang JH, Aparicio AM, Zeghouf M, Davis RW, Cherfils J, Boutry M, Kaiser CA, Cummins CL, Trimble WS, Brown GW, Schimmer AD, Bankaitis VA, Nislow C, Bader GD, Giaever G (2014) Mapping the cellular response to small molecules using chemogenomic fitness signatures. Science 344(6180):208–211. doi:10.1126/science.1250217

Lee JY, Kinch LN, Borek DM, Wang J, Wang J, Urbatsch IL, Xie XS, Grishin NV, Cohen JC, Otwinowski Z, Hobbs HH, Rosenbaum DM (2016) Crystal structure of the human sterol transporter ABCG5/ABCG8. Nature 533(7604):561–564. doi:10.1038/nature17666

Li XC, Jacob MR, Ding Y, Agarwal AK, Smillie TJ, Khan SI, Nagle DG, Ferreira D, Clark AM (2006) Capisterones A and B, which enhance fluconazole activity in *Saccharomyces cerevisiae*, from the marine green alga *Penicillus capitus*. J Nat Prod 69(4):542–546. doi:10.1021/np050396y

Limtrakul P, Anuchapreeda S, Buddhasukh D (2004) Modulation of human multidrug-resistance MDR-1 gene by natural curcuminoids. BMC Cancer 4:13. doi:10.1186/1471-2407-4-13

Liu M, Healy MD, Dougherty BA, Esposito KM, Maurice TC, Mazzucco CE, Bruccoleri RE, Davison DB, Frosco M, Barrett JF, Wang YK (2006) Conserved fungal genes as potential targets for broad-spectrum antifungal drug discovery. Eukaryot Cell 5(4):638–649

Liu XL, Tee HW, Go ML (2008) Functionalized chalcones as selective inhibitors of P-glycoprotein and breast cancer resistance protein. Bioorg Med Chem 16(1):171–180. doi:10.1016/j.bmc.2007.10.006

Liu M, Zeng R, Zhang L, Li D, Lv G, Shen Y, Zheng H, Zhang Q, Zhao J, Zheng N, Liu W (2015) Multiple cyp51A-based mechanisms identified in azole-resistant isolates of *Aspergillus fumigatus* from China. Antimicrob Agents Chemother 59(7):4321–4325. doi:10.1128/AAC.00003-15

Lo HJ, Tseng KY, Kao YY, Tsao MY, Lo HL, Yang YL (2015) Cph1p negatively regulates MDR1 involved in drug resistance in *Candida albicans*. Int J Antimicrob Agents 45(6):617–621. doi:10.1016/j.ijantimicag.2015.01.017

Loftus BJ, Fung E, Roncaglia P, Rowley D, Amedeo P, Bruno D, Vamathevan J, Miranda M, Anderson IJ, Fraser JA, Allen JE, Bosdet IE, Brent MR, Chiu R, Doering TL, Donlin MJ, D'Souza CA, Fox DS, Grinberg V, Fu J, Fukushima M, Haas BJ, Huang JC, Janbon G, Jones SJ, Koo HL, Krzywinski MI, Kwon-Chung JK, Lengeler KB, Maiti R, Marra MA, Marra RE, Mathewson CA, Mitchell TG, Pertea M, Riggs FR, Salzberg SL, Schein JE, Shvartsbeyn A, Shin H, Shumway M, Specht CA, Suh BB, Tenney A, Utterback TR, Wickes BL, Wortman JR, Wye NH, Kronstad JW, Lodge JK, Heitman J, Davis RW, Fraser CM, Hyman RW (2005) The genome of the basidiomycetous yeast and human pathogen *Cryptococcus neoformans*. Science 307(5713):1321–1324

Lohberger A, Coste AT, Sanglard D (2014) Distinct roles of *Candida albicans* drug resistance transcription factors TAC1, MRR1, and UPC2 in virulence. Eukaryot Cell 13(1):127–142. doi:10.1128/EC.00245-13

Lomovskaya O, Bostian KA (2006) Practical applications and feasibility of efflux pump inhibitors in the clinic—a vision for applied use. Biochem Pharmacol 71(7):910–918. doi:10.1016/j.bcp.2005.12.008

Lum PY, Armour CD, Stepaniants SB, Cavet G, Wolf MK, Butler JS, Hinshaw JC, Garnier P, Prestwich GD, Leonardson A, Garrett-Engele P, Rush CM, Bard M, Schimmack G, Phillips JW, Roberts CJ, Shoemaker DD (2004) Discovering modes of action for therapeutic compounds using a genome-wide screen of yeast heterozygotes. Cell 116(1):121–137

Maebashi K, Niimi M, Kudoh M, Fischer FJ, Makimura K, Niimi K, Piper RJ, Uchida K, Arisawa M, Cannon RD, Yamaguchi H (2001) Mechanisms of fluconazole resistance in *Candida albicans* isolates from Japanese AIDS patients. J Antimicrob Chemother 47(5):527–536

Maki N, Dey S (2006) Biochemical and pharmacological properties of an allosteric modulator site of the human P-glycoprotein (ABCB1). Biochem Pharmacol 72(2):145–155. doi:10.1016/j.bcp.2006.04.008

Mamnun YM, Pandjaitan R, Mahe Y, Delahodde A, Kuchler K (2002) The yeast zinc finger regulators Pdr1p and Pdr3p control pleiotropic drug resistance (PDR) as homo- and heterodimers in vivo. Mol Microbiol 46(5):1429–1440

Mansfield BE, Oltean HN, Oliver BG, Hoot SJ, Leyde SE, Hedstrom L, White TC (2010) Azole drugs are imported by facilitated diffusion in *Candida albicans* and other pathogenic fungi. PLoS Pathog 6(9):e1001126. doi:10.1371/journal.ppat.1001126

Mathema B, Cross E, Dun E, Park S, Bedell J, Slade B, Williams M, Riley L, Chaturvedi V, Perlin DS (2001) Prevalence of vaginal colonization by drug-resistant *Candida* species in college-age women with previous exposure to over-the-counter azole antifungals. Clin Infect Dis 33(5):E23–27. doi:10.1086/322600

Maurya IK, Thota CK, Verma SD, Sharma J, Rawal MK, Ravikumar B, Sen S, Chauhan N, Lynn AM, Chauhan VS, Prasad R (2013) Rationally designed transmembrane peptide mimics of the multidrug transporter protein Cdr1 act as antagonists to selectively block drug efflux and chemosensitize azole-resistant clinical isolates of *Candida albicans*. J Biol Chem 288(23):16775–16787. doi:10.1074/jbc.M113.467159

Mellado E, Diaz-Guerra TM, Cuenca-Estrella M, Rodriguez-Tudela JL (2001) Identification of two different 14-alpha sterol demethylase-related genes (cyp51A and cyp51B) in *Aspergillus fumigatus* and other *Aspergillus* species. J Clin Microbiol 39(7):2431–2438

Miyazaki H, Miyazaki Y, Geber A, Parkinson T, Hitchcock C, Falconer DJ, Ward DJ, Marsden K, Bennett JE (1998) Fluconazole resistance associated with drug efflux and increased transcription of a drug transporter gene, PDH1, in *Candida glabrata*. Antimicrob Agents Chemother 42(7):1695–1701

Mohammadi F, Hashemi SJ, Zoll J, Melchers WJ, Rafati H, Dehghan P, Rezaie S, Tolooe A, Tamadon Y, van der Lee HA, Verweij PE, Seyedmousavi S (2016) Quantitative analysis of single-nucleotide polymorphism for rapid detection of TR34/L98H- and TR46/Y121F/T289A-positive *Aspergillus fumigatus* isolates obtained from patients in Iran from 2010 to 2014. Antimicrob Agents Chemother 60(1):387–392. doi:10.1128/AAC.02326-15

Moitra K, Lou H, Dean M (2011) Multidrug efflux pumps and cancer stem cells: insights into multidrug resistance and therapeutic development. Clin Pharmacol Ther 89(4):491–502. doi:10.1038/clpt.2011.14

Monk BC, Goffeau A (2008) Outwitting multidrug resistance to antifungals. Science 321 (5887):367–369. doi:10.1126/science.1159746

Monk BC, Harding DR (2005) Peptide motifs for cell-surface intervention: application to anti-infective and biopharmaceutical development. BioDrugs 19(4):261–278

Monk BC, Niimi K, Lin S, Knight A, Kardos TB, Cannon RD, Parshot R, King A, Lun D, Harding DR (2005) Surface-active fungicidal D-peptide inhibitors of the plasma membrane proton pump that block azole resistance. Antimicrob Agents Chemother 49(1):57–70

Monk BC, Tomasiak TM, Keniya MV, Huschmann FU, Tyndall JD, O'Connell JD 3rd, Cannon RD, McDonald JG, Rodriguez A, Finer-Moore JS, Stroud RM (2014) Architecture of a single membrane spanning cytochrome P450 suggests constraints that orient the catalytic domain relative to a bilayer. Proc Natl Acad Sci USA 111(10):3865–3870. doi:10.1073/pnas.1324245111

Morio F, Loge C, Besse B, Hennequin C, Le Pape P (2010) Screening for amino acid substitutions in the *Candida albicans* Erg11 protein of azole-susceptible and azole-resistant clinical isolates: new substitutions and a review of the literature. Diagn Microbiol Infect Dis 66(4):373–384. doi:10.1016/j.diagmicrobio.2009.11.006

Morschhauser J (2016) The development of fluconazole resistance in *Candida albicans*—an example of microevolution of a fungal pathogen. J Microbiol 54(3):192–201. doi:10.1007/s12275-016-5628-4

Morschhauser J, Barker KS, Liu TT, Bla BWJ, Homayouni R, Rogers PD (2007) The transcription factor Mrr1p controls expression of the MDR1 efflux pump and mediates multidrug resistance in *Candida albicans*. PLoS Pathog 3(11):e164. doi:10.1371/journal.ppat.0030164

Nakamura K, Niimi M, Niimi K, Holmes AR, Yates JE, Decottignies A, Monk BC, Goffeau A, Cannon RD (2001) Functional expression of *Candida albicans* drug efflux pump Cdr1p in a Saccharomyces cerevisiae strain deficient in membrane transporters. Antimicrob Agents Chemother 45(12):3366–3374. doi:10.1128/AAC.45.12.3366-3374.2001

Nascimento AM, Goldman GH, Park S, Marras SA, Delmas G, Oza U, Lolans K, Dudley MN, Mann PA, Perlin DS (2003) Multiple resistance mechanisms among *Aspergillus fumigatus* mutants with high-level resistance to itraconazole. Antimicrob Agents Chemother 47(5):1719–1726

Nierman WC, Pain A, Anderson MJ, Wortman JR, Kim HS, Arroyo J, Berriman M, Abe K, Archer DB, Bermejo C, Bennett J, Bowyer P, Chen D, Collins M, Coulsen R, Davies R, Dyer PS, Farman M, Fedorova N, Fedorova N, Feldblyum TV, Fischer R, Fosker N, Fraser A, Garcia JL, Garcia MJ, Goble A, Goldman GH, Gomi K, Griffith-Jones S, Gwilliam R, Haas B, Haas H, Harris D, Horiuchi H, Huang J, Humphray S, Jimenez J, Keller N, Khouri H, Kitamoto K, Kobayashi T, Konzack S, Kulkarni R, Kumagai T, Lafon A, Latge JP, Li W, Lord A, Lu C, Majoros WH, May GS, Miller BL, Mohamoud Y, Molina M, Monod M, Mouyna I, Mulligan S, Murphy L, O'Neil S, Paulsen I, Penalva MA, Pertea M, Price C,

Pritchard BL, Quail MA, Rabbinowitsch E, Rawlins N, Rajandream MA, Reichard U, Renauld H, Robson GD, Rodriguez de Cordoba S, Rodriguez-Pena JM, Ronning CM, Rutter S, Salzberg SL, Sanchez M, Sanchez-Ferrero JC, Saunders D, Seeger K, Squares R, Squares S, Takeuchi M, Tekaia F, Turner G, Vazquez de Aldana CR, Weidman J, White O, Woodward J, Yu JH, Fraser C, Galagan JE, Asai K, Machida M, Hall N, Barrell B, Denning DW (2005) Genomic sequence of the pathogenic and allergenic filamentous fungus *Aspergillus fumigatus*. Nature 438(7071):1151–1156

Niimi M, Nagai Y, Niimi K, Wada S, Cannon RD, Uehara Y, Monk BC (2002) Identification of two proteins induced by exposure of the pathogenic fungus *Candida glabrata* to fluconazole. J Chromatogr B Analyt Technol Biomed Life Sci 782(1–2):245–252

Niimi K, Harding DR, Parshot R, King A, Lun DJ, Decottignies A, Niimi M, Lin S, Cannon RD, Goffeau A, Monk BC (2004) Chemosensitization of fluconazole resistance in *Saccharomyces cerevisiae* and pathogenic fungi by a D-octapeptide derivative. Antimicrob Agents Chemother 48(4):1256–1271

Niimi K, Maki K, Ikeda F, Holmes AR, Lamping E, Niimi M, Monk BC, Cannon RD (2006) Overexpression of *Candida albicans* CDR1, CDR2, or MDR1 does not produce significant changes in echinocandin susceptibility. Antimicrob Agents Chemother 50(4):1148–1155

Niimi K, Harding DR, Holmes AR, Lamping E, Niimi M, Tyndall JD, Cannon RD, Monk BC (2012) Specific interactions between the *Candida albicans* ABC transporter Cdr1p ectodomain and a D-octapeptide derivative inhibitor. Mol Microbiol 85(4):747–767. doi:10.1111/j.1365-2958.2012.08140.x

Nim S, Rawal MK, Prasad R (2014) FK520 interacts with the discrete intrahelical amino acids of multidrug transporter Cdr1 protein and acts as antagonist to selectively chemosensitize azole-resistant clinical isolates of *Candida albicans*. FEMS Yeast Res 14(4):624–632. doi:10.1111/1567-1364.12149

Nim S, Monico A, Rawal MK, Duarte N, Prasad R, Di Pietro A, Ferreira MU (2016) Overcoming multidrug resistance in *Candida albicans*: macrocyclic diterpenes from euphorbia species as potent inhibitors of drug efflux pumps. Planta Med. doi:10.1055/s-0042-106169

Nishikawa JL, Boeszoermenyi A, Vale-Silva LA, Torelli R, Posteraro B, Sohn YJ, Ji F, Gelev V, Sanglard D, Sanguinetti M, Sadreyev RI, Mukherjee G, Bhyravabhotla J, Buhrlage SJ, Gray NS, Wagner G, Naar AM, Arthanari H (2016) Inhibiting fungal multidrug resistance by disrupting an activator-Mediator interaction. Nature 530(7591):485–489. doi:10.1038/nature16963

O'Brien HE, Parrent JL, Jackson JA, Moncalvo JM, Vilgalys R (2005) Fungal community analysis by large-scale sequencing of environmental samples. Appl Environ Microbiol 71(9):5544–5550. doi:10.1128/AEM.71.9.5544-5550.2005

Onishi J, Meinz M, Thompson J, Curotto J, Dreikorn S, Rosenbach M, Douglas C, Abruzzo G, Flattery A, Kong L, Cabello A, Vicente F, Pelaez F, Diez MT, Martin I, Bills G, Giacobbe R, Dombrowski A, Schwartz R, Morris S, Harris G, Tsipouras A, Wilson K, Kurtz MB (2000) Discovery of novel antifungal (1,3)-beta-D-glucan synthase inhibitors. Antimicrob Agents Chemother 44(2):368–377

Onyewu C, Wormley FL Jr, Perfect JR, Heitman J (2004) The calcineurin target, Crz1, functions in azole tolerance but is not required for virulence of *Candida albicans*. Infect Immun 72(12):7330–7333

Palecek SP, Parikh AS, Kron SJ (2002) Sensing, signalling and integrating physical processes during *Saccharomyces cerevisiae* invasive and filamentous growth. Microbiology 148(Pt 4):893–907

Parker JL, Newstead S (2014) Molecular basis of nitrate uptake by the plant nitrate transporter NRT1.1. Nature 507(7490):68–72. doi:10.1038/nature13116

Parker JE, Warrilow AG, Price CL, Mullins JG, Kelly DE, Kelly SL (2014) Resistance to antifungals that target CYP51. J Chem Biol 7(4):143–161. doi:10.1007/s12154-014-0121-1

Parsons AB, Brost RL, Ding H, Li Z, Zhang C, Sheikh B, Brown GW, Kane PM, Hughes TR, Boone C (2004) Integration of chemical-genetic and genetic interaction data links bioactive compounds to cellular target pathways. Nat Biotechnol 22(1):62–69

Parsons AB, Bussey H, Boone C (2006) Functional genomic approaches to fungal pathogenesis, drug target validation, and antifungal drug discovery. In: Heitman Jea (ed) Molecular principles of fungal pathogenesis

Paul S, Moye-Rowley WS (2014) Multidrug resistance in fungi: regulation of transporter-encoding gene expression. Front Physiol 5:143. doi:10.3389/fphys.2014.00143

Paulsen IT, Skurray RA (1993) Topology, structure and evolution of two families of proteins involved in antibiotic and antiseptic resistance in eukaryotes and prokaryotes—an analysis. Gene 124(1):1–11

Paumi CM, Chuk M, Snider J, Stagljar I, Michaelis S (2009) ABC transporters in *Saccharomyces cerevisiae* and their interactors: new technology advances the biology of the ABCC (MRP) subfamily. Microbiol Mol Biol Rev 73(4):577–593. doi:10.1128/MMBR.00020-09

Payne DJ, Gwynn MN, Holmes DJ, Pompliano DL (2007) Drugs for bad bugs: confronting the challenges of antibacterial discovery. Nat Rev Drug Discov 6(1):29–40

Pedersen BP, Kumar H, Waight AB, Risenmay AJ, Roe-Zurz Z, Chau BH, Schlessinger A, Bonomi M, Harries W, Sali A, Johri AK, Stroud RM (2013) Crystal structure of a eukaryotic phosphate transporter. Nature 496(7446):533–536. doi:10.1038/nature12042

Perea S, Lopez-Ribot JL, Kirkpatrick WR, McAtee RK, Santillan RA, Martinez M, Calabrese D, Sanglard D, Patterson TF (2001) Prevalence of molecular mechanisms of resistance to azole antifungal agents in *Candida albicans* strains displaying high-level fluconazole resistance isolated from human immunodeficiency virus-infected patients. Antimicrob Agents Chemother 45(10):2676–2684

Pfaller MA (2012) Antifungal drug resistance: mechanisms, epidemiology, and consequences for treatment. Am J Med 125(1 Suppl):S3–13. doi:10.1016/j.amjmed.2011.11.001

Pfaller MA, Diekema DJ (2007) Epidemiology of invasive candidiasis: a persistent public health problem. Clin Microbiol Rev 20(1):133–163

Pfaller MA, Diekema DJ (2010) Epidemiology of invasive mycoses in North America. Crit Rev Microbiol 36(1):1–53. doi:10.3109/10408410903241444

Piddock LJ (2006) Clinically relevant chromosomally encoded multidrug resistance efflux pumps in bacteria. Clin Microbiol Rev 19(2):382–402. doi:10.1128/CMR.19.2.382-402.2006

Posteraro B, Sanguinetti M, Sanglard D, La Sorda M, Boccia S, Romano L, Morace G, Fadda G (2003) Identification and characterization of a *Cryptococcus neoformans* ATP binding cassette (ABC) transporter-encoding gene, CnAFR1, involved in the resistance to fluconazole. Mol Microbiol 47(2):357–371

Prasad R, Goffeau A (2012) Yeast ATP-binding cassette transporters conferring multidrug resistance. Annu Rev Microbiol 66:39–63. doi:10.1146/annurev-micro-092611-150111

Prasad R, Rawal MK (2014) Efflux pump proteins in antifungal resistance. Front Pharmacol 5:202. doi:10.3389/fphar.2014.00202

Prasad R, De Wergifosse P, Goffeau A, Balzi E (1995) Molecular cloning and characterization of a novel gene of *Candida albicans*, CDR1, conferring multiple resistance to drugs and antifungals. Curr Genet 27(4):320–329

Quistgaard EM, Low C, Guettou F, Nordlund P (2016) Understanding transport by the major facilitator superfamily (MFS): structures pave the way. Nat Rev Mol Cell Biol 17(2):123–132. doi:10.1038/nrm.2015.25

Ramage G, Bachmann S, Patterson TF, Wickes BL, Lopez-Ribot JL (2002) Investigation of multidrug efflux pumps in relation to fluconazole resistance in *Candida albicans* biofilms. J Antimicrob Chemother 49(6):973–980

Ramage G, Martinez JP, Lopez-Ribot JL (2006) *Candida* biofilms on implanted biomaterials: a clinically significant problem. FEMS Yeast Res 6(7):979–986

Rawal MK, Khan MF, Kapoor K, Goyal N, Sen S, Saxena AK, Lynn AM, Tyndall JD, Monk BC, Cannon RD, Komath SS, Prasad R (2013) Insight into pleiotropic drug resistance ATP-binding cassette pump drug transport through mutagenesis of Cdr1p transmembrane domains. J Biol Chem 288(34):24480–24493. doi:10.1074/jbc.M113.488353

Rawal MK, Shokoohinia Y, Chianese G, Zolfaghari B, Appendino G, Taglialatela-Scafati O, Prasad R, Di Pietro A (2014) Jatrophanes from *Euphorbia squamosa* as potent inhibitors of

Candida albicans multidrug transporters. J Nat Prod 77(12):2700–2706. doi:10.1021/np500756z

Redhu AK, Shah AH, Prasad R (2016) MFS transporters of *Candida* species and their role in clinical drug resistance. FEMS Yeast Research 16. doi:10.1093/fensyr/fow043

Rees DC, Johnson E, Lewinson O (2009) ABC transporters: the power to change. Nat Rev Mol Cell Biol 10(3):218–227

Reynolds TB, Fink GR (2001) Bakers' yeast, a model for fungal biofilm formation. Science 291 (5505):878–881

Ricardo E, Costa-de-Oliveira S, Dias AS, Guerra J, Rodrigues AG, Pina-Vaz C (2009) Ibuprofen reverts antifungal resistance on *Candida albicans* showing overexpression of CDR genes. FEMS Yeast Res 9(4):618–625. doi:10.1111/j.1567-1364.2009.00504.x

Riggle PJ, Kumamoto CA (2006) Transcriptional regulation of MDR1, encoding a drug efflux determinant, in fluconazole-resistant *Candida albicans* strains through an Mcm1p binding site. Eukaryot Cell 5(12):1957–1968. doi:10.1128/EC.00243-06

Roemer T, Jiang B, Davison J, Ketela T, Veillette K, Breton A, Tandia F, Linteau A, Sillaots S, Marta C, Martel N, Veronneau S, Lemieux S, Kauffman S, Becker J, Storms R, Boone C, Bussey H (2003) Large-scale essential gene identification in *Candida albicans* and applications to antifungal drug discovery. Mol Microbiol 50(1):167–181

Rogers PD, Barker KS (2003) Genome-wide expression profile analysis reveals coordinately regulated genes associated with stepwise acquisition of azole resistance in *Candida albicans* clinical isolates. Antimicrob Agents Chemother 47(4):1220–1227

Rognon B, Kozovska Z, Coste AT, Pardini G, Sanglard D (2006) Identification of promoter elements responsible for the regulation of MDR1 from *Candida albicans*, a major facilitator transporter involved in azole resistance. Microbiology 152(Pt 12):3701–3722. doi:10.1099/mic.0.29277-0

Rustad TR, Stevens DA, Pfaller MA, White TC (2002) Homozygosity at the *Candida albicans* MTL locus associated with azole resistance. Microbiology 148(Pt 4):1061–1072. doi:10.1099/00221287-148-4-1061

Rutledge RM, Esser L, Ma J, Xia D (2011) Toward understanding the mechanism of action of the yeast multidrug resistance transporter Pdr5p: a molecular modeling study. J Struct Biol 173 (2):333–344. doi:10.1016/j.jsb.2010.10.012

Sagatova AA, Keniya MV, Wilson RK, Monk BC, Tyndall JD (2015) Structural insights into binding of the antifungal drug fluconazole to *Saccharomyces cerevisiae* lanosterol 14alpha-demethylase. Antimicrob Agents Chemother 59(8):4982–4989. doi:10.1128/AAC.00925-15

Sagatova AA, Keniya MV, Wilson RK, Sabherwal M, Tyndall JD, Monk BC (2016) Triazole resistance mediated by mutations of a conserved active site tyrosine in fungal lanosterol 14alpha-demethylase. Sci Rep 6:26213. doi:10.1038/srep26213

Saini P, Prasad T, Gaur NA, Shukla S, Jha S, Komath SS, Khan LA, Haq QM, Prasad R (2005) Alanine scanning of transmembrane helix 11 of Cdr1p ABC antifungal efflux pump of *Candida albicans*: identification of amino acid residues critical for drug efflux. J Antimicrob Chemother 56(1):77–86. doi:10.1093/jac/dki183

Sanchez-Leon E, Riquelme M (2015) Live imaging of beta-1,3-glucan synthase FKS-1 in *Neurospora crassa* hyphae. Fungal Genet Biol 82:104–107. doi:10.1016/j.fgb.2015.07.001

Sanglard D, Bille J (2002) Current understanding of the modes of action of and resistance mechanisms to conventional and emerging antifungal agents for treatment of *Candida* infections. In: Calderone RA (ed) Candida and Candidiasis. ASM Press, Washington, DC, pp 349–383

Sanglard D, Kuchler K, Ischer F, Pagani JL, Monod M, Bille J (1995) Mechanisms of resistance to azole antifungal agents in *Candida albicans* isolates from AIDS patients involve specific multidrug transporters. Antimicrob Agents Chemother 39(11):2378–2386

Sanglard D, Ischer F, Koymans L, Bille J (1998) Amino acid substitutions in the cytochrome P-450 lanosterol 14alpha-demethylase (CYP51A1) from azole-resistant *Candida albicans* clinical isolates contribute to resistance to azole antifungal agents. Antimicrob Agents Chemother 42(2):241–253

Sanglard D, Ischer F, Calabrese D, Majcherczyk PA, Bille J (1999) The ATP binding cassette transporter gene CgCDR1 from *Candida glabrata* is involved in the resistance of clinical isolates to azole antifungal agents. Antimicrob Agents Chemother 43(11):2753–2765

Sanglard D, Ischer F, Bille J (2001) Role of ATP-binding-cassette transporter genes in high-frequency acquisition of resistance to azole antifungals in *Candida glabrata*. Antimicrob Agents Chemother 45(4):1174–1183

Sanglard D, Ischer F, Marchetti O, Entenza J, Bille J (2003) Calcineurin A of *Candida albicans*: involvement in antifungal tolerance, cell morphogenesis and virulence. Mol Microbiol 48 (4):959–976

Sanguinetti M, Posteraro B, La Sorda M, Torelli R, Fiori B, Santangelo R, Delogu G, Fadda G (2006) Role of AFR1, an ABC transporter-encoding gene, in the in vivo response to fluconazole and virulence of *Cryptococcus neoformans*. Infect Immun 74(2):1352–1359

Saraswathy M, Gong S (2013) Different strategies to overcome multidrug resistance in cancer. Biotechnol Adv 31(8):1397–1407. doi:10.1016/j.biotechadv.2013.06.004

Sauna ZE, Bohn SS, Rutledge R, Dougherty MP, Cronin S, May L, Xia D, Ambudkar SV, Golin J (2008) Mutations define cross-talk between the N-terminal nucleotide-binding domain and transmembrane helix-2 of the yeast multidrug transporter Pdr5: possible conservation of a signaling interface for coupling ATP hydrolysis to drug transport. J Biol Chem 283(50):35010–35022. doi:10.1074/jbc.M806446200

Schmid D, Ecker G, Kopp S, Hitzler M, Chiba P (1999) Structure-activity relationship studies of propafenone analogs based on P-glycoprotein ATPase activity measurements. Biochem Pharmacol 58(9):1447–1456

Schubert S, Barker KS, Znaidi S, Schneider S, Dierolf F, Dunkel N, Aid M, Boucher G, Rogers PD, Raymond M, Morschhauser J (2011a) Regulation of efflux pump expression and drug resistance by the transcription factors Mrr1, Upc2, and Cap1 in *Candida albicans*. Antimicrob Agents Chemother 55(5):2212–2223. doi:10.1128/AAC.01343-10

Schubert S, Popp C, Rogers PD, Morschhauser J (2011b) Functional dissection of a *Candida albicans* zinc cluster transcription factor, the multidrug resistance regulator Mrr1. Eukaryot Cell 10(8):1110–1121. doi:10.1128/EC.05100-11

Schumacher MA, Brennan RG (2003) Deciphering the molecular basis of multidrug recognition: crystal structures of the *Staphlyococcus aureus* multidrug binding transcription regulator QacR. Res Microbiol 154(2):69–77

Selmecki A, Forche A, Berman J (2006) Aneuploidy and isochromosome formation in drug-resistant *Candida albicans*. Science 313(5785):367–370

Seoighe C, Federspiel N, Jones T, Hansen N, Bivolarovic V, Surzycki R, Tamse R, Komp C, Huizar L, Davis RW, Scherer S, Tait E, Shaw DJ, Harris D, Murphy L, Oliver K, Taylor K, Rajandream MA, Barrell BG, Wolfe KH (2000) Prevalence of small inversions in yeast gene order evolution. Proc Natl Acad Sci U S A 97(26):14433–14437

Shahi P, Moye-Rowley WS (2009) Coordinate control of lipid composition and drug transport activities is required for normal multidrug resistance in fungi. Biochim Biophys Acta 1794 (5):852–859

Sharma M, Prasad R (2011) The quorum-sensing molecule farnesol is a modulator of drug efflux mediated by ABC multidrug transporters and synergizes with drugs in *Candida albicans*. Antimicrob Agents Chemother 55(10):4834–4843. doi:10.1128/AAC.00344-11

Sharma M, Manoharlal R, Shukla S, Puri N, Prasad T, Ambudkar SV, Prasad R (2009) Curcumin modulates efflux mediated by yeast ABC multidrug transporters and is synergistic with antifungals. Antimicrob Agents Chemother 53(8):3256–3265. doi:10.1128/AAC.01497-08

Shedletzky E, Unger C, Delmer DP (1997) A microtiter-based fluorescence assay for (1,3)-beta-glucan synthases. Anal Biochem 249(1):88–93. doi:10.1006/abio.1997.2162

Shields DC, Wolfe KH (1997) Accelerated evolution of sites undergoing mRNA editing in plant mitochondria and chloroplasts. Mol Biol Evol 14(3):344–349

Shukla S, Sauna ZE, Prasad R, Ambudkar SV (2004) Disulfiram is a potent modulator of multidrug transporter Cdr1p of *Candida albicans*. Biochem Biophys Res Commun 322 (2):520–525. doi:10.1016/j.bbrc.2004.07.151

Shukla S, Rai V, Saini P, Banerjee D, Menon AK, Prasad R (2007) *Candida* drug resistance protein 1, a major multidrug ATP binding cassette transporter of *Candida albicans*, translocates fluorescent phospholipids in a reconstituted system. Biochemistry 46 (43):12081–12090. doi:10.1021/bi700453e

Silva LV, Sanguinetti M, Vandeputte P, Torelli R, Rochat B, Sanglard D (2013) Milbemycins: more than efflux inhibitors for fungal pathogens. Antimicrob Agents Chemother 57(2):873–886. doi:10.1128/AAC.02040-12

Singh N, Perfect JR (2007) Immune reconstitution syndrome associated with opportunistic mycoses. Lancet Infect Dis 7(6):395–401. doi:10.1016/S1473-3099(07)70085-3

Sionov E, Chang YC, Kwon-Chung KJ (2013) Azole heteroresistance in *Cryptococcus neoformans*: emergence of resistant clones with chromosomal disomy in the mouse brain during fluconazole treatment. Antimicrob Agents Chemother 57(10):5127–5130. doi:10.1128/AAC.00694-13

Sklar LA, Carter MB, Edwards BS (2007) Flow cytometry for drug discovery, receptor pharmacology and high-throughput screening. Curr Opin Pharmacol 7(5):527–534. doi:10.1016/j.coph.2007.06.006

Smriti KS, Dixit BL, Gupta CM, Milewski S, Prasad R (2002) ABC transporters Cdr1p, Cdr2p and Cdr3p of a human pathogen *Candida albicans* are general phospholipid translocators. Yeast 19 (4):303–318. doi:10.1002/yea.818

Snelders E, Huis In 't Veld RA, Rijs AJ, Kema GH, Melchers WJ, Verweij PE (2009) Possible environmental origin of resistance of *Aspergillus fumigatus* to medical triazoles. Appl Environ Microbiol 75(12):4053–4057. doi:10.1128/AEM.00231-09

Snelders E, van der Lee HA, Kuijpers J, Rijs AJ, Varga J, Samson RA, Mellado E, Donders AR, Melchers WJ, Verweij PE (2008) Emergence of azole resistance in *Aspergillus fumigatus* and spread of a single resistance mechanism. PLoS Med 5(11):e219. doi:10.1371/journal.pmed.0050219

Snider J, Hanif A, Lee ME, Jin K, Yu AR, Graham C, Chuk M, Damjanovic D, Wierzbicka M, Tang P, Balderes D, Wong V, Jessulat M, Darowski KD, San Luis BJ, Shevelev I, Sturley SL, Boone C, Greenblatt JF, Zhang Z, Paumi CM, Babu M, Park HO, Michaelis S, Stagljar I (2013) Mapping the functional yeast ABC transporter interactome. Nat Chem Biol 9(9):565–572. doi:10.1038/nchembio.1293

Spitzer M, Griffiths E, Blakely KM, Wildenhain J, Ejim L, Rossi L, De Pascale G, Curak J, Brown E, Tyers M, Wright GD (2011) Cross-species discovery of syncretic drug combinations that potentiate the antifungal fluconazole. Mol Syst Biol 7:499. doi:10.1038/msb.2011.31

Strouse JJ, Ivnitski-Steele I, Njus HM, Foutz TD, Yao T, Weiner WS, Schroeder CE, Simpson DS, Maki BE, Li K, Golden JE, Waller A, Evangelisti AM, Young SM, Perez D, Chavez SE, Garcia MJ, Ursu O, Fara DC, Bologa CG, Carter MB, Salas VM, Tegos GP, Oprea TI, Edwards BS, Larson RS, Aube J, Sklar LA (2010) Selective efflux inhibition of ATP-binding cassette sub-family G member 2. In: Probe reports from the NIH molecular libraries program, Bethesda (MD)

Sun J, Bankston JR, Payandeh J, Hinds TR, Zagotta WN, Zheng N (2014) Crystal structure of the plant dual-affinity nitrate transporter NRT1.1. Nature 507(7490):73–77. doi:10.1038/nature13074

Suzuki K, Saito K, Tsujimura S, Nakayamada S, Yamaoka K, Sawamukai N, Iwata S, Nawata M, Nakano K, Tanaka Y (2010) Tacrolimus, a calcineurin inhibitor, overcomes treatment unresponsiveness mediated by P-glycoprotein on lymphocytes in refractory rheumatoid arthritis. J Rheumatol 37(3):512–520. doi:10.3899/jrheum.090048

Taglicht D, Michaelis S (1998) *Saccharomyces cerevisiae* ABC proteins and their relevance to human health and disease. Methods Enzymol 292:130–162

Tamay-Cach F, Villa-Tanaca ML, Trujillo-Ferrara JG, Aleman-Gonzalez-Duhart D, Quintana-Perez JC, Gonzalez-Ramirez IA, Correa-Basurto J (2016) In silico studies most employed in the discovery of new antimicrobial agents. Curr Med Chem

Tanabe K, Lamping E, Adachi K, Takano Y, Kawabata K, Shizuri Y, Niimi M, Uehara Y (2007) Inhibition of fungal ABC transporters by unnarmicin A and unnarmicin C, novel cyclic peptides from marine bacterium. Biochem Biophys Res Commun 364(4):990–995. doi:10.1016/j.bbrc.2007.10.110

Tanabe K, Lamping E, Nagi M, Okawada A, Holmes AR, Miyazaki Y, Cannon RD, Monk BC, Niimi M (2011) Chimeras of *Candida albicans* Cdr1p and Cdr2p reveal features of pleiotropic drug resistance transporter structure and function. Mol Microbiol 82(2):416–433. doi:10.1111/j.1365-2958.2011.07820.x

Thakur JK, Arthanari H, Yang F, Pan SJ, Fan X, Breger J, Frueh DP, Gulshan K, Li DK, Mylonakis E, Struhl K, Moye-Rowley WS, Cormack BP, Wagner G, Naar AM (2008) A nuclear receptor-like pathway regulating multidrug resistance in fungi. Nature 452(7187):604–609. doi:10.1038/nature06836

Thornewell SJ, Peery RB, Skatrud PL (1997) Cloning and characterization of CneMDR1: a *Cryptococcus neoformans* gene encoding a protein related to multidrug resistance proteins. Gene 201(1–2):21–29

Tong AH, Boone C (2006) Synthetic genetic array analysis in *Saccharomyces cerevisiae*. Methods Mol Biol 313:171–192

Tsao S, Rahkhoodaee F, Raymond M (2009) Relative contributions of the *Candida albicans* ABC transporters Cdr1p and Cdr2p to clinical azole resistance. Antimicrob Agents Chemother 53 (4):1344–1352

Tusnady GE, Sarkadi B, Simon I, Varadi A (2006) Membrane topology of human ABC proteins. FEBS Lett 580(4):1017–1022

van den Brule S, Smart CC (2002) The plant PDR family of ABC transporters. Planta 216(1):95–106

Vandeputte P, Ferrari S, Coste AT (2012) Antifungal resistance and new strategies to control fungal infections. Int J Microbiol 2012:713687. doi:10.1155/2012/713687

Venkatesh M, Wang H, Cayer J, Leroux M, Salvail D, Das B, Wrobel JE, Mani S (2011) In vivo and in vitro characterization of a first-in-class novel azole analog that targets pregnane X receptor activation. Mol Pharmacol 80(1):124–135. doi:10.1124/mol.111.071787

Vermitsky JP, Edlind TD (2004) Azole resistance in *Candida glabrata*: coordinate upregulation of multidrug transporters and evidence for a Pdr1-like transcription factor. Antimicrob Agents Chemother 48(10):3773–3781. doi:10.1128/AAC.48.10.3773-3781.2004

Vermitsky JP, Earhart KD, Smith WL, Homayouni R, Edlind TD, Rogers PD (2006) Pdr1 regulates multidrug resistance in *Candida glabrata*: gene disruption and genome-wide expression studies. Mol Microbiol 61(3):704–722

Verrier PJ, Bird D, Burla B, Dassa E, Forestier C, Geisler M, Klein M, Kolukisaoglu U, Lee Y, Martinoia E, Murphy A, Rea PA, Samuels L, Schulz B, Spalding EJ, Yazaki K, Theodoulou FL (2008) Plant ABC proteins—a unified nomenclature and updated inventory. Trends Plant Sci 13(4):151–159

Wada S, Niimi M, Niimi K, Holmes AR, Monk BC, Cannon RD, Uehara Y (2002) *Candida glabrata* ATP-binding cassette transporters Cdr1p and Pdh1p expressed in a *Saccharomyces cerevisiae* strain deficient in membrane transporters show phosphorylation-dependent pumping properties. J Biol Chem 277(48):46809–46821

Wada S, Tanabe K, Yamazaki A, Niimi M, Uehara Y, Niimi K, Lamping E, Cannon RD, Monk BC (2005) Phosphorylation of *Candida glabrata* ATP-binding cassette transporter Cdr1p regulates drug efflux activity and ATPase stability. J Biol Chem 280(1):94–103

White TC (1997) Increased mRNA levels of ERG16, CDR, and MDR1 correlate with increases in azole resistance in *Candida albicans* isolates from a patient infected with human immunodeficiency virus. Antimicrob Agents Chemother 41(7):1482–1487

White TC, Marr KA, Bowden RA (1998) Clinical, cellular, and molecular factors that contribute to antifungal drug resistance. Clin Microbiol Rev 11(2):382–402

Wirsching S, Michel S, Kohler G, Morschhauser J (2000a) Activation of the multiple drug resistance gene MDR1 in fluconazole-resistant, clinical *Candida albicans* strains is caused by mutations in a trans-regulatory factor. J Bacteriol 182(2):400–404
Wirsching S, Michel S, Morschhauser J (2000b) Targeted gene disruption in *Candida albicans* wild-type strains: the role of the MDR1 gene in fluconazole resistance of clinical *Candida albicans* isolates. Mol Microbiol 36(4):856–865
Wolfe KH, Shields DC (1997) Molecular evidence for an ancient duplication of the entire yeast genome. Nature 387(6634):708–713
Wortman JR, Fedorova N, Crabtree J, Joardar V, Maiti R, Haas BJ, Amedeo P, Lee E, Angiuoli SV, Jiang B, Anderson MJ, Denning DW, White OR, Nierman WC (2006) Whole genome comparison of the *A. fumigatus* family. Med Mycol 44(Suppl):3–7
Yamamoto S, Hiraga K, Abiko A, Hamanaka N, Oda K (2005) A new function of isonitrile as an inhibitor of the Pdr5p multidrug ABC transporter in *Saccharomyces cerevisiae*. Biochem Biophys Res Commun 330(2):622–628. doi:10.1016/j.bbrc.2005.03.009

Chapter 24
Inhibitors of Cancer Cell Multidrug ABC Transporters: Extrapolation to Combat *Candida albicans* Resistance to Antifungals

Attilio Di Pietro

Abstract A number of human cancer cells become resistant to anticancer agents upon overexpression of three main membrane ATP-binding cassette (ABC) multidrug transporters, ABCB1, ABCG2 and ABCC1, of which inhibitors have been identified. Similarly, *Candida albicans* may become resistant to antifungals upon overexpression of the CaCrd1p ABC transporter. Structural and functional analyses of CaCdr1p indicate common features with the three human multidrug ABC transporters: same degenerescence of ATP-binding motifs as ABCC1, same "reverse" topology as ABCG2, and comparable pattern of hydrophobic, non-conjugated, substrates as ABCB1. Assaying representative inhibitors of the different human multidrug ABC transporters indicated that only those targeting ABCB1 were efficient against CaCdr1p transport activity. Potent inhibitors, able to prevent Nile Red efflux from a *Saccharomyces cerevisia* strain transformed to overexpress CaCdr1p and to sensitize its growth to fluconazole, were identified among macrocyclic diterpenes such as jatrophanes and lathyranes, and among other classes of compounds known to interact with ABCB1, but with different structure–activity relationships. Interestingly, other derivatives from the same series and other classes of compounds were able to inhibit CaMdr1p, a non-ABC multidrug transporter of *Candida albicans*, belonging to the Major Facilitator Superfamily (MFS) and overexpressed concomitantly to CaCdr1p in antifungal-resistant strains.

Keywords Multidrug resistance (MDR) · ABC transporters · Cancer cells · Yeast multidrug transporters · Inhibitors · Antifungal efflux

---

A. Di Pietro (✉)
Institute of Protein Biology and Chemistry, MMSB,
UMR 5086 CNRS-University of Lyon, Lyon, France
e-mail: a.dipietro@ibcp.fr

R. Prasad (ed.), *Candida albicans: Cellular and Molecular Biology*,
DOI 10.1007/978-3-319-50409-4_24

24.1 Introduction

At least half of human tumors that are resistant to multiple anticancer drugs overexpress membrane ATP-binding cassette (ABC) transporters which expel cytotoxic drugs out of the cells, using the energy driven by ATP hydrolysis, down to an intracellular concentration below the cytotoxic threshold. They typically acquire a multidrug resistance (MDR) phenotype.

Two multidrug ABC transporters are highly overexpressed in many types of cancer cells, namely ABCB1 (Endicott and Ling 1989) and ABCG2 (Allikmet et al. 1998; Doyle et al. 1998; Miyake et al. 1999). A third one, ABCC1 (Cole et al. 1992), is slightly overexpressed and restricted to a few types of tumors (Cole 2014).

Each functional transporter is constituted of two cytosolic nucleotide-binding domains (NBDs) involved in ATP binding and hydrolysis thanks to "ATP-binding canonical motifs", and two alpha-helical transmembrane domains (TMDs) involved in drug binding and transport. In ABCB1, the topological arrangement is TMD1-NBD1-TMD2-NBD2, from the *N*-teminal to the *C*-terminal ends. ABCC1 includes an additional *N*-terminal TMD (TMD0-TMD1-NBD1-TMD2-NBD2) and partially degenerated ATP-binding motifs, while ABCG2 is a "half-transporter", containing only one TMD and one NBD with a "reverse topology" (NBD-TMD), which needs to at least homodimerize to be functional.

24.2 Inhibitors of Human Multidrug ABC Transporters

Identifying selective and potent inhibitors able to prevent the efflux of multiple drugs by ABCB1 and/or ABCG2 has constituted a major strategy to abolish multidrug resistance of cancer cells with the aim of setting a new chemotherapeutic treatment combining such inhibitors with conventional anticancer agents.

24.2.1 ABCB1 Inhibitors

During the last three decades, many efforts have been spent to identify, characterize, and optimize ABCB1 inhibitors, as illustrated in Fig. 24.1a for verapamil, the cyclosporin A derivative Valspodar/PSC833, and a number of synthetic compounds (Elacridar/GF120918, Tariquidar, Biriquidar, Laniquidar, Ontogen/OC144-093, MS209 and Zosuquidar/LY335979), some of them having reached clinical trials but, unfortunately, with limited efficacy (for a review, see Srivalli and Lakshmi 2012).

Our group has contributed to new ABCB1 inhibitors by screening new classes of compounds (Fig. 24.1b): steroid analogues such as RU49953 (Perez-Victoria et al.

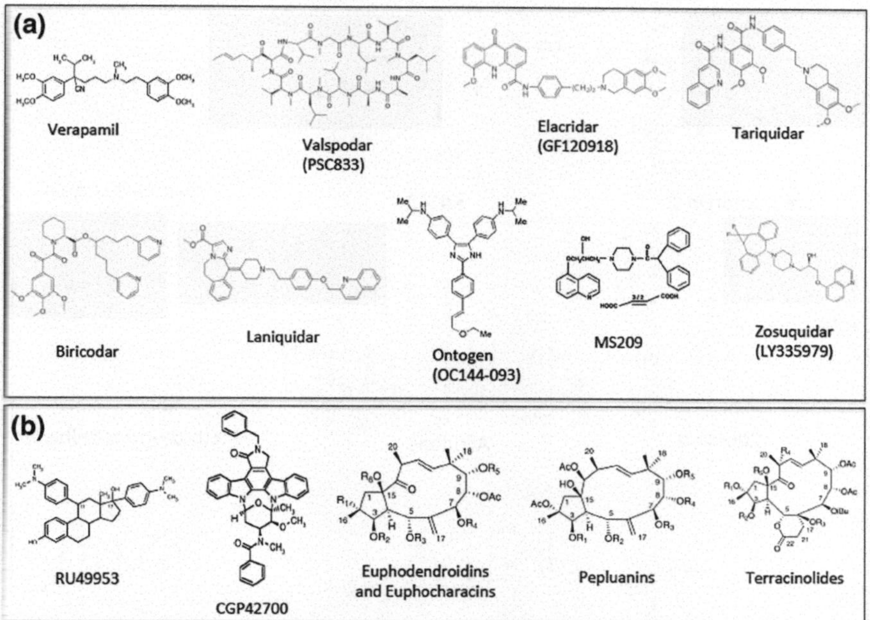

Fig. 24.1 ABCB1 inhibitors. **a** second- and third-generation inhibitors reported in the literature. **b** inhibitors identified within our studies

2003), protein kinase inhibitors such as CGP42700 (Conseil et al. 2001), and natural macrocyclic diterpenes such as Euphodendroidins and Euphocharacins, Pepluanins, and Terracinolides (Corea et al. 2009).

24.2.2 ABCG2 Inhibitors

Less inhibitors are known for more recently discovered ABCG2 (Fig. 24.2). The first-identified selective inhibitor, Fumitremorgin C, was unfortunately highly neurotoxic, and optimization with synthetic analogues led to Ko143 (Allen et al. 2002).

Our main contributions concerned hydrophobic flavones such as 6-prenylchrysin (Ahmed-Belkacem et al. 2006), rotenoids (Macalou et al. 2007), acridones (Boumendjel et al. 2007), methoxy-*trans*stilbenes (Valdameri et al. 2012b), methoxy-chalcones (Gauthier et al. 2012), chromones (Valdameri et al. 2012a) such as MBLII-141 showing preclinical activity (Honorat et al. 2014), and indenoindoles (Jabor Gozzi et al. 2015).

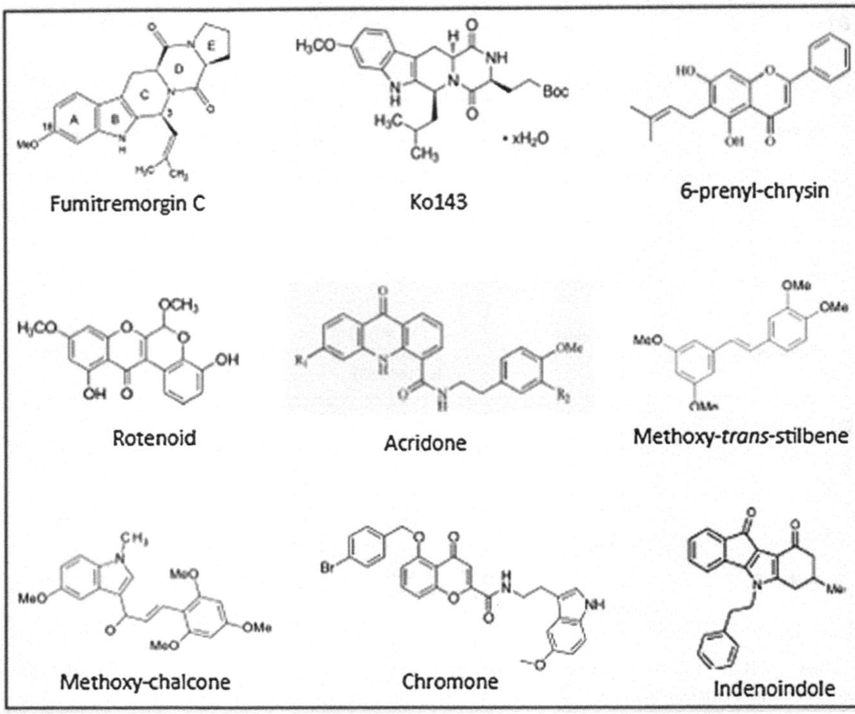

Fig. 24.2 ABCG2 inhibitors optimized from different series of compounds

24.3 Extrapolation to Yeast Multidrug Resistance to Antifungals

Resistance of yeasts such as *Saccharomyces cerevisiae* to multiple cytotoxic drugs including antifungals was characterized by a Pleiotropic Drug Resistance (PDR) phenotype (Balzi and Goffeau 1995). It displays striking similarities with tumor resistance to anticancer agents, especially through the overexpression of multidrug ABC transporters with structural and functional homology to the human cancer cell multidrug transporters.

24.3.1 Main Role of Overexpressed CaCdr1p in Candida albicans *Resistance to Azoles*

A key mechanism of *Candida albicans* resistance to antifungals such as azoles is related to overexpression of two multidrug ABC transporters (Prasad et al. 2015), namely CaCdr1p and CaCdr2p, among which CaCdr1p is the major determinant of resistance.

In addition, a lower but significant role is played by overexpression of CaMdr1p, another type of membrane drug-efflux transporter belonging to the Major Facilitator Superfamily (MFS) and using a proton gradient as energy source, with a different panel of drug substrates when compared to CaCdr1p (Pasrija et al. 2007).

24.3.2 Similarities of CaCdr1p with Human Multidrug ABC Transporters

The similarities are both structural, toward the domains organization and topology, and the canonical sequences involved in ATP binding and hydrolysis, and functional, toward the nature of transport drug substrates.

Structural similarities:

As for other yeast PDR ABC transporters, CaCdr1p displays the same "reverse topology" as human ABCG2, except that is a "full-transporter" (NBD1-TMD1-NBD2-TMD2) (Prasad et al. 2015).

Additionally, CaCdr1p NBDs contain partially degenerated or "deviant" ATP-binding motifs as for ABCC1 and other members of the human ABCC subfamily. This concerns the Walker A and Walker B as well as the Q-loop in NBD1, and the Signature as well as the D-loop in NBD2 (Prasad et al. 2015). The first ATP-binding site is then not catalytic, but it regulates ATP hydrolysis at the second site; the exact mechanism and meaning of such a regulation are only poorly known. It is worthwhile mentioning that such a partial degeneracy is not observed for plant PDR ABC transporters (Crouzet et al. 2006).

Functional similarities:

CaCdr1p transports a wide panel of hydrophobic, non-conjugated, substrates similar to ABCB1. This includes different types of antifungals, as well as herbicides, antibiotics, anticancer drugs, lipids, steroids, and fluorescent dyes (Prasad et al. 2015). These substrates are either cationic or neutral but never anionic, by difference with ABCB1 substrates and some ABCG2 substrates.

CaCdr1p structural and functional similarities with human MDR transporters concern as well *Candida albicans* CaCdr2p, and many other yeast PDR transporters such as *Saccharomyces cerevisiae* ScPdr5p and ScSnq2p (Balzi and Goffeau 1995).

24.4 Experimental Screening of CaCdr1p Inhibitors

The cell-based screening used is described in details in the previous Chap. 23 of this book. A summary is given here below

- The *Saccharomyces cerevisiae* strain AD1-8u^{-}, devoid of endogenous membrane transporters by A. Decottignies and A. Goffeau in Louvain-La-Neuve, Belgium, was transformed to overexpress either CaCdr1p (AD-CDR1, Nakamura et al. 2001) or CaMdr1p (AD-MDR1, Pasrija et al. 2007).
- The transport assay monitored Nile Red (NR) accumulation by flow cytometry. NR was used at 7 μM, and the cells incubated at 30 °C for 30 min with or without of inhibitor at a tenfold excess (70 μM). The starting, untransformed, strain AD1-8u^{-} was used as a control of maximal NR accumulation (no efflux).
- Cytotoxicity and sensitization to fluconazole (FLC). Yeast cells were grown for 48 h at 30°C in either the absence or presence of inhibitors at varying concentrations (0.15–80 μM). The resistance index (RI) was the ratio between the IC_{50} values for AD-CDR1 or AD-MDR1 relative to the AD1-8u^{-} control; a RI value close to 1 indicated a lack of transport whereas higher values suggested an efflux of the cytotoxic compound.

Sensitization by inhibitors to FLC cytotoxicity was evaluated by the checkerboard method recommended by CLSI (formerly NCCLS), and was expressed as the fractional inhibitory concentration index (FICI), as the sum of FICs for each agent (FLC and inhibitors); the FIC of each agent was calculated as the MIC of the agent in combination divided by the MIC of the agent alone (Niimi et al. 2004). FICI values lower than 0.5 indicated a significant sensitization, resulting from synergistic interactions between the inhibitor and FLC.

24.5 Identification of CaCdr1p Inhibitors Among Classes of Compounds Inhibiting Human ABCB1

Representatives of series of compounds known to inhibit either ABCB1 or ABCG2 have been early assayed for their ability to interact with ScPdr5p and inhibit its drug transport activity within enriched membranes. A number of ABCB1 inhibitors, such as hydrophobic flavonoids (Conseil et al. 2000), protein kinase C inhibitors (Conseil et al. 2001) and hydrophobic steroid analogues (Conseil et al. 2003), strongly interacted with purified ScPdr5p and inhibited its drug-efflux activity in yeast membranes, as monitored by rhodamine 6G quenching, and moderately inhibited its ATPase and UTPase activities. However, no inhibition was observed on whole *Saccharomyces cerevisiae* cells overexpressing either ScPdr5p or CaCdr1p (unpublished experiments).

In contrast, macrocyclic diterpenes previously shown to inhibit ABCB1 drug-efflux activity (Corea et al. 2009) also strongly inhibited the drug-efflux activity of CaCdr1p in AD-CDR1 cells, leading to NR intracellular accumulation, and sensitization of yeast growth to FLC cytotoxicity. By difference, ABCG2 inhibitors such as chalcones (Gauthier et al. 2012) were not efficient on AD-CDR1 cells but, unexpectedly, efficiently inhibited the drug-efflux activity of CaMdr1p in the AD-MDR1 strain.

24.5.1 *Jatrophanes and Lathyranes*

Jatrophanes from the Iranian spurge *Euphorbia squamosa*, especially deacetylserrulatin B and euphosquamosin C (Fig. 24.3), strongly inhibited the drug-efflux of CaCdr1p (Kaur Rawal et al. 2014): NR efflux was respectively inhibited by 88% and around 50%, while very low FICI values of 0.03 and 0.06 were obtained for sensitization to FLC. Both inhibitors appeared to be themselves transported by CaCdr1p since high RI values of 11.6 and 24.8 were observed.

Lathyranes such as latilagascenes also strongly inhibited NR efflux (Nim et al. 2016), but in a nonselective way by difference with the jatrophane Euphomelliferene A which was Cdr1p selective (*cf.* below). These compounds appeared not to be transported by CaCdr1p since the RI values were close to unity.

24.5.2 *Other Classes of Compounds*

Other series of compounds known to interact with ABCB1, such as other types of macrocyclic diterpenes (in collaboration with M. Litaudon) and verapamil analogues (in collaboration with E. Teodori), which are under current investigation, also show inhibition capacity toward CaCdr1p drug-efflux activity.

24.6 CaMdr1p Inhibitors

The same series of macrocyclic diterpenes assayed on AD-CDR1 cells in the above Sect. 24.5.1 were also assayed, under the same conditions, on AD-MDR1 cells. Deacetylserrulatin B (Fig. 24.4) also showed a significant activity on CaMdr1p, although less potent than for CaCdr1p, toward both NR efflux and sensitization to FLC cytotoxicity, whereas euphosquamosin C was much less efficient. Both compounds appeared to be also transported by CaMdr1p (Kaur Rawal et al. 2014).

The latilagascenes described above for CaCdr1p were similarly efficient on CaMdr1p, with a strong inhibition of NR efflux. More interestingly, euphomelliferine

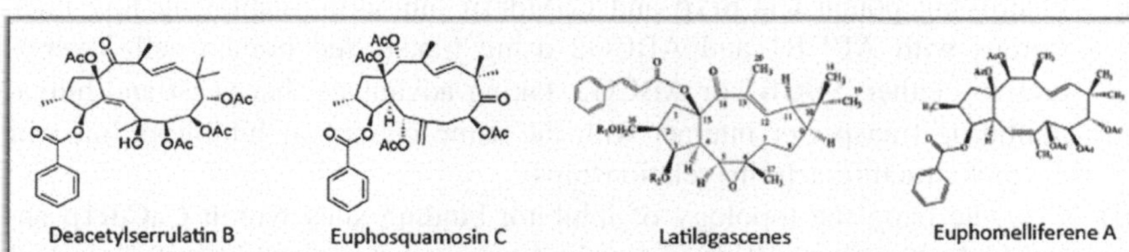

Fig. 24.3 CaCdr1p inhibitors identified among ABCB1-interacting compounds

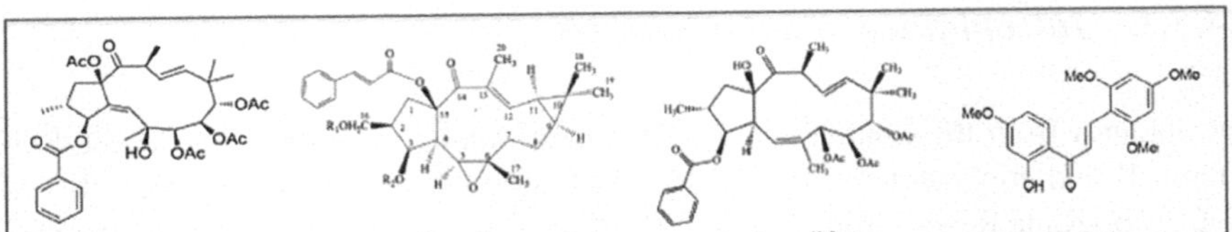

Fig. 24.4 CaMdr1p inhibitors identified among ABCB1- and ABCG2-interacting compounds

(Fig. 24.4) as well as euphomelliferene B induced a strong sensitization to FLC cytotoxicity with FICI values of 0.071 (Nim et al. 2016).

The other types of macrocyclic diterpenes and verapamil analogues currently under investigation also appear to inhibit CaMdr1p drug-efflux activity. Unexpectedly, known ABCG2 such as methoxy-chalcones (Fig. 24.4), which did not modify CaCdr1p, were quite efficient on CaMdr1p (in collaboration with A. Boumendjel). A new series of piperazinyl-pyrrolo-quinoxaline derivatives is also being investigated for inhibition of both CaMdr1p and CaCdr1p (in collaboration with J. Guillon and M. Le Borgne).

In conclusion, a number of macrocyclic diterpenes known to interact with human ABCB1 were found to inhibit CaCdr1p. This is consistent with the fact that both types of transporters share common substrates, and that some inhibitors are themselves transported. Interestingly, various inhibitors interact with the MFS Mdr1p transporter, or with both multidrug transporters, but with different structure–activity relationships. In contrast, ABCG2 inhibitors appeared only to able to inhibit CaMdr1p. Such characteristics might be related to the polyspecificity of all these multidrug transporters.

24.7 Perspectives

The following perspectives may be investigated in future studies:

(i) Screening new series of compounds such as new derivatives of jatrophanes or lathyranes, or other macrocyclic diterpenes, or other new classes of compounds.

(ii) Identifying potent CaCdr1p and CaMdr1p inhibitors displaying low interactions with ABCB1 and ABCG2 using transfected human cells overexpressing either ABCB1 or ABCG2, taking advantage that yeast and human multidrug transporter interact with the same classes of inhibitors but with distinct structure-activity relationships.

(iii) Characterizing the topology of inhibitor-binding sites within CaCdr1p and CaMdr1p by using collections of single-point mutants and identifying which residues are essential for inhibition.

Acknowledgements Funding was provided by the CNRS and University of Lyon (UMR5086). The following collaborators are warmly acknowledged: R. Prasad, M. Kaur Rawal and S. Nim (Amity University Haryana, Gurgaon, India), O. Taglialatela-Scafati (University of Naples, Italy), G. Appendino (University of Novara, Italy), M-J.U. Ferreira (University of Lisbon, Portugal), M. Litaudon (CNRS-ICSN of Gif-sur-Yvette, France), E. Teodori (University of Florence, Italy), A. Boumendjel (University of Grenoble, France), M. Le Borgne (University of Lyon, France) and J. Guillon (University of Bordeaux, France). The work was initiated thanks to interesting discussions with A. Goffeau (University of Louvain-La-Neuve, Belgium) and B.C. Monk (University of Dunedin, New Zealand).

References

Ahmed-Belkacem A, Pozza A, Macalou S, Perez-Victoria JM, Boumendjel A, Di Pietro A (2006) Inhibitors of cancer cell multidrug resistance mediated by breast cancer resistance protein (BCRP/ABCG2). Anticancer Drugs 17:239–243

Allen JD, van Loevezijn A, Lakhai JM, van der Valk M, van Tellingen O, Reid G, Schellens JH, Koomen GJ, Schinkel AH (2002) Potent and specific inhibition of the breast cancer resistance protein multidrug transporter in vitro and in mouse intestine by a novel analogue of fumitremorgin C. Mol Cancer Ther 1:417–425

Allikmets R, Schriml LM, Hutchinson A, Romano-Spica V, Dean M (1998) A human placenta-specific ATP-binding cassette gene (ABCP) on chromosome 4q22 that is involved in multidrug resistance. Cancer Res 58:5337–5339

Balzi E, Goffeau A (1995) Yeast multidrug resistance: the PDR network. J Bioeneg Biomembr 27:71–76

Boumendjel A, Macalou S, Ahmed-Belkacem A, Fortune A, Blanc M, Di Pietro A (2007) Acridone derivatives: design, synthesis and inhibition of breast cancer resistance protein ABCG2. Bioorg Med Chem 15:2892–2897

Cole SP (2014) Targeting multidrug resistance protein 1 (MRP1, ABCC1): past, present and future. Annu Rev Pharmacol Toxicol 54:95–117

Cole SPC, Bhardwaj G, Gerlach JH, Mackie JE, Grant CE, Almquist KC, Stewart AJ, Kurz EU, Duncan AM, Deeley RG (1992) Overexpression of a transporter gene in a multidrug-resistant human lung cancer cell line. Science 258:1650–1654

Conseil G, Decottignies A, Jault J-M, Comte G, Barron D, Goffeau A, Di Pietro A (2000) Prenylflavonoids as potent inhibitors of the Pdr5p multidrug ABC-transporter from Schizosaccharomyces cerevisiae. Biochemistry 39:6910–6917

Conseil G, Pérez-Victoria JM, Jault J-M, Gamarro F, Goffeau A, Hofmann J, Di Pietro A (2001) Protein kinase C effectors bind to multidrug ABC transporters and inhibit their activity. Biochemistry 40:2564–2571

Conseil G, Pérez-Victoria JM, Renoir JM, Goffeau A, Di Pietro A (2003) Potent competitive inhibition of drug binding to the Saccharomyces cerevisiae ABC exporter Pdr5p by the hydrophobic estradiol-derivative RU49953. Biochem Biophys Acta 1614:241–244

Corea G, Di Pietro A, Dumontet C, Fattorusso E, Lanzotti V (2009) Jatrophane diterpenes from Euphorbia spp. as modulators of multidrug resistance in cancer therapy. Phytochem Rev 8:431–447

Crouzet J, Trombik T, Fraysse AS, Boutry M (2006) Organization and function of the plant pleiotropic drug resistance ABC transporter family. FEBS Lett 579:6791–6795

Doyle LA, Yang W, Abruzzo LV, Krogmann T, Gao Y, Rishi AK, Ross DD (1998) A multidrug resistance transporter from human MCF-7 breast cancer cells. Proc Natl Acad Sci USA 95:15665–15670

Endicott JA, Ling V (1989) The biochemistry of P-glycoprotein-mediated multidrug resistance. Annu Rev Biochem 58:171–247

Gauthier C, Valdameri G, Terreux R, Kachadourian R, Day BJ, Winnischofer SMB, Rocha MEM, Frachet V, Ronot X, Di Pietro A, Boumendjel A (2012) Investigation of chalcones as selective inhibitors of the breast cancer resistance protein ABCG2: critical role of methoxy substituents in both inhibition and cytotoxicity. J Med Chem 55:3193–3200

Honorat M, Guitton J, Gauthier C, Bouard C, Lecerf-Schmidt F, Peres B, Terreux R, Gervot H, Rioufol C, Boumendjel A, Puisieux A, Di Pietro A, Payen L (2014) MBL-II-141, a chromone derivative, enhances irinotecan anticancer efficiency in ABCG2-positive xenografts. Oncotarget 5:11957–11970

Jabor Gozzi G, Bouaziz Z, Winter E, Daflon-Yunes N, Aichele D, Nacereddine A, Marminon C, Valdameri G, Zeinyeh W, Bollacke A, Guillon J, Lacoudre A, Pinaud N, Cadena SM, Jose J, Le Borgne M, Di Pietro A (2015) Converting potent indeno[1,2-b]indole inhibitors of protein kinase CK2 into selective inhibitors of the breast cancer resistance protein ABCG2. J Med Chem 58:265–277

Kaur Rawal M, Shokoohinia Y, Chianese G, Zolfaghari B, Appendino G, Taglialatela-Scafati O, Prasad R, Di Pietro A (2014) Jatrophanes from *Euphorbia squamosa* as potent inhibitors of *Candida albicans* multidrug transporter. J Nat Prod 77:2700–2706

Macalou S, Ahmed-Belkacem A, Borelli F, Capasso R, Fattorusso E, Taglialatela-Scafati O, Di Pietro A (2007) Non-prenylated rotenoids, a new class of potent breast cancer resistance protein inhibitors. J Med Chem 50:1933–1938

Miyake K, Mickley L, Litman T, Zhan Z, Robey R, Cristensen B, Brangi M, Greenberger L, Dean M, Fojo T, Bates SE (1999) Molecular cloning of cDNAs which are highly overexpressed in mitoxantrone-resistant cells: demonstration of homology to ABC transport genes. Cancer Res 8–24

Nakamura K, Niimi M, Niimi K, Holmes AR, Yates JE, Decottignies A, Monk BC, Goffeau A, Cannon RD (2001) Functional expression of Candida albicans drug efflux pump Cdr1p in a Saccharomyces cerevisiae strain deficient in membrane transporter. Antimicrob Agents Chemother 45:3366–3374

Niimi K, Harding DR, Parshot R, King A, Lun DJ, Decottignies A, Niimi M, Lin S, Cannon RD, Goffeau A, Monk BC (2004) Antimicrob Agents Chemother 48:1256–1271

Nim S, Monico A, Kaur Rawal M, Duarte N, Prasad R, Di Pietro A, Ferreira MJU (2016) Overcoming multidrug resistance in Candida albicans: macrocyclic diterpenes from Euphorbia species as potent inhibitors of drug-efflux pumps. Planta Med (in press)

Pasrija R, Banerjee D, Prasad R (2007) Structure and function analysis of CaMdr1p, a major facilitor superfamily antifungal efflux transporter protein of *Candida albicans*: identification of amino acid residues critical for drug/H^+ transport. Eukaryot Cell 6:443–453

Perez-Victoria FJ, Conseil G, Munoz-Fernandez F, Perez-Victoria JM, Dayan G, Marsaud V, Castanys S, Gamarro F, Renoir JM, Di Pietro A (2003) RU49953: a non-hormonal steroid derivative that potently inhibits P-glycoprotein and reverts cellular multidrug resistance. Cell Mol Life Sci 60:526–535

Prasad R, Banerjee A, Khandelwal NK, Dhamgaye S (2015) The ABCs of *Candida albicans* multidrug transporter Cdr1. Eukaryot Cell 14:1154–1164

Srivalli KMR, Lakshmi PK (2012) Overview of P-glycoprotein inhibitors: a rational outlook. Braz J Pharm Sci 48:353–367

Valdameri G, Genoux-Bastide E, Peres B, Gauthier C, Guitton J, Terreux R, Winnischofer SMB, Rocha MEM, Boumendjel A, Di Pietro A (2012a) Substituted chromones as highly-potent, nontoxic inhibitors, specific for the breast cancer resistance protein ABCG2. J Med Chem 55:966–970

Valdameri G, Pereira Rangel L, Sapatafora C, Guitton J, Gauthier C, Arnaud O, Ferreira-Pereira A, Falson P, Winnischofer SMB, Rocha MEM, Tringali C, Di Pietro A (2012b) Methoxy stilbenes as potent, specific, untransported and noncytotoxic inhibitors of breast cancer resistance protein. ACS Chem Biol 7:322–330